T0355781

THE BIRDS OF SOUTH AMERICA

VOLUME I **THE OSCINE PASSERINES**

VOLUME I THE OSCINE PASSERINES

JAYS AND SWALLOWS

WRENS, THRUSHES, AND ALLIES

VIREOS AND WOOD-WARBLERS

TANAGERS, ICTERIDS, AND FINCHES

Project support has been provided by
RARE Center for Tropical Bird Conservation,
Massachusetts Audubon Society,
and Pan-American Section
of International Council for Bird Preservation

THE BIRDS OF SOUTH AMERICA

by Robert S. Ridgely and Guy Tudor

with the collaboration of WILLIAM L. BROWN

In association with World Wildlife Fund

UNIVERSITY OF TEXAS PRESS, AUSTIN

Text printed in China
Color plates printed in China

Fifth printing, 2006

Requests for permission to reproduce material
from this work should be sent to:
Permissions
University of Texas Press
P.O. Box 7819
Austin, TX 78713-7819
www.utexas.edu/utpress/about/bpermission.html

LIBRARY OF CONGRESS CATALOGING-IN-PUBLICATION DATA

Ridgely, Robert S., 1946–
The birds of South America : the oscine passerines : jays and swallows, wrens, thrushes, and allies, vireos, and wood-warblers, tanagers, icterids, and finches / by Robert S. Ridgely and Guy Tudor with the collaboration of William L. Brown.—1st ed.
p. cm.
Bibliography: p.
Includes index.
ISBN-13: 978-0-292-70756-6
ISBN-10: 0-292-70756-8
1. Passeriformes—South America. 2. Birds—South America.
I. Tudor, Guy, 1934– . II. Brown, William L. III. Title.
IV. Title: Oscine passerines.
QL696.P2R53 1989 88-20899
598.8098—dc19 CIP

∞ The paper used in this book meets the minimum requirements of ANSI/NISO Z39.48-1992 (R1997) (Permanence of Paper).

To the memory of
EUGENE EISENMANN,
one of the first ornithologists born in the neotropics,
who more than anyone else who ever lived
actively encouraged the serious study of its birds
by both visitors and residents alike.

And to the people of South America,
in whose hands the future of the world's finest avifauna lies.

CONTENTS

FOREWORD

A SUDDEN FLASH of color in the backyard, along a hiking trail, across a farm field—the chance sighting of an unknown and vividly colored bird. The pleasure in that instant of perception is often what begins to transform a mere onlooker into a student of birdlife. It is a feeling that has transformed those unfamiliar with nature into devoted amateur naturalists. It has set others on the path to become scientists dedicated to the formal study and conservation of the world's avifauna.

Whether our interest is professional or personal, our appreciation for birds is strengthened and enriched by a more intimate knowledge of them. In North America, such knowledge has been fostered by the wide availability of a number of excellent field handbooks to birds, guides that have stimulated enduring interest and support for the conservation of birds and their habitat.

As this concern has grown, so has awareness of the avian ties between North America and the New World tropics (over half of the 660 bird species found in the United States spend the winter in Latin America and the Caribbean). This migratory link with birdlife in Latin America has begun to intrigue North Americans, at the same time that interest is spreading within the region itself. Such interest comes as little wonder. Within the cloud forests of the high Andes, throughout the vast Amazon basin, and in many other South American habitats, an impressive 3,100 species, migrant and resident, are found.

Yet many of South America's bird species—and other animals and plants as well—face the peril of extinction. As population pressures and economic aspirations multiply, short-sighted economic development measures are exacting a grim toll from the tropical habitats that allow wildlife to flourish. Commercial logging, mining activities, conversion into farmland and pastureland, slash-and-burn agriculture—all threaten to destroy tropical ecosystems that until now have survived for millions of years.

WWF has worked since 1961 to protect all forms of endangered wildlife and their habitats around the world, particularly in the tropics, the home of most of the world's species. Our methods include helping to establish new parks that safeguard birds along with a wide range of other species, training Latin American biologists and other conservation professionals, and promoting environmentally sound development. In South America alone, we have helped to create or maintain more than sixty protected areas, encompassing more than 14 million hectares of critical wildlands (an area roughly half the size of Ecuador). Manu National Park in Peru is one of these areas; its 1.5 million hectares shelter nearly 10 percent of all bird species on earth.

WWF has protected birds in other ways as well. Concerned with the impact of international trade in particular species, the organization has supported surveys that have provided the evidence necessary to remove such birds as the Hyacinth Macaw from commercial trade. At another level, it has helped to support an international network of sister reserves to protect Western Hemisphere migratory shorebirds.

As a complement to such efforts, WWF has been pleased to support preparation of this volume of *Birds of South America*. Bob Ridgely, one of today's most important and dedicated ornithologists, and Guy Tudor, one of the world's outstanding wildlife illustrators, have created a remarkably thorough handbook, presenting the most current information available on South American avifauna. As the first modern field handbook to cover all the true songbirds (the oscine passerines) of South America, it is indispensable. We are confident that *Birds of South America* will increase awareness of the continent's magnificent and diverse birdlife and concern for its survival.

This volume will ably serve researchers and conservationists and provide new knowledge to bird enthusiasts throughout the Western Hemisphere. Most important, we believe it will provide added inspiration to South Americans themselves, as it is they who ultimately will determine the fate of the extraordinary avian heritage that is the subject of this volume.

RUSSELL E. TRAIN
Chairman of the Board
World Wildlife Fund and The Conservation Foundation

PREFACE

IN 1970 Rodolphe Meyer de Schauensee published his *Guide to the Birds of South America,* in which, for the first time, was assembled in a single volume a description of each species of South American bird. Finally, and despite the absence of useful pictures, field identification of South America's incredibly diverse birdlife was at least feasible—Ridgely well remembers the frustrations he and Eugene Eisenmann felt in 1969 attempting to identify birds around Machu Picchu on the basis of only their English and Latin names! He likewise recalls the satisfying thrill of being able to be positive that that magnificent gnateater mist-netted outside Belém, Brazil, with Tom Lovejoy in 1971 was a male Hooded: Meyer de Schauensee's description "fit," and he was in the correct range.

But "BSA" (as it came to be called), while immensely helpful, proved not an easy book to use in the field. It was written by a museum-based ornithologist, one who knew his birds mostly as museum skins: Meyer de Schauensee had little knowledge of field characters, vocalizations, or the birds' behavior or ecology. But then, back in those days, it must be emphasized, no one knew these birds very well.

BSA changed that or, really, it permitted that to change. With identification at least possible, ornithologists and birders descended on South America in numbers dwarfing anything seen before. We count ourselves fortunate to have been part of that first wave. Local scientists and conservationists also gradually became more active, and all of us were fortunate that, at the same time, the South American travel infrastructure was also improving to the point where access to numerous heretofore difficult or totally inaccessible areas was now feasible (even relatively easy). We made many mistakes along the way, but gradually information accumulated in memories and in notebooks and on tape cassettes, much of it not in the literature and none of it reflected in BSA, which came to be viewed as increasingly obsolete.

Gradually we came to feel that the time had come for a second-generation field-oriented book on the birds of South America, one with accurate pictures and a text which better reflected the current level of knowledge. Our belief that this was timely has in no way been shaken by the appearance in recent years of several regional books (in some of which we have played a role), notably the superb *Guide to the Birds of Colombia* by S. Hilty and W. L. Brown. For what has been needed is a synthesis, an overview of the *entire* continental fauna, one which attempts to bring together the massive amount of information now available. We might point out that this has been an ongoing, continuing process over the past fifteen-plus years, for us and of course others;

in this connection, we should emphasize that where there are differences, everything appearing in our previous efforts should be considered as superseded by this work.

We here present our first volume of a projected four, in this book covering some 750 oscine passerine birds. Our explanation for having departed from the traditional phylogenetic list sequence, commencing work with the passerine groups and not the nonpasserines (in other words, proceeding "backwards"), is simply that we wanted to first tackle the birds which needed attention the most. Work on our next volume, which will cover all of the suboscine passerines, proceeds.

We hope we have managed to organize and present information which is accurate and useful for the professional and semiprofessional, but also palatable and understandable to the layman. Certainly that has been our goal: to make some sense of the great mass of information out there and to convey the excitement, the vitality, pervading all levels of South American ornithology today.

ROBERT S. RIDGELY

GUY TUDOR

ACKNOWLEDGMENTS

IT IS a virtual truism that books of this sort are not created in a vacuum; nonetheless, we sense that in our case this is even more than typically true. We owe a debt of gratitude to a vast throng of people; indeed, as this has become virtually a lifework for us both, it has involved at some level virtually everyone we know professionally, as well as numerous others. Mentioning "everyone" thus becomes something of a problem! But let us begin.

While the idea for *Birds of South America* had been gestating in our minds for several years, it was destined not to come to fruition until after discussions with various staff and board members of World Wildlife Fund-U.S., notably Thomas Lovejoy, Phil Humphrey, and Howard Brokaw. What developed was a commitment of major financial support for what has proved an arduous task, one which has (as usual!) taken longer than was originally anticipated, in part because the scope of our efforts was expanded. For their patience, encouragement, and support we are immensely grateful—and beyond the three individuals mentioned above, we should also include two important subsequent "contacts," Nancy Hammond and Byron Swift. We can only hope that our final product exceeds their expectations.

As the project came to consume more time and energy than had been expected, we gradually felt the need to seek supplemental financial assistance from other sources. This has been forthcoming, and, while not on a scale of WWF's support, it has been substantial, and we are grateful for their confidence and generosity. The following organizations or individuals provided major support: the Pan-American Section of the International Council for Bird Preservation (at the urging of its then-chairman, William Belton), the RARE Center for Tropical Bird Conservation (with the special prompting of its president, Ken Berlin, and one of its vice-presidents, David O. Hill), the Massachusetts Audubon Society (with the special interest of its president, Gerald Bertrand), William L. Brown, and Claudia Wilds.

Since late 1982 one of us (RSR) has been a member of the ornithology staff at the Academy of Natural Sciences in Philadelphia (ANSP). Apart from having one of the world's finest neotropical bird collections at one's fingertips, a more congenial and stimulating place in which to work and write cannot be imagined—at times it was almost too much fun! Its atmosphere is due in no small measure to the dynamic leadership of its chairman, Frank Gill, and I am grateful to him for having provided the opportunity to be part of that team. Others at the academy who have been particularly helpful include George Glenn, J. Pete Myers, Robert Peck, Mark Robbins, and various of the invaluable

VIREO staff. In addition, the academy library, with its superb book and journal collection, has proven to be an invaluable source of information. I remain especially indebted to Elizabeth Kellogg and Mary Dolack for enduring the tedium of the clerical side of the book's production, in particular my seemingly endless stream of minor (but nonetheless important) revisions.

Over the years both of us have spent a great deal of time among the incomparably vast bird collections in the American Museum of Natural History (AMNH) in New York City. Never, however, have we spent more time than during the ongoing research and preparation of *Birds of South America,* and we owe a tremendous debt of thanks to the two recent ornithology departmental chairmen, Lester Short and Wesley Lanyon, for allowing us such complete access. One of us (GT) is especially grateful for permission to borrow virtually any specimen needed; it should be noted that with only one or two exceptions, all the illustrations in this book were prepared through the use of AMNH material. Other staff members there who have been especially helpful or encouraging in various ways over the years include John Bull, Mary LeCroy, and Francois Vuilleumier; the latter's assistance with certain Andean groups was particularly valuable.

An increasingly important center for neotropical bird research has been the Louisiana State University Museum of Zoology (LSUMZ) in Baton Rouge. The collective field experience of the staff there is legendary, and a significant amount of that has been placed at our disposal, not only through access to their marvelous bird collections from Peru and elsewhere in South America but also through their steady flow of important publications and through personal correspondence and conversations with one or the other or both of us. We thus recognize our special debt of gratitude to (in particular) our friends John O'Neill, Ted Parker, J. Van Remsen, and Tom Schulenberg and can only wish them continued success in their efforts to learn more about the neotropical avifauna.

We have also benefited from occasional access (as time permitted) to several other important museum bird collections, in particular those held at Harvard University's Museum of Comparative Zoology (MCZ) and at the Carnegie Museum (CM) in Pittsburgh. Curators of both these institutions were invariably helpful and also permitted the occasional loan of crucial specimen material when needed, for which we are thankful.

Of course by no means all our work has been done in a museum setting—one of us in particular (RSR) has had the great good fortune to have been able to spend significant portions of virtually every year since the mid-1970s in various countries in South America. While there I have been assisted by innumerable individuals, far too many to mention here, from well-known scientists and conservationists to "lowly" woodsmen, all of them contributing to a life which is now replete with fine memories. Some are specifically mentioned elsewhere in other contexts and will not be repeated here; others who have not been, and to whom I feel I owe a very special debt of gratitude, include Felipe

Benavides, Gunnar Hoy, Christoph and Ina Hrdina, Fernando Ortiz-Crespo, William Phelps, Jr., Maurice Rumboll, the late Augusto Ruschi, and Helmut Sick.

Many of the above-named individuals offered us their counsel and opinions on matters taxonomic and nomenclatural. We are deeply appreciative to all and only regret that we cannot, in our own conclusions, agree with everyone all of the time, for these matters are sometimes too divisive to permit unanimous opinions. But we would here be remiss not to especially single out Burt Monroe, who, through a steady and voluminous stream of correspondence dating from 1983, has helped to clarify our thoughts as we (especially RSR) proceeded through the stormy sea that is systematics today.

Much of what is known about South American birds today remains unpublished or, at best, is scattered in a multitude of papers in various languages. We count ourselves fortunate that we have developed, over the years, a veritable host of cooperators who have shared information or valuable photographs or tape recordings with us. Many have been mentioned above (or will be named again below), but we would like to express our special thanks to the following individuals; some may have provided only one piece of information, or one photo, or one tape recording, but it may have been a vital one, and we are thankful to all: Peter Alden, Allan Altman, William Belton, Angelo Capparella, Tom Davis, Tristan Davis, John Dunning, Victor Emanuel, Jon Fjeldså, Crawford Greenewalt, Paul Greenfield, John Gwynne, Heinz Haberyan, Jorge Hernandez, Steve Hilty, Robin Hughes, Michel Kleinbaum, Neils Krabbe, Sid Lipschutz, Jorge Mata, Larry McQueen, Tony Meyer, Charles Munn, Ted Parker, Chris Parrish, Peter Post, Mark Robbins, Rose Ann Rowlett, Michel Sallaberry, Tom Schulenberg, Bruce Sorrie, Doug Stotz, Roberto Straneck, Bret Whitney, and Edwin Willis.

A small number of other people were even more crucial, contributing as they did to building a personal support system around what are all too often stressful and chaotic life-styles. Eugene Eisenmann was not only a close personal friend of ours (he was such, in fact, even before we two knew each other) but also served more than anyone else as our personal mentor and helped get both of us started on our ornithological and conservation careers; we only wish he was alive to see the fruits of our labor. Marc Weinberger provided valued counsel on legal and contractual matters. Victor Emanuel helped provide one of us (RSR) with the opportunity to vastly expand his knowledge of neotropical birds, without which this book could never have come into being; his efforts to promote meaningful bird-oriented tourism in the neotropics are unexcelled. John Dunning's fame as *the* bird photographer of South America is already well established. We have benefited tremendously from his efforts, for John put any photo out of his vast collection at our disposal, thus greatly increasing the accuracy of this book's artwork; further, he provided some financial support as well. Julie Ridgely was RSR's ever-helpful partner during the early research and travel phase of the project, while on the home front my parents,

Beverly and Barbara Ridgely, could always be counted on for encouragement, advice, and love. More recently, Peg Ridgely has been selflessly supportive, in her own special way, of RSR's efforts, while Michelle le Marchant has constantly underscored her commitment to the work through steady administrative assistance to GT; furthermore, she vastly facilitated communications between us. We will always be grateful to each and all.

THE BIRDS OF SOUTH AMERICA

VOLUME I THE OSCINE PASSERINES

ABBREVIATIONS

AMNH	American Museum of Natural History
ANSP	Academy of Natural Sciences of Philadelphia
BSA	*Birds of South America* (Meyer de Schauensee 1970)
C	*A Guide to the Birds of Colombia* (Hilty and Brown 1986)
CM	Carnegie Museum
FM	Field Museum
LSUMZ	Louisiana State University Museum of Zoology
MCZ	Museum of Comparative Zoology
P	*A Guide to the Birds of Panama* (Ridgely 1976)
USNM	Museum of Natural History at the Smithsonian Institution
V	*A Guide to the Birds of Venezuela* (Meyer de Schauensee and Phelps, Jr. 1977)
WWF	World Wildlife Fund

PLAN OF THE BOOK

THE South American avifauna is by far the largest in the world, containing as it does over 3,000 full species or over a third of the world's total. The continent's striking topographic diversity provides part of the explanation, with various species being confined to one or another climatic zone or ecological region, but not the full answer; for example, several well-investigated sites in eastern Ecuador and eastern Peru now boast amazing lists of 450 to even 550 species of birds. These totals are almost twice as large as would be found at comparable sites in tropical Africa or Asia. It's all a complex marvel, wonderful to experience, but its very richness can also be thoroughly intimidating, with accurate bird identification a difficult task indeed.

It is our contention that the most important step in accurate species-level identification is actually determining the correct group to which the organism in question belongs. This having been accomplished, usually determining the correct species is a relatively simple task, for it often can be based on range, habitat, obvious plumage characters, etc. This premise, then, has become the overriding principle of *Birds of South America*.

Both of us have been accumulating information on neotropical birds for twenty years or more, and it becomes ever clearer that this quest for knowledge will be a never-ending one, particularly as this relates to study in the field. Nonetheless, for the purposes of this book we in a sense reverted to an "old-fashioned" approach, at least in order to orient our thinking. We decided to attempt to actually examine each described species and subspecies of bird, comparing them and placing each in the context of what we had already learned. The incomparable neotropical collection at the American Museum of Natural History in New York City provided the basis for much of this work, supplemented by the collections of various other major U.S. museums as necessary (in particular, the Academy of Natural Sciences of Philadelphia, the Louisiana State University Museum of Zoology, the Museum of Comparative Zoology at Harvard, and the Carnegie Museum in Pittsburgh). That we were persistent in our efforts is attested to by the fact that we eventually examined specimens of all but a handful of the species dealt with in this volume, all of those exceedingly rare in collections (*Nemosia rourei, Conothraupis mesoleuca, Sporophila melanops, S. nigrorufa, S. palustris,* and *S. zelichi*), and a very large majority (surely well over 95 percent) of the described subspecies.

It was a tedious but very often illuminating experience, one which has in a very fundamental way shaped our presentation of this book. So central has it been that we would be remiss not to here mention the debt we owe to Charles E. Hellmayr, compiler of the vastly useful (despite the dates when they were written) and seemingly under-appreciated *Catalogue of Birds of the Americas*, which greatly aided in orienting our taxonomic thinking (and provided descriptions of various subspecies not readily available to us). Further, we might also here acknowledge our special debt to certain collectors for the importance and completeness of their collections; we would single out, as being especially helpful to our research, Emil B. Kaempfer (in eastern Brazil and Paraguay; collection in AMNH), M. E. Carriker, Jr. (in the Andes, especially of Colombia, Peru, and Bolivia; collections in ANSP and USNM), and John Weske (in eastern Peru; collection in AMNH).

During our AMNH sessions we made notes on relationships, racial variation and its relative significance, distribution (e.g., unpublished locality data for certain species), and so on, but our primary goal was to establish how the various genera might best be assembled (and, in the case of larger genera, how they might best be broken down into subgroups) on each plate, for it is around each plate that the book has been organized.

We might point out here that the sequence of families, subfamilies, genera, and species which we use here is *not* necessarily an accurate reflection of present taxonomic thinking (though usually it is). We have more or less followed the "standard" phylogenetic sequence as published, for instance, in the 1983 AOU Check-list and in Meyer de Schauensee (1966, 1970), so our "flow" will be relatively familiar to experienced readers. However, it should be emphasized that because of recent changes in taxonomic thinking, there really is no "standard" sequence, and changes, some of them quite drastic, are almost unquestionably in the offing. We have at times had to make some minor modifications in order to associate certain visually similar groups with other, similar-looking but not closely related groups. Thus, for example, the one South American species of lark is placed near the pipits, etc.

One final point that must be made concerns Spanish and Portuguese vernacular names. We considered including these, and would have liked to. However, it seems that as yet there is nothing approaching a consensus concerning a valid list of appropriate names in either language. That being the case, *we* would have been forced to make that choice—and, English being our primary language, we did not feel we alone should do so. It is a complicated issue, complicated especially by the fact that some species have a different name in each of up to a dozen Latin American countries and perhaps yet another in Spain/Portugal. We and the World Wildlife Fund are anxious to see *Birds of South America* translated into Spanish and Portuguese as soon as is feasible; we hope that such a list is available by that time.

THE PLATES AND FACING PAGES

Our basic approach has been to associate similar genera on the same plate together. These are usually, but not always, also closely related to each other; however, note that we have given priority to placing visually similar genera together.

The genus thus becomes our most important grouping, and out of the mass of tanagers, for instance, certain similar genera are placed together on the same plate. Species-rich genera present a special problem, which we have attempted to address by the creation of formalized "groups" within that genus (A, B, C, D, etc.), usually of visually similar species, less often of geographically associated species, occasionally of very distinct or miscellaneous species with no apparent close relatives within the genus. These do not necessarily have any taxonomic standing, though most often they do. Sometimes there are "subgroups" as well, when these have been deemed useful for clarifying purposes; these are marked by a "bullet" (•) where they appear on each facing page.

Our intent, then, for the users of this book when they are confronted with an unusual bird is this: peruse the plates at the beginning of the book (we hope these will already be familiar from previous study). *A member of every genus and at least one member of every visually distinct group (within genera) have been illustrated,* with some very minor exceptions. The actual species you are dealing with may turn out not to have been portrayed, but we expect this should not present an insuperable problem, for the species which *is* portrayed should be sufficiently similar to permit you to get into the right group.

The text on the page facing each plate discusses the salient characteristics of each genus or group and lists each species illustrated, together with the nonillustrated species under their appropriate group. Ranges of all but the most widespread species are briefly indicated, except where this is clear from the species' name. Generally, details concerning species-level identification have not been included on the facing pages; for these points, you must refer to the main text. Note that the sequence and groupings in the main text strictly follow those which have been established on the plates, with a very few minor exceptions. The descriptions in the main text are also oriented back toward the plates; thus, the species and subspecies which are illustrated have been given precedence, with comparisons of other species based on them. This, combined with the text's behavioral and distributional information, should enable one to confirm an identification.

Early on in this project we realized that actually depicting every species and important subspecies would be prohibitively expensive and time-consuming. That having been recognized, what we attempted to do was to intelligently select the species which most required an illustration; in the end, about two-thirds of the species were portrayed. It may be of interest to set out the factors which especially influenced our decisions:

1. At least one member of each genus, and of each group within large genera, should be illustrated. The only exceptions permitted were of certain well-known North American migrants, though it should be noted that a selection of the common migrants has been included.

2. The range of variation within each genus or group must be shown. This frequently resulted in several species within a unit being depicted.

3. A preference should be given to species which are numerous or widespread, all other factors being equal, over those which are relatively scarce or localized.

4. A preference should be given to species found mainly or entirely in eastern or southern South America; these tend to be less well known as compared with birds found in northern and western South America, in part because of the illustrations in *A Guide to the Birds of Venezuela* (Meyer de Schauensee and Phelps, Jr. 1977) and *A Guide to the Birds of Colombia* (Hilty and Brown 1986). We recognize that this latter dictum at times conflicted with the preceding rule of thumb; in such cases, we tended to favor illustrating those species which have seldom or never before been portrayed.

Finally, we might point out that the decision was made to try to illustrate approximately two-thirds of the species within each (large) genus. In certain cases this has meant we have selected more "visually similar" species in one genus than in another; thus, all the *Hylophilus* greenlets look relatively alike (as the genus is quite uniform), whereas there is great superficial variation in *Tangara* tanagers (as almost every species in this genus looks "different"). One benefit of this should be that the more difficult genera (i.e., those in which the various species are quite alike) have been more thoroughly covered than they might otherwise have been.

Furthermore, we decided not to give short shrift to females, as so often is done—these have been depicted as fully as possible. Subspecies, on the other hand, have been dealt with on the plates to a much lesser extent, although a few really divergent ones (some of which could ultimately be split as full species) are included. We must emphasize, however, that *the reader needs to check the main text, as racial variation in many species can be significant* (and thus *very* confusing in the field). *Note that the subspecies which is illustrated is always named on the facing page* (if no subspecies is given, the species is monotypic), but that *the range which is given is that of the species as a whole* (not just the listed and illustrated subspecies).

THE MAIN TEXT

The arrangement of our text is more or less standard; nonetheless, a few explanatory comments are needed.

It should be noted again that the sequence of genera, and of our subgroups within genera, is *precisely the same in the text as it is in the text on the plate facing pages;* even the wording describing our subgroups remains the same. We hope, thereby, to facilitate transferring between plates and text.

A few introductory comments on each genus are included; in some cases, particularly for larger or more complex genera, these comments are more amplified. An attempt is made to define the genus in nontechnical terms, together with comments on its systematic relationships. Usually field characters are *not* given here (though these are included in the genus accounts on the plate facing pages). Note that our higher-level (subfamily and family) taxonomy and sequence basically follow those of the 1983 AOU Check-list, with a few deviations where we felt it necessary.

As always, the bulk of the main text is taken up with the *species accounts*. These are broken into four or five paragraphs, as follows.

Identification

A full description of most species is included. Doing so, rather than merely giving characters useful for recognition in the field, was deemed necessary because, among other things, not all species could be illustrated. Field characters are, of course, emphasized in the usual way, being given *in italics*. Generally we describe soft-parts (eyes, bill, legs) first, if these are significant for our purposes. The actual description commences with the head and proceeds posteriorly on the bird's upperparts, then moves to the throat and does the same on the bird's underparts. The adult male plumage is usually described first, followed by the adult female and then various subadult/immature/juvenile plumages when necessary; this sequence is altered most often in the case of certain North American migratory species (which in South America may be most often seen in such plumages; then these are given first).

For species which have been illustrated, almost always *the race which is depicted is described first,* and then the species' other races are compared back to that, with only the differences noted; obviously this is not necessary in the case of monotypic species. *We only mention races* (= subspecies) *which are noticeably different in the field.* Likewise for species which have *not* been illustrated, *comparison is usually made back to a species which is* (and these unillustrated species are sometimes *not* described fully, their differences merely being noted).

Subspecies are usually named and are always given *in italics*. The nominate race refers to the first described race of that species. What happens is that when the species is first discovered and described it has only two names: the genus and species names. The species is then "monotypic." At some later point, a similar but morphologically distinct form may be described which is obviously related to the original species. This then becomes a subspecies of the original species, and the first described form becomes the "nominate subspecies" (or race). The species has now become "polytypic." Note further that sometimes groups of similar subspecies are discussed together, as a "subspecies group." As an example, *Thraupis palmarum* (the Palm Tanager) was described on the basis of a bird from e. Brazil; that bird became the "type specimen" and that locality (Bahia, Brazil) became the "type locality." The species is now monotypic. Later, other forms were described from other areas in South America; these were clearly all closely related to *Thraupis palmarum* and, thus, were considered subspecies, e.g., *T. p. melanoptera* of Amazonia. The species has now become polytypic, and one may refer to a group of even more closely related races as a group (e.g., the *violilavata* group from west of the Andes).

Species and subspecies not illustrated here but which are depicted in Meyer de Schauensee and Phelps, Jr. (1977), *A Guide to the Birds of*

Venezuela (V); Hilty and Brown (1986), *A Guide to the Birds of Colombia* (C); or Ridgely (1976), *A Guide to the Birds of Panama* (P), are so indicated, together with the appropriate plate number (e.g., V-40, etc.).

The bird's overall length, in both centimeters and inches, is given; we independently measured specimens of every species available to us (from museum study skins).

Most of the terms we have used in describing birds are standard and self-evident and require no explanation. However, some such terms do seem to cause some confusion or are more or less unique to neotropical birds; these are depicted in Figure 1 and briefly described here.

CORONAL PATCH. An often hidden area, usually of a bright color, in the crown; this may be exposed more prominently when the bird is agitated or is vocalizing.

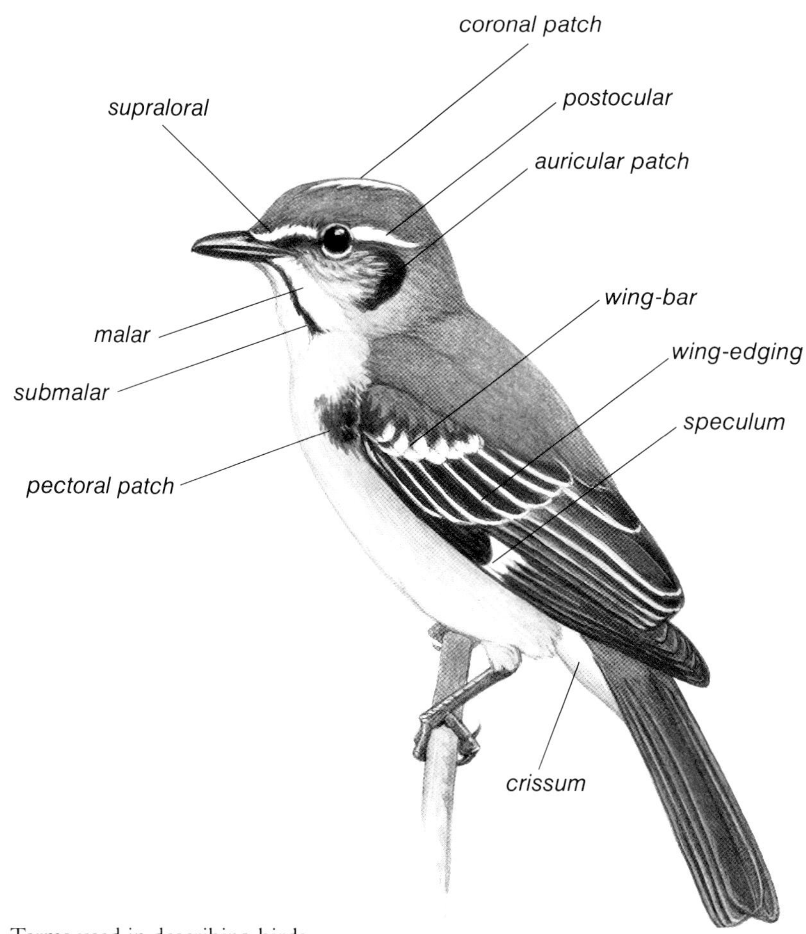

Fig. 1. Terms used in describing birds

SUPRALORAL. A short stripe, usually of a pale color, which extends from the base of the maxilla, over the lores, to just above or in front of the eye (but not extending behind the eye, in which case it becomes a *superciliary*). The supraloral is sometimes joined to an eye-ring to form "spectacles."
POSTOCULAR. Another short stripe, also usually pale, which extends back from the eye; in some birds this is reduced to a postocular spot.
AURICULAR PATCH. An area, typically of a dark color, on the bird's ear-coverts; typically it is oval or crescent-shaped.
WING-BARS. Very often a bird's wing-coverts are tipped with a paler color; these are its wing-bars. Note that there are two sets of covert feathers (and thus often two wing-bars), the lesser wing-coverts being the upper and shorter ones, the greater wing-coverts being the lower and longer ones. A few species show only one wing-bar (or one shows much more strongly)—in these cases it is almost always the greater wing-coverts which are marked.
WING-EDGING. The flight feathers of a bird are almost always dark. Particularly, its inner shorter feathers (the secondaries and tertiaries) are almost always edged with a paler color; most often this is very narrow, in which case it is not mentioned. If the edging is more strongly marked, or if it extends to the bird's longer outer flight feathers (its primaries), it is mentioned. Typically the wing-coverts do not show wing-edging (i.e., along the *edge,* rather than at the *tip,* of the covert feathers); when they do, this too is mentioned.
SPECULUM. Some small birds show a patch of pale color at the base of the primaries in a folded wing; this is termed a *speculum* (note that it is here used in a sense completely different from the "speculum" shown in the secondaries of many flying waterfowl).
MALAR. The area between the bird's cheeks and its throat is termed its *malar area.* This is elsewhere sometimes referred to as the *moustachial area.*
SUBMALAR. Often there is a streak just below the actual malar area; this we term a *submalar streak.*
PECTORAL PATCH. Many birds show an area of contrasting color on the sides of the chest or breast; this we term a *pectoral patch,* and if it extends all the way across the chest it becomes a *pectoral band.*
CRISSUM. We use this term to refer to the most posterior feathers on the underparts, posterior to the bird's vent and underlying its tail feathers. The term *under tail-coverts* is synonymous (but longer!).

A few additional points of clarification. A bird's bill consists of two mandibles; we term the upper one the *maxilla,* the lower one the *lower mandible* (unfortunately a shorter term for it seems not to exist). An *eye-ring* is very narrow and typically consists of only a ring of bright color on the skin surrounding the eye, but sometimes includes a narrow ring of feathers as well; a broader area of color around the eye is called the *ocular area.* Depending on the pattern of the bird, we sometimes use the term *chest* as distinct from *breast,* in which case chest means simply the upper part of what is conventionally called the breast; if not specified, then by breast we mean the entire area. Likewise we distinguish between *chin* and *throat*, the chin being the uppermost part of the throat, but not expressly referred to unless necessary. Finally, by *sides* we mean the sides of the breast, by *flanks* the sides of the belly.

Similar Species

Comparison is here made to other species with which the bird in question is most likely to be confused within its range. For the most part, discussion is here limited to morphological differences (eye, color, pattern, etc.), but at times reference is made to behavioral, vocal, or distributional cues.

Habitat and Behavior

Information on the species' general abundance, habitat preferences, behavior, and vocalizations (in approximately that order) is presented here. In the main, this has been derived from our own (usually RSR's) observations, though to a varying extent data from other sources have been summarized and incorporated, particularly for scarcer species. In cases where our own experience with the bird is limited or nonexistent, we have credited the observations or notes of others, either to an individual (in which case usually the data was conveyed directly to RSR) or to a published reference. References which have been used only once in this work are given a full citation at that point; those referred to more often are cited in the text merely by author and year, with their full citation being given in the bibliography.

It should be noted that the abundance terms used (*common, rare,* etc.) are only relative, especially as compared to similar birds: thus a "common" seedeater is really much more numerous, in absolute numbers, than a "common" hawk ever will be. Further, the abundance terms are tied to the frequency with which an *experienced* observer will record the species, particularly an individual familiar with neotropical bird vocalizations; those less experienced will record various species less often, on average.

A full explanation of the habitat terms used will be found below.

Range

The species' general distribution is encapsulated here, together with information on its seasonal presence in various areas (in the case of migratory species). Disjunct portions of a bird's range are marked by semicolons; where its range is known or presumed to be continuous, this is indicated by commas. Localities from which the species is known only from sightings are usually so indicated. The bird's extralimital range is also given, together with its normal elevation range (given in meters above sea level). Months of occurrence for migratory species are also included. The written range statement is designed to be interpreted in conjunction with the map of the species' range placed adjacent to the species account. An explanation as to how these data were compiled will be found below (see "The Maps").

Note

This final paragraph contains comments on taxonomy and nomenclature; for species for which these are more or less undisputed, this paragraph is omitted. Where our generic or species-level taxonomy differs from that of Meyer de Schauensee (1966, 1970) this is so indicated, together with a brief explanation for our new treatment. A full citation is given for species described since 1970; as noted previously, other works referred to only once are here cited fully, with the citations of those referred to more frequently being given in the bibliography. Numerous other works, in particular the 1983 AOU Check-list and the various volumes of the *Check-list of the Birds of the World* (Peters) and the *Catalogue of Birds of the Americas* (Hellmayr) are referred to repeatedly. However, given our own research and experience and the giant strides toward greater and more comprehensive understanding over the last ten to twenty years, we have *not* felt that we necessarily had to follow the lead of any particular source, and thus a few taxonomic and nomenclatural revisions are here introduced to the semipopular literature. We have attempted to be consistent in our treatment but at the same time to follow the courses suggested by the most recent research, and we well recognize that these sometimes conflict; that is, some groups recently dealt with have been split apart as full species (e.g., the *Diglossa carbonaria* complex), while others with similar biogeographic patterns have either not been dealt with by any recent researcher (e.g., the *Hemispingus melanotis* complex) or have been analyzed and show similar patterns but are at least for now retained as one species (e.g., the *Hemispingus superciliaris* complex). It's all very complicated, and as new data (particularly biochemical) become available species-level taxonomy will inevitably have to be altered. We hope we have been reasonably evenhanded in our treatments, and we keenly anticipate future developments.

The English names of South American birds present another contentious issue. Most such names have become more or less well established since they were introduced to the semipopular literature by Eisenmann and Meyer de Schauensee in the latter's *The Species of Birds of South America* (1966). Eisenmann, however, was the first to admit that many of "his" names were chosen somewhat arbitrarily, and that many reflected virtually no knowledge of the birds in life (and were merely descriptive, based on museum study skins, or translations of the Latin species name). Numerous names could now be "improved," but in the interests of stability it seems desirable not to do so unless there exists a very good reason. Thus, in the vast majority of cases we have employed Meyer de Schauensee's names or those used in the 1983 AOU Check-list (usually giving priority to the latter when there exists a difference, as there surprisingly often does). But anyone familiar with Meyer de Schauensee's books and the AOU Check-list will note that there are some "new" names in this volume, for we have not invariably shied away from innovation (as a few would urge us to do). For while we recognize that stability is desirable, we also feel that a name should at least be accurate and not fundamentally misleading, that it should ideally reflect relationship, and that if at all possible it should be short and memorable.

Some general precepts will perhaps better disclose where we stand on this issue than further discussion. We have at least considered alteration or change under the following conditions:

1. *Taxonomic revision.* This normally forces a change, regardless of whether a "split" or a "lump" is involved. Ideally in such situations the expanded species will have a name different from any of its subunits, but this is not always possible, and in these cases normally the name employed for the more widespread of the subunits is employed.

2. *The international issue.* We agree with those who propose that there be only one English name in worldwide use for the same species and that each name be used for only one species. (This seems simple enough but is often difficult in practice!) As there are relatively few South American oscine passerines which range widely on other continents, this issue has posed few problems in this volume. Two notable exceptions involve *Riparia riparia* (which we "internationalize" to Common Sand-Martin from the familiar, to Americans, Bank Swallow) and the *Turdus* thrush problem (which see).

3. *The group-name issue.* The number of species in certain groups of South American birds causes problems, not only because it has caused difficulties in creating adequate names, but because having so many different species being called the same thing causes confusion in people's minds—think how fundamentally different a Scarlet Tanager is from a Scarlet-and-white Tanager. We grant that it is too late to do much about this now (and we regret it) and thus opt not to advocate radical change—we in fact, for example, would prefer to call all the *Tangara* tanagers "callistes," as the aviculturists do, so as to immediately differentiate them from all other tanagers, but have not gone so far as to do so. However, we have here introduced a few new group names which in these cases we hope will clarify relationships; for example, both the *Coryphospingus* finches are called pileated-finches, rather than have one a Pileated Finch and the other a seemingly unrelated Red-crested Finch.

4. *Consistency.* Ideally all members of the same genus will have the same group name, but consistency is not always easy to achieve, nor in some cases does it seem desirable. Thus, for example, while most *Sicalis* are now well known as yellow-finches, it does not seem necessary to call the familiar Saffron Finch the Saffron Yellow-Finch.

5. *Name shortening.* We tend to favor short names where this is possible. Thus Peruvian Meadowlark seems adequate when all the members of its group have red breasts, Great-billed Seed-Finch seems clearer than Greater Large-billed, and so on.

6. *Other literature.* Where a change is deemed necessary for clarification or accuracy, we often have opted to employ an English name suggested elsewhere. Thus, for *Cyanoloxia,* which is not an indigo blue, we use W. H. Hudson's name adopted straight from the Latin,

and for the Lesser Red-breasted Meadowlark we use L. Short's briefer and more descriptive Pampas Meadowlark, etc.

7. *Geographical names.* Where we opt for a change in name, we have tended to favor geographical epithets where these are possible over yet another descriptive modifier (which we feel have been rather over-employed of late). Thus, for example, the Chestnut-mantled Oropendola, whose mantle is really not very chestnut (hardly differing in this regard from its relative the Black Oropendola), we call the Baudo Oropendola, which highlights its very restricted range; further, the Black-cheeked Mountain-Tanager becomes the Santa Marta Mountain-Tanager (it is found only on the Santa Martas, and all *Anisognathus* have black cheeks), and so on.

8. *Familiarity.* We have tended not to tamper with well-established names of relatively well-known birds, even when these could be "improved." Modifications thus are proposed mostly for species which are little known, in cases where we hope the long-term gain is worth the temporary minor inconvenience.

Having stated all this, it may appear that many English names have been changed, but a quick perusal of the text will reveal that this is not so. We have given the matter much thought, and while we recognize that it may be impossible to please everyone in such matters, we have tried to be reasonable. We would be remiss not to expressly acknowledge here our special debt of gratitude to Dr. Burt Monroe, now chairman of the AOU "Check-list Committee," who over the past several years has provided much insight on matters taxonomical and nomenclatural. He has helped to clarify our thinking on numerous points, but we hasten to add that responsibility for our final conclusions rests with us.

THE MAPS

As so much time and effort were expended on our distribution maps, a few comments on how they were prepared seems in order.

We were fortunate indeed that William L. Brown of Toronto, Canada (coauthor of *A Guide to the Birds of Colombia,* Hilty and Brown 1986) offered his services at an early point. Brown carefully compiled a point-by-point working distribution map for each species (and in the case of certain complex groups, each subspecies) based on the published literature and on other material sent to him by RSR. The maps were sent for review and comments to RSR, who then added numerous unpublished locality records (both specimens and reliable sightings), deleted localities which after careful research proved to be erroneous, assessed the migrant status (if any) of each species, and otherwise refined the plotted distributions based on his knowledge and experience in the field. The revised maps were then returned to Brown for review: Brown then prepared a "final" distribution map for each species (but these continued to be revised up until the last possible moment, based on personal fieldwork, recently published references, and personal communications to RSR and Brown; *no data, however, were incorporated after March 1987*).

A fine young Pennsylvania artist, Tracy Pedersen, had meanwhile been engaged to prepare four endpaper maps and a series of twelve "core maps," one depicting all of South America, the other eleven showing various portions of the continent. She then plotted each species' final distribution on whichever of the twelve maps permitted showing the greatest detail; thus, wide-ranging species are shown on the entire South American map, while a southeastern Brazil endemic is shown on a map depicting only that region. Note that distributions in Panama are *not* shown.

The area in which a species is resident is shown by *shading*, while areas in which it occurs only as a nonbreeding migrant or visitant are enclosed by a *solid black line*. Isolated records are shown by individual locality "dots." We used our judgment in deciding whether to show a species' range as continuous between two recorded locality points or as separated discrete populations; taken into special consideration were the degree of habitat continuity in the intervening area and the degree of morphological variation in the taxa found in the two areas (if any). Where there existed considerable uncertainty, we tended to be conservative, breaking the units into separate areas, sometimes with a query in between. Nonetheless, bear in mind that a species will of course be found within its shaded area *only in its appropriate habitat;* if that is locally distributed, then so will be the bird. Also, be aware that to some extent, especially in montane areas, the distributions shown are only approximations of reality; for instance, many birds restricted to the subtropical zone along the eastern slope of the Andes have much "narrower" distributions than could actually be shown.

Finally, a few comments on the area we consider to be part of "South America" are in order. Essentially we have followed the limits established by Meyer de Schauensee (1970: xii) with a few minor modifications. Thus, all the continental inshore islands *are* included (e.g., Trinidad and Tobago; various small islands off the northern coast of Venezuela; the Netherlands Antilles [Aruba, Bonaire, and Curaçao]; and Fernando de Noronha, off the northeastern coast of Brazil), but islands more properly considered as part of the West Indies (e.g., Grenada) are *not*. To the south, we have opted to include the Falkland Islands (or Islas Malvinas—in referring to them as the Falklands we are *not* making any political statement but merely recognizing that this book is being written in the English language), as their avifauna is really very similar to that of Patagonia and Tierra del Fuego. However, various other islands farther out in the South Atlantic (e.g., South Georgia) are not included except incidentally (e.g., some notes on the endemic South Georgia Pipit have been incorporated). Likewise, the Juan Fernández Islands far off the Chilean coast have not been included (except for incidental comments), nor have the Galápagos Islands, situated even farther off the Ecuadorian coast.

HABITATS

A FIRM understanding of the important South American bird habitats is essential, and these are briefly described here. There are three basic types of *natural* habitats to be considered, each with numerous subdivisions: lowland forest; open and semiopen lowland areas; and montane areas, which may be either forested or nonforested. Each has been degraded or altered by man's activities to varying degrees; these activities result in various secondary habitats, which are briefly referred to later. Bear in mind that under typical conditions none of the primary habitats are entirely discrete from each other—rather, what is usually found is a gradual shift from one to the other (often termed an *ecotone*). Finally, note that the various littoral and marine habitats are not discussed here, as they are of little or no importance to the oscine passerines; these habitats will be described in future volumes.

LOWLAND FORESTS

South America being a relatively humid continent (compared to, for example, Africa), forests dominate most of the lowlands and are found (or until recently were found) over vast expanses at tropical and subtropical latitudes. Trees of a multitude of species form a continuous canopy of varying heights above the ground; often the ground is so effectively shaded that the understory itself is relatively open. Occasional emergent trees, such as various *Bombax* spp., tower over the remainder of the canopy. Forests are not, however, immune to perturbation, for even under natural circumstances treefalls are frequent, resulting in much denser second growth as saplings compete for space in the canopy. There exists a gradation of forest types, depending on the annual rainfall an area receives.

Humid forest remains more or less fully leaved throughout the year (meaning leaves of various trees are dropped at varying times of the year, depending on the species, or are slowly dropped throughout the year); these areas never really dry out, for they receive two to three meters of annual rainfall and there is no pronounced dry season. In the popular conception, this is "rain forest." Vast sections of Amazonia east to the Guianas are carpeted by humid forest, and another major area is (or was) found in the coastal lowlands of eastern Brazil. This latter is geographically very isolated from Amazonia (despite showing some avifaunal similarities) and it supports numerous endemic species; as it also has been to a large extent destroyed by man's activities, the region has a high number of critically endangered birds, with several already apparently extinct. In general it is wettest in western Amazonia (this is where bird species diversity is greatest) and somewhat drier eastward (e.g., in lower Amazonian Brazil, where avian diversity is somewhat lower).

A number of forest bird species seem to be more or less restricted to the relatively stunted and less rich forests which grow on *white sand soils* (sometimes called *sandy-belt forest;* these apparently are remains of ancient ocean shorelines). And it should be emphasized that even on a microgeographic level the Amazonian lowland forests are not uniform. A particularly useful distinction can be made between *terra firme forest,* growing on uplands which are never flooded, and *várzea* and *swampy forests,* which are periodically flooded and thus have waterlogged soil; *transitional forest* is intermediate between them. *Várzea* forest is typically found along watercourses which flood seasonally and overspill their usual banks; swampy forest is found more in low-lying, poorly drained areas, often far from rivers or streams, with stands of *Mauritia* palms frequent in some regions. Because most or all Amazonian rivers flood seasonally or temporarily, the forest found on river islands is almost always *várzea* forest. In some areas (particularly in southeastern Peru and adjacent regions, but also locally elsewhere) are found dense thickets of a spiny bamboo (*Guadua* sp.); these thickets harbor a surprisingly distinctive set of often rare or little-known birds.

Where rainfall is especially high (at least three to four meters per year), *wet forest,* sometimes termed *pluvial forest,* is found; in South America this type of forest is found only in Pacific western Colombia (extending to some extent into Panama) and northwestern Ecuador, and it is very rich in birds, with a high number of endemics.

At the other end of the spectrum, *deciduous* or *dry forest* is found where rainfall is somewhat reduced and where there is a pronounced dry season. In part because it burns more readily, this type of forest has been extensively modified by man for agricultural purposes; it was once, for instance, extensive in northern Colombia and southwestern Ecuador. Extensive deciduous forest is found along the ecotone between the southern edge of humid Amazonian forests in Brazil and the *cerrado* to the south, but here too it is being rapidly cleared.

Farther south (i.e., at a higher latitude) and isolated by open habitats across central Brazil is *subtropical forest,* found in portions of interior southeastern Brazil (where it is contiguous with humid forests found farther north in eastern Brazil), eastern Paraguay, and northeastern Argentina. The term might better be semitropical forest, we admit, but the term *subtropical* seems now so well entrenched that we hesitate to tamper with it; note further that the same term (*subtropical forest*) is also used to refer to montane forests (e.g., on the Andes at moderate elevations; see below). Frosts are possible, though in most places they are not regular. These forests are similar in structure to those found in tropical latitudes but are less diverse, in part because of the cooler conditions. A "southern pine" (*Araucaria* sp.) is frequent in some places, especially in Brazil.

Even farther south, the subtropical forests become lower in stature and more fragmented by ever more extensive areas of grassland and other open habitats; this low woodland is called *monte* and is characteristic of Uruguay and adjacent parts of Brazil and Argentina. Here, and in other areas where forest terrain gradually merges into more open habitats, there often occur strips of forest (or woodland) along watercourses; this is called *gallery forest*. Floristically it is less diverse than more continuous forest, but it nonetheless provides vital cover for many birds which would otherwise not be found in the region. Uniform stands of *Mauritia* palms are found in some such low-lying areas, especially across interior Brazil.

OPEN AND SEMIOPEN LOWLAND HABITATS

Where rainfall is less heavy and soils more permeable, a variety of open habitats is found. Structurally, the basic difference between these and the various forest or woodland habitats is that, even if there are a few trees, these are relatively scattered or low in stature, such that a continuous canopy is not formed. Grass may or may not be found as a ground cover. These habitats may be designated roughly as follows, starting with the more "wooded."

MATORRAL. Low scrub or scrubby woodland found in the coastal lowlands and on lower mountain slopes in central Chile. The resemblance to the chaparral of California is striking, and the two regions enjoy similar climates, with rainy winters and hot, dry summers.

CHACO. An extensive area of semiarid to arid scrub and low woodland found from southern Bolivia to northern Argentina.

CAATINGA. Another area of semiarid to arid scrub and low woodland, but in this case with little or no grass cover, found in interior northeastern Brazil.

SAVANNA. A general term for natural grasslands, in some cases with scattered trees or groves of trees; some of these natural grasslands or savannas are maintained to a greater or lesser degree by fire. Note that over extensive areas man has created what in effect are artificial savannas by clearing forest and planting grasses (usually of species not native to the Americas) as forage for domestic animals. These man-created grasslands are generally not nearly as rich in birdlife as are natural savannas, though some species have been able to adapt to them. Natural savannas have been much modified by man, either by conversion to agriculture or through overgrazing by cattle and too-frequent fires; they are among the more critically imperiled habitats in South America. Several subdivisions are often given separate names; these are as follows:

Pampas. Open, grassy, low-lying plains found in eastern Argentina, Uruguay, and Rio Grande do Sul, Brazil. Now extensively converted to intensive agricultural use, though birdlife remains abundant, as only a few species have failed to adapt well.

Campos. Open, grassy, often rolling plains on the central plateau of interior Brazil. Extensive campos in a natural state are now extremely rare and fragmented. True campos, in which trees are nonexistent or very few, gradually blend into *cerrados*.

Cerrado. A distinctive savanna community found across wide areas of interior Brazil; grass cover is dense (and the grasses very tall, at least where these have not been affected by overgrazing or an excessive frequency of fires) and there is a scattering of low, gnarled trees. Where tree cover is denser, this is often called *cerradão;* the latter gradually blends into deciduous woodland and forest along the southern edge of Amazonia. The *cerrado* has also been extensively modified by man's activities, and some of its most characteristic species are now quite scarce and very localized.

Campina. A specialized term used for small areas of natural savanna, scrub, or *cerrado*-like habitat found scattered in Amazonia, primarily in lower and middle Amazonian Brazil. For the most part *campinas* are found where there is underlying sandy soil, which is relatively poor in nutrients; note that where the vegetation is somewhat better developed, these savannalike habitats merge into sandy-belt forest (referred to above under humid forest).

PATAGONIAN STEPPE. Southward in Argentina from the pampas is found this bleak open expanse of low scrub and grass, now much modified by heavy sheep grazing; this extends south to northern Tierra del Fuego and into adjacent southern Chile. Much the same habitat is found on the various subantarctic islands found offshore, notably on the Falklands.

DESERT AND DESERT SCRUB. True desert, where rainfall is extremely light, is limited in extent in South America, though the continent does boast the world's driest desert, the notorious Atacama of coastal northern Chile and southwestern Peru, where in places rainfall has never been recorded. Truly arid, desertlike conditions are found in only three limited areas in South America: along the Pacific coast of Chile and Peru (also inland to the upper Marañón River valley of northern Peru) and in pockets along the coast of southern Ecuador; in north-central Argentina along the southern periphery of the *chaco;* and in pockets along the Caribbean coast of northeastern Colombia and Venezuela (especially in the west). Each of these supports some endemic elements. Additionally, arid intermontane valleys are found in some places in the Andes; these have been created by a rain-shadow effect, in which nearby cooler ridges receive the most moisture, leaving little to fall in the warmer valleys below. These too may show some biogeographic affinity to more extensive nonforested regions, which may be situated a vast distance away. While absolute desert is virtually or entirely devoid of vegetation, this occurs only in extremely arid regions which receive little or no rain. Far more prevalent, and much more important for birds, are various desert scrub habitats in which a sparse growth of low trees, shrubs, and cactus (especially *Cereus* sp.) is found. The few riparian areas in desert regions support a somewhat lusher vegetation (usually much modified by man's activities), with the tree *Prosopis* sp. being especially important in southwestern Ecuador and northern Peru.

LLANOS AND PANTANAL. These two terms both refer to seasonally inundated grasslands, with associated marshes and ponds and nearby shrubbery. The *llanos* are found in interior Venezuela and northeastern Colombia, whereas the *pantanal* lies in the very extensive floodplain along the upper and middle course of the Paraná River, in southwestern Brazil, eastern Bolivia, Paraguay, and northern Argentina. Justifiably renowned for their fabulous concentrations of waterbirds, both areas also support an abundance of smaller birds.

RIPARIAN AREAS. A general term employed for the shrubby, often early-succession habitats found along rivers and other bodies of water. In particular, along the Amazon and its major tributaries is found a very distinctive community of birds restricted to a short series of discrete habitats which succeed each other temporally (especially on islands); these include *Gynerium* cane, several *Salix* willows, a small tree called *Tessaria,* and *Cecropia* trees. Note that, as employed here, this habitat does grade into *várzea* forest (riparian areas being more open, *várzea* more wooded).

MONTANE AREAS

Areas above about 800 to 1000 meters in elevation are considered, for our purposes, to be *montane.* Average temperature drops with increased elevation; thus, it is possible in montane areas to quickly change climatic zones, and likewise their associated bird faunas will change. A high degree of endemicity is often to be found, in part because of a montane area's frequent isolation from other montane areas (which would often have similar habitats) by a "sea" of intervening lowlands. The following are the more important habitats; bear in mind that because the Andes form by far the largest montane landmass in South America, to some extent the habitat terms are slanted toward them, with comments relating to the other montane areas included as needed.

FOOTHILLS. A relatively narrow zone along the base of the Andes, centered at about 800 to 1200 meters, the foothill or upper tropical zone is usually very humid, and at tropical latitudes it harbors a distinctive avifauna, particularly along its east (or Amazon-facing) slope from Colombia to Bolivia, also in western Colombia and Ecuador. As the terrain at these elevations is so often steep, and as the climate is so often super wet, for the most part this habitat remains relatively little modified.

SUBTROPICS. Basically a mid-elevation zone, the subtropical level is usually considered to lie between 1000 to 1200 meters and 2300 to 2500 meters, but these levels vary with latitude (bear in mind the potential confusion with the latitudinal subtropical forests found in southeastern South America; see above). In many areas the subtropical zone is covered with humid montane forest, which is often exceptionally luxuriant due to rich soils and optimal growing conditions. Larger trees are often heavily laden with epiphytic growth (orchids, bromeliads, ferns, mosses, etc.), and tree ferns (e.g., *Cyathea* spp.) are especially frequent. Typical subtropical forest is not found on the western slope of the Andes south of northern Peru (and there it is patchy and relatively not so humid) or on the eastern slope south of northern Bolivia (extending patchily south to northwestern Argentina). Good subtropical forest is usually difficult to reach: in most areas where there is reasonable access, the original habitat has been removed for various agricultural pursuits. Fortunately, at least for the present there still remain remote regions where substantial tracts of subtropical zone forest remain little modified; some areas are formally protected.

TEMPERATE ZONE. An upper-elevation, forested zone, the temperate zone is usually considered to lie between 2300 to 2500 meters and treeline, which, depending on exposure and precipitation, is found between 3100 and 3500 meters. Basically it is an upward extension of the subtropical zone; diversity is lower, markedly so at higher elevations. Stands of bamboo (especially *Chusquea* sp.) are frequent in many areas (perhaps mostly where there has been some disturbance), with a number of bird species being quite closely associated thereto. Typical temperate zone forest is found locally along the entire eastern slope of the Andes south to northern Argentina (at lower elevations southward), while on the western slope it extends to northern Peru. In La Paz and Cochabamba, Bolivia, the Amazon basin–facing slopes (which here are exceptionally steep) are locally called the *yungas;* this area actually extends down to incorporate subtropical zone forests as well.

CLOUD FOREST. Here employed, in a *restricted* sense, to refer to a montane forest which is frequently enveloped in clouds and hence has a particularly wet, mossy aspect. Often but not always such forest is found on ridges, with its elevation varying, depending on local climate, but always above 600 meters. A number of species appear to be essentially restricted to cloud forests, at whatever level. Note that some other books have used the term more loosely, considering virtually any montane forest as a cloud forest.

ELFIN FOREST. A stunted forest growing on an exposed ridge whose stature is reduced in size by poor ridge-top soils and/or frequent exposure to winds. Typically it is very mossy and virtually impossible to penetrate, with roots growing as a massive tangle some one to two meters above the actual ground; essentially it is a specialized type of cloud forest. It is in such elfin forests on ridges east of the Andes, especially in Peru, that many of the most recently discovered bird species have been discovered.

PÁRAMO. The wet, cold grassland found above timberline in the northern Andes south mainly to Ecuador occurs very locally into Peru, principally near timberline along the eastern slope. In areas which are not intensively grazed or regularly burned the grass can be quite tall. A distinctive tall composite plant, *Espeletia* sp., is characteristic. Habitats which are visually similar to *páramo* but which support totally different plant communities are found above timberline on the mountains (usually referred to as *tepuis*) of southern Venezuela and adjacent areas. The mountains of southeastern Brazil, as well as the Santa Marta Mountains of northern Colombia and the Perijá Mountains on the Colombia-Venezuela border, also support *páramo* vegetation above treeline.

PUNA. A relatively drier, above-timberline grassland which replaces the *páramo* southward; some areas of *puna*-like grassland are found in Ecuador, but it is especially characteristic of the high elevation plain (altiplano) found in the Andes of central and southern Peru, western

Bolivia, northwestern Argentina, and northern Chile. Shrubby areas are found in places, usually where the *puna* is more protected. Essentially this zone is continuous in Argentina with the Patagonian steppe region (see above) and numerous birds have dispersed north within it.

MONTANE SCRUB. A rather variable habitat, but one that is found mainly or entirely on arid Andean slopes at various elevations (but principally high). Cactus (*Cereus* and *Opuntia* spp.) are frequent in some areas, but usually the dominant vegetation is low trees and shrubby growth.

POLYLEPIS. A very distinctive genus of tree or low shrub which grows in monospecific (or nearly so) groves above timberline from Ecuador south to western Bolivia and northern Chile, usually isolated from other forest by *páramo* or *puna* grassland. *Polylepis* is easily recognized by its gnarled form and flaky reddish bark. Several other shrubs, notably *Gynoxys* spp., are often found associated with the trees. The habitat supports a small but characteristic and specialized avifauna, with some species being rarely or never found away from it. The habitat itself is much imperiled by cutting for firewood; there is evidence that in the past its distribution was far more extensive than at present.

NOTHOFAGUS (SOUTHERN BEECH) FOREST. A magnificent tall forest dominated by several species of *Nothofagus* trees is found on the lower slopes of the Andes from southern Chile and Argentina south to Tierra del Fuego. Totally isolated from other forests, it harbors a depauperate but highly distinctive avifauna; there are relatively few oscine passerines.

SECONDARY AND HUMAN-INFLUENCED HABITATS

Most of these terms should be self-explanatory; a garden is a garden the world around. Worth noting is our use of the term *plantation* to refer to an agricultural area where the crop (typically cacao or coffee) is shaded by tall trees; these are often surprisingly rich in birdlife. On the other hand, we have *not* employed the term *capoeira* (referred to by Meyer de Schauensee 1970: xiii), as that term does not seem to be used outside of Brazil. Further, the simple fact that secondary succession does normally occur is vital; it is the birds which can adapt to the mosaic of forest and woodland edge, agricultural fields, and regenerating shrubby areas which will in the long run have the greatest chance of survival over wide areas (i.e., outside of protected areas). These are the birds which inevitably will become the most familiar to the greatest number of people, and it is worth emphasizing that given a little encouragement, a quite large range of species can adapt to man-altered environments, though they will fail to persist in areas in which any vestige of natural habitat has been essentially obliterated.

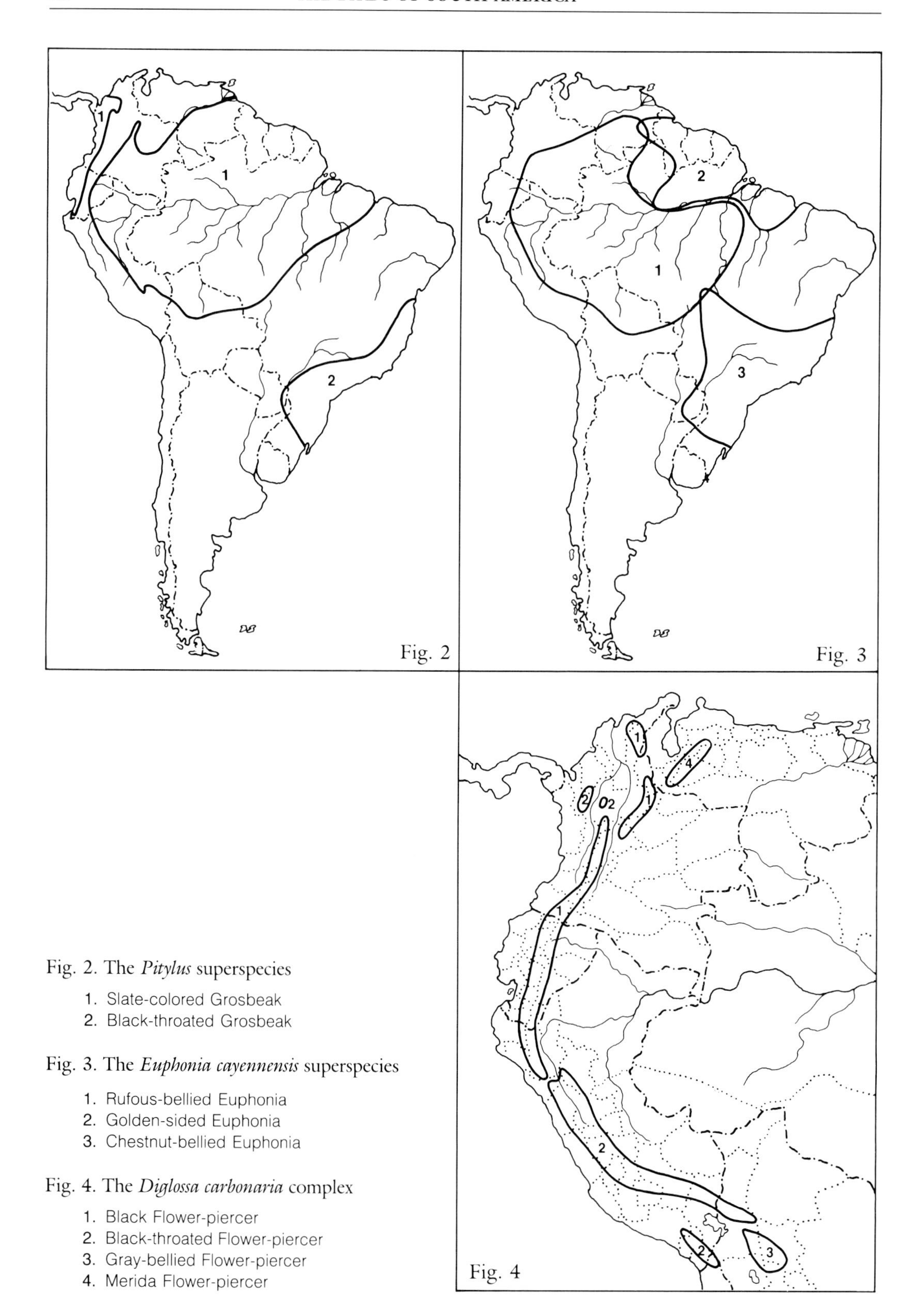

Fig. 2. The *Pitylus* superspecies

1. Slate-colored Grosbeak
2. Black-throated Grosbeak

Fig. 3. The *Euphonia cayennensis* superspecies

1. Rufous-bellied Euphonia
2. Golden-sided Euphonia
3. Chestnut-bellied Euphonia

Fig. 4. The *Diglossa carbonaria* complex

1. Black Flower-piercer
2. Black-throated Flower-piercer
3. Gray-bellied Flower-piercer
4. Merida Flower-piercer

BIOGEOGRAPHY

CENTRAL to an understanding of South American bird distribution is some familiarity with the "endemic center" or "forest refugia" concept as it applies to the present ranges of a large percentage of South America's birds. Further, the concept provides one of the primary explanations for the extraordinary diversity of the South American avifauna. While a detailed analysis is clearly beyond the scope of the present work, it is hoped that the following overview will provide some basis for understanding the concept and its implications. We should emphasize here our debt to J. Haffer, who was the first to formulate this hypothesis as it relates to birds (Haffer 1974); Cracraft (1985) further developed the idea, stressing centers of endemism rather than forest refugia per se.

Why is one species found in one area, while another species, obviously closely related, is found in similar habitat in another area? Explaining such patterns is difficult, but one approach lies in examining the historical record. Based on evidence from pollen deposition analysis, it is evident that the climates of various parts of South America have not been stable through recent geologic time. In particular, there have been relatively arid and relatively humid periods, largely correlated (at least over the last million years or so) with glacial advances and retreats in both polar regions; a glacial advance would tie up much available water and result in a comparatively dry tropical climate, with conditions gradually becoming more humid during glacial retreats. The implications for various habitats were profound indeed: though difficult to actually track, it seems safe to presume that with increasing aridity a once-continuous forest cover would become progressively drier and might actually become restricted to certain regions which were comparatively wet, due perhaps to orthographic rainfall (proximity to mountains) or some other cause. Savanna or some other semi-open habitat would replace the forest in intervening areas. Conversely, as global conditions became warmer and wetter, forests would slowly advance at the expense of the savannas, until ultimately they were once again more or less continuous. We are, it should be noted, presently in an interglacial period, with a relatively mild and damp global climate as compared with conditions which were prevalent at various times in the recent past.

Birds (and of course all other animals) are fundamentally tied to, and dependent on, the habitats they live in. A species adapted to forest cannot live in a grassland, and vice versa. The expansion and contraction of forest cover through the Pleistocene would thus have had massive influences on birds' ranges: these too would have shifted back and forth, expanding into newly available habitat, gradually being extirpated from unsuitable areas. From a biogeographic standpoint, the most important situation which might develop concerns the potential fragmentation of formerly continuous habitat into discrete units. A species, once continuously distributed but now isolated into these "refuges" of suitable habitat, might then begin to diverge genetically (as no gene interchange would be possible). In time, this might proceed to the point where interbreeding would become unlikely or impossible. The opportunity for this to be tested might occur subsequent to a return to continuous habitat, with the forms once again in contact. At that point, two (or more) species might have been created out of what formerly was but one. These would then be termed *semispecies* (or *allospecies*), the units of a *superspecies,* whose members are more closely related to each other than they are to any other in their genus and whose ranges generally are *parapatric* or *allopatric* (i.e., they replace each other geographically, never or only very locally occurring together). This pattern has repeated itself over and over again in South America more (because of topography) than in any other continent and provides one major reason for why there are so many birds in South America. In some cases not enough differentiation would have developed for two forms to achieve reproductive separation when they once again came into contact; if they were morphologically somewhat distinct, these would then be called *subspecies.* And while the concept was originally applied to birds found in humid forest (and still is usually thought of in such terms), it applies equally well to birds found in semiopen, nonforest habitats (whose ranges were expanding and contracting at the same time as those of the forests). Birds found in montane areas have also been influenced by these climatological events, with more montane-adapted birds enjoying wider ranges during cooler periods and having their ranges fragmented during warmer ones.

Three examples should serve to illustrate the superspecies concept. Figure 2 shows the ranges of the Slate-colored and Black-throated Grosbeaks (*Pitylus grossus* and *P. fuliginosus*): the northern species is widespread and little differentiated, while the quite distinct southeastern semispecies is being isolated from the genus' main range by the unsuitable dry open *cerrado* and *caatinga* regions. Presumably at one time forest cover between the two regions was continuous. Figure 3 depicts a similar situation, but in this case two Amazonian forms have differentiated to full species status (Rufous-bellied and Golden-sided Euphonias, *Euphonia rufiventris* and *E. cayennensis*), in addition to having a third species (Chestnut-bellied Euphonia, *E. pectoralis*) isolated in southeastern South America. Finally, Figure 4 shows an Andean superspecies, the *Diglossa carbonaria* complex, with its four component forms arrayed essentially north to south.

Numerous South American bird species are not widely distributed but rather are restricted to a certain region or habitat. To a large extent their ranges have been defined by the historical factors described above, of course. We speak of these species with relatively restricted distributions as being endemic to such and such an area, and to a large extent these ranges are not randomly distributed. Rather, a group of species tends to have ranges which are more or less congruent with each other, and such an area can be termed an *endemic center*. To a large extent these endemic centers occupy the same areas as Haffer's (1974) postulated "forest refugia," at least insofar as they relate to lowland forest (and not open or montane habitats).

Figures 5 and 6 depict the major lowland and montane endemic centers, respectively. A brief description of each follows; bear in mind that these do not necessarily refer to forested areas.

Lowland Endemic Centers

CHOCÓ. An area of very wet lowland forest (with the lower western slopes of the Western Andes also often included) extending from western Colombia into western Ecuador, with some species reaching eastern Panama; it is exceptionally rich, with numerous very distinct species and several endemic genera.

NECHÍ/MAGDALENA/MARACAIBO. An area of humid lowland forest extending locally across northern Colombia to northwestern Venezuela in the Maracaibo basin; it shows numerous affinities with Middle America.

GUAJIRA. An arid zone found principally near the Caribbean coast in northeastern Colombia and northern Venezuela.

LLANOS. A savanna-dominated area with much standing water in the rainy season and extensive gallery forests found in interior Venezuela and northeastern Colombia.

TUMBESIAN. An arid zone with desert vegetation and some deciduous woodland in the Pacific lowlands of southwestern Ecuador and northwestern Peru.

MARAÑÓN. A small arid zone in the upper Marañón River valley of northern Peru; Andean slopes above are also often included.

PERUVIAN COASTAL. Another arid zone, found in the Pacific lowlands of western Peru and northern Chile; most of the birds found here actually occur in riparian areas of river valleys.

NAPO (OR UPPER AMAZONIAN). A very important, largely forested region in the upper Amazon basin, with numerous endemics and the richest overall diversity of any area in South America (or the world); diversity becomes progressively lower as one proceeds eastward through the other Amazon basin centers (following).

MADEIRA. Another important forested area in upper Amazonia, lying to the east of the previous area but showing numerous avifaunal similarities.

TAPAJÓS. Yet another forested area in Amazonia, situated south of the Amazon and to the east of the Madeira.

PARÁ (BELÉM). The area near the mouth of the Amazon, on its south side; as with the three previous areas, it is dominated by forest and shows certain avifaunal similarities with the other Amazonian centers, though with numerous endemics as well.

GUIANAN. Numerous species are confined to the area roughly defined by the lower Río Negro to the west and the Amazon to the south; some range up on the lower slopes of the ancient mountains of the

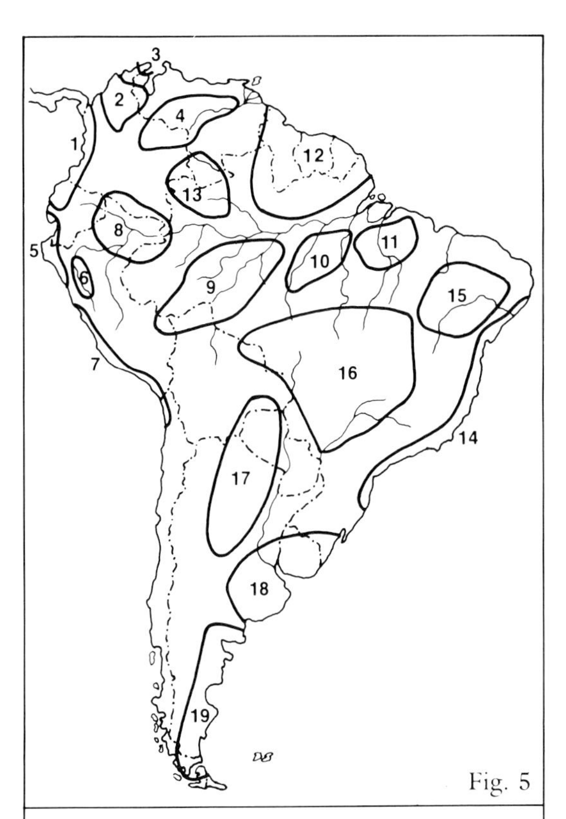

Fig. 5. Lowland endemic centers

1. Chocó
2. Nechí/Magdalena/Maracaibo
3. Guajira
4. Llanos
5. Tumbesian
6. Marañón
7. Peruvian Coastal
8. Napo
9. Madeira
10. Tapajós
11. Pará (Belém)
12. Guianan
13. Upper Rio Negro
14. Brazilian Coastal
15. Caatinga
16. Cerrado
17. Chaco
18. Pampas
19. Patagonian Steppe

Fig. 6. Montane endemic centers

1. Santa Marta
2. Perijá
3. Venezuelan Coastal
4. Paria
5. Mérida Andean
6. North Andean
7. Peruvian Andean
8. Argentinian Andean
9. Austral
10. Tepui
11. Serra do Mar

Guianan shield, and a few have managed to gain a foothold on the south bank of the Amazon.

UPPER RIO NEGRO. The area drained by this important Amazon tributary is dominated by infertile white-sand soils. Its forests are floristically impoverished, while the rivers flow with relatively clear, dark "black water."

BRAZILIAN COASTAL. An extremely important lowland forest zone stretching along the eastern coast of Brazil, with numerous species also extending west into eastern Paraguay and northeastern Argentina; these latter species are sometimes considered to more properly belong to another (but closely related) center, the Paraná center. Note that we here expressly exclude a more montane element found in this area, its endemic species being considered to form the Serra do Mar center (see below). Perhaps no other region in South America has as large a number of endemic birds; a connection with the Amazonian fauna is, nonetheless, indicated (at least in the distant past), with many species being found in both but not in intervening, more open areas.

CAATINGA. A very arid zone of dry scrub and woodland in interior northeastern Brazil.

CERRADO. An open area of savanna with at most widely scattered trees (though there is some gallery forest) found on the central plateau of interior Brazil.

CHACO. Dense thorny scrub and woodland zone of interior south-central South America; surprisingly few endemics, at least at the species level.

PAMPAS. Low-lying grassland with scattered patches of *monte* woodland in eastern Argentina, Uruguay, and extreme southeastern Brazil.

PATAGONIAN STEPPE. Wind-swept and cold scrubland found across vast stretches of southern Argentina and adjacent Chile.

Montane Endemic Centers

SANTA MARTA. An isolated but very tall mountain in northern Colombia.

PERIJÁ. Chain of mountains straddling the Colombia/Venezuela border between the Santa Martas and the Andes; most differentiation has occurred only at the subspecies level.

PARIA. Rather isolated range of mountains found in northeastern Venezuela, the northeasternmost extension of the "Andes"; Trinidad lies just offshore and forms the ultimate outlier.

VENEZUELAN COASTAL. Range of mountains of moderate height extending along most of the northern coast of Venezuela.

MÉRIDA ANDEAN. The northernmost part of the "true" Andes, situated in western Venezuela, separated from the northern mountains by lowlands in Lara and from the Andes to the south by the Táchira Depression.

NORTH ANDEAN. Encompasses essentially the Andes of Colombia and Ecuador, extending to extreme northern Peru north of the Marañón Depression; subcenters in the Western Andes and the Eastern Andes can be discerned.

PERUVIAN ANDEAN. Basically encompasses the Andes of Peru (north to the Marañón Depression) and adjacent western Bolivia. Important avifaunal differences exist between their (more arid) west and (more humid) east slopes.

ARGENTINIAN ANDEAN. The eastern slopes of the Andes in southern Bolivia and northwestern Argentina form this zone; forests are relatively limited in extent.

ALTIPLANO. Here considered to be the center encompassing the dry high altiplano zone of the Andes from central Peru to northern Chile and northwestern Argentina, largely covered by grasslands or sparse shrubbery.

AUSTRAL. The humid-forested center of the far southern Andes, extending south all the way to Tierra del Fuego.

TEPUI. A region of isolated, often flat-topped mountains found in southern Venezuela and adjacent areas, supporting a very distinct avifauna.

SERRA DO MAR. The montane equivalent of the lowland Brazilian coastal center and almost equally as distinct; to some extent, of course, the two blend into each other, but each has its own unique components.

Finally, note that we sometimes use the terms "trans-Andean" and "cis-Andean" when describing bird distributions; "trans-Andean" refers to birds ranging in the lowlands *west* of the Andes, "cis-Andean" to birds found *east* of them. Often they are members of a superspecies pair. The terms seem to have been introduced to the ornithological literature by Haffer (1974).

MIGRATION

LONG-DISTANCE migration in South American birds has been an under-recognized phenomenon until very recently. Particularly unappreciated was the extent to which a number of species breeding in southern South America vacate their nesting grounds during the austral winter and move north into Amazonia or even beyond. Far more generally known is the extent to which North American breeding birds move south during the northern winter, though in fact a majority of the species and individuals involved do not actually reach South America—for most species Middle America (and to a lesser extent the West Indies) is a more important wintering ground.

The situation is very complex, with each migratory species' range and habitat requirements being essentially unique to it, such that generalizations are difficult and certainly beyond the scope of this work. However, we felt it might at least be useful to list the oscine passerines which are migratory to or within South America. Table 1 lists the austral migrants now known to be long-distance migrants, at least to some extent; note that this list is surely conservative (due to inadequate data) and that other species will almost certainly be shown to be migratory (among them various *Sporophila* seedeaters and certain other Emberizine finches). Table 2 tabulates the northern migrants, many of which, as noted above, occur in South America only marginally; species marked with an asterisk pass the northern winter mainly (or entirely) here.

TABLE 1.
Austral Migrants

Brown-chested Martin	Lined Seedeater
Gray-breasted Martin	Double-collared Seedeater
Southern Martin	Black-bellied Seedeater
White-rumped Swallow	Chestnut Seedeater
Chilean Swallow	Marsh Seedeater
Blue-and-white Swallow	Carbonated Sierra-Finch
Creamy-bellied Thrush	Gray-hooded Sierra-Finch
Andean Slaty-Thrush	Patagonian Sierra-Finch
Eastern Slaty-Thrush	Common Diuca-Finch
Patagonian Mockingbird	Cinnamon Warbling-Finch
White-banded Mockingbird	Grassland Yellow-Finch
Correndera Pipit	Yellow-rumped Siskin
Red-eyed Vireo	

TABLE 2.
Northern Migrants

NOTE: This table excludes the casual migrants found in Appendix 1.

Purple Martin *
Caribbean Martin *
Tree Swallow
Sand-Martin *
Cliff Swallow *
Barn Swallow
Swainson's Thrush
Gray-cheeked Thrush *
Veery *
Red-eyed Vireo *
Yellow-green Vireo
Black-whiskered Vireo *
Yellow-throated Vireo
Golden-winged Warbler
Tennessee Warbler
Yellow Warbler
Blackpoll Warbler *
Bay-breasted Warbler
Cerulean Warbler *
Chestnut-sided Warbler
Blackburnian Warbler *
Cape May Warbler
Black-throated Green Warbler
Black-and-white Warbler
American Redstart
Black-throated Blue Warbler
Prothonotary Warbler
Northern Waterthrush
Louisiana Waterthrush
Canada Warbler *
Hooded Warbler
Mourning Warbler
Connecticut Warbler *
Kentucky Warbler
Common Yellowthroat
Scarlet Tanager *
Summer Tanager
Bobolink *
Northern Oriole
Orchard Oriole
Rose-breasted Grosbeak

CONSERVATION

WE count ourselves as fortunate, for we live at perhaps the ideal moment to see birds. Quite apart from optical improvements and so on, prior to the 1960s and 1970s access to most parts of South America was so difficult that most of its birds could not be seen without a tremendous amount of effort. Now, a transportation infrastructure has built up to the point that virtually every South American bird can be found within at least a reasonable striking distance of a road or hotel. But improved access has come about at a substantial price, that of deteriorating, declining habitat. We ourselves, in the past twenty years, can bear testimony to the fact that many areas we have known have greatly deteriorated from a wildlife perspective and that numerous birds have declined drastically during the same period.

Essentially we view this book as an opportunity to help prevent the situation from deteriorating further. Perhaps we are being naively over-optimistic, but it is our hope and belief that by presenting some of what we know about South America's birds and by documenting the overwhelming abundance of the South American avifauna, a vastly increased number of people are going to begin to really care what happens to it; people care about what they can name and know something about. Furthermore, birds are perhaps the most highly visible indicators of the healthiness of an environment. Engaging, often colorful or songful, they have a unique appeal. And once awakened to the appeal of birds, people will start to want to have lots of birds around them and will want to protect rarer ones in more remote places, too.

We do not need to return to the pre–Spanish Conquest era. It is our firm belief that sound, sustainable development and rational use of resources can take place which will not thwart efforts toward species protection. Probably no species included in this volume requires large areas of inviolate, undisturbed habitat; in fact, rather few birds, even among the large nonpasserines, do. A mosaic of natural, more or less undisturbed areas interspersed with areas under cultivation will suffice for all but the rarest and most "sensitive" of birds. What cannot, however, be allowed to proceed for any longer is a continuation of the massive and total destruction of vast areas of natural habitat which has in recent decades taken place. Development of this sort creates what in effect are "biological deserts," and regions dominated by such areas are, sadly, becoming all too frequent. What is perhaps most discouraging about such large areas where natural vegetation has been removed is that even after they are abandoned (as all too often most soon are) they only very slowly, if ever, revert to their natural state, remaining merely static, overrun with weeds, a near-perpetual colossal mistake.

Our fervent hope is that gradually the shortsightedness of this approach will come to be more generally recognized and that more rational, sustainable development will come to take precedence. This, in conjunction with the reservation of adequate samples of all the habitats in a region, should suffice to protect some populations of almost all bird species adequately. (One must, of course, add the caveat that the ultimate cause of most environmental stress, human population increase, will also have to be dealt with and resolved, ideally soon.)

Despite the preceding relatively optimistic comments, there are a number of oscine passerine birds in South America which do appear to have declined substantially and whose status must be regarded as a cause of some concern. Their difficulties have in most cases been noted under their respective species accounts. Nonetheless, in addition it would seem appropriate to here list all the species which we believe may be at some risk, together with some comments on how their vulnerable situations developed. Most are birds with relatively small ranges whose habitats have been substantially affected by man's activities, but we have *not* included a species simply because its range is small if the habitats there are not known to be at risk. Thus, none of the "Santa Marta Mountain endemics" are listed; though each species is restricted to the slopes of that one mountain, habitats found there are largely intact, and in addition much now has formal protected status. On the other hand, certain species which have relatively large ranges but which are rare have been included, even if no reason for their rarity seems evident; thus, two Amazonian birds with extensive distributions but few actual records, White-bellied Dacnis and Dotted Tanager, are listed. While admittedly a somewhat personal view, we hope that this summary will prove useful—if it helps to prevent the further slide of even one species toward extinction, its inclusion will have been worthwhile.

BEAUTIFUL JAY, *Cyanolyca pulchra*
Never numerous in its limited range (lower elevations on the western slope of the Andes in southwestern Colombia and northwestern Ecuador), this jay seems to have declined for inexplicable reasons—this despite minimal recent modification of its forest habitat.

AZURE JAY, *Cyanocorax caeruleus*
For reasons which remain obscure, this jay has largely if not entirely disappeared from the better part of its range; it remains numerous only in Rio Grande do Sul, Brazil.

NICEFORO'S WREN, *Thryothorus nicefori*
Still known from only a few specimens taken in the 1940s at the type locality in northeastern Colombia, this species (a close relative of the much more widespread Rufous-and-white Wren) is obviously local; it could be at risk if its habitat is being impacted by man's activities.

APOLINAR'S WREN, *Cistothorus apolinari*
Endemic to marsh vegetation in the Andes of northeastern Colombia, numbers of this wren have declined drastically due to drainage of much of its former habitat.

RUFOUS-THROATED DIPPER, *Cinclus schulzi*
This distinctive dipper has a limited range in northwestern Argentina and adjacent Bolivia, an area which has only a few streams suitable for dippers; many of these have been altered by man's activities, especially siltation and diversion or damming.

CHACO PIPIT, *Anthus chacoensis*
A very rare pipit, unknown in life, which has been recorded from a few localities in southeastern Paraguay and northern Argentina; certainly very local and may be at risk from habitat modification.

OCHRE-BREASTED PIPIT, *Anthus nattereri*
Rare and very local in southeastern South America, with few specimens and very few recent reports; unquestionably declining due to overgrazing or modification of natural grasslands.

GRAY-HEADED WARBLER, *Basileuterus griseiceps*
Endemic to the coastal mountains of northeastern Venezuela, this warbler is scarce in what little remains of its montane forest habitat; deforestation has surely had a severe impact on its overall numbers. Two other species which share its limited range, Paria Redstart (*Myioborus pariae*) and Venezuelan Flower-piercer (*Diglossa venezuelensis*), are probably also threatened by deforestation; neither, however, is a forest understory species (unlike the Gray-headed Warbler) and thus they likely are not at such grave risk.

GRAY-THROATED WARBLER, *Basileuterus cinereicollis*
Very poorly known, this warbler is known only from the lower Andean slopes of western Venezuela and northeastern Colombia, a zone that has suffered from extensive deforestation; fortunately it also occurs on the Perijá Mountains, which are more remote and undisturbed.

GIANT CONEBILL, *Oreomanes fraseri*
This very distinctive large conebill is confined to groves of *Polylepis* trees in the high Andes. Because of cutting for firewood, these groves become more restricted each year; hence, the conebill becomes ever scarcer and more localized.

TAMARUGO CONEBILL, *Conirostrum tamarugense*
Only recently described from northern Chile (and since found at one location in southern Peru), this species remains known from only a handful of localities; its distribution remains poorly understood, and its extremely limited oasis and scrub habitat could be threatened by clearance.

BLACK-LEGGED DACNIS, *Dacnis nigripes*
Endemic to the lowlands of southeastern Brazil, this species is rarely seen and is likely at risk because of deforestation over most of its range.

TURQUOISE DACNIS, *Dacnis hartlaubi*
This very rare dacnis has been recorded from a few lower montane localities in western Colombia; deforestation has been especially severe on these lower slopes (which are particularly desirable for agriculture).

SCARLET-BREASTED DACNIS, *Dacnis berlepschi*
This colorful dacnis has been found only in a limited area in southwestern Colombia and northwestern Ecuador; even here it appears to be rare, for unknown reasons, and thus it must be considered potentially threatened by the deforestation which is impacting on increasingly large portions of its range.

WHITE-BELLIED DACNIS, *Dacnis albiventris*
This dacnis seems inexplicably rare across all of its relatively large Amazonian range; though perhaps not actually at risk, the reasons for its rarity should be investigated.

CHERRY-THROATED TANAGER, *Nemosia rourei*
Perhaps already extinct, this tanager has been recorded only twice in the lowlands of southeastern Brazil, the more recent occurrence being as long ago as 1941; massive deforestation over its small range would appear to have been the cause.

BLACK-BACKED TANAGER, *Tangara peruviana*
This poorly known tanager of southeastern Brazil has been found principally on the lower slopes of the Serra do Mar; it is probably threatened by deforestation.

BLUE-WHISKERED TANAGER, *Tangara johannae*
Despite apparently not being a true "forest" tanager (rather, being found in second growth and edge), this species is surprisingly rare, local, and infrequently seen, especially in the Ecuadorian portion of its range (it is also found in southwestern Colombia); the reasons for its obvious scarcity remain unknown.

DOTTED TANAGER, *Tangara varia*
This rather dull *Tangara* tanager enjoys a fairly large range in northeastern South America, but it everywhere appears to be very rare; its forest habitat remains little modified over most of its range, and thus its evident rarity is a puzzle.

SEVEN-COLORED TANAGER, *Tangara fastuosa*
Endemic to the coastal forest of northeastern Brazil, this gaudy tanager has been much affected by widespread deforestation and its numbers depleted by bird trappers; it is now very rare and extremely localized in the few remaining pockets of extensive forest.

GREEN-CHINNED EUPHONIA, *Euphonia chalybea*
This distinctive euphonia is found in southeastern Brazil and adjacent Paraguay and Argentina, where it is now very local due to widespread clearance of its forest habitat. Even in areas where forest does still exist, this species seems to be decidedly scarce, so much so that its future may well be in jeopardy.

SLATY-BACKED HEMISPINGUS, *Hemispingus goeringi*
Apparently rare in montane forest in its limited range in the Andes of western Venezuela, this hemispingus has been found very infrequently in recent decades and is almost certainly threatened by deforestation. The sympatric *H. reyi*, also endemic to the Andes, is more numerous.

GOLD-RINGED TANAGER, *Bangsia aureocincta,* and BLACK-AND-GOLD TANAGER, *B. melanochlamys*

These two tanagers are both endemic to certain parts of the Western Andes of Colombia; always rare and local, they have gone almost unreported in recent decades, and it is believed they are both imperiled by forest destruction.

CONE-BILLED TANAGER, *Conothraupis mesoleuca*

Known from only one specimen, a male taken in the 1930s in central Mato Grosso, Brazil. Recent brief searches have failed to turn it up, but this species may always have been rare and even nomadic (like its congener in western Peru). As much of this part of Brazil remains only lightly modified by man's activities, we presume that it will eventually be relocated.

PAMPAS MEADOWLARK, *Sturnella defilippii*

Conversion of its natural pampas grasslands for more intensive agricultural pursuits has greatly reduced the numbers of this formerly common meadowlark, endemic to eastern Argentina (rarely in Uruguay and extreme southern Brazil); it is not yet endangered, but its future status will need to be monitored.

SAFFRON-COWLED BLACKBIRD, *Agelaius flavus*

Found in southeastern South America, this attractive blackbird's range seems to have undergone a general contraction in recent decades and it is now rare or absent from all parts of it except Rio Grande do Sul, Brazil. The reasons for its decline remain uncertain, but probably involve drainage and conversion of its meadow and marsh habitat.

FORBES' BLACKBIRD, *Curaeus forbesi*

This blackbird of eastern Brazil is known from very few specimens and its habitat requirements are unknown; though apparently rare, it is possible that to some extent it has been overlooked due to its similarity to other blackbirds.

RED-BELLIED GRACKLE, *Hypopyrrhus pyrohypogaster*

This spectacular grackle is endemic to Andean slopes in Colombia; as it requires montane forest in order to survive and as it is found at elevations (subtropical) which have been severely deforested, the species may be gravely threatened.

COLOMBIAN MOUNTAIN-GRACKLE, *Macroagelaius subalaris*

Closely related to another species found on the *tepuis,* this grackle is found only in montane Andean forests of northeastern Colombia; very little has been seen of it in recent decades, and it likely is as threatened as the Red-bellied Grackle for similar reasons.

BAUDO OROPENDOLA, *Psarocolius cassini*

Known from a mere handful of specimens, this oropendola is restricted to a very small area in northwestern Colombia; no one has found it in recent decades, and the species could be becoming scarce as a result of the extensive deforestation which is sweeping northwestern Colombia.

YELLOW CARDINAL, *Gubernatrix cristata*
This lovely cardinal of southeastern South America has declined greatly due to excessive trapping for the cage bird market. It is now extirpated or rare over much of its former range, despite abundant remaining suitable habitat.

BLUE FINCH, *Porphyrospiza caerulescens*
This small finch of Brazil's *cerrado* seems to be very scarce and local and is surely declining due to excessive burning, overgrazing, and outright destruction of much of its native habitat. Two other Emberizine finches found widely but now very locally across the Brazilian *cerrado* region are the Black-masked Finch (*Coryphaspiza melanotis*) and Coal-crested Finch (*Charitospiza eucosma*); both are being affected by the same factors.

HOODED SEEDEATER, *Sporophila melanops*
Known from only one specimen taken in the nineteenth century in central Brazil, this species remains of uncertain status; it may not be a valid taxon.

BUFFY-FRONTED SEEDEATER, *Sporophila frontalis*
Endemic to forests of southeastern Brazil and adjacent Paraguay and Argentina, this very distinctive seedeater is now much reduced in overall numbers due to a combination of deforestation and (in Brazil) depletion due to excessive trapping for the cage bird traffic. Temminck's Seedeater (*S. falcirostris*), with much the same range, is also now rare (and never seems to have been as numerous), though it was never in demand as a cage bird.

TUMACO SEEDEATER, *Sporophila insulata*
Never found anywhere other than on Tumaco Island, just off the coast of southwestern Colombia, this species has not been seen in recent years and may have been extirpated because of heavy settlement on the island; there remains the hope that it will be found on the adjacent mainland.

BLACK-AND-TAWNY SEEDEATER, *Sporophila nigrorufa*
A very rare seedeater recorded from only the upper Paraguay River basin in Mato Grosso, Brazil, and eastern Bolivia, this species has hardly ever been seen in life and is known from only a handful of specimens; no explanation for its rarity is apparent.

MARSH SEEDEATER, *Sporophila palustris*
Cage bird trafficking has apparently reduced numbers of this always scarce and localized seedeater of south-central South America. In fact, a number of *Sporophila* species found in this region (where seedeater diversity is very high) are now rare and very local, including Rufous-rumped (*S. hypochroma*), Black-bellied (*S. melanogaster*), Chestnut (*S. cinnamomea*), Dark-throated (*S. ruficollis*), and Narosky's (*S. zelichi*) Seedeaters; of these apparently the Marsh and Narosky's Seedeaters are the most threatened. Habitat modification may have also affected numbers.

OLIVE-HEADED BRUSH-FINCH, *Atlapetes flaviceps*
This brush-finch has been found in only a limited area on the eastern slope of Colombia's Central Andes, where it appears to be threatened by massive deforestation (despite not being a true forest bird).

PALE-HEADED BRUSH-FINCH, *Atlapetes pallidiceps*
With a minute range in one semiarid valley of southern Ecuador, this brush-finch is gravely imperiled by woodland destruction; recent brief searches have failed to find any at all, though small populations probably do yet exist.

TANAGER FINCH, *Oreothraupis arremonops*
This very distinct monotypic genus is found only in the Western Andes of western Colombia and northwestern Ecuador, where it is extremely scarce and very local (seemingly absent from numerous "suitable" areas). Considerable amounts of its forest habitat still remain; the reasons for its rarity remain unknown.

BLACK-THROATED FINCH, *Melanodera melanodera*
Until recently this species appears to have been quite numerous on the Fuegian grasslands to which it is restricted, as indeed it still is on the Falkland Islands; however, on the mainland Black-throated Finches have declined precipitously of late. The reasons for this decline remain uncertain, but we suspect that it may be correlated with severe overgrazing by sheep. The Yellow-bridled Finch (*M. xanthogramma*), with a similar range but mostly nesting in treeline scrub, seems also to be a very scarce bird, but as far as we are aware there is no evidence that its numbers have dropped.

CINEREOUS WARBLING-FINCH, *Poospiza cinerea*
Endemic to interior southern Brazil, for some reason this warbling-finch seems to have gone virtually unrecorded in recent years. While some of its *cerrado*/woodland habitat may have been converted for agriculture, this seems an inadequate explanation for its present rarity.

COCHABAMBA MOUNTAIN-FINCH, *Poospiza garleppi*
A rare bird found only in *Polylepis*-dominated groves of Cochabamba, Bolivia; never numerous, it may be quite threatened by severe overcutting for firewood. The Rufous-breasted Warbling-Finch (*P. rubecula*) of western Peru may be similarly threatened, but the Tucuman Mountain-Finch (*P. baeri*) of northwestern Argentina seems relatively safe (though it is far from numerous) because of that area's much lower population density.

GRASSHOPPER SPARROW, *Ammodramus savannarum*
Here at the southern end of their extensive range, Grasshopper Sparrows are now very rare and local in South America, known only from western Colombia and western Ecuador. Overgrazing and other modification of their always limited habitat are the cause.

SAFFRON SISKIN, *Carduelis siemeradzkii*
This siskin has an extremely limited range in southwestern Ecuador; apparently never very numerous, it now seems decidedly rare and is almost certainly declining as a result of severe and extensive woodland destruction.

RED SISKIN, *Carduelis cucullata*
The plight of this attractive and unusual siskin of northern Venezuela and adjacent Colombia is relatively well known: its numbers have been tremendously depleted by the activities of bird trappers, who sell the bird to canary breeders desirous of the siskin's infusion of red pigment. As a result the species has been rendered almost extinct in the wild. The Yellow-faced Siskin (*C. yarrellii*) of northeastern Brazil may also have been put at risk by the activities of bird trappers, though presumably not to as extreme a degree.

THE PLATES

FACING PAGE TERMS

Note that, under more complex genera, major Groups (A, B, C, etc.) have been employed to associate visually similar species. Additional subsets within the Groups have been broken out where deemed useful; these are preceded by • (see "Plates and Facing Pages," p. 4).

Numbers represent a full species; modifying letters (e.g., 1a, 2b, etc.) refer to different subspecies, an age stage, or some other variation. If the gender symbol (♂, ♀) is lacking, sexes are identical or nearly so in plumage. The subspecies illustrated here is specified; if none is, then the species is monotypic; i.e., it shows no racial variation.

Species *not illustrated* are listed in each Group under their most visually similar relatives; these are preceded by "*Also*." The species sequence *within* each Group may not correspond exactly with that of the main text, although the Group sequences do.

Species-level *field marks* are not usually given; rather, we describe its genus or group and expect the reader to refer to the main text for most identification criteria.

Ranges are given (in very encapsulated form) for species which are not widespread; note that these refer to the entire species and are not merely the range of the listed subspecies. The following abbreviations are employed (apart from "standard" ones such as "se." for southeastern, "cen." for central, etc.):

se. S. Am. = southeastern South America (i.e., se. Brazil and adjacent e. Paraguay and ne. Argentina)
Amaz. = Amazonian
Arg. = Argentina
Bol. = Bolivia
Col. = Colombia
Ecu. = Ecuador
Par. = Paraguay
Uru. = Uruguay
Venez. = Venezuela
spp. = refers to several species in a general way (e.g., *Chlorothraupis* spp.)
ssp. = subspecies
♂♂, ♀♀ = plural

Text page numbers are given for each genus and for major Groups within genera.

All birds on the same plate are drawn to *scale* unless divided by a solid line, which indicates that two different-size scales have been employed. Occasionally an inset is drawn to smaller scale; this is normally obvious but nonetheless is specified in a note at the bottom of that facing page.

PLATE 1: JAYS

Cyanolyca Jays — PAGE 40

Beautiful, sleek jays found in *montane Andean forests*. All 4 have a *black mask* (with 3 also showing a narrow pectoral collar of either black or white), not the black-bibbed effect of the *Cyanocorax* jays.

1. **BLACK-COLLARED JAY**, *C. armillata meridana* (Venez. to Ecu.)
2. **WHITE-COLLARED JAY**, *C. viridicyana cyanolaema* (Peru and Bol.)

Also: Turquoise Jay, *C. turcosa* (mainly Ecu.)
Beautiful Jay, *C. pulchra* (w. Col. and nw. Ecu.; lacks pectoral collar)

Cyanocorax Jays — PAGE 42

Somewhat heavier-bodied jays found in *lowlands* (only the unique Green Jay being montane). All have *black extending from sides of head over chest* (forming a "bib"), but otherwise they may be uniform or boldly patterned in shades of violet blue and white. Usually bold, noisy, and conspicuous. Only 1 species occurs in many regions (at most 2 are ever sympatric, usually 1 from each Group), so *note ranges carefully*.

A. *Iris dark;* no facial markings; all but 1 *without* white tail tips. — PAGE 42

- Body color blue to purplish mauve; *tail all dark.*

3. **PURPLISH JAY**, *C. cyanomelas* (mainly Bol. to Par. and n. Arg.)

Also: Azure Jay, *C. caeruleus* (mainly se. Brazil)
Violaceous Jay, *C. violaceus* (w. Amazonia to Venez.)

- *Belly and terminal tail white.*

4. **CURL-CRESTED JAY**, *C. cristatellus* (s.-cen. Brazil)

B. *Iris pale;* blue to white nape and facial markings, tail tipped white. — PAGE 45

- *Belly mauve.*

5. **AZURE-NAPED JAY**, *C. heilprini* (sw. Venez. and adjacent areas)

- *Belly creamy to white.*

6. **PLUSH-CRESTED JAY**, *C. c. chrysops* (s.-cen. S. Am.)
7. **WHITE-TAILED JAY**, *C. mystacalis* (sw. Ecu. and nw. Peru)

Also: White-naped Jay, *C. cyanopogon* (e. Brazil)
Black-chested Jay, *C. affinis* (n. Col. and nw. Venez.)
Cayenne Jay, *C. cayanus* (ne. S. Am.)

C. Distinctive mainly *green and yellow* coloration. — PAGE 48

8. **GREEN JAY**, *C. y. yncas* (Andes and n. Venez.)

1
2
3
1
4
5
6
7
8
TUDOR

PLATE 2: SWALLOWS & MARTINS

Phaeoprogne Martins PAGE 49
Brown-backed. Widespread in lowlands, but not as gregarious as *Progne.*

1. BROWN-CHESTED MARTIN, *P. tapera fusca*

Progne Martins *Large* swallows with forked tails. *Violet blue at least above,* with some showing *white or whitish below;* ♀♀ duskier. *Identification often tricky.* Typically most numerous around towns. All migratory to some extent. PAGE 50

2. GRAY-BREASTED MARTIN, *P. c. chalybea* (widespread in lowlands)
3. SOUTHERN MARTIN, *P. modesta elegans* (breeds s. S. Am. and coastal Peru)

Also: Purple Martin, *P. subis* (N. Am. breeder, wintering locally)
Caribbean Martin, *P. dominicensis* (breeds Tobago; winters S. Am.?)

Tachycineta Swallows Basically *bicolored,* glossy blue or blue-green above, *white below;* usually a *white rump.* Young birds brownish above. Found in open country, often near water; several species highly migratory. PAGE 53

4. WHITE-WINGED SWALLOW, *T. albiventer* (tropical lowlands)
5. WHITE-RUMPED SWALLOW, *T. leucorrhoa* (s.-cen. S. Am.)
5a. Adult; 5b. Immature

Also: Mangrove Swallow, *T. albilinea* (nw. Peru)
Chilean Swallow, *T. leucopyga* (s. S. Am.)
Tree Swallow, *T. bicolor* (rump *not* white; rare N. Am. migrant)

Notiochelidon Swallows Rather small; note *dark crissum,* somewhat forked tail. Mostly in Andes, Blue-and-white also in lowlands. PAGE 56

6. BROWN-BELLIED SWALLOW, *N. murina* (Venez. to Bol.)
7. BLUE-AND-WHITE SWALLOW, *N. c. cyanoleuca*

Also: Pale-footed Swallow, *N. flavipes* (locally from Col. to Bol.; pink-buff throat)

Atticora Swallows PAGE 58
Like *Notiochelidon,* but *tail long and deeply forked.* Found near water in lowlands.

8. WHITE-BANDED SWALLOW, *A. fasciata* (Amazonia to Guianas)
9. BLACK-COLLARED SWALLOW, *A. melanoleuca* (e. Amazonia locally to s. Brazil)

Neochelidon Swallows *Small, dark, forest-based* swallow; *white thighs* diagnostic but hard to see. Humid tropical lowlands, often scarce and local. PAGE 60

10. WHITE-THIGHED SWALLOW, *N. tibialis griseiventris*

Stelgidopteryx Swallows PAGE 60
Brown-backed with *cinnamon buff throat* (at least). Widespread in lowlands.

11. SOUTHERN ROUGH-WINGED SWALLOW
11a. *S. ruficollis uropygialis;* 11b. *S. r. ruficollis*
12. TAWNY-HEADED SWALLOW, *S. fucata* (s.-cen. S. Am., locally northward)

Riparia Sand-Martins Small; brown back and *chest band.* N. Am. migrant. PAGE 62
Not illustrated: Sand-Martin, *R. riparia*

Hirundo Swallows Variable in appearance (generic limits now broader). Three groups in S. Am.: Andean Swallow (*"Haplochelidon"*), a dull-plumaged swallow of high Andes; the Cliff/Cave Swallow group (*"Petrochelidon"*), with almost square tails and buff rumps; and familiar "true" *Hirundo* group, with characteristic tail streamers. PAGE 63

13. ANDEAN SWALLOW, *H. a. andecola* (Peru and Bol.)
14. CHESTNUT-COLLARED SWALLOW, *H. r. rufocollaris* (sw. Ecu. and nw. Peru)
15. BARN SWALLOW, *H. rustica erythrogaster,* nonbreeding plumage (widespread as N. Am. migrant)

Also: Cliff Swallow, *H. pyrrhonota* (N. Am. migrant)

NOTE: Flying birds (3, 5a, 9, 11a, 15) drawn to smaller scale.

2

1

2♂

3♂

5a

4

5b

6

8

7

9

10

12

15

13

11a

11b

G. TUDOR

14

PLATE 3: DONACOBIUS & LARGER WRENS

Donacobius PAGE 67

Unmistakable and conspicuous, *mostly dark brown and buff* with yellow eye and white in wings and tail (visible mainly in flight). *Grassy areas near water* in tropical lowlands.

1. BLACK-CAPPED DONACOBIUS, *D. a. atricapillus*

Campylorhynchus Wrens Notably large, conspicuous, and noisy wrens. Patterns also bold, most species *strongly barred, striped, or spotted.* Basically arboreal, some being found in arid regions, others at forest borders in more humid regions, mainly in nw. S. Am. Songs very loud and rhythmic, in most species harsh and guttural (Thrush-like more musical). PAGE 68

A. Two dissimilar, large species; *not* boldly banded above. PAGE 68

2. BICOLORED WREN, *C. griseus minor* (Col. and Venez.)
3. THRUSH-LIKE WREN
   3a. *C. turdinus hypostictus* (Amazonia)
   3b. *C. turdinus unicolor* (Bol. and sw. Brazil)

B. *Mantle boldly banded or striped* and *spotted below* (except divergent *C. albobrunneus*); the *C. zonatus* superspecies. All basically *allopatric.* PAGE 70

4. FASCIATED WREN, *C. fasciatus* (sw. Ecu. and nw. Peru)

Also: Stripe-backed Wren, *C. nuchalis* (n. Col. and Venez.)
Band-backed Wren, *C. zonatus* (n. Col. and nw. Ecu.)
White-headed Wren, *C. albobrunneus* (w. Col.; mostly *white,* mantle plain brown)

Thryothorus Wrens A large group of mid-size wrens, widespread across all but s. S. Am., though diversity greatest in north and west. All more or less brown or rufous above, whitish to buff below, sometimes with spotting; tail usually barred, in some also on wings. Most species difficult to see clearly in their typically dense undergrowth habitat; all far more often heard than seen, with characteristic, loud, usually rollicking and musical songs. PAGE 72

A. Three distinctive, *relatively boldly patterned* wrens of *nw. S. Am.* PAGE 72

5. BLACK-BELLIED WREN, *T. f. fasciatoventris*
6. BAY WREN, *T. nigricapillus schotti*

Also: Sooty-headed Wren, *T. spadix* (mainly chestnut, head blackish)

B. The *"moustached"* group; all with chestnut mantles, *unbarred wings* (breast may be spotted). PAGE 74

7. PLAIN-TAILED WREN, *T. e. euophrys* (Andes from s. Col. to n. Peru)
8. MOUSTACHED WREN, *T. g. genibarbis* (mainly Brazil)
9. CORAYA WREN, *T. c. coraya* (Amazonia)

Also: Inca Wren, *T. eisenmanni* (Andes of s. Peru; recalls 7)
Whiskered Wren, *T. mystacalis* (montane areas of Venez. to Ecu.; recalls 8)

C. Miscellaneous *small* wrens of *n. and w. S. Am.* PAGE 77

10. STRIPE-THROATED WREN, *T. leucopogon*
11. SPECKLE-BREASTED WREN, *T. sclateri paucimaculatus*
12. RUFOUS-BREASTED WREN, *T. r. rutilus*

D. *Relatively simply patterned; buff to whitish below,* with *barred wings.* PAGE 79

13. BUFF-BREASTED WREN, *T. leucotis bogotensis*
14. LONG-BILLED WREN, *T. longirostris bahiae* (e. Brazil)
15. SUPERCILIATED WREN, *T. superciliaris baroni* (sw. Ecu. and nw. Peru)

Also: Fawn-breasted Wren, *T. guarayanus* (n. Bol. and sw. Brazil)
Rufous-and-white Wren, *T. rufalbus* (ne. Col. and n. Venez.)
Niceforo's Wren, *T. nicefori* (local in n. Col.)

E. Distinctive; *small, short-tailed, gray.* PAGE 83

16. GRAY WREN, *T. griseus* (w. Amaz. Brazil)

1
2
3b
3a
4
5
6
7
8
10
9
11
12
13
16
15
14
G Tudor

PLATE 4: WRENS, GNATWRENS, & GNATCATCHERS

Odontorchilus Wrens — Pair of aberrant, arboreal wrens with very disjunct distributions. *Gnatcatcherlike in shape and overall coloration* (aside from barred tail). PAGE 83

1. GRAY-MANTLED WREN, *O. b. branickii* (Andes)

Also: Tooth-billed Wren, *O. cinereus* (Amaz. Brazil)

Troglodytes Wrens — Small, "standard" wrens, basically brown with few field marks other than superciliary. All except the widespread House Wren range in montane forest. PAGE 84

2. HOUSE WREN, *T. aedon musculus*
3. MOUNTAIN WREN, *T. solstitialis macrourus* (Andes)

Also: Santa Marta Wren, *T. monticola;* Tepui Wren, *T. rufulus*

Cistothorus Wrens — *Small and buffy,* with *streaked backs.* Local in *grasslands and marshy/boggy areas.* Unless singing, hard to see. PAGE 87

4. GRASS WREN 4a. *C. p. platensis;* 4b. *C. platensis aequatorialis*

Also: Apolinar's Wren, *C. apolinari* (Col. Andes)
Merida Wren, *C. meridae* (Venez. Andes)

Cinnycerthia Wrens — Fairly large, *uniform rufous or rufous-brown* wrens of *montane Andean forest,* usually in dense, low cover. Often in large groups. Adults may have white on face. PAGE 89

5. SEPIA-BROWN WREN, *C. p. peruana* (Col. to Bol.)
5a. Adult; 5b. Variant Adult

Also: Rufous Wren, *C. unirufa* (mainly Col. and Ecu.)

Henicorhina Wood-Wrens PAGE 91

Small, with *stubby,* cocked tails and *boldly streaked cheeks.* Mainly in forest tangles.

6. GRAY-BREASTED WOOD-WREN, *H. l. leucophrys* (montane)

Also: Bar-winged Wood-Wren, *H. leucoptera* (n. Peru mts.)
White-breasted Wood-Wren, *H. leucosticta* (lowlands)

Cyphorhinus Wrens — Chunky, short-tailed; bill heavy at base with ridged culmen. *Contrasting orange-rufous foreparts.* Small groups on forest floor. Note *very distinctive songs.* PAGE 93

7. MUSICIAN WREN (Amazonia to Guianas)
7a. *C. aradus modulator;* 7b. *C. aradus faroensis*

Also: Song Wren, *C. phaeocephalus* (west of Andes; *large* bare ocular area)
Chestnut-breasted Wren, *C. thoracicus* (Andes)

Microcerculus Wrens — Small, *stub-tailed* wrens of humid forest floor. Bills rather long; *faces plain.* Most immatures scaly below. Difficult to see, but have *lovely, far-carrying songs.* PAGE 96

8. SOUTHERN NIGHTINGALE-WREN, *M. m. marginatus* (mainly scaled below west of Andes)
9. WING-BANDED WREN, *M. bambla albigularis* immature (Guianas to Amazonia)

Also: Flutist Wren, *M. ustulatus* (*tepui* region; uniform rufous-brown)

Microbates Gnatwrens PAGE 99

Note rather long bill and short tail (*constantly flipping*). Forest lower growth.

10. COLLARED GNATWREN, *M. c. collaris* (Guianas to Amazonia)

Also: Tawny-faced Gnatwren, *M. cinereiventris* (pattern between 10 and 11; Col. to Peru)

Ramphocaenus Gnatwrens PAGE 100

Much like *Microbates,* but bill and tail both longer. Favors lighter woodland.

11. LONG-BILLED GNATWREN, *R. melanurus trinitatis* (widespread)

Polioptila Gnatcatchers — Small, slender, and long-tailed, with *gray predominating.* Much more arboreal than gnatwrens; easier to see. PAGE 101

A. ♂♂ with *glossy black cap.* PAGE 101

12. TROPICAL GNATCATCHER, *P. p. plumbea* (tropical lowlands)

Also: Creamy-bellied Gnatcatcher, *P. lactea* (se. S. Am.)

B. Lacks black cap (but may have *mask*). PAGE 103

13. GUIANAN GNATCATCHER, *P. g. guianensis* (ne. S. Am.)
14. MASKED GNATCATCHER, *P. d. dumicola* (s.-cen. S. Am.)

Also: Slate-throated Gnatcatcher, *P. schistaceigula* (nearest 13; w. Col. and w. Ecu.)

4

1
3
4a
2
4b
5a
6
5b
7a
8
9
7b
10
11
12♀
13♂
12♂
14♂
14♀

G TUDOR

PLATE 5: THRUSHES I

Myadestes Solitaires — PAGE 105

With relatively somber plumage, these solitaires are clad mainly in brown and gray; *silver in wing and tail* flashes in flight. Inconspicuous in montane forest; heard far more often than seen.

1. ANDEAN SOLITAIRE, *M. ralloides plumbeiceps* (Venez. to Bol.)

Also: Varied Solitaire, *M. coloratus* (extreme nw. Col.)

Cichlopsis Solitaires — PAGE 106

An essentially *uniform brown* solitaire, formerly placed in *Myadestes*. Found *very* locally inside humid lower montane forest.

2. RUFOUS-BROWN SOLITAIRE, *C. leucogenys gularis*

Entomodestes Solitaires — PAGE 107

Striking, large, *mainly black* solitaires with *white cheeks;* white in wings and tail conspicuous in flight. Local in Andean forests; note weird, ringing songs.

3. WHITE-EARED SOLITAIRE, *E. leucotis* (Peru and Bol.)

Also: Black Solitaire, *E. coracinus* (w. Col. and nw. Ecu.)

Catharus Nightingale-Thrushes (residents) and Thrushes (migrants) — PAGE 108

Fairly small thrushes, generally with rather subdued coloration, though the 3 resident species have *bill, eye-ring, and legs contrastingly orange*. Residents are primarily Andean; the 3 n. migrants overwinter widely across the n. two-thirds of S. Am. All tend to remain hidden inside forest and woodland; lovely songs of the 3 residents are often heard.

A. *Resident* species; *bill and legs orange*. — PAGE 109

• *Dark gray or olive above,* head blackish.

4. SPOTTED NIGHTINGALE-THRUSH, *C. dryas maculatus* (w. Venez. to nw. Arg.)

Also: Slaty-backed Nightingale-Thrush, *C. fuscater* (w. Venez. to Bol.; unspotted, with *pale* iris)

• Mantle brown.

5. ORANGE-BILLED NIGHTINGALE-THRUSH, *C. aurantiirostris birchelli* (Col. and Venez.)

B. N. *migrants;* bill and legs dull, underparts spotted. — PAGE 110

6. SWAINSON'S THRUSH, *C. ustulatus swainsoni*

Also: Gray-cheeked Thrush, *C. minimus;* Veery, *C. fuscescens*

Platycichla Thrushes — PAGE 112

Similar to *Turdus*. Strongly sexually dimorphic, dingy ♀♀ being *very* like certain *Turdus* thrushes. Both species arboreal in montane forest.

7. YELLOW-LEGGED THRUSH, *P. f. flavipes* (locally in n. and e. S. Am.)

Also: Pale-eyed Thrush, *P. leucops* (Andes south to Bol. and *tepuis;* ♂ black, *iris white*)

Turdus Thrushes — PAGE 114

A well-known group of thrushes of worldwide distribution, well represented in S. Am. They occupy a variety of wooded or semiopen habitats. Some of the forest species are very shy; this combined with their often dull plumage (especially in certain ♀♀) makes them difficult to identify.

We have divided the genus into 3 basic groups, and on this plate present the first, Group A, a set of species which are either relatively large and uniformly colored or have black on head. *Most of these are found in the Andes.* For the second 2 groups, see Plate 6.

A. *Large and uniform,* or *strongly patterned with black on head* (♂); ♀, when different, is brown with *dingy crissum* (as in ♀ *Platycichla,* but *unlike* Group B). — PAGE 114

• *Large to very large; ashy brown to black* with *yellow/orange bill and legs.*

8. GLOSSY-BLACK THRUSH, *T. serranus fuscobrunneus* (Venez. to Arg.)

9. GREAT THRUSH, *T. fuscater gigantodes* (Venez. to Bol.)

Also: Chiguanco Thrush, *T. chiguanco* (Ecu. to Chile and Arg.)

• *Black-headed/hooded effect* with *yellow bill* (♂); lower underparts may be *rufous.*

10. AUSTRAL THRUSH, *T. falcklandii magellanicus* (Chile and s. Arg.)

11. BLACK-HOODED THRUSH, *T. olivater sanctamartae* (locally in n. and e. S. Am.)

Also: Chestnut-bellied Thrush, *T. fulviventris* (Venez. to n. Peru)

1
3
6
5
2
4
7♂
8♂
7♀
9
10♂
11♂
G TUDOR

PLATE 6: THRUSHES II

Turdus Thrushes

PAGE 114

We here continue coverage of the *Turdus* thrushes; cf. Plate 5 for the first group in the genus. Included here are the more "generalized" or "typical" thrushes, many being very similar in appearance, difficult to categorize, and confusing, together with 2 distinctive species with no apparent close relatives. Most of these thrushes are brown above (the *tone varying*) and have the throat more or less streaked and the *crissum usually whitish*. Sexes are similar in most, and when they differ it is never by as much as in some of those found in Group A. They are *found mostly in the lowlands and foothills* (a few up into the subtropical zone). Habitats vary from lower growth to forest canopy, with a few species in semiopen areas or clearings. Behavior also variable, some species being shy and skittish, while others may boldly hop on lawns.

B. "Generalized" *Turdus* thrushes. PAGE 119

- *Lower underparts entirely bright rufous.*
  1. RUFOUS-BELLIED THRUSH, *T. r. rufiventris* (s. and e. S. Am.)
- *Mantle grayish olive brown;* most very dull.
  2. BARE-EYED THRUSH, *T. n. nudigenis* (mainly Venez. and Guianas)
  3. CREAMY-BELLIED THRUSH, *T. amaurochalinus* (s. and e. S. Am.)

  Also: Ecuadorian Thrush, *T. maculirostris* (w. Ecu. and nw. Peru)
  Clay-colored Thrush, *T. grayi* (n. Col.)
  Unicolored Thrush, *T. haplochrous* (n. Bol.)
  Black-billed Thrush, *T. ignobilis* (n. S. Am. to Amazonia)
- *Mantle warm brown.*
  4. PALE-BREASTED THRUSH, *T. l. leucomelas* (widespread)
  5. WHITE-NECKED THRUSH, *T. albicollis phaeopygus* (widespread; se. races have rufous flanks, yellow bill, etc.)
  6. HAUXWELL'S THRUSH, *T. hauxwelli* (w. Amazonia)

  Also: White-throated Thrush, *T. assimilis* (w. Col. and nw. Ecu.; closest to 5)
  Pale-vented Thrush, *T. obsoletus* (w. Col. and nw. Ecu.; similar to 6)
  Cocoa Thrush, *T. fumigatus* (n. and e. S. Am.; similar to 6)
- *Mantle dark olive brown; bright yellow bill* and eye-ring (♂); *distinctive voice.*
  7. LAWRENCE'S THRUSH, *T. lawrencii* (w. Amazonia)
- *Mantle gray* (♂♂); brown-backed ♀♀ resemble *T. albicollis* or *T. ignobilis* (above).
  8. ANDEAN SLATY-THRUSH, *T. nigriceps* (Andes from Peru to Arg.)

  Also: Eastern Slaty-Thrush, *T. subalaris* (se. S. Am.)

C. Two distinctive endemics of w. Ecu./nw. Peru. PAGE 129

9. PLUMBEOUS-BACKED THRUSH, *T. reevei*
10. MARANON THRUSH, *T. maranonicus*

1
2
3♂
4
5
6
7♂
8♂
9
10
TUDOR

PLATE 7: DIPPERS, MOCKINGBIRDS, PIPITS, ETC.

Cinclus Dippers — PAGE 130

A pair of distinctive, aquatic birds—no other passerine birds are so closely tied to *swift-flowing Andean streams.* The 2 S. Am. species are easily distinguished.

1. WHITE-CAPPED DIPPER, *C. l. leucocephalus*
2. RUFOUS-THROATED DIPPER, *C. schulzi* (nw. Arg. and s. Bol.)

Mimus Mockingbirds — PAGE 132

Slender, long-tailed birds found in open or scrub country throughout much of S. Am., with most species being found in the s. half of the continent (Tropical is only one found widely *north* of the Amazon). Basically brown or grayish, with *wings and/or tails strikingly patterned in black and white;* note also their facial patterns. Most mockingbirds are well known and familiar, with some fearlessly hopping about on lawns or around ranch yards. All are vocally gifted, being not only superb mimics but also having fine, strong songs of their own.

A. *Typical mockingbirds.* — PAGE 132

3. TROPICAL MOCKINGBIRD, *M. gilvus melanopterus*
4. CHALK-BROWED MOCKINGBIRD, *M. saturninus modulator*
5. WHITE-BANDED MOCKINGBIRD, *M. triurus*

Also: Patagonian Mockingbird, *M. patagonicus* (Arg. and adjacent s. Chile; most like 4)
Brown-backed Mockingbird, *M. dorsalis* (w. Bol. and nw. Arg.; most like 5)

B. *Larger* mockingbirds with *bold, black "whisker"; Pacific slope only.* — PAGE 136

6. LONG-TAILED MOCKINGBIRD, *M. longicaudatus albogriseus* (w. Ecu. and w. Peru)

Also: Chilean Mockingbird, *M. thenca*

Margarops Thrashers — PAGE 137

Coarse-looking, brown, vaguely thrushlike mimid found (in our area) only on *Bonaire.*

Not illustrated: Pearly-eyed Thrasher, *M. fuscatus*

Anthus Pipits — PAGE 138

Slender, mostly streaked small birds found in *grasslands,* reaching maximum diversity in s. S. Am. Pipits can be distinguished from other grassland-inhabiting birds by their slender bills and *short notched tails with white outer rectrices.* Note, further, that they *walk* and do not hop (as do finches, etc.). The various species are often not easy to identify, but the task is not impossible. Concentrate on noting ground color of underparts, extent of streaking on underparts (and whether fine or coarse), pattern of upperparts (whether distinctly streaked or more mottled), presence or absence of a malar streak, and size (whether "standard" or small). Displaying ♂♂ have *characteristic songs;* thus, these are considerably easier.

A. *Smaller pipits;* lowlands only; both *very* similar. — PAGE 138

7. YELLOWISH PIPIT, *A. l. lutescens* (*only* pipit in lowlands of n. and cen. S. Am.)

Also: Chaco Pipit, *A. chacoensis* (n. Arg. and s. Par.; rare)

B. *Larger pipits;* Andes and s. lowlands. — PAGE 139

8. SHORT-BILLED PIPIT, *A. f. furcatus*
9. HELLMAYR'S PIPIT, *A. h. hellmayri*
10. CORRENDERA PIPIT, *A. c. correndera*
11. PARAMO PIPIT, *A. b. bogotensis* (only in Andes)

Also: Ochre-breasted Pipit, *A. nattereri* (se. Brazil and adjacent Arg. and Par.; rare)

Eremophila Larks — PAGE 144

Only 1 species in S. Am., an isolated race of a common Holarctic species. Not likely to be confused in its limited range, the *E. Andes of Col.*

12. HORNED LARK, *E. alpestris peregrina*

7
1
2
3
4
5
6
7
8
9
10
11
12
G TUDOR

PLATE 8: VIREOS & WOOD-WARBLERS (in part)

Cyclarhis Peppershrikes PAGE 145

Chunky with *heavy bill;* note "brows." Sluggish; tireless songsters in woodland.

1. RUFOUS-BROWED PEPPERSHRIKE
   1a. *C. g. gujanensis;* 1b. *C. gujanensis virenticeps*

Also: Black-billed Peppershrike, *C. nigrirostris* (Col. and Ecu.)

Vireolanius Shrike-Vireos PAGE 147

More brightly colored than *Cyclarhis.* Canopy of humid forest; sing tirelessly.

2. SLATY-CAPPED SHRIKE-VIREO, *V. leucotis simplex* (local, w. Col. to Amazonia)

Also: Yellow-browed Shrike-Vireo, *V. eximius* (Col. and w. Venez.; bright green)

Vireo Vireos PAGE 148

Dull plumaged, but most show a rather distinct superciliary. Larger than wood-warblers or greenlets, with heavier bills. Most nonillustrated species are N. Am. migrants.

3. RED-EYED VIREO, *V. olivaceus chivi*
4. BROWN-CAPPED VIREO, *V. l. leucophrys* (mainly Andes)

Also: Yellow-green Vireo, *V. flavoviridis*
Noronha Vireo, *V. gracilirostris* (off ne. Brazil)
Black-whiskered Vireo, *V. altiloquus*
Yellow-throated Vireo, *V. flavifrons* (spectacles and wing-bars)

Hylophilus Greenlets A confusing group of drab, somewhat warblerlike vireos, differing from *Vireo* in smaller size, shorter tails, and more pointed, *often flesh-colored* bills; none shows wing-bars. Arboreal in forest, woodland, and shrubbery. Pairs or groups often follow mixed flocks; they vocalize frequently and *these songs often aid in identification.* PAGE 152

A. *Relatively simple songs.* Typically *pale iris,* more conical bill; usually in scrub or edge. PAGE 152

• Distinctive *head pattern;* se. S. Am. forests.

5. RUFOUS-CROWNED GREENLET, *H. p. poecilotis*

• *"Typical"* of group; most with pale iris.

6. SCRUB GREENLET, *H. f. flavipes* (Col.; iris *dark* in most of Venez.)
7. GRAY-CHESTED GREENLET, *H. s. semicinereus* (e. Amazonia)
8. ASHY-HEADED GREENLET, *H. pectoralis* (Guianas to cen. Brazil)

Also: Lemon-chested Greenlet, *H. thoracicus* (local; Amazonia, se. Brazil)
Olivaceous Greenlet, *H. olivaceus* (e. Ecu. and e. Peru foothills)
Tepui Greenlet, *H. sclateri*

B. *More complex songs.* All with *dark iris,* more slender bill. All but 1 in *forest canopy.* PAGE 156

9. GOLDEN-FRONTED GREENLET, *H. aurantiifrons saturatus* (Col. and Venez.); *scrub, edge*
10. DUSKY-CAPPED GREENLET, *H. hypoxanthus flaviventris* (Amazonia)
11. BUFF-CHEEKED GREENLET, *H. m. muscicapinus* (s. Venez. to cen. Brazil)
12. LESSER GREENLET, *H. decurtatus minor* (w. Col. and w. Ecu.)

Also: Brown-headed Greenlet, *H. brunneiceps* (upper Rio Negro area)
Rufous-naped Greenlet, *H. semibrunneus* (Andes, mainly Col.)

C. *Humid forest understory.* Distinctive song; iris color varies geographically. PAGE 159

13. TAWNY-CROWNED GREENLET
    13a. *H. ochraceiceps ferrugineifrons;* 13b. *H. ochraceiceps luteifrons*

Resident *Parula* and *Dendroica* Warblers PAGE 161

Both distinctive. Parula fairly widespread, but resident Yellows strictly coastal.

14. TROPICAL PARULA, *P. p. pitiayumi*
15. YELLOW WARBLER, *D. petechia peruviana*

North American Wintering Warblers PAGE 165

Many N. Am. breeding warblers pass the n. winter in S. Am. (nearly all in north or west). A sampling of most common, widespread species shown; 16 more species in text.

16. BLACKPOLL WARBLER, *Dendroica striata* (nonbreeding plumage)
17. BLACKBURNIAN WARBLER, *D. fusca*
18. CANADA WARBLER, *Wilsonia canadensis*
19. AMERICAN REDSTART, *Setophaga ruticilla*
20. NORTHERN WATERTHRUSH, *Seiurus noveboracensis*

8
1a
1b
2
3
4
5
6
7
8
9
10
11
12
13a
13b
14♂
15♂
16
17♀
18♀
19♀
20
G TUDOR

PLATE 9: RESIDENT WOOD-WARBLERS

Granatellus Chats 1. ROSE-BREASTED CHAT, *G. p. pelzelni* (Guianas and e. Amaz.) PAGE 175

Myioborus Redstarts Active, *montane* warblers, all with *conspicuous white in tail;* head patterns vary. Typically 2 species present in an area: Slate-throated at lower elevations, a member of either *ornatus* or *brunneiceps* groups above it. PAGE 176

A. Redstarts with *slaty throat* (yellow in all others). PAGE 177
2. SLATE-THROATED REDSTART, *M. miniatus verticalis*

B. *Ornatus/melanocephalus* group. Andes from Santa Martas to n. Bol.; *all allopatric.* PAGE 177
3. GOLDEN-FRONTED REDSTART, *M. ornatus chrysops* (mainly Col.)
4. SPECTACLED REDSTART, *M. m. melanocephalus* (s. Col. to Bol.)
5. WHITE-FRONTED REDSTART, *M. albifrons* (w. Venez.)
Also: Yellow-crowned Redstart, *M. flavivertex* (Santa Marta Mts.)

C. *Brunniceps* group. S. Andes and Venez. (especially *tepuis*); *all allopatric.* PAGE 180
6. TEPUI REDSTART, *M. c. castaneocapillus*
Also: Brown-capped Redstart, *M. brunniceps* (Andes of Bol. and n. Arg.)
Paria Redstart, *M. pariae* (ne. Venez.)
Guaiquinima Redstart, *M. cardonai;* White-faced Redstart, *M. albifacies*

Geothlypis Yellowthroats 7. MASKED YELLOWTHROAT, *G. aequinoctialis velata* PAGE 182
Also: Olive-crowned Yellowthroat, *G. semiflava* (w. Col. and w. Ecu.)
Common Yellowthroat, *G. trichas* (rare n. migrant)

Basileuterus Warblers *Dull plumaged* warblers of lower growth in woodland and forest, especially Andes. Identification tricky: *head and facial patterns* important, voice also crucial. Some tanagers (e.g., *Hemispingus*) quite similar. PAGE 184

A. The "*citrine*" group (olive and yellow with *no* lateral crown striping); all in Andes. PAGE 185
8. CITRINE WARBLER (w. Venez. to Bol.)
8a. *B. l. luteoviridis;* 8b. *B. luteoviridis euophrys*
Also: Pale-legged Warbler, *B. signatus* (s. Peru to nw. Arg.)
Black-crested Warbler, *B. nigrocristatus* (Venez. to n. Peru)

B. The "*gray-headed*" group (sides of head decidedly gray, usually with coronal stripe); mainly in Andes. PAGE 187
- *Entirely yellow below;* note restricted ranges.
9. GRAY-AND-GOLD WARBLER, *B. f. fraseri* (arid w. Ecu. and nw. Peru)
10. GRAY-HEADED WARBLER, *B. griseiceps* (ne. Venez.)
- *Throat* (at least) *dingy white;* lateral head stripes.
11. RUSSET-CROWNED WARBLER (w. Venez. to Bol.)
11a. *B. coronatus regulus;* 11b. *B. coronatus castaneiceps*
Also: White-lored Warbler, *B. conspicillatus* (Santa Marta Mts.)
Gray-throated Warbler, *B. cinereicollis* (w. Venez. and ne. Col.)

C. The "*stripe-headed*" group (prominent coronal or lateral head striping); widespread. PAGE 189
- *Bold auricular patch,* or *yellowish buff below,* or both.
12. THREE-STRIPED WARBLER, *B. tristriatus auricularis* (Venez. to Bol.)
Also: Santa Marta Warbler, *B. basilicus*
Pirre Warbler, *B. ignotus* (nw. Col.)
- *Bright* olive and yellow, *including superciliary.*
13. TWO-BANDED WARBLER, *B. b. bivittatus* (s. Peru to Arg. and on *tepuis*)
Also: Golden-bellied Warbler, *B. chrysogaster* (sw. Col. and w. Ecu.; cen. Peru)
- *Mantle grayish olive; superciliary pale gray to whitish.*
14. GOLDEN-CROWNED WARBLER, *B. culicivorus auricapillus* (widespread)
15. WHITE-BELLIED WARBLER, *B. hypoleucus* (s. Brazil)
Also: Three-banded Warbler, *B. trifasciatus* (s. Ecu. and nw. Peru)
- Crown and ear-coverts *brick red.*
16. RUFOUS-CAPPED WARBLER, *B. rufifrons mesochrysus* (n. Col.)

D. The "*Phaeothlypis*" subgenus; mainly in *lowlands;* semiterrestrial. PAGE 194
- *Plumage* like "citrine" group's but *behavior* of "*Phaeothlypis.*"
Not illustrated: Flavescent Warbler, *B. flaveolus* (drier lowlands)
- Underparts *grayish to buffy whitish.*
17. WHITE-STRIPED WARBLER, *B. leucophrys* (s. Brazil)
18. WHITE-RIMMED WARBLER, *B. leucoblepharus* (se. Brazil area)
19. BUFF-RUMPED WARBLER, *B. f. fulvicauda* (w. Col. to w. Amazonia)
Also: River Warbler, *B. rivularis* (lacks tail pattern of 19; e. lowlands)

9
1♀
1♂
2
4
5
3
6
7♂
7♀
8a
8b
9
10
11a
11b
12
13
14
15
16
17
18
19
G Tudor

PLATE 10: FLOWER-PIERCERS, ANDEAN CONEBILLS, ETC.

Diglossa Flower-piercers *Unmistakable upturned and hooked bills* (in all but 1 species). *Predominantly blue to gray to black,* several species with some rufous to chestnut below. Sexes usually alike. Most species in *Andes* (others in n. Venez. mts. and on *tepuis*), primarily at high elevations, mainly in shrubby areas and forest borders. Species-level taxonomy is much disputed; here a relatively narrow species concept is employed. PAGE 198

A. "Blue" and *tepui Diglossa.* PAGE 198

- Entirely varying *shades of blue.*
  1. BLUISH FLOWER-PIERCER, *D. caerulescens pallida*
  2. MASKED FLOWER-PIERCER, *D. c. cyanea*
  3. DEEP-BLUE FLOWER-PIERCER, *D. g. glauca*

  Also: Indigo Flower-piercer, *D. indigotica* (w. Col. and nw. Ecu.)
- *Tepuis* of s. Venez. and adjacent areas.

  Not illustrated: Greater Flower-piercer, *D. major*
  Scaled Flower-piercer, *D. duidae*

B. *Lafresnayii* and *carbonaria* groups (complex pair of superspecies arranged visually here). PAGE 201

- *Mainly black,* with or without gray shoulders.
  4. GLOSSY FLOWER-PIERCER, *D. lafresnayii* (w. Venez. to n. Peru)

  Also: Black Flower-piercer, *D. humeralis* (Col. to n. Peru)
- *Belly chestnut;* with or without a moustache.
  5. BLACK-THROATED FLOWER-PIERCER, *D. brunneiventris* (Peru and Bol.)

  Also: Merida Flower-piercer, *D. gloriosa* (w. Venez.)
  Chestnut-bellied Flower-piercer, *D. gloriossisima* (local in n. Col.)
- *Belly gray;* with no moustache.
  6. GRAY-BELLIED FLOWER-PIERCER, *D. carbonaria* (Bol.)
- *Belly black; prominent white to rufous moustache,* usually a pectoral band.
  7. MOUSTACHED FLOWER-PIERCER, *D. mystacalis unicincta* (Peru and Bol.)

C. *Albilatera* group; sexes *differ.* PAGE 205

- Gray to blackish with *white tuft at sides* (latter echoed in brown ♀).
  8. WHITE-SIDED FLOWER-PIERCER, *D. a. albilatera*

  Also: Venezuelan Flower-Piercer, *D. venezuelensis* (ne. Venez.)
- Gray above, *rusty* below (♀ *streaky* below).
  9. RUSTY FLOWER-PIERCER, *D. sittoides decorata*

Oreomanes Conebills PAGE 206

Large; *white on face;* sharply pointed bill. *Polylepis* groves in high Andes.

10. GIANT CONEBILL, *Oreomanes fraseri sturninus*

Nephelornis Parduscos PAGE 207

Plain brownish tanager of *timberline* woodland. Very local in *cen. Peru.*

11. PARDUSCO, *Nephelornis oneillei*

Xenodacnis Dacnises PAGE 208

Active tanager of *high Andean shrubbery* in Peru and s. Ecu., in or near *Polylepis* woodland.

12. TIT-LIKE DACNIS, *Xenodacnis parina petersi*

Conirostrum Conebills Small warblerlike tanagers with *slender, sharply pointed bills.* Andean forests and shrubbery (*Conirostrum* of lowlands are on Plate 11). PAGE 208

A. "Typical" conebills; sexes similar. PAGE 209

- Grayish above, *prominent L-shaped wing-mark.*
  13. CINEREOUS CONEBILL, *C. c. cinereum*

  Also: Tamarugo Conebill, *C. tamarugense* (sw. Peru and n. Chile; rufous brow and throat)
- Gray above and *all rufous below.*
  14. WHITE-BROWED CONEBILL, *C. ferrugineiventre* (Peru and Bol.)

  Also: Rufous-browed Conebill, *C. rufum* (Col.)
- *Mainly blue above, contrasting dark chest.*
  15. BLUE-BACKED CONEBILL, *C. sitticolor intermedium*

B. *All dark* with blue or white *crown* (♀ olive with bluish cap). PAGE 211

16. CAPPED CONEBILL, *C. albifrons atrocyaneum*

10
1
3♂
2
4
7
8♂
6
5
9♀
9♂
10
11
12♂
12♀
15
13
16♀
14
16♂
G TUDOR

PLATE 11: HONEYCREEPERS & SMALL LOWLAND TANAGERS

Dacnis Dacnises — Small warblerlike tanagers with *short, sharply pointed bills,* sexes differ, ♂♂ brighter and usually with *much blue,* ♀♀ *dull and often confusing. All arboreal in humid forest. Only 3 species common and widespread (Blue, Black-faced, Yellow-bellied); others either rare or range-restricted.* PAGE 212

Only ♂♂ diagnosed here; for ♀♀, see text.

A. "Standard" dacnises; typically black and turquoise. PAGE 212

- *Entirely black underparts* (in ♂).

1. **SCARLET-THIGHED DACNIS,** *D. venusta fuliginata* (w. Col., nw. Ecu.)

- *Black lores* and (usually) black throat.

2. **BLUE DACNIS,** *D. cayana paraguayensis* (nw. races deeper blue)

Also: Black-legged Dacnis, *D. nigripes* (se. Brazil; rare)
Viridian Dacnis, *D. viguieri* (nw. Col.; greener, no black throat)

- *Broad black mask;* iris yellow.

3. **BLACK-FACED DACNIS,** *D. lineata aequatorialis* (belly white *east* of Andes)

Also: Turquoise Dacnis, *D. hartlaubi* (w. Col.; stubby bill, black throat)

B. Miscellaneous other dacnises (all distinctive). PAGE 216

4. **SCARLET-BREASTED DACNIS,** *D. berlepschi* (sw. Col., nw. Ecu.)
5. **WHITE-BELLIED DACNIS,** *D. albiventris* (Amazonia)
6. **YELLOW-BELLIED DACNIS,** *D. flaviventer* (Amazonia)

Cyanerpes Honeycreepers — Slender *decurved bills* and *brightly colored legs. ♂♂ dark blue or purple,* ♀♀ green and *streaky below.* Arboreal, mostly in humid forest. PAGE 217

7. **PURPLE HONEYCREEPER,** *C. c. caeruleus*
8. **SHORT-BILLED HONEYCREEPER,** *C. nitidus*

Also: Shining Honeycreeper, *C. lucidus* (nw. Col.)
Red-legged Honeycreeper, *C. cyaneus* (red legs; ♂ with turquoise crown)

Chlorophanes Honeycreepers PAGE 220

Larger than *Cyanerpes,* with *stouter yellow bill.* Widespread in humid forest.

9. **GREEN HONEYCREEPER,** *C. s. spiza*

Coereba Bananaquits — *Short decurved bill.* Variable; all mainland races dark gray above with eyestripe, yellow below with gray throat; some island races all black. Widespread in lowlands, more numerous coastally. PAGE 221

10. **BANANAQUIT,** *C. flaveola intermedia*

Conirostrum Conebills — Small and warblerlike, with *slender pointed bills. Mainly bluish gray,* paler below. *Found in lowlands* (*Conirostrum* of Andes are on Plate 10), either in deciduous or gallery woodlands or in coastal areas and river-edge habitats; *not* in humid forest. PAGE 222

11. **BICOLORED CONEBILL,** *C. b. bicolor* (coastal and along Amazon)
12. **CHESTNUT-VENTED CONEBILL,** *C. s. speciosum*

Also: Pearly-breasted Conebill, *C. margaritae* (along Amazon; local)
White-eared Conebill, *C. leucogenys* (w. Col. and nw. Venez.; ♂ has black crown, white cheeks)

Chrysothlypis and *Hemithraupis* Tanagers PAGE 224

Small tanagers with slender pointed bills. *Strongly patterned or colored* ♂♂ distinctive (though dissimilar), olive and yellow-whitish ♀♀ confusing (see text). Arboreal, mostly in humid forest.

13. **SCARLET-AND-WHITE TANAGER,** *C. salmoni* (w. Col. and nw. Ecu.)
14. **GUIRA TANAGER,** *H. g. guira*
15. **YELLOW-BACKED TANAGER,** *H. flavicollis peruana*

Also: Rufous-headed Tanager, *H. ruficapilla* (se. Brazil; resembles 14)

Nemosia Tanagers PAGE 227

Distinctive, *gray and white* with white lores, yellow irides. Widespread in lighter woodland.

16. **HOODED TANAGER,** *N. pileata caerulea*

Also: Cherry-throated Tanager, *N. rourei* (se. Brazil, perhaps extinct; red bib)

1♀
2♂
3♂
2♀
3♀
4♂
5♂
6♀
7♂
8♂
6♂
7♀
9♂
9♀
10
12♀
12♂
13♂
11
14♂
15♂
16♂
14♀
16♀
G TUDOR

PLATE 12: MONTANE *TANGARA* & SMALL ANDEAN TANAGERS

Thlypopsis Tanagers — Contrasting *orange/rufous heads or crowns;* sexes alike. PAGE 228
Most are arboreal in *Andean woodlands,* but 1 species widespread in lowlands east of Andes.

A. *Grayish mantle; no* yellow below. PAGE 228

1. ORANGE-HEADED TANAGER, *T. sordida chrysopis* (e. lowlands)
2. BUFF-BELLIED TANAGER, *T. inornata* (nw. Peru)
3. RUFOUS-CHESTED TANAGER, *T. ornata media* (s. Col. to s. Peru)

Also: Fulvous-headed Tanager, *T. fulviceps* (n. Venez., Col.)
Brown-flanked Tanager, *T. pectoralis* (cen. Peru)

B. *Olive mantle; all-yellow underparts.* PAGE 230

4. RUST-AND-YELLOW TANAGER, *T. ruficeps* (s. Peru to nw. Arg.)

Chlorochrysa Tanagers PAGE 231

Brilliantly colored and (most species) *intricately patterned;* ♀♀ duller, but most show ♂'s pattern. Humid subtropical Andean forests.

5. MULTICOLORED TANAGER, *C. nitidissima* (w. Col.)
6. ORANGE-EARED TANAGER (e. slope)
   6a. *C. c. calliparaea;* 6b. *C. calliparaea bourcieri*

Also: Glistening-green Tanager, *C. phoenicotis* (sw. Col., w. Ecu.; vivid green)

Montane *Tangara* Tanagers (including *Iridophanes*) PAGE 233

Complex, widespread, and numerous genus. We here deal with the montane (primarily Andean) *Tangara;* those found in lowlands are treated on the following plate (there is some overlap, especially in foothills). Most *Tangara* are *ornately patterned and colorful;* sexes usually alike. Arboreal, active, and gregarious, often among the most prominent members of Andean mixed-species flocks.

A. *Striking blue head,* mostly black body. PAGE 233

7. BLUE-NECKED TANAGER, *T. cyanicollis caeruleocephala*

B. Montane *Tangara* with *crown typically brighter.* PAGE 234

- *Crown yellow to orange;* underparts variable.

8. GOLDEN TANAGER, *T. arthus pulchra*
9. SAFFRON-CROWNED TANAGER, *T. xanthocephala lamprotis* (yellow crown in n. races)

Also: Flame-faced Tanager, *T. parzudakii* (red foreface, opalescent below)

- *Boldly spangled* with opalescent crown.

10. BERYL-SPANGLED TANAGER, *T. n. nigroviridis*

- Predominantly *cobalt blue.*

11. BLUE-AND-BLACK TANAGER
    11a. *T. v. vassorii;* 11b. *T. vassorii atrocaerulea*

C. Montane *Tangara* with *crown typically black.* PAGE 237

- *Mostly opalescent green to bluish* with buff on belly.

12. GOLDEN-EARED TANAGER, *T. chrysotis* (e. slope)
13. GOLDEN-NAPED TANAGER, *T. r. ruficervix*

Also: Metallic-green Tanager, *T. labradorides*
Blue-browed Tanager, *T. cyanotis* (e. slope; black head, blue brow)

- Similar to above, but *crown green, cheeks rufous.*

Not illustrated: Rufous-cheeked Tanager, *T. rufigenis* (n. Venez.)

- Black crown and *streaky metallic gorget;* ♀♀ mostly green (*T. heinei* superspecies).

14. SILVER-BACKED TANAGER, *T. v. viridicollis* (s. Ecu., Peru)
15. STRAW-BACKED TANAGER, *T. a. argyrofenges* (local in e. Peru, Bol.)

Also: Black-capped Tanager, *T. heinei* (mostly Venez. to Ecu.)
Sira Tanager, *T. phillipsi* (very local in e. Peru)

D. *Contrasting black head* and *mostly opalescent body;* ♀♀ mostly olive. Note longer bill of *Iridophanes.* PAGE 241

16. GOLDEN-COLLARED HONEYCREEPER, *Iridophanes p. pulcherrima*

Also: Black-headed Tanager, *T. cyanoptera* (Venez. and adjacent areas)

12
1
2
3
4
5♂
6b♀
6a♂
7
8
9
10
11a
11b
12
13
14♂
14♀
15♂
16♂
16♀
TUDOR

PLATE 13: LOWLAND & FOOTHILL *TANGARA* TANAGERS

Lowland *Tangara* Tanagers

PAGE 243

An exceptionally large genus (no other in S. Am. is as big). We have divided it approximately in half, with Andean members on the previous plate; comments here apply only to this lowland subset. Note that there is some overlap along either base of the Andes. Best known for their *bright colors* and *bold patterns,* though there are some duller species. Sexes usually alike (♀♀ may be washed out), but the "Scrub Tanager complex" is much less patterned. Most favor humid forest, foraging mainly in the canopy, often in mixed flocks. A few species (including most of the "Scrub Tanager complex") range in clearings, gallery woodland, or scrub.

E. Miscellaneous *n. Col. and Pacific slope (south to w. Ecu.) Tangara;* some more widespread *Tangara* (other Groups) also range to Pacific slope. PAGE 243

1. GRAY-AND-GOLD TANAGER, *T. palmeri*
2. RUFOUS-THROATED TANAGER, *T. rufigula*
3. SILVER-THROATED TANAGER, *T. i. icterocephala*

Also: Plain-colored Tanager, *T. inornata* (small, mostly gray)

F. "Scrub Tanager complex." *Opalescent* greenish blue to buff, with *rufous crown* (usually) and *black mask or face.* PAGE 245

4. SCRUB TANAGER, *T. vitriolina* (Col. and n. Ecu.)
5. BURNISHED-BUFF TANAGER, *T. cayana flava* (lacks median stripe in north and west)
6. CHESTNUT-BACKED TANAGER, *T. preciosa* (se. Brazil and nearby)

Also: Black-backed Tanager, *T. peruviana* (se. Brazil only)
Green-capped Tanager, *T. meyerdeschauenseei* (very local in s. Peru; nearest 4)

G. Predominantly *"golden green" Tangara.* PAGE 247

- *Chestnut red head.*

7. BAY-HEADED TANAGER, *T. g. gyrola*

Also: Rufous-winged Tanager, *T. lavinia* (w. Col., nw. Ecu.)

- *Bold black ear-patch.*

8. GREEN-AND-GOLD TANAGER, *T. s. schrankii* (Amazonia)

Also: Emerald Tanager, *T. florida* (w. Col., nw. Ecu.)
Blue-whiskered Tanager, *T. johannae* (w. Col., nw. Ecu.)

- *Spotted or speckled* (mainly Amaz.).

9. SPOTTED TANAGER, *T. p. punctata*
10. YELLOW-BELLIED TANAGER, *T. x. xanthogastra*

Also: Speckled Tanager, *T. guttata*
Dotted Tanager, *T. varia* (speckling minute; rare)

H. *Predominantly blue and black Tangara* (mainly Amaz.). PAGE 252

- *Contrasting pale blue or golden hood.*

11. MASKED TANAGER, *T. nigrocincta*

Also: Golden-hooded Tanager, *T. larvata* (w. Col., w. Ecu.)

- Blue face and breast; *belly yellow to white.*

Not illustrated: Turquoise Tanager, *T. mexicana*

- *Mostly deep blue;* bill more pointed than other *Tangara.*

12. OPAL-CROWNED TANAGER, *T. callophrys*
13. OPAL-RUMPED TANAGER, *T. velia cyanomelaena* (Amaz. races purplish below)

I. *Multicolored Tangara;* all but one species *mainly e. Brazil.* PAGE 255

14. PARADISE TANAGER, *T. chilensis coelicolor* (Amazonia)
15. GREEN-HEADED TANAGER, *T. seledon*
16. RED-NECKED TANAGER, *T. c. cyanocephala*
17. BRASSY-BREASTED TANAGER, *T. desmaresti*

Also: Seven-colored Tanager, *T. fastuosa* (most like 15; ne. Brazil)
Gilt-edged Tanager, *T. cyanoventris* (like 17 but head golden, breast blue)

1
4♂
2
5♂
7♂
3♂
6♂
10
8♂
9
11♂
13♂
12
14
15♂
16♂
17
GTUDOR

PLATE 14: EUPHONIAS, CHLOROPHONIAS, ETC.

Euphonia Euphonias — PAGE 258

Small, short-tailed tanagers with stubby bills. Strong sexual dimorphism in most species: ♀♀ generally olive above, yellowish or grayish below (and often difficult to identify), ♂♂ typically *steely blue above* (often with *yellow forecrown*) and yellow to ochraceous below, *with or without a dark throat.* A few are less dimorphic, ♂♂ being more or less "hen-feathered." Widespread arboreal birds, most euphonias range in or near forest of some type, though a few are most frequent in savannas or clearings. Almost all inhabit lowlands, only the Orange-bellied and Golden-rumped being common in the Andes. *Only ♂♂ are diagnosed here;* for ♀♀ see text.

A. "Typical" euphonias with *throat entirely or mostly yellow.* — PAGE 259

- Throat entirely yellow.
  1. THICK-BILLED EUPHONIA, *E. laniirostris zopholega* (n. and w. S. Am.)

  Also: Violaceous Euphonia, *E. violacea* (e. S. Am.)
- *Dark chin* and *narrow* yellow forehead.
  2. GREEN-CHINNED EUPHONIA, *E. chalybea* (se. Brazil and adjacent areas)

B. "Typical" euphonias with *throat dark.* — PAGE 261

- *Underparts yellow to ochraceous yellow.*
  3. WHITE-VENTED EUPHONIA, *E. m. minuta*
  4. PURPLE-THROATED EUPHONIA, *E. c. chlorotica*
  5. ORANGE-BELLIED EUPHONIA
     5a. *E. xanthogaster brevirostris*

  Also: Trinidad Euphonia, *E. trinitatis* (n. S. Am.)
  Velvet-fronted Euphonia, *E. concinna* (local in w. Col.)
  Tawny-capped Euphonia, *E. anneae* (forecrown rufous; nw. Col.)
- *Lower* underparts *deep ochraceous to rufous.*
  5. ORANGE-BELLIED EUPHONIA
     5b. *E. xanthogaster ruficeps*
  6. ORANGE-CROWNED EUPHONIA, *E. saturata* (w. Col. to nw. Peru)

  Also: Finsch's Euphonia, *E. finschi* (ne. S. Am.)
  Fulvous-vented Euphonia, *E. fulvicrissa* (w. Col., nw. Ecu.)

C. "Atypical" euphonias. — PAGE 266

- *Bright blue crown;* throat dark.
  7. GOLDEN-RUMPED EUPHONIA, *E. cyanocephala aureata*
- *No* crown patch, *golden tuft on sides; E. cayennensis* superspecies.
  8. RUFOUS-BELLIED EUPHONIA, *E. rufiventris* (Amazonia)
  9. CHESTNUT-BELLIED EUPHONIA, *E. pectoralis* (se. S. Am.)

  Also: Golden-sided Euphonia, *E. cayennensis* (ne. S. Am.; most like 9)
- *Above gray or olive* (not steel blue).
  10. PLUMBEOUS EUPHONIA, *E. plumbea* (ne. S. Am.)
  11. WHITE-LORED EUPHONIA, *E. c. chrysopasta* (Amazonia to Guianas)

  Also: Bronze-green Euphonia, *E. mesochrysa* (Andes; most like 11 but forecrown yellow, no white on face)

Chlorophonia Chlorophonias — PAGE 270

Larger, plumper, even shorter-tailed relative of *Euphonia.* Predominantly *bright green.* Arboreal, mostly in humid subtropical forests.

12. CHESTNUT-BREASTED CHLOROPHONIA, *C. pyrrhophrys* (Andes of Venez. to Peru)
13. BLUE-NAPED CHLOROPHONIA, *C. cyanea longipennis*

Also: Yellow-collared Chlorophonia, *C. flavirostris* (small with yellow collar, orange bill, etc.; w. Col., nw. Ecu.)

Tersina Tanagers — PAGE 272

Distinctive, with broad flat bill. Gregarious and conspicuous, often sallying after insects. Locally common at forest borders; hole nester. Formerly in separate family.

14. SWALLOW TANAGER, *Tersina v. viridis*

1♂
2♂
3♂
4♂
1♀
5a♂
4♀
5a♀
5b♂
7♀
6♂
7♂
11♂
8♂
8♀
10♂
14♀
9♂
13♀
12♂
13♂
14♂
GTUDOR

PLATE 15: BUSH-TANAGERS, HEMISPINGUSES, ETC.

Catamblyrhynchus Plushcap — PAGE 273

Dark gray and chestnut with *prominent yellow forecrown*. Stubby bill. Lower growth in Andean forest, especially with *bamboo*. Formerly in separate family.

1. PLUSHCAP, *C. d. diadema*

Urothraupis Bush-Tanagers — PAGE 274

Basically *black above*, white and grayish below. Recalls an *Atlapetes*. *Shrubbery near timberline*, Andes of s. Col. and Ecu.

2. BLACK-BACKED BUSH-TANAGER, *U. stolzmanni*

Chlorospingus Bush-Tanagers — PAGE 274

Stocky, *mainly olive* tanagers; some very drab. *Iris often pale* (varies), some with white postocular spot or stripe or with yellow across chest or on throat. Sexes alike. Mostly found in upper tropical and subtropical *Andean forests*, where usually numerous and conspicuous.

A. *"Capped" appearance* (cap usually gray or brown), with yellow pectoral band. — PAGE 274

3. COMMON BUSH-TANAGER
   3a. *C. ophthalmicus peruvianus*
   3b. *C. ophthalmicus nigriceps*; 3c. *C. ophthalmicus argentinus*
4. ASHY-THROATED BUSH-TANAGER, *C. canigularis signatus*

B. *Nondescript* species, either olive or grayish below. W. Col. and w. Ecu. — PAGE 276

5. DUSKY BUSH-TANAGER, *C. s. semifuscus*

Also: Tacarcuna Bush-Tanager, *C. tacarcunae*
Yellow-green Bush-Tanager, *C. flavovirens*

C. Grayish below with *contrasting bright yellow on throat*. Col. to Bol. — PAGE 277

6. YELLOW-WHISKERED BUSH-TANAGER, *C. parvirostris medianus* (e. slope)

Also: Yellow-throated Bush-Tanager, *C. flavigularis*

Cnemoscopus Bush-Tanagers — PAGE 278

Contrasting gray hood; slender (usually pink) bill. Note *tailwagging*. Andean forests.

7. GRAY-HOODED BUSH-TANAGER, *C. r. rubrirostris* (bill dark in Peru)

Hemispingus Hemispinguses — PAGE 279

More slender and warblerlike than *Chlorospingus,* with thinner bills; some closely resemble certain *Basileuterus* warblers. Variably but usually simply patterned; often with *bold superciliary* (white, yellow, or rufous). Sexes alike. Range mainly in *temperate zone of Andean forests,* foraging primarily in lower growth and edge. Most species usually in groups.

A. *Underparts yellow to olive* (except in 9b). — PAGE 279

- *Prominent superciliary.*

8. BLACK-CAPPED HEMISPINGUS, *H. atropileus auricularis*
9. SUPERCILIARIED HEMISPINGUS
   9a. *H. s. superciliaris*
   9b. *H. superciliaris leucogaster* (n. and cen. Peru)

Also: Orange-browed Hemispingus, *H. calophrys* (Bol., s. Peru)
Parodi's Hemispingus, *H. parodii* (superciliary yellow; s. Peru)

- Superciliary *indistinct or lacking.*

10. OLEAGINOUS HEMISPINGUS, *H. f. frontalis*

Also: Gray-capped Hemispingus, *H. reyi* (*no* superciliary; w. Venez.)

B. *Underparts pale grayish* (but see also 9b); *iris pale.* — PAGE 282

11. BLACK-HEADED HEMISPINGUS, *H. verticalis* (mainly Col. and Ecu.)
12. DRAB HEMISPINGUS, *H. xanthophthalmus* (Peru, nw. Bol.)

C. *Underparts ochraceous to rufous.* — PAGE 283

13. THREE-STRIPED HEMISPINGUS, *H. trifasciatus* (s. Peru, nw. Bol.)
14. BLACK-EARED HEMISPINGUS
   14a. *H. m. melanotis*; 14b. *H. melanotis piurae* (nw. Peru)
15. RUFOUS-BROWED HEMISPINGUS, *H. rufosuperciliaris* (Peru)

Also: Slaty-backed Hemispingus, *H. goeringi* (w. Venez.)

15
1
2
3a
3b
3c
4
5
6
7
8
9a
9b
10
11
12
13
14a
14b
15
G TUDOR

PLATE 16: ANDEAN TANAGERS

Pipraeidea Tanagers PAGE 285

Basically blue (*brighter on crown*) and buff tanager with black mask and small bill. Arboreal at edge of montane forests and in se. S. Am.

1. FAWN-BREASTED TANAGER, *P. melanonota venezuelensis*

Delothraupis Mountain-Tanagers PAGE 286

Blue and rufous tanager with black mask and *submalar streak*. Higher montane forests in *Andes of Peru and Bol.*

2. CHESTNUT-BELLIED MOUNTAIN-TANAGER, *D. castaneoventris*

Creurgops Tanagers PAGE 287

Simply patterned, mostly gray or *gray and rufous* tanagers of subtropical zone Andean forests. Bill rather heavy. Both rather uncommon.

3. SLATY TANAGER, *C. dentata* (se. Peru and nw. Bol.)
4. RUFOUS-CRESTED TANAGER, *C. verticalis* (w. Venez. to Peru; ♀ similar)

Iridosornis Tanagers PAGE 288

Richly colored and boldly patterned tanagers of *Andean forest lower growth. Blue or purplish* usually predominates. Bill quite stubby.

A. *Extensive yellow throat.* PAGE 288

5. YELLOW-THROATED TANAGER, *I. analis* (mainly e. Ecu. and e. Peru)

Also: Purplish-mantled Tanager, *I. porphyrocephala* (w. Col. and nw. Ecu.)

B. *Yellow on head.* PAGE 289

6. GOLDEN-COLLARED TANAGER, *I. jelskii* (e. Peru, nw. Bol.)
7. GOLDEN-CROWNED TANAGER, *I. r. rufivertex* (mainly Col. and Ecu.)

Also: Yellow-scarfed Tanager, *I. reinhardti* (e. Peru)

Anisognathus Mountain-Tanagers PAGE 290

Boldly patterned, colorful tanagers of *Andean forests and shrubby areas*. All but 1 *some shade of rich yellow below;* note also facial markings and/or crown stripes. Arboreal and usually conspicuous, often in mixed flocks.

A. *Yellow median crown stripe,* more conical bill ("*Compsocoma*" subgenus). PAGE 291

8. BLUE-WINGED MOUNTAIN-TANAGER, *A. flavinucha somptuosa* (Venez. to Bol.)

Also: Black-chinned Mountain-Tanager, *A. notabilis* (back yellow-olive, less blue on wing; sw. Col. and w. Ecu.)

B. Stubbier bill ("true" *Anisognathus*). PAGE 292

- *Some shade of yellow below;* "teardrop" below eye.

9. LACRIMOSE MOUNTAIN-TANAGER, *A. lachrymosus caerulescens* (Venez. to Peru)

Also: Santa Marta Mountain-Tanager, *A. melanogenys* (crown blue, lacks postauricular spot)

- *Underparts bright red.*

10. SCARLET-BELLIED MOUNTAIN-TANAGER, *A. igniventris ignicrissa* (Col. to Bol.)

Bangsia Tanagers PAGE 293

Chunky and short-tailed tanagers of foothills and lower Andean slopes; *all found in w. Col.,* 2 spilling over into adjacent Ecu. Colors and patterns vary, but all share *yellow chest patch.*

A. Mainly dark green with *contrasting color on face;* mandible flesh color. PAGE 293

11. MOSS-BACKED TANAGER, *B. edwardsi*

Also: Gold-ringed Tanager, *B. aureocincta* (yellow encircles cheeks; rare)

B. *Dark blue or black with contrasting yellow.* PAGE 294

12. GOLDEN-CHESTED TANAGER, *B. rothschildi*

Also: Black-and-gold Tanager, *B. melanochlamys* (yellow to median belly; rare)

Wetmorethraupis Tanagers PAGE 295

Rare, little-known tanager reminiscent of *Bangsia*. Foothill forests in n. Peru.

13. ORANGE-THROATED TANAGER, *W. sterrhopteron*

1♀
1♂
2
3♀
3♂
4♂
5
6
7
8
11
12
9
10
13
G TUDOR

PLATE 17: ANDEAN TANAGERS & *THRAUPIS* TANAGERS

Dubusia Mountain-Tanagers PAGE 296

Large tanager with black bib, *buff across chest,* and at least some *frosty pale blue streaking above* (usually just on head). Lower growth in temperate zone Andean forest and shrubbery, usually in pairs, not with flocks.

1. BUFF-BREASTED MOUNTAIN-TANAGER, *D. t. taeniata*

Buthraupis Mountain-Tanagers PAGE 296

Large, strikingly patterned tanagers of *high-elevation* Andean forests. Heavier-bodied than *Anisognathus* mountain-tanagers; colors vary quite dramatically. Hooded is gregarious, noisy, and conspicuous, but the others are more retiring and *much* scarcer.

2. HOODED MOUNTAIN-TANAGER, *B. montana cyanonota*
3. BLACK-CHESTED MOUNTAIN-TANAGER, *B. eximia chloronota* (mainly Col. and Ecu.)
4. GOLDEN-BACKED MOUNTAIN-TANAGER, *B. aureodorsalis* (very local near timberline in Peru)

Also: Masked Mountain-Tanager, *B. wetmorei* (olive above, yellow outlining black face; local near timberline, s. Col. to n. Peru)

Chlorornis Tanagers PAGE 299

Mostly vivid green; face chestnut, bright *red bill and legs.* Heavy build much like a *Buthraupis* mountain-tanager's. Gregarious and often conspicuous in higher-elevation Andean forests.

5. GRASS-GREEN TANAGER, *C. riefferii celata*

Sericossypha Tanagers PAGE 299

Very large, spectacular black tanager with *showy snow white crown* and *crimson throat* (darker in ♀, black in immature). Conspicuous but rare and wide-ranging in flocks in canopy of humid Andean forest; arresting, jaylike calls.

6. WHITE-CAPPED TANAGER, *S. albocristata*

Thraupis Tanagers PAGE 300

The *typical, average* "tanager": difficult to generalize. Usually *subdued coloration* (♂ Blue-and-yellow being the exception). Mostly in *semiopen habitats,* edge and scrub. Active and conspicuous birds, with squeaky call notes and songs.

A. *Montane or temperate zone species.* PAGE 300

7. BLUE-AND-YELLOW TANAGER, *T. bonariensis composita* (back olive in ♂♂ of Andes from Ecu. to n. Chile and Bol.)
8. BLUE-CAPPED TANAGER, *T. c. cyanocephala*

B. *Lowland species.* PAGE 302

- *Dull* olive to bluish with *contrasting* pattern on wings.

9. PALM TANAGER, *T. palmarum melanoptera*
10. GOLDEN-CHEVRONED TANAGER, *T. ornata* (se. Brazil)

- Some shade of (usually pale) *grayish blue;* often white or blue on shoulder.

11. BLUE-GRAY TANAGER, *T. e. episcopus*
12. SAYACA TANAGER, *T. s. sayaca*

Also: Glaucous Tanager, *T. glaucocolpa* (arid Caribbean)
Azure-shouldered Tanager, *T. cyanoptera* (se. Brazil)

1
2
3
4
5
6♀

8
9
10♂
7♂
7♀
11
12
G TUDOR

PLATE 18: "RED" TANAGERS

Piranga Tanagers PAGE 305

Typical, "classic" tanagers: arboreal, ♂♂ *mainly red,* ♀♀ *mainly olive or yellow. Wings often contrastingly darker.* Some species have rich caroling song.

A. *Bold white wing-bars* (both sexes); small size. PAGE 305

1. WHITE-WINGED TANAGER, *P. leucoptera venezuelae* (mts., Venez. to Bol.)

B. *Predominantly rosy red to scarlet* (♂♂) or *olive and yellow* (♀♀). PAGE 306

2. HEPATIC TANAGER, *P. flava saira*

Also: Summer Tanager, *P. rubra* (N. Am. migrant)
Scarlet Tanager, *P. olivacea* (black or dusky wings; N. Am. migrant)

C. Unmistakable *scarlet hood;* sexes similar. PAGE 309

3. RED-HOODED TANAGER, *P. rubriceps* (Andes, Col. to Peru)

Ramphocelus Tanagers PAGE 309

Obvious *pale silvery on bill,* with lower mandible typically swollen (especially ♂♂). ♀♀ duller (except 7). Shrubby habitats at edge and in clearings, often near water; widespread in lowlands. Most species conspicuous and commonly seen.

A. Both sexes with *bright contrasting rump* (vermilion to yellow); ♂ otherwise black. PAGE 309

4. FLAME-RUMPED TANAGER, *R. f. flammigerus* (w. Col. and w. Ecu.)

B. *Blackish maroon to crimson* with black wings and tail. PAGE 310

- *R. carbo* superspecies; ♀ pinkish brown.

5. SILVER-BEAKED TANAGER, *R. c. carbo*
6. BRAZILIAN TANAGER, *R. b. bresilius* (e. Brazil)

Also: Crimson-backed Tanager, *R. dimidiatus* (w. Col. and nw. Venez.)
Huallaga Tanager, *R. melanogaster* (local in e. Peru)

- *Black mask and back;* sexes similar.

7. MASKED CRIMSON TANAGER, *R. nigrogularis* (w. Amazonia)

Calochaetes Tanagers PAGE 313

Sexes alike. Arboreal in subtropical zone forests on *e. slope of Andes from Col. to Peru.*

8. VERMILION TANAGER, *C. concinneus*

Habia Ant-Tanagers PAGE 314

Rather inconspicuous tanagers of forest and woodland undergrowth. Only 1 species (Red-crowned) across most of S. Am., others being entirely *Colombian.* Often in small groups; loud, scratchy calls may attract attention.

A. Lack obvious crests; both sexes with *paler throats.* PAGE 314

9. RED-CROWNED ANT-TANAGER, *H. rubica peruviana*

Also: Red-throated Ant-Tanager, *H. fuscicauda*

B. *Conspicuous scarlet crest;* sexes similar PAGE 316

10. SOOTY ANT-TANAGER, *H. gutturalis*

Also: Crested Ant-Tanager, *H. cristata* (all reddish)

Rhodinocichla Thrush-Tanagers PAGE 316

Aberrant, *semiterrestrial,* shy tanager of dry woodland undergrowth. Note mimidlike bill. Underparts and eyestripe ochraceous in ♀. N. Col. and Venez.

11. ROSY THRUSH-TANAGER, *R. rosea harterti*

1♀
2♀
1♂
2♂
3♂
4♀
5♂
6♂
7♂
5♀
8
10♂
9♂
9♀
11♂
G TUDOR

PLATE 19: LOWLAND TANAGERS

Chlorothraupis Tanagers — PAGE 317

Rather *drab, dark-looking* tanagers of forest lower growth, mainly in foothills. *Olive* usually predominates; heavy bill. *W. Col. and w. Ecu.* (except for Olive Tanager).

1. OLIVE TANAGER, *C. carmioli frenata* (mostly e. slope of Andes)
2. OCHRE-BREASTED TANAGER, *C. stolzmanni dugandi*

Also: Lemon-spectacled Tanager, *C. olivacea*

Mitrospingus Tanagers — PAGE 319

Mainly *olive and gray* tanagers of humid forest and borders. Behavior of the 2 species very different (see text).

3. OLIVE-BACKED TANAGER, *M. o. oleagineus* (*tepui* area)

Also: Dusky-faced Tanager, *M. cassinii* (dark face, pale iris; w. Col. and w. Ecu.)

Eucometis Tanagers — PAGE 320

Bright olive and yellow tanager with contrasting *gray head*. Excitable, with *expressive bushy crest*. In pairs in woodland/forest undergrowth, east of Andes often near water.

4. GRAY-HEADED TANAGER, *E. p. penicillata*

Heterospingus Tanagers — PAGE 321

Striking tanager of humid forest canopy in *w. Col. and nw. Ecu.* Even mostly gray ♀ easily known: *yellow rump, white pectoral patch.*

5. SCARLET-BROWED TANAGER, *H. x. xanthopygius*

Lanio Shrike-Tanagers — PAGE 322

Mid-size tanagers of forest canopy in Amazonia, often leading flocks of other tanagers, etc. Despite name, bill only slightly heavier and more hooked than its near-relative, *Tachyphonus*. Strong sexual dimorphism: ♂♂ *boldly patterned,* ♀♀ *much duller* (and quite similar to certain ♀ *Tachyphonus*).

6. WHITE-WINGED SHRIKE-TANAGER, *L. v. versicolor* (*south* of Amazon)

Also: Fulvous Shrike-Tanager, *L. fulvus* (*north* of Amazon; lacks white in wing)

Cyanicterus Tanagers — PAGE 323

Fairly large, heavy-bodied tanager with notably stout bill. ♂'s blue of a deeper, more cobalt hue, extending over head and chest. Forest canopy in *ne. S. Am.;* usually in mixed flocks.

7. BLUE-BACKED TANAGER, *C. cyanicterus*

Tachyphonus Tanagers — PAGE 324

Widespread group of small to mid-size tanagers. ♂♂ show some pale leaden blue on mandible (less in ♀♀). ♂♂ *mainly black*, often with flat crest, and with *varying amount of white on inner shoulder and under wing;* ♀♀ duller, typically more olive or rufescent. Range mostly in canopy of humid forest and woodland, often with mixed flocks. A minority (those in Group C) favor semiopen or edge habitats. Only ♂♂ diagnosed below; for ♀♀ see text.

A. Small, with *prominent white shoulder.* — PAGE 324

8. WHITE-SHOULDERED TANAGER, *T. l. luctuosus* (humid forested lowlands)

B. *Bright* (usually prominent) *coronal crest.* — PAGE 325

- *Yellow or buff rump.*

9. FLAME-CRESTED TANAGER, *T. cristatus cristatellus* (mainly Amazonia)
10. YELLOW-CRESTED TANAGER, *T. rufiventer* (w. Amazonia)

Also: Fulvous-crested Tanager, *T. surinamus* (nearest 9; Amazonia)

- *Without* buff rump (plumage mainly black).

Not illustrated: Tawny-crested Tanager, *T. delatrii* (w. Col. and w. Ecu.)

C. Mainly *all black*, with *no obvious crest.* — PAGE 327

11. RED-SHOULDERED TANAGER, *T. phoeniceus* (locally in savannas)
12. RUBY-CROWNED TANAGER, *T. coronatus* (se. S. Am.)

Also: White-lined Tanager, *T. rufus* (nearest to 12; semiopen areas in lowlands)

19
2
1
3
4
6♀
7♀
5♂
6♂
8♂
10♂
9♀
9♂
8♀
12♂
11♀
12♀
G TUDOR

PLATE 20: MISCELLANEOUS TANAGERS

Tanagers on this plate are in distinct, usually monotypic genera (in 2 cases, a pair of species is involved). Most are very distinct, but 4 groupings of genera, based on distribution and appearance, can be discerned. Sexes similar unless otherwise noted.

A. Forest endemics found mainly in *se. Brazil* and adjacent areas.

Orchesticus Tanagers — Remarkably *foliage-gleaner-like* but for its *robust bill.* Gleans and probes in foliage and among epiphytes, usually well above ground in montane forests. PAGE 330

1. BROWN TANAGER, *O. abeillei*

Stephanophorus Tanagers PAGE 330

Dark blue with contrasting *crown patch.* Lighter woodlands and forest borders.

2. DIADEMED TANAGER, *S. diadematus*

Orthogonys Tanagers — Large *olive* tanager with fairly long slender bill. Usually in groups in canopy of montane forests. PAGE 331

3. OLIVE-GREEN TANAGER, *O. chloricterus*

Trichothraupis Tanagers — *Contrasting black "goggles" and wings;* ♀ lacks goggles. Behavior recalls *Habia* ant-tanagers': in pairs or small groups in forest and woodland lower growth. Ranges also in Peruvian and Bolivian Andes. PAGE 331

4. BLACK-GOGGLED TANAGER, *T. melanops*

Pyrrhocoma Tanagers — An inconspicuous tanager of forest borders and lower growth. Usually in pairs apart from flocks. PAGE 332

5. CHESTNUT-HEADED TANAGER, *P. ruficeps*

B. *Open country* tanagers found mainly in *interior Brazil.*

Schistochlamys Tanagers PAGE 333

Note black on face and stout, mostly bluish bills. In pairs, usually not with flocks.

6. BLACK-FACED TANAGER, *S. melanopis aterrima* (fairly widespread)
7. CINNAMON TANAGER, *S. ruficapillus* (only Brazil)

Neothraupis Tanagers — Distinctive *gray, black, and white* tanager found in *cerrado* of Brazil and adjacent areas. Note remarkable plumage convergence with certain shrikes (*Lanio* spp.). PAGE 334

8. WHITE-BANDED TANAGER, *N. fasciata*

Cypsnagra Tanagers PAGE 334

Mostly *black and white* tanager; *cerrado* of Brazil and adjacent areas.

9. WHITE-RUMPED TANAGER, *C. h. hirundinacea*

Compsothraupis Tanagers PAGE 335

Aberrant large, *mostly black* tanager (♂ with *red throat*) of *ne. Brazil.*

10. SCARLET-THROATED TANAGER, *C. loricata*

C. Miscellaneous "*black and white*" tanagers.

Conothraupis Tanagers — Striking ♂♂ basically *all black and white* (with seedeaterlike pattern), ♀♀ (unknown in 1 species) much duller, basically olive. Deciduous woodland in semiarid areas. Still poorly known and perhaps not even tanagers. PAGE 336

11. BLACK-AND-WHITE TANAGER, *C. speculigera* (w. Ecu. and Peru)

Also: Cone-billed Tanager, *C. mesoleuca* (Mato Grosso, Brazil; rare)

Lamprospiza Tanagers — Unmistakable *pied black and white plumage,* with *red bill.* Small groups with flocks in canopy of Amaz. forests. ♂ with back black. PAGE 336

12. RED-BILLED PIED TANAGER, *L. melanoleuca*

Cissopis Tanagers — *Very large black and white* tanager with *notably long graduated tail.* Pairs or small groups are conspicuous at forest borders and in clearings; widespread. PAGE 337

13. MAGPIE TANAGER, *C. l. leveriana*

1
2
3
4♂
4♀
5♀
5♂
7
6
8
9
11♂
10♂
11♀
12♀
13
G TUDOR

PLATE 21: MEADOWLARKS, MARSH BLACKBIRDS, ETC.

Sturnella Meadowlarks and Blackbirds — PAGE 338

Bill varies from being quite long and very pointed to rather blunt and shorter (in the *Leistes* subgenus). *Underparts always brightly colored* (either red or yellow), but ♀♀ and (in some species) nonbreeding ♂♂ duller, with bright colors subdued. Gregarious icterids of *open grassy country,* where they walk about (often in the open) or perch on fences or clumps of grass.

A. *Underparts yellow* with black crescent; sexes alike. — PAGE 338

1. EASTERN MEADOWLARK, *S. magna praticola* (n. S. Am.)

B. *Underparts red;* ♀♀ duller, streakier below. — PAGE 339

2. LONG-TAILED MEADOWLARK, *S. l. loyca* (Arg. and Chile)

Also: Pampas Meadowlark, *S. defilippii* (se. S. Am.)
Peruvian Meadowlark, *S. bellicosa* (w. Ecu. to n. Chile)

C. Similar to Group B, but blacker and *bill shorter* ("*Leistes*" subgenus). — PAGE 341

3. WHITE-BROWED BLACKBIRD, *S. superciliaris* (s. S. Am.)

Also: Red-breasted Blackbird, *S. militaris* (♂ lacks eyebrow; n. S. Am.)

Dolichonyx Bobolinks — PAGE 342

Highly gregarious migrant from N. Am.; *favors ricefields. Spiky tail.* Usually mostly buffy with *prominent head striping;* nuptial plumage of ♂ gradually attained during northward passage.

4. BOBOLINK, *D. oryzivorus* (prenuptial plumage)

Pseudoleistes Marshbirds — PAGE 343

Large and heavy-bodied, with *entirely brown and yellow plumages;* sexes alike. Gregarious in marshes and adjacent grassy terrain in *se. S. Am.*

5. YELLOW-RUMPED MARSHBIRD, *P. guirahuro*

Also: Brown-and-yellow Marshbird, *P. virescens* (lacks yellow rump)

Gymnomystax Blackbirds — PAGE 344

Large and conspicuous blackbird which is oriolelike only in its *bright yellow and black* coloration; sexes alike. Open country and along rivers in n. S. Am.

6. ORIOLE BLACKBIRD, *G. mexicanus*

Amblyramphus Blackbirds — PAGE 345

Unmistakable *bright red hood* marks this large, long-billed blackbird of *reedbeds in s. S. Am.* Sexes alike (but cf. immature).

7. SCARLET-HEADED BLACKBIRD, *A. holosericeus*

Agelaius Blackbirds — PAGE 345

Medium-size, simply patterned blackbirds with straight pointed bills. ♀♀ duller and streaky (in all but one species). All species associated with *marshes,* though they may forage in nearby open country. Most species gregarious and conspicuous in appropriate habitat, often noisy.

A. *Bright yellow underparts or hood* (♂). — PAGE 345

8. SAFFRON-COWLED BLACKBIRD, *A. flavus* (se. S. Am.)

Also: Yellow-hooded Blackbird, *A. icterocephalus* (♀ with yellow only on throat; n. S. Am.)

B. Predominantly or entirely black (♂); mostly e. or s. S. Am. — PAGE 347

9. CHESTNUT-CAPPED BLACKBIRD, *A. r. ruficapillus*
10. UNICOLORED BLACKBIRD, *A. c. cyanopus*

Also: Yellow-winged Blackbird, *A. thilius* (both sexes with yellow shoulder)
Pale-eyed Blackbird, *A. xanthophthalmus* (sexes alike, all black; very local in e. Ecu. and e. Peru)

1
3♂
2♂
4♂
3♀
7
5
6
10♀
10♂
8♂
8♀
9♂
G TUDOR

PLATE 22: BLACKBIRDS, COWBIRDS, GRACKLES, & ORIOLES

Lampropsar Grackles PAGE 349
Slender, long-tailed; plush forecrown hard to see. *Water edges* in Amazon and Orinoco basins.
1. VELVET-FRONTED GRACKLE, *L. t. tanagrinus*

Gnorimopsar, Oreopsar, Curaeus, and *Dives* Blackbirds PAGE 350
Four closely related genera, differing slightly in *bill shape* and *degree of shiny shaft streaking on head.* They mostly separate out by *range.* All are conspicuous, often gregarious icterids of open or semiopen country; loud, arresting vocalizations.
2. CHOPI BLACKBIRD, *Gnorimopsar c. chopi* (s.-cen. and e. S. Am.)
3. AUSTRAL BLACKBIRD, *Curaeus c. curaeus* (s. Arg., Chile)
Also: Bolivian Blackbird, *Oreopsar bolivianus* (arid Bol. highlands)
Forbes' Blackbird, *Curaeus forbesi* (e. Brazil; rare)
Scrub Blackbird, *Dives warszewiczi* (w. Ecu. and w. Peru)

Molothrus Cowbirds Rather small, with conical bills. Most ♂♂ have *lustrous sheen;* ♀♀ duller, especially Shiny. Gregarious in open country; most spp. are *brood parasites.* PAGE 353
A. ♂♂ *glossed;* ♀♀ usually duller. PAGE 353
4. SCREAMING COWBIRD, *M. rufoaxillaris* (n. Arg. and adjacent areas; sexes alike)
5. SHINY COWBIRD, *M. bonariensis venezuelensis* (widespread)
Also: Bronzed Cowbird, *M. aeneus* (local in n. Col.)
B. Both sexes "hen-plumaged" with *rufous wings.* PAGE 355
6. BAY-WINGED COWBIRD, *M. b. badius* (s. and e. S. Am.)

Scaphidura Cowbirds Large, wide-ranging cowbird of more humid lowlands. PAGE 356
Neck ruff prominent in ♂. Flight shape and style distinctive (see text); *brood parasite.*
7. GIANT COWBIRD, *S. o. oryzivora*

Quiscalus Grackles Note bill shape, *pale irides, keel-shaped tails;* ♀♀ smaller and duller. Common in n. S. Am., often near coast. PAGE 356
8. CARIB GRACKLE, *Q. l. lugubris*
Also: Great-tailed Grackle, *Q. mexicanus*

Macroagelaius Mountain-Grackles PAGE 358
Slender, long-tailed grackles of restricted range in montane forest.
9. TEPUI MOUNTAIN-GRACKLE, *M. imthurni* (*tepui* region)
Also: Colombian Mountain-Grackle, *M. subalaris* (ne. Col.)

Hypopyrrhus Grackles Spectacular large grackle of *Col. Andes.* Rare in montane forests. PAGE 359
10. RED-BELLIED GRACKLE, *H. pyrohypogaster*

Icterus Orioles and Troupials Rather slender icterids with sharply pointed bills. *Predominantly orange or yellow and black,* sexes alike except in the two n. migrants. Arboreal in pairs or family groups, most often not with mixed flocks. All are good songsters. PAGE 359
A. *Mainly black.* PAGE 359
11. MORICHE ORIOLE, *I. chrysocephalus* (n. Amazonia)
12. EPAULET ORIOLE, *I. cayanensis pyrrhopterus* (Guianas to n. Arg.; n. races have yellow epaulets)
B. *N. Am. migrants* (only to Col. and Venez.); rather small, ♂♂ black-headed. PAGE 361
Not illustrated: Northern Oriole, *I. galbula;* Orchard Oriole, *I. spurius*
C. *Large and bright orange;* always with "*shaggy*" *bib* and *pale iris.* PAGE 362
13. TROUPIAL 13a. *I. icterus croconotus;* 13b. *I. icterus ridgwayi*
D. Typical "*black-bibbed*" orioles. PAGE 363
• *Yellow back.*
14. YELLOW-BACKED ORIOLE, *I. chrysater giraudii* (Col. and Venez.)
Also: Yellow Oriole, *I. nigrogularis* (n. S. Am.; white wing-bar)
• *Black back.*
15. WHITE-EDGED ORIOLE, *I. graceannae* (w. Ecu. and nw. Peru)
Also: Yellow-tailed Oriole, *I. mesomelas* (w. Venez. to nw. Peru)
Orange-crowned Oriole, *I. auricapillus* (n. Col. and Venez.)

1 2 3 4 5♀ 5♂ 6 7♂ 8♂ 9 10 11 12 13a 13b 14 15

G TUDOR

PLATE 23: CACIQUES & OROPENDOLAS

Amblycercus Caciques — PAGE 366

Similar to *Cacicus*, but note *yellow iris*. Unlike most of them, this species *skulks in dense undergrowth*.

1. YELLOW-BILLED CACIQUE, *A. holosericeus australis* (n. S. Am. and in Andes)

Cacicus Caciques — PAGE 367

Mid-size to large icterids, *mainly black*, most with *yellow on rump and/or wing-coverts* or *red on rump*. Sexes basically alike; ♀♀ sootier in some. Most arboreal in lowland forest and woodland, a few in montane forest, one (Solitary) skulking in lower growth; generally conspicuous and very vocal. Some species *highly colonial as nesters*.

A. *All black.* — PAGE 367

2. SOLITARY CACIQUE, *C. solitarius*

Also: Ecuadorian Cacique, *C. sclateri* (e. Ecu. and ne. Peru)

B. *Rump scarlet-red.* — PAGE 368

3. SCARLET-RUMPED CACIQUE, *C. u. uropygialis* (Andes; w. Col. and w. Ecu.)

Also: Red-rumped Cacique, *C. haemorrhous* (e. lowlands; *large* rump patch)

C. *Rump yellow* (often a wing-patch as well). — PAGE 370

- Lowlands.

4. YELLOW-RUMPED CACIQUE, *C. c. cela*

5. GOLDEN-WINGED CACIQUE, *C. chrysopterus* (s. S. Am.)

Also: Selva Cacique, *C. koepckeae* (no yellow on wing; rare in e. Peru)

Andes; size of 4.

Not illustrated: Mountain Cacique, *C. leucoramphus*

Ocyalus Oropendolas — PAGE 373

Small, distinctive oropendola found *locally along Amazon* and some of its major tributaries. Rather "intermediate" toward *Cacicus*.

6. BAND-TAILED OROPENDOLA, *O. latirostris*

Psarocolius Oropendolas — PAGE 374

Large to very large icterids, basically dark (varying from olive to chestnut and black) with *contrasting yellow outer tail feathers*, conspicuous in flight. Bill long and pointed, often swollen onto forehead (even as a casque). ♂♂ much larger than ♀♀. Arboreal, mostly in humid forest of lowlands and adjacent clearings; some species occur mainly on lower Andean slopes. Their nesting colonies are a characteristic sight.

A. *Smaller* oropendolas, with swollen casque on bill. — PAGE 374

7. CASQUED OROPENDOLA, *P. oseryi* (e. Ecu. and e. Peru)

Also: Chestnut-headed Oropendola, *P. wagleri* (w. Col. and nw. Ecu.)

B. Typical *large* oropendolas. — PAGE 375

- Mostly *black*.

8. CRESTED OROPENDOLA, *P. d. decumanus*

- Predominantly *olive green and chestnut*.

9. RUSSET-BACKED OROPENDOLA, *P. angustifrons alfredi* (Andes to Amazonia; nominate race black-billed)

10. DUSKY-GREEN OROPENDOLA, *P. atrovirens* (s. Peru and Bol.)

Also: Green Oropendola, *P. viridis* (Guianas to Amazonia; bill red-tipped)

C. *Very large* oropendolas, mainly black and chestnut (except 11b); note *bare facial skin*. The "*Gymnostinops*" subgenus. — PAGE 378

11. OLIVE OROPENDOLA
   11a. *P. b. bifasciatus*
   11b. *P. bifasciatus yuracares*

Also: Black Oropendola, *P. guatimozinus* (n. Col.)
Baudo Oropendola, *P. cassini* (nw. Col.; rare)

NOTE: Inset of displaying and nesting Crested Oropendolas (8) drawn to smaller scale.

23
1
2
3
4♂
5
6
8♀
8♂
8
7
11a
11b
9
10
GTUDOR

PLATE 24: SALTATORS & FOREST GROSBEAKS

Periporphyrus Grosbeaks PAGE 381

Black hood in conjunction with *mostly red* (♂) *or olive* (♀) *plumage*. In pairs inside forest; usually inconspicuous. Ne. S. Am.

1. RED-AND-BLACK GROSBEAK, *P. erythromelas*

Pitylus Grosbeaks PAGE 381

Gray grosbeaks with *massive red bills;* ♀♀ duller. Found in pairs in humid forest, widespread; rich, musical song.

2. SLATE-COLORED GROSBEAK, *P. g. grossus*

Also: Black-throated Grosbeak, *P. fuliginosus* (e. Brazil)

Caryothraustes Grosbeaks PAGE 382

Mostly olive with *distinctive* (but very different) *facial patterns*. Both found in humid lowland forest, but behavior differs (see text).

3. YELLOW-SHOULDERED GROSBEAK, *C. humeralis* (locally in Amazonia)
4. YELLOW-GREEN GROSBEAK, *C. c. canadensis* (e. S. Am.)

Saltator Saltators PAGE 384

Large finches with stout bills. Olive or gray above, typically with *black-bordered throat patch,* and often a *white superciliary* of varying length and breadth. Some species have orange or red bills. Sexes basically alike. Arboreal and usually conspicuous in forest edge and scrub; loud, often musical songs.

A. Duller, typical saltators. PAGE 384

- *Streaked underparts* (except in part of arid Pacific); olive above.

5. STREAKED SALTATOR (drier regions from Venez. to w. Peru)
   5a. *S. albicollis striatipectus;* 5b. *S. albicollis flavidicollis*

- *White or buff throat patch margined black;* olive or gray above with white eyestripe of varying length.

6. GRAYISH SALTATOR, *S. coerulescens azarae*
7. THICK-BILLED SALTATOR, *S. maxillosus* (se. S. Am.)

Also: Green-winged Saltator, *S. similis* (se. S. Am.)
Buff-throated Saltator, *S. maximus*

B. *Sides of head broadly black,* bordered by (except in Black-cowled) *light eyebrow and throat.* PAGE 387

- Note color and *contrasting black wings*.

8. BLACK-WINGED SALTATOR, *S. a. atripennis* (Col. and Ecu.)

- *Gray above and buffy below* (adult); details vary.

9. GOLDEN-BILLED SALTATOR (*postocular only* and *orange bill;* Andes and se. S. Am.)
   9a. *S. aurantiirostris albociliaris;* 9b. *S. a. aurantiirostris,* immature

Also: Black-cowled Saltator, *S. nigriceps* (entire hood black; s. Ecu. and n. Peru)
Orinocan Saltator, *S. orenocensis* (full superciliary; Venez., ne. Col.)

C. Three *distinctive* saltators. PAGE 389

- *Black face* with orange bill, brown upperparts.

10. BLACK-THROATED SALTATOR, *S. atricollis* (mostly cen. Brazil; *cerrado*)

- Mainly *gray* with *rufous belly.*

11. RUFOUS-BELLIED SALTATOR, *S. rufiventris* (highlands of Bol. and nw. Arg.)

- *Lacks* eyebrow; note *habitat* and *unusual tail.*

12. MASKED SALTATOR, *S. cinctus* (e. Andean forests of Ecu. and Peru; rare and local)

24
1♀
1♂
2♂
3
4
5a
5b
6
7♀
8
9a♂
9b
10
11
12
G Tudor

PLATE 25: CARDINALS, GROSBEAKS, SEED-FINCHES, ETC.

Paroaria Cardinals *Boldly patterned* finches with *red heads* (usually). PAGE 390
Immatures brown above. Shrubbery and scrub. Most species common and conspicuous.

A. Smaller species; *black above. Near water.* PAGE 390

1. RED-CAPPED CARDINAL, *P. g. gularis* (Amazon and Orinoco drainages)
   1a. Adult; 1b. Immature
2. CRIMSON-FRONTED CARDINAL, *P. baeri xinguensis* (local in cen. Brazil)

Also: Yellow-billed Cardinal, *P. capitata* (upper Paraguay River basin)

B. Larger species; *gray above.* PAGE 391

3. RED-CRESTED CARDINAL, *P. coronata* (s. S. Am.)

Also: Red-cowled Cardinal, *P. dominicana* (ne. Brazil)

Cardinalis Cardinals PAGE 392
Unmistakable with *long pointed crest.* ♂ *red;* ♀ mostly sandy brown. Desert scrub, *Caribbean coast.*

4. VERMILION CARDINAL, *C. phoeniceus*

Gubernatrix Cardinals PAGE 393
Unmistakable with *bold pattern showing much yellow, long crest.* Scrub and woodland. Mostly Arg.

5. YELLOW CARDINAL, *G. cristata*

Pheucticus Grosbeaks Large, *thick-set* finches with heavy bills. All have *bold white wing-patches;* the 2 resident spp. are basically yellow and black (♀♀ duller, more mottled) and are found in the Andes and *chaco,* while the n. migrant is quite different, ♂ mostly black and white, ♀ brown and white with head striping. PAGE 393

6. BLACK-BACKED GROSBEAK
   6a. *P. aureoventris uropygialis;* 6b. *P. a. aureoventris*

Also: Southern Yellow-Grosbeak, *P. chrysogaster* (much more yellow)
Rose-breasted Grosbeak, *P. ludovicianus* (n. migrant)

Cyanocompsa and *Cyanoloxia* Grosbeaks PAGE 395
Medium-size to fairly large, usually with *heavy bill.* ♂♂ *blue* (the shade varying), ♀♀ brown (often rich). Usually inconspicuous, but with good melodic songs. Humid lowland forest (Blue-black) to lighter woodland and scrub.

7. ULTRAMARINE GROSBEAK, *Cyanocompsa brissonii sterea*
8. GLAUCOUS-BLUE GROSBEAK, *Cyanoloxia glaucocaerulea* (se. S. Am.)

Also: Blue-black Grosbeak, *Cyanocompsa cyanoides*

Porphyrospiza Finches *Slender yellow bill. Open* cerrado *in Brazil.* ♀ *streaky brownish.* PAGE 398

9. BLUE FINCH, *P. caerulescens*

Amaurospiza Seedeaters PAGE 399
♂♂ *dark* blue, ♀♀ *tawny* brown. Both spp. local, furtive inside humid montane forest.

10. BLACKISH-BLUE SEEDEATER, *A. moesta* (se. S. Am.)

Also: Blue Seedeater, *A. concolor* (n. Andes)

Haplospiza Finches PAGE 400
Sharply pointed bill. ♂♂ *uniform gray,* ♀♀ *streaked below.* Forest undergrowth.

11. UNIFORM FINCH, *H. unicolor* (se. S. Am.)

Also: Slaty Finch, *H. rustica* (Andes)

Oryzoborus Seed-Finches *Enormous squared-off bill* is larger than any seedeater's, but cf. *Cyanocompsa* grosbeaks (especially ♀♀). ♂♂ *predominantly or all black,* ♀♀ *rich dark brown.* In pairs (*not flocks*) in shrubby clearings, often near water. PAGE 401

12. LESSER SEED-FINCH, *O. angolensis torridus*
13. GREAT-BILLED SEED-FINCH, *O. maximiliani magnirostris*

Also: Large-billed Seed-Finch, *O. crassirostris*

1a
1b
2
3
4♀
5♀
5♂
6a♂
6b♀

7♀
9♂
8♂
7♂
10♀
10♂
12♂
13♂
12♀
11♂
11♀
13♀
G TUDOR

PLATE 26: SEEDEATERS & GRASSQUITS

Volatinia Grassquits — Abundant in grassy areas and roadsides. Note *pointed bill.* — PAGE 403

1. BLUE-BLACK GRASSQUIT, *V. j. jacarina*

Tiaris Grassquits — PAGE 404

Bill somewhat narrower and more pointed than *Sporophila's*. Open areas in lowlands.

2. SOOTY GRASSQUIT, *T. fuliginosa fumosa* (local in Col., Venez., e. Brazil)
3. DULL-COLORED GRASSQUIT, *T. o. obscura* (lower Andean slopes; local)

Also: Black-faced Grassquit, *T. bicolor* (Caribbean lowlands)
Yellow-faced Grassquit, *T. olivacea* (mainly Col.)

Dolospingus Seedeaters — PAGE 407

Somewhat larger, more conical bill than *Sporophila's*. Sandy soil areas of s. Venez. region.

4. WHITE-NAPED SEEDEATER, *D. fringilloides*

Sporophila Seedeaters — Numerous genus of small finches found in open to semiopen areas of lowlands, most diverse in s.-cen. S. Am.; a few favor wooded habitats. Often in large mixed flocks when not breeding. *Bill thick and stubby.* ♀♀ hard to identify; *only* ♂♂ *diagnosed here.* — PAGE 407

A. *Boldly patterned in black and white (or rusty);* bill black. — PAGE 408

5. LINED SEEDEATER, *S. l. lineola*
6. VARIABLE SEEDEATER, *S. americana murallae* (south to Amazonia)
7. RUSTY-COLLARED SEEDEATER, *S. collaris melanocephala* (s. S. Am.)

Also: Lesson's Seedeater, *S. bouvronides* (Guianas to Amazonia; like 5)

B. "*Hooded*" group; *lack* face pattern; *bill bluish.* — PAGE 411

8. YELLOW-BELLIED SEEDEATER, *S. n. nigricollis*

Also: Dubois' Seedeater, *S. ardesiaca* (se. Brazil)
Hooded Seedeater, *S. melanops* (s.-cen. Brazil; very rare)
Black-and-white Seedeater, *S. luctuosa* (Andes; all black above)

C. "*Collared*" group; *gray upperparts; bill yellowish.* — PAGE 413

9. DOUBLE-COLLARED SEEDEATER, *S. c. caerulescens* (s.-cen. S. Am.)
10. PARROT-BILLED SEEDEATER, *S. p. peruviana* (arid Pacific)

Also: White-throated Seedeater, *S. albogularis* (ne. Brazil)

D. *Predominantly gray to olive;* bill yellow or black. — PAGE 414

11. BUFFY-FRONTED SEEDEATER, *S. frontalis,* subadult (se. Brazil area)
12. SLATE-COLORED SEEDEATER, *S. schistacea longipennis* (local)
13. PLUMBEOUS SEEDEATER, *S. p. plumbea* (local in savannas)

Also: Gray Seedeater, *S. intermedia* (nearest 12; n . S. Am.)
Temminck's Seedeater, *S. falcirostris* (like 12; se. S. Am.)
Drab Seedeater, *S. simplex* (like 3 with wing-bars; w. Peru and sw. Ecu.)

E. *Sharply bicolored* (Bol. race *black* above); *bill yellow.* — PAGE 418

14. WHITE-BELLIED SEEDEATER, *S. l. leucoptera* (s.-cen. S. Am.)

F. *Both sexes streaked above* and with white at base of tail. — PAGE 418

15. CHESTNUT-THROATED SEEDEATER, *S. telasco* (Pacific lowlands)

Also: Tumaco Seedeater, *S. insulata* (mostly rufous below; sw. Col.; rare)

G. Small; *cinnamon to chestnut (or black) below.* Many local or rare. — PAGE 419

- *Black cap;* otherwise mostly cinnamon (whiter in young birds).

16. CAPPED SEEDEATER, *S. b. bouvreuil* (savannas of e. S. Am.)

- *Gray* (usually) *or black above;* tawny to rufous below *and on rump.*

17. TAWNY-BELLIED SEEDEATER, *S. hypoxantha* (s.-cen. S. Am.)
18. RUFOUS-RUMPED SEEDEATER, *S. hypochroma* (s.-cen. S. Am.; rare)

Also: Ruddy-breasted Seedeater, *S. minuta* (n. S. Am.; much like 17)
Black-and-tawny Seedeater, *S. nigrorufa* (e. Bol., w. Mato Grosso; rare)

- *Entirely* gray above; chestnut or black on *median* underparts.

19. BLACK-BELLIED SEEDEATER, *S. melanogaster* (se. Brazil)

Also: Chestnut-bellied Seedeater, *S. castaneiventris* (Amazonia)

- "Chestnut" and "Marsh" groups; note *gray cap, contrasting throat,* or both.

20. CHESTNUT SEEDEATER, *S. cinnamomea* (s.-cen. S. Am.; rare)
21. MARSH SEEDEATER, *S. palustris* (s.-cen. S. Am.; rare)
22. DARK-THROATED SEEDEATER, *S. ruficollis* (s.-cen. S. Am.)

Also: Narosky's Seedeater, *S. zelichi* (ne. Arg.; very rare)

NOTE: ♂ White-naped Seedeater shown at smaller scale.

26
1♂
1♀
2♂
3
4♀
4♂
5♂
5♀
6♂
7♂
7♀
8♂
9♂
9♀
10♂
11♂
12♂
12♀
13♂
14♂
15♂
15♀
16♂
17♂
17♀
18♂
19♂
20♂
21♂
22♂
G TUDOR

PLATE 27: BRUSH-FINCHES & ALLIES

Lysurus Finches

PAGE 425

Smaller than most *Atlapetes; dark.* Secretive in forest undergrowth on lower Andean slopes.

1. OLIVE FINCH, *L. castaneiceps*

Also: Sooty-faced Finch, *L. crassirostris* (nw. Col.)

Atlapetes Brush-Finches

PAGE 426

Medium-size finches of *lower growth and shrubbery, mostly at forest borders;* a few are terrestrial inside humid forest. Most range at *various levels in the Andes.* Forage mostly as pairs or small groups, usually remaining independent of mixed flocks (though a few species regularly accompany them); they feed mostly by flicking with bill. Many are excitable and can be lured into view by squeaking.

A. *Underparts yellow to olive* (only a yellow *throat* in one).

PAGE 426

- Mainly *dark olive* with *white orbital area.*

2. WHITE-RIMMED BRUSH-FINCH, *A. leucopis* (sw. Col. and Ecu.)

- *Broad black mask* with *contrasting crown* (except Santa Marta).

3. RUFOUS-NAPED BRUSH-FINCH
3a. *A. rufinucha latinuchus;* 3b. *A. rufinucha melanolaimus*

4. TRICOLORED BRUSH-FINCH, *A. tricolor crassus* (locally Col. to Peru)

5. PALE-NAPED BRUSH-FINCH, *A. p. pallidinucha* (mainly Col. and Ecu.)

Also: Santa Marta Brush-Finch, *A melanocephalus* (near 3b but no rufous)
Moustached Brush-Finch, *A. albofrenatus* (w. Venez. and ne. Col.)
Yellow-throated Brush-Finch, *A. gutturalis* (white below throat; Col.)

- *Head all blackish or yellow-olive* (no crown stripe); Col.

6. DUSKY-HEADED BRUSH-FINCH, *A. fuscoolivaceus*

Also: Olive-headed Brush-Finch, *A. flaviceps* (rare)

- *Head striped black and yellow;* nw. Arg.

7. YELLOW-STRIPED BRUSH-FINCH, *A. citrinellus*

- *Head all cinnamon to chestnut; A. fulviceps* superspecies.

8. OCHRE-BREASTED BRUSH-FINCH, *A. semirufus denisei* (Venez. and ne. Col.)

9. FULVOUS-HEADED BRUSH-FINCH, *A. fulviceps* (Bol. and nw. Arg.)

Also: Tepui Brush-Finch, *A. personatus*

B. *Underparts white to gray* (no yellow).

PAGE 432

- *Broad black mask with rufous crown.*

10. SLATY BRUSH-FINCH, *A. schistaceus taczanowskii*

11. WHITE-WINGED BRUSH-FINCH, *A. l. leucopterus* (Ecu. and nw. Peru)

Also: Bay-crowned Brush-Finch, *A. seebohmi* (s. Ecu. and nw. Peru)

- *Head variable,* but *no rufous;* s. Ecu. and w. Peru.

12. WHITE-HEADED BRUSH-FINCH, *A. albiceps*

13. RUSTY-BELLIED BRUSH-FINCH, *A. n. nationi*

Also: Pale-headed Brush-Finch, *A. pallidiceps* (rare)

- *Head mostly rufous;* cen. Peru.

14. RUFOUS-EARED BRUSH-FINCH, *A. r. rufigenis*

- *Olive mantle* and (in most) *black pectoral band.*

15. CHESTNUT-CAPPED BRUSH-FINCH, *A. brunneinucha frontalis*

16. STRIPE-HEADED BRUSH-FINCH
16a. *A. torquatus poliophrys;* 16b. *A. torquatus borelli*

Also: Black-headed Brush-Finch, *A. atricapillus* (no pectoral band; Col.)

Oreothraupis Finches

PAGE 437

Recalls a *large* brush-finch. Humid forest undergrowth in W. Andes of Col. and nw. Ecu.

17. TANAGER FINCH, *O. arremonops*

1
2
3a
3b
4
5
6
7
8
9
10
11
12
13
14
15
16a
16b
17
G TUDOR

PLATE 28: SIERRA-FINCHES & MONTANE ALLIES

Catamenia Seedeaters

PAGE 438

Stubby yellow to pink bills. Mainly *gray* (♂♂) or *streaky brown* (♀♀ and immatures). Primarily *Andean,* where often common in open grassy or shrubby terrain at higher elevations. Forage mostly on the ground; frequently in groups when not breeding.

1. PARAMO SEEDEATER, *C. h. homochroa*
2. BAND-TAILED SEEDEATER, *C. a. analis*

Also: Plain-colored Seedeater, *C. inornata*

Idiopsar Finches

PAGE 439

Fairly large *all-gray* finch of *high rocky Andean grasslands from s. Peru to Arg. Long* pointed bill. In pairs or alone (not in flocks).

3. SHORT-TAILED FINCH, *I. brachyurus*

Phrygilus Sierra-Finches

PAGE 440

Small to medium-size, almost entirely *Andean* in distribution, especially at higher elevations. Plumage patterns vary, but *gray predominates* in most. Largely terrestrial in open grassy or shrubby terrain. Often in small groups, frequently with other species.

A. *Mostly plain gray;* ♀♀ brown and streaky.

PAGE 440

4. PLUMBEOUS SIERRA-FINCH, *P. unicolor inca* (Venez. to Patagonia)
5. ASH-BREASTED SIERRA-FINCH, *P. p. plebejus* (Ecu. to n. Chile and Arg.)

B. *Yellow bill,* with *gray to black underparts;* ♀♀ streaky.

PAGE 441

6. BAND-TAILED SIERRA-FINCH, *P. alaudinus venturii* (Ecu. to cen. Chile and w. Arg.)
7. MOURNING SIERRA-FINCH, *P. f. fruticeti* (Peru to Chile and Arg.)

Also: Carbonated Sierra-Finch, *P. carbonarius* (Arg.)

C. *Hooded effect* (gray to black); ♀♀ duller.

PAGE 443

8. BLACK-HOODED SIERRA-FINCH, *P. atriceps* (s. Peru to n. Chile and n. Arg.)
9. GRAY-HOODED SIERRA-FINCH, *P. gayi caniceps* (Chile and Arg.)

Also: Peruvian Sierra-Finch, *P. punensis*
Patagonian Sierra-Finch, *P. patagonicus*

D. Large; *gray above, white below,* with back rufous or gray. Sexes alike.

PAGE 446

10. RED-BACKED SIERRA-FINCH, *P. dorsalis* (nw. Arg. and adjacent Chile and Bol.)

Also: White-throated Sierra-Finch, *P. erythronotus* (sw. Bol. and adjacent Peru and Chile)

Diuca Diuca-Finches

PAGE 447

Very close to the last *Phrygilus* grouping (thus *essentially gray and white*). *Conspicuous white in tail* (and in 1 species, also wing). Open or shrubby country. Behavior much like *Phrygilus*'s, sometimes flocking with them.

11. WHITE-WINGED DIUCA-FINCH, *D. s. speculifera* (high Andes, mainly Peru and Bol.)

Also: Common Diuca-Finch, *D. diuca* (mainly Arg. and Chile)

Melanodera Finches

PAGE 448

Pair of finches restricted to *Patagonia.* Both boldly patterned with *black throats,* ♀♀ much duller. Open grassy or shrubby country, in flocks when not breeding; both species scarce.

12. BLACK-THROATED FINCH, *M. melanodera princetonia*

Also: Yellow-bridled Finch, *M. xanthogramma* (less yellow in wing)

28
1♀
2♂
3
1♂
4♀
2♀
4♂
7♂
5♂
6♂
7♀
9♀
9♂
8♂
10
11
12♀
12♂
TUDOR

PLATE 29: WARBLING-FINCHES, INCA-FINCHES, ETC.

Xenospingus Finches — PAGE 450

Sleek *gray* finch of *arid Pacific slope. Long slender yellow bill.*

1. SLENDER-BILLED FINCH, *X. concolor* (w. Peru, n. Chile)

Piezorhina Finches — PAGE 450

Hefty gray finch of *arid plains* in nw. Peru. *Massive yellow bill.*

2. CINEREOUS FINCH, *P. cinerea*

Incaspiza Inca-Finches — PAGE 451

Colorfully patterned finches of arid scrub in *w. Peru,* often quite high on Andean slopes. Note their *complex facial and dorsal patterns; conspicuous white outer tail feathers.* ♀♀ slightly duller.

A. Larger species. — PAGE 451

3. GREAT INCA-FINCH, *I. pulchra*
4. RUFOUS-BACKED INCA-FINCH, *I. personata*

Also: Gray-winged Inca-Finch, *I. ortizi*

B. Smaller species. — PAGE 452

5. BUFF-BRIDLED INCA-FINCH, *I. laeta*

Also: Little Inca-Finch, *I. watkinsi*

Poospiza Warbling-Finches and Mountain-Finches — PAGE 453

Widespread and rather diverse genus of finches found mostly in the *s. Andes or in lowlands of s. S. Am.* All favor woodland or scrub. Patterns vary, but most are colorful and quite strongly marked; many flash *white in outer tail feathers in flight.* Except for Collared and Ringed, sexes basically alike. Most are quite active, gleaning in foliage and on smaller branches, but several Andean species are larger and behave somewhat differently (see text).

A. *Lowland* and *southeastern* warbling-finches. — PAGE 453

- Gray and white, with *black pectoral band* (♂♂).

6. COLLARED WARBLING-FINCH, *P. hispaniolensis* (sw. Ecu. and w. Peru)

Also: Ringed Warbling-Finch, *P. torquata* (s.-cen. S. Am.)

- Basically *bicolored,* gray and white.

7. BLACK-CAPPED WARBLING-FINCH, *P. melanoleuca* (s.-cen. S. Am.)

Also: Cinereous Warbling-Finch, *P. cinerea* (head gray; local in s. Brazil)

- Mainly gray above *with at least some rufous below.*

8. RED-RUMPED WARBLING-FINCH, *P. lateralis cabanisi* (se. S. Am.)
9. CINNAMON WARBLING-FINCH, *P. ornata* (Arg.)
10. BAY-CHESTED WARBLING-FINCH, *P. thoracica* (se. Brazil)

Also: Black-and-rufous Warbling-Finch, *P. nigrorufa* (nearest 14; se. S. Am.)

B. *Andean* warbling-finches and mountain-finches. — PAGE 457

- Basically gray above with *rufous superciliary;* usually *rufous to chestnut below* (rufous and gray in 12).

11. RUSTY-BROWED WARBLING-FINCH, *P. e. erythrophrys* (Bol., nw. Arg.)
12. TUCUMAN MOUNTAIN-FINCH, *P. baeri* (nw. Arg.)

Also: Rufous-breasted Warbling-Finch, *P. rubecula* (black face; w. Peru)
Cochabamba Mountain-Finch, *P. garleppi* (Bol.; local)

- *White below* with *rufous usually confined to sides;* upperparts grayish brown, superciliary white.

13. RUFOUS-SIDED WARBLING-FINCH, *P. hypochondria affinis* (Bol., nw. Arg.)

Also: Plain-tailed Warbling-Finch, *P. alticola* (w. Peru)
Bolivian Warbling-Finch, *P. boliviana* (also rufous pectoral band)

- *Mainly deep chestnut below.*

14. BLACK-AND-CHESTNUT WARBLING-FINCH, *P. w. whitii* (Bol., nw. Arg.)

- Somewhat *larger* than congeners; *broad chestnut pectoral band.*

15. CHESTNUT-BREASTED MOUNTAIN-FINCH, *P. caesar* (local in s. Peru)

1
3♂
4♂
2
5♂
6♂
7
8
6♀
9♂
10
11
14
13
12
15
TUDOR

PLATE 30: SPARROWS & OPEN COUNTRY FINCHES

Arremon Sparrows PAGE 461

Head black striped gray or white (all black in one); below white with *pectoral band* (usually). Often a *colorful bill.* Woodland undergrowth in *lowlands* (not montane like larger *Atlapetes*).

1. PECTORAL SPARROW (locally Venez. to e. Brazil)
   1a. *A. t. taciturnus;* 1b. *A. taciturnus semitorquatus*
2. SAFFRON-BILLED SPARROW, *A. flavirostris polionotus* (s. S. Am.)
3. GOLDEN-WINGED SPARROW, *A. s. schlegeli* (n. Col. and Venez.)

Also: Orange-billed Sparrow, *A. aurantiirostris* (Col. to n. Peru)
Black-capped Sparrow, *A. abeillei* (sw. Ecu. and nw. Peru)

Arremonops Sparrows — Dull, less patterned than *Arremon.* Favor arid scrub, clearings. PAGE 463

4. BLACK-STRIPED SPARROW, *A. c. conirostris* (Venez. to w. Ecu.; local)

Also: Tocuyo Sparrow, *A. tocuyensis* (w. Venez. and ne. Col.)

Zonotrichia Sparrows — *Familiar and widespread* in open to semiopen terrain. PAGE 464

5. RUFOUS-COLLARED SPARROW, *Z. capensis subtorquata*

Ammodramus Sparrows PAGE 465

Small, *streaky* sparrows; *yellow lores* at least. Mostly in *open grasslands.*

6. GRASSLAND SPARROW, *A. h. humeralis*

Also: Yellow-browed Sparrow, *A. aurifrons* (grassy borders in Amazonia)
Grasshopper Sparrow, *A. savannarum* (Col. and n. Ecu.)

Aimophila Sparrows PAGE 467

Larger than *Ammodramus,* with *complex facial patterns.* Shrubby areas with grass.

7. STRIPE-CAPPED SPARROW, *A. s. strigiceps* (n. Arg., w. Par.)

Also: Tumbes Sparrow, *A. stolzmanni* (sw. Ecu. and nw. Peru)

Spiza Dickcissels — Gregarious N. Am. migrant; sparrow-like. PAGE 468

Not illustrated: Dickcissel, *S. americana (local in open country in n. S. Am.)*

Embernagra Pampa-Finches and Serra-Finches PAGE 469

Large, *rather stout* finches of *s. grasslands.* Mostly olive/gray with *orange bill.*

8. GREAT PAMPA-FINCH, *E. p. platensis* (Bol. to cen. Arg. and s. Brazil)

Also: Pale-throated Serra-Finch, *E. longicauda* (e. Brazil; local)

Emberizoides Grass-Finches PAGE 470

Streaked, with eye-ring and *long pointed tail.* Tall grass in open country.

9. WEDGE-TAILED GRASS-FINCH, *E. herbicola sphenurus*

Also: Lesser Grass-Finch, *E. ypiranganus* (local in se. S. Am.)
Duida Grass-Finch, *E. duidae* (s. Venez.; rare)

Coryphaspiza Finches — Local in *tall grass savannas. Graduated tail showing white.* PAGE 471

10. BLACK-MASKED FINCH, *C. m. melanotis* (s.-cen. S. Am.)

Donacospiza Reed-Finches PAGE 472

Buffy brown finch of *grassy areas in and near marshes. Rather long slender tail.*

11. LONG-TAILED REED-FINCH, *D. albifrons* (se. S. Am.)

Saltatricula Chaco-Finches — *Colorfully patterned* finch, common in *dry chaco.* PAGE 473

12. MANY-COLORED CHACO-FINCH, *S. multicolor*

Lophospingus Finches PAGE 473

Mostly gray finches with *long upstanding crests. White tail corners* conspicuous in flight.

13. BLACK-CRESTED FINCH, *L. pusillus* (mainly n. Arg.)
14. GRAY-CRESTED FINCH, *L. griseocristatus* (mainly Bol.)

Charitospiza Finches — Small finch of *Brazil's cerrado.* Note laid-back *crest in both sexes.* PAGE 474

15. COAL-CRESTED FINCH, *Charitospiza eucosma*

Coryphospingus Pileated-Finches PAGE 475

Flat red, black-bordered crests in ♂♂; *white eye-ring* in both sexes.

16. RED PILEATED-FINCH, *C. cucullatus rubescens* (interior s. S. Am.)
17. GRAY PILEATED-FINCH, *C. pileatus brevicaudus* (Col., Venez., e. Brazil)

1a♂
1b♀
2♂
3
4
5
6
7
8
9
10♂
11♂
12
13♂
14♂
15♂
15♀
16♂
16♀
17♂
G TUDOR

PLATE 31: YELLOW-FINCHES, SISKINS, ETC.

Sicalis Yellow-Finches — PAGE 476

Predominantly *yellow* finches of *open country;* some species more olive or brownish, ♀♀ may be streaky. Most are quite *gregarious;* usually feed on ground, often around buildings.

A. *Arid Pacific slope* or *widespread in lowlands* (some locally higher). — PAGE 477

- *Massive bill;* mostly white below.
  1. **SULPHUR-THROATED FINCH**, *S. taczanowskii* (sw. Ecu. and nw. Peru)
- *Bright yellow* with *orange forecrown;* ♀♀ and/or immatures much duller.
  2. **SAFFRON FINCH**
     2a. *S. f. flaveola;* 2b. *S. flaveola pelzelni*
     2c. *S. f. flaveola,* immature
  Also: Orange-fronted Yellow-Finch, *S. columbiana*
- Mostly *olive above* and *yellow below;* range also locally in highlands.
  3. **STRIPE-TAILED YELLOW-FINCH**, *S. citrina browni*
  4. **GRASSLAND YELLOW-FINCH**, *S. luteola luteiventris*
  Also: Raimondi's Yellow-Finch, *S. raimondii* (w. Peru)

B. *Andes* (Peru south) and/or *Patagonia.* — PAGE 480

- *Contrastingly patterned in gray and citron yellow.*
  5. **BRIGHT-RUMPED YELLOW-FINCH**, *S. uropygialis connectens* (Peru to nw. Arg.)
  Also: Citron-headed Yellow-Finch, *S. luteocephala* (local in Bol.)
- *Mostly greenish yellow* with *relatively little pattern.*
  6. **PUNA YELLOW-FINCH**, *S. lutea* (s. Peru to nw. Arg.)
  7. **GREENISH YELLOW-FINCH**, *S. o. olivascens* (Peru to n. Chile and w. Arg.)
  Also: Greater Yellow-Finch, *S. auriventris* (mainly Chile)
  Patagonian Yellow-Finch, *S. lebruni*

Carduelis Siskins and Goldfinches — PAGE 484

♂♂ usually *boldly patterned in black, yellow, and olive;* ♀♀ much duller. Favor *open or semiopen country,* but a few Andean species more associated with forest. *Gregarious.*

A. "*Hooded*" siskins. — PAGE 484

- "Typical"; body *olive and yellow.*
  8. **HOODED SISKIN**, *C. m. magellanica*
  9. **THICK-BILLED SISKIN**, *C. c. crassirostris* (high Andes)
  Also: Saffron Siskin, *C. siemeradzkii* (sw. Ecu.)
  Olivaceous Siskin, *C. olivacea* (e. slope Andes)
- Body *red* (♂) or with *salmon patches* (♀); now very rare.
  10. **RED SISKIN**, *C. cucullata* (mainly Venez.)

B. "*Capped*" and "*black-mantled*" siskins. — PAGE 487

- *Black cap.*
  11. **YELLOW-FACED SISKIN**, *C. yarrellii* (ne. Brazil)
  12. **BLACK-CHINNED SISKIN**, *C. barbata* (s. Arg. and Chile)
  Also: Andean Siskin, *C. spinescens* (n. Andes)
- *Black mantle and bib;* yellow below.
  13. **YELLOW-RUMPED SISKIN**, *C. uropygialis* (high Andes; mainly Chile and Arg.)
  Also: Yellow-bellied Siskin, *C. xanthogastra* (local; Venez. to Bol.)
- *Underparts mostly black.*
  14. **BLACK SISKIN**, *C. atrata* (high Andes; Peru to nw. Arg. and n. Chile)
- *Bicolored* (all yellow below), with *white* wing-patch.
  Not illustrated: Lesser Goldfinch, *C. psaltria* (Venez. to nw. Peru)

Rhodospingus Finches — PAGE 483

♂ *unique;* ♀ seedeaterlike but tail shorter, *bill more pointed.* Gregarious.

15. **CRIMSON FINCH**, *R. cruentus* (shrubby areas, w. Ecu. and nw. Peru)

Introduced Waxbills, Sparrows, and Finches — PAGE 490

Only 1 widespread (House Sparrow); all associated with urban/suburban areas. *All distinctive.*

16. **COMMON WAXBILL**, *Estrilda astrild* (e. Brazil)

Not illustrated: House Sparrow, *Passer domesticus*
European Goldfinch, *Carduelis carduelis* (Uru.)
European Greenfinch, *C. chloris* (Uru.)

1
2a♂
3♂
3♀
2b♀
2c
4♂
5♂
6♂
7♂
7♀
8♂
9♂
11♂
12♂
8♀
10♀
12♀
13♂
15♂
15♀
14♂
16
G TUDOR

The Oscines

ORDER PASSERIFORMES

ALL the species included in this volume of *Birds of South America* are now considered to belong in the Oscine suborder of the Order Passeriformes, which comprises what are known as the songbirds. Almost half the total number of S. Am. Passeriform species will be found here; the other half, the so-called suboscines, will be dealt with in our next volume, now in preparation. The 2 suborders differ mainly in their syrinx morphology, with the oscines containing perhaps the world's finest singing birds.

Taxonomy within the Oscine suborder is very controversial and will likely soon be made even more confusing by the gradual adoption of the findings now emerging from biochemical and DNA research. We have attempted to steer a relatively conservative course in such matters so as not to further confuse an already confusing situation. Essentially, we have followed most of the "higher order" changes made in the 1983 AOU Check-list. As a result there are some significant departures from the taxonomy of Meyer de Schauensee (1966, 1970). A brief summary of the 2 major "problem areas" follows:

1. The 1983 AOU Check-list has included within its enlarged Family Muscicapidae 2 groups which were formerly (including by Meyer de Schauensee 1966, 1970) considered families in their own right, the Sylviidae (Old World Warblers) and the Turdidae (Thrushes). These were there considered subfamilies; however, recent biochemical evidence is beginning to indicate that they may not be so closely related after all, and we therefore maintain both as full families.

2. The "9-primaried assemblage" comprises a very large group of overwhelmingly New World birds, the vast majority of which are found in South America. Up until recently its various component groups (wood-warblers, tanagers, etc.) were treated as full families (e.g., in Meyer de Schauensee 1966, 1970). Recent evidence, however, shows that they are more closely related to each other than this would seem to indicate; as a result, the 1983 AOU Check-list has combined them all into 1 huge family, the Emberizidae. We have followed this course here, but, so as not to let the complex become too large and unwieldy, have separated its component groups as subfamilies. Thus, our arrangement will not look all that "unfamiliar," but a few alterations should be listed here: Virtually all the Honeycreeper family (Coerebidae) has been subsumed into the Tanager subfamily, the 1 exception being the Bananaquit, which is maintained in its own subfamily (Coerebinae). The Swallow Tanager was formerly placed in its own family (Tersinidae) but is now considered to be merely a tribe in the Tanager subfamily. The Plush-capped "Finch" was formerly considered to represent a monotypic family (Catamblyrhychidae), but it too has more recently been treated as a subfamily in the Emberizidae, probably not particularly close to the "finches." Finally, the only S. Am. members left in the Fringillidae (which formerly included all the "finches," seedeaters, brush-finches, etc.) are the dozen-odd members of the genus *Carduelis*.

We must emphasize that the sequence of families, subfamilies, genera, and species which we employ *is not necessarily an indication of any relationship*. For us the association of various visually similar groups together took precedence, though as a general rule we have tried to stay close to the "normal" sequence, if only for familiarity's sake.

Corvidae

JAYS

OF this varied and widespread family, only jays are found in South America, no true crows being found south of Nicaragua (a rather inexplicable absence, given their abundance in tropical areas elsewhere). Jays in South America are not especially conspicuous or numerous in most areas. We have placed this family first in our sequence of the oscines to reflect recent DNA evidence which indicates that the Corvidae is the most divergent family of oscine passerines in the New World (and is part of a basically Old World radiation).

Cyanolyca Jays

Beautifully patterned, mainly blue jays found in the Andes and Middle America. Smaller and slenderer than *Cyanocorax* jays (mostly lowland birds), never showing much white on underparts or tail.

Cyanolyca armillata

BLACK-COLLARED JAY

PLATE: 1

Other: Collared Jay (in part)

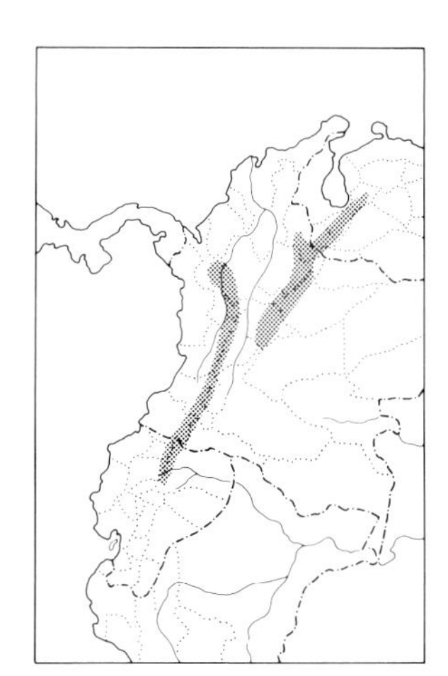

IDENTIFICATION: 33 cm (13″). *Andes from w. Venezuela to n. Ecuador. Mostly rich dark blue,* crown paler; forehead and broad mask across sides of head black, *extending down narrowly on sides of neck to encircle blue throat patch;* underside of tail black. N. races including *meridana* (Venezuela to E. Andes of ne. Colombia) more violet blue than is *quindiuna* (Cen. and W. Andes of Colombia south into Ecuador).

SIMILAR SPECIES: Note montane range and overall blue coloration. Range narrowly overlaps that of rather similar Turquoise Jay, but latter is shorter-tailed and more greenish blue generally and has markedly paler blue throat and crown; voices also differ. Note wide hiatus between range of this species and that of White-collared Jay to the south.

HABITAT AND BEHAVIOR: Rare to locally fairly common in humid montane forest and forest borders. Usually not very conspicuous, hopping and peering about in foliage and along branches, principally at middle and upper levels. Usually in small groups. Has an unusually large vocal repertoire (J. W. Hardy), though many of its calls are not particularly loud or arresting, and they are usually short and disjointed, almost seeming to be given at random; most frequent are a rising "shrwee?," a rapid chatter (e.g., "jet-jtjtjtjt"), a low, guttural "wor" or soft "craah," and several high, thin notes. Black-collared Jays seem most numerous in w. Venezuela (e.g., in Táchira on the Queniquea road), less common in Colombia (even where adequate habitat remains), and known from only 2 records in Ecuador.

RANGE: Andes of w. Venezuela (north to Trujillo), Colombia (where local, and known from W. Andes only at their n. end), and extreme n. Ecuador (on e. slope of Andes in extreme e. Carchi and nw. Napo). Mostly 1800–3000 m.

NOTE: The black-collared n. birds are here considered a species distinct from the white-collared nominate group, following Goodwin (1976). Not only are the morphological differences sharply defined, but they also appear to differ vocally. Goodwin (1976) calls *C. armillata* the Collared Jay, but we feel that name is best reserved for the enlarged species and thus favor calling *armillata* the Black-collared Jay.

Cyanolyca viridicyana

WHITE-COLLARED JAY

PLATE: 1

Other: Collared Jay (in part)

IDENTIFICATION: 34 cm (13¼"). *Andes of Peru and Bolivia. Mostly greenish blue.* Forehead and broad mask across sides of head black, barely contrasting with dark greenish blue throat; mask bordered above by *narrow milky white band and conspicuous milky white forecrown, throat bordered below by narrow but conspicuous white collar.* Underside of tail black. *Jolyaea* (Peru from Amazonas south to Junín) rather different, being more purplish blue generally, with throat essentially concolor with rest of underparts (and thus contrasting more with black mask), and having narrower and less conspicuous white pectoral collar and less crisply defined milky white area on crown.

SIMILAR SPECIES: Does not overlap with any of the other montane, blue *Cyanolyca* jays (all of which are found to the north); in any case, none shows the white collar.

HABITAT AND BEHAVIOR: Uncommon to fairly common in humid montane forest and forest borders. Behavior similar to that of Black-collared Jay, but its voice seems quite different and is markedly less varied (in this resembling more the geographically intermediate Turquoise Jay), with loud repeated notes being most frequently given (e.g., "jeet-jeet-jeet" or "jho-jho-jho-jho").

RANGE: Andes of e. Peru (north to Amazonas) and nw. Bolivia (La Paz and Cochabamba). Mostly 2000–3000 m.

Cyanolyca turcosa

TURQUOISE JAY

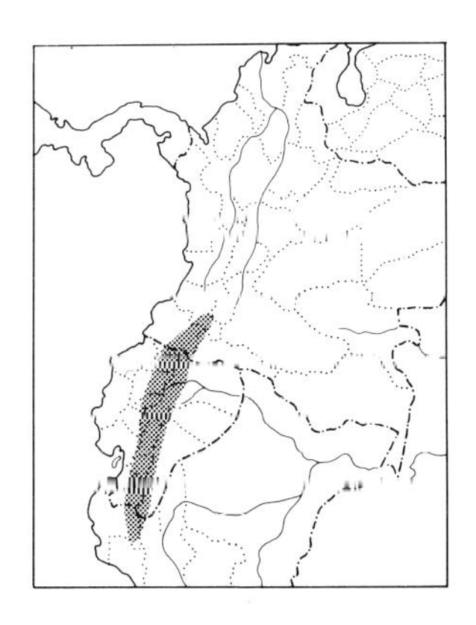

IDENTIFICATION: 32 cm (12½"). *Andes of Ecuador and adjacent Colombia and Peru.* Closely resembles *quindiuna* race of Black-collared Jay (reportedly sympatric, at least locally), but *more greenish blue generally* and *throat and crown notably brighter and paler blue* (almost turquoise). Tail of Turquoise Jay is proportionately shorter. See C-12.

SIMILAR SPECIES: Cf. Black-collared Jay; remember that *quindiuna* race of that species is *less* purplish in color than is illustrated *meridana*.

HABITAT AND BEHAVIOR: Fairly common to locally common in humid montane forest, forest borders, and better-developed secondary growth. General behavior similar to that of Black- and White-collared Jays, but seems better able to tolerate somewhat disturbed or partially cutover areas, perhaps especially where there are extensive stands of alders

(*Alnus* sp.). By far the most often heard call is a loud arresting "jeeyr! jeeyr! jeeyr!," usually given in series but sometimes singly; a variety of other, less raucous, more "conversational" (or contact) notes is also given. Readily found along various roads west of Quito in Pichincha, Ecuador (e.g., the Chiriboga road and the Nono-Mindo road).
RANGE: Andes of extreme s. Colombia (both slopes of Nariño), Ecuador, and extreme n. Peru (n. Piura and nw. Cajamarca). 1500–3000 m.

Cyanolyca pulchra

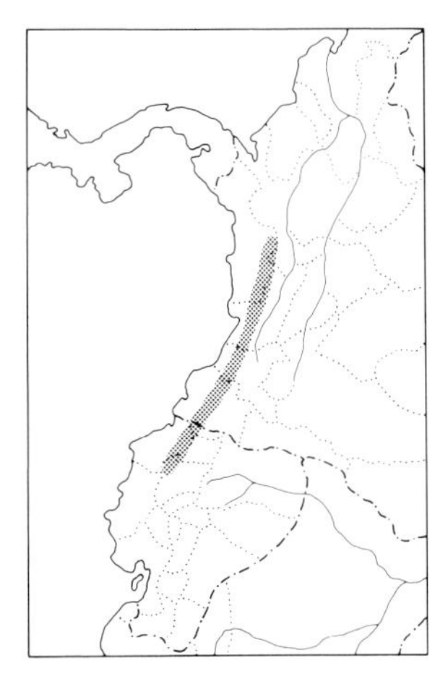

BEAUTIFUL JAY

IDENTIFICATION: 27 cm (10½"). *W. Andes in w. Colombia and nw. Ecuador.* Mostly purplish blue, but strongly suffused with brown across mantle and on breast (especially in females and immatures). *Crown contrastingly milky bluish white,* with forehead and broad mask across sides of head black. See C-12.
SIMILAR SPECIES: Black-collared and Turquoise Jays lack the whitish crown and show a prominent black pectoral collar outlining the throat.
HABITAT AND BEHAVIOR: Rare and local in very humid forest and forest borders. Usually forages singly or in pairs, generally remaining inconspicuous unless calling; tends to forage more in lower growth than its S. Am. congeners. Its vocalizations include a fairly soft "chee-chee-chee," sometimes becoming louder and more emphatic (e.g., "chewp-chewp-chewp"), as well as other harsh, somewhat strident notes. Seems to have declined in nw. Ecuador over the past 10 to 15 years for reasons unknown.
RANGE: Pacific slope of W. Andes in w. Colombia (north to extreme s. Chocó) and nw. Ecuador (south to Pichincha). Mostly 900–1800 m.

Cyanocorax Jays

The *Cyanocorax* jays are, as a group, widely distributed through the lowlands of South America, though never are more than 2 found together and in most areas only 1 is found. A single species, the Green Jay, is more montane. All but the last are predominantly blue, most species being contrastingly patterned with blue or white facial markings and white on the underparts and tail. All are noisy, garrulous, and highly social birds. Though most remain unstudied from a behavioral perspective, those that have been (e.g., the Green Jay in Colombia) are known to remain in groups of related individuals throughout the year and to nest in a cooperative manner (immatures helping the single pair of breeding adults); it seems likely that the others will be found to exhibit similar behavior once they are better known.

GROUP A.

Iris dark; no facial markings; all but 1 *without* white tail tips; 4 jays, probably comprising a superspecies; all have dark irides, no pale markings on the face, and similar harsh voices. The Curl-crested Jay diverges dramatically in superficial appearance (white lower underparts, white on tail, etc.) but otherwise seems clearly to fall within this group.

Cyanocorax cyanomelas

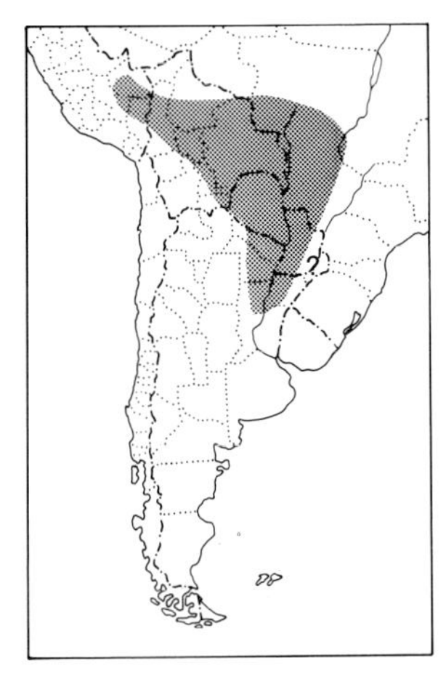

PURPLISH JAY

PLATE: I

IDENTIFICATION: 37 cm (14½"). *S.-cen. South America. A drab, purplish brown jay with no obvious field characters.* Forecrown, sides of head, throat, and chest sooty black, becoming dull purplish brown on mantle and dull brownish purple on belly; tail violet blue (black from below).
SIMILAR SPECIES: The dullest S. Am. jay, showing little pattern except in strong light. Azure Jay is much brighter blue generally, with none of the brown tone. Violaceous Jay nearly overlaps this species in se. Peru (though Purplish is essentially restricted to valleys near the base of Andes, while Violaceous occurs out in Amaz. lowlands); Violaceous Jay is paler and brighter generally, with conspicuous pale band on hindneck (lacking in Purplish).
HABITAT AND BEHAVIOR: Fairly common to common in deciduous and gallery woodland and adjacent scrub. Noisy and conspicuous birds, they usually are found in groups of up to 6 to 8 individuals, regularly in the company of Plush-crested Jays. They are bold and often inquisitive, coming in to investigate squeaking; though basically arboreal, they also drop to the ground. Its flight often appears labored, with slow wing-strokes, usually ending with a long upward glide to its perch. Its calls are loud and often raucous (especially when excited), a repeated "jar-jar-jar-jar . . ." or "craa-craa-craa-craa . . ." being most frequent. Numerous at various points around Santa Cruz and Cochabamba, Bolivia, and outside Asunción, Paraguay.
RANGE: Se. Peru (locally in Cuzco and Madre de Dios), n. and e. Bolivia, Paraguay (virtually throughout, but perhaps not in extreme east or in driest parts of *chaco*), sw. Brazil (w. Mato Grosso), and n. Argentina (commonly in e. Formosa, e. Chaco, n. Santa Fe, and Corrientes; less frequent in Misiones). To about 2000 m (in Bolivia).

Cyanocorax caeruleus

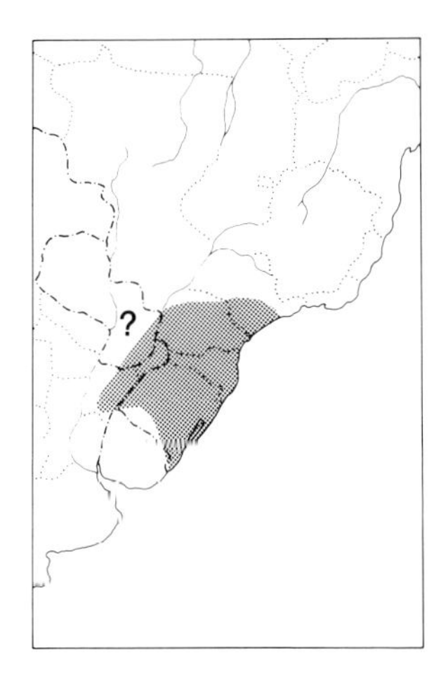

AZURE JAY

IDENTIFICATION: 38 cm (15"). *Se. South America. Mostly purplish blue;* head, throat, and chest contrastingly black. What is apparently another color phase is more *greenish blue* instead of purplish blue; it is recorded from throughout the species' range but always seems less numerous.
SIMILAR SPECIES: Somewhat similar to Purplish Jay, which it replaces eastward, but obviously much bluer. The only jay across all but a small portion of its range.
HABITAT AND BEHAVIOR: Rare to locally fairly common in forest, secondary woodland, and borders; at least in Brazilian portion of its range seems most numerous where *Araucaria* predominates. Behavior and voice very similar to Purplish Jay's. Overall numbers have declined greatly due to deforestation across much of its former range; now seems most numerous in se. Brazil (though even there very local), with population perhaps greatest in Rio Grande do Sul. Also may no longer occur in w. portion of its range (see below).
RANGE: Se. Brazil (s. São Paulo south to Rio Grande do Sul), e. Paraguay, and ne. Argentina. To about 1000 m. Distribution in Para-

guay and Argentina uncertain. Old records exist west to the Paraguay River around Asunción, Paraguay, and in e. Formosa and e. Chaco, Argentina, but recent work has revealed the presence of only Purplish Jays there. These old records may represent misidentifications (as Short 1975: 288 suggests), but it is also possible that the Purplish Jay is now outcompeting the Azure in these areas for reasons which remain obscure. Even in extreme e. Paraguay and ne. Argentina the Azure Jay seems to have declined substantially (and this is the case even where suitable habitat remains): there is only 1 recent report of the Azure Jay from e. Paraguay (that of a pair seen west of Puerto Stroessner in Aug. 1977; RSR), and only a very few recent sightings exist from Iguazú Nat. Park in Misiones, Argentina (*fide* M. Rumboll).

Cyanocorax violaceus

VIOLACEOUS JAY

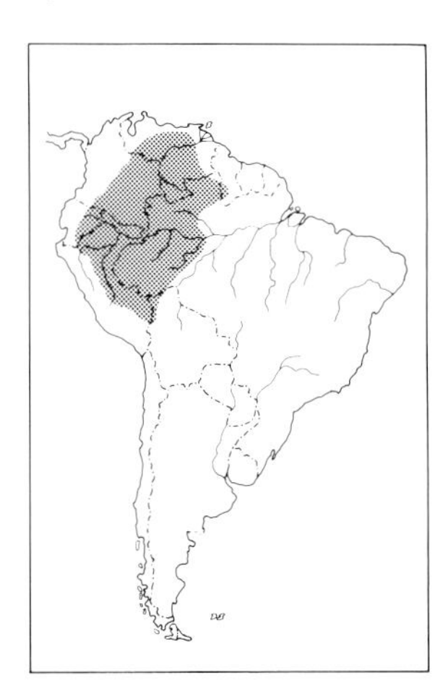

IDENTIFICATION: 37 cm (14½"). *W. Amazonia to Venezuela. Mostly dull violaceous lavender,* darker above than below, brightest on wings and tail. *Head, throat, and chest contrastingly black,* with *band on hindcrown and sides of neck milky bluish white.* Often looks somewhat crested, feathers of crown often being raised, those of nape always laid flat. See V-10, C-12.

SIMILAR SPECIES: The only jay across most of its upper Amazon and upper Orinoco range. In sw. Venezuela and adjacent Colombia and Brazil cf. Azure-naped Jay, and in se. Peru, Purplish Jay.

HABITAT AND BEHAVIOR: Fairly common to common in *várzea* forest and forest borders, clearings, and along margins of rivers or lakes; rarely or never in continuous *terra firme* forest. Noisy and conspicuous birds, trooping about in flocks of up to a dozen or so individuals; perhaps most numerous in stands of *Cecropia* along the Amazon and its tributaries, there frequently seen flying 1 by 1 from a shore or island to another. Almost always remain well above the ground. The commonest call is a loud, harsh screaming "jyeeer! jyeeer! . . ." uttered either singly or in (sometimes long) series.

RANGE: Venezuela (north through the llanos, but absent from the extreme east), s. Guyana, e. Colombia, e. Ecuador, e. Peru, extreme n. Bolivia (Pando), and n. and w. Brazil (east only to the Rio Purús and upper Rio Negro area). To about 1000 m.

Cyanocorax cristatellus

CURL-CRESTED JAY

PLATE: 1

IDENTIFICATION: 34.5 cm (13½"). *Cerrado of s.-cen. Brazil.* Long *upstanding frontal crest* apparent at all times (never laid flat). Rather short-tailed. Head, neck, and chest sooty black, becoming dull violet brown on hindneck and upper back. Otherwise dull violet blue above, brightest on wings and tail. *Terminal half of tail white* (all white from below). *Lower underparts white.*

SIMILAR SPECIES: All the other jays with white lower underparts have some pale blue or whitish markings on face, a pale iris, and narrower

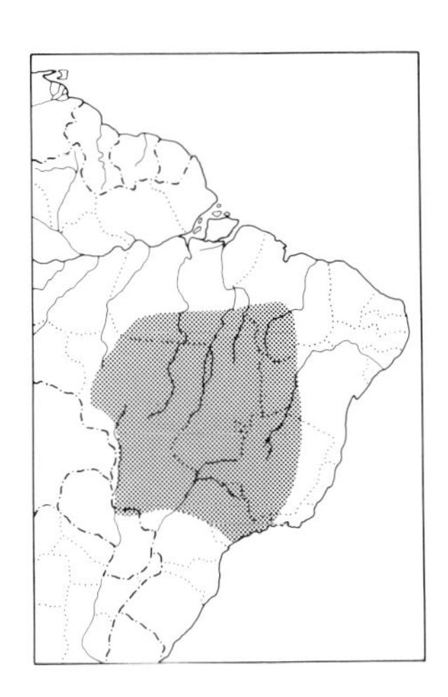

white tail tipping; the frontal crest is, in any case, diagnostic. Compare especially to Plush-crested and White-naped Jays, both of which are partially sympatric with Curl-crested, though always occurring in more wooded terrain; calls also differ.

HABITAT AND BEHAVIOR: Fairly common to locally common in *cerrado* and edges of gallery woodland. Generally occurs in small straggling groups; bold, inquisitive, and easy to see (though less so when nesting). Never seems to actually flock with other species of jays. The loud calls are very similar to Purplish Jay's, typically a sharp "kyaar!," often repeated several times. Easily seen around Brasilia.

RANGE: Interior cen. and s. Brazil (sw. Pará, s. Maranhão, and s. Piauí south through Mato Grosso and Goiás to Minas Gerais and São Paulo) and extreme ne. Paraguay (Concepción at San Luis de la Sierra; Laubmann 1939). To about 1100 m.

GROUP B.

Iris pale; blue to white nape and facial markings, tail tipped white; 6 jays with largely allopatric ranges across the lowlands of South America. They are united by their pale irides, conspicuous white or blue facial markings, and overall pattern (including white in tail); voices are also similar, though distinctively different from the previous group's.

Cyanocorax heilprini

AZURE-NAPED JAY

PLATE: 1

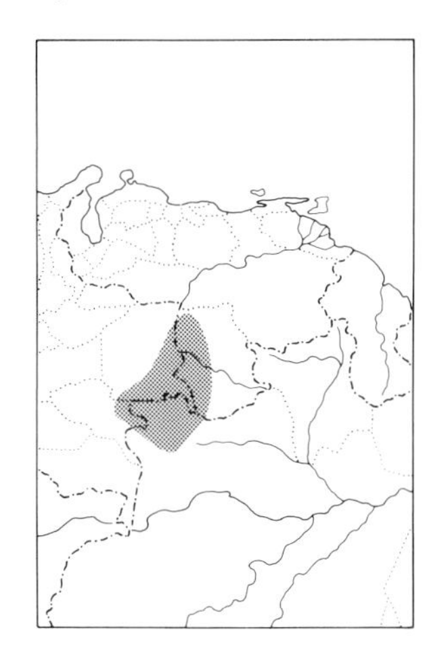

IDENTIFICATION: 33.5 cm (13¼"). *Limited area in sw. Venezuela and adjacent Colombia and Brazil.* Forecrown, sides of head, throat, and chest black, with feathers of forehead stiff and upstanding (imparting a slight bushy-crested look); hindcrown, nape, and *short moustache milky bluish.* Otherwise mostly grayish brown glossed with bluish purple; crissum white, and *tail relatively narrowly tipped white.*

SIMILAR SPECIES: Violaceous Jay lacks the white tail-tipping. Cf. also Cayenne Jay (no known overlap).

HABITAT AND BEHAVIOR: Recorded from forest borders and light woodland in sandy soil areas. Very little has been recorded about this jay. Its voice seems likely to resemble that of the other pale-eyed jays, though J. Borrero (in Hilty and Brown 1986) reports that it resembles the Violaceous Jay(?).

RANGE: Sw. Venezuela (w. Amazonas), extreme e. Colombia (e. Guainía and e. Vaupés), and extreme nw. Brazil (upper Rio Negro drainage). Below 250 m.

Cyanocorax chrysops

PLUSH-CRESTED JAY

PLATE: 1

IDENTIFICATION: 35.5 cm (14"). *S-cen. South America. Crown feathers stiff and plushlike,* extending to rear as short crest. Head, sides of neck, throat, and chest black; spot above eye bright silvery blue, with spot below eye and short moustache deep blue. *Band on hindneck milky bluish white,* becoming violet blue on upper back; otherwise dark violet blue above, brightest on wings and tail, with tail broadly tipped white.

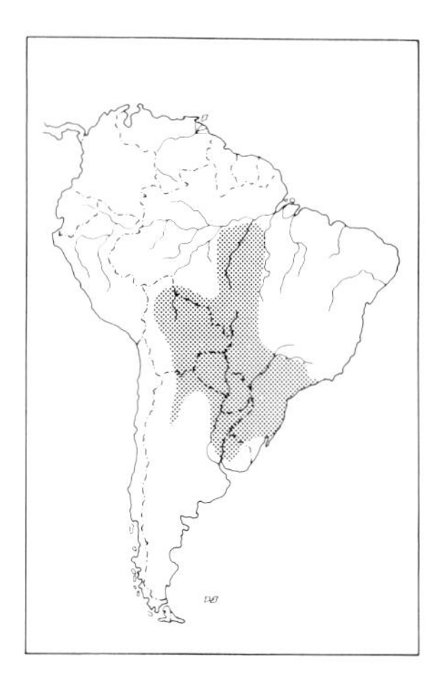

Lower underparts creamy white. *Diesingii* (locally on lower Rio Madeira and lower Rio Tapajós in Amaz. Brazil) has facial markings much reduced and lacks the pale band on hindneck.

SIMILAR SPECIES: White-naped Jay has different head shape, lacking crest on hindcrown, broader *white* band on nape, and essentially brown upperparts. Cf. also Curl-crested Jay.

HABITAT AND BEHAVIOR: Fairly common to common in forest, woodland, and borders, occasionally venturing out into scrub or groves in agricultural areas. Range in groups of up to 10 to 12 individuals, often accompanied by the larger-bodied Purplish Jay; they forage actively at all levels, hopping and peering about on branches and in foliage. Usually not at all shy. Among their variety of loud and arresting calls (which often serve to draw attention to the birds), the most frequent is a ringing "cho-cho-cho" (typically trebled, but sometimes not); they sometimes imitate other birds.

RANGE: N. and e. Bolivia (west to Beni and Cochabamba), cen. and s. Brazil (Pará south of the Amazon from lower Rio Madeira and lower Rio Tapajós south through Mato Grosso to São Paulo and Rio Grande do Sul), Paraguay, Uruguay, and n. Argentina (south to Tucumán, Catamarca, Chaco, n. Santa Fe, and Entre Ríos). The supposed race, *interpositus,* of Plush-crested Jay found in Pernambuco, Brazil, is now considered (J. W. Hardy, *Condor* 71: 365, 1969) to represent molting or worn subadult *C. cyanopogon.* To about 2800 m (in s. Bolivia), but mostly below 1500 m.

Cyanocorax cyanopogon

WHITE-NAPED JAY

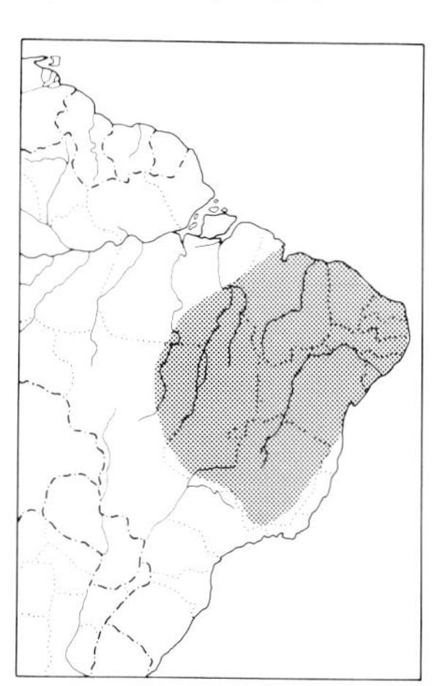

IDENTIFICATION: 34.5 cm (13½"). *Ne. Brazil.* Feathers of forecrown stiff and plushlike (but those of hindcrown are *not* noticeably so). Head, sides of neck, throat, and chest black; spot above eye bright silvery blue, with spot below eye and short moustache deep blue. *Broad area on hindneck white, becoming dull brown on remaining upperparts;* tail broadly tipped white. Lower underparts white.

SIMILAR SPECIES: Plush-crested Jay is similar but always shows obvious stiff crest and narrower and more bluish band on hindneck, is bluer (not so brown) above. The 2 are not known to be sympatric. Cf. also Curl-crested Jay (with which broadly sympatric, though normally in a different habitat).

HABITAT AND BEHAVIOR: Fairly common in deciduous woodland, gallery woodland, and dense scrub. Behavior and vocalizations much like Plush-crested Jay's. At least in Bahia seems to have declined in recent years, due perhaps to cagebird trappings.

RANGE: E. Brazil (se. Pará, Maranhão, and Ceará south through Goiás and much of Bahia to Minas Gerais and e. Mato Grosso). Earlier recorded specimens of *C. cyanopogon* from São Paulo and Paraná are actually *C. chrysops* (see Haffer 1975: 148); though their ranges approach each other closely, these 2 species do not seem to have ever been actually taken at the same locality. To about 1100 m.

Cyanocorax affinis

BLACK-CHESTED JAY

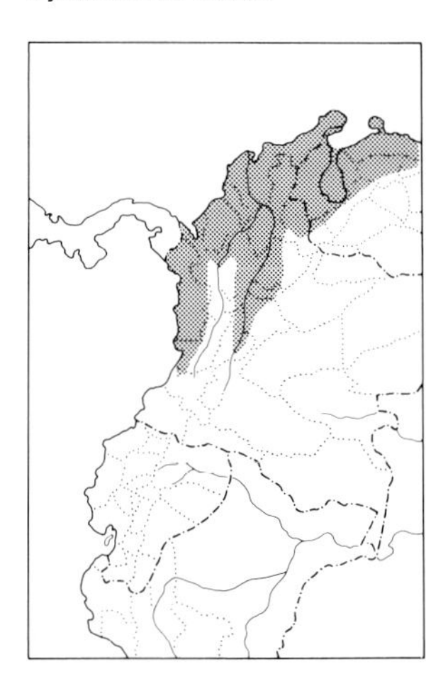

IDENTIFICATION: 35.5 cm (14"). *N. Colombia and nw. Venezuela.* Head, sides of neck, throat, and chest black; spots above and below eye and short moustache deep blue. Mantle violet brown, more violet blue on wings and tail; *tail broadly tipped white. Lower underparts creamy white.* See V-10, P-11.

SIMILAR SPECIES: The only jay in its range with white belly and tail-tipping.

HABITAT AND BEHAVIOR: Uncommon to locally common in a variety of forest types (both humid and rather dry), but occurs mainly at forest edge and in second growth and overgrown clearings. Active and alert birds, Black-chested Jays move about in groups of up to 6 to 8 individuals, foraging at all levels, though usually not too high above ground. Their loud ringing calls ("cho! cho!" or "chowng-chowng" is perhaps the most frequent) regularly draw attention to them, though they may remain mostly hidden in heavy cover for protracted periods.

RANGE: N. Colombia (south in Pacific lowlands to Valle and in Magdalena valley south to Huila) and nw. Venezuela (Maracaibo basin east through Falcón). Also Costa Rica and Panama. Mostly below 1500 m (but recorded to 2200 m in Colombia).

Cyanocorax mystacalis

WHITE-TAILED JAY

PLATE: 1

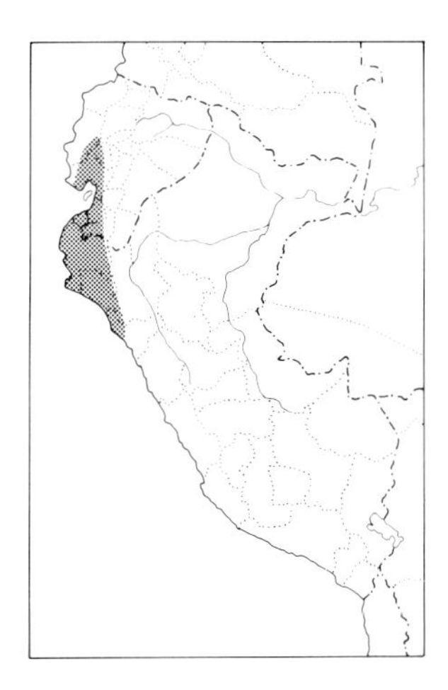

IDENTIFICATION: 32.5 cm (12¾"). *Sw. Ecuador and nw. Peru.* Head, sides of neck, throat, and chest black; small spot above eye and broad moustache (which arches up to reach eye) white. *Large area on hindneck and upper back white;* upperparts otherwise dark blue, brightest on wings; *tail mostly white,* with only central feathers blue. Lower underparts pure white.

SIMILAR SPECIES: This exceptionally attractive jay cannot be confused in its limited range, where it is the only jay found.

HABITAT AND BEHAVIOR: Fairly common to common in dry woodland, arid scrub, and cactus-dominated desert vegetation; absent from humid areas. Behavior similar to that of other *Cyanocorax* jays, though tends to forage more on the ground and to be unusually conspicuous (in part due to its open or semiopen habitats). Its vocal repertoire is quite limited (J. W. Hardy and RSR), with a fast, rather dry repetition, "cho-cho-cho . . ." or "chah-chah-chah . . ." being most typical. Numerous just west of Guayaquil, in the Chongon Hills and along the first part of the highway west toward Salinas.

RANGE: Sw. Ecuador (Guayas, El Oro, and w. Loja) and nw. Peru (south to w. La Libertad). To 1200 m.

Cyanocorax cayanus

CAYENNE JAY

IDENTIFICATION: 33 cm (13"). *Ne. South America.* Resembles geographically distant White-tailed Jay, but tail mostly violet blue (only tipped white) and mantle brown glossed with violet, brightest on

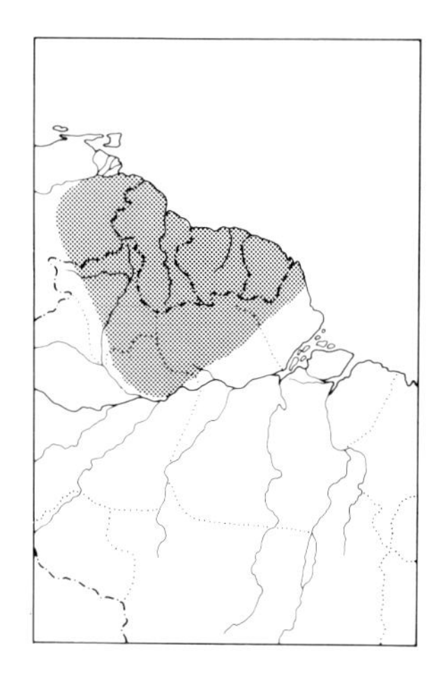

wings; moustache slightly narrower, and with discrete small white spot below eye. See V-10.

SIMILAR SPECIES: *The only jay found in most of its range.* Compare to more uniform-looking Violaceous Jay (which lacks any facial markings or tail tips) and to Azure-naped Jay (which is not mainly white below).

HABITAT AND BEHAVIOR: Fairly common in forest borders, secondary woodland, and woodland and adjacent scrub in sandy soil regions; essentially absent from continuous forest. Behavior much like that of other jays, with its most frequent call being (like all the others in this group) a repetitive loud ringing "choh-choh-choh" or some variant thereof.

RANGE: Guianas, se. Venezuela (Bolívar), and ne. Brazil (relatively few records, though known south to Manaus area near lower Rio Negro and from n. Amapá). To 1100 m (in Venezuela).

GROUP C. Distinctive mainly *green and yellow* coloration.

Cyanocorax yncas

GREEN JAY

PLATE: 1

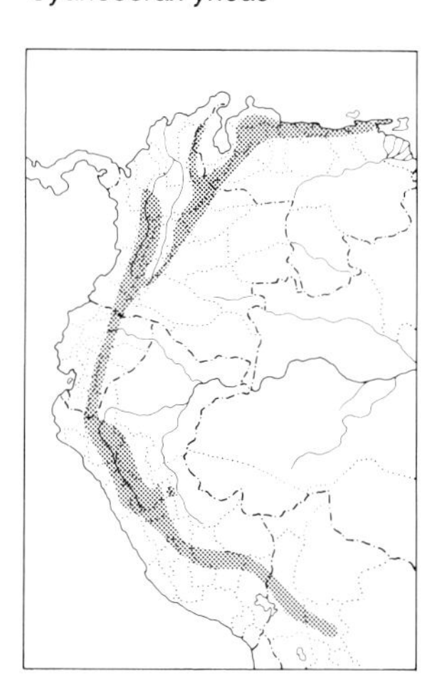

IDENTIFICATION: 29.5–30.5 cm (11½–12"). *An unmistakable, mostly green and yellow jay of the Andes and Venezuelan mountains.* Stiff frontal crest of blue feathers, longest in Venezuelan races, shorter (but still quite evident) in birds from s. Colombia south. Mostly green above and yellow below, with *outer tail feathers also yellow* (very conspicuous in flight); sides of head and neck and throat and chest black, with blue spot above eye and broad blue moustache arching up to touch lower eye. *Crown and nape bluish white in most of its range,* but blue in races from E. Andes of Colombia north into Venezuela (see V-10).

HABITAT AND BEHAVIOR: Fairly common to common in humid montane forest and forest borders and secondary woodland. Also regularly ventures out into trees and shrubbery in clearings; in fact, in many regions it tolerates a good deal of habitat disturbance and, unlike so many other of the larger montane forest birds, it often maintains sizable populations in such areas. Green Jays are almost always conspicuous, moving about and foraging at all levels in groups of up to about 10 birds. Though quiet when actually feeding, they are noisy at other times, with a variety of loud calls being given, both as alarm and contact notes. The most frequent is a usually trebled, nasal "nyaa-nyaa-nyaa" or "quin-gun-gun" (from which its local Colombian name of "Quinquin" is derived); also has a distinctive metallic clicking call as well as a "clee-op" (often trebled), etc.

RANGE: Mts. and foothills of n. Venezuela (east to Sucre) and Andes of w. Venezuela, Colombia (not south of Cauca in W. Andes), e. Ecuador, e. Peru, and nw. Bolivia (La Paz and Cochabamba); Perijá Mts. on Venezuela-Colombia border. Also Texas and Mexico to Honduras. Mostly 1400–2600 m, but recorded down almost to sea level in Venezuela.

Hirundinidae

SWALLOWS AND MARTINS

A distinctive and familiar group of aerial feeding birds, the swallows differ from the superficially alike swifts in their more maneuverable wings and flight and in their habit of perching on branches or phone wires (swifts can only cling to vertical surfaces; they cannot grasp). A high proportion of the S. Am. swallows are migratory and most are very gregarious; though most species are common, a few have restricted ranges or are poorly known (or under-recorded) migrants.

Phaeoprogne Martins

This monotypic genus is very close to *Progne* structurally and in appearance (aside from being much browner) but differs in its solitary nesting habits. The 2 genera should perhaps be merged.

Phaeoprogne tapera

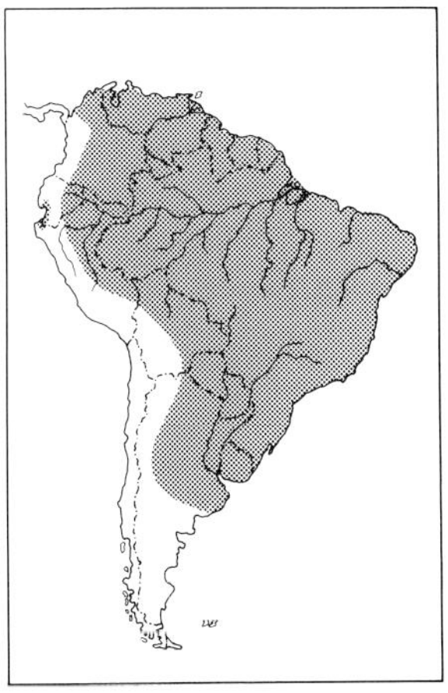

BROWN-CHESTED MARTIN PLATE: 2

IDENTIFICATION: 17.5 cm (7"). *Dull grayish brown above,* duskier on wings and tail. Below white with *distinct grayish brown breast band,* continuing down on median breast as band of droplike spots. White crissum feathers long and silky and often visible from above along edge of rump and base of tail. Foregoing applies to *fusca,* the race breeding in s. South America but migrating northward during austral winter, some birds even reaching the Caribbean. Nominate race (resident south to Amazonia and ne. Brazil) similar but with less well defined breast band merging into grayish white throat and lacking the droplike spots on the median breast.

SIMILAR SPECIES: Though color pattern is reminiscent of Sand Martin's, this species is so much larger that confusion is unlikely. None of the *Progne* martins, even immatures, are so brown above: all show a noticeable violet blue gloss. Flight profile and even color and pattern are at times remarkably like Solitary Sandpiper's (*Tringa solitaria*).

HABITAT AND BEHAVIOR: Generally common in semiopen or open areas with scattered trees, often near water, especially when not breeding (at which time it may gather in large flocks and associate with other swallows). The s. *fusca* race nests as scattered pairs which settle in the vicinity of hornero nests. If the nest is vacant, the martins quickly move in, but if the horneros are still in residence a protracted struggle ensues, with the martins apparently nearly always the victors by sheer persistence. The n.-breeding nominate race reportedly nests in cavities, including arboreal termitaries (F. Haverschmidt), but perhaps also uses hornero nests when these are available. Foraging Brown-chested Martins have a swift, graceful flight similar to that of other martins, but on the whole this species often seems sluggish, resting for long periods on branches at the edge of woods or near water (less often on wires). Especially when nesting (but also at other times), they often make very slow, fluttery, circular flights with wings characteristically bowed downward. Unlike the *Progne* martins, they rarely seem to fly at any height above the ground.

RANGE: Widespread south to cen. Argentina (south to Mendoza, La Pampa, and Buenos Aires); west of the Andes only in nw. Venezuela, n. Colombia (Caribbean lowlands and south in Magdalena valley to Huila), and sw. Ecuador (north to Guayas) and nw. Peru (Tumbes); vagrants to coastal sw. Peru in Arequipa (R. A. Hughes). *Fusca* breeds from s. Bolivia and s.-cen. Brazil southward, where it is present ca. Sept.–Apr. Also Costa Rica and Panama (in austral winter). To about 3000 m (as austral migrant in Colombia).

Progne Martins

The martins are a group of large swallows with comparatively broad, somewhat triangular wings and a fairly long, forked tail. They alone among the American swallows exhibit quite marked sexual dimorphism. Identification points are often subtle, though fortunately they tend to be rather tame and conspicuous, often perching in large groups on wires. All are essentially allopatric on their breeding ranges but often mix when on migration. Their taxonomy is chaotic, with as many as 8 species being recognized (e.g., by the 1983 AOU Check-list) or as few as 4 (e.g., *Birds of the World,* vol. 9), but regardless of the total number clearly all belong in one superspecies. We adopt an intermediate approach but admit to remaining uncertain just what represents the best course. E. Eisenmann and F. Haverschmidt (*Condor* 72 [3]: 368–369, 1970) suggest that the dark forms breeding in tropical latitudes (*modesta* on the Galápagos and *murphyi*) probably evolved separately from temperate-breeding *subis* and *elegans.*

Progne chalybea

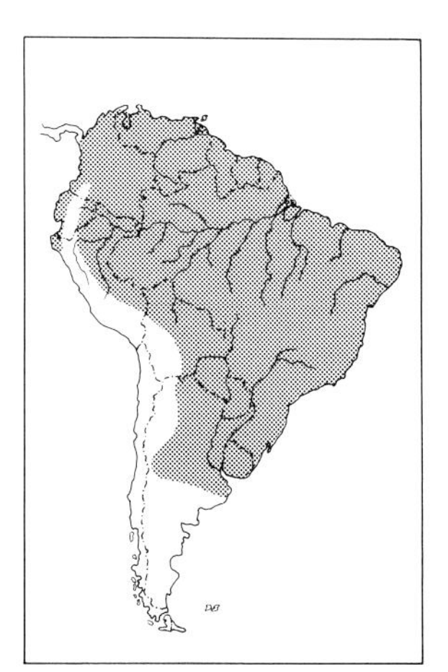

GRAY-BREASTED MARTIN

PLATE: 2

IDENTIFICATION: 18–19 cm (7–7½″), larger in s.-breeding *domestica* (e. Brazil and cen. Bolivia southward, but migrating north in austral winter as far as Venezuela and Suriname). Male *glossy steely blue above,* blacker on wings and tail. Throat and breast grayish brown, sometimes with some glossy blue on chest (this has been viewed as a very rare melanistic phase); lower underparts white, sometimes with dark streaks. Female similar but less strongly glossed blue above. Immature browner above and showing less contrast below.

SIMILAR SPECIES: *By far the most numerous and widespread Progne in South America;* learn it well as the basis of comparison with other martins. This is the *least* sexually dimorphic species. When searching for the others, look for birds with pale foreheads or collars (these are female Purples), all-dark underparts (these are female Southerns), or sharply demarcated dark-and-white underparts (these may be Caribbeans); see also respective males of these species. Brown-chested Martin is much browner above and has a pale throat (never brownish, as in all Gray-breasts).

HABITAT AND BEHAVIOR: Locally very common in semiopen areas and (especially) in towns and inhabited areas; most numerous near water, but also often frequent away from it. A gregarious and familiar martin, often nesting under eaves of buildings; away from buildings smaller numbers nest in holes of dead snags. Though absent from generally forested areas, does colonize new clearings rapidly. Their flight tends to be higher than that of most other swallows and often seems very leisurely, the birds circling far up in the sky for protracted periods. Males have a rich, liquid gurgling song, given especially near where they are nesting, both in flight and when perched; both sexes also regularly give various short rich calls (e.g., "chu" or "chu-chu").

RANGE: Widespread in lowlands south to n. Argentina (south to Mendoza, Córdoba, and Buenos Aires); present in s. part of breeding range only during austral summer (mostly Sept.–Mar.); on Pacific slope south to nw. Peru; Trinidad, casual on Tobago. In Argentina breeds mainly east of range of Southern Martin, though there is some overlap; the 2 species occur together regularly on migration. Also Middle America. To about 1200 m.

Progne modesta

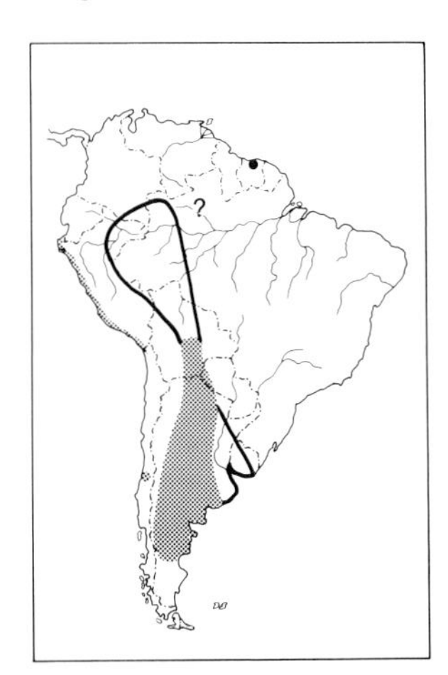

SOUTHERN MARTIN

PLATE: 2

IDENTIFICATION: 18–19.5 cm (7–7¾"). *Breeds locally in s. and w. South America,* migrating northward. Male *glossy steely blue,* blacker on wings and tail. Female not as strongly glossed blue above and sometimes shows a grayish brown forecrown. *Mostly dusky brown below,* feathers often edged pale, giving scaly effect; belly usually only slightly (if at all) paler than remainder of underparts. *Murphyi* (coastal Peru and n. Chile) notably smaller than *elegans* (breeding in s. South America, migrating to Amazonia), and males are never as highly glossed with blue.

SIMILAR SPECIES: Female Southern Martins are the only martin with essentially *uniform* underparts; compare to female Purple (best distinguished by its usually prominent partial collar). Male Southern closely resembles male Purple Martin (the 2 could occur together during the austral summer in n. Argentina or possibly elsewhere on migration), and the 2 are usually not safely distinguished in the field; Southern is slightly larger and has a slightly longer and more deeply forked tail.

HABITAT AND BEHAVIOR: Fairly common around towns, buildings in settled or agricultural areas, and cliffs along coast; wintering birds in w. Amazonia roost and feed along rivers and may gather in exceptionally large concentrations: a quarter-million or more roosted in downtown Iquitos, Peru, during the late 1970s (D. Oren, *Condor* 82 [3]: 344–345, 1980). General behavior much like Gray-breasted Martin's and often flocking with that species on its wintering grounds.

RANGE: Pacific coastal Peru (where resident locally from Piura south) and extreme n. Chile (where perhaps only casual); *elegans* breeds locally from w. Bolivia (north to Cochabamba and w. Santa Cruz) south across much of Argentina (south locally to Chubut; absent from much of the northeast) and is also recorded in Uruguay: *elegans* passes the austral winter (ca. Apr.–Oct.) mostly in w. Amazonia, whence it is recorded from ne. Peru, se. Colombia, and w. Brazil, with one record (of 5 specimens collected) from coastal Suriname. Vagrant *elegans* have overshot their normal wintering range on occasion, hence recorded from Panama and Florida, and have also strayed to cen. Chile and the Falkland Is. Recorded to over 2000 m in Bolivia, but mostly in lowlands.

NOTE: Following most recent authors (but not the 1983 AOU Check-list), we consider the 3 allopatric forms in this species as conspecific; if they are deemed full species, then *Progne modesta* becomes the Galapagos Martin, *P. elegans* the Southern Martin, and *P. murphyi* the Peruvian Martin. We unite them in part because males of *murphyi* appear intermediate in terms of size and gloss between *elegans* and the small, dull *modesta*, but it is possible they are not so closely related. A few hybrids between *P. m. elegans* and *P. chalybea* have been taken in nw. Argentina (E. Eisenmann and F. Haverschmidt, *Condor* 72 [3]: 368–369, 1970).

Progne subis

PURPLE MARTIN

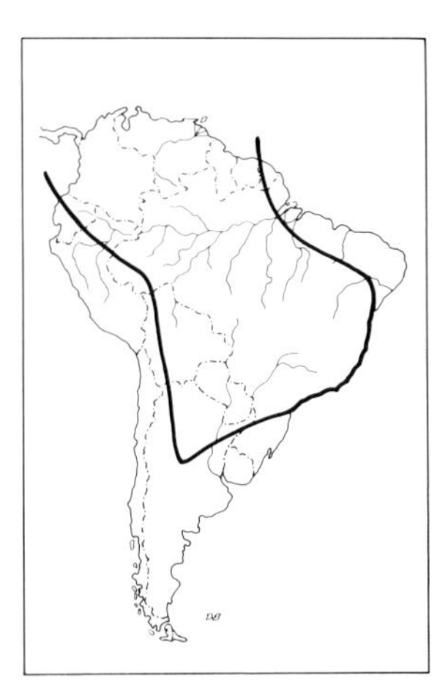

IDENTIFICATION: 18–19 cm (7–7½"). Male closely resembles male Southern Martin (*elegans*) and is not safely distinguished in the field; it is very slightly smaller and its tail is shorter and not quite so deeply forked. Female and immature sooty brown with slight blue gloss on crown and mantle and *distinct grayish or whitish forecrown* (giving hoary effect), usually showing a *fairly obvious whitish semicollar on sides of neck*. Throat and breast pale brownish gray, becoming whitish on belly; entire underparts usually with *streaked* effect (sometimes heavy).

SIMILAR SPECIES: Female Southern Martin lacks the hoary whitish forecrown and collar (forecrown absent in a few Purples, but if present definitely *not* Southern) and is much darker below (typically with scaly rather than streaked effect). Female Purple also resembles female Gray-breasted Martin, but latter too never shows the pale forecrown and collar.

HABITAT AND BEHAVIOR: A n.-winter resident in South America (mostly Sept.–Mar.), but details of migratory routes, etc., remain imperfectly known. Main wintering area appears to be in s. Brazil, especially in w. São Paulo and s. Mato Grosso, in both of which very large concentrations have been found. Passage through n. South America appears to be rapid, though from the number of specimens collected along the lower Amazon one suspects this is a major staging area. Purple Martins are often noted in steady migratory flight, frequently accompanying wintering N. Am. swallows: they are usually seen over semiopen country, typically near water. Roosting birds often congregate in city plazas and parks.

RANGE: Nonbreeding transient and winter visitor mostly in lowlands east of the Andes south to n. Argentina (recorded only Córdoba and Formosa) and s. Brazil (south to São Paulo and Rio de Janeiro); not recorded French Guiana or Paraguay, but probably occurs in both. Breeds North America, occurring as transient through Middle America. When migrating has been recorded at over 3000 m in Colombian and Venezuelan Andes, but mostly in lowlands.

NOTE: The closely related Cuban Martin (*P. cryptoleuca*), considered a full species by the 1983 AOU Check-list (though many others have considered it a race of *P. subis* or of *P. dominicensis*), leaves Cuba during the n. winter. Where it goes remains unknown, but South America seems likely; specimens of *cryptoleuca* have been collected on Curaçao in Sept.–Oct. In any case it would be extremely difficult to identify unless collected; male differs from male Purple only in having a few white feathers on lower belly, while female seems essentially identical to female Gray-breasted.

Progne dominicensis

CARIBBEAN MARTIN

Other: Snowy-bellied Martin

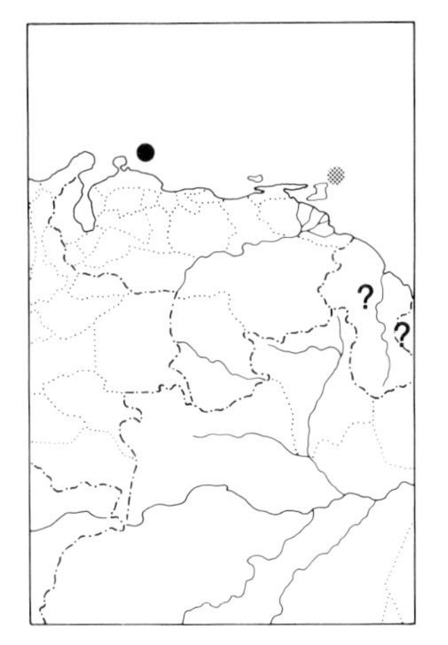

IDENTIFICATION: 18 cm (7″). *Nests on Tobago,* leaving during nonbreeding season for an unknown destination (but presumably somewhere in South America). Male resembles male Southern or Purple Martin but has *sharply demarcated white area on median lower breast, belly, and crissum.* Female has much the same pattern as male, but underparts are grayish brown where male is glossy steel blue, with brown extending down on flanks.

SIMILAR SPECIES: Sharply patterned males should be relatively easy; no other martin shows such a contrast below. Female, however, can be confused with female Gray-breasted Martin (though latter lacks the brown on the flanks). Cf. also female Purple Martin (with pale forehead and partial collar, as well as streaky effect below, etc.).

HABITAT AND BEHAVIOR: Fairly common over semiopen areas, particularly along the coast; also occurs inland in smaller numbers, especially near water. General behavior much like other martins'. Seen entering holes and crevices in cliffs overlooking ocean; presumably this is where they nest. Evidently present on Tobago mostly (or only) Feb.–Sept.

RANGE: Nests on Tobago, leaving during the nonbreeding season, when presumably somewhere in South America; also breeds on many of the West Indies (not Cuba), and likely it is these birds which have been recorded as transients on Curaçao in Oct. and May. There are also possible sightings from Guyana (see Snyder 1966: 228).

NOTE: Yet another martin, *P. sinaloae* (Sinaloa Martin), also probably occurs as a winter visitant in South America but is so far unrecorded (doubtless at least in part because of identification difficulties). It breeds in w. Mexico and has been recorded as a transient in Guatemala. It resembles the Caribbean Martin (and has often been considered conspecific with that bird) but has more white below, with blue-black on flanks more restricted; females of the 2 are very similar. A specimen would certainly be needed in order to confirm its presence in South America.

Tachycineta Swallows

A familiar group of swallows with greenish blue to blue upperparts and white underparts; only 1 species does not have a prominent white rump, and all have notched tails. In fresh (nonbreeding) plumage, several species tend to be greener above and smudgier grayish on breast and rump and have tertials more broadly edged white. Immatures of all species are ashy brown above. Typically seen in low swooping flight over open country, especially near water; the 3 migratory species may gather in large flocks when they are not nesting.

Tachycineta albiventer

WHITE-WINGED SWALLOW

PLATE: 2

IDENTIFICATION: 13.5 cm (5¼"). Glossy bluish green above (slightly bluer when breeding) with white rump; wings more blackish, with *large patch of white on inner flight feathers and upper wing-coverts* (conspicuous both at rest and in flight), though in very worn plumage the white is abraded and often not so obvious. Below pure white.

SIMILAR SPECIES: White-rumped and Chilean Swallows both lack white in wing. Cf. also Mangrove Swallow (no overlap) and Tree Swallow (which lacks the white rump and also has no white in wing).

HABITAT AND BEHAVIOR: Fairly common to common around larger rivers and lakes. Usually in pairs or small groups, perching prominently on sticks or branches out over the water. Often allows a close approach, then slips off its perch to fly leisurely low over the water; only occasionally does it fly high, and it is almost never seen away from water (though sometimes it forages over adjacent pastures, etc.). Regularly occurs with White-banded Swallow across much of Amazonia, though that species tends to be more tied to forest borders and is less often fully in the open.

RANGE: Widespread in lowlands south to n. Argentina (south to Jujuy, Salta, Chaco, Corrientes, and Misiones) and s. Brazil (south to Rio Grande do Sul); west of Andes only in n. Colombia and nw. Venezuela; Trinidad. To about 500 m.

Tachycineta albilinea

MANGROVE SWALLOW

IDENTIFICATION: 13 cm (5"). *Nw. Peru. Glossy bluish green above with whitish rump and narrow inconspicuous whitish loral streak;* wings and tail blacker, with some whitish edging on tertials. Below grayish white, vaguely streaked dusky on breast.

SIMILAR SPECIES: There is no other really similar swallow in this species' limited S. Am. range. Blue-and-white Swallow lacks white rump and is bluer above, etc.

HABITAT AND BEHAVIOR: Rare to uncommon around coastal lagoons and in irrigated agricultural land. Behavior similar to White-winged Swallow's, which this species replaces in nw. Peru and in Middle America.

RANGE: Nw. Peru (Tumbes south to La Libertad). Also Middle America. Below 100 m.

NOTE: The Peruvian race, *stolzmanni*, is very isolated from the rest of the range of *T. albilinea*, Middle America south to e. Panama. For this reason, and because of certain morphologic differences (including smaller bill, less if any white on supraloral, grayish not white under wing-coverts, and grayer underparts), some have felt it deserved full species status. If the latter course is taken, probably the best English name (none seems to have been recently suggested) would be West Peruvian Swallow.

Tachycineta leucorrhoa

WHITE-RUMPED SWALLOW

PLATE: 2

IDENTIFICATION: 13.5 cm (5¼"). *S. South America, migrating northward during austral winter. Mostly dark glossy blue above* with *narrow inconspicuous white supraloral streak* (often extending over base of bill) and *white rump;* wings and tail blacker. Below white. Nonbreeding birds are more greenish blue above.

SIMILAR SPECIES: Chilean Swallow is *very* similar (and at any distance cannot be distinguished) but lacks the supraloral streak and always seems to be truly blue above (never with the green sheen shown by many White-rumps).

HABITAT AND BEHAVIOR: Fairly common to common in semiopen areas and at edge of woodland, particularly near water; nests under eaves of houses or in holes found in dead snags, fence posts, etc. When breeding, pairs are found scattered across the countryside, often hovering around their nest site, the male giving its pretty musical chortling song. At other seasons it may gather in very large flocks. Particularly common and conspicuous in the pampas of Argentina and Uruguay, but at least partially migratory in these areas, with few or no birds being present May–Aug.

RANGE: Breeds in s. Brazil (north to cen. Mato Grosso, Goiás, and Minas Gerais), Paraguay, Uruguay, and n. Argentina (south to La Pampa and Buenos Aires); during austral winter more s. breeders move northward and are then recorded north to n. Mato Grosso, Brazil, as well as in n. Bolivia and se. Peru (where rare; recorded north to Junín). N. limit of breeding uncertain with possible nesting indicated in n. Bolivia (Beni at Chatarona; M. A. Carriker) and in cen. Brazil (Goiás on Bananal Is.; RSR). To about 1000 m.

NOTE: See comments under Chilean Swallow.

Tachycineta leucopyga

CHILEAN SWALLOW

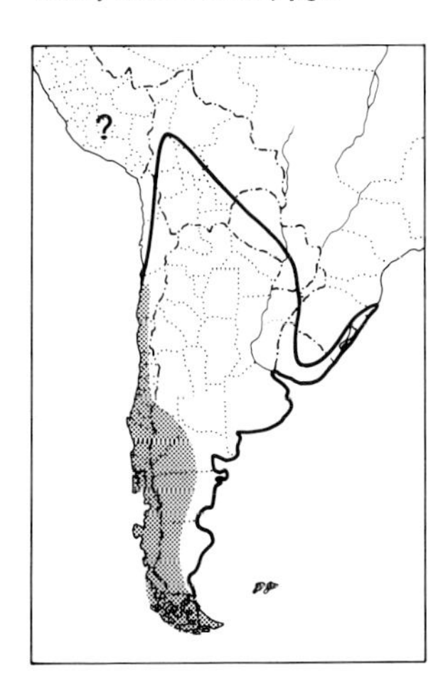

IDENTIFICATION: 13.5 cm (5¼"). *Extreme s. South America* (nesting *south* of range of White-rumped), *migrating northward during austral winter.* Closely resembles White-rumped Swallow. Differs in generally *lacking that species' white supraloral and narrow frontlet* (beware: an occasional Chilean does show a supraloral, but it *never* extends across forehead) and in being *bluer above* (even during nonbreeding season never showing the green sheen shown by White-rumps at that time). The distinctions are so subtle that many birds cannot be identified with certainty in the field.

HABITAT AND BEHAVIOR: Fairly common to common in semiopen areas and woodland borders (latter especially when breeding). Behavior very similar to White-rumped Swallow's, though more a bird of wooded areas when nesting. During nonbreeding season occurs in large flocks and is strongly migratory (more so than White-rumped), with entire population leaving s. part of breeding range.

RANGE: Breeds in cen. and s. Chile (north to Atacama) and s. Argentina (north to Neuquén and w. Río Negro); during austral winter migrates north and east across remainder of Argentina to Uruguay, Paraguay, s. Brazil (Rio Grande do Sul), and Bolivia (only 1 definite record, a specimen taken from a flock of Blue-and-whites in June 1981 on the upper Río Beni in La Paz); Falkland Is. Migrants may reach s. Peru (sightings from Ayacucho and Arequipa, but in no case has it been possible to critically distinguish it from White-rumped Swallow, also a possibility), and 1 sighting from Netherlands Antilles (Curaçao; confusion with White-rumped again possible). Mostly below 1000 m.

NOTE: It has been suggested that *T. leucopyga* and *T. leucorrhoa* are only subspecifically distinct. They have largely if not entirely allopatric breeding ranges, are very similar morphologically, and have identical vocalizations. Intergradation has not been shown to date.

Tachycineta bicolor

TREE SWALLOW

IDENTIFICATION: 13.5 cm (5¼"). *A rare migrant to n. South America.* The only S. Am. *Tachycineta* swallow *without a white rump.* Glossy dark greenish blue above (bluest when breeding); wings blacker, often with some white edging on tertials. Pure white below. Female similar but duller. Immature browner above and not as pure white below, often with smudgy brown area across breast.

SIMILAR SPECIES: Pattern superficially like that of Blue-and-white Swallow, but note that Tree lacks latter's black crissum. All the other *Tachycineta* swallows have white rumps. Cf. also Sand-Martin.

HABITAT AND BEHAVIOR: Rare, irregular, and local n.-winter visitant (recorded only Jan.–Apr., but probably occurs earlier) to semiopen or open areas near water, most often near Caribbean coast. Was perhaps overlooked until recently, but now birders are encountering flocks with fair regularity (though may not occur annually in South America).

RANGE: Nonbreeding visitant to Colombia (mostly along Caribbean coast, but also one record from Nariño highlands), Venezuela (several sightings from Falcón), Guyana (1 record), and off Trinidad. Breeds North America, wintering south, increasingly irregularly, to Panama. Mostly coastal, but has been recorded to 2800 m (in Colombia).

Notiochelidon Swallows

A small group of 3 swallows with slightly forked tails; the crissum is dark and the feet some shade of flesh pink in all three. Immatures are more brownish above. All are essentially birds of montane areas (though the Blue-and-white also occurs out into the lowlands, especially as a migrant), with 2 being restricted to the Andes. The genus is probably close to *Atticora.* The 1983 AOU Check-list used the genus name *Pygochelidon* for *N. cyanoleuca;* we prefer, however, to follow *Birds of the World* and Meyer de Schauensee (1966, 1970) in leaving it in *Notiochelidon.*

Notiochelidon murina

BROWN-BELLIED SWALLOW

PLATE: 2

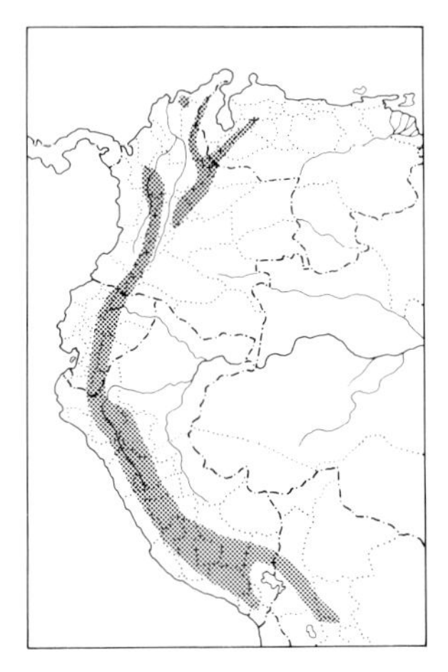

IDENTIFICATION: 13.5 cm (5¼"). *A rather slim, dark swallow of the Andes from Venezuela to Bolivia.* Above dark steely bluish green, duskier on wings and tail; *tail quite noticeably forked. Below entirely smoky grayish brown,* with blackish crissum.

SIMILAR SPECIES: Most likely confused with Andean Swallow (found only from Peru to Bolivia and Chile), but slenderer with longer, more forked tail (though the 2 are about the same overall length) and uniformly dark below (not with white belly). Brown-bellieds often fly with Blue-and-white Swallows, from which they are easily distinguished by their overall dark appearance.

HABITAT AND BEHAVIOR: Fairly common to common over open and semiopen country, often grassy, in the highlands. Usually seen in small, loose groups flying rapidly and gracefully at varying heights. Nest in small colonies in crevices on cliffs or roadcuts, also sometimes under eaves of buildings (though less apt to be around human habitation and to perch on wires than is Blue-and-white).

RANGE: Andes of w. Venezuela (north to Trujillo), Colombia (though absent from much of the W. Andes), Ecuador, Peru (south on w. slope to Arequipa), and nw. Bolivia (La Paz and Cochabamba); Santa Marta Mts. of Colombia and Perijá Mts. on Colombia-Venezuela border. Mostly 2500–4000 m.

Notiochelidon cyanoleuca

BLUE-AND-WHITE SWALLOW

PLATE: 2

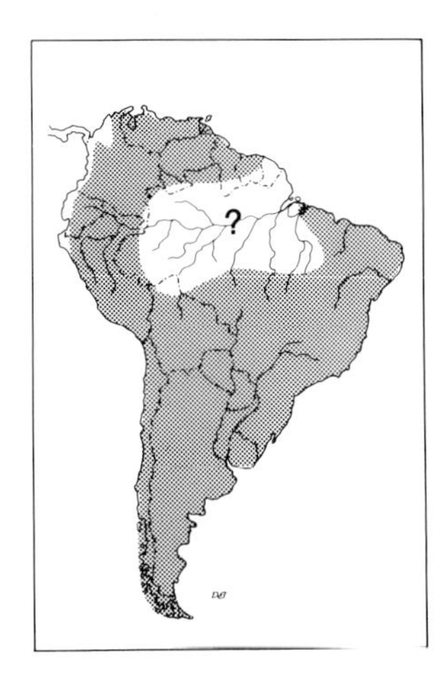

IDENTIFICATION: 12–12.5 cm (4¾–5"). *Entirely glossy steel blue above.* White below, often with a few dusky spots on chest, and with *black crissum.* Tail slightly forked. *Patagonica* (breeding Chile and s. Argentina, migrating to n. South America during austral winter) similar to nominate group but slightly larger with gray (not blackish) under wing-coverts and black on crissum more restricted to sides.

SIMILAR SPECIES: Tree Swallow has white crissum. Cf. also the very similar Pale-footed Swallow.

HABITAT AND BEHAVIOR: Common and widespread (especially in the highlands and the south temperate zone, in tropical lowlands more locally and mostly as an austral migrant) in semiopen and open country, with numbers usually greatest in towns or agricultural regions. Spreads rapidly into deforested areas. Regularly in large flocks, flying at all heights but most often high; frequently perches on phone wires. A familiar, confiding swallow, nesting both as separate pairs or in groups (depending on availability of suitable sites); nests under eaves or in other holes in houses, in crevices of cliffs or roadcuts, even sometimes in niches in the bark of date palms.

RANGE: Widespread virtually throughout South America, though there seem to be no records from across much of cen. and e. Amazonia (it should occur here as a migrant, however), and occurring in tropical lowlands mainly or only as an austral migrant; population of *patagonica* breeding in s. Argentina and s. Chile present there only during austral summer (mostly Oct.–Apr.); Trinidad. Also Costa Rica and Panama, with *patagonica* wintering regularly north to Panama and casually to Mexico. To about 3500 m.

Notiochelidon flavipes

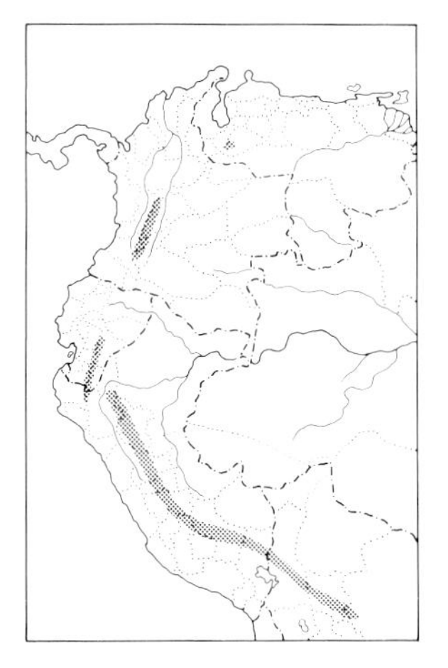

PALE-FOOTED SWALLOW

IDENTIFICATION: 12 cm (4¾"). *Local in temperate zone forests on e. slope of Andes.* Entirely glossy steel blue above. *Throat and upper chest pinkish buff;* remaining underparts mostly white, with *sides and flanks sooty brown* and crissum dark steel blue (looks black at a distance). See C-40.
SIMILAR SPECIES: Closely resembles much more common Blue-and-white Swallow and is likely to be overlooked unless you are specifically looking for it. Against the sky the cinnamon throat is usually impossible to discern, so look for Pale-footed's smaller size and especially its dark sides and flanks (which are often quite conspicuous). The 2 species often fly together (though Pale-footed is much more of a forest bird in general), in which case Pale-footed's faster, more direct flight may be apparent. Note that *both* species have pale pinkish feet.
HABITAT AND BEHAVIOR: Locally not uncommon over temperate zone forest and forest borders. Usually seen flying in groups of 5 to 15 over forest canopy and in clearings and along roads; occasionally larger flocks are noted, and it sometimes flies with Blue-and-white Swallows. Rarely seen perched, then usually on dead snags. May nest in cavities found in large clumps of moss on trees (*fide* T. Parker), but has also been seen perching on roadcuts and inspecting crevices. Up until the 1970s this swallow was believed very rare (and was known from only 3 specimens), but recent fieldwork has shown that its range may be more or less continuous in appropriate habitat; the species is difficult to collect, and it was doubtless much overlooked among the abundant Blue-and-whites until its specialized habitat requirements were first discerned by several keen workers from the LSUMZ. Perhaps the most accessible place where Pale-footed Swallows are regularly found is along the Abra Malaga road in Cuzco, Peru.
RANGE: Andes of w. Venezuela (recent sightings from Mérida near La Azulita; K. Kaufmann et al.), w. Colombia (Cen. Andes in Caldas, Tolima, and Cauca and E. Andes in Nariño), e. Ecuador (Sangay Nat. Park), e. Peru (recorded Cajamarca, Amazonas, San Martín, La Libertad, Huánuco, Pasco, Junín, Cuzco, and Puno), and nw. Bolivia (La Paz, Cochabamba, and w. Santa Cruz); almost all these records are recent, with to date no specimens from Venezuela, Ecuador, or Bolivia. Mostly 2400–3000 m, rarely down to 2135 m or up to over 3200 m.

Atticora Swallows

A pair of attractive swallows with bold blue-black and white patterns. Both have long, deeply forked tails (though they are often held closed in a point) and are restricted to rivers in the lowlands east of the Andes.

Atticora fasciata

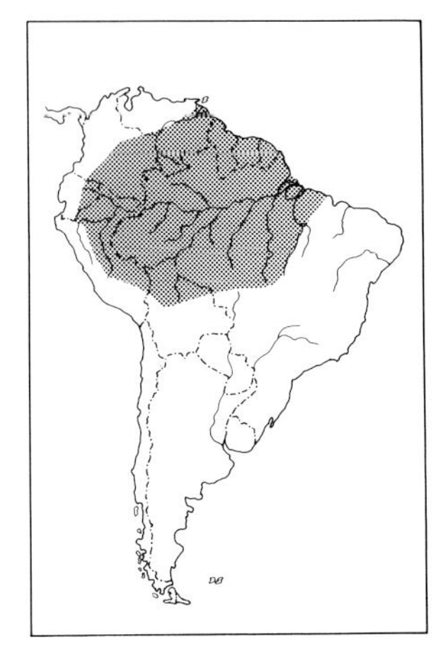

WHITE-BANDED SWALLOW

PLATE: 2

IDENTIFICATION: 14.5 cm (5¾"). *A pretty, dark steel blue swallow of Amazonia with an unmistakable white breast band;* thighs also white (hard to see in the field). Immature sootier or browner, particularly below, but already with adult's pattern.

HABITAT AND BEHAVIOR: Uncommon to locally common (always conspicuous where it occurs) along rivers and larger streams bordered with forest. Generally most numerous along rivers with clear or black water, scarcer or absent in "white" (or muddy) rivers; rarely or never on lakes. Usually in small groups, often perching on branches protruding from or hanging over the water. At such times they will frequently not be noticed until they abruptly flush as a group (sometimes looking remarkably like the bats which frequent much the same places) and then fly rapidly, at times almost darting, low over the water. Though sometimes with White-winged Swallows, White-bandeds tend not to favor as open areas as that species.

RANGE: E. Colombia (north to Meta and Guainía), s. Venezuela (Amazonas, Bolívar, and Delta Amacuro), Guianas, e. Ecuador, e. Peru, n. Bolivia (Beni and La Paz), and Amaz. Brazil (south to n. Mato Grosso and east to e. Pará). To about 1000 m.

Atticora melanoleuca

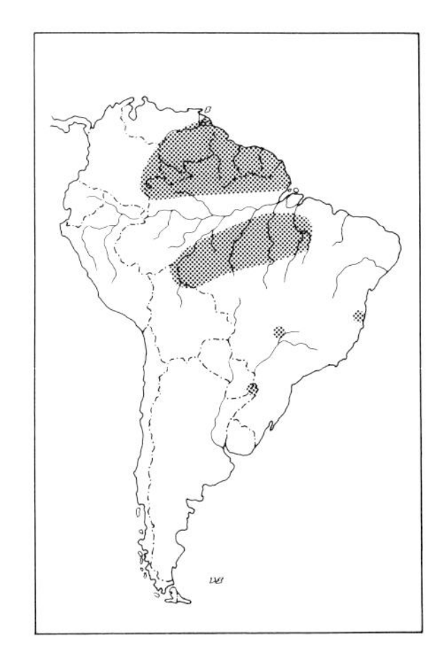

BLACK-COLLARED SWALLOW

PLATE: 2

IDENTIFICATION: 14.5 cm (5¾"). *Local near rapids in rivers of e. South America. Steely blue-black above, blacker on wings and tail. Mostly white below with conspicuous blue-black band across breast* and blue-black crissum.

SIMILAR SPECIES: No other swallow has the white underparts with the black breast band. At times the breast band can be hard to discern, especially when flying against the sky, in which case somewhat resembles Blue-and-white Swallow (but note very different tail shape, etc.).

HABITAT AND BEHAVIOR: Locally uncommon to fairly common along rivers and larger streams where there are waterfalls or rocky rapids. Apparently most numerous and widespread in blackwater areas of Guianas and upper Rio Negro drainage; much less common elsewhere (and its presence is strictly tied to that of rocky torrents). Behavior much like that of White-banded Swallow, but seems more prone to circle high over rivers and adjacent forest, often with other swallows. Often perches on rocks out in the middle of rapids.

RANGE: Guianas, s. Venezuela (Amazonas, Bolívar, and Delta Amacuro), extreme e. Colombia (e. Vichada, e. Guainía, and e. Vaupés), and locally in Amaz. and s. Brazil (north of Amazon mainly in upper Rio Negro area, also in Amapá; not along Amazon itself; south of Amazon along drainages of Rios Madeira, Tapajós, Xingú, and Tocantins; isolated records from sc. Bahia and s. Goiás on the upper Rio Paraná; small numbers resident at Iguaçu Falls on the Rio Iguaçu on the border between Paraná and Misiones, Argentina, where they were first noted around 1976). To about 300 m.

Neochelidon Swallows

A monotypic genus whose sole representative is a small, dark, forest-based swallow with characteristic white thighs. Found locally in humid tropical lowlands.

Neochelidon tibialis

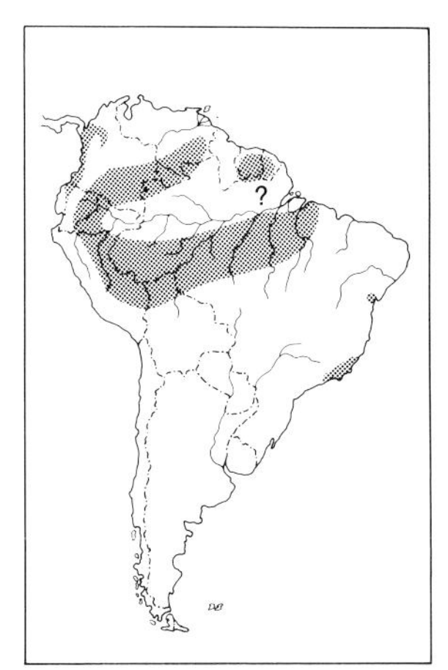

WHITE-THIGHED SWALLOW PLATE: 2

IDENTIFICATION: 11–11.5 cm (4¼–4½"). *A small, uniform-looking dark swallow whose white thighs are diagnostic but hard to see.* Above dark brown, very slightly glossed green, with rump a paler ashy brown. Below uniform pale ashy brown with blackish crissum. *Minimus* (west of Andes in Colombia and Ecuador) differs from nominate group (with *griseiventris*) found east of the Andes in being slightly smaller, shorter-tailed, and darker and sootier generally (especially below) with rump barely contrasting if at all; see C-40.

SIMILAR SPECIES: Most likely confused with Southern Rough-winged Swallow, especially east of the Andes (where *both* are pale-rumped); Rough-wing's underparts are, however, quite different (paler generally, with cinnamon buff throat and yellowish median belly) and it is larger.

HABITAT AND BEHAVIOR: Locally uncommon to fairly common along forest borders or in small clearings, often along roads; seems more numerous and widespread west of the Andes. Usually in small groups, most often not with other swallows (sometimes accompanying Rough-wings), often perching on dead branches, less frequently wires. Its usual flight is fast and erratic. In Panama birds have been seen entering holes in dead trees, where they presumably nest.

RANGE: N. and w. Colombia (more humid Caribbean lowlands east to middle Magdalena valley, Pacific lowlands) and w. Ecuador (south to e. Guayas); locally east of Andes in s. Venezuela (Bolívar and Amazonas), Guianas (not recorded from Guyana but should occur), se. Colombia (north to Meta and Vaupés), e. Ecuador, e. Peru, n. Bolivia (Pando), and Amaz. Brazil (mostly south of the Amazon, east to e. Pará; also north of Manaus); coastal e. Brazil (Bahia, Espírito Santo, and Rio de Janeiro; recently found in e. São Paulo, *fide* E. O. Willis). Also Panama. To about 1000 m.

Stelgidopteryx Swallows

A small group of basically grayish brown swallows found in open country of the lowlands. Both S. Am. species are marked by their cinnamon buff throats. Both nest in holes in banks, but unlike *Riparia* they are not especially gregarious.

Stelgidopteryx ruficollis

SOUTHERN ROUGH-WINGED SWALLOW PLATE: 2

Other: Rough-winged Swallow

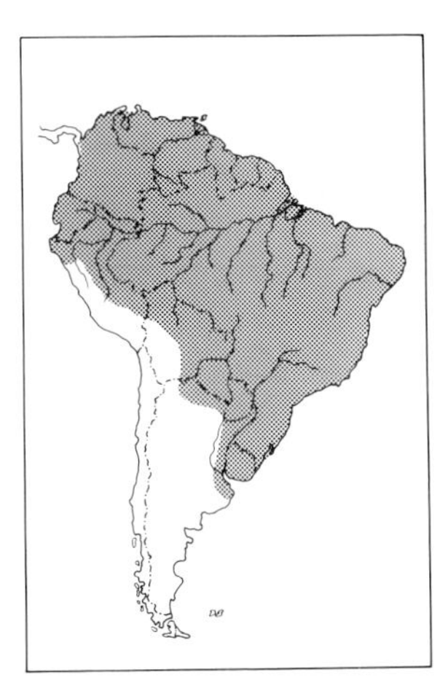

IDENTIFICATION: 13 cm (5″). Above grayish brown, slightly paler on rump (but with little contrast, especially southward and eastward); wings and tail more blackish, with tertials edged whitish. *Throat cinnamon buff,* becoming pale grayish brown on breast and sides and pale yellowish on belly and crissum. Races west of Andes and in Venezuela and on Trinidad (including *uropygialis*) have *rump conspicuously whitish* and are usually not so yellowish on belly.

SIMILAR SPECIES: This widespread but rather drab, brownish swallow is easily known in the w. and n. parts of its range by its pale rump and elsewhere by its rather uniform appearance in conjunction with its buff throat. Cf. Tawny-headed Swallow and Sand-Martin.

HABITAT AND BEHAVIOR: Generally common and widespread in open areas and clearings, with numbers usually largest near water. Often in small groups, perching on dead branches or on wires. Flight is rapid and usually direct, and it generally does not fly very high above the ground. Its call is an upslurred "djreeet." Nests in holes in banks, sometimes in loose colonies, typically along rivers or dug into roadcuts. The outer web of the outer primary of adult males is indeed rough (serrated), though its adaptive significance is unknown.

RANGE: Widespread in lowlands south to n. Argentina (south to Salta, Formosa, Entre Ríos, and n. Buenos Aires); west of Andes south to nw. Peru (south to Cajamarca); southernmost breeders are evidently migratory to Amazonia during austral winter, though details of this are sketchy, as this area is also inhabited by resident birds of the same race; Trinidad. Also Costa Rica and Panama. Mostly below 1000 m, but in smaller numbers and locally up to just over 2000 m.

NOTE: Does not include the Northern Rough-winged Swallow (*S. serripennis*), recently shown to be a distinct species (cf. F. G. Stiles, *Auk* 98 [2]: 282–293, 1981). The Northern Rough-wing is migratory and may occur in South America, though as yet there are no documented records from the mainland; it is regular and not uncommon as a n. winterer as far south as Panama (breeding as far south as Costa Rica) and has been reported (1 sighting) from Bonaire in the Netherlands Antilles. There is also 1 specimen in the AMNH (#500993) of *serripennis* which is labeled as being from "Brazil" but has no further data. Northern Rough-wings are much drabber than the s. species, being essentially uniform grayish brown, paler below, especially on belly (no buff throat, yellow on belly, pale rump, etc.).

Stelgidopteryx fucata

TAWNY-HEADED SWALLOW PLATE: 2

IDENTIFICATION: 12 cm (4¾″). *Mainly in s. South America,* very local northward. *Superciliary and nuchal collar cinnamon merging into buff on throat and chest;* crown brown, feathers edged tawny. Otherwise grayish brown above, duskier on wings and tail, with tertials edged whitish. Lower underparts white.

SIMILAR SPECIES: Flight style and overall appearance similar to Southern Rough-winged Swallow's, but smaller and never shows pale rump; its tawny head is obvious, given reasonable light.

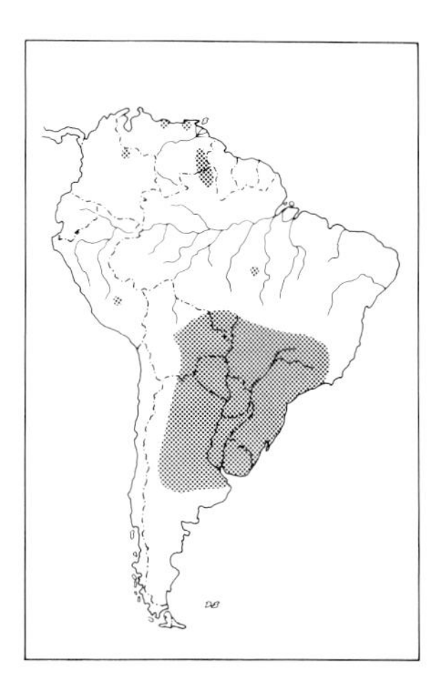

HABITAT AND BEHAVIOR: Uncommon to fairly common in open, often grassy country; generally most numerous in vicinity of water. Usually in pairs or small groups, tending not to associate with other swallows. General behavior similar to that of Southern Rough-winged Swallow and like that species nests in holes in banks (W. Belton). Numerous on the Gran Sabana of s. Bolívar, Venezuela, and also in *cerrado* grasslands of s. Goiás, Brazil (e.g., at Brasilia).
RANGE: E. Bolivia (north to Cochabamba and Santa Cruz), s. Brazil (north to Mato Grosso, s. Goiás, and Minas Gerais), Paraguay, Uruguay, and n. Argentina (south to Mendoza and Buenos Aires). Status northward uncertain, with some records perhaps representing migrants (breeders in s. part of range are known to depart during austral winter, but perhaps mostly go no farther than the n. part of the main breeding range), though breeding-condition birds have been taken in Venezuela; recorded from s. Peru (Cuzco), Amaz. Brazil (se. Pará on Serra do Cachimbo), ne. Colombia (Arauca), and Venezuela (Distrito Federal, Sucre, and e. Bolívar) and adjacent n. Brazil (Roraima). To 1600 m.

NOTE: Formerly placed in the monotypic genus *Alopochelidon* (e.g., by Meyer de Schauensee 1966, 1970); Short (1975) suggested merging it into *Stelgidopteryx*, a treatment which seems preferable. However, we note that males do lack the serrated outer web of outermost primary found in true *Stelgidopteryx*.

Riparia Sand Martins

Only 1 species of *Riparia* swallow occurs in the New World, that a nearly cosmopolitan one. It superficially resembles the *Stelgidopteryx* swallows but is more gregarious; in South America it occurs mainly with other migratory swallows.

Riparia riparia

SAND MARTIN

Other: Bank Swallow

IDENTIFICATION: 12 cm (4¾"). *Uniform grayish brown above,* wings and slightly notched tail duskier. Throat whitish, contrasting with *grayish brown chest band* (often extending down as a line of spots on mid-breast); remaining underparts white. See C-40.
SIMILAR SPECIES: This small brown and white swallow is not likely to be confused because of its distinctive pattern below, though the chest band can sometimes be hard to see on flying birds. Brown-chested Martin (especially its s. race) has much the same pattern but is so much larger, with different flight style, that confusion is improbable. Cf. also Southern Rough-winged Swallow.
HABITAT AND BEHAVIOR: Locally fairly common to common n.-winter resident (mostly Sept.–Apr.) in open agricultural terrain and grassy areas. Often seen in fairly large groups, usually associating with large numbers of other swallows (especially Barns and Cliffs). Has a distinctive fast fluttery flight, quite different from the more languid,

deeper wing-strokes of many other swallows. Its usual call is a dry, gravelly "drrt," sometimes given in series.

RANGE: Nonbreeding transient and winter visitor mostly in lowlands south to n. and cen. Chile (sightings from Arica and Valdivia) and n. Argentina (recorded south only to Tucumán, Córdoba, and Buenos Aires, but probably a few straggle farther); most birds range east of the Andes, but a few also regularly migrate along the Pacific coast; the number of birds reaching s. South America appears to be small, but it remains uncertain where the largest concentrations of overwinterers occur. Breeds in North America, wintering mainly in South America; also occurs widely in the Old World. Mostly below 1000 m, though recorded to 3700 m as a transient.

NOTE: Known as the Bank Swallow in virtually all American literature. However, it is elsewhere universally known as the "Sand Martin" or "Common Sand-Martin," with its 3 Old World congeners (*cincta, paludicola,* and *congica*) also known as various "sand-martins." Though there is nothing intrinsically "wrong" about the name "Bank Swallow," it seems time for Americans to synchronize with the rest of the world on this point.

Hirundo Swallows

A diverse group of swallows found around the world; 4 species are recorded from South America, with an additional fifth likely also occurring. Generic limits in this complex of swallows are uncertain; *Petrochelidon* was recently merged into it (e.g., by the 1983 AOU Check-list). One result is that generalizations are difficult, either for appearance (though all are basically blue to blue-black above) or for nesting (though most are colonial).

Hirundo andecola

ANDEAN SWALLOW

PLATE: 2

IDENTIFICATION: 13.5 cm (5¼"). *High Andes of Peru, Bolivia, and n. Chile.* Tail notched. Above dull glossy blue-black, with *rather contrastingly paler dusky brown rump and tail. Throat and chest ashy brown, gradually becoming whitish on lower underparts.*

SIMILAR SPECIES: In its *puna* zone habitat it is most likely confused with Brown-bellied and Blue-and-white Swallows. Brown-bellied is, however, uniformly dark below, has a more deeply forked tail, and lacks the paler rump; Blue-and-white is pure white below (aside from its black crissum) and much bluer above.

HABITAT AND BEHAVIOR: Locally common over open *puna* grasslands and rocky slopes and around towns and habitations. Usually in small groups, sometimes associating with other swallows, though it ranges at higher elevations than any of them. Often feeds around herds of grazing animals (cattle, sheep, llamas, even vicuña where these survive). Nesting birds have been seen entering holes in cliffs and roadcuts and under eaves (much like a *Notiochelidon* swallow); no structure similar to that of the other *Hirundo* swallows has been described.

RANGE: High Andes of cen. and s. Peru (north to Lima and Junín), w. Bolivia (La Paz south to Potosí and Tarija), and n. Chile (Tarapacá). Mostly 2500–4400 m.

NOTE: Formerly placed in the genus *Petrochelidon*, now lumped into *Hirundo*. It is, however, with some reservation that we ally this species with the *Hirundo/Petrochelidon* complex. Its Andean distribution is unusual for that group and its nesting behavior unique (unless, as seems unlikely, it constructs a mud nest inside the burrow). A case could be made for resurrecting the monotypic genus, *Haplochelidon*, in which it was formerly placed, but in light of reported voice similarities to other *Petrochelidon* (*fide* T. Parker), at least for now we continue to associate it with that group.

Hirundo rufocollaris

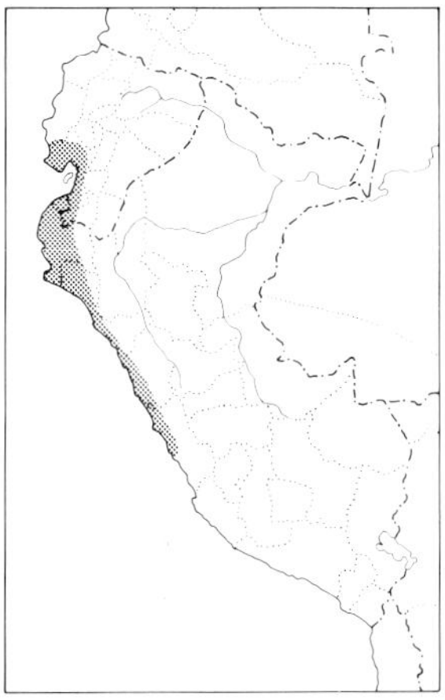

CHESTNUT-COLLARED SWALLOW

PLATE: 2

Other: Cave Swallow

IDENTIFICATION: 12 cm (4¾"). *Sw. Ecuador and w. Peru.* Tail essentially square. Forehead dull chestnut; crown and back glossy blue-black, latter streaked grayish, with *conspicuous intervening rufous nuchal collar. Rump also rufous;* wings and tail blackish. Throat and cheeks whitish, contrasting with *rufous-chestnut chest and sides;* lower underparts buffy whitish. *Aequatorialis* (Ecuadorian portion of range) similar but with throat and cheeks more strongly tinged buff; has deeper chestnut on chest and sides.

SIMILAR SPECIES: In its limited range (where Cliff Swallow is a very scarce migrant) it is easily known by the pale rump, rufous collar, and rufous-chestnut below—a very handsome swallow. Cliff Swallow differs in its *dark* cheeks and throat, and it lacks the pectoral-collar effect of this species; Cliff's forehead is usually conspicuously whitish.

HABITAT AND BEHAVIOR: Locally fairly common to common over agricultural land (especially where irrigated) and around towns and habitations. Rather a gregarious swallow, feeding in low swooping flocks over fields and around houses; frequently forages over water when this is present. Nests in colonies on cliffs or under eaves on the sides of buildings, perhaps especially those made of adobe. Nests are constructed of mud and are ball- or retort-shaped with a side entrance which is often extended out as a "neck" (T. Parker); their shape thus resembles those of the Cliff Swallow more than the open cup nests (also of mud) of the Cave Swallows of Texas and the West Indies.

RANGE: Sw. Ecuador (recorded only from Guayaquil and Loja) and nw. Peru (Piura south to Lima). To about 1300 m.

NOTE: Here considered a species distinct from the Cave Swallow (*H. fulva*); both were formerly placed in the genus *Petrochelidon*, now merged by most authorities into *Hirundo*. Apart from the rather striking morphological differences (as great as the differences between *H. fulva* and *H. pyrrhonota*), nest shape is also distinctive, that of *rufocollaris* resembling the more distantly related Cliff and not the Cave. The Cave Swallows breeding in Texas and Mexico are known to be migratory and are presumed to spend the n. winter in South America, though as yet there are no records. One specimen, identified as *pallida* (breeding in n. Mexico), was taken in Oct. 1952 on Curaçao, and others have been seen on migration in Panama (note that the West Indian breeding races are not definitely known to be long-distance migrants). Cave differs from Chestnut-collared Swallow in having entire cheek, throat, and chest area cinnamon buff, with *no* deep rufous or chestnut below, and in its brighter forehead. It would most likely be found among Cliff Swallows.

Hirundo pyrrhonota

CLIFF SWALLOW

IDENTIFICATION: 13–13.5 cm (5–5¼"). Tail essentially square. Mostly dull, steely blue-black above with *prominent buffy whitish forehead,* buffy grayish nuchal collar, and *conspicuous cinnamon-rufous rump;* back with a few narrow whitish streaks. *Sides of head and throat dark chestnut,* with black patch on center of lower throat and upper chest; breast grayish, fading to white on belly. One race (*melanogaster,* breeding in sw. United States and w. Mexico) has forehead dark chestnut. Immatures (frequently seen until at least Nov.–Dec.) are much duller but already show most of adult's pattern (including the rump patch), though throat and breast are buffy grayish and forehead patch is lacking.
SIMILAR SPECIES: Most likely confused with Barn Swallow, and the 2 species are often found together; bear in mind that Barn Swallow's tail *lacks* streamers during much of its S. Am. sojourn, though it is always forked (never squared off). Barn Swallow lacks the pale rump patch and never has the pale forehead shown by most Cliffs. Cf. also Chestnut-collared Swallow.
HABITAT AND BEHAVIOR: Locally common n.-winter resident (mostly Sept.–Apr.) in open agricultural areas and over grassy savannas. Often in large flocks, particularly when actually migrating. General behavior similar to Barn Swallow's, though it seems to associate less with sugarcane fields.
RANGE: Nonbreeding transient and winter visitor south to n. Chile (Tarapacá) and n. Argentina (south regularly to Córdoba and Buenos Aires; C. Olrog records it south to Tierra del Fuego); most individuals apparently move south into s.-cen. South America during middle of n. winter; at least so far not recorded from e. South America (e. Venezuela and the Guianas to ne. Brazil). Breeds in North America, wintering mainly in South America. Mostly below 1000 m, though recorded to 3700 m as a transient.

Hirundo rustica

BARN SWALLOW

PLATE: 2

IDENTIFICATION: 14–16.5 cm (5½–6½"), depending on length of tail streamers. *Long deeply forked tail with band of white spots on inner webs;* tail much shorter in immature plumage and in adults during their southward passage (gradually lengthening Jan.–Mar.). Above steely blue with chestnut forehead. Throat chestnut to buff, with *remaining underparts cinnamon buff to pale buff* (chestnut and cinnamon buff in breeding plumage males), often with some black demarcating the shift in colors. Immature similar but much duller, not as blue above and lacking the chestnut forehead; throat dull buffyish, with remaining underparts whitish.

SIMILAR SPECIES: Birds seen in South America often look rather ragged, lacking the glossy crisp pattern of breeders in North America, but are usually readily recognized by the long forked tail and rusty to buff underparts. Immatures, with their short tails and mainly whitish underparts, are most likely to cause confusion; cf. especially their fellow migrants, the Bank and Cliff Swallows.

HABITAT AND BEHAVIOR: Locally very common n.-winter resident (mostly Aug.–May, though scattered individuals linger throughout the year) in open country. Shows a particular predilection for sugarcane fields, in which vast concentrations often roost and then spread out to forage over nearby areas. By far the most numerous and widespread swallow which is migratory from North America. The Barn Swallow's flight is exceptionally graceful and buoyant; it swoops around rapidly, often describing a roughly circular course, usually close to the ground. Like all the other migrant swallows it is usually silent on its wintering grounds. One of the most surprising and interesting recent discoveries in S. Am. ornithology concerns the nesting of this species in Argentina. M. M. Martínez (*Neotropica* 29: 83–86, 1983) discovered in 1980 near Mar Chiquita in e. Buenos Aires province 6 pairs nesting in Jan. of that year, 5 in Jan.–Feb. of 1981, and 11 between Nov. 1981 and Mar. 1982. Most nests were placed under highway bridges and many produced young. Barn Swallows are present here only Oct.–Apr., posing the interesting question of where these individual swallows go during the remainder of the year. Observers should be alert to document other nestings: it seems quite possible that the habit might spread, at least across temperate zone South America.

RANGE: Nonbreeding transient and n.-winter visitor to lowlands throughout South America, though numbers are very small south of cen. Chile and n. Argentina (casual vagrants have occurred south to Tierra del Fuego and a few occur annually on the Falkland Is.); recently a few have been found breeding in Argentina (see above). Breeds in North America, wintering mainly in South America, also occurs widely in the Old World. Mostly below 1000 m, though recorded to 3700 m as a transient.

Troglodytidae

WRENS

BASICALLY an American family (only one species in the Old World), the wrens reach their maximum diversity in Middle America and the tropical parts of South America. Somberly colored, mostly in brown and rufous, often with black wing- and tail-barring, wrens are best known for their marvelous songs, some of them ranking in anyone's list of "best bird singer" (for example, the Song and Musician Wrens). Most are rather skulking in behavior, creeping in dense, low vegetation or hopping on the ground; a few, notably *Campylorhynchus,* are more arboreal and conspicuous.

Donacobius

The unique Donacobius is an easily recognized, mostly dark brown and buff bird of marshes or grassy areas near water. Its relationships have been debated, but recent evidence (e.g., R. Kiltie and J. W. Fitzpatrick, *Auk* 101 [4]: 804–811, 1984) indicates that it is a wren (and not, as previously supposed, a mimid). The Donacobius frequently duets and engages in cooperative breeding, both traits being frequent in the Troglodytidae and absent or infrequent in the Mimidae. Its open cup nest is, however, very atypical for a wren.

Donacobius atricapillus

BLACK-CAPPED DONACOBIUS

PLATE: 3

Other: Black-capped Mockingthrush

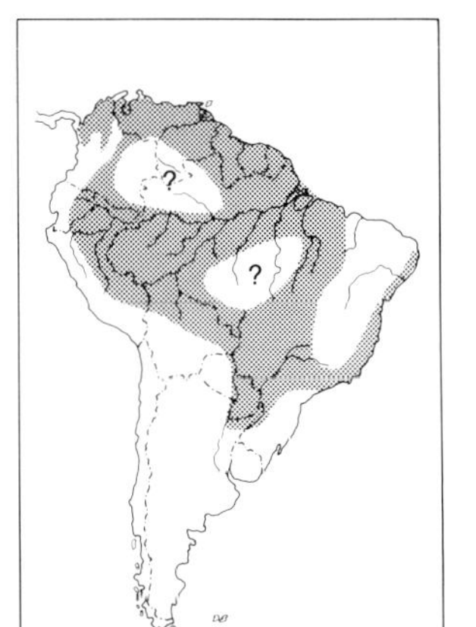

IDENTIFICATION: 21.5–22 cm (8½″). *An unmistakable sleek, long-tailed bird of tropical marshes and lake borders. Iris conspicuously yellow.* Head and neck black, becoming chocolate brown on remaining upperparts, tawnier on rump. White patch at base of primaries and *tail feathers broadly tipped white. Below rich buff,* often with some fine black barring on sides; a bare orange patch on sides of neck is usually concealed but can be expanded during display (see below). *Albovittatus* of n. Bolivia, as well as immatures of all races, show a narrow but conspicuous white postocular streak.

HABITAT AND BEHAVIOR: Locally common in grass and other marshy vegetation bordering lakes and sluggish rivers and in damp pastures or marshes; particularly associated with floating mats of vegetation around oxbow lakes in Amazonia. Pairs or small groups are usually conspicuous, perching at the top of clumps of grass or low bushes or flying weakly across small bodies of water. Very vocal, giving a large repertoire of calls, including a loud "quoit-quoit-quoit-quoit," a harsh "jeeeyaa" (sometimes repeated in series), and a low "chirru." Displaying pairs are frequently seen, and their delightful performance is worth studying. Typically they perch a few inches apart, one slightly above the other and usually facing the other way; they bob their heads, jerkily wag their partially fanned tails asynchronously, and call antiphonally, the upper bird giving a harsh "chrrr," the lower bird a loud "kweéa." Often the orange skin on the sides of the neck is then exposed.

RANGE: Widespread in lowlands south to n. and e. Bolivia, e. Paraguay, ne. Argentina (south to Corrientes and e. Chaco), and s. Brazil (south to w. Paraná and São Paulo); west of Andes south only to w. Colombia; seemingly absent from the blackwater region of the upper Rio Negro drainage, from much of interior ne. Brazil, and perhaps from the upper drainages of the Rios Xingú and Tapajós. Also e. Panama. Mostly below 600 m.

NOTE: A number of names have been attached to this species. Because it neither mocks nor is a thrush, "mockingthrush" seems poor and "mockingwren" seems only marginally better. We opt to follow the 1983 AOU Check-list and Ridgely (1976) in calling it the Black-capped Donacobius, though an argument could be made that since there is only one member of the genus, simply calling it the Donacobius would suffice.

Campylorhynchus Wrens

The largest wrens, usually boldly patterned, often strongly barred or spotted. Iris color is variable, usually reddish brown (except in Stripe-backed); bills look pale in most species, though darker in Bicolored. *Campylorhynchus* wrens are conspicuous (compared to the rest of their generally furtive family) and noisy, with calls scratchy and guttural in most species, more musical in the Thrush-like and Bicolored; all species duet frequently. They exhibit all levels of sociality, some species occurring in pairs, others remaining in large groups. Large, globular, often untidy nests with a side entrance are built, with the more social species exhibiting "helper at the nest" behavior. These nests are also often used as dormitories. Diversity is greatest in n. and w. South America (and in Middle America), with only 1 species (Thrush-like) extending farther.

GROUP A

Two dissimilar, large species; *not* boldly banded above.

Campylorhynchus griseus

BICOLORED WREN

PLATE: 3

IDENTIFICATION: 21–22 cm (8¼–8½"). *A large wren, basically dark brown above and white below, with a bold white superciliary.* Outer tail feathers blackish, broadly banded or tipped white (visible mainly from below). *Minor* (much of n. Venezuela) is rufescent across the mantle, while mantle of nominate race (s. Venezuela and adjacent Guyana and Brazil) is more grayish brown. *Albicilius* (n. Colombia to nw. Venezuela) has blackish nape extending over upper back, while *bicolor* (Magdalena valley in Colombia) is larger and darker generally, more or less uniform sooty above.

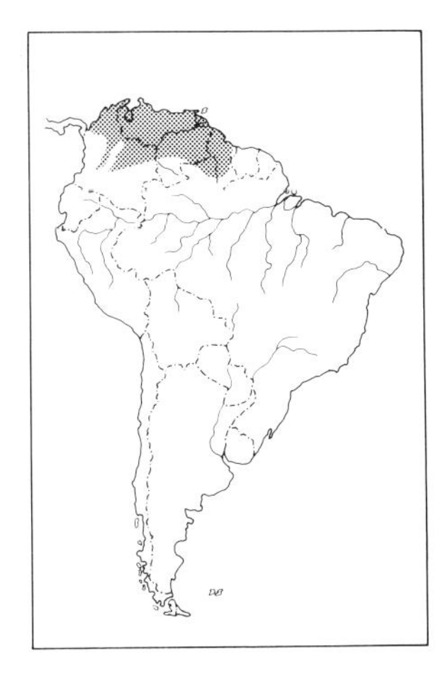

SIMILAR SPECIES: All other *Campylorhynchus* wrens in or near its range are either boldly barred above (Stripe-backed and Band-backed) or boldly spotted below (Thrush-like).

HABITAT AND BEHAVIOR: Fairly common to common in arid and semiarid regions, both in cactus-dominated scrub and semiopen light woodland; sometimes in the trees around houses and towns. Conspicuous and bold birds, they move about deliberately on branches at all heights and also regularly drop to the ground, hopping with tail held high. Palms are especially favored in many areas. Bicolored Wrens are often noisy and have a variety of loud, usually guttural calls; these are often given as a rhythmic duet by a pair (e.g., "owk-cha-chok, owk-cha-chok . . ." or "owk-cha-cha-chok, owk-cha-cha-chok . . .").

RANGE: N. Colombia (more arid Caribbean lowlands, south in Magdalena valley to Huila, and east of Andes south of Meta and Vichada), Venezuela (semiopen areas except in south), interior Guyana, and extreme n. Brazil (Roraima). To 2000 m (in Colombia).

Campylorhynchus turdinus

THRUSH-LIKE WREN

PLATE: 3

IDENTIFICATION: 20.5 cm (8″). Dull grayish brown above, feathers centered darker, giving somewhat scaly effect; vague white superciliary. Wings and tail grayish brown. *Below white thickly spotted with grayish brown* except on throat; lower belly and crissum tinged ochraceous. Nominate race (Brazil in interior northeast and in coastal southeast) less thickly spotted below than is *hypostictus* (all of Amazonia). *Unicolor* (n. and e. Bolivia, w. Mato Grosso, adjacent Paraguay) looks quite different, however; it is *paler and grayer above,* has a more pronounced superciliary, and is *either virtually or entirely unspotted below.*

SIMILAR SPECIES: Does not occur with any other *Campylorhynchus* wren. Virtually unmistakable otherwise, and really not very "thrush-like."

HABITAT AND BEHAVIOR: Fairly common to common in canopy and borders of humid forest (both *terra firme* and *várzea*) and forest borders, also regularly out into clearings and second growth provided a few tall trees are left standing. Usually found in family groups of up to 6 to 8 birds, which hop about on larger limbs and in viny tangles, generally well above the ground. Does not often associate with mixed flocks and on the whole is not too often seen, especially relative to the large numbers which are heard; tends to remain in dense epiphytic growth. Their frequently heard, very loud choruses are startling in their sudden explosiveness. Usually given as a duet, the phrases are complex and variable (e.g., "chookadadoh, choh, choh" or "choh-do-do-chit," often repeated several times) and are often cued by a few harsh, scratchy notes. In their musical quality these vocalizations are unlike those of any other S. Am. *Campylorhynchus,* and they represent one of the most characteristic and memorable sounds of upper Amazonia. The song of the so different-looking *unicolor* race is seemingly identical.

RANGE: Se. Colombia (north to Meta), e. Ecuador, e. Peru, n. and e. Bolivia, Amaz. Brazil (east to n. Goiás and cen. Maranhão); sw. Brazil (upper Paraguay River basin in w. Mato Grosso) and adjacent n. Paraguay (sightings along Río Apa in July 1977; RSR); coastal e. Brazil (e. Bahia and Espírito Santo); heard once in Rio Grande do Sul, Brazil (H. Sick). To 1100–1200 m along base of Andes.

GROUP B

Mantle boldly banded or striped and *spotted below* (except divergent *C. albobrunneus*); the *C. zonatus* superspecies. All basically *allopatric*.

Campylorhynchus fasciatus

FASCIATED WREN

PLATE: 3

IDENTIFICATION: 19 cm (7½"). *Sw. Ecuador and nw. Peru. Above boldly banded blackish and pale brownish gray to whitish,* somewhat more uniform grayish on head, with narrow whitish superciliary. Below whitish, *boldly and thickly spotted with grayish brown.*

SIMILAR SPECIES: The only *Campylorhynchus* wren in its range. Band-backed Wren is found in more humid nw. Ecuador just to the north of Fasciated's range; Fasciated is slightly larger and "dustier"-looking (less crisply patterned) overall (e.g., spots below are larger and blurrier) and lacks ochraceous tinge to lower underparts.

HABITAT AND BEHAVIOR: Common in scrub, thorny low woodland, agricultural areas with hedgerows and scattered trees, and desert vegetation with cactus predominating. Found only in arid or semiarid regions. Conspicuous, occurring in small, noisy groups. Its calls are loud and nonmusical (though often rhythmic), consisting of harsh, scratchy phrases, often rapidly repeated several times; as in other *Campylorhynchus* wrens, pairs often duet. Numerous in arid areas west of Guayaquil, Ecuador.

RANGE: Sw. Ecuador (Guayas, El Oro, and Loja) and nw. Peru (Pacific slope south to Ancash, rarely to n. Lima). Mostly below 1500 m (locally to over 2500 m in Peru).

Campylorhynchus nuchalis

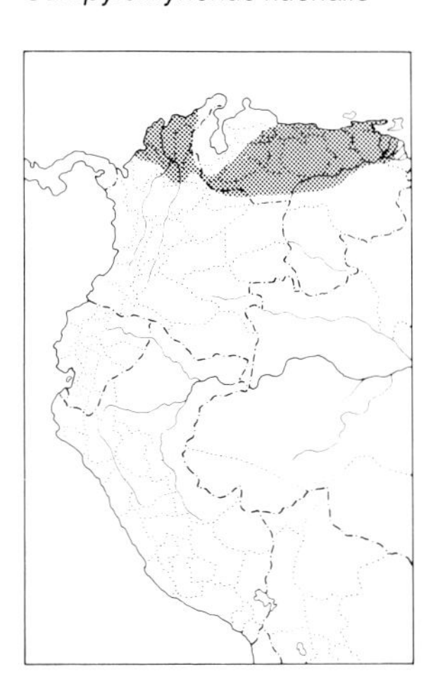

STRIPE-BACKED WREN

IDENTIFICATION: 17.5 cm (7"). *Venezuela and n. Colombia.* General form resembles that of Fasciated Wren. Iris pale yellow. Crown light grayish brown spotted black, with white superciliary and faint rufescent nuchal collar; *neck and back broadly striped black and white, wings and tail broadly banded black and white.* Below white, *spotted black* (fewest spots in *pardus* of Colombia). See V-31, C-42.

SIMILAR SPECIES: Band-backed Wren is banded (not striped) on back and has ochraceous tone on belly; it tends to be found in more humid areas.

HABITAT AND BEHAVIOR: Fairly common to common in dry woodland, scattered trees, and shrubbery in agricultural areas and around rural houses and in gallery woodland. Usually in small groups of 4 to 8 birds, with general behavior much like Fasciated Wren's; its rhythmic vocalizations are also similar, with a variety of harsh, scratchy notes,

none of them musical. Reported to appropriate old nests of other birds, usually large, untidy structures with a side entrance, such as those of *Myiodynastes* flycatchers and *Pitangus* kiskadees.
RANGE: N. Colombia (more arid Caribbean lowlands west to Córdoba, along lower Río Sinú; also east of Andes in llanos of Arauca) and n. Venezuela (but absent from the northwest). To 800 m.

Campylorhynchus zonatus

BAND-BACKED WREN

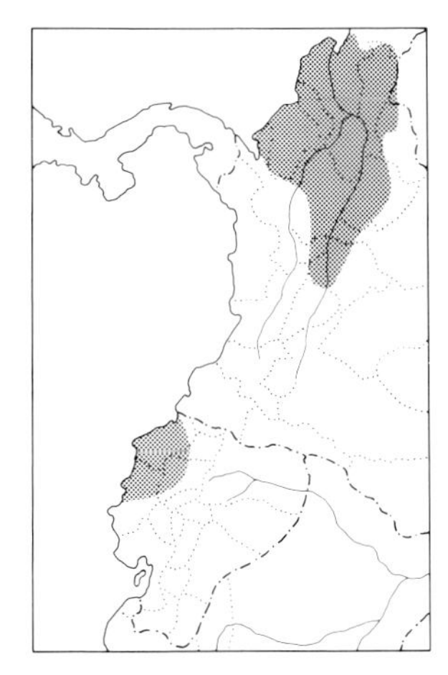

IDENTIFICATION: 18.5 cm (7¼"). *N. Colombia and* (disjunctly) *nw. Ecuador.* General form resembles Fasciated Wren's. Crown brownish gray irregularly spotted blackish, with vague whitish superciliary and somewhat more rufescent on nape. *Above otherwise boldly banded black and buffy whitish. Below mostly whitish boldly spotted black,* becoming *ochraceous on belly* with some black barring on flanks. See C-41. S. Am. birds are considerably duller than those found in Middle America (with tawny belly, etc.).
SIMILAR SPECIES: Slightly smaller than Fasciated Wren (found in drier areas to the south of this species' range in Ecuador), not so grayish overall and with ochraceous belly (lacking in Fasciated); Bandback's pattern is neater, more crisply delineated. Stripe-backed Wren is broadly striped (not banded) on back and lacks the ochraceous belly. See also White-headed Wren.
HABITAT AND BEHAVIOR: Fairly common in canopy and borders of humid forest and in clearings with scattered tall trees remaining. Found mostly in humid regions (in any case only in areas with tall forest). Usually in small groups, clambering about actively among epiphytes and in viny tangles, generally remaining well above the ground but sometimes lower in clearings. Its harsh, scratchy, rhythmical vocalizations are similar to Fasciated Wren's.
RANGE: N. Colombia (humid Caribbean lowlands east to the Santa Marta Mts. and south to middle Magdalena valley in s. Cundinamarca); nw. Ecuador (Esmeraldas and Pichincha to Manabí). Also Mexico to w. Panama. To about 1000 m (locally to 1600 m in Colombia).

Campylorhynchus albobrunneus

WHITE-HEADED WREN

IDENTIFICATION: 18.5 cm (7¼"). *Pacific w. Colombia.* Unmistakable, with *entire head and underparts white;* back, wings, and tail dark brown. See C-41. Birds from sw. Nariño (described as a subspecies, *aenigmaticus,* but perhaps hybrids between White-headed and Band-backed Wrens; see below) are variable but have head more or less brownish and breast more or less spotted and are tinged buff on belly with brown barring on flanks.
SIMILAR SPECIES: Nariño birds differ from Band-backed Wrens in their solid brown upperparts (with only a trace of barring) and more diffuse pattern below.
HABITAT AND BEHAVIOR: Uncommon and seemingly local in canopy and (especially) borders and breaks in humid forest. Behavior and vo-

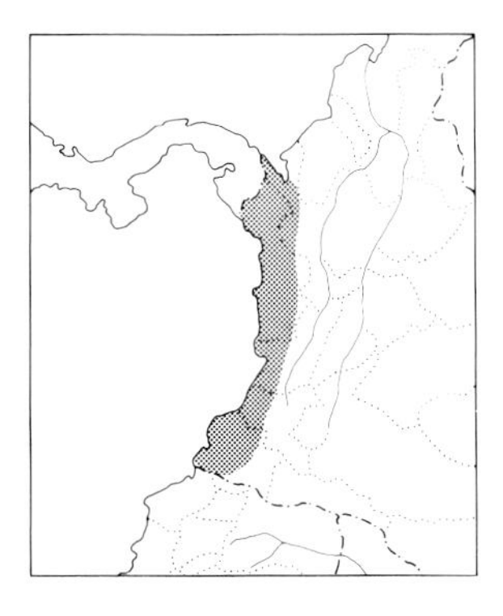

calizations very similar to Band-backed and Fasciated Wrens'. Fairly regular along lower Buenaventura road in Colombia.

RANGE: Pacific w. Colombia (Chocó to Nariño). Also e. Panama. Mostly below 1000 m, occasionally to 1500 m.

NOTE: *Aenigmaticus* (see above) was believed by Meyer de Schauensee (1966) to be intermediate between *C. albobrunneus harterti* (n. Pacific Colombia) and Amaz. *C. turdinus,* and he therefore considered *C. albobrunneus* conspecific with *C. turdinus.* This we believe to be totally implausible: their voices are completely different and their ranges separated by the Andes. We concur with Haffer (1975) that what it must represent is a hybrid population between *C. zonatus* and *C. albobrunneus.* This would fit geographically, and these 2 are similar ecologically and vocally. Further work will perhaps demonstrate that these 2 should be considered conspecific.

Thryothorus Wrens

A rather large genus of medium-size wrens found mainly in n. and w. South America (and Middle America), with only 2 species extending to e. Brazil and none to Argentina. Various shades of brown and rufous predominate, paler (often whitish at least in part) below, and virtually all have streaked cheeks and a white superciliary. They tend to skulk in lower growth of woodland and forest borders (not usually deep inside humid forest, however), usually foraging apart from mixed flocks. All are heard far more often than seen, and their wonderfully varied musical songs are notable for their antiphonal qualities, typically given as a duet by members of a pair. Nests are dome-shaped with a side entrance.

GROUP A

Three distinctive, *relatively boldly patterned* wrens of *nw. South America.*

Thryothorus fasciatoventris

BLACK-BELLIED WREN

PLATE: 3

IDENTIFICATION: 15 cm (6″). *N. and w. Colombia.* Chestnut brown above with narrow whitish superciliary and blackish mask through eyes; tail barred black and wings vaguely barred. *Throat and chest white, contrasting sharply with black breast and belly;* breast and belly very narrowly barred white.

SIMILAR SPECIES: No other wren shows the sharp contrast below.

HABITAT AND BEHAVIOR: Uncommon to locally fairly common in dense thickets and undergrowth at edge of forest and woodland and in overgrown clearings; particularly fond of low-lying areas near water. A difficult wren to observe, heard much more often than seen. The song is one of the finest in the genus, with a very rich, low gurgling quality; each phrase is repeated several times before going on to the next, and many end with a characteristic upslurred "wheeowheét" or "cream-of-wheat." Quality of song similar to that of Whiskered Wren.

RANGE: N. Colombia (south in the Pacific lowlands to Valle and south in the Magdalena valley to Tolima). Also Costa Rica and Panama. To 1000 m.

Thryothorus nigricapillus

BAY WREN

PLATE: 3

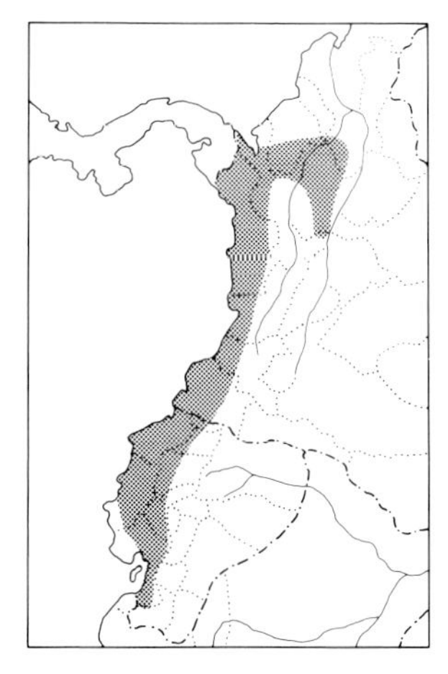

IDENTIFICATION: 14.5 cm (5¾"). *W. Colombia and w. Ecuador. Schotti* (nw. Colombia south to Valle) has *head and neck mostly black,* with narrow white superciliary and *white patch on ear-coverts;* otherwise bright rufous-chestnut above, wings and tail barred black. *Entire underparts narrowly barred black and white,* flanks and lower belly tinged rufous. Nominate race (w. Ecuador) has *throat and breast white,* with barring narrower and restricted to sides, flanks, and belly. *Connectens* (sw. Colombia north to Cauca) is intermediate.

SIMILAR SPECIES: Variable below, but no other wren of its range has the contrasting black head with the white ear-spot.

HABITAT AND BEHAVIOR: Common in undergrowth at edge of forest and woodland and in second-growth and thickly overgrown clearings; especially fond of *Heliconia* thickets and tangles in damp low-lying areas, and usually absent from undergrowth of continuous forest (except in openings such as treefalls). A lively and fast-moving wren which, though it generally keeps to dense cover, often is inquisitive, even appearing briefly in full view on occasion. Very vocal and thus heard more often than seen; the song is rich and musical in quality, consisting of a variety of fast, loud, ringing phrases (e.g., "see-me, how-wet-I-am"), each repeated several times, then abruptly on to the next. Also gives several loud, distinctive calls, including a "heetowíp" and a chirring.

RANGE: W. Colombia (Pacific lowlands and east along humid n. base of Andes to middle Magdalena valley in Bolívar) and w. Ecuador (south to El Oro). Also Nicaragua to Panama. Mostly below 1200 m, occasionally higher.

Thryothorus spadix

SOOTY-HEADED WREN

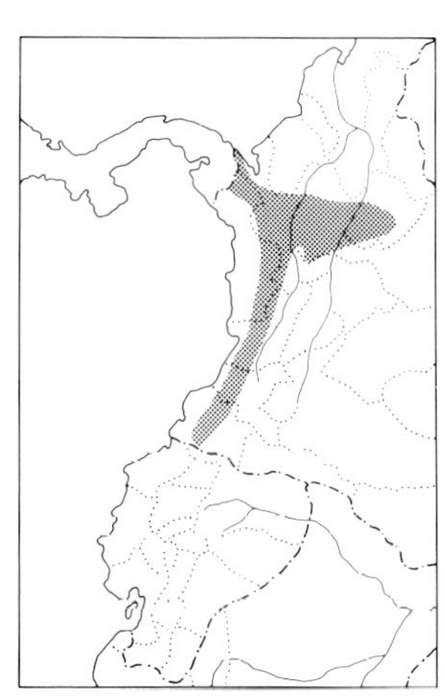

IDENTIFICATION: 14.5 cm (5¾"). *W. Colombia. Crown sooty, becoming black on sides of head and throat,* with fine white streaks on ear-coverts and trace of superciliary; *otherwise mainly chestnut brown above,* becoming brighter chestnut on breast and duller brown on belly, and barred black on tail; wings unbarred. See C-41, P-32.

SIMILAR SPECIES: The blackish head and throat in contrast to the remaining rufous-chestnut is distinctive; Bay Wren shows white on throat and has solid white ear-patch and barred wings. Sooty-headed can look very dark in the dim light of forest undergrowth.

HABITAT AND BEHAVIOR: Uncommon to locally common in undergrowth of mossy forest, forest borders, and overgrown clearings; particularly favors dense viny tangles. Usually in pairs which forage independently of mixed flocks, often probing into curled-up dead leaves. Generally shy and difficult to see. Its song consists of a series of fast, loud phrases, in quality much like that of many others in the genus.

RANGE: W. Colombia (Pacific lowlands south to s. Nariño and east along n. base of Andes to middle Magdalena valley in Santander). Also e. Panama. Mostly 800–1800 m.

GROUP B

The "*moustached*" group; all with chestnut mantles, *unbarred wings* (breast may be spotted).

Thryothorus euophrys

PLAIN-TAILED WREN

PLATE: 3

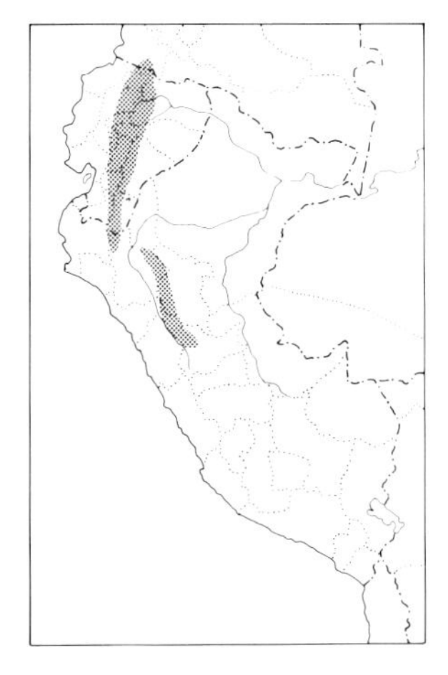

IDENTIFICATION: 16 cm (6¼″). *Andes of s. Colombia, Ecuador, and n. Peru.* Crown mostly brownish gray with long white superciliary surmounted by black; face mostly black and gray with white malar and black submalar streaks. *Upperparts otherwise bright rufous (wings and tail unbarred).* White throat and grayish chest *coarsely spotted black;* lower underparts grayish buff, buffiest on flanks and crissum. *Longipes* (e. Ecuador) resembles nominate race (w. Ecuador and adjacent Colombia) but more sparsely spotted below; *atriceps* (n. Peru north of the Marañón) has blackish crown and likewise is less spotted below. *Schulenbergi* (n. Peru south of the Marañón) is the dullest race, with browner (less rufescent) upperparts, grayish superciliary and throat, and relatively little spotting below.
SIMILAR SPECIES: High Andean range in conjunction with large size, unbarred wings and tail, and spotting below render confusion unlikely. In w. Ecuador compare to Whiskered Wren (which ranges up well into the subtropical zone, coming close to lower limit of Plain-tailed); it lacks spotting below, has barred tail, etc.
HABITAT AND BEHAVIOR: Fairly common in undergrowth of temperate zone forest and woodland, especially at borders and where there are dense stands of *Chusquea* bamboo. Usually in pairs and, though often heard, exceptionally difficult to see (without use of tape playback), remaining in dense cover and rarely accompanying mixed flocks. The song is a loud, fast, rollicking duet performed by mated pairs, perfectly antiphonal, with a common phrase (in w. Ecuador) being "worry chee-chee, worry chee-choo"; though musical, it lacks the rich liquid quality of the Whiskered Wren. Plain-tails are frequently heard (and with effort seen) at various localities in the W. Andes near Quito, Ecuador (e.g., the upper Chiriboga road and the Nono-Mindo road).
RANGE: Andes of extreme s. Colombia (on w. slope in w. Nariño), Ecuador (north on e. slope to Napo, south on w. slope to Chimborazo), and n. Peru (Cajamarca, Amazonas, and San Martín). Mostly 2000–3300 m.

Thryothorus eisenmanni

INCA WREN

IDENTIFICATION: 16 cm (6¼″). *Andes of Cuzco, Peru.* In general form resembles Plain-tailed Wren but *much more spotted below.* Head pattern similar but crown gray, in striking contrast with even brighter rufous upperparts. *Throat and breast white boldly spotted black,* spots also extending to whitish upper belly; sides, flanks, and lower belly buff.
SIMILAR SPECIES: This very handsome wren is unlikely to be confused in its limited range. The spotting below continues much farther down on the underparts than it does in Plain-tailed Wren (its replacement species, found well to the north in the Andes).

HABITAT AND BEHAVIOR: Locally common at the borders of montane forest and woodland, especially where there are dense and extensive stands of *Chusquea* bamboo. Behavior similar to Plain-tailed Wren's, though seems much less prone to remain in heavy cover and in fact is easily seen in the scrubby woodland borders at the ruins of Machu Picchu. The song is also basically similar but is faster and usually more jumbled-sounding. It seems surprising that this wren went so long before being collected (first in 1974, though seen at Machu Picchu as early as 1965); possibly it increased after the clearing of forest around the ruins, followed gradually by regeneration with much bamboo.
RANGE: Andes of s. Peru (Cuzco from Cordillera Vilcabamba south to valleys of Río Urubamba and Río Santa María). 1830–3350 m.

NOTE: A newly described species: T. A. Parker, III and J. P. O'Neill, *Neotropical Ornithology,* AOU Monograph no. 36, pp. 9–15, 1985.

Thryothorus genibarbis

MOUSTACHED WREN

PLATE: 3

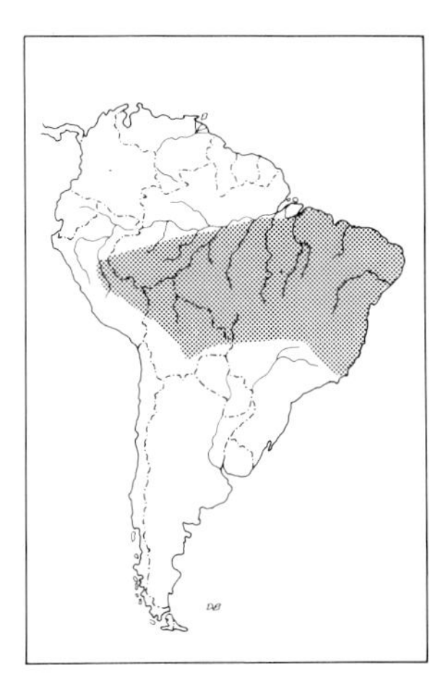

IDENTIFICATION: 15.5 cm (6"). *S. Amazonia east to e. Brazil* (*mystacalis* of Venezuela to Ecuador considered a distinct species; see following). *Crown and nape grayish brown,* becoming rufous-brown on upperparts (with little contrast); long superciliary white, ear-coverts black with white streaks, and *prominent white malar and narrow black submalar streak* (the "moustache"). Throat white becoming dull buffy grayish on breast and dull buff on belly (deepest in *bolivianus* of n. Bolivia). Tail boldly barred with black, with little or no barring on wings.
SIMILAR SPECIES: Less sharply patterned than the geographically distant Whiskered Wren, lacking the contrasty gray band across the upper back, etc. More likely to be confused with Coraya Wren (with which it is sympatric in parts of Amazonia), but latter always looks essentially black-faced (though in some races there are a few white streaks on ear-coverts).
HABITAT AND BEHAVIOR: Locally fairly common to common in undergrowth of forest and woodland borders and in w. Amazonia, especially in dense thickets (often with bamboo, *Bambusa* sp.) along the margins of lakes and rivers. Pairs work through lower growth and are difficult to see under most circumstances. In se. Peru its song is a series of fast, rollicking phrases, almost always given as an antiphonal duet by members of a pair; often the song consists of a phrase followed by a fast "chochocho." Its whining, querulous calls (e.g., "jeeyr") are also frequently given and sometimes incorporated into the song. Overall quality of song is more like that of Coraya Wren than the slower, rich, mellow gurgling of Whiskered Wren.
RANGE: Se. Peru (north to Ucayali), n. and e. Bolivia (south to Santa Cruz), and Amaz. and e. Brazil (only south of the Amazon; east to the Atlantic coast, and south to Mato Grosso, Goiás, Minas Gerais, and Rio de Janeiro). To 1500 m (on slopes of Andes in Bolivia).

NOTE: Does not include the *T. mystacalis* group (Whiskered Wren), considered conspecific by Meyer de Schauensee (1966, 1970). The latter differs in its mainly montane range and its very different voice, as well as morphology; their ranges are entirely allopatric, but it seems doubtful that they would interbreed were they to be in contact. We include in *T. genibarbis* the races *juruanus, intercedens,* and *bolivianus.*

Thryothorus mystacalis

WHISKERED WREN

Other: Moustached Wren (in part)

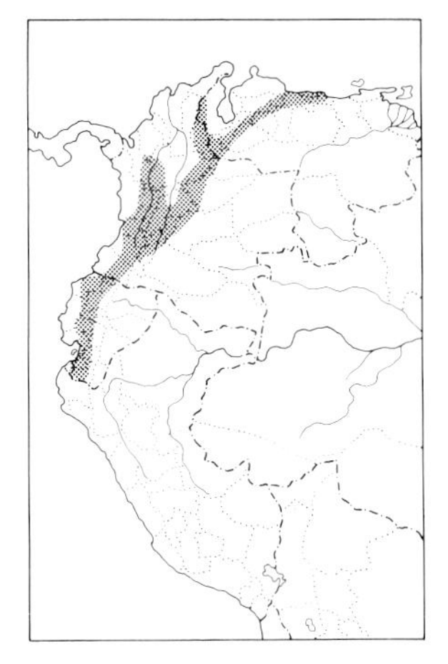

IDENTIFICATION: 16 cm (6¼"). *Montane areas from Venezuela to s. Ecuador;* here considered a species distinct from true Moustached Wren, *T. genibarbis* (see preceding). *Crown to upper back gray, contrasting with otherwise rufous-chestnut upperparts;* long superciliary white, ear-coverts streaked black and white, and *prominent white malar and bold black submalar streak* (the "whisker"). Throat white, becoming gray on breast, buffier on belly. Tail rufous-chestnut barred black, but wings essentially unbarred. See V-31, C-42. *Ruficaudatus* (n. Venezuela) has rufous, unbarred tail and its throat and the malar streak are buff.
SIMILAR SPECIES: This large wren with its distinctly bicolored back should not be confused in its mostly montane range. Could overlap with the even higher-ranging Plain-tailed Wren in w. Ecuador, but latter has spotted breast, unbarred tail, etc. Range does not approach that of Moustached Wren.
HABITAT AND BEHAVIOR: Uncommon to fairly common in dense undergrowth of forest and woodland borders and openings and in heavily overgrown clearings; infrequent actually inside continuous forest. Perhaps the most difficult *Thryothorus* wren to see, but behavior otherwise similar to the others'. Quality and pace of its song are reminiscent of Black-bellied Wren's, being exceptionally rich and musical, often with a slow, gurgling quality and usually with long pauses between phrases; 1 frequent phrase is a "tee-to, to-whít."
RANGE: Coastal mts. of n. Venezuela (east to Miranda) and Andes of w. Venezuela, Colombia (south on e. slope apparently only to Cundinamarca), and w. Ecuador (where also recorded locally from lowlands; south to El Oro); Perijá Mts. on Venezuela-Colombia border. Mostly 800–2400 m, but locally lower (to almost sea level) in w. Ecuador.

NOTE: See comments under Moustached Wren. We include in *T. mystacalis* the races *macrurus, amaurogaster, saltuensis, yananchae, consobrinus, ruficaudatus,* and *tachirensis;* some of these may not be worthy of recognition.

Thryothorus coraya

CORAYA WREN

PLATE: 3

IDENTIFICATION: 14.5 cm (5¾"). Crown dusky brown, becoming chestnut brown on remaining upperparts, with tail barred black; wings unbarred. *Face mainly black, extending down over sides of throat;* vague superciliary, some streaking on cheeks, and lower eyelid white. Throat white, becoming dingy grayish buff on remaining underparts. Two races on *tepuis* of Bolívar, Venezuela, and adjacent Guyana (*obscurus* and *ridgwayi*) are much more rufescent on breast and belly; *cantator* (Junín, Peru) and *herberti* (lower Amaz. Brazil south of the Amazon) lack streaking on cheeks.

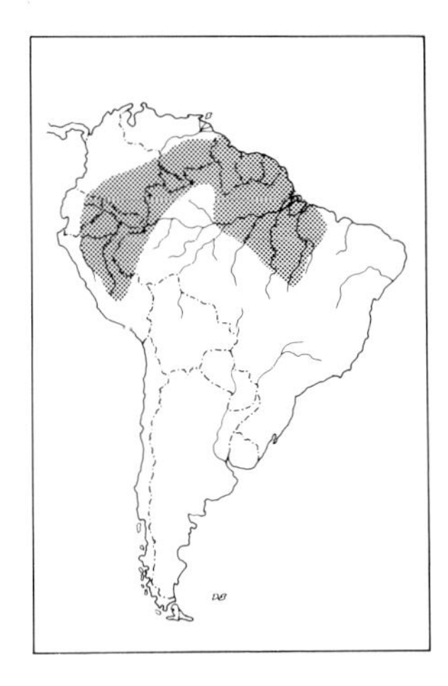

SIMILAR SPECIES: Though a broad area in the malar region usually looks darker than the cheeks themselves, nonetheless the *essentially black-faced look* is unlike that of any other wren. Never shows the black and white malar streaks of Moustached and Whiskered Wrens. Cf. the widely sympatric Buff-breasted Wren.

HABITAT AND BEHAVIOR: Fairly common to common in undergrowth of humid forest (both *terra firme* and *várzea*) and secondary woodland. Most frequent near water. Behavior much like other similar wrens', but more apt to join mixed flocks and quite often forages higher above ground, regularly up to mid-levels in viny tangles at edge, along with Gray Antbirds, Pygmy Antwrens, etc.; partly as a result, this is a relatively easy *Thryothorus* wren to see. Its loud and frequently given vocalizations often draw attention to it; calls include a distinctive "chidip-chidip, choopu" or a "whoo-oor, wheeeer, chuchuchuchuchu," either often repeated rapidly, while its fast-phrased song is reminiscent of that of numerous other *Thryothorus*. Both calls and songs are almost always given as a duet by a pair.

RANGE: Se. Colombia (north to Meta and Guainía), e. Ecuador, ne. Peru (south to n. Cuzco), s. Venezuela (Amazonas and Bolívar), Guianas, and locally in Amaz. Brazil (upper Rio Negro region and along lower Amazon from lower Rio Negro east, also south to n. Goiás and east to w. Maranhão). Mostly below 1000 m (but recorded to 2400 m in Venezuela).

GROUP C

Miscellaneous *small* wrens of *n. and w. South America.*

Thryothorus leucopogon

STRIPE-THROATED WREN

PLATE: 3

IDENTIFICATION: 12 cm (4¾"). *W. Colombia and w. Ecuador.* Iris dull yellow. Above dull olive brown, wings and tail barred black. Narrow superciliary white, with *sides of head and throat streaked black and white.* Remaining underparts pale cinnamon brown (more grayish brown in *grisescens* of n. Chocó, Colombia).

SIMILAR SPECIES: Superficially similar to the larger Buff-breasted Wren but with obviously streaked throat and not as warm buff below; behavior also differs markedly (see below). Cf. also Rufous-breasted Wren (no known overlap; latter has much brighter breast, spotted throat, unbarred wings, etc.).

HABITAT AND BEHAVIOR: Uncommon and seemingly local in borders of secondary woodland and forest. Not well known in S. Am. portion of its range. In Panama this is a relatively easy wren to observe, foraging in pairs mostly some 3 to 10 m above the ground, often in viny tangles; unlike most other *Thryothorus* wrens it regularly accompanies mixed flocks. Its most commonly given call is a distinctive, fast, 3-noted "chu, ch-chu" or "chu, ch-chu, ch-chu."

RANGE: W. Colombia (Pacific lowlands and along n. base of Andes east to s. Córdoba) and w. Ecuador (south to Manabí). Also e. Panama. To 900 m.

NOTE: We follow the 1983 AOU Check-list and several other recent authors in considering the *leucopogon* group as a species distinct from *T. thoracicus* (Stripe-breasted Wren) of s. Central America.

Thryothorus sclateri

SPECKLE-BREASTED WREN

PLATE: 3

Other: Spot-breasted Wren

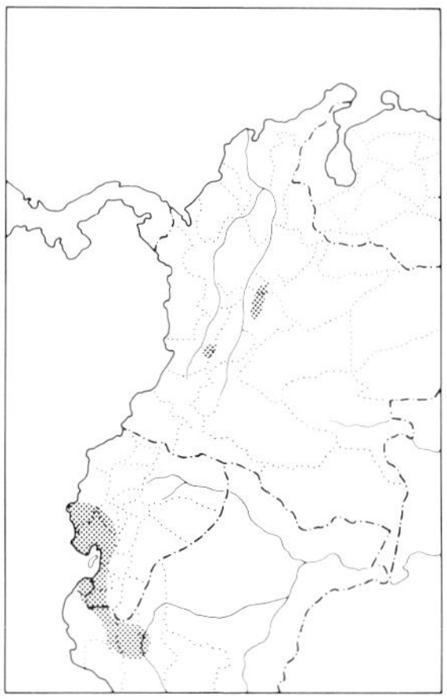

IDENTIFICATION: 13.5–14 cm (5¼–5½"). *Sw. Ecuador to nw. Peru and very locally in w. Colombia.* Brown above, slightly more rufescent on crown; tail barred blackish (but wings unbarred). Narrow superciliary white, with *sides of head and neck streaked or speckled black and white, Paucimaculatus* (sw. Ecuador and adjacent Peru) has throat white, breast and upper belly pale grayish *speckled with black;* flanks tinged rufous. *Sclateri* (Marañón valley of n. Peru) and *columbianus* (w. Colombia) *even more boldly speckled black below* (with effect almost of wavy black barring), with this extending from throat to belly; *sclateri* has a slightly longer bill and is a bit larger than the other races.
SIMILAR SPECIES: Note the small size and spotting or barring below. No other really similar wren in range, but from above Stripe-throated Wren is rather similar (the 2 overlap only in w. Ecuador). Not known to occur with Rufous-breasted Wren.
HABITAT AND BEHAVIOR: Locally fairly common to common in thickets and undergrowth (often viny) of deciduous forest and woodland. Usually in pairs, foraging independently of flocks most of the time; for a *Thryothorus* wren, generally not too difficult to see. Its song is a series of fast, repeated phrases which, while attractive, is not as musical as that of many of its congeners. A frequently given call is uncannily similar to the sound made by rubbing one's finger against the tines of a comb.
RANGE: Locally in w. Colombia (known only from w. slope of Cen. Andes above Palmira and Valle and from several localities on w. slope of E. Andes in Cundinamarca); sw. Ecuador (north to Manabí) and nw. Peru (on Pacific slope in Tumbes and Piura and in upper Marañón valley of Cajamarca). 1300–2000 m in Colombia, to about 1600 m in Ecuador and Peru.

NOTE: The taxonomy of the *T. rutilus* group is disputed; some would recognize but 1 polytypic species (e.g., *Birds of the World,* vol. 9), others 3 (e.g., the 1983 AOU Check-list), others 2 (by merging the 2 widely disjunct spot- or speckle-breasted groups; e.g., Meyer de Schauensee 1966). Spot- or speckle-breasted types occupy either end of the overall range, with the rufous-breasted group occurring in the center. None of the 3 main forms is presently in contact; the 2 which come closest (*T. sclateri columbianus* and *T. rutilus hypospodius* on either side of the E. Andes in Cundinamarca, Colombia) do not show any signs of intergradation, and it is of interest that *T. rutilus laetus,* which looks "intermediate," does not (at least now) occur anywhere near the range of *T. sclateri.* Vocally and behaviorally all are similar. For now we prefer to maintain all 3 as full species.

Thryothorus rutilus

RUFOUS-BREASTED WREN

PLATE: 3

IDENTIFICATION: 14 cm (5½"). *Ne. Colombia and n. Venezuela.* Brown above, slightly more rufescent on crown; tail barred with blackish.

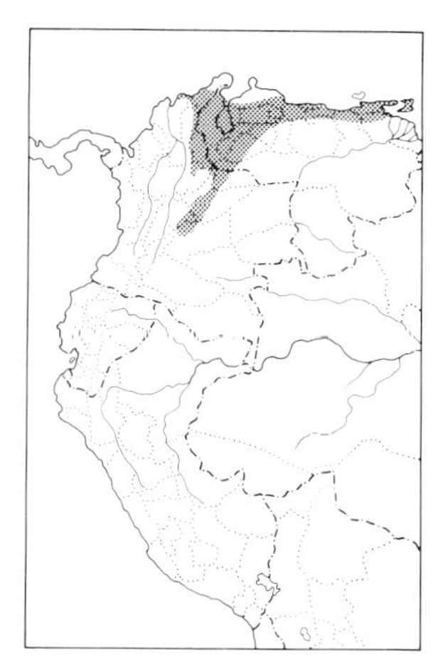

Narrow superciliary white, with *sides of head and throat speckled black and white, in sharp contrast to bright orange-rufous breast;* center of belly whitish, flanks more or less tawny-rufous. *Laetus* (lower slopes of Santa Marta and Perijá mts. in Colombia and Venezuela and on Guajira Peninsula) has breast sparsely spotted black, while *intensus* (lower Andean slopes of w. Venezuela) is the most richly colored on breast and lacks the white mid-belly.

SIMILAR SPECIES: The contrast between the speckled face and the bright rufous breast is unique among the wrens.

HABITAT AND BEHAVIOR: Fairly common to common in thickets and undergrowth of deciduous woodland forest borders; favors viny tangles, where it often moves about well above the ground and therefore is less difficult to see than many other wrens. Song and calls are similar to Speckle-breasted Wren's; as in that species they are often given as a duet by members of a pair (e.g., a repeated "wee-chee, weecheewee-ee-yee"). Most numerous in foothill areas (especially 500–1200 m), except on Trinidad and Tobago, where it is more widespread.

RANGE: Ne. Colombia (Santa Marta area south on w. slope of E. Andes to Santander, on e. slope of Andes to Meta) and n. Venezuela (east to the Paria Peninsula); Trinidad and Tobago. Also Costa Rica and Panama. To 1900 m.

NOTE: See comments under Speckle-breasted Wren.

GROUP D

Relatively simply patterned; buff to whitish below, with *barred wings.*

Thryothorus leucotis

BUFF-BREASTED WREN

PLATE: 3

IDENTIFICATION: 14–14.5 cm (5½–5¾"). Above rufous-brown to grayish brown, wings and tail barred black. Narrow superciliary white, sides of head streaked blackish and white. Throat white, *gradually deepening to pale buff on breast and cinnamon buff to rufous on belly. Rufiventris* of interior s. Brazil is slightly larger and more rufescent below, while *venezuelanus* (n. Colombia and w. Venezuela) is paler below.

SIMILAR SPECIES: Cf. Long-billed and Fawn-breasted Wrens. Coraya, Moustached, and Whiskered Wrens all show much more black on sides of face, have unbarred wings, and are less rufescent below. Stripe-throated is smaller and has obviously streaked throat and a pale eye.

HABITAT AND BEHAVIOR: Uncommon to locally common (seemingly less numerous and more local in w. Amazonia) in thickets and undergrowth of forest and woodland borders, overgrown clearings, and even mangroves; most numerous near water, and in Amazonia largely confined to *várzea* shrubbery and borders. Usually in pairs, foraging actively through lower growth, remaining in heavy cover and generally not easy to see. The song is loud, vigorous, and often quite musical, typically a phrase repeated over and over, then changing abruptly to another (e.g., "choreéwee, choreéwee, choreéwee . . . wheeooreéticktweeoo wheeooreétickweeoo . . ."). Calls include a sharp "chit-cho" or "chit, cho-cho."

RANGE: Widespread east of Andes south to extreme n. Bolivia (Pando) and interior s. Brazil (south to Mato Grosso, n. São Paulo, and Minas Gerais; does not extend to e. coastal lowlands); west of Andes only in w. Colombia (Caribbean lowlands and south in Magdalena valley to Huila); absent from part of n. Venezuela. Also Panama. To 950 m.

NOTE: See comments under Long-billed and Fawn-breasted Wrens.

Thryothorus longirostris

LONG-BILLED WREN

PLATE: 3

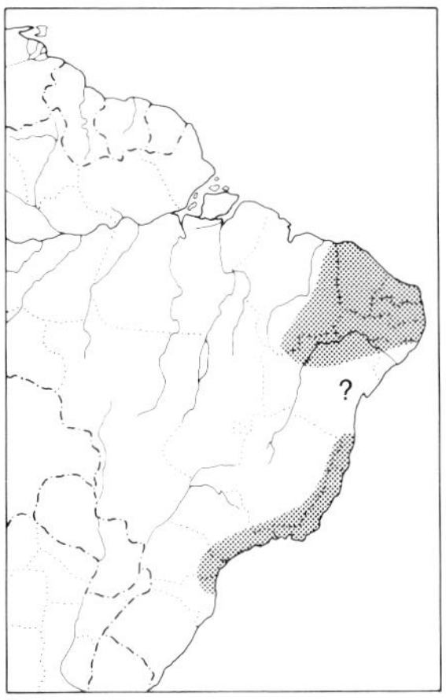

IDENTIFICATION: 15 cm (6″). *E. Brazil. Strikingly long-billed* (about 1 cm or ½″ longer than in Buff-breasted), but otherwise quite similar to Buff-breasted Wren (the 2 may be conspecific; see below). Rather rufescent brown above (especially when compared to the quite grayish brown–backed race of Buff-breasted with which it is most nearly in contact, *rufiventris*). *Bahiae* (ne. Brazil) is somewhat paler generally than more s. nominate race, with whiter, less streaked cheeks. Several specimens from s. Piauí have bills of intermediate length.

SIMILAR SPECIES: Plumage duplicated by various races of Buff-breasted Wren; best to go by range and by its obviously longer bill. Moustached Wren has shorter bill, strong black and white stripes in malar region, unbarred wings, etc.

HABITAT AND BEHAVIOR: Uncommon to fairly common in thickets and undergrowth at edge of forest and woodland and in shrubby clearings; seems less tied to water than Buff-breasted Wren usually is. General behavior and vocalizations much like Buff-breasted Wren's. Quite readily found in Tijuca Nat. Park above Rio de Janeiro.

RANGE: E. Brazil (northeast from Ceará and Pernambuco south to s. Piauí and n. Bahia and in coastal southeast from s. Bahia to e. Santa Catarina). To about 900 m.

NOTE: At least 2 specimens in AMNH taken in s. Piauí by E. Kaempfer are intermediate in bill length between *T. leucotis rufiventris* and *T. longirostris bahiae* (1 each from Parnaguá and Gilbues). Of the 3 other Piauí specimens which we could locate, 1 from Parnaguá appears typical of *longirostris bahiae*, while 2 from Therezinha and Rio Parnaiba (Belo Horizonte) seem typical of *T. leucotis rufiventris*. These intermediate specimens were first noted by K. C. Parkes, Jr., and could argue for treating the 2 forms as conspecific (the name *T. longirostris* would have priority). We suspect that this will ultimately by shown to be the case but are reluctant to do so without a more thorough analysis of the situation. One cautionary note is that bill length in *both* races of *longirostris* varies appreciably.

Thryothorus guarayanus

FAWN-BREASTED WREN

IDENTIFICATION: 13.5 cm (5¼″). *N. Bolivia and adjacent Brazil.* Closely resembles Buff-breasted Wren (the 2 may be conspecific). Compared to *T. leucotis peruanus* (the race of Buff-breasted which comes closest in s. Peru), Fawn-breasted is slightly more grayish (less rufescent) above, has more strongly streaked cheeks, and its throat is buff-tinged (not as whitish). Compared to *T. leucotis rufiventris* (the race of Buff-breasted

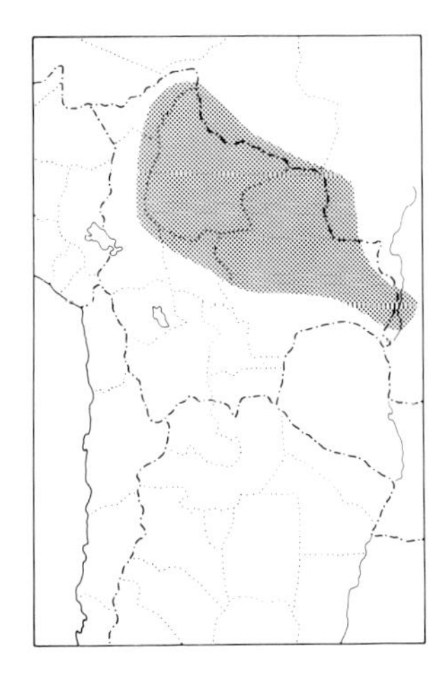

which comes closest in sw. Brazil), Fawn-breasted is smaller but otherwise similar in color.

SIMILAR SPECIES: So close in appearance to Buff-breasted Wren that the 2 can probably be told apart in the field only by range. Moustached Wren has strong black and white stripes in malar region, unbarred wings, etc.

HABITAT AND BEHAVIOR: Uncommon to fairly common in thickets near water and undergrowth in woodland and forest borders. General behavior and vocalizations much like Buff-breasted Wren's. Numerous at various localities around Corumbá in sw. Mato Grosso, Brazil.

RANGE: N. Bolivia (Pando and Beni south to La Paz and Cochabamba and east across all of Santa Cruz) and adjacent sw. Mato Grosso, Brazil. To about 400 m.

NOTE: Probably conspecific with *T. leucotis,* of which it would be the smallest race; bill length, color of upperparts, boldness of cheek striping, and color of underparts can all be matched by one or another of the races of *T. leucotis.* Perhaps the only factor mitigating against our doing so (over 50 years ago it also caused Hellmayr, in *Birds of the Americas,* vol. 13, to do likewise) is that no intergradation has been shown between *guarayanus* and the appreciably larger race of *T. leucotis, rufiventris,* found in Brazil immediately to the north. An alternate treatment is to consider *guarayanus* as a race of *T. leucotis* while raising *T. rufiventris* to species rank; this treatment was advocated by M. E. Carriker (*Proc. Acad. Nat. Sci. Phil.* 87: 337, 1935). Clearly, a thorough analysis of this complex is needed, as is more fieldwork, and until such is completed we believe it best not to alter the generally accepted arrangement. If *guarayanus* is deemed conspecific with *T. leucotis,* then the former name would have priority; if *longirostris* is also included, then its name has priority. Regardless, Buff-breasted Wren would seem the best English name for the expanded species.

Thryothorus superciliaris

SUPERCILIATED WREN

PLATE: 3

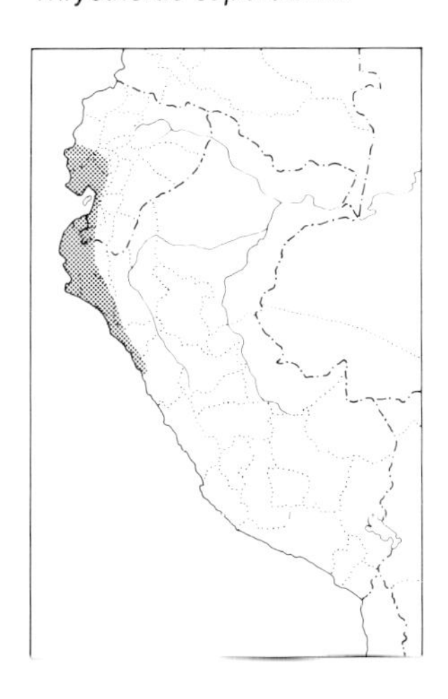

IDENTIFICATION: 14.5 cm (5¾"). *Sw. Ecuador and nw. Peru.* Above bright rufescent brown, slightly duller on crown; wings and tail barred black. *Long broad superciliary white; sides of head mostly white (unstreaked).* Mostly white below, slightly deepening to buff on flanks and lower belly. *Baroni* (El Oro, Ecuador, south into Peru) is slightly brighter above.

SIMILAR SPECIES: Recalls Buff-breasted Wren but does not occur with it (part of same superspecies); its superciliary is fuller and more conspicuous, its cheeks are whiter and essentially unstreaked, and it is whiter below than any race of Buff-breasted.

HABITAT AND BEHAVIOR: Common in undergrowth of drier woodland, arid scrub, and agricultural regions with scattered trees and hedgerows. Not nearly as tied to water as Buff-breasted Wren, occurring far from water over most of its range and in fact avoiding humid areas; partly because of its semiopen habitat it is much easier to see than that species. The song is fast and very rollicking, with phrases so run together that the effect is sometimes quite warbled.

RANGE: Sw. Ecuador (north to Manabí) and coastal w. Peru (south to Ancash). To about 500 m.

Thryothorus rufalbus

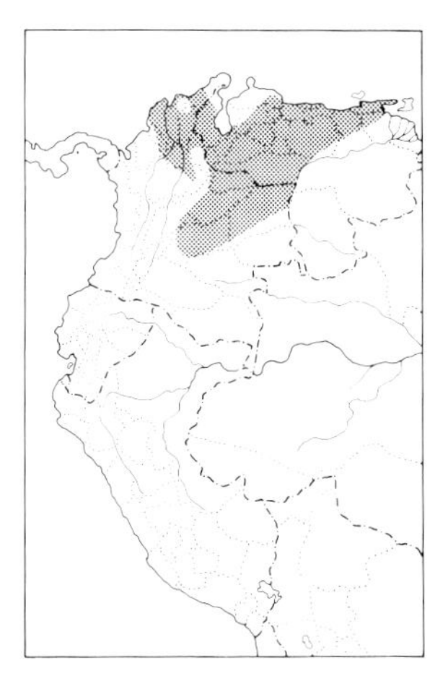

RUFOUS-AND-WHITE WREN

IDENTIFICATION: 14.5 cm (5¾"). *Colombia and Venezuela.* General form recalls Buff-breasted Wren, but *much whiter below. Rich rufous above;* wings and tail barred black. Long superciliary white, sides of head and neck white streaked black. *White below,* tinged brownish on sides and flanks; crissum barred black. See V-31, P-22.

SIMILAR SPECIES: A basically *bicolored* wren, with rich rufous upperparts and white underparts. Buff-breasted Wren is generally much buffier below, though its palest race, *venezuelensis,* is somewhat similar (but is much duller rufous-brown above and more generally tinged with buff on breast and, especially, belly). Cf. also Niceforo's Wren.

HABITAT AND BEHAVIOR: Fairly common to common in undergrowth and borders of deciduous woodland and gallery forest, locally into more humid woodland and forest borders. Very shy and furtive, tending to move about slowly in thick lower growth, sometimes even hopping on ground, rarely fully in the open. Its song is unmistakable, very unlike any other in the genus; it consists of 4 to 5 slow, low-pitched "hooting" whistles, often preceded and ending with a higher note; 1 recurring phrase is a "wheee, hoo-hoo-hoo-hoo-whít." Usually sings from heavy cover.

RANGE: N. Colombia (Santa Marta area and lower Magdalena valley south to s. Cesar and s. Bolívar; llanos region of northeast south to Meta and Vichada) and w. and n. Venezuela (south to Apure, Guárico, Anzoátegui, and the Paria Peninsula). Also Mexico to Panama. To 1500 m.

Thryothorus nicefori

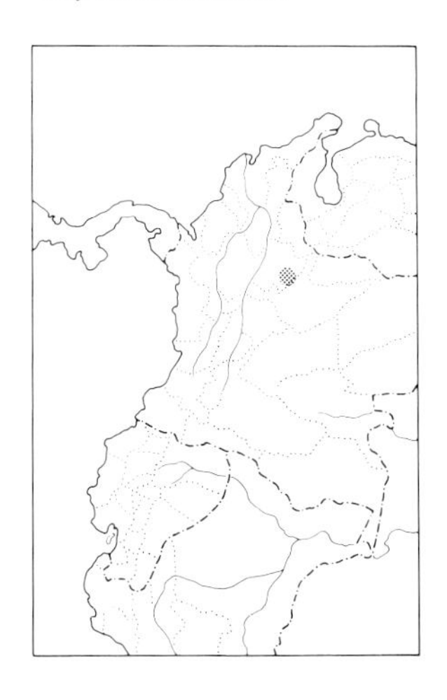

NICEFORO'S WREN

IDENTIFICATION: 14.5 cm (5¾"). *Very locally in E. Andes of n. Colombia.* Resembles Rufous-and-white Wren (and is perhaps only a race of that species). *Brown above,* becoming somewhat more rufescent on lower back and rump; wings and tail barred black. Long superciliary white, sides of head white streaked black. Below white, *tinged grayish on sides and flanks;* crissum barred black.

SIMILAR SPECIES: Rufous-and-white Wren is much more rufous above and its sides and flanks are tinged brownish (not grayish); both species have usually concealed subapical white spots on rump, but these are larger in Niceforo's (and possibly more conspicuous?). Cf. also Buff-breasted Wren.

HABITAT AND BEHAVIOR: Unknown in life; recorded from only a few specimens. It would be of particular interest to determine the quality and pattern of its song.

RANGE: N. Colombia (San Gil, on w. slope of E. Andes in Santander). 1100 m.

NOTE: Perhaps only a well-marked race of *T. rufalbus;* however, note that races of the latter in South America show only minimal variation.

GROUP E

Distinctive; *small, short-tailed, gray.*

Thryothorus griseus

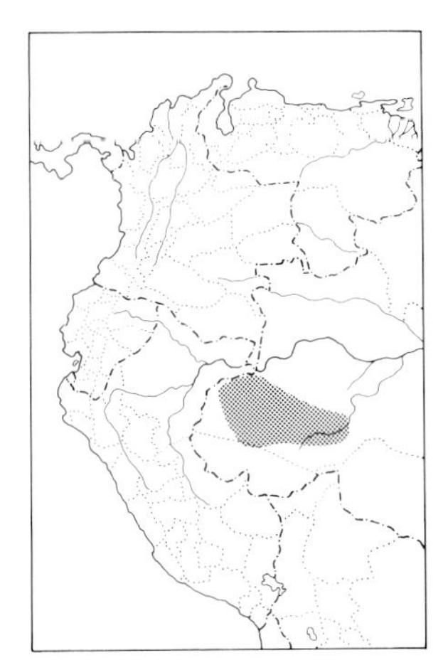

GRAY WREN

PLATE: 3

IDENTIFICATION: 11.5 cm (4½"). *Very locally in w. Amaz. Brazil. Very short tail* (proportionately much shorter than in other *Thryothorus* wrens) *Above brownish gray* with inconspicuous short whitish superciliary; cheeks only lightly streaked. Tail barred gray and dusky. *Below uniform pale smoky gray.*

SIMILAR SPECIES: Short-tailed and basically gray, the Gray Wren should be easily distinguished from other wrens. All other *Thryothorus* wrens are much browner or more rufescent and have longer tails. Tooth-billed Wren is also gray above but is whitish below, and its shape with long tail is utterly different (being very similar to the Gray-mantled Wren of the Andes; see below). Shape and color of Gray Wren are also superficially similar to certain male *Myrmotherula* antwrens.

HABITAT AND BEHAVIOR: Apparently locally common in tangled undergrowth at edge of *várzea* forest and woodland and in overgrown clearings. Forages at varying levels in viny tangles, moving in pairs or small groups. The song is rhythmic and melodic, consisting of a phrase repeated 5 to 8 times, then on to the next (e.g., "tor-chílip, tor-chílip . . . ," or "fiddle-dip, fiddle-dip . . . ," or "chur-dúrt, chur-dúrt . . ."), typically getting stronger as it goes along (S. Hilty). Has been seen along the Rio Javarí just south of Leticia, Colombia.

RANGE: W. Amaz. Brazil (Amazonas along Rio Javarí, upper Rio Juruá, and upper Rio Purús). 200 m.

NOTE: A very divergent *Thryothorus* wren, with unique gray coloration and short tail. Possibly it deserves to be generically separated.

Odontorchilus Wrens

A pair of closely related wrens, very unusual in their gnatcatcherlike form and even behavior. One is Andean, the other occurs in the central Amazon basin. They have been considered close to *Campylorhynchus* and do agree in their arboreal behavior, but differ markedly in size and pattern and are much less vocal.

Odontorchilus branickii

GRAY-MANTLED WREN

PLATE: 4

IDENTIFICATION: 12 cm (4¾"). Above dull bluish gray, browner on crown and palest on forecrown. Vague whitish superciliary and sides of head streaked gray and whitish. *Below whitish. Long tail gray barred* with black. *Minor* (w. slope of Andes in sw. Colombia and nw. Ecuador) has central tail feathers uniform gray with no barring.

SIMILAR SPECIES: *Gnatcatcherlike shape* (even to the long cocked tail) is unique among the wrens except for the congeneric Tooth-billed (see below). No gnatcatcher has a similar pattern.

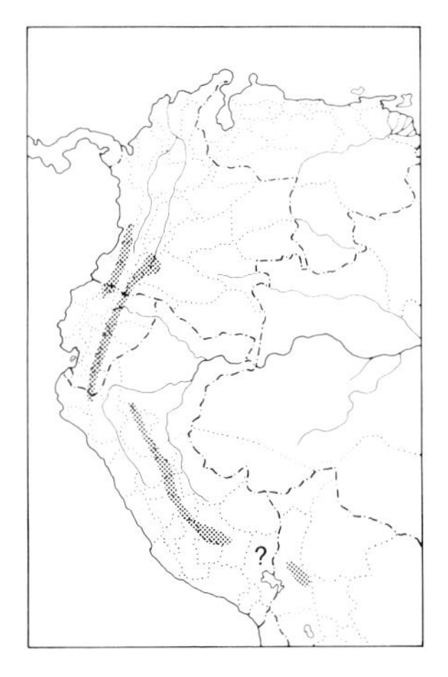

HABITAT AND BEHAVIOR: Uncommon and local (perhaps merely overlooked) in canopy and borders of humid forest. An active arboreal bird which usually cocks its long barred tail, giving it a distinctive silhouette. Generally found singly or in pairs, almost invariably while accompanying mixed flocks of tanagers, etc. in the canopy and subcanopy. Often forages by hopping out on thick horizontal branches, inspecting epiphytes and peering underneath. Generally quiet, but in e. Ecuador has been seen to mount to an exposed perch in early morning and there repeatedly give what is probably its song, a fairly short, thin, dry trill. Has been recently found at a number of e. slope localities (seems to be an easier bird to see than it was to collect); numerous at Coca Falls on the road to Lago Agrio, Ecuador.

RANGE: W. slope of W. Andes in sw. Colombia (Valle in the upper Anchicayá valley) and nw. Ecuador (Esmeraldas at Paramba); slopes above upper Magdalena valley in Huila, Colombia; and e. slope of E. Andes in s. Colombia (sighting in w. Caquetá), e. Ecuador, e. Peru, and nw. Bolivia (La Paz). Mostly 1400–2200 m, occasionally higher; on Pacific slope 800–1100 m.

Odontorchilus cinereus

TOOTH-BILLED WREN

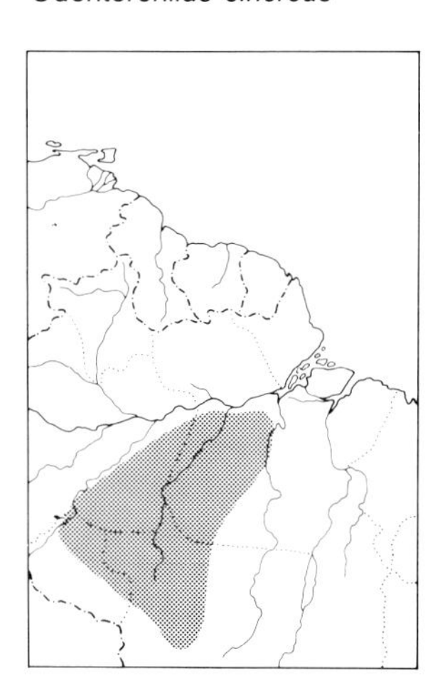

IDENTIFICATION: 12 cm (4¾"). Very closely resembles Gray-mantled Wren but ranges are far separated: Tooth-billed found only in *cen. Amaz. Brazil.* Bill somewhat heavier; mantle somewhat duller, more brownish gray (less blue); throat and breast tinged buff.

SIMILAR SPECIES: Should be distinctive in its range. Gray Wren is uniform gray with short tail. Compare also to Guianan and female Tropical Gnatcatchers (neither of which has a barred tail, etc.).

HABITAT AND BEHAVIOR: Nothing is recorded, but presumably very much like Gray-mantled Wren's.

RANGE: Amaz. Brazil south of the Amazon in n. and cen. Mato Grosso (upper Rio Madeira south to Serra dos Parecís nw. of Cuiabá) and e. Pará (lower Rio Tapajós and lower Rio Xingú area).

NOTE: Were it not for their far-separated ranges at different elevations in different habitats, we would be inclined to consider the 2 presently recognized species in *Odontorchilus* as conspecific; they are very similar morphologically. *Cinereus* would have priority, but we would suggest that "Gray-mantled" would be a better name for the enlarged species as opposed to "Tooth-billed" (the "tooth" near the tip of the maxilla is impossible to see in the field and indeed is hard to discern in the hand).

Troglodytes Wrens

Small, "typical" wrens, basically brown and, compared to most other members of the family, dully patterned except for a bold superciliary (and even that is lacking in the House Wren). Tails are often held cocked. The familiar House Wren is a virtual commensal of man, occurring almost throughout South America, but the other 3 S. Am. *Troglodytes* are restricted to montane forest and shrubbery.

Troglodytes aedon

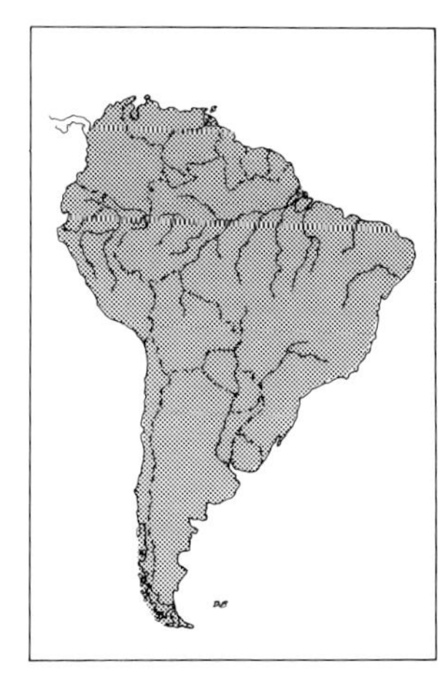

HOUSE WREN

PLATE: 4

IDENTIFICATION: 11.5–12 cm (4½–4¾"). *A small brown and buff wren without obvious features.* Brown to buffy brown above, with weak buffy whitish superciliary and indistinct dusky barring on wings and tail. Below whitish to buffy whitish, usually palest on throat and belly. *Cobbi* of the Falkland Is. is the largest race, while *tobagensis* (Tobago) has the whitest underparts and *puna* (altiplano of Peru south to La Paz, Bolivia) is the most rufescent below (with belly solid rufous); *tecellatus* (coastal sw. Peru and adjacent Chile) has broken black barring across back.

SIMILAR SPECIES: Perhaps the most wide-ranging passerine bird in South America (few settled areas are without them), House Wrens are best recognized by their lack of obvious field marks and familiar behavior. Mountain Wren is much more of a forest bird and has a more prominent buff eyestripe. Grass Wren has streaked back, etc.

HABITAT AND BEHAVIOR: Common and very widespread, occurring in virtually all open or semiopen habitats, even quickly appearing in clearings recently made in forested regions. Though usually most numerous near human habitation, it also occurs in shrubby areas away from man's direct influence. A nervous and high-strung but at the same time often a tame little bird, constantly scolding but usually not moving very far when disturbed. The ebullient song is given throughout the year, a cheerful, gurgling warble whose pattern and quality should at once be familiar to N. Am. observers. A common call, at least in Ecuador, is a nasal "jeeyáh," repeated at intervals, very different from N. Am. birds.

RANGE: Widespread virtually throughout South America; Trinidad and Tobago and Falkland Is. Also North and Middle America and the Lesser Antilles. Up to over 4000 m in the Andes.

Troglodytes solstitialis

MOUNTAIN WREN

PLATE: 4

IDENTIFICATION: 10.5–11 cm (4–4¼"). *Andes from Venezuela to Argentina. Rufous-brown above,* barred blackish on wings and tail; *prominent long superciliary buff* in n. birds (*solitarius* and nominate), buffy whitish in *macrourus* (most of e. Peru), and whitish in *frater* and *auricularis* (Puno, Peru, south to Argentina). Below pale to deep buff, brightest on breast and whitest on median belly; crissum whitish barred with black.

SIMILAR SPECIES: A more rufescent, shorter-tailed version of the House Wren, with much more conspicuous superciliary; much less familiar than that species, being more or less arboreal in montane forest and woodland. Pattern reminiscent of *fulva* race of Sepia-brown Wren (found in extreme s. Peru and nw. Bolivia) but much smaller, etc.

HABITAT AND BEHAVIOR: Fairly common to common in humid montane forest and woodland, mostly at edge or in mid-level tangles around openings. Forage mostly at middle levels but sometimes go higher or move through undergrowth; often they are noted hopping actively along moss-covered limbs (sometimes even trunks), inspecting crevices and epiphytes. Usually encountered with mixed flocks of tanagers, warblers, etc. The short song is high-pitched, fast, and not very musical, "peet, twee-titititeteh," fading at the end; not very loud, it is apt to be overlooked.

RANGE: Andes of w. Venezuela (north to s. Lara), Colombia, Ecuador, nw. and e. Peru (south on w. slope to Cajamarca), Bolivia, and nw. Argentina (south to Tucumán and adjacent Catamarca); Perijá Mts. on Venezuela-Colombia border. Mostly 1500–3500 m, but lower in Argentina (regular down to at least 700 m).

NOTE: Does not include *T. monticola* (Santa Marta Wren) or Ochraceous Wren (*T. ochraceus*) of Costa Rica and Panama. Latter has been recorded on Cerro Tacarcuna virtually on Colombian border and may range into Colombian territory; it resembles the Mountain Wren but is more uniformly ochraceous below, etc.

Troglodytes monticola

SANTA MARTA WREN

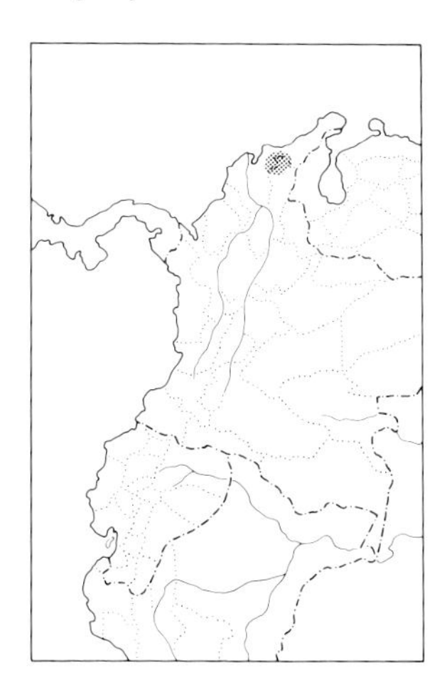

IDENTIFICATION: 11.5 cm (4½"). *High elevations on Colombia's Santa Marta Mts.;* often considered a race of Mountain Wren but here regarded as a full species. Somewhat larger than Mountain Wren, also with a buff superciliary. Darker, much less rufescent brown above; wings and tail more conspicuously barred with blackish, and *flanks and crissum boldly barred blackish.*

SIMILAR SPECIES: Does not occur with Mountain Wren. See also House Wren (though latter may not occur as high on the Santa Marta Mts. as this species is found).

HABITAT AND BEHAVIOR: Not well known, but reportedly found in low, thick shrubbery at timberline and in sheltered spots higher up in the *páramo* zone; it is shy, silent, and easily overlooked (Todd and Carriker 1922). We are not aware of any recent reports of this bird, as it is apparently not found as low as the San Lorenzo ridge, the only part of the Santa Marta massif now generally visited.

RANGE: Santa Marta Mts. of n. Colombia. Mostly 3200–4600 m.

NOTE: Here considered a species distinct from Andean *T. solstitialis,* following Hellmayr (*Birds of the Americas,* vol. 13) and the 1983 AOU Check-list but not Meyer de Schauensee (1966, 1970). Morphologically it stands apart strikingly from all the races of *T. solstitialis* (indeed, *T. ochraceus* of Costa Rica and w. Panama seems closer to *solstitialis*), and its very high elevational preference is also quite different, though what "prevents" it from occurring lower remains unknown.

Troglodytes rufulus

TEPUI WREN

IDENTIFICATION: 12 cm (4¾"). *Tepui region of s. Venezuela and adjacent Brazil.* A large and intensely colored relative of the Mountain Wren; also somewhat longer-tailed. *Chestnut brown above* with blackish barring on wings and tail and *ochraceous buff superciliary. Below mostly pale rufous-brown,* more whitish on median breast and belly. See V-31. Two races found on the *tepuis* of Amazonas and adjacent w. Bolívar, Venezuela, are more grayish below (almost entirely gray in *wetmorei* of

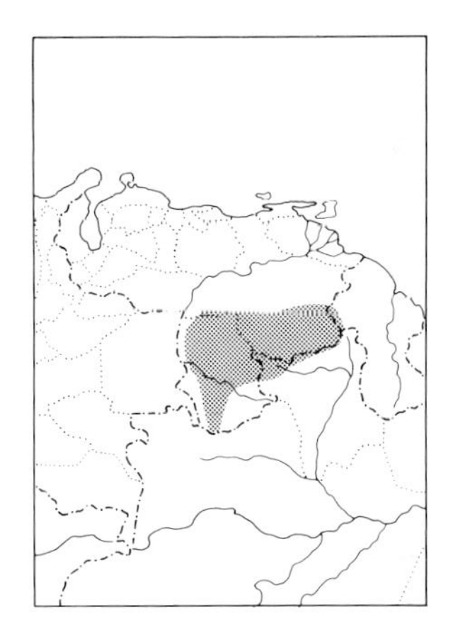

Cerro de la Neblina). Juvenal generally more blackish, with dusky scaling below (unique in this genus).

SIMILAR SPECIES: Does not occur with Mountain Wren. See also the much more grayish House Wren.

HABITAT AND BEHAVIOR: Not well known. Recorded from forest edge and shrubbery in semiopen areas. Apparently tends to remain low, sometimes even hopping on the ground (Meyer de Schauensee and Phelps 1978), which seems quite different from typical behavior of Mountain Wren. The song is a fast jumbled high-pitched twittering, similar to but rather longer than voice of Mountain Wren (A. Altman recording).

RANGE: *Tepuis* of s. Venezuela (Bolívar and Amazonas) and adjacent n. Brazil (Roraima). 1000–2800 m.

Cistothorus Wrens

Small buffy wrens characterized by their streaked backs and typically rather short bills. All are found in more or less open places, usually in tall grass near water, and are secretive and hard to see except in the breeding season. The number of species is uncertain, but we suspect that the traditional arrangement will prove incorrect, with the wide-ranging *C. platensis* being separated into several allospecies.

Cistothorus platensis

GRASS WREN

PLATE: 4

IDENTIFICATION: 9.5–11 cm (3¾–4¼"), *polyglottos* of se. Brazil being smallest, *falklandicus* of Falkland Is. largest. Variable, but can be generally known by its *small size and streaked back in conjunction with its grass/sedge habitat.* Nominate group (Argentina and s. Brazil) has crown streaked black and buff, back and rump broadly streaked black and buffy whitish, and vague whitish superciliary; wings and tail rufous-brown barred black. Mostly whitish below, buff on flanks and crissum. *Aequatorialis* group (Andes from Colombia to nw. Bolivia) is larger and longer-billed, with brown and essentially unstreaked crown and buff superciliary; only the back is streaked (rump plain brown); underparts are more buffy ochraceous. *Alticola* (with oddly disjunct range on the *tepuis,* locally in the n. coastal mts. and Andes of Venezuela, and on Santa Marta Mts. of Colombia) is also quite different, being more grayish brown above with streaks narrower and confined to back; its superciliary is faint or lacking and underparts are mostly whitish.

SIMILAR SPECIES: This small but not particularly short-tailed wren can be difficult even to glimpse unless singing, but the streaked back (readily seen even as the bird quickly flushes) together with its small size, behavior, and habitat should preclude confusion with other wrens (House Wren is much plainer). In Andes of Colombia and Venezuela see the Apolinar's and Merida Wrens, however. Grass Wren could be confused with several of the streak-backed canasteros, some of which occur sympatrically with it; look for the wren's barred wings and shorter (not pointed) barred tail.

HABITAT AND BEHAVIOR: Locally common in tall grassy or sedgy situations, sometimes with scattered bushes or low trees but typically only where these do not predominate. Most prevalent in damp areas (at least in Argentina sometimes even in actual marshes), but in s. Brazil (elsewhere?) also found in dry *cerrado* grasslands. Inconspicuous and apt to be overlooked unless singing or calling (when it may mount to the top of a bush or clump of grass), tending to creep around on or close to the ground. Often cocks tail very sharply. The songs vary between groups (quite possibly there is more than 1 species in South America) but are always complex, consisting of sometimes musical, sometimes buzzy phrases, often introduced by 1 to 2 shorter notes; the overall effect is often quite finchlike. Andean birds (e.g., *aequatorialis* in Colombia and Ecuador) tend to give short, discrete phrases (e.g., "sisisisi, trrr, chee-ee-ee-ee-ee") followed by discrete, often lengthy pauses, whereas Argentinian birds (nominate race) deliver a more or less continuous series of phrases, most of them either 4 or 6 repetitions of a single note or a trill. Its calls also often draw attention, an oft-repeated, scolding "tchew-tchew-tchew-tchew-tchew" in the Andes or a harsher, drier "jtt, jtt, jtt, jtt" in the s. lowlands.

RANGE: Locally in coastal mts. and Andes of Venezuela and more widely in Andes of Colombia (where also recorded locally from e. lowlands in Meta), Ecuador, Peru, and w. Bolivia (south to w. Santa Cruz and in Tarija; also locally in lowlands in Beni); lowlands (mostly) of Chile (north to Coquimbo), Argentina (north to Jujuy, Córdoba, and Corrientes), and s. Brazil (north to Goiás and Minas Gerais) south to Tierra del Fuego; *tepui* region of se. Venezuela (Bolívar) and adjacent Guyana; Falkland Is. Seems almost certain to occur in Uruguay (but as yet unrecorded) and perhaps also in Paraguay (though no definite records known). Also North America. To about 4000 m in the Andes.

NOTE: As mentioned above, more than 1 species seems likely to be involved in South America, but more study is needed. Note that birds of North America, the *stellaris* group (which also very likely merits species status), are now known as the Sedge Wren (appropriate there, but certainly not so in South America).

Cistothorus apolinari

APOLINAR'S WREN

Other: Apolinar's Marsh-Wren

IDENTIFICATION: 12 cm (4¾"). *Marshes in the E. Andes of Colombia.* Resembles Grass Wren of Colombia (*aequatorialis* group), but larger, with *short superciliary grayish* (not buff), rump as well as mantle streaked black and buffy whitish, and *underparts dingier and grayer,* tinged dull buff on flanks and crissum. See C-42.

SIMILAR SPECIES: Grass Wren is not known to occur in the same *marsh vegetation* as this species inhabits; rather, it is found on grassy slopes and *páramos* primarily at higher elevations.

HABITAT AND BEHAVIOR: Locally fairly common in tall marsh vegetation, principally cattails but also bullrushes. Reclusive and not easy to see unless singing (though it will respond to squeaking), creep-

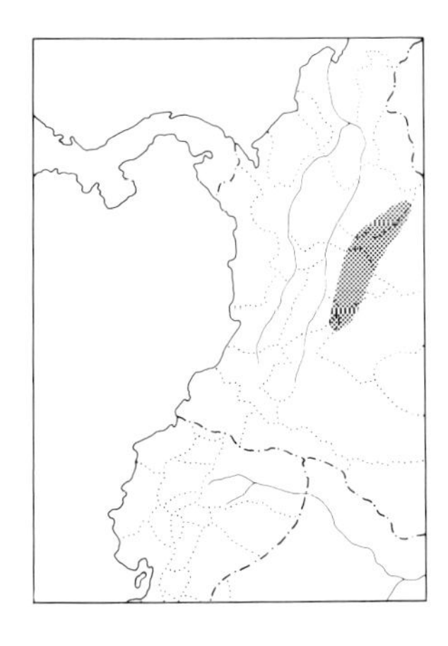

ing through dense vegetation near water level. The song is a lively, fast, rhythmic "toou-cheecheechee, toou-cheecheechee, toou-cheechee-chee . . . ," the first note characteristically low and gravelly; it is usually delivered from a semiexposed perch. The Apolinar's Wren has declined greatly due to the drainage of wetlands across most of its limited range; a substantial population remains, however, in the marshes fringing Tota Lake in s. Boyacá, and some persist at Parque La Florida outside Bogotá near the international airport.

RANGE: E. Andes of Colombia (Boyacá and Cundinamarca). 2500–3500 m.

NOTE: Calling this species a "marsh-wren," as it formerly was, implies that it in some way is allied to the Marsh Wren (*C. palustris*) of North America, which it is not. We prefer to call it simply Apolinar's Wren.

Cistothorus meridae

MERIDA WREN

Other: Paramo Wren

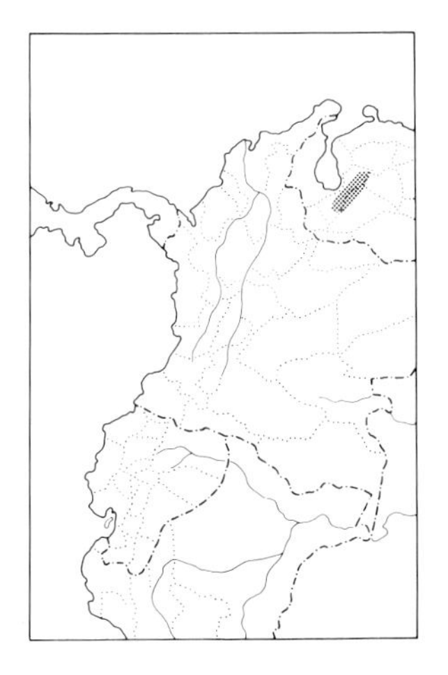

IDENTIFICATION: 10 cm (4"). *W. Andes of Venezuela.* Closely resembles Grass Wren, 1 form of which (*alticola*) occurs in virtual if not actual sympatry with it in Mérida. With care distinguished by its *bolder and whiter superciliary* (*alticola*'s superciliary is faint or lacking), more extensive streaking on upperparts (reaching to rump), and *blackish barring on flanks* (lacking in Grass Wren). See V-31.

HABITAT AND BEHAVIOR: Locally common in damp, boggy areas in *páramo,* often where there is much *Espeletia.* General behavior similar to Grass Wren's, but its song is distinctly different, being much less complex, merely a series of simple musical phrases (e.g., "ts-ts-tseee-ts-tseeeu" or "tseee, teeeeu, tee-ee-ee-ee-ee?"); also a long, chattering scold note (T. Meyer). Numerous in appropriate habitat in the Sierra Nevada Nat. Park of Mérida.

RANGE: Andes of w. Venezuela (Trujillo and Mérida). 3000–4100 m.

NOTE: As so many of the races of Grass Wren found in the Andes are characteristic of the *páramo* zone, we find it misleading and inappropriate for *meridae* to be singled out as the "Paramo Wren." We prefer to emphasize its very restricted range, almost entirely encompassed by the state of Mérida.

Cinnycerthia Wrens

A pair of rather short-billed, uniform-looking, fairly large wrens found in Andean forests; both species are uniform in appearance. Notable for moving about in large groups and for their strikingly melodious songs.

Cinnycerthia peruana

SEPIA-BROWN WREN

PLATE: 4

IDENTIFICATION: 14.5–16 cm (5¾–6¼"). *Mostly dull rufous-brown,* brightest on head and palest on throat, with inconspicuous grayish superciliary; *wings and tail narrowly but distinctly barred black.* What are apparently breeding adults have *variable amounts of white on face,* usually just on forehead and around eye but occasionally encompassing

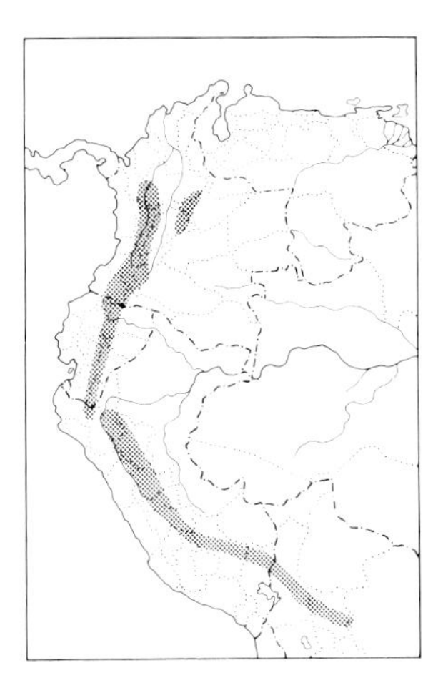

entire foreface; this is seen mainly in nominate *peruana* (found in Peru except in extreme north and south). N. birds (found in Colombia and Ecuador) are similar but larger and slightly less rufescent, and they lack the grayish superciliary (though in some birds, perhaps mostly young females, entire head and throat are washed with gray); they do not show as much facial white as do adult *peruana,* and when present it seems restricted to forehead; see C-41. *Fulva* (Bolivia and Puno, Peru) looks quite different, with *conspicuous long buffy whitish superciliary,* dark lores, and paler underparts.

SIMILAR SPECIES: In area of overlap with Rufous Wren (Colombia south to extreme n. Peru) Sepia-brown is most likely to be confused with that species, but it is a duller rufous-brown with more pronounced black barring on wings and tail. Pattern of *fulva* race recalls that of much smaller Mountain Wren.

HABITAT AND BEHAVIOR: Uncommon to locally common in undergrowth of humid montane forest, forest borders, and well-developed second-growth woodland, especially where *Chusquea* bamboo is prevalent. Usually in small groups (almost never alone, occasionally up to a dozen or so together) which rummage in tangles and thick foliage, sometimes dropping to ground. Frequently with mixed flocks, often seeming to be their nuclear species. Though often bold and even inquisitive, their quick, nervous movements and tendency to remain in heavy cover make them difficult to see well. The call is a fast, gravelly chattering, "ch-d-d-dt, ch-d-d-dt . . . ," often given as a group. Often in the midst of calling 1 bird will gradually begin to give its fine, loud song: this consists of a series of very varied phrases, all with the same rich musical quality (e.g., "pur-tee, turturturturtur; teur, teee, turturturturtur; pur-teeyr, curp, tutututututu" and so on).

RANGE: Andes of Colombia (north to s. Santander in E. Andes), nw. and e. Ecuador (on w. slope recorded south to Pichincha), e. Peru, and nw. Bolivia (La Paz and Cochabamba). Mostly 1500–2500 m, but locally higher (to 3000–3300 m), especially in Peru and Bolivia (perhaps an instance of competitive release, as here *C. unirufa* is absent), and down to 900 m on Pacific slope of Colombia.

Cinnycerthia unirufa

RUFOUS WREN

IDENTIFICATION: 16.5 cm (6½"). General form resembles that of Sepia-brown Wren, but *wing- and tail-barring are so faint as to be difficult to see. Uniform deep rufous-brown,* palest on throat; lores blackish; wings and tail obscurely barred blackish. See C-41. Nominate race (sw. Venezuela and E. Andes of Colombia south to Cundinamarca) is an even brighter and paler rufous; see V-31, C-42. Occasional birds have white on forehead and chin. *Chakei* (Perijá Mts. on Venezuela-Colombia border), also bright rufous, reportedly has a white or pale gray iris (eye dark in other races).

SIMILAR SPECIES: Sepia-brown Wren is a duller brown than this species, lacks the dusky lores, and has more prominent wing- and tail-barring; it is slightly smaller and is generally found at lower elevations,

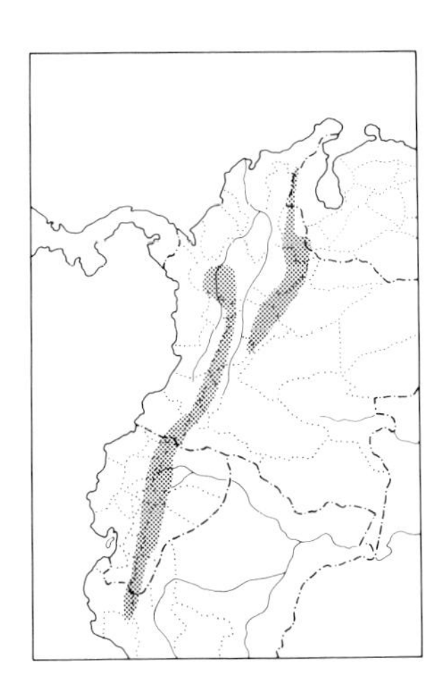

though there is local overlap. Rufous Wren is superficially much like Rufous Spinetail (with which it is sympatric), but latter has longer, more pointed tail, lacks all wing- and tail-barring, and doesn't troop about in large family groups.

HABITAT AND BEHAVIOR: Fairly common to common in undergrowth of humid montane forest and forest borders, ranging up to shrubbery at and slightly above timberline. Behavior generally similar to Sepia-brown Wren's, though flock size is regularly larger (up to 15 to 20 birds). Has a marvelous song, often given as a duet (the 2 birds sitting side by side), sometimes with other members of the group contributing their usual "swit, swit" call notes; the primary performer gives a series of loud and often long, sometimes modulated warbles and trills, all the while accompanied by the other birds' backdrop of softer "wort-wort-wort-wort . . ." notes.

RANGE: Andes of extreme sw. Venezuela (sw. Táchira), Colombia, nw. and e. Ecuador (on w. slope not recorded south of Pichincha), and extreme n. Peru (Cerro Chinguela on Piura-Cajamarca border); Perijá Mts. on Venezuela-Colombia border. 2200–3800 m.

Henicorhina Wood-Wrens

Small wrens with stubby, cocked tails and crisply streaked cheeks; their short tails immediately distinguish them from all *Thryothorus* wrens (some of which are otherwise somewhat similar, with streaked cheeks, etc.). Found in forest undergrowth, with 2 species in the Andes and 1 occurring widely in the lowlands of n. South America.

Henicorhina leucophrys

GRAY-BREASTED WOOD-WREN

PLATE: 4

IDENTIFICATION: 11 cm (4¼″). *Montane areas from Venezuela to Bolivia.* Mostly rufous-brown above, crown with blackish sides; long narrow white superciliary, with sides of head black prominently streaked white; wings and tail lightly barred black. Upper throat whitish, lower throat usually streaked with black; *remaining underparts mostly gray,* with flanks and lower belly rufous. *Hilaris* (sw. Ecuador) is palest gray below, while *brunneiceps* (nw. Ecuador north into sw. Colombia on Pacific slope) is darkest gray below and has brown crown nearly concolor with back. On the Santa Marta Mts. the situation is complicated, with 2 races replacing each other altitudinally: the upper one (*anachoreta,* down to about 2000 m) is fairly typical of the species, but the lower one (*bangsi,* up to about 2000 m) is rather different, with *pale grayish white throat and breast* and *no* streaking on lower throat.

SIMILAR SPECIES: Closely resembles White-breasted Wood-Wren, but latter is mostly white (not gray) below, lacks streaking on lower throat, and typically occurs at lower elevations (though often sympatric in w. Colombia). In n. Peru see also Bar-winged Wood-Wren.

HABITAT AND BEHAVIOR: Common in undergrowth of humid montane forest and forest borders and in secondary woodland and densely overgrown clearings. Usually in pairs, creeping about in dark tangles usually very close to the ground; often inquisitive and thus not too difficult to see if one patiently remains quiet near where they are moving about and scolding. Heard far more often than seen; its song is one of *the* commonly heard bird sounds of the Andean forest; the wrens sing through the day and apparently year-round. The song is often produced by a duetting pair and consists of several repetitions of a fairly long phrase (e.g., "cheerooeecheé—cheeweé—cheerooweechee") with many variations, then on to the next phrase; phrases are longer but not so melodic as those of the White-breasted. Birds from the Pacific slope of sw. Colombia (w. Cauca and w. Nariño) give a strikingly different series of tinkling notes with a variable (not so phrased) pattern (S. Hilty). The call of all races is a dry, gravelly "chrrr," often given in a series or stuttered.

RANGE: Coastal mts. of n. Venezuela (east to Miranda) and Andes of w. Venezuela, Colombia, Ecuador, e. Peru, and nw. Bolivia (south to w. Santa Cruz); Perijá Mts. on Venezuela-Colombia border and Santa Marta Mts. of n. Colombia. Also Mexico to Panama. Mostly 1500–3000 m, but regularly lower (to 500–700 m) in w. Colombia and w. Ecuador.

NOTE: More than one species could be involved. The situation on the Santa Martas, where *bangsi* and *anachoreta* replace each other altitudinally, warrants further investigation: the usual pattern is for such forms to turn out to be full species when they are carefully studied. In this case, however, apparent intergrades are known between *bangsi* and *anachoreta,* these having been taken at more or less intermediate elevations (K. C. Parkes, pers. comm.). The reported different song type in sw. Colombia (in at least part of the range of *brunneiceps*) is also intriguing.

Henicorhina leucoptera

BAR-WINGED WOOD-WREN

IDENTIFICATION: 11 cm (4¼"). *Known only from several remote, isolated mountain ranges in n. Peru.* Resembles a pale Gray-breasted Wood-Wren, but with *2 obvious white wing-bars* contrasting with black coverts. Throat and breast white flammulated with gray especially on sides, becoming rufous-brown on lower underparts (especially on flanks). Tail heavily barred black.

SIMILAR SPECIES: The only wood-wren with wing-bars. This, combined with its restricted range, should make identification easy. Sympatric Gray-breasteds are much grayer below and show streaking on lower throat.

HABITAT AND BEHAVIOR: Locally common in undergrowth of thick, mossy cloud forest, especially in stunted forest along exposed ridges. Behavior, including voice, similar to Gray-breasted Wood-Wren's. Apparently represents a relict form whose present distribution seems to have been restricted by ecological pressures from *H. leucophrys,* its closest relative. The Bar-winged is only known from ornithologically depauperate mountain ranges isolated east of the main Andes, with its numbers being greatest near the top of these ranges.

RANGE: N. Peru (Cordillera del Condor in Cajamarca; below Abra Patricia, and ne. of Jirillo on trail to Balsapuerto, both in San Martín;

above Utcubamba in La Libertad); as the Cordillera del Condor forms part of the border between Peru and Ecuador, will probably be found in the latter country as well. 1350–2450 m.

NOTE: A newly described species: J. W. Fitzpatrick, J. W. Terborgh, and D. Willard, *Auk* 94 (2): 195–201, 1977.

Henicorhina leucosticta

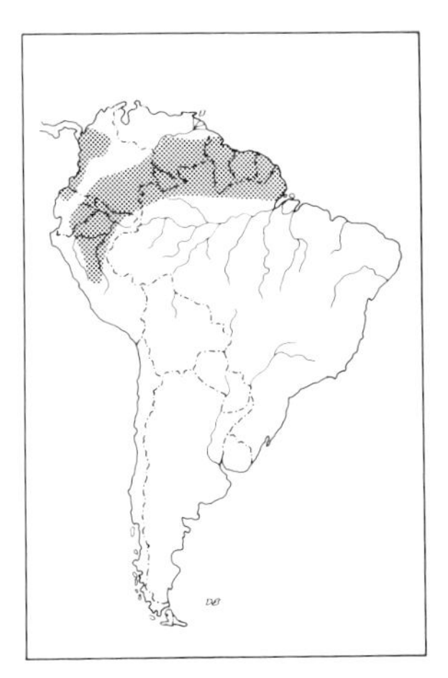

WHITE-BREASTED WOOD-WREN

IDENTIFICATION: 11 cm (4¼"). Resembles Gray-breasted Wood-Wren *but generally replaces it at lower elevations;* tail even shorter. Nominate group (with *hauxwelli* and *dariensis,* in all of Amaz. portion of range as well as nw. Colombia) differs strikingly in having *crown black,* contrasting with chestnut brown back; *throat and breast obviously white,* becoming grayish on sides and rufous-brown on flanks. However, the group of races found in w. Colombia and nw. Ecuador have brown crowns and thus more closely resemble Gray-breasted. They are best distinguished by their basically white throats and breasts, *without any streaking on lower throat;* in some races (especially *inornata* of w. Colombia and nw. Ecuador) the white is dingier, causing potential confusion, but it never is as gray as in that species. See P-22, V-31.

SIMILAR SPECIES: See Gray-breasted Wood-Wren. Problems are likely to be greatest in w. Colombia, where the 2 species approach each other altitudinally and are closest morphologically (both with brown crowns); helpful is the fact that while the race of White-breasted there (*inornata*) is quite dirty whitish below, Gray-breasteds there (*brunneiceps*) have the darkest gray breast of any race of that species.

HABITAT AND BEHAVIOR: Fairly common to common in undergrowth of humid forest, forest borders, and advanced second-growth woodland. Particularly fond of tangles around treefalls or logs and damp forested ravines; most numerous in hilly areas. General behavior similar to Gray-breasted Wood-Wren's. Voice, too, is reminiscent but on the whole richer and more melodic, with individual phrases tending to be shorter but repeated more often (up to 4 to 6 times); common phrases are "churry-churry-cheer" (or "pretty-pretty-bird"), "turytuoree? tury-tuoree?," and "tsee, chury-chury-chury."

RANGE: W. and n. Colombia (Pacific lowlands and east along humid n. base of Andes to w. Cundinamarca) and nw. Ecuador (south to Pichincha); east of Andes in se. Colombia (north to Meta and Vaupés), e. Ecuador, and ne. Peru (south to Huánuco) east across s. Venezuela (Amazonas and Bolívar) and n. Brazil (recorded only well north of the Amazon, with the exception of Amapá) to the Guianas. Also Mexico to Panama. Mostly below 1100 m, but recorded higher (to 1600–1800 m) in w. Colombia and s. Venezuela.

Cyphorhinus Wrens

A distinctive group of wrens of forest undergrowth and floor, characterized by their thick bills with a high, ridged culmen. Each of the

3 species has contrasting orange-rufous foreparts. Rather furtive birds, they are most often recorded from their marvelous songs.

Cyphorhinus aradus

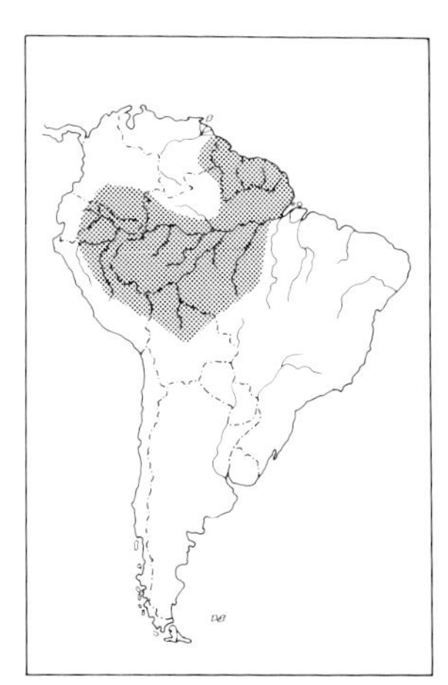

MUSICIAN WREN

PLATE: 4

IDENTIFICATION: 12.5 cm (5″). Small bare area around eye grayish blue (not very conspicuous). *Modulator* group (w. Amazonia) has *fore-crown, superciliary, and throat and breast dark orange-rufous.* Otherwise dull brown, wings and tail barred black. Nominate group (with *faroensis;* Guianas to lower Amaz. Brazil north of the Amazon) is slightly smaller and has brighter, paler orange-rufous foreparts, contrasting more with paler belly, and *conspicuous partial collar of white streaks* on a black background extending from sides of neck to upper back. *Griseolateralis* (small range south of the Amazon in Brazil in Rio Tapajós drainage) resembles nominate group but has collar reduced to a whitish superciliary and is grayer above and (especially) on belly.
SIMILAR SPECIES: Does not overlap with Song Wren (found west of the Andes); cf. also Chestnut-breasted Wren of Andes. Might briefly be mistaken for a number of female antbirds (many of which share the orange-rufous color), and *modulator* group (especially) could be confused with any of the rufous-throated leaftossers (*Sclerurus* sp.).
HABITAT AND BEHAVIOR: Fairly common in undergrowth of humid forest, mostly in *terra firme* but also in transitional or *várzea* forest. Usually found in pairs or small family groups, hopping about on or near the ground, often pausing on a low branch or log to look the observer over, then moving on. They forage mostly in leaf litter and most often move apart from mixed flocks, though they do occasionally follow antswarms. Musician Wrens tend to be wary, but they often give away their presence by their chuckling "churrs," given especially when disturbed. Their lovely song is unusual, several gutteral "churrs" followed abruptly by a beautiful, sometimes complex phrase of clear, melodic notes that may range way up and down in pitch; the same phrase is usually repeated over and over, then suddenly the bird switches to another, all the while with occasional low, guttural notes in syncopation. The overall effect is charming and memorable.
RANGE: Se. Colombia (north to Caquetá and Vaupés), e. Ecuador, e. Peru, and n. Bolivia (south to Cochabamba and w. Santa Cruz) east across Amaz. Brazil (though lacking from the blackwater region of the upper Rio Negro drainage and absent south of the Amazon east of the lower Rio Tapajós around Santarém) to the Guianas and se. Venezuela (e. Bolívar only). Mostly below 600 m, locally to 1000 m along base of Andes and in the *tepui* region.

NOTE: Often suggested is the possibility that the *modulator* group of w. Amazonia is a species distinct from the true *aradus* group (with a collar of white streaks on neck, etc.). However, when all the races of the complex are laid out, it seems clear that the geographically intermediate races of w. Amaz. Brazil (*transfluvialis* and *interpositus*) are also morphologically intermediate between the *modulator* group and *griseolateralis* (of cen. Amaz. Brazil), with the latter in turn being intermediate toward the *aradus* group. Thus a stepped clinal situation seems to exist, and this combined with the close similarity in songs of all groups compels us to regard them as conspecific. On the other hand, the trans-Andean *phaeocephalus* group does stand apart (large bare ocular area, etc.), and thus we separate it as a distinct species; the 1983 AOU Check-list concurs.

Cyphorhinus phaeocephalus

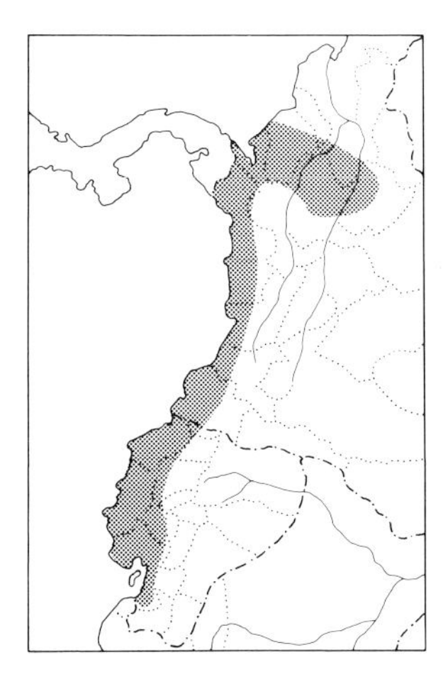

SONG WREN

IDENTIFICATION: 13 cm (5"). *W. Colombia and w. Ecuador.* Resembles the *modulator* group of Musician Wren, but with *conspicuous pale blue bare ocular area* (broadest behind eye, in some birds extending back as a short streak). Nominate group (w. Ecuador north in Pacific Colombia to s. Chocó) mostly dark brown, becoming quite dusky on crown; *sides of head, throat, and chest deep rufous;* wings and tail barred black; see C-41. *Lawrencii* group (n. Colombia) similar but with paler brighter rufous on head and foreneck (forecrown not duskier), contrasting with grayish upper belly; see P-22. Some birds (immatures?) in latter group have a white patch on chin.

SIMILAR SPECIES: Several antbirds have a similar pale blue ocular area, but none of these has barred wings; cf. especially female Chestnut-backed and Bare-crowned Antbirds.

HABITAT AND BEHAVIOR: Fairly common in undergrowth of humid forest and tall second-growth woodland. Behavior very similar to Musician Wren's, as is song, though latter usually is not quite as complex or elaborate.

RANGE: W. Colombia (Pacific lowlands and east along humid n. base of Andes to middle Magdalena valley to e. Antioquia) and w. Ecuador (south to El Oro). Also Honduras to Panama. Mostly below 700 m (occasionally to 1000 m).

NOTE: See comments under Musician Wren.

Cyphorhinus thoracicus

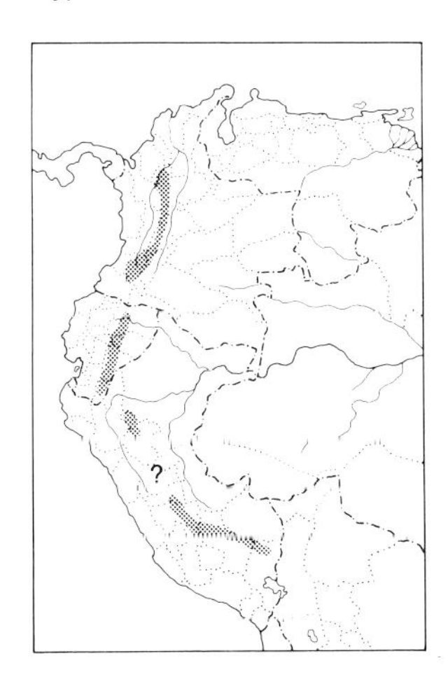

CHESTNUT-BREASTED WREN

IDENTIFICATION: 15 cm (6"). *Andes from Colombia to Peru.* Dark brown above, duskier on crown and around eyes; *wings and tail unbarred. Throat and breast orange-rufous,* extending up over sides of neck to cheeks; lower underparts dark brown. See C-41.

SIMILAR SPECIES: Found at higher elevations than the other, smaller *Cyphorhinus* wrens, though a narrow zone of contact (or even overlap) is possible. Chestnut-breasted is only one with *unbarred* wings. On w. slope, Song Wren shows distinctive and conspicuous bare pale blue ocular area; on e. slope note that Chestnut-breasted has rufous of breast extending up over sides of head and neck (*leaving a dusky-capped look*), while Musician has a rather different rufous superciliary and brownish cheeks.

HABITAT AND BEHAVIOR: Uncommon to locally common in undergrowth of humid montane forest, including mossy cloud forest. Behavior similar to that of other *Cyphorhinus* wrens but seems even more difficult to see, in part because so much of its habitat is dense and inaccessible. Chestnut-breasted's song is, however, quite different. It lacks the interspersed "churring" of other *Cyphorhinus* wrens and is

much simpler, usually consisting of a repetition of the same minor-key note or pair of notes, the second a half-tone either higher or lower (e.g., "peer-pur, peer-pur, peer-pur, peer-pur . . ." or "tur-lee, tur-lee, tur-lee, tur-lee . . ."); the repetitions (which sometimes have up to 4 notes) may go on for several minutes with barely a pause. The overall quality is haunting and ethereal, difficult to describe but never forgotten once recognized and somehow perfectly appropriate to its mountain haunts. Seems especially numerous in various areas above Cali, Colombia (e.g., at the "West Crest" and at Pichindé).
RANGE: Andes of Colombia (not recorded from E. Andes), e. Ecuador, and e. Peru (south to Puno). Mostly 1300–2300 m, lower (to 700 m) in w. Colombia.

Microcerculus Wrens

A group of small, dark, obscure wrens of the humid forest floor. Bills are long and slender, tails are very stubby, and (as befits their semiterrestrial habits) legs are long. Songs of all species are far-carrying and beautiful and quickly become accustomed background sounds; however, these wrens are among the more difficult of all neotropical birds to see.

Microcerculus marginatus

SOUTHERN NIGHTINGALE-WREN PLATE: 4

Other: Nightingale Wren

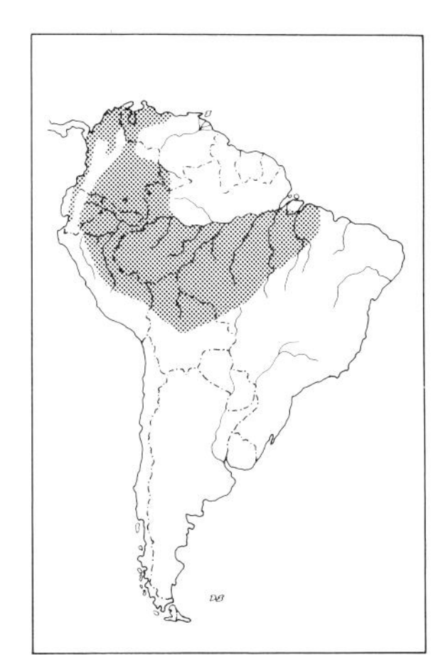

IDENTIFICATION: 11 cm (4¼"). Nominate race (widespread east of Andes) uniform dark brown above, with wings and tail essentially unbarred. *Below mostly white,* with sides, flanks, and lower belly solid to mottled brown. Immatures have white feathers scaled dusky, especially on breast. West of Andes the situation is complex and nomenclature confused (see below). *Occidentalis* (Pacific lowlands of Colombia and nw. Ecuador) is very similar to nominate but slightly darker, more rufescent brown above. To the south in w. Ecuador, all birds (for which we revive the name *taeniatus*) are *boldly scaled below,* dusky brown on a white background, with scaling least on the throat. Birds from n. Colombia and nw. Venezuela (*squamulatus, antioquensis,* and *corrasus*) are likewise boldly scaled below in much the same way; see V-31. *No* birds in the range of *taeniatus* or the *squamulatus* group are ever white-breasted below.
SIMILAR SPECIES: This furtive, small dark wren of the forest floor should not be confused on the all too infrequent occasions when it is seen, despite the confusing variation in its plumages. Wing-banded Wren always shows a very bold wing-bar, while Flutist Wren is more or less solidly rufous-brown; White-breasted Wood-Wren has a superciliary, boldly streaked cheeks, etc. Cf. also the Collared Gnatwren.
HABITAT AND BEHAVIOR: Fairly common in undergrowth of humid forest, with a particular preference for ravines or thick tangled places

along streams. Very furtive and difficult to see, walking quietly on or very near the ground, almost constantly teetering its rear end; usually solitary and never with flocks. Heard far more often than it is seen. Two basic song types are known in South America, and both are unforgettable once learned. All trans-Andean birds and those from w. Amazonia (Colombia, Ecuador, and ne. Peru south at least to Huánuco; east of the Río Ucayali the song shifts, apparently) sing the same song as birds from Panama. After a short, fast, irregular opening phrase (often not heard at a distance), there follows a long, very slow series of almost sibilant pure-toned whistles (usually 10 or more), each about a half-tone lower than the preceding and given at progressively longer intervals (toward the end pauses may last more than 10 seconds, with an almost tantalizing effect). In easternmost Amaz. Brazil the song is similar but the pauses do *not* become progressively longer. However, in e. Peru and n. Bolivia a very different song is heard; this consists of a series of pure, clear notes seemingly given almost at random, with changes in length (of both the notes and the pauses), pitch, and loudness.

RANGE: W. Venezuela (east along the coastal mts. to Distrito Federal and along e. base of Andes in Táchira and w. Apure), n. and w. Colombia and w. Ecuador (south to Guayas); e. Colombia (north to Meta and Vaupés and along base of Andes to Norte de Santander), e. Ecuador, e. Peru, n. Bolivia, and Amaz. Brazil (mostly south of the Amazon, where it extends east to nw. Maranhão; north of it only in upper Rio Negro area). Also Costa Rica and Panama. Mostly below 1000 m, in smaller numbers higher (to 1800 m in Venezuela).

NOTE: F. G. Stiles (*Wilson Bull.* 95 [2]: 169–183, 1983) has recently published evidence for regarding the birds of n. Middle America as a distinct species, *M. philomela*, based mainly on their very different songs. However, we do not believe the evidence yet justifies his regarding *M. luscinia* as a species distinct from *marginatus*. As noted above (in part), birds he called *luscinia* (ranging only in Panama and s. Costa Rica), as well as the trans-Andean scaly and white-breasted forms and birds from a large part of the nominate *marginatus*' range, *all* sing what appears to be the *same song*. The striking shift in song type which does occur (involving the same subspecies, nominate *marginatus*, across Amazonia) seems comparable to what Stiles found in Costa Rica, though a correlated morphological change has not yet been found. We thus believe it preferable to continue to regard all the S. Am. forms (together with those of Panama and s. Costa Rica) as conspecific under the name *M. marginatus*, while recognizing that further study may alter this. We also believe it unwise to shift English names given the present uncertainty regarding true species limits; we thus call *M. marginatus* the Southern Nightingale-Wren and would favor calling *M. philomela* the Northern Nightingale-Wren. The 35th Supplement to the AOU Check-list (*Auk* 102 [3]: 683, 1985) suggested calling *M. marginatus* the Scaly-breasted Wren, an unfortunate choice, as birds from most of its range do *not* have scaly breasts.

Microcerculus bambla

WING-BANDED WREN

PLATE: 4

IDENTIFICATION: 11.5 cm (4½"). Above dark brown, wings blackish *crossed by conspicuous single white band.* Throat and breast gray, flanks and lower belly dark brown, flanks obscurely barred dusky. Immatures

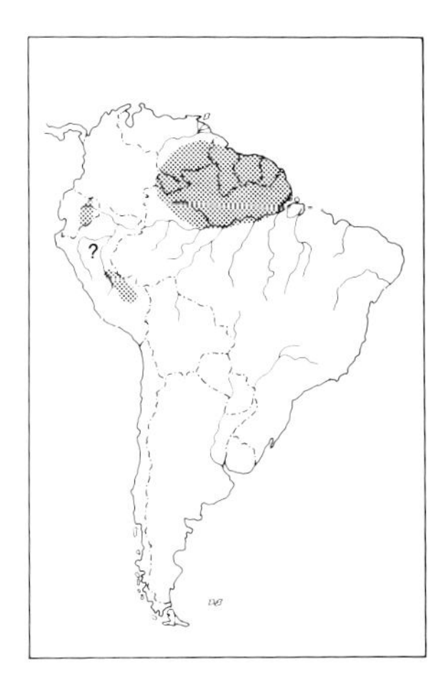

are more or less scaled below, with usually whiter throat; they already show the wing-band.

SIMILAR SPECIES: General form and behavior similar to Southern Nightingale-Wren's (the 2 are largely allopatric, though at least in e. Ecuador they may occur together); even in dim forest light the white wing-band stands out prominently. Male Banded Antbird has a similar strong white wing-band.

HABITAT AND BEHAVIOR: Rare to locally fairly common in undergrowth of humid forest. Behavior much like Southern Nightingale-Wren's, which it seems to replace in ne. South America, but much outnumbered by that species in e. Ecuador (where it seems to occur mainly at elevations *above* that species) and e. Peru. The song is similar to that of the widespread n. type of Southern Nightingale, consisting of a series of high, clear notes, all on the same pitch (perhaps slightly falling), at first separated, then gradually accelerating into a crescendo, followed by a distinct pause before another series.

RANGE: S. Venezuela (Bolívar and Amazonas), Guianas, and Amaz. Brazil north of the Amazon (west to the Rio Negro drainage); e. Ecuador (Napo and Pastaza); se. Peru (Huánuco on Cerros del Sira; Madre de Dios in Manu Nat. Park; Cuzco in Cosñipata valley). To about 1100 m (but recorded to 1500 m in Venezuela).

Microcerculus ustulatus

FLUTIST WREN

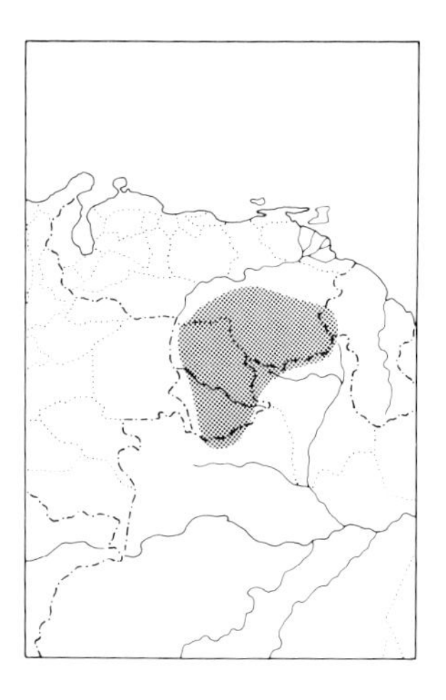

IDENTIFICATION: 11.5 cm (4½"). *Tepuis of s. Venezuela and adjacent areas.* Above dark chestnut brown. Below rufescent brown, becoming darker, more chestnut brown on belly; *faintly scaled throughout with dusky.* See V-31. Birds from remote *tepuis* of w. Bolívar and Amazonas are grayer on throat and more prominently scaled below.

SIMILAR SPECIES: A very uniform-looking dark wren of the forest floor. Even immature Southern Nightingale-Wrens (with scaling below) are whiter on the throat; they occur at lower elevations. Tepui Wren is superficially rather similar (and also essentially uniform rufous-brown) but has a longer tail; it shows an eyebrow and barring on wings and tail and lacks scaling below.

HABITAT AND BEHAVIOR: Fairly common on or near the ground in undergrowth of humid montane forest, usually in dense, tangled thickets. Behavior similar to other *Microcerculus* wrens', likewise extremely furtive and heard far more often than seen. The song is a beautiful glissando: after 1 or 2 brief introductory notes, it gradually and slowly slides upward, the whole song often taking 10 to 20 or even more seconds. A variant is similar but the notes are more clipped and gradually drop in pitch. Can be heard (and, with tape recorder and patience, seen) along the upper part of the Escalera road in Bolívar, Venezuela.

RANGE: *Tepuis* of s. Venezuela (Bolívar and Amazonas) and adjacent n. Brazil (Roraima) and Guyana (Mt. Twek-quay). 860–2100 m.

Sylviidae

OLD WORLD WARBLERS (GNATWRENS, GNATCATCHERS)

THE gnatcatchers and gnatwrens are the only members of this vast assemblage to occur in South America; otherwise the Sylviids are basically Old World in distribution. Some have considered them to form a family of their own, the Polioptilidae (they at least form a subfamily, the Polioptilinae), while others regard the Sylviids to be part of an expanded Muscicapidae. Gnatcatchers are slender, mainly gray birds with long, usually cocked tails that inhabit woodland and scrub and are basically arboreal. Gnatwrens, on the other hand, are predominantly brown and found in forest and woodland lower growth; their bills are longer.

Microbates and *Ramphocaenus* Gnatwrens

Distinctive, small, wrenlike birds of forest and woodland undergrowth. The 2 gnatwren genera are closely related and are here combined for convenience. All 3 species have long, slender, rather pale bills (especially *Ramphocaenus*), and their tails are usually held cocked and often flipped about to odd angles; tails of *Microbates* are short and stubby, while those of *Ramphocaenus* are considerably longer.

Microbates collaris

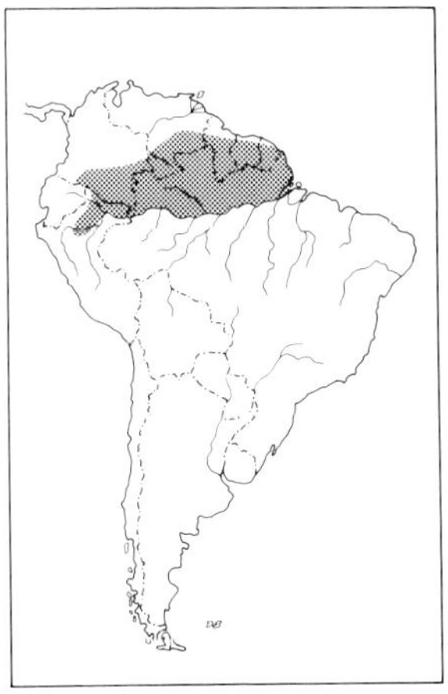

COLLARED GNATWREN PLATE: 4

IDENTIFICATION: 10.5 cm (4″). Long bill; short stubby tail. Above dark olivaceous to rufescent brown; *long narrow white eyestripe,* bordered below by a blackish stripe, then a wider white patch across lower cheeks, and finally a *broad black malar stripe.* Mostly white below with a *broad black pectoral band,* grayish on sides and flanks.

SIMILAR SPECIES: Vaguely wrenlike, but combination of long bill, boldly striped facial pattern, and pectoral band should be distinctive. Cf. Tawny-faced Gnatwren (with which it possibly overlaps in se. Colombia) and Banded Antbird.

HABITAT AND BEHAVIOR: Uncommon and seemingly local (but inconspicuous and probably often overlooked; locally more numerous) in lower growth of *terra firme* forest, rarely at forest edge. Usually found singly or in pairs, moving actively through undergrowth; often with mixed flocks, occasionally even in attendance at antswarms. Because of its near constant motion, it is often hard to see well for any length of time. The call is a series of soft thin notes repeated at 3-to-4-second intervals, "peeeee . . . peeeee . . . peeeee . . ."; also has a harsh, scolding note.

RANGE: Guianas (not recorded Guyana, but must occur), s. Venezuela (Bolívar and Amazonas), se. Colombia (Caquetá and Vaupés south to Putumayo and Amazonas), ne. Peru (Loreto and n. San Martín), and n. Amaz. Brazil (north of the Amazon only). To about 500 m.

Microbates cinereiventris

TAWNY-FACED GNATWREN

Other: Half-collared Gnatwren

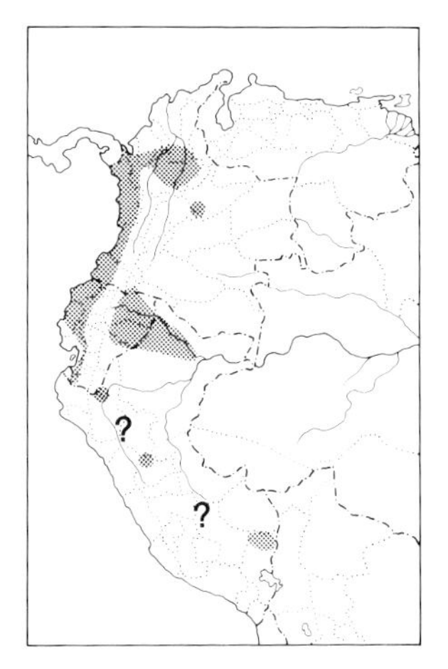

IDENTIFICATION: 10.5 cm (4"). Long bill; short stubby tail. Dark brown above with *sides of head and neck contrastingly bright tawny.* Throat white bordered on sides by broad black malar stripe and below by *partial collar of black streaks across chest; lower underparts gray,* more whitish on mid-breast and mid-belly; see P-24. *Cinereiventris* (w. Colombia and w. Ecuador) has *prominent blackish postocular streak;* see C-42. *Peruvianus* (e. Peru) more uniform and darker gray below.
SIMILAR SPECIES: In dim light of forest interior even the bright pale face can be hard to make out and then might be confused with various wrens, especially Nightingale. Long-billed Gnatwren lacks the gray below and the partial collar and is longer-tailed. Cf. Collared Gnatwren (similar form and behavior but very differently patterned).
HABITAT AND BEHAVIOR: Uncommon to fairly common in lower growth of humid forest, rarely at forest edge. Perky, active behavior is very similar to Collared Gnatwren's (see above). Voice also similar, with a nasal, complaining scold "nyeeeh" being most often heard; neither species is, however, particularly vocal (especially for a forest interior bird).
RANGE: W. and n. Colombia (Pacific lowlands and north around humid n. base of Andes to middle Magdalena valley in e. Antioquia), w. Ecuador (south to El Oro and w. Loja); se. Colombia (in lowlands along e. base of Andes north to e. Cundinamarca), e. Ecuador, and e. Peru (n. Loreto and very locally along e. base of Andes from Amazonas to Puno). Also Nicaragua to Panama. To about 1000 m.

NOTE: We follow the 1983 AOU Check-list and Ridgely (1976) in calling this species the Tawny-faced rather than the Half-collared Gnatwren.

Ramphocaenus melanurus

LONG-BILLED GNATWREN

PLATE: 4

IDENTIFICATION: 12 cm (4¾"). *Very long slender bill; tail long* and rather slender. Uniform brown above; tail dusky, with outer feathers narrowly tipped white. Below whitish to pale grayish, with sides broadly tinged cinnamon buff in some races (e.g., *trinitatis*). *Rufiventris* group (west of Andes from Venezuela to Ecuador) has grayer back, contrasting more with brown crown, its cheeks and most of its underparts (aside from whitish throat) are uniform cinnamon buff, and its white tail-tipping is more prominent; see P-24.
SIMILAR SPECIES: Compare to Collared and Tawny-faced Gnatwrens (both much more true forest species than is Long-billed). Among the wrens, only the Long-billed of Brazil has anywhere near so long a bill, but it is a very different bird.

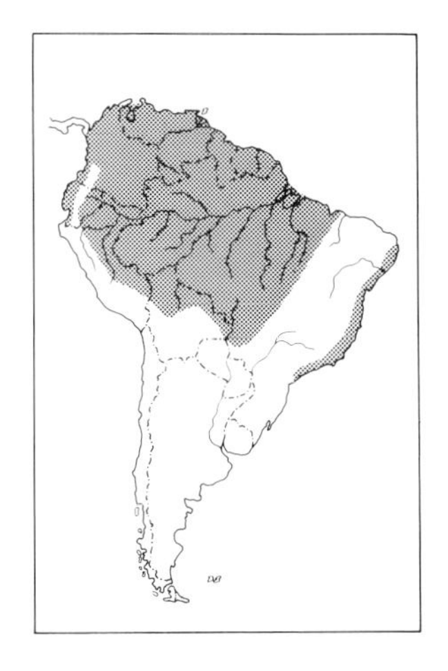

HABITAT AND BEHAVIOR: Uncommon to fairly common (tending to be overlooked until its distinctive song is learned) in undergrowth of deciduous and secondary woodland, tangled forest borders, and thickets in clearings; in se. Brazil and on Trinidad also quite often inside humid forest. Usually found singly or in pairs as they hop through viny tangles, etc.; often hard to get an unobstructed view. Sometimes with mixed flocks of other undergrowth insectivorous birds, but at least as often encountered away from them. Its pretty call is a clear, musical trill, sometimes on 1 pitch but more often slightly rising and becoming slightly louder.

RANGE: Widespread in lowlands south to n. Bolivia (La Paz) and cen. Brazil (south to s. Mato Grosso, Goiás, and Maranhão); coastal e. Brazil (Pernambuco south to São Paulo); west of Andes south to sw. Ecuador (El Oro and w. Loja); Trinidad. Also Mexico to Panama. Mostly below 1500 m.

Polioptila Gnatcatchers

The gnatcatchers are a distinctive group of very small slender birds with long narrow tails constantly flipped around in an animated fashion. Most species are predominantly bluish gray, often patterned with black on head; in most tails are black with white outer feathers. They glean actively in foliage, in all but one species (Masked) normally well above the ground.

GROUP A

Males with *glossy black cap;* generally whiter below.

Polioptila plumbea

TROPICAL GNATCATCHER

PLATE: 4

IDENTIFICATION: 11–11.5 cm (4¼–4½"). Male mainly bluish gray above with *glossy black crown* (sharply cut off from gray back); wings blackish, with inner flight feathers edged white (in some races broadly so); tail black, with outer feathers white. Below white to pale grayish, grayest on breast. *Bilineata* (w. and n. Colombia east to Santa Marta Mts. and w. Ecuador and Pacific nw. Peru) similar but has white of throat *arching up onto cheeks and encircling eyes;* see C-42, P-24. Female like male but black of head replaced by gray. Female *atricapilla* (ne. Brazil) does, however, have a black streak on ear-coverts, while female of *maior* (upper Marañón valley of n. Peru) is very distinct in that *like male it has a glossy black cap,* with white lores and a broken eye-ring (interrupted behind eye); *maior* is also a darker bluish gray above (in male as well).

SIMILAR SPECIES: By far the most numerous and widespread gnatcatcher across n. South America; cf. Guianan and Slate-throated Gnatcatchers. Female of *atricapilla* rather resembles female Masked Gnatcatcher (both have black streak on ear-coverts) but Masked is larger and distinctly grayer on underparts. Color pattern of both sexes vaguely recalls Hooded Tanager.

HABITAT AND BEHAVIOR: Uncommon to locally common in light or secondary woodland, forest borders, clearings with scattered trees, arid scrub, and mangroves; in Amazonia also found in lower *várzea* growth on islands and along major rivers, but in general more numerous and widespread in drier regions. An active little bird which constantly twitches its usually cocked tail. Gleans from foliage and smaller twigs; often accompanies mixed flocks of other insectivorous birds. Frequently gives a somewhat nasal "nyeeah" and also has a thin but fairly musical high song consisting of a series of simple notes, "seet, weet, weet-weet, weet-pitee, weet-pitee" or some such variation, often fading in intensity.

RANGE: Widespread south to e. Peru (Madre de Dios) and Amaz. and ne. Brazil (south to Piauí and e. Bahia); west of Andes south to Lima, Peru. Also Mexico to Panama. Mostly below 1200 m, locally higher in arid montane valleys (e.g., to 2500 m or more in *maior* of n. Peru).

NOTE: More than one species is probably involved. *Maior* (Marañon Gnatcatcher) of the upper Marañón River valley in n. Peru was long regarded as a species and seems particularly distinct on account of its virtual lack of sexual dimorphism. It also differs vocally from the other forms (*fide* T. Parker). Also, the white-faced *bilineata* group of Middle America and nw. South America differs notably from cis-Andean nominate *plumbea* group: it differs as much from the nominate *plumbea* group as the *bilineata* group does from Middle American *P. albiloris* (and the latter 2 do occur sympatrically). For now, however, it seems best to continue to regard all forms as conspecific. A study of the presumed contact zone between *bilineata* and the fully black-capped *plumbiceps* race found in the Santa Marta area of n. Colombia would be of interest.

Polioptila lactea

CREAMY-BELLIED GNATCATCHER

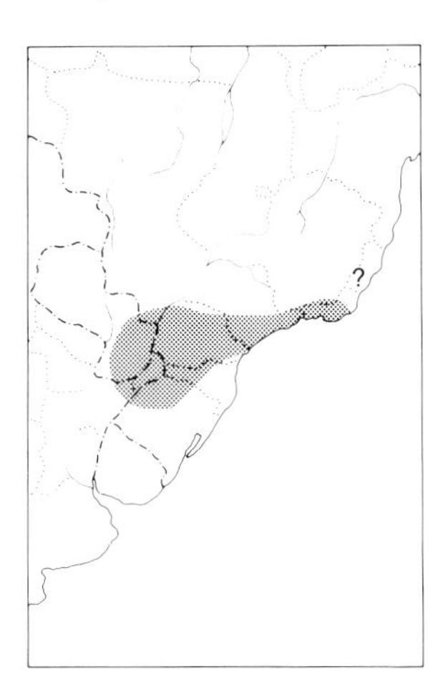

IDENTIFICATION: 11.5 cm (4½"). *Se. Brazil and adjacent Paraguay and Argentina.* Resembles Tropical Gnatcatcher (note their widely disjunct ranges), especially the *bilineata* group (with *white face*). Somewhat darker bluish gray dorsally and *distinctly creamy yellowish below.*

SIMILAR SPECIES: Masked Gnatcatcher has a very different facial pattern, occupies more open scrubbier habitat, etc. At a distance Creamy-bellied Gnatcatcher is perhaps most likely to be confused with similarly shaped Bay-ringed Tyrannulet.

HABITAT AND BEHAVIOR: Uncommon in canopy and borders of humid forest and woodland. Behavior much like that of Tropical Gnatcatcher but seems more forest-associated. The song is a simple, fast repetition of a high, thin note, "weet-weet-weet-weet" (up to about 6 "weets"). Can be found in small numbers at Iguazú Falls, particularly on the Argentinian side.

RANGE: Se. Brazil (Rio de Janeiro south to n. Rio Grande do Sul; perhaps north to Espírito Santo), e. Paraguay (Canendiyu to Itapúa), and ne. Argentina (Misiones and Corrientes). Below 400 m.

GROUP B Lacks black cap (but may have *mask*); generally grayer below.

Polioptila guianensis

GUIANAN GNATCATCHER

PLATE: 4

IDENTIFICATION: 11 cm (4¼"). Local in *ne. South America.* Male *mostly gray,* with narrow white eye-ring and white belly. Wings dusky, with inner flight feathers edged gray; tail black, with outer feathers white. Female similar but slightly paler on facial area.

SIMILAR SPECIES: Both sexes rather resemble female Tropical Gnatcatcher but are grayer on breast and lack Tropical's contrast between its gray crown and white lower cheek; note also that female Tropical has white (not gray) edging to inner flight feathers.

HABITAT AND BEHAVIOR: Rare and seemingly local in canopy and borders of humid forest. Seen singly or in pairs, usually when accompanying large mixed-species flocks moving through the canopy and subcanopy. Can be seen at Suriname's Brownsberg Reserve, but not numerous.

RANGE: Guianas; sw. Venezuela (Amazonas) and extreme nw. Brazil (upper Rio Negro); lower Amaz. Brazil (near Manaus and from lower Rio Tapajós east to Belém area). To about 500 m.

Polioptila schistaceigula

SLATE-THROATED GNATCATCHER

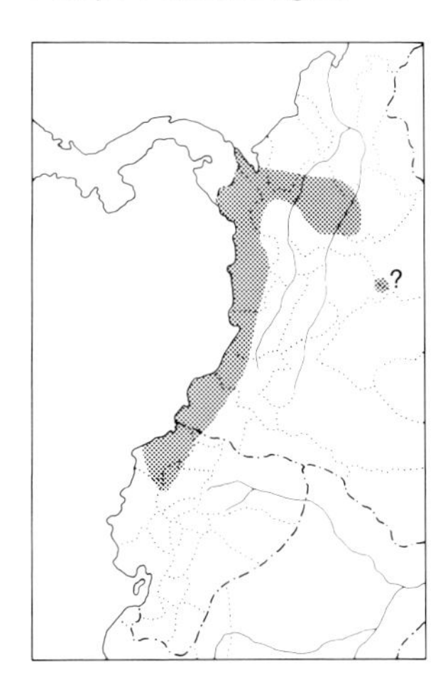

IDENTIFICATION: 11 cm (4¼"). *W. Colombia and nw. Ecuador. Mostly slaty gray* with white lores and narrow white eye-ring. *Breast and belly contrastingly white.* Wings and tail blackish, outer tail feathers very narrowly fringed white (*not all white,* as in other gnatcatchers). See C-42. Sexes are apparently alike, but birds from nw. Colombia and Panama (an undescribed race?) are darker, essentially blackish slate where nominate is slaty gray, but with same pattern; see P-24. The paucity of specimen material precludes a more definitive statement.

SIMILAR SPECIES: Nothing really similar in range. Pattern rather resembles geographically far-removed Guianan Gnatcatcher, but Slate-throated is a much darker bird; note virtual absence of white in its tail. See also Gray-mantled Wren.

HABITAT AND BEHAVIOR: Rare in canopy and borders of humid forest. Found singly or in pairs, usually as they are accompanying mixed foraging flocks in the canopy and subcanopy, sometimes coming lower at forest edge. Look for it in the lowlands around Buenaventura, Colombia, particularly on the road to Bajo Calima.

RANGE: W. and n. Colombia (Pacific lowlands and in humid lowlands along humid n. base of Andes east to middle Magdalena valley in w. Santander; also a questionable record from e. base of E. Andes in Cundinamarca, perhaps mislabeled) and nw. Ecuador (Esmeraldas and Pichincha). Also e. Panama. To about 1000 m.

Polioptila dumicola

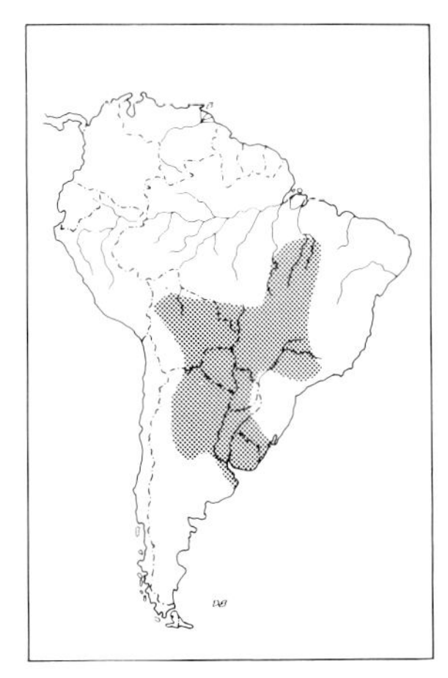

MASKED GNATCATCHER

PLATE: 4

IDENTIFICATION: 12.5 cm (5″). Male *blue-gray above* with *wide black mask* bordered below by thin white stripe. Wings blackish, inner flight feathers edged white; tail black, with outer feathers white. *Below pure gray* (paler than upperparts), fading to whitish on belly. Female similar but with much *narrower black area on ear-coverts,* with lower cheeks as well as lores and a narrow eye-ring whitish. *Berlepschi* (interior Brazil, south to s. Mato Grosso, west through Beni lowlands of n. Bolivia) is quite different: paler (and not as blue) gray above and mostly whitish below (merely tinged gray across breast) and has *narrower black mask* (white of throat extends up over lower cheeks); female like female of nominate race but paler, both above and below.

SIMILAR SPECIES: The only gnatcatcher over virtually all of its range. Cf. Creamy-bellied Gnatcatcher. More likely confused with female Tropical where their ranges nearly meet in interior ne. Brazil (no recorded overlap, though possible), especially as female of that race of Tropical (*atricapilla*) also has black area on ear-coverts. Masked differs, however, in its larger size, grayer breast, and white on lores and eye-ring.

HABITAT AND BEHAVIOR: Uncommon to very common in lighter woodland and scrub (including *monte*), semiopen *cerrado,* and gallery woodland; most numerous in the *chaco,* where it is one of the commonest birds. Conspicuous, active, and excitable, pairs or small groups of this attractive gnatcatcher are most often seen with mixed flocks (especially during the nonbreeding season). Foraging is much like that of other gnatcatchers, though Masked tends to remain much lower. Its soft, musical song is very variable but usually consists of only a few short sweet notes or phrases; a common one is "suwee, tu-tu-tu-tu . . . suwee, tu-tu-tu-tu." At times it is given almost interminably, while the bird continues to forage.

RANGE: Cen. and sw. Brazil (north to Mato Grosso, se. Pará, and n. Goiás and south to Rio Grande do Sul, but absent from coastal region), n. and e. Bolivia (west to Beni lowlands and Cochabamba highlands), Paraguay (except in the far east), Uruguay, and n. Argentina (south to San Luis and Buenos Aires). To about 2400 m (in Bolivia).

NOTE: Two species could be involved. The 2 races comprising the species are quite distinct morphologically and may differ vocally as well (though songs of both are variable). However, until their presumed contact zone in Brazil and Bolivia has been analyzed, we prefer to maintain them as conspecific. So far as we know, no intermediate specimens have been taken.

Turdidae

SOLITAIRES, THRUSHES, AND ALLIES

THE thrushes are a cosmopolitan group, most diverse in the Old World but with members on every continent; like the Sylviids, some have considered the thrushes to merely form a subfamily in an expanded Muscicapidae family, but we prefer to maintain them as a full family. Thrushes are widespread in South America, particularly in the genus *Turdus;* solitaires and nightingale-thrushes are more montane and localized in distribution. Rather subdued in color, though often handsomely patterned and with brightly colored soft-parts, the thrushes are best known for their often superb songs.

Myadestes Solitaires

A small genus of thrushes with relatively short, broad bills. The solitaires are mainly arboreal, though unlike the *Turdus* thrushes they place their cup nest of moss on or near the ground (often on a bank or the side of a cliff). They are perhaps best known for their magnificent ethereal songs.

Myadestes ralloides

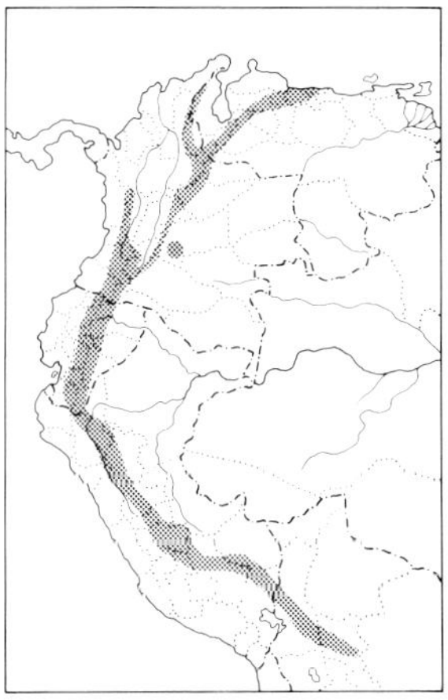

ANDEAN SOLITAIRE

PLATE: 5

IDENTIFICATION: 17–18 cm (6¾–7″). Bill rather short, dusky above and yellow below (but all blackish in nominate race, from n. Peru south). *Mostly warm rufescent brown above,* brightest across mantle, grayer on crown (especially forecrown). *Below leaden gray.* Wings with *silvery gray band along base of primaries,* tail dark brown with *outer feathers silvery gray,* both normally hidden except in flight. Immature mostly rufous-brown heavily spotted with buff; these are regularly seen.

SIMILAR SPECIES: Much slimmer than any of the *Turdus* thrushes and tends to perch much more upright than they do. Though general coloration subdued, the silvery in the wings and tail flashes conspicuously in flight. Cf. Varied Solitaire (no overlap).

HABITAT AND BEHAVIOR: Fairly common to common in montane forest, borders, and shady secondary woodland. A rather shy and unobtrusive bird, usually seen singly or in pairs, often perching motionless, almost trogonlike, for long periods. Its presence is most often made known by its beautiful song of clear, flutelike liquid notes, sometimes interspersed with more guttural or gurgling ones; the delivery is leisurely and the pauses between phrases are long (e.g., "tleee . . . leedle-lee . . . lulee . . . turdelee . . . treelee . . . teul-teul . . ."). The song is quite loud and carries well, even over the roar of mountain streams, the vicinity of which the bird so often frequents; it is, however, ventriloquial, and as the singing bird usually remains motionless on an often well-concealed perch, it is difficult to track down to its source.

RANGE: Coastal mts. of n. Venezuela (east to Distrito Federal) and Andes of w. Venezuela, Colombia, Ecuador, nw. and s. Peru (south on w. slope to Cajamarca), and n. Bolivia (La Paz and Cochabamba); Perijá Mts. on Venezuela-Colombia border. Mostly 1200–2700 m, lower (to 800 m) on Pacific slope.

Myadestes coloratus

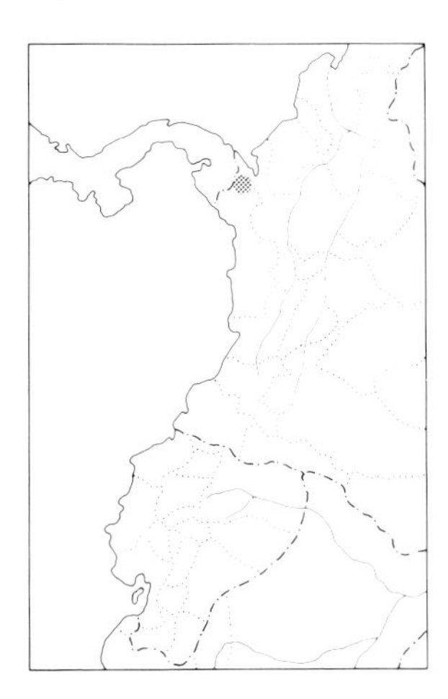

VARIED SOLITAIRE

IDENTIFICATION: 17–18 cm (6¾–7"). *Extreme nw. Colombia.* Resembles Andean Solitaire (no overlap), but *bill and legs yellow-orange, forehead and face black* in contrast to gray of rest of head, and outer tail feathers grayer (thus not flashing as conspicuously as in Andean). See P-32.

HABITAT AND BEHAVIOR: Fairly common to common in montane cloud forest. Only recently found in Colombia (Rodríguez 1982) and better known in the Panama portion of its range. In the latter it has recently been found to be numerous above 1350 m on Cerro Pirre. As with the Andean Solitaire, the Varied is recorded mostly by its leisurely song, which resembles that of the Andean; as in that species singing birds are very difficult to locate. Most singing occurs at dawn and dusk and also when low clouds create foggy conditions.

RANGE: Extreme nw. Colombia (nw. Chocó on slopes of Cerro Tacarcuna). Also e. Panama. 1100–1500 m.

NOTE: We follow the 1983 AOU Check-list in considering *coloratus* as a species distinct from *M. ralloides* and *M. melanops* (of Middle America). However, as the similarity in their voices indicates, they are closely related and it may ultimately prove best to merge them; "intermediate" by range, *coloratus* combines characters of both the other forms (rufous mantle, black facial area, etc.).

Cichlopsis Solitaires

Though recently associated with the genus *Myadestes,* we believe that *leucogenys* is sufficiently distinct to justify resurrecting the genus *Cichlopsis.* Without *leucogenys, Myadestes* was a uniform genus of small thrushes with short, rather wide bills, typically with gray predominating and silvery or pale gray in the tail and an oblique, dull wing-band. *Leucogenys* is very different, being predominantly rufous-brown, larger, and with a longer bill and no wing- or tail-markings. Furthermore, its usual posture differs from the characteristic upright stance of *Myadestes,* the one nest known (found in Brazil in Sept. 1977; H. Sick and RSR) was located well above the ground in the crotch of a tree (rather than on or near the ground), and its voice is very different. The species remains little known, with a seemingly relict distribution.

Cichlopsis leucogenys

RUFOUS-BROWN SOLITAIRE

PLATE: 5

IDENTIFICATION: 20.5–21 cm (8–8¼"). Maxilla blackish, *mandible yellow-orange*. Uniform rufous-brown above and on breast, with *median throat ochraceous tawny* set off by darker malar area; belly dull buffy grayish, becoming ochraceous buff on crissum. Nominate race of se. Brazil and *gularis* of the *tepuis* are similar. *Peruvianus* (locally on e. slope of Andes in Peru) similar, but with ochraceous tawny supraloral as well as *entire* throat, and with dull ochraceous (not so gray) lower underparts and *bright orange-ochraceous crissum. Chubbi* of nw. Ecuador and s. Colombia is patterned like *peruvianus* but its *throat and supraloral are rich reddish chestnut;* it has a narrow yellow eye-ring; see C-43.
SIMILAR SPECIES: In general coloration similar (especially the duller Brazilian and *tepui* races) to various mostly brown *Turdus* and female *Platycichla* thrushes, but it has a shorter, obviously bicolored bill and *no streaking on the throat* (rather, with quite prominent tawny on at least the middle of the throat). Andean Solitaire is very different, with gray underparts, silvery on wing and tail, etc.
HABITAT AND BEHAVIOR: *Very* local, but in a few localities not uncommon; inhabits humid lower montane forest (at least in the W. Andes often mossy). An inconspicuous, solitary bird of the forest interior. Tends to perch more horizontally (more like a *Turdus*) than do *Myadestes* solitaires. The songs of the nominate race (se. Brazil) and *chubbi* (nw. Ecuador) are similar and completely unlike that of the *Myadestes* solitaires, being a series of complex, rapidly uttered phrases, quite musical in quality though with some chattering or even twittering notes (e.g., "tleeowít-tsiii-trrrrr-tr-tr-teeo"), with much variation, each phrase generally very unlike the next. Usually it sings from a perch in mid-canopy, often fluttering its wings while doing so. Regular in the Nôva Lombardia Reserve above Santa Teresa in Espírito Santo, Brazil, and on the upper Escalera road in e. Bolívar, Venezuela.
RANGE: W. slope of W. Andes in sw. Colombia (1 record from Anchicayá valley of w. Valle) and nw. Ecuador (very local in Esmeraldas and Pichincha); e. slope of Andes in cen. Peru (Río Perené in Junín; Cerros del Sira in Huánuco); *tepuis* of se. Venezuela (Bolívar) and adjacent Guyana, and recently reported from Suriname as well (Donahue and Pierson 1982); se. Brazil (s. Bahia and Espírito Santo). 550–1300 m.

Entomodestes Solitaires

A pair of large, mainly black solitaires, both species with white in wing and tail flashing very conspicuously in flight. Larger than *Myadestes* solitaires, with proportionately longer tails. Each species has a limited range in the Andes.

Entomodestes leucotis

WHITE-EARED SOLITAIRE

PLATE: 5

IDENTIFICATION: 24 cm (9½"). *A strikingly patterned solitaire of the e. slope of the Andes in Peru and Bolivia.* Bill black above, orange below;

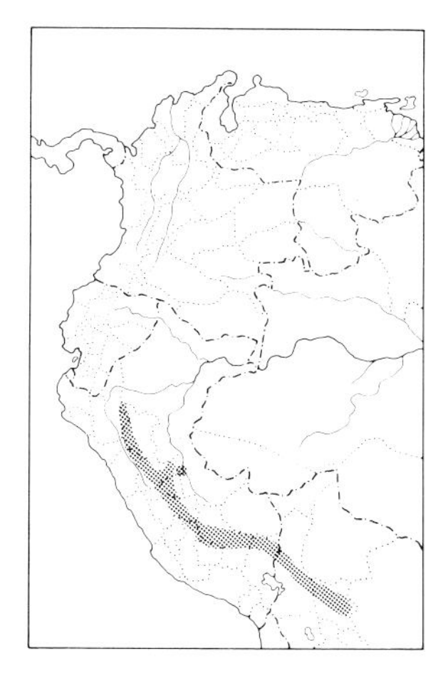

iris brown. Head mostly black but with *very broad white patch below eye.* Upperparts otherwise mostly bright rufous-chestnut; wings and tail mostly black, with *wide white band at base of primaries* (not visible on closed wing) *and outer tail feathers mostly white* (both very conspicuous in flight). *Below mainly black,* with white pectoral patch (usually hidden by wing) and some rufous-chestnut on flanks. Female has crown duskier.

SIMILAR SPECIES: Can hardly be confused; cf. the smaller Andean Solitaire.

HABITAT AND BEHAVIOR: Fairly common in montane forest and forest borders. Active and often seemingly quite nervous and shy—as a result frequently seen only briefly or in flight. Usually seen singly or in pairs, at times following forest flocks or congregating in small groups at fruiting trees. The song is a strange, ringing, almost nasal "wreeeeeeenh," weak but at the same time surprisingly far-carrying; singing birds are, however, difficult to locate as the song often almost seems to "float on air" and is usually given from a hidden or obscured perch. Perhaps most readily seen along the Chapare road north of Cochabamba (city), Bolivia.

RANGE: E. slope of Andes in cen. and s. Peru (north to Amazonas) and n. Bolivia (La Paz and Cochabamba). Mostly 1500–2800 m.

Entomodestes coracinus

BLACK SOLITAIRE

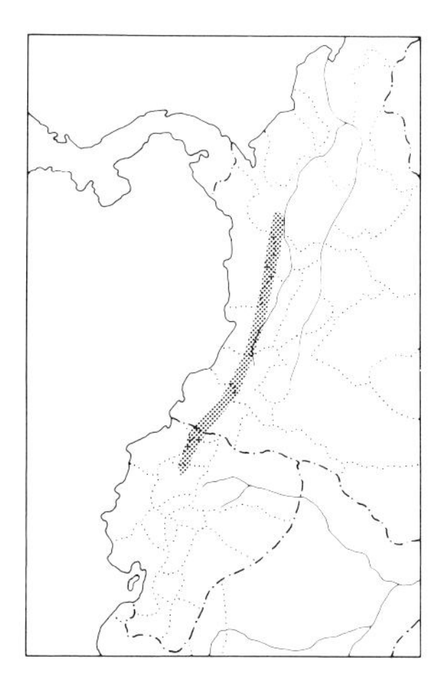

IDENTIFICATION: 23 cm (9"). *W. Colombia and nw. Ecuador.* Resembles White-eared Solitaire (no overlap) but *lacks all rufous-chestnut* in plumage; thus, all jet black aside from white on head, pectoral patch, wings, and tail. Iris red rather than brown. See C-43.

SIMILAR SPECIES: Essentially unmistakable in its limited range.

HABITAT AND BEHAVIOR: Uncommon to locally fairly common in wet, mossy cloud forest and forest borders; seems absent, or at most seasonal, from areas where mossy conditions do not prevail. Behavior very much like White-eared Solitaire's (see above). Black Solitaire is in general a much less often encountered bird; about the only accessible locality where the species is numerous is above the town of Junín, along the road to Tumaco in w. Nariño, Colombia.

RANGE: W. slope of W. Andes in Colombia (north to s. Chocó) and nw. Ecuador (very local in Esmeraldas and Pichincha). Mostly 600–1600 m.

Catharus Nightingale-Thrushes (resident spp.) and Thrushes (migrant spp.)

Half of the 6 S. Am. members of this genus are migratory from their breeding grounds in North America, while the other 3 are resident in montane areas of the n. and w. parts of the continent, mainly on Andean slopes. The 3 residents are characterized by their orange bills and legs (dark in juvenals) and they remain low to the ground; the 3 migrants are duller generally and tend to be more arboreal (especially the Swainson's). All are shy and inconspicuous birds, with the residents being heard much more often than seen. Nests are bulky cups placed low to the ground.

GROUP A

Resident species; *bill and legs orange.*

Catharus dryas

SPOTTED NIGHTINGALE-THRUSH

PLATE: 5

IDENTIFICATION: 17 cm (6¾"). *Bill and legs orange;* narrow but conspicuous eye-ring orange-yellow. Head black; remaining upperparts olive. *Below apricot yellowish, conspicuously spotted with dusky,* flanks washed with grayish. In older museum specimens upperparts fade to gray, underparts to whitish.

SIMILAR SPECIES: Slaty-backed Nightingale-Thrush has white eye and lacks the spotting below. None of the migrant *Catharus* thrushes have brightly colored soft-parts or the black head.

HABITAT AND BEHAVIOR: Locally fairly common in undergrowth of humid montane forest. Shy and inconspicuous, generally encountered hopping singly on forest floor, but much more often heard than seen. Its song is a series of beautiful short phrases, the notes with a pure, liquid quality (e.g., "tru-lee? . . . chee-lolee . . . troloweé . . . chee-trelelee . . . troloweé . . ."), with rather short pauses between each phrase; the overall effect is a little more hurried than in the Slaty-backed Nightingale-Thrush. Singing birds are very hard to track down, the bird usually moving off at your approach, though often continuing to sing, and the overall effect is that the song "floats on the air."

RANGE: Locally in Andes of w. Venezuela (north to Lara), Colombia (e. slope of E. Andes south to Meta and at head of Magdalena valley in Huila), sw. and e. Ecuador (on w. slope from Pichincha to El Oro and Loja and on entire e. slope), e. Peru, w. Bolivia, and extreme nw. Argentina (Jujuy and Salta); Perijá Mts. on Colombia-Venezuela border and Macarena Mts. in Colombia. Also Mexico to Honduras. 700–2300 m.

Catharus fuscater

SLATY-BACKED NIGHTINGALE-THRUSH

IDENTIFICATION: 18 cm (7"). Similar to Spotted Nightingale-Thrush but *lacks spotting below; iris whitish* (dark in juvenal). *Bill and legs orange;* narrow but *fairly conspicuous orange eye-ring. Above mostly dark slaty gray,* head somewhat more blackish (especially in nominate race of Venezuela, most of Colombia, and Ecuador). Below mostly grayish olive, but median breast and belly pale yellow. See V-32, C-43. *Sanctaemartae* (of Colombia's Santa Marta Mts.) and *mentalis* (southernmost Peru and Bolivia) are smokier gray below.

SIMILAR SPECIES: Most apt to be confused with Spotted Nightingale-Thrush, which, however, is spotted below and olive (not gray) above and has a dark eye. Cf. also male Pale-eyed Thrush (a larger, arboreal bird which is all black and lacks the eye-ring).

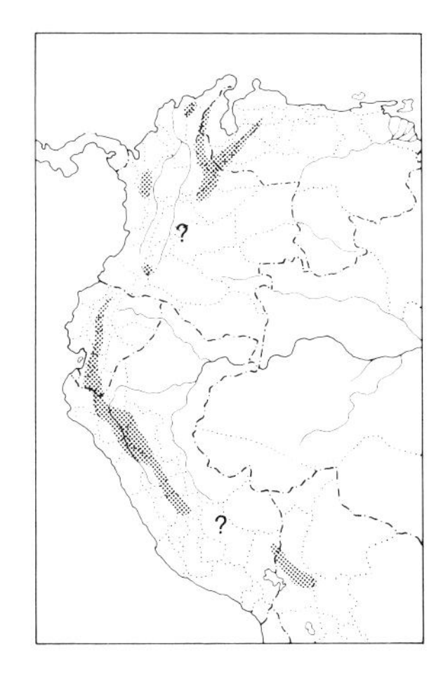

HABITAT AND BEHAVIOR: Locally fairly common in undergrowth of humid montane forest. Behavior much like Spotted Nightingale-Thrush's and is equally shy and difficult to see—the 2 species seem rarely if ever actually to occur together, rather, replacing each other in an as yet not understood way. The song is also rather similar, though simpler and delivered in a more leisurely way (e.g., "toh-toh-tee, tee-toh" or "tlee-to-tleedelee, to-wee-tlee?" repeated slowly over and over). In Middle America Slaty-backed has been regularly seen at small swarms of *Labidus* army ants, and this may also occur in South America.
RANGE: Locally in Andes of w. Venezuela (north to Trujillo), Colombia (n. end of E. Andes south to Santander and Boyacá, head of Magdalena valley in Huila, and n. end of W. Andes), w. Ecuador (Pichincha south to El Oro and Loja), nw. and e. Peru (south on w. slope to Cajamarca), and nw. Bolivia (La Paz); Perijá Mts. on Colombia-Venezuela border and Santa Marta Mts. in Colombia. Also Costa Rica and Panama. Mostly 800–2300 m.

Catharus aurantiirostris

ORANGE-BILLED NIGHTINGALE-THRUSH PLATE: 5

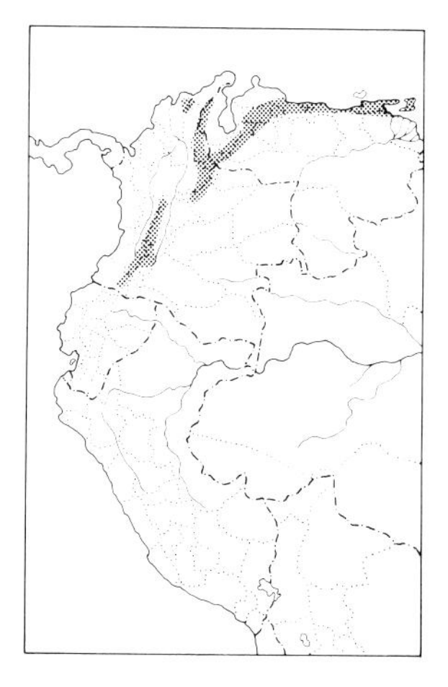

IDENTIFICATION: 16.5 cm (6½"). *Venezuela and Colombia. Bill and legs orange,* narrow orange eye-ring. *Above olivaceous to rufescent brown.* Below mostly pale gray, fading to whitish on throat and median belly. *Phaeopleurus* (sw. Colombia) has decidedly grayer head; see C-43.
SIMILAR SPECIES: Neither of the other S. Am. nightingale-thrushes has a brown back. Migrant *Catharus* thrushes from North America have bill and legs dull.
HABITAT AND BEHAVIOR: Fairly common to common in undergrowth of forest borders, lighter woodland, regenerating clearings, and coffee plantations. Generally hard to see, keeping to dense vegetation, though it may hop into the open along a trail. For a nightingale-thrush its song is poor and unmusical, consisting of short, jumbled phrases with a mostly squeaky or unmusical quality. Often a bird continues to sing for long periods without a pause, usually remaining nearly motionless and thus hard to see. The call is a distinctive nasal "waaa-a-a-a," very like one of the Veery's calls.
RANGE: Coastal mts. of n. Venezuela (east to Paria Peninsula) and Andes of w. Venezuela and Colombia (south to Nariño); Perijá Mts. on Colombia-Venezuela border and Santa Marta Mts. of Colombia; Trinidad. Also Mexico to Panama. 600–2200 m.

GROUP B

N. *migrants;* bill and legs dull, underparts spotted.

Catharus ustulatus

SWAINSON'S THRUSH PLATE: 5

IDENTIFICATION: 18 cm (7"). *Buff lores and bold buff eye-ring* (often imparting a spectacled effect); otherwise uniform olivaceous brown above, with cheeks buffy brownish. Below whitish washed with buff

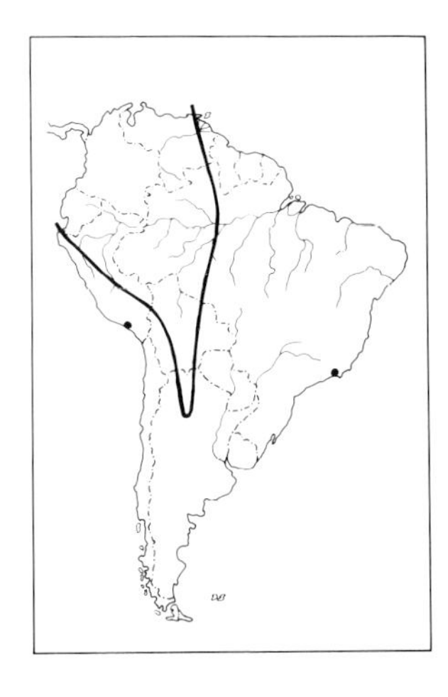

across breast, where also spotted with blackish. Apparently the more rufescent (nominate *ustulatus*) or grayer (*almae*) w.-breeding races of Swainson's Thrush do not reach South America (they seem to over-winter mainly or entirely in n. Middle America).

SIMILAR SPECIES: The obvious buff on the face will usually distinguish this commonest of the 3 migrant N. Am. *Catharus,* but note that both Veery and Gray-cheeked Thrushes do sometimes show a narrow inconspicuous eye-ring. The 3 S. Am. nightingale-thrushes all have brightly colored bills and legs. All the *Turdus* thrushes, many of which are also basically brown, are larger and none (except for the range-restricted Maranon Thrush) show spotting below when adult.

HABITAT AND BEHAVIOR: Fairly common to common transient and n.-winter resident (mostly Oct.–Apr.) in lower and middle growth of forest, forest borders, secondary woodland, and adjacent clearings. Commonest in montane areas (especially on Andean slopes), less numerous in lowlands (where it principally occurs as a transient). Tends to keep to dense cover and is usually quite shy, thus often hard to see. Primarily frugivorous on its wintering grounds, where it only infrequently descends to the ground. Regularly gives a faint version of its song, a lovely series of upward-rolling phrases, especially while on northward passage in Mar.–Apr.; its liquid short call note, "whit," is also frequently heard.

RANGE: Nonbreeding visitor to Colombia, Venezuela, Ecuador, e. Peru, n. and e. Bolivia, and nw. Argentina (south to Tucumán and w. Córdoba); a few records from w. Amaz. Brazil and from Arequipa in coastal sw. Peru; seen once in Rio de Janeiro, Brazil. Breeds North America, wintering from Mexico to South America (smaller numbers in the West Indies). Mostly below 2000 m, higher as transient.

Catharus minimus

GRAY-CHEEKED THRUSH

IDENTIFICATION: 18 cm (7″). Resembles Swainson's Thrush but *never shows the prominent buff lores and eye-ring* of the Swainson's. The Gray-cheeked has at most a narrow whitish eye-ring (sometimes not apparent). Upperparts are usually a dull olivaceous brown (a few birds on southward passage show a somewhat warmer, more rufescent tone); *cheeks usually look grayish*. Generally not as buffy on breast, with sides and flanks often tinged decidedly grayish.

SIMILAR SPECIES: The Veery always is more rufescent above and shows less spotting on its breast. Often a good look is required to distinguish this species from the usually commoner Swainson's Thrush, especially during southward passage when the spectacled effect of Swainson's may be obscured. In general, Gray-cheeked is a duller and grayer-looking bird (not so buffy), particularly on head, neck, and breast—but be careful!

HABITAT AND BEHAVIOR: Uncommon transient and n.-winter resident (Sept.–May) in lower growth of forest, forest borders, and secondary woodland. Main wintering area may be in s. Venezuela (based on number of specimens taken there) and the w. Amazon basin. Its habits are similar to Swainson's Thrush's but it seems even more reclusive, tending to remain closer to the ground in denser undergrowth. Singing in South America is apparently infrequent, perhaps most often while on spring passage, but Gray-cheeked's "veer" call is regularly heard.

RANGE: Nonbreeding visitor to Guyana, Venezuela, Colombia, e. Ecuador, e. Peru, and w. Amaz. Brazil; one sighting from Suriname (Raleigh Falls); Trinidad. Breeds in n. North America, wintering in South America. Mostly below 1500 m, higher as a transient.

Catharus fuscescens

VEERY

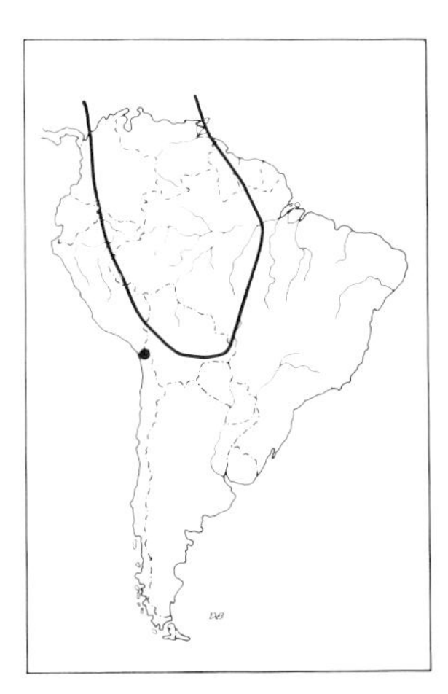

IDENTIFICATION: 18 cm (7"). Resembles Swainson's Thrush but *never shows the prominent buff lores and eye-ring* of the Swainson's. The Veery has at most a narrow whitish eye-ring (usually not apparent). *Color of upperparts varies from bright to rather dark rufous-brown.* Throat and chest washed with buff; *only indistinctly spotted with dusky on chest* (spots both smaller and fewer than those of Swainson's).

SIMILAR SPECIES: Much less spotted below than either Gray-cheeked or Swainson's Thrush and more rufescent above. Shows less of an eye-ring than either, but the dark eye often does seem large and prominent.

HABITAT AND BEHAVIOR: Uncommon transient and n.-winter resident (Sept.–Apr.) in lower growth of forest, forest borders, and secondary woodland. Even more than the Gray-cheeked, the Veery is shy and reclusive and rarely seen in South America, where in fact it is markedly under-recorded relative to its high numbers when in North America. Where most of them overwinter is a mystery yet to be solved. Regularly gives its distinctive down-slurred call, "pheuw," and its more nasal "waaa-a-a-a," but seems not to have been heard singing in South America.

RANGE: Nonbreeding visitor to Guyana, Venezuela, n. and e. Colombia (rather few records), w. Amaz. Brazil (east to the lower Rio Tapajós and south to nw. Mato Grosso), and n. Bolivia (Cochabamba and Santa Cruz); a single sighting from e. Peru (Madre de Dios) and accidental in n. Chile (Arica). Breeds in North America, wintering in South America. Mostly below 1500 m, higher as a transient.

Platycichla Thrushes

A pair of thrushes found in montane areas of South America. The genus is weakly differentiated from *Turdus* (plumage, vocalizations, and behavior are similar); it should probably be sunk.

Platycichla flavipes

YELLOW-LEGGED THRUSH

PLATE: 5

IDENTIFICATION: 22 cm (8½"). *Bill and legs bright yellow in male;* female with legs yellowish and bill variable but usually dusky, often with yellowish on ridge or at base; *narrow yellow eye-ring and dark iris in both*

sexes. Male mostly black with *contrasting gray back, rump, and belly. Melanopleura* of ne. Venezuela and Trinidad similar but with underparts mostly black (*only flanks and crissum gray*), while *xanthoscelus* (restricted to Tobago) is *all* black. Females of all races similar and *dingy:* olivaceous brown above; below buffy brownish, *streaked dusky on throat, paler* (more whitish) *on median belly,* but dingy on crissum. Tonalities seem to vary (with age?), some being more rufescent above and buffier below, others more olivaceous above and grayer below.

SIMILAR SPECIES: Male easily known by its pied black and gray pattern (on Tobago there is no other all-black thrush). Female, however, is considerably more difficult. From the similar Pale-eyed Thrush it is best told by its yellow eye-ring and dark (not pale grayish) iris and its throat streaking. Female Black-hooded Thrush is also very similar but is larger and more uniform tawny-buff below (not paler on mid-belly); it often shows the shadowy effect of the dark hood and usually a full yellow bill. Female Glossy-black Thrush is larger and is darker and more uniform below, with little throat streaking. Be aware, however, that under normal field conditions many of these thrushes will be seen only poorly or briefly; as a result many may have to be left unidentified.

HABITAT AND BEHAVIOR: Fairly common in canopy and borders of forest and secondary woodland and in adjacent clearings and coffee plantations where tall trees provide shading. Essentially arboreal; because of this and a rather shy nature these birds are often hard to see well except when they congregate at fruiting trees. Males are often located by tracking down their songs. This song varies in quality; sometimes it is quite musical and pleasant, at others quite squeaky. It consists of a series of phrases, each often repeated several times, then a pause and on to the next. Mimicking of other birds is frequent in e. Brazil (though fidelity is often poor, such that it is often hard to tell which species is being imitated).

RANGE: Mts. of ne. Colombia (Santa Marta Mts. and south in E. Andes to Norte de Santander) and w. and n. Venezuela (east to Paria Peninsula); *tepuis* of s. Venezuela (Bolívar) and adjacent Guyana and extreme n. Brazil (Roraima); e. Brazil (s. Bahia south to n. Rio Grande do Sul and in Paraíba) and adjacent e. Paraguay (no recent records) and ne. Argentina (Misiones); Trinidad and Tobago. To about 2000 m, but most numerous 500–1500 m.

Platycichla leucops

PALE-EYED THRUSH

IDENTIFICATION: 21.5 cm (8½″). Male with bright yellow bill and legs and *bluish white iris* (but no distinct eye-ring); female has blackish bill, brownish yellow legs, and grayish brown iris (with *no eye ring*). Male *all lustrous black.* Female dark brown above; somewhat paler brown on sides of head, throat, and breast, *becoming more brownish gray on belly.* See C-43.

SIMILAR SPECIES: The male is the only all-black S. Am. thrush with whitish irides (and these stand out very prominently); compare to Yellow-legged Thrush (the 2 overlap only in Venezuela and the *tepui*

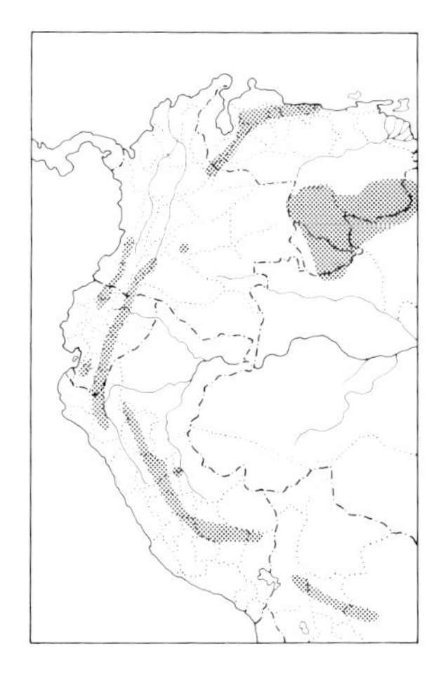

region). Female Yellow-legged similar but has narrow yellow eye-ring and dusky throat streaking. Female Glossy-black Thrush is larger and has a narrow yellow eye-ring, as well as being uniform brown below (not becoming grayer on belly). Female Black-hooded Thrush is also larger and uniform tawny-buff below, as well as often having a suggestion of the male's dusky hood.

HABITAT AND BEHAVIOR: Uncommon and seemingly local in canopy and borders of humid montane forest. Usually found singly or in pairs, though larger numbers will congregate at fruiting trees. Shy and generally not very often seen. The song is a series of short phrases, some of them musical but most of them quite high and thin (almost squeaky), usually with long intervening pauses (such that the pauses are usually longer than the songs); one frequent phrase will remind N. Am. observers of the Swainson's Thrush.

RANGE: Locally in coastal mts. of n. Venezuela (Lara and Miranda) and in Andes of sw. Venezuela, s. Colombia (recorded only from Valle, the head of the Magdalena valley in Huila, and w. Meta), w. and e. Ecuador (on w. slope south to Pichincha and in El Oro), e. Peru, and nw. Bolivia (south to w. Santa Cruz); *tepuis* of s. Venezuela (Bolívar and Amazonas) and adjacent Guyana and n. Brazil (Roraima). Mostly 900–2000 m, occasionally wandering lower into adjacent lowlands; also a few (questionable?) records from up to 3100 m.

Turdus Thrushes

The thrushes in this genus include many widespread, familiar birds of semiopen country and woodland or forest borders, but some, particularly several in Group B, are much more reclusive and less often seen. Though easily recognized as thrushes, identification to species level is often tricky (female *Platycichla* thrushes further complicate the issue, as do the frequently seen spotted immature plumages). Foraging behavior varies, with some species habitually hopping on the ground, searching for insects, while others are exclusively arboreal. *Turdus* songs also vary, though the lilting, caroling pattern of distinct phrases is common to many (their musical qualities may differ markedly). All *Turdus* build substantial cup nests, often lined with mud, usually well above the ground. The genus is found virtually throughout the world, and group names (e.g., "thrush" or "robin" or "ouzel" or whatever) present a problem. The 1983 AOU Check-list calls some species "thrushes" and others "robins," which has the disadvantage of being inconsistent and obscuring their relationship. Here we follow Meyer de Schauensee (1966, 1970) in calling all the South American *Turdus* "thrushes" and would suggest that all the American *Turdus* be given the "thrush" group name (with the exception of the familiar but certainly "misnamed" American Robin, *T. migratorius*).

GROUP A

Large and uniform, or *strongly patterned with black on head* (male); female, when different, is brown with *dingy crissum* (as in female *Platycichla,* but *unlike* Group B).

Turdus serranus

GLOSSY-BLACK THRUSH

PLATE: 5

IDENTIFICATION: 25 cm (9¾"). Yellow-orange bill and legs, narrow orange eye-ring in male; bill and legs dull yellowish brown, narrow orange-yellow eye-ring in female (bill yellower in birds from Venezuela); *iris dark* in both sexes. Male *all lustrous black*. Female uniform rufescent brown, only slightly paler below (and with *no* gray on belly). *Atrocericeus* (most of Venezuelan range) more olivaceous brown above, grayish brown below; *cumanensis* (ne. Venezuela) rather different, *more or less uniform sooty with fully yellow bill.*

SIMILAR SPECIES: Great Thrush is apt to be confused with male, particularly in w. Colombia, n. Ecuador, and se. Peru, where its races are very dark; but it is obviously larger and not as deep a black and its habitat and behavior differ. Male Pale-eyed Thrush has pale iris, *lacks* eye-ring. Female resembles female Yellow-legged, Pale-eyed, and Black-hooded Thrushes, but combination of eye-ring (which eliminates Pale-eyed) and uniform and rather dark brown underparts (lacking paler median belly of Yellow-legged and paler ochraceous tone of Black-hooded) will identify it under favorable circumstances. Note that Yellow-legged and Black-hooded Thrushes do not occur in the Andes south of Colombia.

HABITAT AND BEHAVIOR: Fairly common in montane forest, forest borders, and advanced secondary woodland. Though it occurs regularly at forest edge, it never ventures far into the open except occasionally to fly between patches of woodland (Great Thrush *very* different in this respect). Usually shy and inconspicuous, and on the whole not often seen except at fruiting trees and when males are singing (though even then they are often hidden). The very fast song is tirelessly repeated, with short intervals between phrases; in quality it is often quite shrill, usually rising slightly (e.g., "tee-do-do-eét").

RANGE: Coastal mts. of n. Venezuela (east to Sucre) and Andes of w. Venezuela, Colombia, Ecuador, nw. and e. Peru (on w. slope south to Cajamarca), w. Bolivia, and nw. Argentina (Salta and Jujuy); Perijá Mts. on Colombia-Venezuela border. Mostly 1500–2800 m (lower on Pacific slope of Colombia and Ecuador and in Argentina).

Turdus fuscater

GREAT THRUSH

PLATE: 5

IDENTIFICATION: 33 cm (13"). *By far the largest S. Am. thrush.* Bill orange, legs yellow-orange, *eye-ring bright yellow* (last may be lacking in some presumed immatures). *Uniform dark grayish brown,* darkest on wings and tail (almost blackish), paler on belly. Racial variation is substantial: *ockendoni* (se. Peru in Cuzco and Puno) is the darkest, being sooty blackish; *quindio* (Cen. and W. Andes of Colombia and all but southernmost Ecuador) is the next darkest (sooty above, brownish below) and *gigantodes* (s. Ecuador and Peru south to Junín) is only slightly paler; while *cacozelus* (Colombia's Santa Marta Mts.) is much the palest, being buffy olivaceous below. Nominate *fuscater* of Bolivia has head markedly darker than rest of body, with face and lores blackish, and male has throat quite prominently streaked dusky.

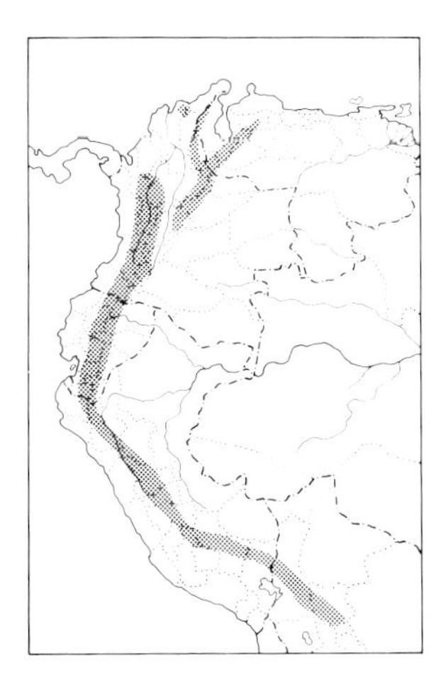

SIMILAR SPECIES: Great Thrush is so much larger than other *Turdus* thrushes that it can usually be known on that basis alone. Cf. Glossy-black Thrush (a far more secretive, forest-based species). In area of overlap with Chiguanco Thrush (s. Ecuador to nw. Bolivia) it is most apt to be confused with that species, but note that in this area Chiguanco *does not show an eye-ring.*

HABITAT AND BEHAVIOR: Common to very common in cleared areas with hedgerows and patches of low woodland and at borders of forest and secondary woodland. Absent from extensively forested regions, though it quickly colonizes new clearings. This thrush has doubtless increased substantially in response to widespread deforestation, and in many areas it is now one of the most frequently seen birds. This is especially the case from Ecuador northward: to the south, where the highlands tend to be more arid, this thrush is more or less restricted to humid areas on the e. slope. Often seen hopping about on the ground, especially on short-grass pastures or roadsides; also regularly visits fruiting trees. It is so large that in flight it often is not recognized as a thrush (looking almost like a small hawk, etc.). Not only is the Great Thrush conspicuous, it also is noisy, with frequently heard loud calls, most often an arresting "keeyert!" or "kurt!-kurt!-kurt!-kurt!" regularly given in flight or when flushed. Its song is, however, infrequently heard; it is rather varied, consisting of a series of rapidly uttered musical phrases.

RANGE: Andes of w. Venezuela (north to Lara and Trujillo), Colombia, Ecuador, nw. and e. Peru (on w. slope south to Cajamarca), and nw. Bolivia (La Paz and Cochabamba); Colombia's Santa Marta Mts. and Perijá Mts. on Colombia-Venezuela border. Mostly 1800–4000 m.

Turdus chiguanco

CHIGUANCO THRUSH

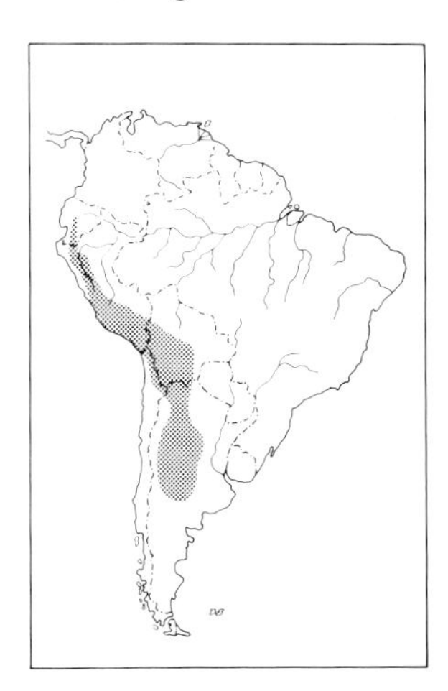

IDENTIFICATION: 27–28 cm (10½–11"). Bill yellow to orange-yellow; narrow yellow eye-ring in males of *anthracinus* only (*no* eye-ring in n. pair of races, a helpful point of distinction from Great Thrush); legs yellowish to dull orange. In general form resembles the larger Great Thrush. In n. part of range (south to Peru and adjacent Bolivia) *pale ashy brown,* slightly darker above. *Anthracinus* (north to Cochabamba, Bolivia) is *notably darker:* essentially sooty in male, slightly paler sooty brownish in female.

SIMILAR SPECIES: Easily confused with Great Thrush, and the 2 are sympatric (or sometimes nearly so, Chiguanco tending to favor drier areas) in some regions. In view of the geographic variation in both species, it is often best to go by size and (in Ecuador and Peru) by the presence or absence of an eye-ring. From Ecuador south to cen. Peru they are about the same grayish brown color, but in se. Peru the Great is darker, while in Bolivia the reverse is true, Chiguanco being darker.

HABITAT AND BEHAVIOR: Fairly common to common in agricultural areas with scattered trees and hedgerows, gardens, and light, scrubby woodland. Found primarily in arid regions but usually near watercourses or in irrigated areas. Hops about on ground like many other thrushes of semiopen country and often becomes quite tame around habitations. The song (in Argentina) is weak and relatively unmusical, consisting merely of a short phrase repeated over and over.

RANGE: Andes of s. Ecuador (north to Chimborazo), Peru, w. Bolivia, n. Chile (south to Antofogasta), and nw. Argentina (south to Mendoza and La Pampa); absent from e. slope in Ecuador, Peru, and nw. Bolivia. Mostly 1500–4000 m, but lower (nearly to sea level) in Peru and Chile.

Turdus falcklandii

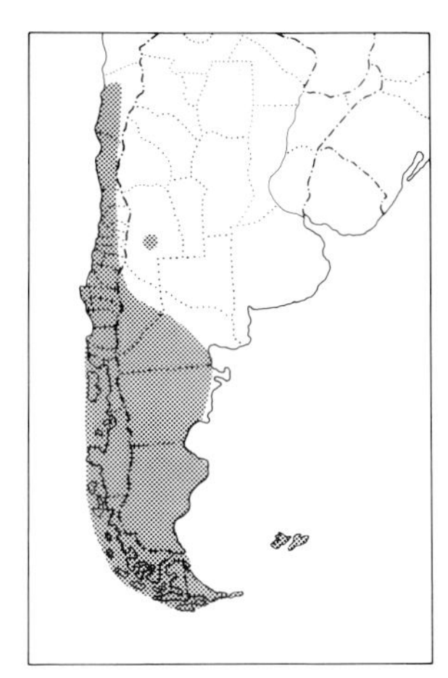

AUSTRAL THRUSH

PLATE: 5

IDENTIFICATION: 25.5–26.5 cm (10–10½"). *Chile and s. Argentina.* Bill and legs golden yellow. *Head blackish,* contrasting with olive brown of remaining upperparts; flight feathers and tail duskier. Throat whitish, *sharply streaked blackish;* breast pale brownish, becoming ochraceous buff on belly. Female somewhat duller, with head not as black and grayer or more olivaceous yellow. Nominate race (Falkland Is.) slightly larger.

SIMILAR SPECIES: In most of its range it is the only thrush present. Creamy-bellied Thrush overlaps marginally with Austral in s. Argentina; it lacks the contrastingly dark head.

HABITAT AND BEHAVIOR: Common in gardens and agricultural areas with scattered trees and hedgerows and in lighter woodland and forest borders. Feeds mostly on the ground, hopping about on grassy lawns and pastures, but also eats much fruit. Its song in Chile is a rich and measured caroling, with individual phrases often repeated several times.

RANGE: Cen. and s. Chile (north to Atacama) and s. Argentina (north to Neuquén and Río Negro, with 1 austral winter record from Mendoza); Falkland and Juan Fernández Is. To about 2000 m (in cen. Chile).

Turdus olivater

BLACK-HOODED THRUSH

PLATE: 5

IDENTIFICATION: 23–24 cm (9–9½"). Male's bill and narrow eye-ring yellow, legs yellow to dull brownish yellow; female's bill variable, yellow (usually?) to dusky, narrow eye-ring yellow, and legs dull brownish to brownish yellow. Male has *entire hood and chest black,* contrasting with dull olive brown upperparts and ochraceous (varying from dull to rather bright) lower underparts. *Tepui* races have the brightest (most rufescent) underparts and black on bib is reduced to mottled streaking (especially in *kemptoni* of Cerro de la Neblina). *Caucae* (sw. Colombia) is *quite different:* it has black only on head, blackish streaking on throat, and sandy buff underparts (without the ochraceous tone). Female has black of male's hood replaced by dull brown, giving faint echo of the pattern, but otherwise has plumage much like male's (including the *uniform dull ochraceous underparts*).

SIMILAR SPECIES: Male's obvious black hood sets it apart from all other S. Am. thrushes, but female is tricky. It is distinguished from both of the female *Platycichla* thrushes by its uniform lower underparts (not paler on median belly) and further from female Pale-eyed by its eye-ring. Female Glossy-black also *very* similar but darker and more rufescent below. Cf. also other relatively dull *Turdus* (e.g., Black-billed and Clay-colored).

HABITAT AND BEHAVIOR: Fairly common to common in canopy and borders of humid forest and secondary woodland and in coffee plantations where these are shaded by tall trees. Behavior similar to that of other forest-based, arboreal *Turdus,* but perhaps somewhat less shy than most others. Its song is also typical of the genus, a loud and rather musical series of phrases, rather slowly delivered, sometimes with higher and thinner notes intervening. Numerous on the San Lorenzo ridge road in the Santa Martas and at the Escalera in the *tepuis* of se. Venezuela; there are, however, no recent reports of *caucae,* and it may be endangered by widespread deforestation in the upper Cauca valley, Colombia.

RANGE: Coastal mts. of n. Venezuela (east to Guárico and Miranda) and locally in Andes of w. Venezuela and ne. Colombia (Norte de Santander) and in sw. Colombia (slopes above upper Cauca valley in Cauca); Colombia's Santa Marta Mts. and Perijá Mts. on Colombia-Venezuela border; *tepui* region of s. Venezuela (Bolívar and Amazonas) and adjacent Guyana and n. Brazil (Roraima), with a recent report from Suriname (Donahue and Pierson 1982). 800–2600 m.

NOTE: *Caucae* is so distinct in appearance and disjunct in range that we question whether it is actually conspecific with *T. olivater.* However, as it remains poorly known, we leave it as a subspecies of *T. olivater.* Should it be split, Cauca Thrush would be an appropriate name.

Turdus fulviventris

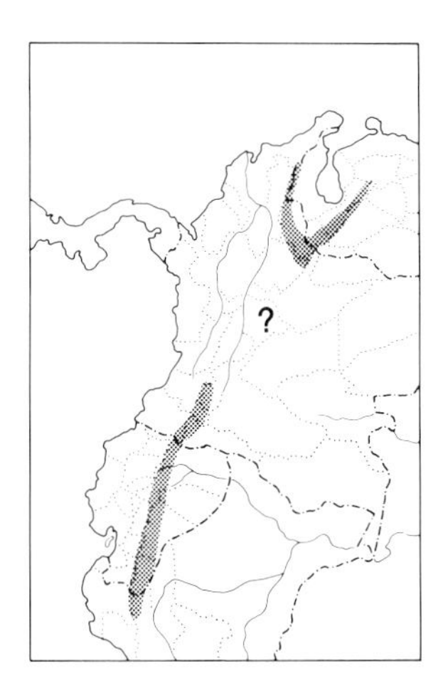

CHESTNUT-BELLIED THRUSH

IDENTIFICATION: 24 cm (9½"). Bill yellow, legs yellowish brown, narrow but conspicuous eye-ring orange. *Head and throat black,* latter somewhat streaked gray. Above otherwise dark gray. Breast slightly paler gray; *belly bright rufous,* becoming whitish on crissum. Female similar but slightly duskier on head (hence less contrast with gray upperparts). See V-32, C-43.

SIMILAR SPECIES: The only thrush in the n. Andes with an orange-rufous belly. Observers from North America will at once be struck by its similarity to the American Robin (*T. migratorius*).

HABITAT AND BEHAVIOR: Uncommon and rather local in humid montane forest, forest borders, and adjacent small clearings and roadsides; often in mossy cloud forest. Usually arboreal but also sometimes seen hopping on the ground in shrubby clearings or along roads (when it looks uncannily like an American Robin). The song, not too often heard, is not one of the better ones in the genus. It is a series of clipped, mimidlike phrases with interspersed harsh or buzzy notes; generally the bird sings from the canopy. Does not occur in many readily accessible localities, but one is along the road north of the Universidad de los Andes Forest beyond La Azulita in Mérida, Venezuela.

RANGE: Andes of w. Venezuela (north to Trujillo), very locally in Colombia (recorded only from E. Andes in Norte de Santander, head of Magdalena valley in Huila, and on e. slope of E. Andes in w. Putumayo and e. Nariño), e. Ecuador, and extreme n. Peru (Cajamarca); Perijá Mts. on Colombia-Venezuela border. Mostly 1400–2600 m.

GROUP B

"Generalized" *Turdus* thrushes. Some shade of brown above, often paler below; throat streaking usually present and crissum usually whitish. Mostly in lowlands (a few up into lower subtropics).

Turdus rufiventris

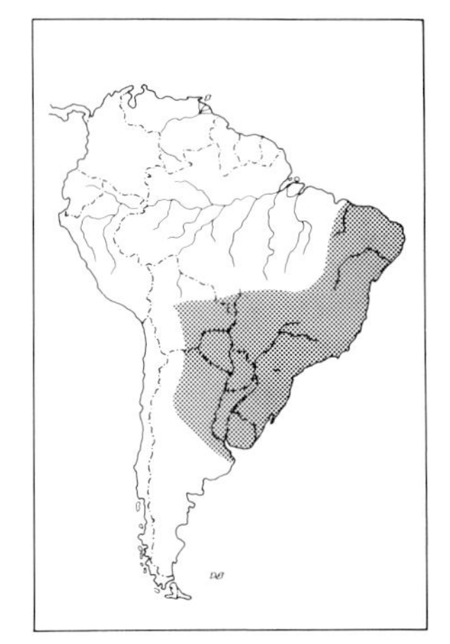

RUFOUS-BELLIED THRUSH

PLATE: 6

IDENTIFICATION: 24.5 cm (9¾"). Bill olive yellow; narrow eye-ring yellow-orange. Above uniform grayish olive brown. Throat white, *prominently streaked with brown;* breast pale grayish brown, becoming *orange-rufous on belly and crissum.* Female slightly paler below. *Juensis* (ne. Brazil) is somewhat paler generally in both sexes.

SIMILAR SPECIES: The only thrush in its range with a rufous belly; does not overlap with Austral Thrush (latter found farther south).

HABITAT AND BEHAVIOR: Common in lighter woodland, forest borders, and clearings and gardens; seems less numerous across much of ne. Brazil. Often seen hopping about on the ground but usually not far from cover, and in dry regions it is more or less restricted to areas near water. As usual with S. Am. *Turdus,* it is quite shy in most places (rarely as familiar as the American Robin, *T. migratorius,* which it recalls in plumage and mannerisms). Its song is a fast, rich caroling, generally given from a hidden perch; in quality it ranks with the best of the S. Am. *Turdus.* The querulous, upslurred call recalls that of the Bare-eyed/Ecuadorian/Clay-colored group.

RANGE: E. Bolivia (Santa Cruz south and east), Paraguay, n. Argentina (south to Córdoba and Buenos Aires), Uruguay, and w. and e. Brazil (Pernambuco, Bahia, and cen. Mato Grosso south and east). To about 2200 m.

Turdus nudigenis

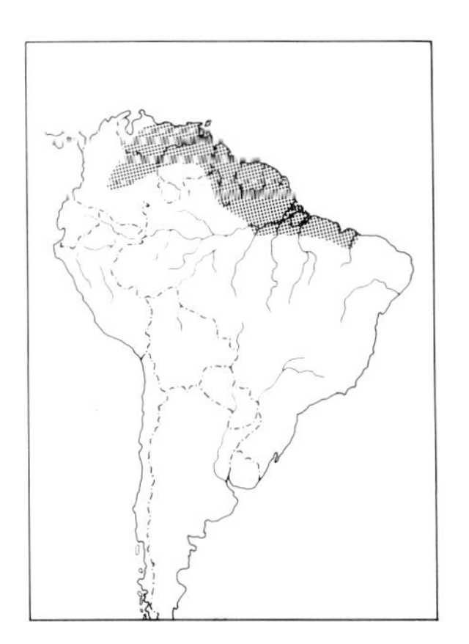

BARE-EYED THRUSH

PLATE: 6

IDENTIFICATION: 23 cm (9"). Bill yellowish, somewhat paler at tip; *very wide bare eye-ring flesh yellow.* Dull grayish olive brown above, paler grayish brown below with center of belly and crissum white; throat whitish streaked brown.

SIMILAR SPECIES: No other thrush has as conspicuous an eye-ring.

HABITAT AND BEHAVIOR: Uncommon to fairly common in semiopen and scrubby areas, borders of forest and woodland, and gardens. Perhaps most numerous on Trinidad and Tobago. Mainly arboreal but at times drops to ground and hops about on lawns, etc. The song is a musical caroling, consisting of a series of short phrases, the call a distinctive querulous "queeow."

RANGE: Ne. Colombia (south to Meta and Vichada), Venezuela (except in the northwest and far south), Guianas, and ne. Brazil (east locally to around Belém and in n. Maranhão; perhaps spreading); Trinidad and Tobago. Also the Lesser Antilles. Mostly below 1000 m (but recorded to 1800 m in Venezuela).

NOTE: See comments under Ecuadorian Thrush.

Turdus maculirostris

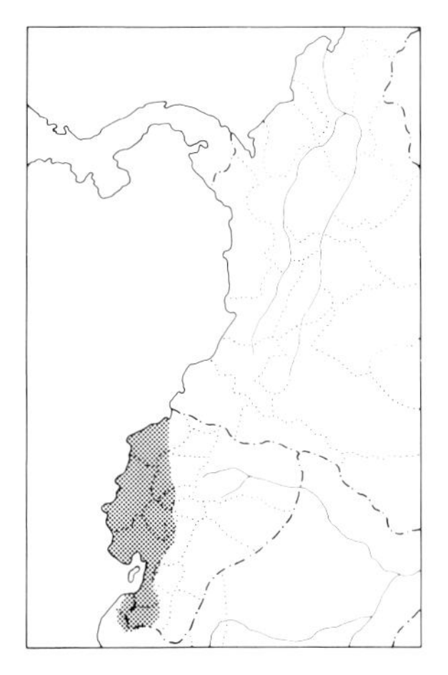

ECUADORIAN THRUSH

IDENTIFICATION: 23 cm (9″). *W. Ecuador and adjacent Peru.* Bill olive yellow, somewhat paler at the tip; *narrow* eye-ring yellow to yellow-orange. Resembles Bare-eyed Thrush (and long considered a disjunct race of that species; see below) but *lacks* the wide eye-ring: eye-ring of *maculirostris* is of comparable width as that of numerous other *Turdus*.
SIMILAR SPECIES: Does not occur with either Bare-eyed or Clay-colored Thrushes. Cf. female Pale-eyed and Glossy-black Thrushes, both of which mostly occur at a higher elevation than this species; both are more uniform below and lack this species' throat streaking.
HABITAT AND BEHAVIOR: Uncommon to fairly common in forest, forest borders, and adjacent clearings with scattered trees. Behavior similar to that of many other *Turdus* thrushes but seems more a forest-based bird than either Bare-eyed or Clay-colored Thrush. Whining, querulous call is similar to both of theirs.
RANGE: W. Ecuador (north to coastal Esmeraldas) and adjacent nw. Peru (Tumbes). To at least 1400 m.

NOTE: *T. maculirostris* is here considered a species distinct from *T. nudigenis*: it lacks the wide bare ocular area so distinctive in *T. nudigenis* (the eye-ring of *maculirostris* is comparable in width to that of *T. grayi*), differs in its forest-based habitat, and has a widely disjunct range.

Turdus grayi

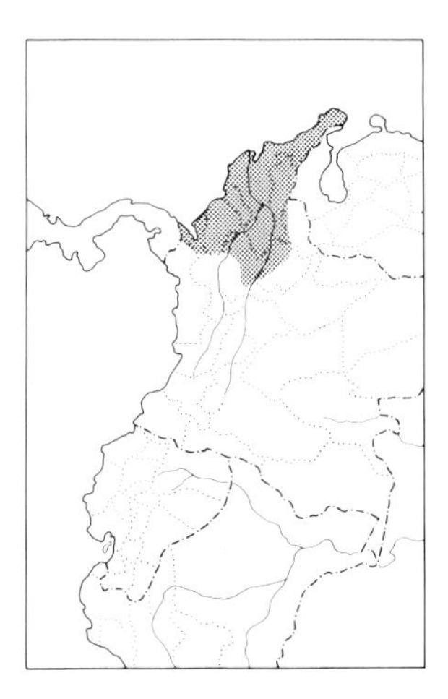

CLAY-COLORED THRUSH

IDENTIFICATION: 23 cm (9″). *N. Colombia.* Bill greenish or olive yellow. Uniform dull olive brown above; *uniform dull sandy to fulvous-brown below* (not paler on mid-belly), with throat lightly streaked brown. See C-43, P-23.
SIMILAR SPECIES: Does not overlap with *very* similar Unicolored Thrush (of Bolivia) or with Bare-eyed or Ecuadorian Thrushes. Does occur with Pale-breasted Thrush (which has contrasting grayish head); cf. also female Black-hooded Thrush (mainly a highland bird).
HABITAT AND BEHAVIOR: Uncommon to fairly common in lighter woodland, clearings, and gardens; principally in more arid regions. Considerably less conspicuous and less numerous in Colombia than it is, for instance, in Panama. Its behavior is similar to many other thrushes', as is its excellent musical caroling song, given mostly early in the morning (often before first light) and at dusk; also often heard is its querulous, slurred call.
RANGE: N. Colombia (n. Chocó east to the Guajira Peninsula and south in middle Magdalena valley to s. Bolívar). Also Mexico to Panama. To about 300 m.

Turdus haplochrous

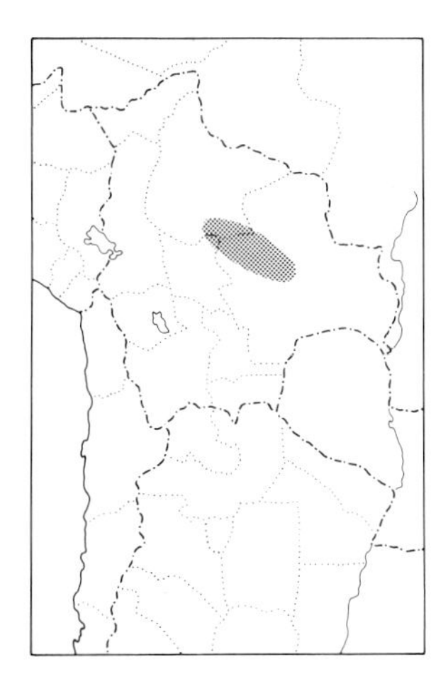

UNICOLORED THRUSH

IDENTIFICATION: 23 cm (9″). *N. Bolivia.* Uniform dull olive brown above; uniform dull sandy brown below (not paler on mid-belly), with throat lightly streaked brown.

SIMILAR SPECIES: In its range probably most apt to be confused with Hauxwell's Thrush, but it lacks that species' whitish on lower belly and crissum. Soft-part colors also probably differ, but these are not known with certainty for Unicolored, which remains a very poorly known bird: bill color probably like Clay-colored's (yellowish green), to which this species bears a strong resemblance.

HABITAT AND BEHAVIOR: Nothing appears to be on record. Probably similar to Bare-eyed Thrush, to which this species is probably allied. Known from only 3 specimens.

RANGE: N. Bolivia (Beni at Río Mamoré; w. Santa Cruz at Palmarito, Río San Julián). 250–350 m.

Turdus amaurochalinus

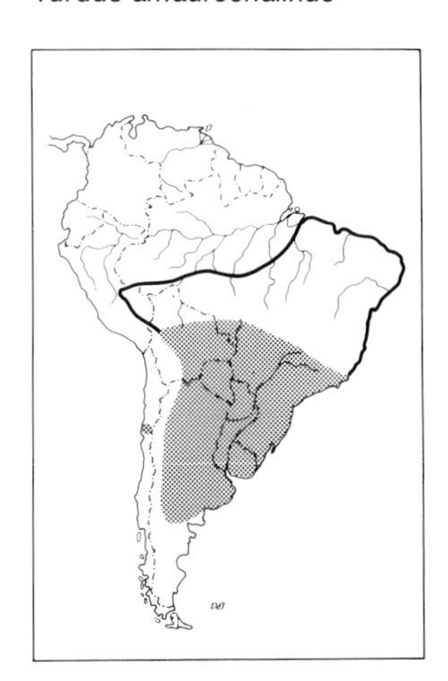

CREAMY-BELLIED THRUSH

PLATE: 6

IDENTIFICATION: 23 cm (9″). Bill bright yellow, sometimes with dusky tip (mostly dusky with a little yellow on mandible in female). Olive brown above with *contrasting blackish lores.* Throat white, *sharply and heavily streaked blackish;* breast pale brownish gray, becoming white on center of belly and crissum. Some birds are buffier on throat and belly.

SIMILAR SPECIES: Pale-breasted Thrush has contrasting gray head and less distinctly streaked throat, and it lacks the blackish lores. Female and immature Creamy-bellieds, whose dusky bills may show little or no yellow, also resemble Black-billed Thrush, but that species does not have the blackish lores and shows only vague streaking on throat. Cf. also White-necked Thrush.

HABITAT AND BEHAVIOR: Fairly common to common at forest borders and in lighter woodland, clearings, and gardens. To some extent migratory, and nonbreeding or transient birds are sometimes encountered in small flocks but usually are more or less solitary and shy like other thrushes. Characteristically quivers its tail upon alighting on a branch or the ground. The song is a typical *Turdus* caroling, which, while musical, usually lacks the spirited quality of various others'. Also often gives a sharp "pok" call note.

RANGE: Breeds e. Bolivia (La Paz east to Santa Cruz and southward), s. Brazil (Mato Grosso, Minas Gerais, and Rio de Janeiro southward), Paraguay, Uruguay, and n. and cen. Argentina (south to Neuquén and Río Negro); during austral winter, some birds move north into se. Peru (Puno, Madre de Dios, and se. Ucayali) and across cen. and ne. Brazil north to s. and e. Pará and n. Maranhão (mostly June–Oct.); casual in Chile (Atacama). N. limit of breeding uncertain (perhaps breeds farther north than given above?), as is proportion of Argentinian birds which migrate (the majority apparently leave, except perhaps along the n. border). To 2100 m (in Bolivia).

Turdus ignobilis

BLACK-BILLED THRUSH

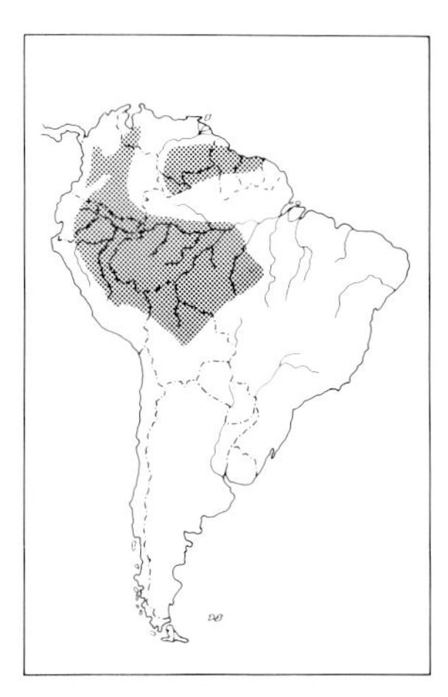

IDENTIFICATION: 21.5–23 cm (8½–9″). *A dingy thrush with a blackish bill.* Widespread *debilis* of w. Amazonia olive brown above; throat white streaked brown, with fairly prominent white patch on upper chest; breast pale grayish brown, gradually fading to white on median belly and crissum. Nominate race and *goodfellowi* (w. Colombia) have less distinct throat streaking and no white patch on upper chest; browner (less grayish) on breast. *Tepui* and Guianan races intermediate. See C-43.

SIMILAR SPECIES: A dull thrush, basically devoid of obvious field marks but often identifiable on that basis alone. Creamy-bellied Thrush basically similar but with sharper throat streaking, blackish lores, and yellow bill (at least in males); the 2 overlap only in s. part of *debilis'* range and only during austral winter. White-necked Thrush is widely sympatric but has narrow yellow eye-ring (*no* eye-ring in Black-billed) and much more prominent throat streaking with larger white throat patch, and it is browner above and grayer below (less dingy overall); it is a true forest bird, unlike Black-billed. Cf. also Pale-breasted Thrush (which always shows grayish head and yellowish bill but is otherwise quite similar) and more uniform and browner Cocoa and Hauxwell's Thrushes; in Colombia's Cauca valley also female Black-hooded Thrush.

HABITAT AND BEHAVIOR: Fairly common to common in clearings, semiopen areas, savannas with gallery woodland, and gardens and around towns. In general a numerous and quite conspicuous thrush, especially in w. Colombia, where it is regularly seen hopping about on lawns or pastures (though it seems to come to ground much less often in Amazonia and the Guianas). The song of *debilis* is a typical *Turdus* series of phrases, often delivered rather softly; it is given especially in predawn darkness but sometimes also at midday. N. Am. observers will recall a striking resemblance to the song of the American Robin (*T. migratorius*).

RANGE: Colombia (Cauca and Magdalena valleys and widely through e. lowlands except in ne. llanos), w. Venezuela (north to Mérida and Barinas), e. Ecuador, e. Peru, n. Bolivia (south to Cochabamba and w. Santa Cruz), and w. and cen. Amaz. Brazil (east to lower Rio Negro and upper Rio Tapajós); s. Venezuela (Amazonas and Bolívar), Guyana, and Suriname. Mostly below 2000 m.

Turdus leucomelas

PALE-BREASTED THRUSH

PLATE: 6

IDENTIFICATION: 23 cm (9″). Bill olive yellowish. Above mostly brown with *gray head and nape;* wings often more rufescent. Throat white streaked with brown; breast pale grayish brown, becoming dingy grayish white on belly (whitest on mid-belly and crissum). N. races (south to lower Amazon area) are slightly more strikingly gray-headed than nominate (some nominate birds, especially females/immatures, have gray essentially restricted to nape).

SIMILAR SPECIES: The gray-headed look is usually distinctive; Pale-breasted is warmer brown above than many other somewhat similar

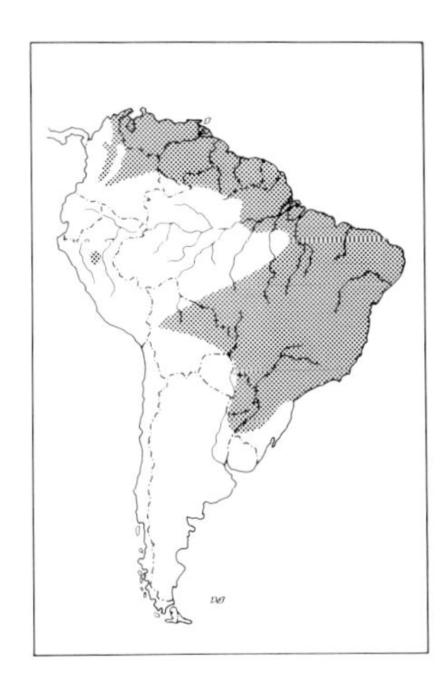

thrushes (e.g., Black-billed, Creamy-bellied). Cf. also White-necked Thrush.

HABITAT AND BEHAVIOR: Fairly common to common in a variety of semiopen habitats, ranging from borders of humid forest and adjacent clearings to deciduous woodland and savannas with gallery woodland. Behavior similar to many other thrushes', though usually less shy than most; often terrestrial. The song is a caroling similar to that of many other thrushes—a series of melodic phrases, with individual notes often repeated a few to many times, usually given from a hidden perch in lower growth. The call is a distinctive guttural, often trebled "wert-wert-wert."

RANGE: N. Colombia (Santa Marta region south into upper Magdalena valley in Tolima and in e. lowlands south to Meta and Vichada), Venezuela, Guianas, most of Brazil (absent from cen. and w. Amazonia and e. Santa Catarina and Rio Grande do Sul), ne. Argentina (Misiones and Corrientes), e. Paraguay, and n. and e. Bolivia (Santa Cruz and La Paz); ne. Peru (recorded only from around Moyobamba in San Martín). Mostly below 1500 m, locally to 2000 m.

Turdus albicollis

WHITE-NECKED THRUSH

PLATE: 6

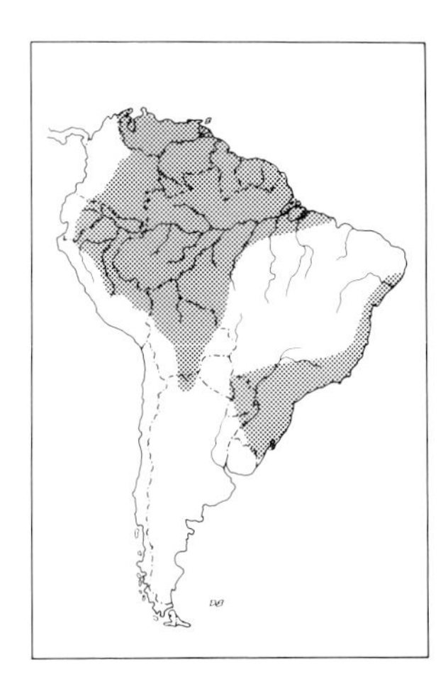

IDENTIFICATION: 20.5–23 cm (8–9"). Bill blackish; narrow eye-ring yellow or orange-yellow. Above deep brown. Throat white, *thickly streaked with dark brown,* and bordered below by *white crescent.* Remaining underparts pale grayish, becoming white on median belly and crissum. *Paraguayensis* and the nominate race (se. Brazil, e. Paraguay, and ne. Argentina) are larger and have bill mostly yellow (only the ridge is black), with back brighter brown and *rufous on flanks. Contemptus* (*yungas* of Bolivia and in nw. Argentina) and *crotopezus* (ne. Brazil) both resemble the widespread n. group (typified by *phaeopygus*) but have lower mandible yellowish.

SIMILAR SPECIES: Combination of contrast between brown upperparts and gray breast, yellow eye-ring, and boldly streaked throat with white crescent below will identify this relatively reclusive thrush of humid forest. Cf. White-throated Thrush (which replaces it west of the Andes).

HABITAT AND BEHAVIOR: Fairly common to common in lower growth of humid forest and advanced secondary woodland. Rarely stray far from cover, though in a few places (e.g., at Rancho Grande in Venezuela) they sometimes emerge to feed on the ground at roadsides or on pastures or lawns. Even shyer than most other S. Am. thrushes, usually just glimpsed, often hopping on forest floor, then retreating back into heavy cover. Occasionally rises to feed in mid-story at fruiting trees and sometimes is seen at army ant swarms. It is far more often heard than seen, and its song is a rather monotonous, almost tired-sounding series of short phrases which carries for long distances but is very hard to track down to its source. Unlike many other thrushes, it regularly sings through midday.

RANGE: N. and e. Colombia (west of Andes only in Santa Marta region, the Perijá Mts., and Norte de Santander), Venezuela (absent from nw. lowlands and the llanos region), Guianas, Amaz. Brazil (east to around Belém and n. Maranhão, south to nw. Mato Grosso), e. Ecuador, e. Peru, n. and e. Bolivia, and nw. Argentina (Salta); coastal e. Brazil (Alagoas south to Rio Grande do Sul), e. Paraguay, and ne. Argentina (Misiones and Corrientes); Trinidad and Tobago. Mostly below 1500 m.

NOTE: Here Middle American and trans-Andean *T. assimilis* is considered a species distinct from S. Am. *T. albicollis,* following the 1983 AOU Check-list. Furthermore, birds from se. South America, the nominate group *sensu strictu* (with *paraguayensis* and perhaps *crotopezus*), differ morphologically from birds of remainder of range, and it has been suggested that they should be separated as a full species (Rufous-flanked Thrush would seem a good English name). However, voices of both groups are similar, and *crotopezus* is somewhat intermediate, so at least for now we unite under *T. albicollis* all forms found east of the Andes.

Turdus assimilis

WHITE-THROATED THRUSH

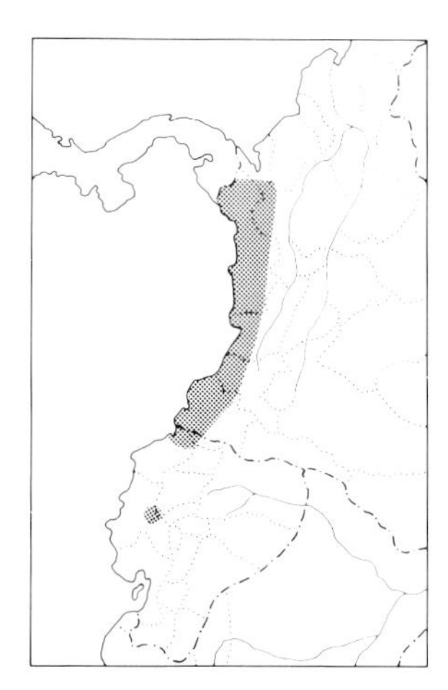

IDENTIFICATION: 21.5 cm (8½"). *W. Colombia and nw. Ecuador.* Bill mostly yellowish (duskier toward base); eye-ring bright yellow. Similar to White-necked Thrush (latter found only *east* of the Andes and in ne. Colombia, with no overlap). *Underparts below the white chest crescent are dull brown* (not the gray of White-necked), becoming white on median belly and crissum. See P-23.
SIMILAR SPECIES: No other thrush in its limited S. Am. range shows such a conspicuous white crescent across its upper chest.
HABITAT AND BEHAVIOR: Uncommon in humid forest canopy and borders, secondary woodland, and adjacent clearings. Primarily a foothill bird. In Middle America (where it seems more numerous) its behavior is similar to White-necked Thrush's, but it is much more arboreal and more easily seen than that species. Song in Panama is quite different from White-necked's, being much louder and fuller (lacking White-necked's "weak" quality), with a more mimidlike quality. It also has a characteristic odd guttural or nasal "enk" or "urrk" call, almost froglike in quality.
RANGE: W. slope of W. Andes in w. Colombia (Chocó south to w. Nariño) and w. Ecuador (Esmeraldas; also 1 specimen from Quevedo, Los Ríos). Also Mexico to Panama. To 900 m.

NOTE: Often considered conspecific with cis-Andean *T. albicollis,* but behavior and vocalizations differ markedly (at least from Middle American birds; *daguae,* the race of *T. assimilis* in South America, remains less well known).

Turdus hauxwelli

HAUXWELL'S THRUSH

PLATE: 6

IDENTIFICATION: 23 cm (9"). *W. Amazonia.* Bill dark gray to dark brown; some birds show a narrow olive eye-ring. *Mostly warm brown,* more rufescent above, paler and more olivaceous below; throat whitish lightly streaked dusky; *median belly and crissum white,* occasionally washed with buff.
SIMILAR SPECIES: Closely resembles Pale-vented and Cocoa Thrushes (neither of which occurs with Hauxwell's except at several central

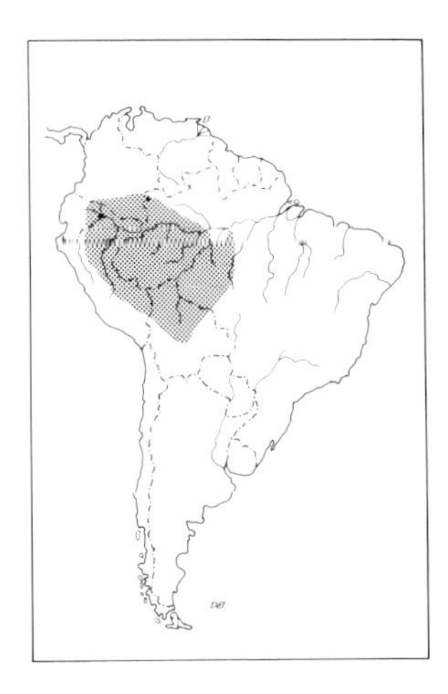

Amaz. localities where both Cocoa and Hauxwell's have been reported taken); cf. under those species. Black-billed Thrush is dingier and more olive generally; it occurs in clearings and semiopen areas and is not basically a forest thrush, as is Hauxwell's. Female Lawrence's Thrush (in which eye-ring is much less obvious than in male) is darker and more grayish brown above, dingier and more olivaceous below.

HABITAT AND BEHAVIOR: Uncommon to fairly common in humid forest, mostly in lower growth and subcanopy, rarely coming to edge; found in both *terra firme* forest and *várzea* (including forest on river islands). An inconspicuous and shy thrush, not often seen in the dense terrain it inhabits, perhaps most likely at fruiting trees.

RANGE: Se. Colombia (north to Vaupés), e. Ecuador, e. Peru, n. Bolivia (south to La Paz and nw. Santa Cruz), and w. Amaz. Brazil (east to the Rio Madeira and almost to the lower Rio Negro). To about 800 m.

NOTE: The *Turdus fumigatus/hauxwelli/obsoletus* complex is a difficult one, with species limits not yet fully established. *T. hauxwelli* and *T. fumigatus* are very close and were considered conspecific by Hellmayr (*Birds of the Americas,* vol. 7). Gyldenstolpe (1945a) later provided evidence indicating that they were apparently sympatric at several localities in the Brazilian Amazon. In *Birds of the World,* vol. 10, these 2 were thus considered full species; there, however, the *obsoletus* group was left as part of *T. fumigatus* rather than of *T. hauxwelli* (to which it is more allied). Meyer de Schauensee (1966, 1970) continued to lump *obsoletus* and *hauxwelli,* leaving *fumigatus* as a species. The latest person to analyze the group was D. Snow (*Bull. B.O.C.* 105 [1]: 30–37, 1985). He was particularly concerned with the *fumigatus/hauxwelli* zone of contact, where he found considerable morphological variation, though "individuals of either type tend to birds of their own type." Despite this, he recommended treating all forms as conspecific (including, by extension, *obsoletus* as well). We feel, however, that the evidence does not yet support this treatment and favor leaving all 3 as allospecies. *T. obsoletus* differs strikingly from both *hauxwelli* and *fumigatus* in its montane habitat and elevation preferences; we consider *columbianus,* together with *parambanus,* to be races of *T. obsoletus. T. hauxwelli* remains a monotypic species of upper Amazonia, with *T. fumigatus* replacing it eastward and northward; we consider *orinocensis* to be a race of *T. fumigatus,* in which it was originally described (though it has since usually been assigned to *T. obsoletus,* as in Meyer de Schauensee and Phelps 1978).

Turdus obsoletus

PALE-VENTED THRUSH

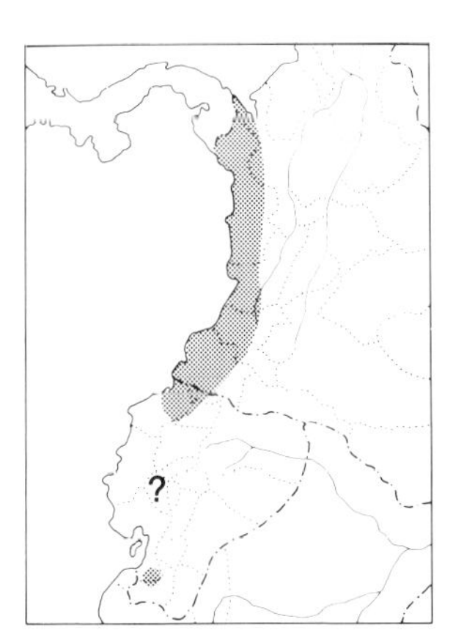

IDENTIFICATION: 23 cm (9"). *W. Colombia and w. Ecuador.* Closely resembles Hauxwell's Thrush (no overlap) but has duller and somewhat darker brown upperparts, and *crissum is always pure white.* See C-43, P-23.

SIMILAR SPECIES: Does not occur with either Hauxwell's or Cocoa Thrush. No other thrush in Pale-vented's range is such a dark brown with a contrasting white lower belly and crissum; cf. especially Black-billed Thrush.

HABITAT AND BEHAVIOR: Rare to uncommon in lower and middle growth inside humid foothill forest; infrequently seen at forest edge. Shy and not often seen. The song in nw. Ecuador is a fairly fast melodic caroling usually delivered from a hidden perch high in the forest canopy.

RANGE: W. Colombia (mostly on w. slope of W. Andes, spilling over to e. slope locally in Valle) and w. Ecuador (Esmeraldas and Pichincha, with 1 recent record from El Oro). Also Costa Rica and Panama. Mostly 500–1500 m.

NOTE: See comments under Hauxwell's Thrush.

Turdus fumigatus

COCOA THRUSH

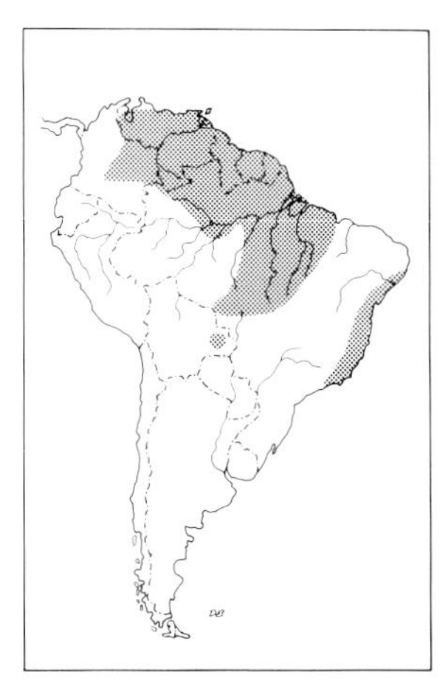

IDENTIFICATION: 23 cm (9"). *N. and e. South America.* Resembles Hauxwell's Thrush, but *warmer and more rufescent brown* generally (especially across breast) and typically with much buffier median belly and crissum (though there is variation, and some Hauxwell's Thrushes can have this area quite strongly washed with buff, while in some Cocoas it can be quite whitish). The 2 overlap only marginally, so far as known. See V-32.
SIMILAR SPECIES: The most uniformly rufescent thrush in its range. White-necked Thrush, which regularly occurs with it, is much grayer below, has a sharply streaked throat, etc.
HABITAT AND BEHAVIOR: Uncommon to locally common in lower and middle growth of forest and forest borders and in adjacent clearings with scattered trees. Found especially near water—near streams, in swampy areas, or in *várzea* forest. Like the other members of this group, it is generally a shy and not often seen bird, quick to retreat to cover when disturbed; seems tamer and more conspicuous on Trinidad. Its song is loud and of good quality: the phrasing more or less typical of the genus (often with the same note or phrase repeated a number of times), the quality usually rich and musical. When breeding, singing is more prevalent, and toward dusk at that season one can sometimes hear lovely choruses of Cocoa Thrushes, all of them remaining invisible in the gathering gloom.
RANGE: Extreme ne. Colombia (Norte de Santander), n. and e. Venezuela (Maracaibo basin east to Delta Amacuro and Bolívar), Guianas, and e. Amaz. Brazil (west to lower Rio Negro and Rio Madeira and south to cen. Mato Grosso and Goiás); e. coastal Brazil (Pernambuco to Rio de Janeiro); Trinidad. Questionably recorded from ne. Bolivia (Santa Cruz). Also some of Lesser Antilles. Mostly below 1000 m, locally higher (in Venezuela recorded to 1800 m).

NOTE: See comments under Hauxwell's Thrush.

Turdus lawrencii

LAWRENCE'S THRUSH

PLATE: 6

IDENTIFICATION: 23 cm (9"). *W. Amazonia.* Male has *bill bright orange-yellow,* sometimes tipped dusky, and *wide eye-ring* (much more conspicuous than in other *Turdus* except for Bare-eyed) *also bright orange-yellow.* Female has bill blackish and narrower eye-ring. Above dark brown, duskier on wings and tail. Below mostly paler, more olivaceous brown; throat whitish streaked with blackish, and median belly and crissum whitish.

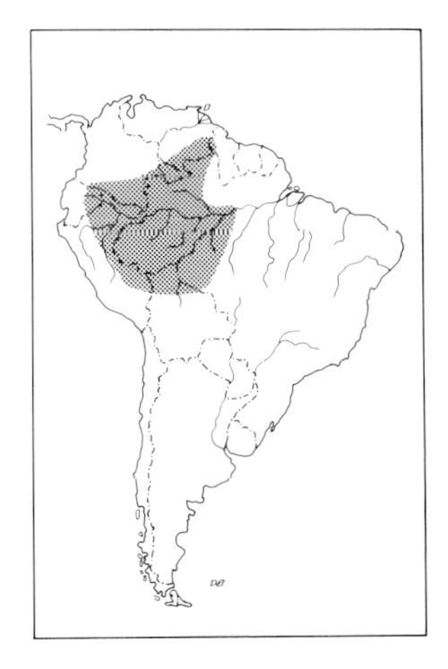

SIMILAR SPECIES: Most apt to be confused with Hauxwell's Thrush, with which it often occurs, but Hauxwell's is more rufescent and lacks the bold, obvious eye-ring. Eye-ring also should distinguish female (with dark bill) from Black-billed Thrush; note also Lawrence's *lack* of a white patch on upper chest.
HABITAT AND BEHAVIOR: Uncommon to fairly common in humid forest, both *terra firme* and *várzea;* will be much overlooked until its remarkable song is recognized (see below). Sometimes feeds on the ground in damp leaf litter near swamps or forest streams, but usually stays well above the ground and, like so many other forest thrushes, is very shy. To T. Parker must go the credit for discovering that Lawrence's Thrush is one of the most remarkable bird mimics of the world. At places such as the Explorer's Inn in se. Peru, males are heard day after day singing from favored perches high in the canopy (though even then they are frustratingly hard to spot unless they chance to fly and land in the open). The song is a near-endless series of near-perfect imitations of portions (usually short) of the songs and call notes of many other birds, ranging from tinamous and parrots to antbirds and grosbeaks. Parker has noted that a single individual may give the vocalizations of up to 35 other species, though most birds seem to give fewer. Some phrases of its own are also included. Certain individuals sing virtually all day, even during the hottest midday hours. Now that its song is known, Lawrence's Thrushes are being found in far greater numbers elsewhere in w. Amazonia (it is still a rare bird in collections); its fame as an astonishingly accomplished mimic will doubtless also soon begin to spread.
RANGE: S. Venezuela (Bolívar and Amazonas), se. Colombia (north to w. Putumayo and Vaupés), e. Ecuador, e. Peru, n. Bolivia (Pando), w. Amaz. Brazil (east to just below the lower Rio Negro and to nw. Mato Grosso). Below 500 m.

Turdus nigriceps

ANDEAN SLATY-THRUSH

PLATE: 6

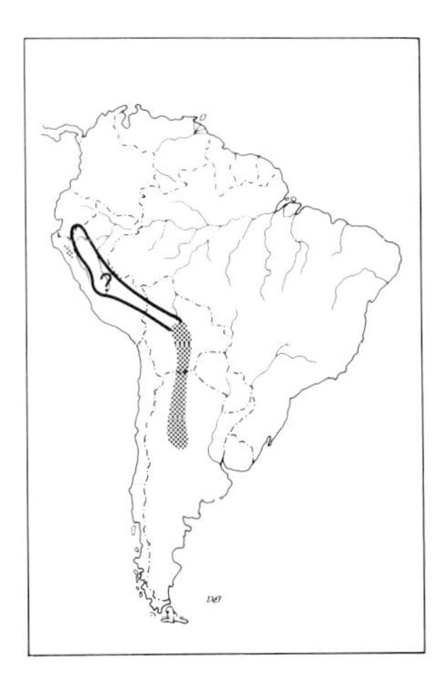

IDENTIFICATION: 21.5 cm (8½"). *A predominantly gray thrush of the Andes from s. Ecuador to nw. Argentina.* Bill, legs, and narrow eye-ring yellow in male; bill blackish, legs more brownish, and eye-ring also yellow in female. *Dark gray above,* shading to black on crown. Throat white sharply streaked black; *breast and flanks gray* (paler than upperparts), becoming white on median belly and crissum. Female patterned much like male but basically *brown* rather than gray, more olivaceous brown above, with brownish breast but belly and flanks grayish.
SIMILAR SPECIES: In its range the gray male is distinctive, but the female can be more of a problem. Cf. especially female Pale-eyed and Glossy-black Thrushes, both of which can be sympatric with this species. White-necked Thrush is grayer below (especially across breast) than female Andean Slaty and has prominent white crescent on upper chest (lacking or at most vague in Andean Slaty).
HABITAT AND BEHAVIOR: Locally common in canopy and borders of humid montane forest. Much less numerous northward but locally

abundant, at least during austral spring (Oct.–Dec.) in nw. Argentina (e.g., along the road to Tafí del Valle in Tucumán and at El Rey Nat. Park in Salta). An arboreal thrush, exceptionally shy and difficult to see, even when singing. Its song is a series of rather high, jumbled phrases, some of the notes quite high-pitched, given quite rapidly but then often followed by a long pause before the next delivery; it is *very* ventriloquial.

RANGE: E. slope of Andes in Peru (north to Piura, Lambayeque, and Amazonas), Bolivia, and nw. Argentina (south to n. Catamarca and Córdoba); 1 old record from s. Ecuador (no recent reports). Recent evidence (*fide* T. Schulenberg) indicates that records from north of about Cochabamba, Bolivia, pertain at least primarily to austral migrants (though there is evidently a breeding population in nw. Peru). Mostly 500–2000 m, perhaps occasionally lower in austral winter.

NOTE: We believe that *T. subalaris* of se. South America deserves to be specifically separated from *T. nigriceps* of the Andes. Obviously well isolated geographically, they also differ in plumage and have rather different songs, that of Andean being more jumbled and musical (not so squeaky). The English names we employ reflect the fact that *T. nigriceps* and *T. subalaris* are clearly allies (probably best considered a superspecies); both are thus best called Slaty-Thrushes, and a geographical modifier then seems appropriate.

Turdus subalaris

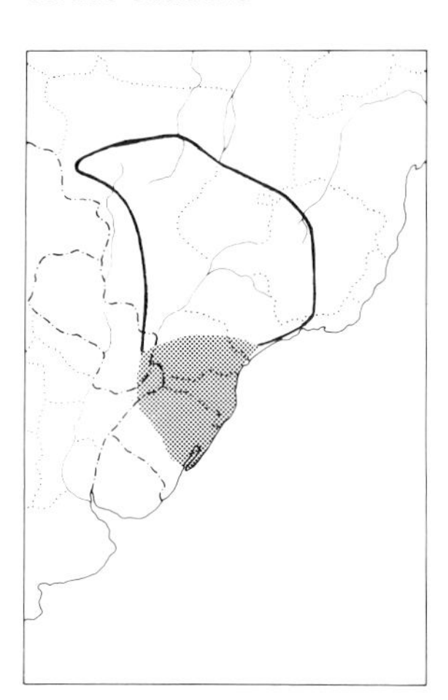

EASTERN SLATY-THRUSH

IDENTIFICATION: 21.5 cm (8½"). *S. Brazil and adjacent areas.* In general coloration and pattern resembles Andean Slaty-Thrush but paler. Male has orange-yellow bill, sometimes with blackish tip, legs yellowish brown to dusky, narrow yellow eye-ring; female like male but with bill yellowish brown. Compared to Andean Slaty, male is paler gray above, often washed olivaceous, and crown is concolor (not black); it also has a prominent white crescent on upper chest below the throat streaking (lacking in Andean), median belly is more extensively white, and it has white (not gray) under wing-coverts. Female Eastern Slaty is grayish brown above; its throat streaking is reduced and not as dark, emphasizing a *prominent white crescent on upper chest;* below mostly pale brownish gray, becoming mostly white on crissum.

SIMILAR SPECIES: The gray male should be distinctive in its range. Female Eastern Slaty-Thrush closely resembles Black-billed Thrush of Amazonia (no overlap known), but bill color differs. Sympatric races of White-necked Thrush show rufous flanks but are otherwise rather similar to this species (with same prominent chest crescent, though throat streaking is less dense). Cf. also Creamy-bellied Thrush (larger, with blackish lores, typically different habitat, etc.).

HABITAT AND BEHAVIOR: Locally common in woodland and forest canopy, gardens with scattered large trees, and parklike areas and plantations (even of eucalyptus sometimes). Apparently has extended its breeding range southward, having not been found in Rio Grande do Sul, Brazil, prior to the 20th century; now it is numerous and widespread in that state. Behavior is similar to that of other basically arboreal, forest-based thrushes; not as shy and wary as the Andean Slaty-

Thrush. Its distinctive song is rather weak, consisting of a short series of high-pitched notes with an oddly squeaky, bell-like quality; it is typically delivered from a hidden perch in the canopy.

RANGE: S. Brazil (north to cen. Mato Grosso, Goiás, and w. Minas Gerais), e. Paraguay (Alto Paraná; no recent records), and ne. Argentina (Misiones). Apparently does not breed in São Paulo or northward, occurring here only during austral winter (though some linger into Oct. and even sing), but exact n. limit of breeding range uncertain. To about 1000 m.

GROUP C

Two distinctive endemic thrushes of w. Ecuador/nw. Peru.

Turdus reevei

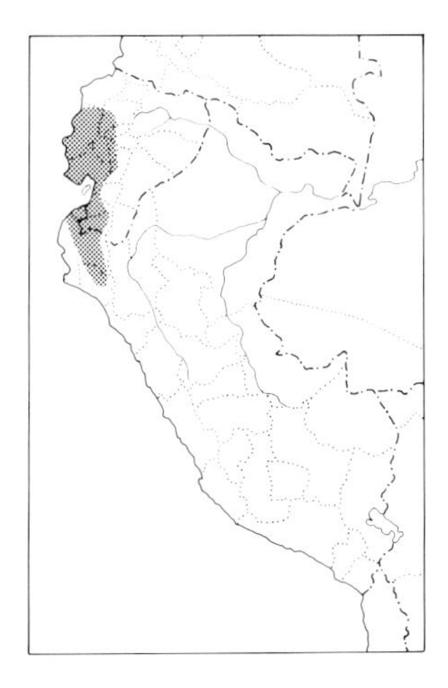

PLUMBEOUS-BACKED THRUSH

PLATE: 6

IDENTIFICATION: 23 cm (9″). *W. Ecuador and nw. Peru.* Bill and legs yellow; *iris bluish white. Above blue-gray,* dullest on head. Throat white, sharply streaked blackish; *remaining underparts pale grayish white,* washed creamy-buff on flanks, becoming white on median belly and crissum.

SIMILAR SPECIES: This handsome, essentially bicolored thrush with its very contrasty pale eye is virtually unmistakable.

HABITAT AND BEHAVIOR: Uncommon to locally (perhaps seasonally) common in forest and woodland, both deciduous and humid, and in borders and adjacent cutover terrain. Mostly arboreal, often gathering in sizable concentrations in fruiting trees. Numerous in the Chongon Hills west of Guayaquil, but even commoner farther south in Ecuador (e.g., in El Oro); its seasonal movements and shifts in abundance are marked but are still not fully worked out (in drier areas primarily during rainy season in the first half of the year).

RANGE: W. Ecuador (Manabí and Pichincha south) and nw. Peru (south to Lambayeque). Mostly below 1500 m (but recorded to 2300 m in Ecuador).

Turdus maranonicus

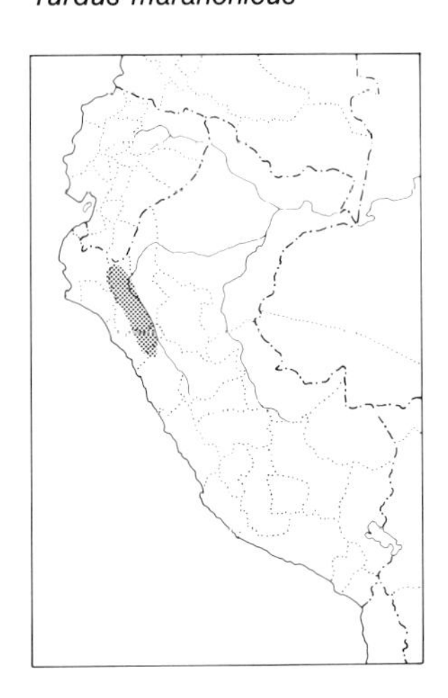

MARANON THRUSH

PLATE: 6

IDENTIFICATION: 21.5 cm (8½″). *Marañón valley of nw. Peru.* Bill olive brown; legs bluish gray; narrow eye-ring bluish gray. Brown above, feathers of crown and back indistinctly edged darker (giving vague scaly effect). White below, *profusely and boldly spotted and scaled with brown* except on crissum.

SIMILAR SPECIES: No other thrush at all resembles this striking Marañón endemic. When searching for it do remember that all thrushes are spotted when young (though none really matches the pattern of this species, being *unspotted above* and so heavily spotted below).

HABITAT AND BEHAVIOR: Apparently not uncommon in deciduous woodland and arid scrub, but virtually no behavioral information is on record.

RANGE: Interior nw. Peru in drainage of upper Marañón River (mostly in Cajamarca, also La Libertad and Piura). Mostly 200–2000 m.

Cinclidae

DIPPERS

TWO SPECIES of dippers are found in South America, where their presence is strictly tied to swiftly flowing montane streams and rivers. Other species are found in Eurasia and North America.

Cinclus Dippers

A pair of distinctive, small, chunky birds, members of the only truly aquatic passeriform family. They are entirely restricted to rocky streams and show a number of characters which clearly are adaptations to this habitat: large feet, dense plumage, an unusually large oil gland, and a very well developed nictitating membrane. Other, better studied members of the family are known to both walk and swim underwater.

Cinclus leucocephalus

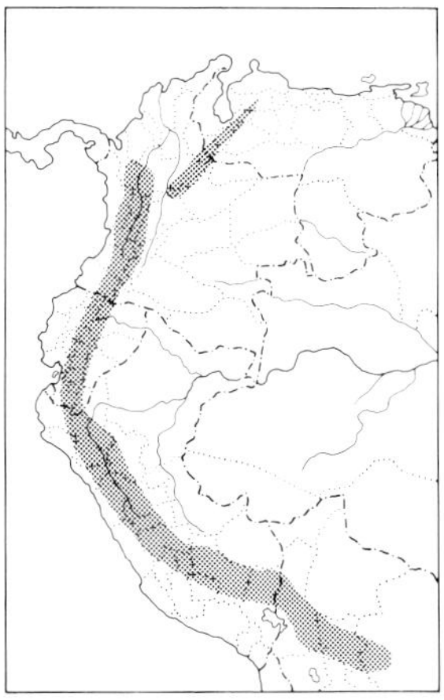

WHITE-CAPPED DIPPER

PLATE: 7

IDENTIFICATION: 15.5 cm (6"). *A boldly patterned, dark and white dipper of the Andes south to Bolivia.* Crown and nape, throat and chest white; otherwise sooty black. *Leuconotus* (Andes south to Ecuador) is quite different, being predominantly white below (blackish confined to sides of chest and lower belly) and with large patch of white on center of back; see C-p. 529. *Rivularis* (Colombia's Santa Marta Mts.), despite its n. range, resembles the s. nominate race in pattern but is grayer (not so black).

SIMILAR SPECIES: Essentially unmistakable in its habitat (fast-flowing streams and rivers). Cf. Rufous-throated Dipper of Argentina and southernmost Bolivia (no known overlap).

HABITAT AND BEHAVIOR: Fairly common along rushing mountain streams and rivers with boulders; sometimes found along surprisingly small rivulets. Occurs in both forested and mostly open areas (e.g., in *páramo*). Usually noted as they perch at water's edge on a rock out in the middle of the torrent or on the shoreline. Presumably they plunge underwater to feed, as do dippers elsewhere, but this seems to be reported infrequently at best; they also pick at insects at the surface. Flight is low over the water, with wings beating rapidly. Infrequently heard song is a loud, musical trill, often hard to hear over the roar of the rushing water.

RANGE: Andes of w. Venezuela (north to Lara), Colombia, Ecuador, Peru (south on w. slope to Arequipa), and nw. Bolivia (south to w. Santa Cruz); Santa Marta Mts. of Colombia and Perijá Mts. on Colombia-Venezuela border. Mostly 1000–3400 m.

Cinclus schulzi

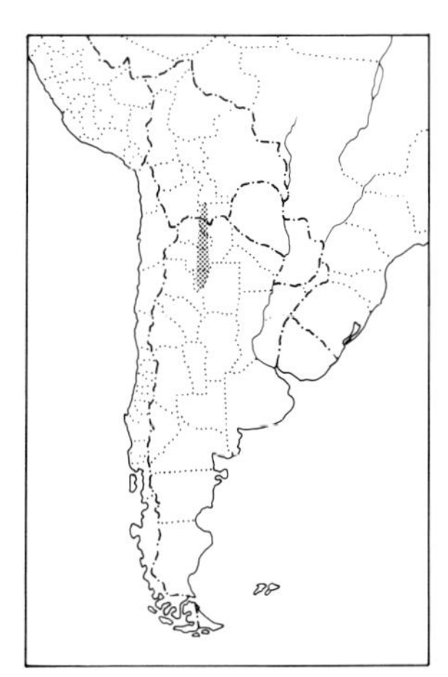

RUFOUS-THROATED DIPPER

PLATE: 7

IDENTIFICATION: 15.5 cm (6"). *Andes of nw. Argentina and adjacent Bolivia.* Mostly *dark leaden gray* with *pale rufous throat patch.* Shows a white wing-patch in flight.

SIMILAR SPECIES: Unmistakable in its habitat and limited range. Shows none of the white of White-capped Dipper (found to the north).

HABITAT AND BEHAVIOR: Rare to uncommon and very local along rushing mountain streams and rivers. Not well known or especially numerous; may be declining due to despoliation of many streams and rivers caused by mining and overgrazing. Can be seen along the road to Tafí del Valle, Tucumán, and along the Río Yala in Jujuy. It is difficult to comprehend why this or some other species of dipper has not extended to the seemingly ideal and abundant habitat found in the Andes from cen. Chile and Argentina south to Tierra del Fuego.

RANGE: E. slope of Andes in s. Bolivia (Tarija) and nw. Argentina (south to Catamarca and Tucumán). It remains to be determined which dipper species inhabits the Andes between Tarija and Santa Cruz, Bolivia; as appropriate habitat exists in this region, presumably one or the other does. 800–2500 m.

NOTE: The correct spelling of the species name (as determined by checking the original description) is *schulzi,* not *schultzi* (as in Meyer de Schauensee 1966, 1970) or *shulzii* (as in *Birds of the World,* vol. 9).

Mimidae

MOCKINGBIRDS, THRASHERS, AND ALLIES

THE mockingbirds are rather thrushlike, long-tailed birds found in South America mainly in the s. part of the continent; no true *Toxostoma* thrashers are found south of Mexico, with others (*Margarops*) in the West Indies and Netherlands Antilles. Long thought most closely related to the wrens and the thrushes, but recent DNA evidence suggests they may, surprisingly, be even closer to the starlings (Sturnidae).

Mimus Mockingbirds

The mockingbirds comprise a group of slender, long-tailed birds which reaches its highest development in s. South America, though 1 species (*M. polyglottos*) is familiar across much of the United States. Despite being somberly colored with various shades of brown, gray, and white (often with bold white areas in the wings and tail), they tend to be well-known birds, mostly because of their renown as superb songsters and mimics of other birds. Irides of adults are usually yellow, darker in juvenals; juvenals of all species are spotted below and often streaked on flanks. It might be pointed out that the Pacific slope species (*M. longicaudatus* and *M. thenca*) seem quite close to the complex of mockingbirds found on the Galápagos Is., currently still separated in the genus *Nesomimus* (a separation perhaps not justified, or if maintained then *longicaudatus* and *thenca* could belong there).

GROUP A

Typical mockingbirds.

Mimus gilvus

TROPICAL MOCKINGBIRD

PLATE: 7

IDENTIFICATION: 24 cm (9½"). Mostly *pale gray above* with dusky patch through eye surmounted by white superciliary. Wings blackish, feathers edged whitish, and with 2 narrow wing-bars; tail blackish, broadly tipped white. Pale grayish white below.

SIMILAR SPECIES: *The only mockingbird in n. South America,* but range nearly contacts that of Chalk-browed in the Guianas and the 2 species overlap in e. Brazil. Chalk-browed should, however, be readily known by its decidedly browner upperparts and more prominent wider superciliary (especially behind the eye).

HABITAT AND BEHAVIOR: Fairly common to common in semiopen country, savannas, gardens and parks in towns, and arid coastal scrub. Seems especially conspicuous and numerous around habitations, where it is often quite tame, hopping about unconcernedly on lawns, etc.

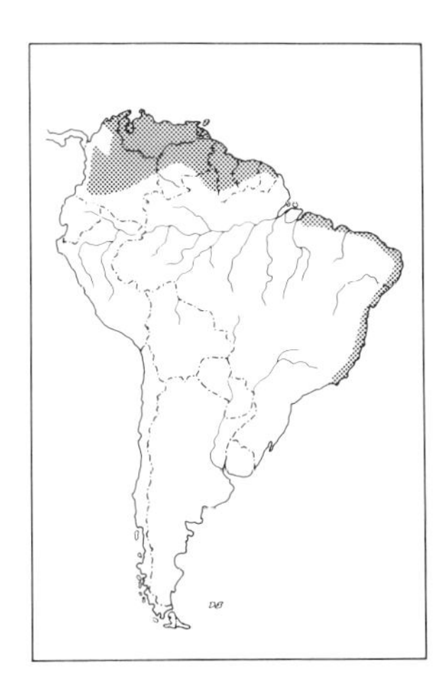

Often lifts and partially spreads its wings: termed "wing-flashing," the function of this display is uncertain. Has a fine song consisting of a series of usually musical phrases, each often repeated several times, together with a few clucks or wheezy notes; unlike the well-known Northern Mockingbird (*M. polyglottos*), only rarely does it mimic other species. Sometimes sings at night.

RANGE: W. and n. Colombia (west of the Andes south to upper Río Patía valley in s. Cauca and to upper Magdalena valley in Huila; east of them south to Meta and Vichada), most of Venezuela (not in Amazonas), Guianas, and extreme n. Brazil (Roraima); coastal e. Brazil (from south of the mouth of the Amazon in e. Pará south to Rio de Janeiro); recently (1987) reported from nw. Ecuador in Pichincha (P. Greenfield); Trinidad and Tobago. Also Mexico to Honduras; introduced to Panama. Mostly below 1500 m, but ranging to 2500 m in cleared areas of Colombian Andes.

Mimus saturninus

CHALK-BROWED MOCKINGBIRD

PLATE: 7

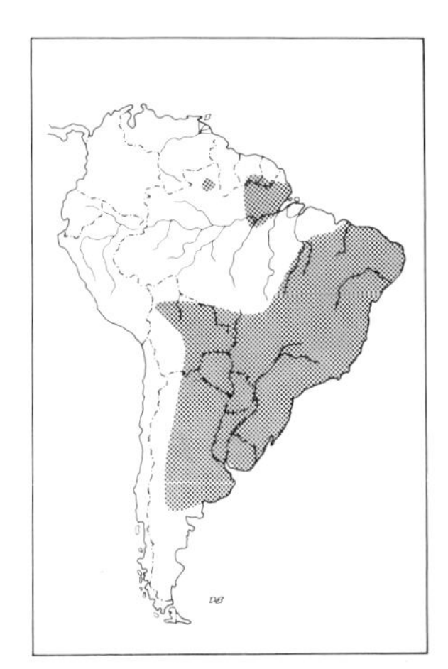

IDENTIFICATION: 26 cm (10¼"). Mostly grayish brown above (more rufescent on rump) with *prominent broad white superciliary* (widest behind the eye) and narrow blackish patch through eye; crown and back feathers edged paler, usually imparting a streaky effect (especially in fresh plumage). Wing feathers edged whitish (often with effect of 2 narrow wing-bars); tail dusky, with outer feathers broadly tipped white, forming *conspicuous white tail corners* (especially prominent in flight). Below dingy grayish white to buffy whitish, often with some vague streaking on flanks. Some birds are strongly tinged rufescent, apparently as a result of dust bathing.

SIMILAR SPECIES: Tropical Mockingbird similar overall but notably grayer (not so brown) and smoother-looking (lacking the streaky effect) above. Cf. also Patagonian, White-banded, and Brown-backed Mockingbirds.

HABITAT AND BEHAVIOR: Common in semiopen areas and around habitations; absent from regions entirely devoid of trees and from forested areas, but present just about anywhere else (even moving into recently deforested areas, e.g., in Espírito Santo). Bold and often inquisitive, these mockingbirds forage mostly on the ground but also regularly perch in bushes and low trees; its tail is often held raised over its back, especially soon after alighting on a perch. During the breeding season males may sing for long periods from an exposed perch; the song is exceptionally variable, consisting of numerous phrases usually without a recognizable pattern, though many notes or phrases are often repeated several times. Portions of the song seem to be its own while others seem to be imitations of other species (often so modified as to be difficult to recognize). Occasionally a singing bird almost seems to be overcome by exuberance, and it bursts several meters into the air and then drops back onto its perch with slow butterflylike flaps, all the while continuing to sing. Its frequently heard call is a sharp "chert."

RANGE: Ne. Brazil along lower Amazon and in Amapá, and in s. Suriname (Sipaliwini); e. and s. Brazil (north to Maranhão, se. Pará, and cen. Mato Grosso), Bolivia (west to Beni), Paraguay, Uruguay, and n. Argentina (south to Mendoza and Río Negro). Mostly below 1000 m (but ranging up to at least 2500 m in dry intermontane valleys of nw. Argentina).

Mimus patagonicus

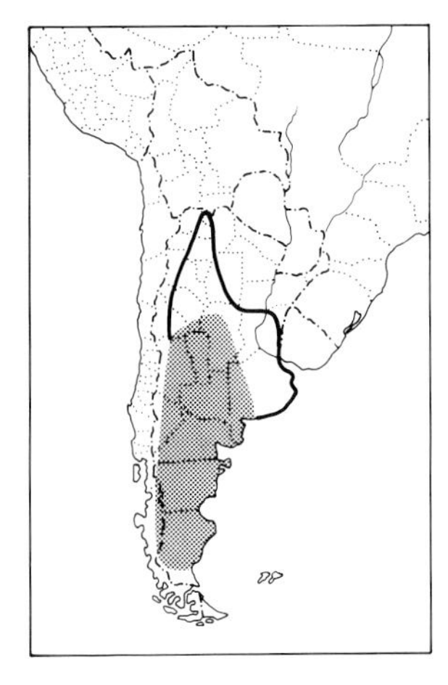

PATAGONIAN MOCKINGBIRD

IDENTIFICATION: 25 cm (9¾″). *Mainly Argentina* (also adjacent Chile). Resembles Chalk-browed Mockingbird (and locally sympatric with it, particularly during austral winter). Best distinguished by less prominent white superciliary, more or less uniform (*no streaking*) brownish gray mantle, bolder white wing-bars and pale wing-edgings, narrow blackish malar streak, *distinct grayish buff wash across breast,* and buffier lower underparts. Patagonian is also slightly smaller and proportionately shorter-tailed.
SIMILAR SPECIES: Besides the Chalk-browed, cf. also White-banded Mockingbird (with much more white in wing and tail, etc.).
HABITAT AND BEHAVIOR: Fairly common to common in bushy Patagonian steppes; tends not to be found around settlements (unlike Chalk-browed). General behavior similar to Chalk-browed Mockingbird's but tends to drop to the ground less often; most often seen perched on top of a bush, quietly surveying its surroundings. Vocally it is also similar to the Chalk-browed, though its song is usually not as loud or forceful.
RANGE: Breeds in cen. and s. Argentina (from Córdoba and s. and w. Buenos Aires south to Santa Cruz) and in adjacent s. Chile (Aysén and Magellanes, casually to n. Tierra del Fuego); in austral winter ranges north in Argentina to Entre Ríos, Salta, and Jujuy. Mostly below 500 m.

Mimus triurus

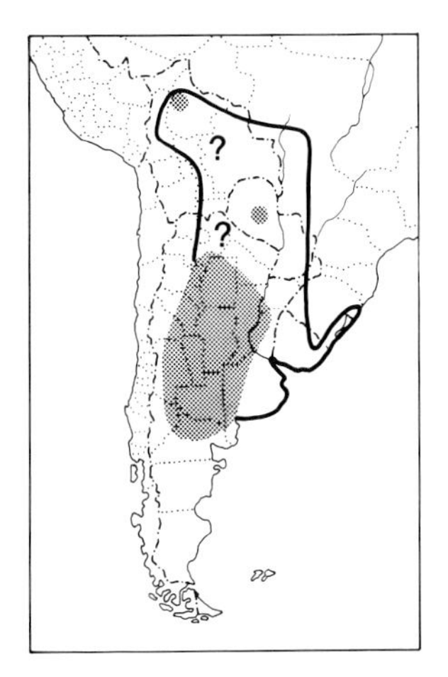

WHITE-BANDED MOCKINGBIRD PLATE: 7

IDENTIFICATION: 23.5 cm (9¼″). Grayish brown above, becoming quite rufescent on rump; long white superciliary, narrow dusky patch through eye, and mottled whitish area below eye. Wings mostly black, but with *long, broad white stripe extending from shoulder to tip of outer secondaries* (obvious even at rest and very conspicuous in flight, when *middle part of wing mostly white*); *tail mostly white* except for black middle pair of feathers. Mostly whitish below, washed with grayish buff across breast and tinged ochraceous on flanks and crissum.
SIMILAR SPECIES: Chalk-browed and Patagonian Mockingbirds both have very much less white in wing and tail. White-banded most resembles Brown-backed Mockingbird; their overall plumage patterns are similar but wings differ: instead of a long white stripe, Brown-backed has just a big *patch* on primary coverts and base of primaries. The 2 overlap at most marginally (only during austral winter, in any case).

HABITAT AND BEHAVIOR: Fairly common to common in low woodland, Patagonian steppe shrubbery, and semiopen country with scattered bushes and trees; at least when not breeding, also often around habitations (in situations similar to those of the Chalk-browed). General comportment typical of other mockingbirds, but its song on the breeding grounds apparently has few rivals (W. H. Hudson refers to it as a "diamond among stones"): it is a long-continued stream of strong, pure, melodic notes interspersed with near-perfect imitations of various other songbirds, not only species with which it breeds in s.-cen. Argentina, but also of various others which it evidently learns during its winter sojourn to the north. A singing bird most often remains hidden inside shrubbery, but on occasion, as if overcome by enthusiasm, it may suddenly mount high into the air and then gradually descend with slow, languid, butterflylike flapping, all the while continuing to sing. When not nesting, however, it only gives 1 or 2 short phrases of its incomparable song.

RANGE: Apparently breeds mainly in cen. and n. Argentina (Tucumán and Entre Ríos south to Río Negro; exact n. breeding limits uncertain), with 1 breeding record from w. Paraguay (50 km south of Orloff) and a few reports from presumed breeding season (Nov.–Dec.) in Beni, Bolivia (J. V. Remsen); in austral winter ranges north through remainder of Argentina through Uruguay and Paraguay to s. Brazil (Rio Grande do Sul and w. and s. Mato Grosso) and Bolivia (north to s. Beni, Cochabamba, and Santa Cruz). Mostly below 500 m, though higher during austral winter northward (to 1950 m in Bolivia).

Mimus dorsalis

BROWN-BACKED MOCKINGBIRD

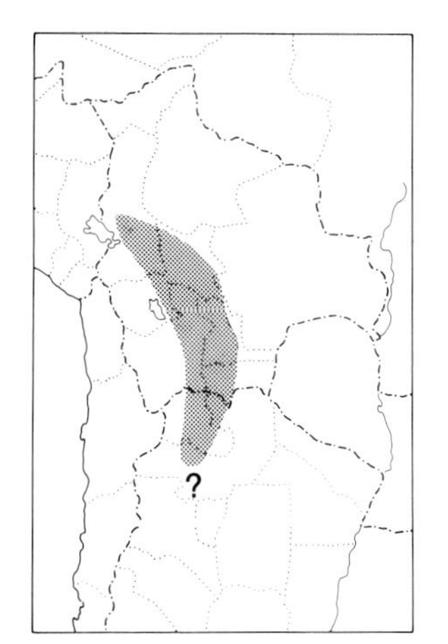

IDENTIFICATION: 25.5 cm (10"). *Highlands of Bolivia and nw. Argentina.* Resembles White-banded Mockingbird but larger and with markedly longer and heavier bill; *more uniformly rufescent brown above* (lacking grayish tone to White-banded's crown and upper back); different wing pattern: *white restricted to large patch on primary coverts and base of primaries* (not extending as long diagonal stripe).

SIMILAR SPECIES: As Brown-backed and White-banded Mockingbirds are not likely to occur together, Brown-backed will most likely be confused with Chalk-browed Mockingbird (which at least in Argentina ranges locally up into the habitat of this species). Chalk-browed differs, however, in its tail pattern (only the corners being white, not the *entire* outer rectrices); more streaked or mottled (not so smooth-looking) upperparts; and in having much less white in wing (only narrow edging and wing-bars; no large white patch).

HABITAT AND BEHAVIOR: Fairly common in arid montane scrub (often with tall cactus) and in agricultural areas with hedgerows and scattered bushes and trees. Behavior similar to other mockingbirds', but so far as known its song does not come close to equaling White-banded's, lacking its strength and variety. Readily seen in Jujuy, Argentina, along the road to Huancabamba.

RANGE: Highlands of Bolivia (La Paz south to Potosí and Tarija) and nw. Argentina (Jujuy and Salta; uncertainly recorded from Tucumán). Recently reported seen once (in 1986) at Putre in extreme n. Chile (Fjeldså 1987). Mostly 2300–3500 m.

GROUP B

Larger mockingbirds with *bold black "whisker"; Pacific slope only.*

Mimus longicaudatus

LONG-TAILED MOCKINGBIRD

PLATE: 7

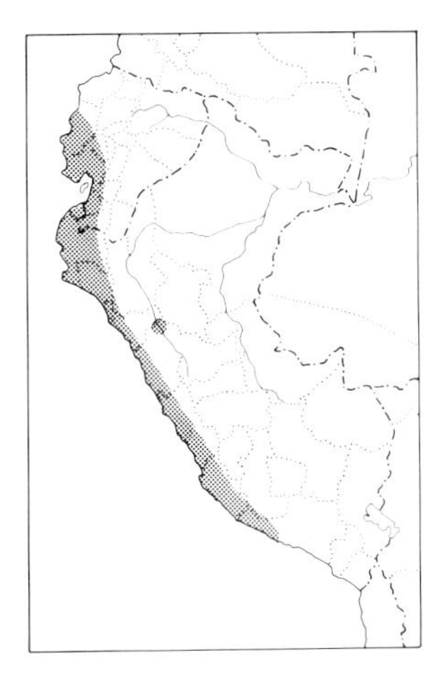

IDENTIFICATION: 29.5 cm (11½"). *W. Ecuador and w. Peru* (the only mockingbird in this area). Basically brownish gray above, feathers centered darker (giving streaked or mottled effect); long, broad superciliary white, with broad blackish eyestripe extending back to behind ear-coverts. Wings dusky brown with paler feather edging and small white patch on primary coverts; *exceptionally long tail* also dusky, all but central feathers broadly tipped white (forming conspicuous white "corners" in flight). Mostly white below, with breast brownish gray (feathers edged paler, often giving scaly effect) and usually some brown streaking on flanks; conspicuous but narrow black submalar streak, becoming scaly as it merges into chest. Nominate race (Pacific slope of Peru from La Libertad south) has smaller white tail corners.
SIMILAR SPECIES: Essentially unmistakable in its restricted range; compare to Chilean Mockingbird, found farther south.
HABITAT AND BEHAVIOR: Common in open arid woodland and shrubby areas (often where cactus is present) and in agricultural regions with hedgerows, brushy areas, or groves of trees. Regularly found in small groves or patches of sparse vegetation very isolated in an otherwise barren coastal desert. General behavior similar to other mockingbirds', though due to the open nature of its habitat, its large size (especially the long tail), and its proclivity for perching in exposed situations it usually seems more conspicuous than the others. Its flight style is, however, somewhat unusual in that it often glides low over the ground for surprisingly long distances, wings outstretched and tail spread. The song is loud and long-continued, with many chuckling and gurgling notes and little obvious pattern; often given right through the middle of the day.
RANGE: Sw. Ecuador (north to Manabí and inland to s. Loja; strictly avoids humid areas) and w. Peru (south to Arequipa on Pacific slope, also in the upper Marañón valley in La Libertad). A mockingbird, either Long-tailed or Chilean, was seen in 1987 near Arica in extreme n. Chile (Fjeldså 1987). To about 2000 m.

Mimus thenca

CHILEAN MOCKINGBIRD

IDENTIFICATION: 27 cm (10½"). *Chile.* Resembles Long-tailed Mockingbird but smaller, shorter-billed, and markedly shorter-tailed. *Browner and more uniform above* (not as gray, and lacking the streaked effect and boldly pied head pattern) and *buffier and more uniform below*

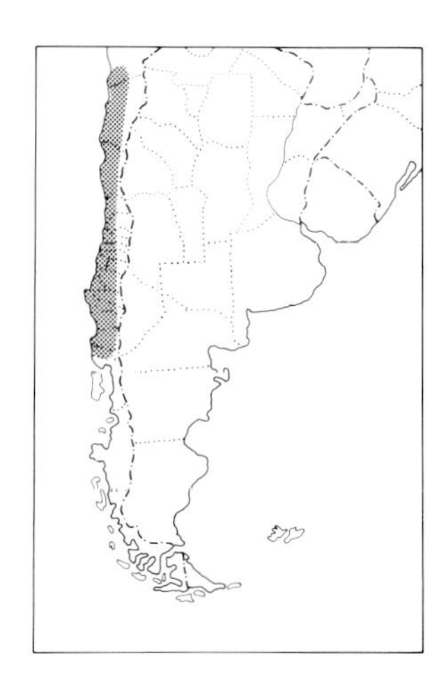

(lacking the scaly effect on breast); brown flank streaking more conspicuous; lacks the white wing-patch.
SIMILAR SPECIES: The only numerous, widespread mockingbird in Chile. Patagonian Mockingbird (found in far south, only as occasional winter visitor in cen. provinces) is smaller, shorter-tailed, and grayer above; has more white on wing and less conspicuous malar streak. The very different White-banded Mockingbird is also an occasional winter visitor.
HABITAT AND BEHAVIOR: Fairly common to common in shrubbery and matorral; tends not to be found (or in smaller numbers) in cultivated areas. General behavior similar to other mockingbirds', particularly the Long-tailed. Like the others, it has an excellent song (consisting not only of its own phrases but also imitations of other birds), which, though not of the quality of the White-banded's, is nonetheless considered to be the finest of any Chilean bird (A. W. Johnson).
RANGE: Cen. Chile (Atacama to Valdivia). Mostly below about 700 m.

Margarops Thrashers

One member of this West Indian genus extends to 2 islands off the n. coast of South America, on 1 of which it may be extinct. Unlikely ever to occur on the mainland.

Margarops fuscatus

PEARLY-EYED THRASHER

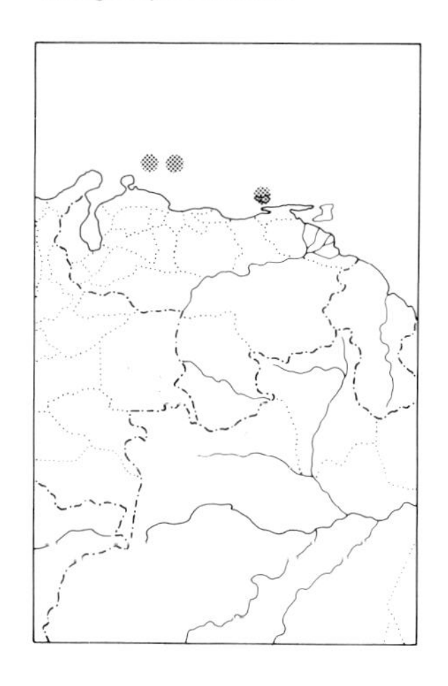

IDENTIFICATION: 27 cm (10½"). Only on *Bonaire.* Iris milky white; long heavy bill, yellowish flesh tipped dusky. *Brown above,* obscurely streaked darker; *outer tail feathers tipped white,* as are the tertials. Below whitish, *coarsely streaked and mottled with brown* except on middle of belly.
SIMILAR SPECIES: Nothing really similar on Bonaire. Compare to the accidental Brown Thrasher (see Appendix).
HABITAT AND BEHAVIOR: Uncommon in dense scrub and low woodland, apparently concentrating near water during dry periods. Tends to skulk in foliage and often is not easy to see well, though elsewhere in its range it is bolder and much more aggressive, even eating eggs and killing young of various other birds. Has a variety of harsh, raucous calls as well as a loud, pleasant thrushlike song (K. H. Voous).
RANGE: Netherlands Antilles (Bonaire; also once seen on Curaçao); formerly on La Orquilla Is. off n. Venezuela (where apparently not recorded since 1908, the reasons for its having disappeared being uncertain). Found mainly in West Indies.

Motacillidae

PIPITS AND WAGTAILS

OF THIS FAMILY, only pipits occur in most of the Americas, and in South America they basically are birds of the s. part of the continent, with several species extending north in the Andes. Only one, the Yellowish, occurs more widely.

Anthus Pipits

The pipits are slim, brown, predominantly streaked terrestrial birds found in grasslands. They are further characterized by their slender bills (which will distinguish them from the many streaked emberizid finches) and their notched tails with pale (usually white) outer feathers, which should serve to distinguish them from the many species of streaked canasteros. Many pipits show an eye-ringed effect. Identification to species level is often difficult, with problems compounded by the variation shown depending on molt stage; fortunately, subspecific variation is so trivial as to not further complicate the situation. Their legs are pale-colored, typically flesh or yellowish, and the hindclaw is notably long, the latter doubtless an adaptation to their characteristic walking habit. The behavior of all the S. Am. pipits is basically similar and will be described here so as to avoid repetition. They usually occur in pairs, even when not breeding being found in at most small, very loose groups, and are usually hard to see when on the ground. Once located they are often tame, sometimes even walking about unconcernedly almost at one's feet. Approached, they tend to crouch, then flush abruptly, often with some variant of a thin "ts-lik" call; usual flight is slow and undulating. All the neotropical species engage in display flights, and not only are these enjoyable performances for the observer but they also serve as a highly useful identification clue.

GROUP A

Smaller pipits; lowlands only; both *very* similar.

Anthus lutescens

YELLOWISH PIPIT

PLATE: 7

IDENTIFICATION: 13 cm (5"). A *small* pipit, *widespread in open tropical lowlands.* Above dark brown streaked with buff and blackish; tail dusky, with outer feathers white (conspicuous in flight). *Below buffy yellowish white,* with band of dark brown streaking across chest. *Peruvianus* of coastal Peru is less tinged with buffy yellow below. Birds of all races in worn plumage are more whitish below (lacking the buffy yellow tone). SIMILAR SPECIES: The only pipit over most of its range, and appreciably smaller than any other (except for the Chaco Pipit). The yellow cast

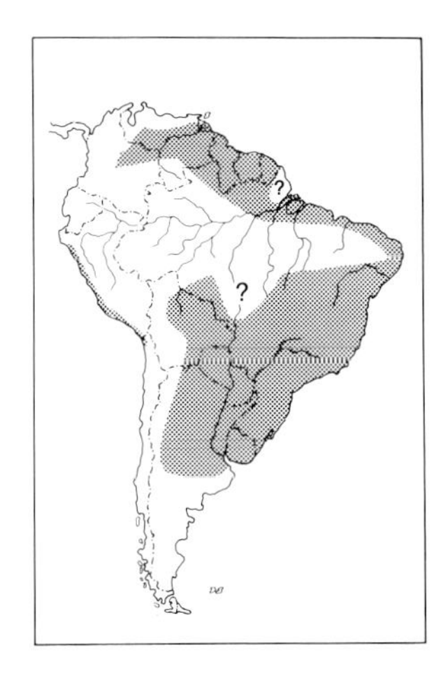

to the underparts is also distinctive, but cf. the even more yellow-ochraceous Ochre-breasted Pipit.

HABITAT AND BEHAVIOR: Locally fairly common to common in grasslands (generally where the grass is quite short) and adjacent barren areas; seems loosely colonial, even when breeding. Most in evidence when males are singing (which they seem to do for much of the year): its distinctive song is usually given in flight, initially a series of "tsits" as the bird ascends 10 to 20 m up into the air, then a long slurred "dzeeeeeeeeeeeeu" as it slowly glides back onto the ground. Also gives a shorter "tsitsirrit" from the ground.

RANGE: Coastal w. Peru (north to Lambayeque) and extreme n. Chile (Arica); ne. Colombia (south to Meta and Vichada), Venezuela (though absent from n. coastal areas and also from s. Amazonas), Guianas, e. and s. Brazil (up the Amazon to lower Rio Tapajós area), e. Bolivia (north and east to Beni), Paraguay, Uruguay, and n. Argentina (south to Mendoza, La Pampa, and Buenos Aires). Also Panama. To 1300 m.

Anthus chacoensis

CHACO PIPIT

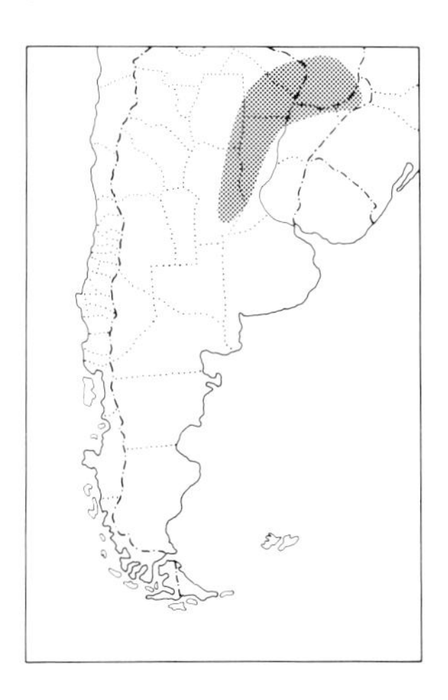

IDENTIFICATION: 13 cm (5"). *Very locally in s. Paraguay and n. Argentina. Closely* resembles Yellowish Pipit and probably not safe to distinguish in the field, at least on presently known characters; may prove not to be a valid species. As described, it differs in its slightly stronger and whiter (less buffy) streaking above, lower underparts with no trace of yellow, and more pronounced flank streaking. In the hand, Chaco also differs *very* slightly from Yellowish in its less extensive white in the outer tail feathers and in its shorter hindclaw.

HABITAT AND BEHAVIOR: Unknown, but probably differs little from Yellowish Pipit. Perhaps the best way to search for this enigmatic bird, and possibly even to confirm its identity as a separate species, would be to locate a Yellowish Pipit–type bird singing a song *differing* from the song so typical of that species.

RANGE: Locally in se. Paraguay (recorded only from Colonia Nueva Italia, south of Asunción) and n. Argentina (recorded from Formosa and Chaco, and there are AMNH specimens taken in Misiones at Barra Concepción and in Córdoba at Leones; the latter 5 specimens were collected in Jan. so were presumably breeding locally).

GROUP B

Larger pipits; Andes and s. lowlands.

Anthus furcatus

SHORT-BILLED PIPIT

PLATE: 7

IDENTIFICATION: 14.5 cm (5¾"). Above streaked and mottled blackish and buffy brown; tail dusky, with outer feathers white or whitish. Below mostly whitish, but breast and flanks strongly tinged ochraceous *with discrete area of broad, bold, blackish streaking across breast* (but streaking does *not* extend down the flanks); rather prominent black malar streak. In worn plumage more grayish below with breast streaking not as bold.

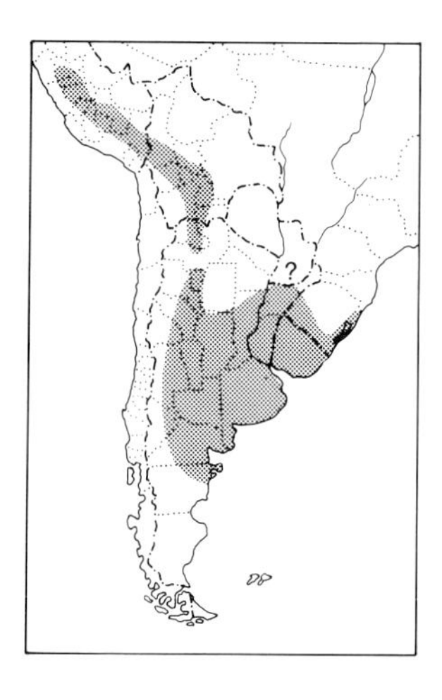

SIMILAR SPECIES: Resembles several other pipits, particularly Hellmayr's. Best distinguishing points include the following (bear in mind that all are subtle): Hellmayr's bolder back streaking (lacking the scaled or mottled effect usually evident in Short-billed), its sparse streaking on flanks (absent in Short-billed), and its narrower breast streaking; Short-billed's bill is slightly shorter and its malar streak is more obvious. Short-billed is also often confused with Correndera Pipit, but the latter is considerably more boldly streaked above (with pair of obvious longitudinal streaks down either side of back) and is also boldly streaked on flanks. Cf. also the rare Ochre-breasted Pipit (slightly smaller and much more orange-ochraceous below).

HABITAT AND BEHAVIOR: Uncommon to locally common in pastures and fields with short grass and in *puna* grasslands. Often in small groups when not breeding, but at that season pairs separate out (though a number may concentrate in favorable habitat). Seems most numerous in the Patagonian steppes of Río Negro, Argentina (*much* less numerous than the Correndera in the pampas of Buenos Aires). Displaying males have a fine, musical (though repetitious) song ("gleeeeeeeu, teedeleh-tleetleetlee" or some variation), given incessantly as the bird hovers high overhead, often almost invisible to the naked eye. W. Belton watched 1 bird which sang continuously for 55 minutes until a tape playback brought the pipit back to earth. This song and display will remind N. Am. observers of the Sprague's Pipit (*A. spragueii*), to which Short-billed is probably related.

RANGE: Andes of cen. and s. Peru (north to Huánuco), w. Bolivia, and nw. Argentina (Jujuy); lowlands and lower Andean slopes in n. and cen. Argentina (Tucumán, Mendoza, Río Negro, and n. Chubut east locally to Buenos Aires and Entre Ríos), Uruguay, and extreme s. Brazil (Rio Grande do Sul); questionably recorded from se. Paraguay. To at least 4000 m (in Peru and Bolivia).

Anthus hellmayri

HELLMAYR'S PIPIT

PLATE: 7

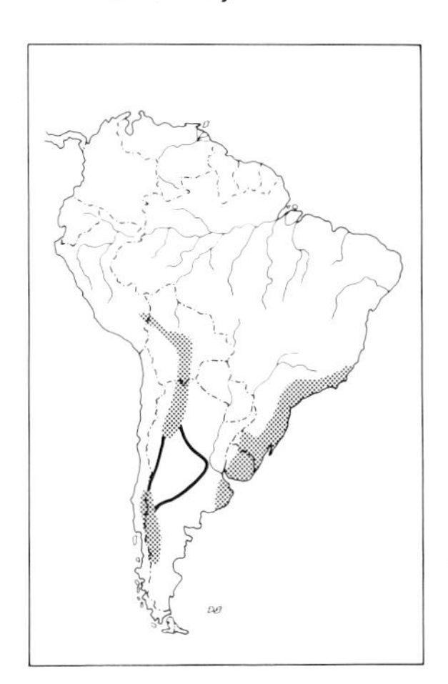

IDENTIFICATION: 14.5 cm (5¾"). Above boldly streaked blackish and light brown; tail dusky with outer feathers dull buff to buffy whitish. Below mostly dull buffy whitish, buffiest across breast; *band of rather narrow, sparse blackish streaking across breast,* with some sparse streaking also extending down flanks. In worn plumage becomes more grayish below, and outer tail feathers are duller and browner.

SIMILAR SPECIES: Hellmayr's has the *finest* breast streaking of any S. Am. pipit (but beware of some worn Short-billeds, which may approximate this, and also the Paramo Pipit, which has only sparse breast spotting). It closely resembles Short-billed Pipit, but back streaking is *bolder* and breast streaking *narrower,* and it has some flank streaking (lacking in Short-billed); also *lacks* a prominent malar streak. Hellmayr's also resembles Correndera Pipit, but latter is more broadly streaked on breast and has more pronounced streaking on flanks; streaking above is bolder, with a distinctive pair of longitudinal stripes down either side of back usually evident.

HABITAT AND BEHAVIOR: Locally uncommon to fairly common in grasslands, most often on dry and often rocky hillsides but also sometimes in moister, flatter terrain. Behavior similar to that of other pipits but seems as likely to sing from a perch (e.g., a rock or fence post) as during short display flights. Song in Brazil described as a short "tu-tee-deée-tu" (W. Belton). Perhaps most numerous in Rio Grande do Sul, Brazil, and also found above treeline on the Itatiaia Mts.; on the other hand, seems scarce or local in the Andes (possibly overlooked due to confusion with other pipits?).
RANGE: Andes of s. Peru (Puno), w. Bolivia, w. Argentina (south locally to Río Negro and w. Chubut, with perhaps isolated records of migrants in Córdoba and Entre Ríos), and s. Chile (Cautín); se. Brazil (Espírito Santo south to Rio Grande do Sul), Uruguay, and e. Argentina (Buenos Aires). Status in Argentina still poorly understood. To about 3700 m (in Peru and Bolivia); e. race (*brasilianus*) to about 2200 m.

Anthus nattereri

OCHRE-BREASTED PIPIT

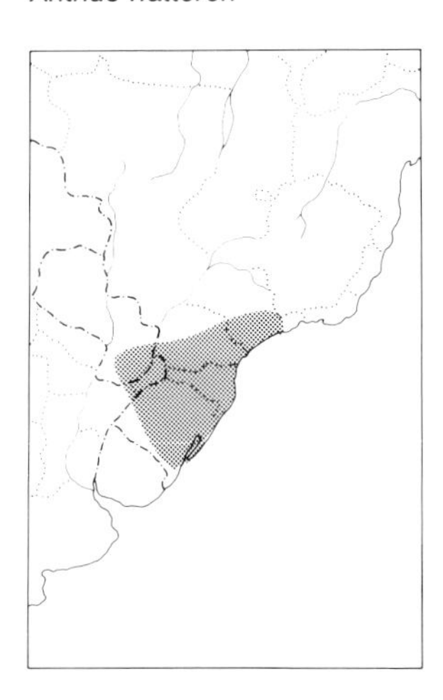

IDENTIFICATION: 13.5 cm (5¼"). A *rare* pipit of *se. South America.* Above boldly streaked blackish and golden-ochraceous; tail dusky with outer feathers white. Throat whitish, *becoming rich golden-ochraceous on breast* (fresh-plumaged birds have an almost orange tone), *breast with fairly heavy black streaking;* belly whitish.
SIMILAR SPECIES: Readily recognized by the strong, bright yellow-orange ground color on breast. Yellowish Pipit is merely tinged yellow below (and especially on belly) and is much less boldly streaked on breast.
HABITAT AND BEHAVIOR: Rare and very local (few specimens recorded) in dry pastures and fields and in rolling grasslands. Not well known, and ecologic relationships with other pipits still imperfectly understood (as well as reason for its rarity). Displaying birds have been recently recorded in se. Paraguay (south of San Patricio, Misiones, in Aug. 1977; RSR) and on 2 occasions in Rio Grande do Sul, Brazil (W. Belton). On all occasions the displays have been relatively short (for a pipit), the bird flying up to some 20 to 25 m above the ground and giving a musical and complex song, ending with a slurred "eeeeeeeur"; at the end of the performance the bird drops rapidly and gives several nasal notes. Possibly threatened by overgrazing or other forms of disturbance to its grassland habitat; there are very few recent reports of the species.
RANGE: Se. Brazil (s. São Paulo to Rio Grande do Sul), se. Paraguay, and ne. Argentina (Misiones and ne. Corrientes).

Anthus correndera

CORRENDERA PIPIT

PLATE: 7

IDENTIFICATION: 15 cm (6"). Above boldly streaked blackish and ochraceous, with a *pair of longitudinal whitish stripes down either side of back* usually evident (and often conspicuous); tail dusky, with outer

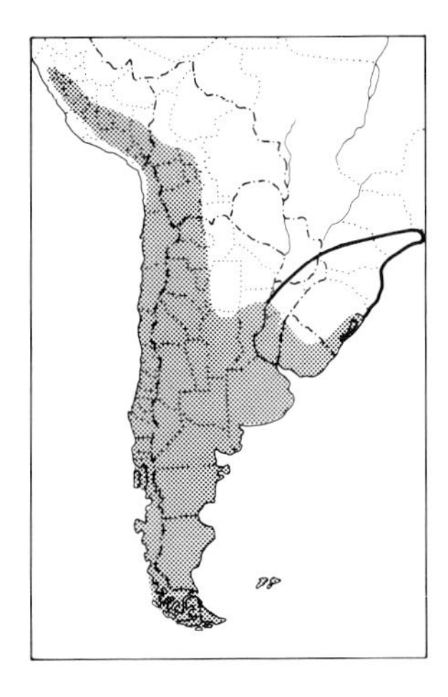

feathers white to whitish. Below buffy whitish to whitish (buffiest across breast and down flanks when in fresh plumage), with *bold blackish spotting across breast, extending down flanks as bold prominent streaking;* usually shows a narrow blackish malar streak. Andean races (*calcaratus* and *catamarcae*) tend to be more rufescent on rump.

SIMILAR SPECIES: Among the S. Am. pipits, the Correndera is the most boldly patterned and (in fresh plumage) the most richly colored: the back pattern is very contrasty, and the breast markings are larger and have more an effect of spots than streaks. Further, it has the longest hindclaw of any S. Am. pipit (occasionally helpful on perched birds). Short-billed is best distinguished by the absence of streaking on its flanks and its more mottled or scaly pattern above, Hellmayr's by its fine, discrete breast streaking. Cf. also Ochre-breasted Pipit (smaller and much more orange-ochraceous below).

HABITAT AND BEHAVIOR: Locally common in pastures, fields, and *puna* grasslands; on the pampas of Buenos Aires, Argentina, where it is very numerous; also frequently occurs in grassy roadside verges. Regularly occurs with Short-billed Pipit in an area of sympatry, tending to be found in lusher areas with taller grass. The Correndera's song is one of *the* sounds of the Buenos Aires pampas; often numerous males can be heard at once. Displaying birds rapidly mount some 20 to 40 m up and there hover into the wind or gradually glide back almost to the ground before mounting back up again; while hovering or descending they constantly repeat a simple but pleasant phrase, typically something like "glishawa-gleeeer, glishawa-gleeeer, glishawa-gleeeer . . ." (the drawled last note being characteristic).

RANGE: Andes of cen. and s. Peru (north to Junín and Lima), w. Bolivia, Chile, and Argentina south to Tierra del Fuego; in latter 2 countries it also is found down to sea level, and in Argentina it occurs east to Entre Ríos and Buenos Aires and on into Uruguay and se. Brazil (Rio Grande do Sul); Falkland Is. Apparently some migration occurs, with a few records from Corrientes, Argentina, and in Brazil north to São Paulo, believed to pertain to austral migrants. To about 4000 m (in the Andes).

NOTE: A related but quite distinct species of pipit is endemic to South Georgia Is., *A. antarcticus* (South Georgia Pipit). It resembles the Correndera but is larger (16.5 cm; 6½") and *much more heavily streaked below,* including throat and flanks; it seems to *lack* Correndera's pair of prominent whitish back stripes. Recent reports indicate that it is not particularly numerous and that it may breed mostly or entirely on smaller islets lying just offshore from South Georgia itself.

Anthus bogotensis

PARAMO PIPIT

PLATE: 7

IDENTIFICATION: 15 cm (6"). Only in the *Andes* (and the *only pipit there north of cen. Peru*). Above boldly streaked blackish and ochraceous; tail dusky with outer feathers buffy whitish to dull buff. *Below rather uniform dull, dark buff* with a few blackish spots or streaks across chest, sometimes (young birds?) extending down flanks as sparse streaking. Birds in worn plumage are much more grayish below.

SIMILAR SPECIES: The virtually uniform dull buffyish or buffy grayish underparts are usually characteristic. In very worn plumage (when dull pale grayish below) it can resemble Hellmayr's Pipit, especially as both have buffyish outer rectrices, but that species shows more streaking on breast (though the streaks are narrow) and always shows at least a little on flanks.

HABITAT AND BEHAVIOR: Locally fairly common in fields and pastures and in *páramo* and *puna* grasslands. Occurs both in rather dry and open as well as wet (e.g., in *pajonal* grasslands near treeline in Peru and Bolivia) habitats. Singing males perform both from the ground (e.g., a rather simple, thin "tseedle-tseedle-tslee") and in a low display flight (when the song is more musical, exuberant, and much more involved).

RANGE: Andes of w. Venezuela (north to Trujillo), Colombia, Ecuador, Peru, nw. Bolivia (south to Cochabamba), and in nw. Argentina (Tucumán). Mostly 2500–4000 m.

Alaudidae

LARKS

THE Horned Lark is the only member of its family found in the Americas, where it maintains an outpost in the Colombian Andes, otherwise not occurring closer than Mexico. The family is much more widespread and diverse in the Old World. Recent DNA evidence suggests that the larks may be fairly closely related to the pipits and wagtails.

Eremophila Larks

Only the Horned Lark has reached the New World, where it is widespread in North America and reappears as a very isolated but barely divergent population in the Colombian Andes.

Eremophila alpestris

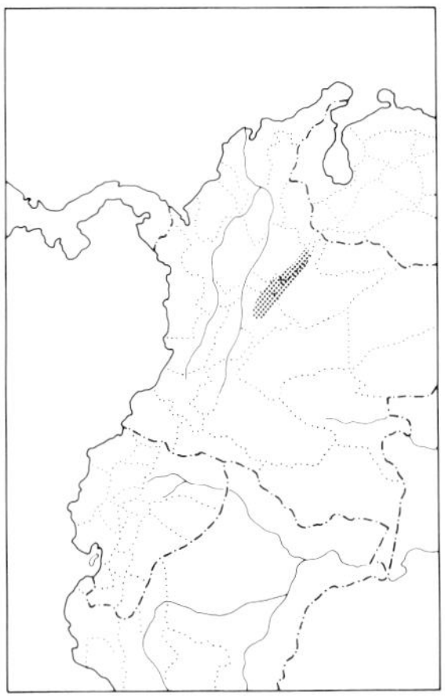

HORNED LARK

PLATE: 7

IDENTIFICATION: 15 cm (6"). *A terrestrial bird of Colombia's E. Andes; complex facial pattern unmistakable* (though duller in female). Above mostly light pinkish brown streaked dusky, most rufescent on nape and sides of neck; *outer web of outer tail feather white* (conspicuous in flight). Face and throat mostly white tinged yellow, with black extending back from forecrown as narrow stripe over superciliary (terminating in the barely visible "horns") and broad black "moustache" below eye. *Broad band across chest black,* with lower underparts whitish, washed pinkish on sides of breast. Female similar but with facial pattern much obscured.

SIMILAR SPECIES: No bird of its range and habitat shares the facial pattern, breast band, and white in the tail.

HABITAT AND BEHAVIOR: Rather uncommon and local on fields and pastures with very short grass (sometimes almost barren areas); *not* found in *páramo* grasslands. Exclusively terrestrial, walking or running about in pairs or small groups. Not especially well known in Colombia: even its (in North America) familiar weak, tinkling song, often delivered in flight, seems not to have been reported, though presumably it is given. Small numbers can sometimes be found around the Bogotá airport.

RANGE: E. Andes of Colombia (Cundinamarca and Boyacá); widespread in North America south to Mexico and in Eurasia. Mostly 2500–3000 m.

Vireonidae

VIREOS

STRICTLY American, the vireos form a discrete group of predominantly olive, arboreal birds found widely in forested and wooded regions. Most are inconspicuous birds, some being much better known from their tirelessly repeated songs. Several members of the genus *Vireo* are strongly migratory. The *Hylophilus* greenlets (possibly more than one genus deserves to be recognized) are more active in behavior and thus are usually easy enough to see, but they are confusing enough that identification to species level is often difficult. The relationships of the Vireo family are obscure; despite their superficial similarity, recent DNA evidence points to their not being especially close to the Emberizidae.

Cyclarhis Peppershrikes

A pair of thickset, bull-headed vireos with heavy, hooked bills. Arboreal, with rather sluggish and inconspicuous behavior, heard more often than seen. Formerly considered a separate family, now the genus *Cyclarhis* is given subfamily rank within the Vireonidae.

Cyclarhis gujanensis

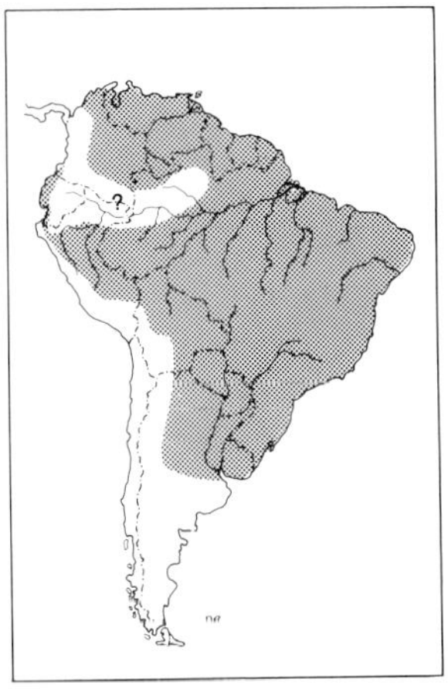

RUFOUS-BROWED PEPPERSHRIKE — PLATE: 8

IDENTIFICATION: 14.5–16 cm (5¾–6¼"). Bill pale brownish (often more leaden below); iris typically hazel; legs pinkish. Variable, best known by *very stout bill;* the *rufous brow* is obvious in all but a few (range-restricted) races. Typical birds olive above, with contrasting gray head and neck and *rufous forehead and superciliary.* Throat and chest pale yellow to olive yellow (*often with the effect of a broad pectoral band*), becoming whitish (sometimes tinged buff) on belly. *Cearensis* (e. Brazil south to n. São Paulo) and *viridis* (Paraguayan and Argentinian *chaco*) have crown tinged brown, with rufous brow still apparent, while *ochrocephala* (se. Brazil and e. Paraguay south to ne. Argentina) may show a distinctly brownish crown with rufous restricted to lores. Most divergent is the *virenticeps* group (w. Ecuador and nw. Peru): *virenticeps* is yellowish olive above (extending onto crown) and *brow is chestnut,* while *contrerasi* and *saturatus* (Peru's middle and upper Marañón valley) are variable but can look very different, with *entire crown rich chestnut* (but usually only mixed olive and chestnut).
SIMILAR SPECIES: Confusingly variable in plumage, but general pattern with the brow and breast band usually apparent. Because of the distinctive bill not likely to be mistaken over most of its large range, but in Colombia and Ecuador cf. Black-billed Peppershrike.
HABITAT AND BEHAVIOR: Fairly common to locally common in a variety of wooded to semiopen habitats; in regions with extensive humid

forest usually restricted to edge or (in Amazonia) *várzea*-associated habitats, though also at times extending into forest canopy. Typically most numerous (certainly most often seen) in arid regions, as vegetation here is lower and foliage is often sparser; also frequent in gallery forest and shade trees over houses. Behavior is sluggish, moving deliberately in foliage; typically in pairs, regularly associating with mixed-bird flocks. Much more often heard than seen, with song repeated tirelessly, often as the bird forages: it is a short, rich musical phrase, rather grosbeaklike (e.g., *Pitylus, Pheucticus*), usually repeated for several minutes before switching to another phrase, which is then also repeated, and so on.

RANGE: Widespread east of Andes south to n.-cen. Argentina (south to La Rioja, San Luis, Córdoba, and n. Buenos Aires), Uruguay, and s. Brazil; seemingly absent from part of upper Amazonia (e.g., e. Ecuador, ne. Peru, nw. Brazil); west of Andes only in sw. Colombia (w. Nariño), w. Ecuador, and nw. Peru; Trinidad. Also Mexico to Panama. Mostly below 1500 m (but regularly higher, even to well over 2000 m, in valleys of e. Peru and n. Bolivia).

Cyclarhis nigrirostris

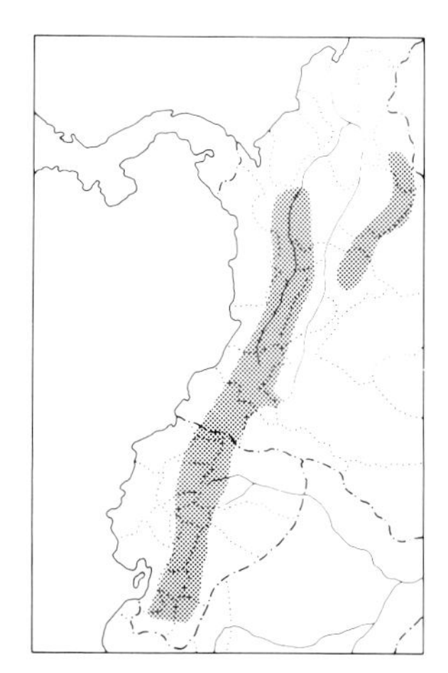

BLACK-BILLED PEPPERSHRIKE

IDENTIFICATION: 14.5–15 cm (5¾–6"). *Colombia and Ecuador. Bill black;* legs bluish gray; iris greenish or grayish yellow. Resembles much more widespread Rufous-browed Peppershrike, especially the *virenticeps* group of the latter. Above mostly olive (*including crown*), with lores and short superciliary deep chestnut. *Below mostly gray* (somewhat paler in nominate race of Colombia except Nariño), center of belly whitish, with olive on sides of chest (sometimes extending across as narrow pectoral band). See C-46.

SIMILAR SPECIES: Rufous-browed Peppershrike's bill is always pale (never black) and its underparts are never as gray as in this species. Note that the facial pattern of Black-billed is not dissimilar from that of *virenticeps* race of Rufous-browed, found in w. Ecuador (where the latter is found exclusively in lowlands, Black-billed on slopes above). Black-billed differs more from other races of Rufous-browed which approach its range (lacking the gray on head and neck, etc.).

HABITAT AND BEHAVIOR: Uncommon to locally fairly common in canopy and borders of humid montane forest at subtropical elevations. Behavior similar to Rufous-browed Peppershrike's, though more strictly a forest bird. Likewise often difficult to see, though single birds or pairs regularly accompany mixed flocks of tanagers, etc. Song very like Rufous-browed's.

RANGE: Andes of Colombia (north in E. Andes to Norte de Santander; Cen. Andes; north in W. Andes to Antioquia) and Ecuador (south on w. slope to El Oro, on e. slope to Zamora-Chinchipe). 1300–2700 m.

Vireolanius Shrike-Vireos

A pair of chunky, large-headed vireos with fairly stout hooked bills, rather resembling the peppershrikes (*Cyclarhis*) except for their notably brighter plumage. Behavior is also similar, and likewise they are more often heard than seen. Formerly considered a separate family, the shrike-vireos are now given subfamily rank within the Vireonidae. Following *Birds of the World* (vol. 14) and the 1983 AOU Check-list, we have merged the genus *Smaragdolanius,* in which the 2 S. Am. species were formerly placed, into *Vireolanius.*

Vireolanius leucotis

SLATY-CAPPED SHRIKE-VIREO

PLATE: 8

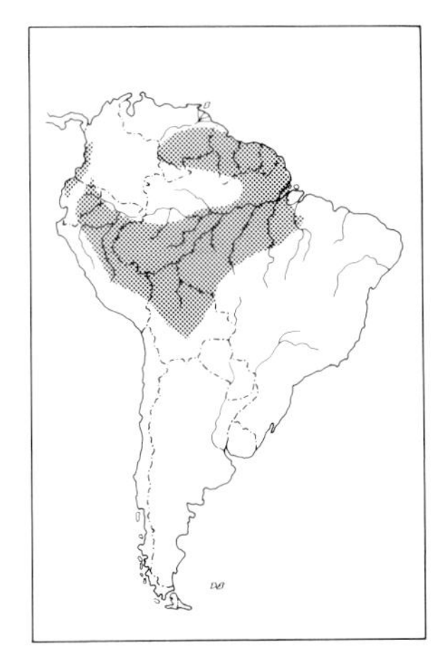

IDENTIFICATION: 14 cm (5½"). Iris lime green; heavy bill black above, bluish gray below; legs usually bluish gray but pink in *mikettae* (of w. Colombia and nw. Ecuador). *Bold face pattern* distinctive in all races. Head gray with *prominent bright yellow superciliary* and spot below eye; otherwise olive green above. *Below mostly bright yellow,* sides and flanks tinged olive. Nominate race (with seemingly disjunct range from s. Venezuela and Guianas to ne. Brazil north of Amazon, in se. Colombia, e. Ecuador, and n. Peru south to San Martín) has a *white cheek stripe* lacking in all other (including *simplex*) races; see V-33.
SIMILAR SPECIES: Cf. Yellow-browed Shrike-Vireo (lacks gray on head, etc.).
HABITAT AND BEHAVIOR: Uncommon to locally fairly common in canopy of humid forest, occasionally coming lower at edge when it is accompanying mixed flocks. Deliberate behavior and habit of remaining high in trees make observation difficult. Presence most often made known by its persistent, far-carrying song, a monotonously repeated "tyeer . . . tyeer . . . tyeer . . ." given at a rate of about 1 note per second; singing birds are also hard to see, but they may eventually respond to tape playback. Usually found in pairs, often associated with mixed flocks of tanagers, etc.
RANGE: Pacific w. Colombia (s. Chocó southward) and nw. Ecuador (south to Pichincha); s. Venezuela (Bolívar and Amazonas), Guianas, much of Amaz. Brazil (east to lower Rio Tocantins and south to n. Mato Grosso; seemingly absent from the northwest), n. Bolivia (south to La Paz, Cochabamba, and w. Santa Cruz), e. Peru, e. Ecuador, and extreme s. Colombia (e. Nariño). Mostly below 1800 m, occasionally higher.

Vireolanius eximius

YELLOW-BROWED SHRIKE-VIREO

Other: Green Shrike-Vireo

IDENTIFICATION: 13.5 cm (5¼"). *N. Colombia and nw. Venezuela.* In general form resembles Slaty-capped Shrike-Vireo but much more brightly colored overall. *Bright emerald green above* with *blue crown and nape* and prominent bright yellow superciliary and small spot below

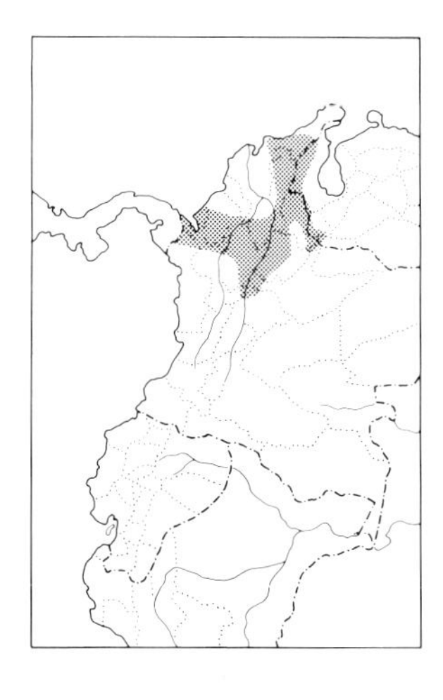

eye. Throat and chest bright yellow, becoming pale yellowish green on remainder of underparts. Female somewhat duller.

SIMILAR SPECIES: Slaty-capped Shrike-Vireo has considerable gray on head and is a much duller, more olive green above. Female Green Honeycreeper is also mostly bright green but lacks the head pattern, is more slenderly built, has a different bill, etc. Blue-naped Chlorophonia, of the same vivid green, might occasionally wander down into the range of this species.

HABITAT AND BEHAVIOR: Uncommon in canopy of humid forest, apparently mostly in foothill zone. Its lethargic behavior and mostly green coloration make observation difficult much of the time, but occasionally 1 or 2 will be seen accompanying mixed flocks of tanagers, etc. In e. Panama the song has recently been found to be very similar to that of the Green Shrike-Vireo (*V. pulchellus*) of Middle America, a tirelessly repeated phrase of 3 or 4 notes, "peer-peer-peer," faster-paced than Slaty-capped Shrike-Vireo's.

RANGE: N. Colombia (more humid parts of Caribbean lowlands and foothill slopes, south in middle Magdalena valley to e. Antioquia; also east of the Andes south to n. Boyacá) and w. Venezuela (slopes of Perijá Mts. and in s. Táchira). Also e. Panama. Mostly below 800 m, occasionally higher.

NOTE: Here considered a species distinct from Middle American *V. pulchellus* (Green Shrike-Vireo), following the 1983 AOU Check-list. No intergrades between the 2 forms have as yet been recorded, but given their similarity in vocalizations and appearance they may ultimately be best considered conspecific.

Vireo Vireos

Rather dull-plumaged, predominantly olive and whitish arboreal birds; most S. Am. species show a more or less pronounced superciliary (in 1 species replaced by spectacles), and wing-bars are lacking in all but 1 species. Rather sluggish, inconspicuous birds of forest and woodland canopy; however, their monotonous, often tirelessly repeated songs do draw attention. Four of the 6 S. Am. species are migratory at least in part.

Vireo olivaceus

RED-EYED VIREO

PLATE: 8

Other: Chivi Vireo

IDENTIFICATION: 14–15 cm (5½–6"). *Iris red to brown,* depending on age and race, with n. *olivaceus* group tending to be a brighter ruby red and perhaps at most a dull carmine in s. *chivi* group. *Crown and nape gray* with *prominent white superciliary bordered above and below by black to dusky lines;* otherwise olive above. Below whitish, often tinged greenish yellow on sides, flanks, and crissum.

SIMILAR SPECIES: Cf. Yellow-green Vireo and the much less common Black-whiskered Vireo. Brown-capped Vireo is smaller with distinctly brown cap, has less distinct head striping, and is more uniform yellow across belly.

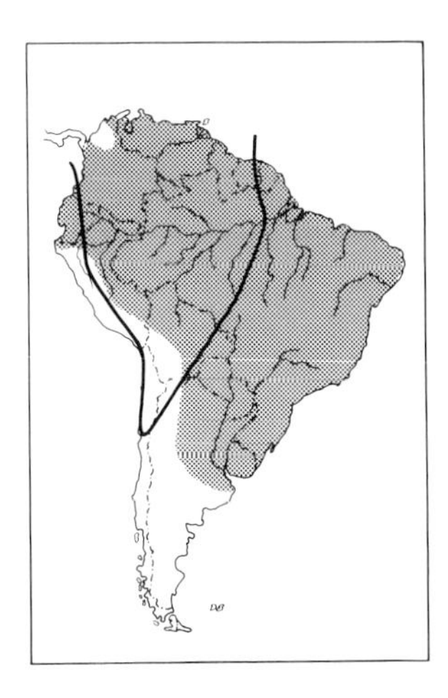

HABITAT AND BEHAVIOR: Common in forest borders, lighter woodland (including various *várzea* and river island habitats), shrubby clearings, and gardens with scattered tall shade trees. A widespread and often familiar bird, especially in edge and semiopen situations, but status of various migratory groups still poorly understood. When not breeding all groups apparently are mostly frugivorous, at other times gleaning for insects among foliage; frequently found with mixed-bird flocks. N. migrants sing rarely or not at all while on their wintering grounds, and wintering austral migrants of the *chivi* group (*chivi* and *diversus*) also may be silent. When breeding the latter have a song consisting of a series of short phrases, less variable but with cadence recognizably similar to that of n. birds (e.g., "cheeweewee, cheewee, cheeweewi, chirchirchir, cheeweewee . . ."). Tropical resident forms (numerous subspecies south to across the Amazon basin), on the other hand, give only a simple 2- or 3-note song (e.g., "cheewit") repeated more or less continuously (and sounding much like a greenlet).
RANGE: Widespread as breeder and/or migrant south to n.-cen. Argentina (south to La Rioja, San Luis, and Buenos Aires), Uruguay, and s. Brazil; west of Andes south to nw. Peru; Trinidad. Also breeds North America (the *olivaceus* group), passing n. winter (mainly Sept.–Apr.) primarily in Amazonia. Red-eyed Vireos breeding in s. South America (north to Bolivia and s. Brazil; *chivi* and apparently also *diversus*) vacate this region during austral winter (mainly Apr.–Aug.) and move to Amazonia and s. Venezuela, where other races are resident. The breeding areas and movements of some races are still not well understood. Mostly below 1500 m, but transients at times occur considerably higher.

NOTE: *V. chivi* (Chivi Vireo), comprising all the forms breeding in South America, has sometimes been accorded full species rank, separate from true *V. olivaceus* breeding in North America. Despite their wide geographic separation when breeding, the *olivaceus* and *chivi* groups are very similar morphologically and have recently been shown to be biochemically close as well (N. Johnson and R. Zink, *Wilson Bull.* 97 [4]: 421–435, 1985). What still remain to be checked, however, are the resident tropical forms, at least some of which have markedly different vocalizations; the significance of this variation, if any, remains to be determined.

Vireo flavoviridis

YELLOW-GREEN VIREO

IDENTIFICATION: 14–15 cm (5½–6"). Very similar to Red-eyed Vireo. Best points to look for are duller, more olive crown (not as gray) with *head striping more obscure* (bordering dark lines inconspicuous or even absent); brighter olive upperparts with yellow or yellowish edging on flight feathers (not greenish or grayish); and *rather bright lemon yellow sides, flanks* (sometimes extending across breast), *and crissum.* Note that some races of the Red-eyed Vireo (e.g., *griseobarbatus* of w. Ecuador and nw. Peru) have almost equally bright yellow sides, flanks, and crissum. See P-24.
SIMILAR SPECIES: It will often not be possible to distinguish between Red-eyed and Yellow-green Vireos with certainty; in any case, a good look will be required. Cf. also the quite similar casual Philadelphia Vireo (Appendix).

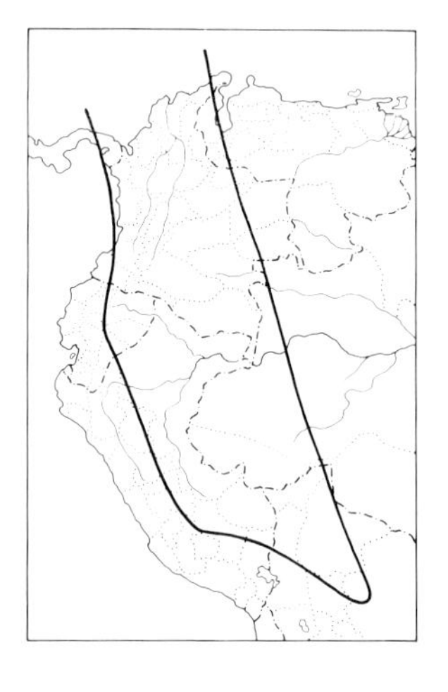

HABITAT AND BEHAVIOR: Uncommon to fairly common n. winter resident (mainly Sept.–Feb.) in canopy and borders of forest, lighter woodland, and clearings. Status in South America not very clear, due in large part to confusion with Red-eyed Vireo; Yellow-green possibly may breed in nw. Colombia. Behavior very similar to the Red-eyed's and, like the n. forms of that species, Yellow-greens only very infrequently sing while in South America.
RANGE: Nonbreeding visitor to Colombia, e. Ecuador, e. Peru, and nw. Bolivia (south to Cochabamba); probably also occurs in w. Amaz. Brazil, but as yet unrecorded. Apparently in Colombia and Ecuador mainly (or entirely?) as a transient. Nests in Middle America, passing nonbreeding season in nw. South America. Mostly below 1500 m but doubtless higher as a transient.

NOTE: Here *V. flavoviridis* is regarded as a full species, based on recent biochemical evidence that it is quite isolated from the *V. olivaceus/chivi* complex (N. Johnson and R. Zink, *Wilson Bull.* 97 [4]: 421–435, 1985). The 1983 AOU Check-list as well as Meyer de Schauensee (1966, 1970) considered it conspecific with *V. olivaceus.*

Vireo gracilirostris

NORONHA VIREO

IDENTIFICATION: 14 cm (5½″). *Restricted to Fernando de Noronha,* an island off the coast of ne. Brazil. *A much duller version of the Red-eyed Vireo.* Bill longer and *very much more slender;* tail also longer. Dingy grayish olive above with narrow but distinct buffy whitish superciliary; eyestripe *not* bordered above by narrow blackish line. Below uniform dull buffy whitish.
SIMILAR SPECIES: Neither Red-eyed nor Black-whiskered Vireo is known to occur on Fernando de Noronha; this species is notably drabber than either.
HABITAT AND BEHAVIOR: Recorded as abundant in all parts of the island to which it is endemic. It forages at all levels, from the treetops to actually hopping about on the ground, and is apparently almost fearless (D. C. Oren, *Bol. Mus. Par. Emi. Goeldi* 1 [1]: 36–37, 1984).
RANGE: Fernando de Noronha Is., off the coast of ne. Brazil.

NOTE: Though since Hellmayr (*Birds of the Americas,* vol. 13, part 8) this form has been considered merely an insular race of *V. olivaceus,* we feel that it stands apart on so many counts that it amply deserves recognition as a full species. Oren (see above) agreed.

Vireo altiloquus

BLACK-WHISKERED VIREO

IDENTIFICATION: 15 cm (6″). Closely resembles Red-eyed Vireo. Bill slightly heavier and longer. Best known by its *narrow blackish malar streak.* Somewhat duller above, with crown usually not as contrastingly gray. Nominate race (breeding in most of the Greater Antilles) somewhat buffier about head and slightly buffier or yellower below than is *barbatulus* (breeding in Florida, Cuba, etc.); both races are migratory to South America.

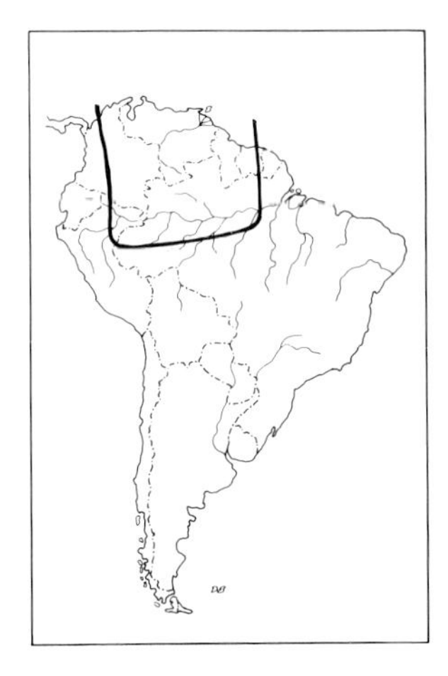

SIMILAR SPECIES: Though the malar streak is diagnostic, one must beware molting (or even wet) Red-eyeds, which frequently seem to show a similar stripe.

HABITAT AND BEHAVIOR: Rare to uncommon n.-winter resident (mostly Sept.–Apr.) in lighter woodland, forest borders, and clearings; common in scrub and mangroves on islands off Venezuela where it is a breeder. Behavior similar to Red-eyed Vireo's, though often seems much tamer. N. migrants apparently never sing on their wintering grounds, but birds resident on islands off Venezuelan coast do sing during their breeding season; this song is a tirelessly repeated series of phrases, much like the Red-eyed's but hoarser and somewhat more abrupt or clipped.

RANGE: Nonbreeding transient and visitor south to ne. Peru, Amaz. Brazil, and Guianas; Trinidad. Breeds in Florida, on West Indies, and on various islands off the coast of Venezuela, including Netherlands Antilles, withdrawing southward from more n. areas. Mostly below 1000 m.

Vireo leucophrys

BROWN-CAPPED VIREO

PLATE: 8

Other: Warbling Vireo (in part)

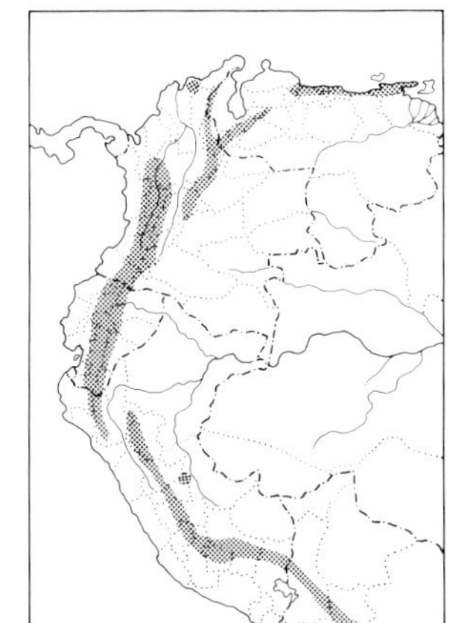

IDENTIFICATION: 12.5 cm (5″). Mostly olive above, but with *distinctly brown crown* and quite prominent *whitish superciliary*. Throat and chest whitish, becoming pale yellowish to buffy yellowish on lower underparts.

SIMILAR SPECIES: This is the only *Vireo* with a brown cap. Several greenlets have rufous on crown (e.g., Rufous-naped), but none of these also shows the eyestripe, and greenlets generally behave quite differently; most greenlets are found in the *lowlands*. Cf. the very rare and quite similar n.-migrant Philadelphia Vireo (Appendix).

HABITAT AND BEHAVIOR: Fairly common to common in humid montane forest, forest borders and tall second growth, and adjacent clearings; generally most numerous in somewhat opened-up, edge situations. Forages deliberately in foliage at varying levels (usually high, but does come lower at edge), often fully in the open for considerable periods; frequently with mixed flocks of tanagers, etc. Song is a short, fast, musical warble, rising at the end, vaguely reminiscent of N. Am. Warbling Vireo (*V. gilvus*) but much shorter. Also often heard is its distinctly rising, buzzy "zreeee" call note, sometimes repeated several times in succession.

RANGE: Coastal mts. of n. Venezuela (east to Paria Peninsula) and Andes of w. Venezuela, Colombia, Ecuador, nw. and c. Peru (south on w. slope to s. Cajamarca), and nw. Bolivia (La Paz and Cochabamba); Perijá Mts. on Venezuela-Colombia border, Santa Marta Mts. of n. Colombia. Also Middle America. Mostly 1300–2500 m, locally to 1000 m in n. Venezuela.

NOTE: We follow the 1983 AOU Check-list in considering the montane Middle and S. Am. *leucophrys* group as a species distinct from the *gilvus* groups (Warbling Vireo) of North America. Though the songs are fairly similar, other vocalizations as well as habitat and plumage differ.

Vireo flavifrons

YELLOW-THROATED VIREO

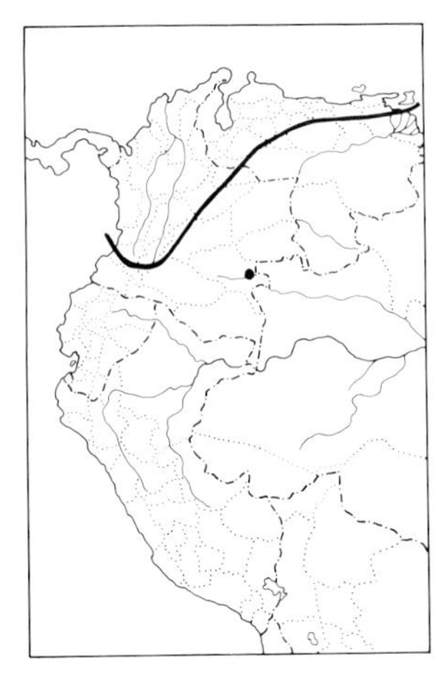

IDENTIFICATION: 14 cm (5½"). Mostly bright olive green with gray lower back and rump and *conspicuous yellow lores and eye-ring* (forming prominent "*spectacles*"); wings and tail blackish with *2 bold white wing-bars. Throat and breast bright yellow,* lower underparts white.

SIMILAR SPECIES: Clean-cut and brightly patterned for a vireo, more so than any other found in South America. Pattern somewhat reminiscent of various tyrannids, but posture and behavior very different, etc.

HABITAT AND BEHAVIOR: Rare to uncommon n.-winter resident (Nov.–Apr.) in forest borders and second-growth woodland. Usually seen singly as they hop lethargically through middle and upper strata of trees, often accompanying mixed flocks of tanagers, wintering warblers, etc. Unlike most other n. migrants, this species gives snatches of its song fairly regularly, a deliberate series of short, rich, husky phrases. Perhaps most numerous in Colombia's Santa Marta Mts.

RANGE: Nonbreeding visitor mainly to mts. of Colombia (south to Valle, once east of Andes in Vaupés) and w. and n. Venezuela (east to Sucre); Trinidad and Tobago, Curaçao (once). Breeds e. North America, wintering mainly in Middle America. To about 1800 m.

Hylophilus Greenlets

A very confusing group of small, warblerlike vireos; many are difficult to identify, and indeed a number of taxonomic problems within the genus (which quite possibly should be split apart) are yet to be resolved. Compared to typical *Vireo* vireos, they are smaller and markedly more active when foraging; note also their pointed bills, which are typically pale (flesh to brownish), at least on the lower mandible.

For field purposes we have divided the genus into 3 groups, but bear in mind that some species are "exceptions" to one degree or another.

1. A usually pale-eyed, scrub- or edge-inhabiting group with simple repetitive songs.

2. A dark-eyed, usually forest canopy–inhabiting group with song (so far as known) consisting of a single but more complex phrase, often tirelessly repeated.

3. The distinctive Tawny-crowned Greenlet (the only species in the forest understory, has a very different song, etc.).

GROUP A

Most characterized by their *pale irides* (but note 2 exceptions: Rufous-crowned and Ashy-headed) and their somewhat more conical, pointed bills. Further, a majority are found in *scrub or edge habitats* (but again there are exceptions: the difficult-to-place Rufous-crowned in forest; the Gray-chested in canopy and borders; and the Lemon-chested in forest borders or scrub). All have a *relatively simple song,* consisting of several repetitions of a single note (e.g., "swee-swee-swee-swee"), in one (Ashy-headed) with an additional trill at the end. Somewhat less active in their behavior than the following group and more often found singly or in pairs than in small groups; all are often with mixed flocks.

Hylophilus poicilotis

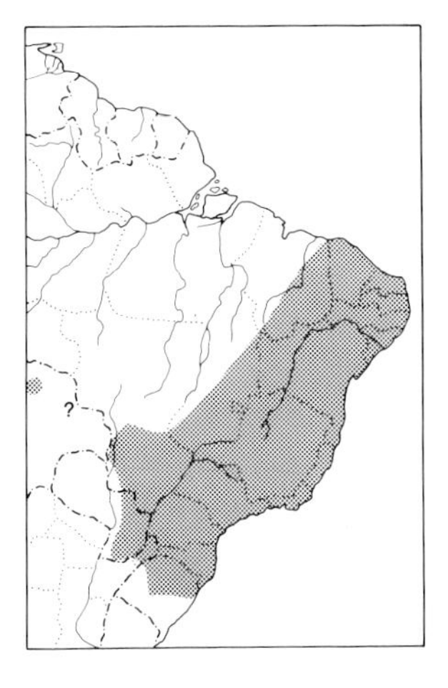

RUFOUS-CROWNED GREENLET

PLATE: 8

IDENTIFICATION: 12.5 cm (5"). *A brightly patterned greenlet of se. South America. Crown bright rufous,* contrasting with rather bright olive upperparts. Sides of head and throat pale grayish, with *conspicuous dusky patch on ear-coverts.* Underparts buffy yellowish, yellowest on flanks. *Amaurocephalus* (ne. Brazil south to s. Bahia) has auricular patch obscure or lacking and has buffier lower underparts.

SIMILAR SPECIES: In its range there is no other even vaguely similar greenlet. Compare to female Plain Antvireo (also with rufous crown but lacking auricular patch, has shorter tail, etc.).

HABITAT AND BEHAVIOR: Fairly common in humid forest and forest borders, second-growth woodland, and (in ne. Brazil) in better-developed *caatinga* woodland and scrub. Found singly or in pairs, foraging at all levels (but mostly 4 to 8 m above ground); frequently with mixed flocks. Often partially cocks its tail (which is rather long for a greenlet). The song is a fairly loud, emphatic (almost ringing) "sweee-sweee-sweee-sweee" (sometimes 5 "sweees"), rather fast.

RANGE: N. Bolivia (recorded only from Beni at San Juan), e. Paraguay (widespread in humid forest), ne. Argentina (Misiones), and s. and e. Brazil (s. Mato Grosso east to Atlantic coastal mts., where recorded south to n. Rio Grande do Sul and north to Piauí, Ceará, and Paraíba). To about 1800 m.

Hylophilus flavipes

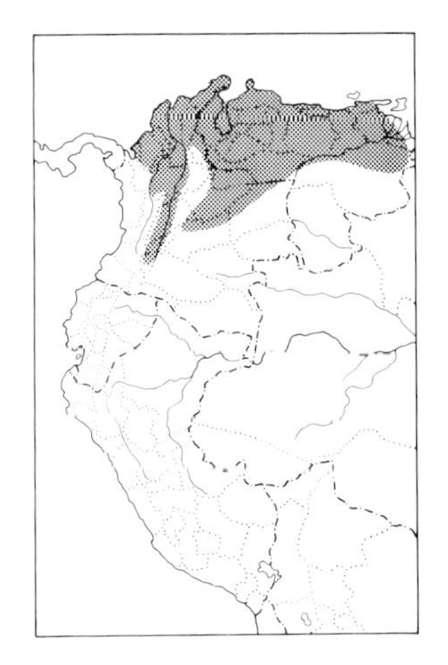

SCRUB GREENLET

PLATE: 8

IDENTIFICATION: 11.5 cm (4½"). *N. Colombia and Venezuela. Bill mostly pinkish* with dusky culmen and tip; legs also pinkish; iris whitish in Colombia but dark brown in Venezuela (except the far west) and on Tobago. *Very plain.* Dull olive to brownish olive above. Throat dull whitish, remaining underparts pale buffy yellowish.

SIMILAR SPECIES: Rather closely resembles Golden-fronted Greenlet, especially across n. Venezuela; here *both* species have dark irides. Apart from their characteristically different vocalizations (see below), note Golden-fronted's brownish cap and yellower forehead as well as its brighter yellow underparts.

HABITAT AND BEHAVIOR: Fairly common to common in arid scrub and lighter woodland. Usually inconspicuous, foraging deliberately in foliage, often fairly low, but also may cling titlike (*Parus* sp.) upside-down to a leaf so as to inspect its underside. More often heard than seen, its song being an easily recognized series of up to 15 to 20 repeats of "tuwee, tuwee, tuwee, tuwee" or "peer, peer, peer, peer" notes, usually given quite rapidly.

RANGE: N. Colombia (south in Magdalena valley to Huila and east of Andes to Meta) and n. Venezuela (generally north of the Orinoco and south of it in n. Bolívar); Tobago, Margarita Is. off n. Venezuela. Also Costa Rica and Panama. To about 1000 m.

NOTE: Despite the striking shift in iris color (pale in nominate race and *galbanus* east to w. Venezuela, dark in *acuticauda* across most of Venezuela and in *insularis* of Tobago), there as yet seems no compelling reason to consider all the S. Am. forms as other than conspecific; vocally all are very similar. Isolated *viridiflavus* of Middle America is more divergent (being yellower below, etc.) and the cadence of its song is markedly slower; it may deserve full species status.

Hylophilus olivaceus

OLIVACEOUS GREENLET

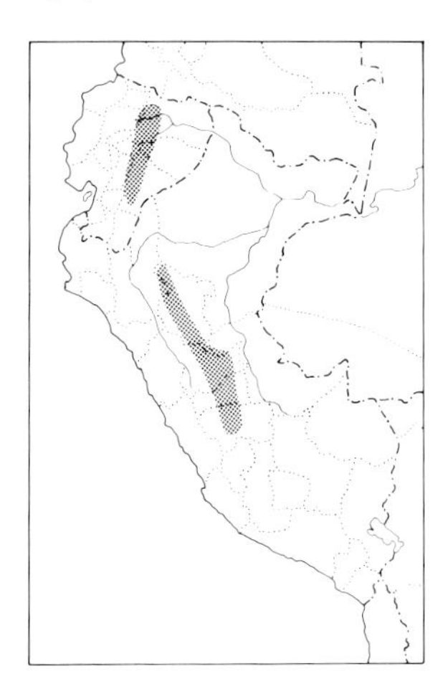

IDENTIFICATION: 12 cm (4¾″). *Lower e. slopes of Andes in Ecuador and Peru.* Bill and legs pale brownish to pinkish; iris whitish to pale yellow. Above dull olive, with forehead more yellowish. Below rather uniform yellowish *olive,* yellowest on belly.

SIMILAR SPECIES: Overall appearance (including soft-part colors), behavior, habitat, and voice similar to Scrub Greenlet's (with which it has been considered conspecific), but no overlap; much more uniformly yellowish olive below than that species. Tawny-crowned Greenlet has much more rufous on crown, browner wings and tail, and pale gray throat.

HABITAT AND BEHAVIOR: Locally uncommon to fairly common in shrubby overgrown clearings and forest borders. Difficult to see in its often almost impenetrable habitat and most often recorded only by voice; its song is a loud, fast, ringing "twee-twee-twee-twee-twee-twee," with up to 10 to 12 "twee" notes. Seems especially numerous in foothills of e. Ecuador (e.g., above Archidona).

RANGE: Lower slopes and e. base of Andes in e. Ecuador (north to Napo) and e. Peru (south to Junín). 600–1600 m.

Hylophilus semicinereus

GRAY-CHESTED GREENLET

PLATE: 8

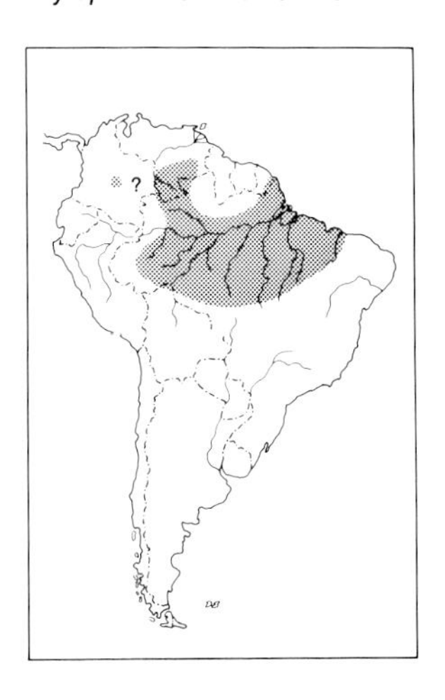

IDENTIFICATION: 12 cm (4¾″). Bill grayish flesh with dark culmen and tip; legs gray; *iris whitish* to pale gray. *Most of crown and nape gray,* contrasting with olive forehead and remaining upperparts. *Mostly gray below,* with a little yellow-olive on sides of chest; whiter on center of belly.

SIMILAR SPECIES: Resembles better-known Lemon-chested Greenlet but much grayer below and lacking latter's broad olive yellow breast band.

HABITAT AND BEHAVIOR: Fairly common to common in canopy and borders of second-growth woodland and humid forest. Not an easy bird to see well as it usually forages in dense foliage and viny tangles; pairs often accompany mixed flocks of tanagers and other birds. Often recorded only by voice, its song is similar to that of Scrub and Lemon-chested Greenlets but is more a repetition of a single slurred note, "seeur-seeur-seeur-seeur. . . ."

RANGE: E. Colombia (recorded only from w. Meta, near Villavicencio, but probably more widespread), s. Venezuela (Amazonas and s. Bolívar), widely in Amaz. Brazil (south to n. Mato Grosso and east to n. Maranhão), and French Guiana; not recorded from Bolivia but likely occurs. To about 400 m.

Hylophilus thoracicus

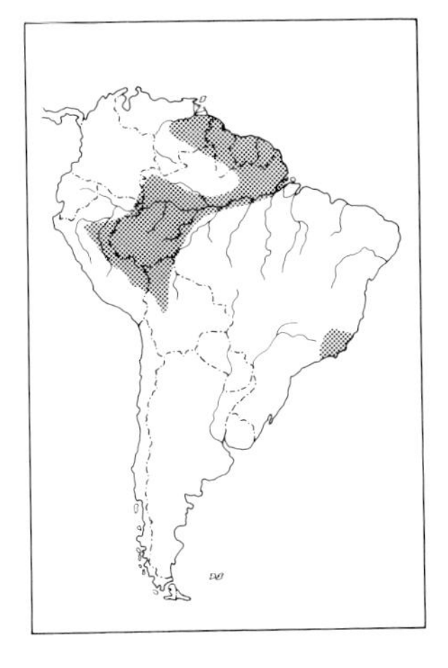

LEMON-CHESTED GREENLET

IDENTIFICATION: 12 cm (4¾"). Bill pinkish brown with dusky culmen; legs pinkish gray; *iris white to pale yellow.* Mostly bright olive above *with gray rearcrown.* Below grayish white to gray with *broad olive yellow band across breast.* See V-33, C-46.
SIMILAR SPECIES: Ashy-headed Greenlet is quite similar but has dark iris, mostly gray head (not just nape), differently patterned song (see below). Gray-chested Greenlet is much more uniformly gray below (lacking the effect of a yellowish pectoral band). Cf. also Tepui Greenlet.
HABITAT AND BEHAVIOR: Uncommon to fairly common in canopy and middle levels of humid forest (both *terra firme* and *várzea*) and forest borders. Nominate race of se. Brazil is also found in scrub and low woodland. Usually found singly or in pairs, foraging quite actively, often accompanying mixed flocks. Its song is a rapidly uttered series, "peeer-peeer-peeer-peeer-peeer," similar to Ashy-headed's but without latter's trilled ending.
RANGE: Se. Venezuela (Bolívar), Guianas, locally in Amaz. Brazil, e. Peru (San Martín south locally to Madre de Dios and Puno), and n. Bolivia (La Paz and Beni); se. Brazil (Rio de Janeiro, Espírito Santo, and s. Minas Gerais). Mostly below 600 m.

Hylophilus pectoralis

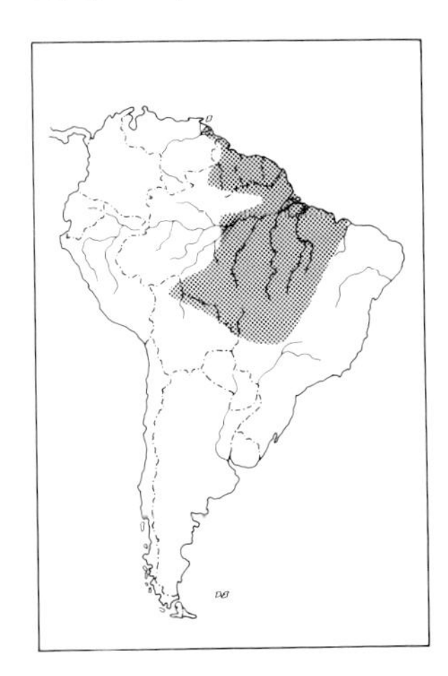

ASHY-HEADED GREENLET

PLATE: 8

IDENTIFICATION: 12 cm (4¾"). Bill and legs grayish flesh, former with dusky culmen; *iris dark reddish brown. Entire head and nape gray,* contrasting with otherwise olive upperparts. Throat whitish, *breast broadly dull yellow,* belly also whitish.
SIMILAR SPECIES: Resembles Lemon-chested Greenlet in pattern, but note the dark iris (not conspicuously whitish); entire head is gray (not just the nape). Tepui Greenlet also similar, though it too has a pale (not dark) eye; further, note its obviously gray wings.
HABITAT AND BEHAVIOR: Fairly common in deciduous and gallery woodland, thickets and overgrown clearings, and occasionally even in gardens or (in the Guianas) at the edge of mangrove forests. Usually quite conspicuous, foraging in a spritely manner among foliage and on outer twigs; often with small mixed flocks of various tyrannids, etc. Presence often made known by its persistent song, a rapidly delivered "peeer-peeeer-peeer, pr-e-e-e-e-e," the number of initial "peeers" varying but always with the characteristic terminal trill. Readily found in coastal Suriname.
RANGE: Extreme e. Venezuela (Delta Amacuro), Guianas, lower Amaz. and cen. Brazil (north of the Amazon west to area north of

Manaus; south of the Amazon south to cen. Mato Grosso and east to Goiás and Maranhão; the Alagoas record is in error, *fide* E. O. Willis), and n. Bolivia (Pando). Below about 400 m.

Hylophilus sclateri

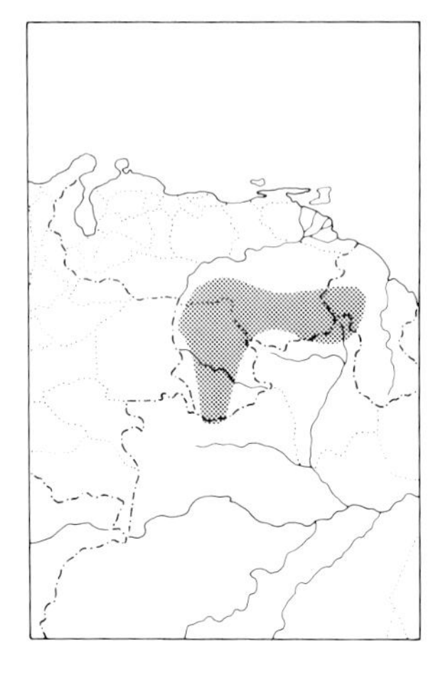

TEPUI GREENLET

IDENTIFICATION: 12 cm (4¾"). *Tepuis of s. Venezuela, etc. Head and nape gray* except for buffyish lores; otherwise olive above but with *contrastingly gray wings and tail.* Dingy grayish white below, with dull yellowish band across breast. See V-33.

SIMILAR SPECIES: This *tepui* endemic combines the overall pattern of the Ashy-headed Greenlet (with the addition of gray wings and tail) with the soft-part colors of the Lemon-chested. Apparently no other similar greenlet occurs with it.

HABITAT AND BEHAVIOR: Uncommon in thickets, low woodland, and forest borders. Behavior similar to other greenlets'; often associates with small flocks which include the Roraiman Antwren and Tepui Redstart. What is apparently its song is a musical "suweé seeu," repeated intermittently, with the quality of a *Sporophila* seedeater (T. Meyer). Small numbers can be found in more open areas along the Escalera road in e. Bolívar, Venezuela.

RANGE: *Tepuis* of s. Venezuela (Bolívar and Amazonas) and adjacent Guyana and extreme n. Brazil (Roraima). 600–2000 m.

GROUP B

These all have *dark* irides and rather more warblerlike bills (less conical, more attenuated), and all but one (Golden-fronted) are found in *forests or forest edge,* typically quite high (often high in canopy). Here they forage actively, often clinging upside-down while searching the undersides of leaves; all are frequent members of mixed flocks of insectivorous birds. Their songs consist of a rapid but often complex phrase (sometimes resembling a single phrase in the Red-eyed Vireo's song), often repeated interminably; the voice of the Brown-headed Greenlet is unrecorded but likely is similar.

Hylophilus aurantiifrons

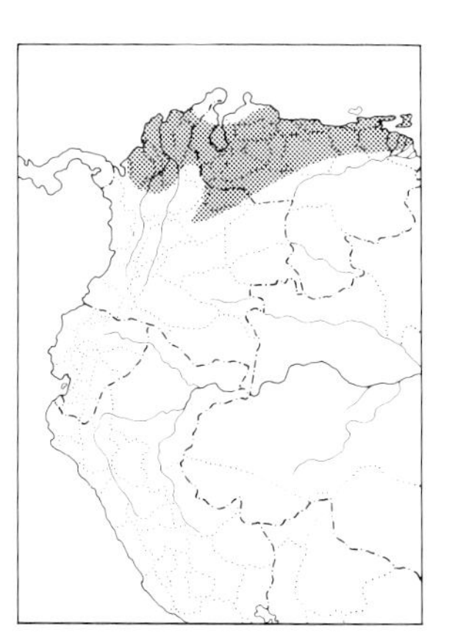

GOLDEN-FRONTED GREENLET

PLATE: 8

IDENTIFICATION: 11.5 cm (4½"). *N. Colombia and Venezuela.* Bill dusky above, flesh below; legs bluish gray; *iris dark. Forehead dull tawny-gold,* becoming *pale brown on crown;* remaining upperparts pale olive. Throat whitish, becoming pale yellowish on remaining underparts; often a tinge of buffy yellow across breast and on sides.

SIMILAR SPECIES: In its range and habitat most likely confused with Scrub Greenlet, which is duller and lacks the yellowish and brown on the crown; note that across most of Venezuela both species have dark irides (Scrub's being pale only in Colombia and w. Venezuela). Their calls are unmistakably different. Cf. also Tawny-crowned Greenlet (very different humid forest habitat, etc.).

HABITAT AND BEHAVIOR: Uncommon to locally fairly common in dry scrub and woodland and in shrubby clearings and woodland borders in somewhat more humid areas. Usually remains fairly low, actively gleaning from foliage like a warbler. Calls frequently, often a rapid phrase "cheetsacheéyou" or sometimes a simpler "cheevee," and also has a more nasal scolding note.

RANGE: N. Colombia (Caribbean lowlands and east of the Andes in Arauca and Casanare) and n. Venezuela (widely north of the Orinoco); Trinidad. Also Panama. To about 1300 m in Venezuela, but usually below 700 m in Colombia.

Hylophilus hypoxanthus

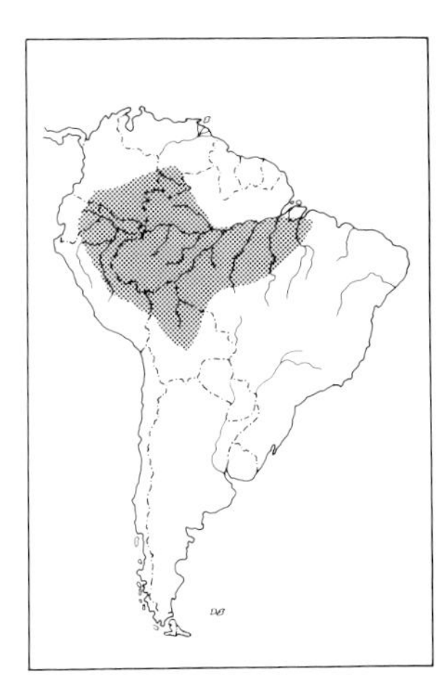

DUSKY-CAPPED GREENLET

PLATE: 8

IDENTIFICATION: 11.5–12 cm (4½–4¾"). Bill dusky above, pinkish below; iris dark. *Crown brown, becoming bronzy brownish olive on back;* olive on rump, wings, and tail. Throat grayish white, becoming *pale yellow on remaining underparts.* Nominate race (upper Rio Negro drainage and s. Venezuela) similar to upper Amazon birds (including *flaviventris*) but uniformly duskier (not so brown) above. *Inornatus* (Amaz. Brazil south of the Amazon between Rios Tapajós and Tocantins) similar but paler yellow below (yellow mainly on belly and crissum), with breast buffy whitish.

SIMILAR SPECIES: *This is the only Amaz. greenlet with basically yellow underparts.* Brown-headed Greenlet is fairly similar but is basically grayish below.

HABITAT AND BEHAVIOR: Fairly common to common (though generally overlooked until its vocalizations are recognized) in canopy and middle levels of *terra firme* forest. Almost invariably found with mixed flocks of tanagers, etc., foraging actively in foliage. Its calls are bright, spritely, and fast, "itsochuwéet" or "purcheechoweér" being typical.

RANGE: S. Venezuela (Amazonas), se. Colombia (north to Putumayo and Guainía), e. Ecuador, e. Peru, n. Bolivia (south to La Paz, Cochabamba, and w. Santa Cruz), and Amaz. Brazil (north of the Amazon east to lower Rio Negro, south of it east to lower Rio Tocantins). To about 500 m.

NOTE: The race *inornatus* has long been confused. Originally described as a full species, it was first considered a disjunct race of *H. brunneiceps* by Hellmayr (*Birds of the Americas,* vol. 13, part 8). This course has been followed ever since, though J. T. Zimmer (*Am. Mus. Novitates* 1160, 1942) registered his doubts. We believe, however, that it actually represents the easternmost race of *H. hypoxanthus. Inornatus* agrees with *H. hypoxanthus* in its rather uniform brown tone to the upperparts, including all of the mantle (quite different from the contrasting pure olive mantle of *H. brunneiceps*) and its mainly yellow underparts (which are very different from the entirely grayish underparts of *H. brunneiceps*). Part of the difficulty seems to have been that the long AMNH series of these greenlets from the Santarém/lower Rio Tapajós area were incorrectly assigned to *H. h. albigula* when, in fact, they are identical to a large series of *inornatus* in the CM from the same region (some even from precisely the same localities); the CM fortunately also had true *albigula* for comparison. True *H. h. albigula* is thus found *only* in the region just to the west of the Tapajós area (in the lower Rio Madeira drainage, etc.); it is yellower-bellied than *H. h. inornatus* and is in fact only weakly differentiated from the other upper Amaz. races of *Hylophilus hypoxanthus.*

Hylophilus brunneiceps

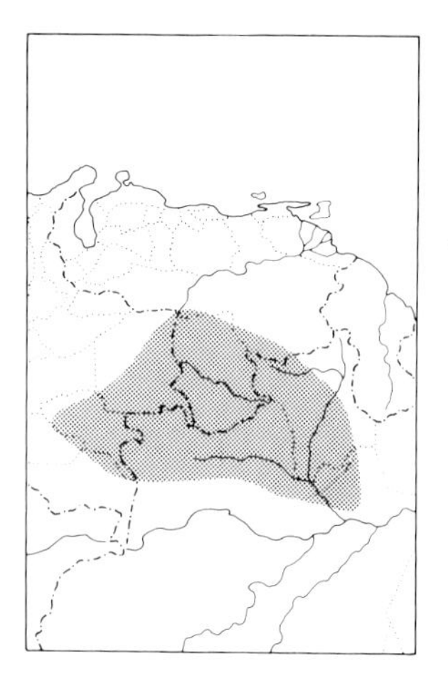

BROWN-HEADED GREENLET

IDENTIFICATION: 11.5 cm (4½"). *Sw. Venezuela and adjacent Colombia and Brazil.* Recalls Dusky-capped Greenlet, and the 2 are sympatric in some areas. The *mantle is olive* (lacking the brown tone of Dusky-capped) and contrasting with the brown crown; *underparts are mostly grayish white* (not mostly yellow), with throat more buffyish. See V-33, C-46.

SIMILAR SPECIES: Cf. Dusky-capped Greenlet.

HABITAT AND BEHAVIOR: Poorly known, but presumably similar to Dusky-capped Greenlet's. Seemingly confined to blackwater areas and their associated sandy-belt forests.

RANGE: S. Venezuela (Amazonas), extreme se. Colombia (e. Guainía and Vaupés), and nw. Brazil (Rio Negro drainage south to north of Manaus at WWF site; *fide* D. Stotz). To about 400 m.

NOTE: See comments under Dusky-capped Greenlet, *H. hypoxanthus.* We now consider *H. brunneiceps* to be a monotypic species confined to the blackwater region of the upper Rio Negro drainage.

Hylophilus muscicapinus

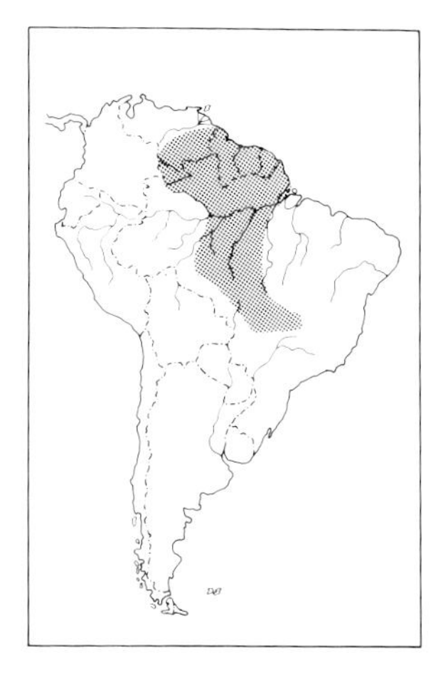

BUFF-CHEEKED GREENLET PLATE: 8

IDENTIFICATION: 11.5 cm (4½"). Iris dark; bill dusky above, grayish flesh below; legs pale gray. Crown mostly gray contrasting with *bright buff lores and sides of head;* remaining upperparts olive. Throat and breast whitish *washed buff,* becoming whitish on belly.

SIMILAR SPECIES: No other greenlet shows that obvious buff on the face and breast. Cf. Brown-headed Greenlet (more uniformly grayish below).

HABITAT AND BEHAVIOR: Fairly common in canopy and middle levels of *terra firme* forest. Behavior similar to that of the other dark-eyed, canopy-inhabiting greenlets of Amazonia and like them regularly with mixed flocks. Its commonest calls are a fast, snappy "pitcheechiweer" or "pitcheeweeu."

RANGE: S. Venezuela (Amazonas and Bolívar), Guianas, and locally in Amaz. Brazil (south from Amapá and n. Pará to cen. Mato Grosso and s. Goiás). To about 600 m.

Hylophilus semibrunneus

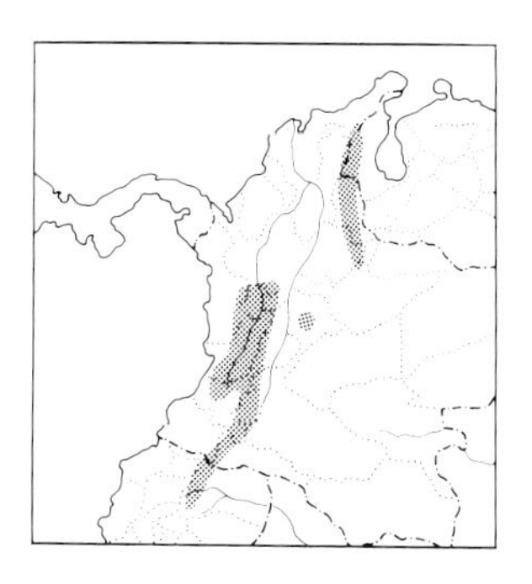

RUFOUS-NAPED GREENLET

IDENTIFICATION: 12.5 cm (5"). *Andes of Colombia and nearby areas.* Bill blackish above, paler flesh below; *iris dark. Crown and nape rich rufous,* contrasting with otherwise rather bright olive upperparts. Lores, lower face, and underparts mainly whitish to grayish white, with patch of rufous on either side of chest; crissum tinged yellowish. See V-33, C-46.

SIMILAR SPECIES: Not likely to be confused in its subtropical Andean range (occurring higher than other nearly overlapping greenlets). Cf. Tawny-crowned and Golden-fronted Greenlets.

HABITAT AND BEHAVIOR: Uncommon in middle levels and border of montane forest and second-growth woodland. Usually found singly or

in pairs, gleaning actively in foliage and on outer twigs; often with mixed flocks of tanagers, warblers, etc. The song is a weak, fast, warbled "wacheera-ditit" (S. Hilty). Seems most numerous in various areas above Cali, in Valle, Colombia (but even there far from common).
RANGE: Locally in Andes of Colombia and on e. slope of Andes in ne. Ecuador (Napo; no recent reports); Perijá Mts. on Venezuela-Colombia border. Mostly 1000–2100 m, to 450 m on Venezuelan side of the Perijá.

Hylophilus decurtatus

LESSER GREENLET

PLATE: 8

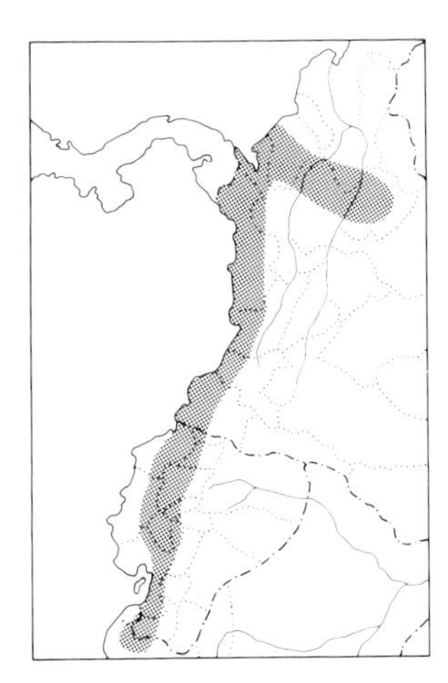

IDENTIFICATION: 10 cm (4"). *W. Colombia to nw. Peru. A plump, puffy-headed, short-tailed greenlet.* Iris dark; bill dusky above, grayish flesh below. Yellowish olive above, somewhat darker on head and with *narrow whitish eye-ring. Grayish white below,* tinged greenish yellow on sides and flanks.
SIMILAR SPECIES: Rather differently shaped from other greenlets; no other *Hylophilus* in its nw. range is likely to be confused with it. In color somewhat reminiscent of Tennessee Warbler (but it too is differently shaped, has a whitish superciliary instead of an eye-ring, etc.).
HABITAT AND BEHAVIOR: Uncommon to fairly common in forest, second-growth woodland, and borders. For the most part found in humid areas but also recorded from drier deciduous forest (e.g., in nw. Peru). A very active bird, often in small groups, foraging mostly at middle and upper levels; frequently accompanies mixed flocks of various insectivorous birds. Its call, a rapid, musical phrase somewhat suggestive of a single phrase of the Red-eyed Vireo's song, "deedereét" or "itsacheét," is constantly repeated.
RANGE: W. Colombia (Pacific lowlands and east to middle Magdalena valley in w. Santander), w. Ecuador, and extreme nw. Peru (Tumbes). Also Mexico to Panama. To about 1000 m.

NOTE: The *minor* group (with *dariensis*) of e. Panama and nw. South America was formerly considered a distinct species but is now generally regarded as conspecific with Middle American *decurtatus* (which when split was called the Gray-headed Greenlet); intergradation occurs in cen. Panama.

GROUP C

A distinct species of greenlet, the only one found in the *understory* of humid forest, where it is most often found with mixed flocks of antwrens, etc. Its song also differs markedly.

Hylophilus ochraceiceps

TAWNY-CROWNED GREENLET

PLATE: 8

IDENTIFICATION: 11.5 cm (4½"). *Iris usually yellowish white to pale gray,* but dark brown in *rubrifrons* group (including *luteifrons* and *lutescens*) of ne. South America (e. Venezuela and Guianas south to lower Amaz. Brazil, west to the Rio Madeira). Over most of its range *crown is tawny* (brightest on forecrown); otherwise olive above, becoming olive brownish on wings and (especially) tail. Below mostly grayish, with ill-defined dull olive band across breast. *Bulunensis* (w. Colombia and

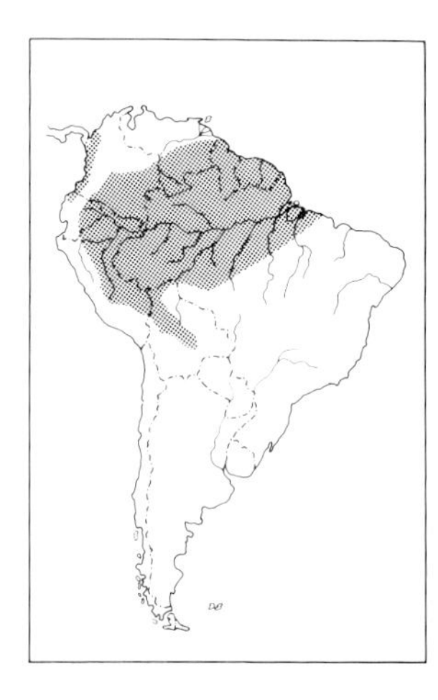

nw. Ecuador) is more uniformly olive below except on throat, while *viridior* (se. Peru and n. Bolivia) has an olive tail (lacking the brown tone of all other races). The *rubrifrons* group has tawny on crown much reduced or (in *luteifrons*) lacking and tends to be buffier below (not so gray), especially on breast.

SIMILAR SPECIES: Often confused, though in most of its range the pale eye in conjunction with the tawny crown is distinctive; most like female Plain Antvireo (and the 2 often forage in the same flock), but note latter's *dark* eye. Even more confusion is created by the dull *rubrifrons* group (possibly a distinct species; see below). With their dark irides and less tawny on crown, these resemble a number of other species (e.g., especially various female antwrens such as Long-winged), but note Tawny-crowned's distinct brownish tinge to tail and its olive tone overall.

HABITAT AND BEHAVIOR: Uncommon in lower growth of humid forest and adjacent second-growth woodland; a forest interior bird, rarely or never seen at edge and never in canopy (unique for a forest-based greenlet). Though wide-ranging, in general not a frequently encountered bird. Most often in pairs, sometimes small groups of 3 to 5 birds, usually accompanying flocks of small insectivorous birds such as antwrens, small flycatchers, etc. The most frequently heard call is a constantly repeated, harsh scolding "nya, nya"; much less often heard is its (presumed) song, a loud, clear, penetrating "teee-yeeé," the second note rising (seemingly much the same in various areas, e.g., the Guianas, Panama, e. Peru, etc.).

RANGE: W. Colombia (Pacific lowlands) and nw. Ecuador (Esmeraldas); Guianas, s. Venezuela (Bolívar and Amazonas), se. Colombia (north to Meta and Guainía), e. Ecuador, e. Peru, n. Bolivia (south to La Paz, Cochabamba, and nw. Santa Cruz), and Amaz. Brazil (south to n. Mato Grosso and east to e. Pará in the Belém area). Also Mexico to Panama. Mostly below 800 m, rarely to 1500 m (in Venezuela).

NOTE: It has been suggested that the dark-eyed *rubrifrons* group represents a species distinct from *H. ochraceiceps*. However, in view of apparent similarities in behavior and, so far as known, vocalizations, it seems best to continue to regard all forms as conspecific, pending detailed study.

Emberizidae

EMBERIZIDS

WE HERE start progressing through the vast 9-primaried assemblage, all now in the Emberizidae family. Because the number of species is so large, we have divided the family into its component subfamilies.

Parulinae

WOOD-WARBLERS

THERE are 2 major groups of warblers in South America: the migrants from North America, of which over 20 species regularly occur, primarily in the n. and w. sectors of the continent; and a large group of resident species, dominated by the obscure and difficult *Basileuterus* genus. The former are mainly arboreal, while the latter favor forest and woodland undergrowth; most are frequent members of mixed-bird flocks. Some *Basileuterus* closely resemble certain *Hemispingus,* a genus now placed in the Thraupinae, but to some extent assignment to 1 genus has been arbitrary (and helps to emphasize the very close relationship of the 2 subfamilies). Forming as they do the first part of the 9-primaried Emberizine assemblage, the warblers of the Americas are *not* closely related to the warblers of the Old World, the Sylviidae. The Parulinae have a strictly New World distribution.

Migrant Wood-Warblers

A diverse assemblage of wood-warblers breeds in North America but passes the n. winter to the south, mainly in Middle America and the West Indies but with a number of species also reaching South America. Most of these are fully treated in the following 2 sections: "arboreal wood-warblers" and "wood-warblers of lower growth"; both sections contain several genera. However, it should be noted that migrant warblers whose occurrence in South America appears to be purely casual—a total of 8 species—are treated only in the Appendix, together with the other very rare migrants. In addition, 2 resident species (at least in part) in these primarily migratory genera are included here.

GROUP A

Various arboreal wood-warblers (mostly migratory from North America).
GENERA: *Parula, Vermivora, Dendroica, Mniotilta,* and *Setophaga*
The warblers in this group (the terms *wood-warblers* and *warblers* are used interchangeably within the N. Am. parulid group) are active, small birds with narrow bills; they regularly occur in flocks during their sojourn in South America, associating with various resident birds of various families. Two species, the Tropical Parula and some races of the Yellow Warbler, are resident. A majority are found most often in woodland and forest borders, but they may occur almost anywhere while actually migrating. Most food is obtained by gleaning for insects in foliage and to a lesser extent on other substrates. They rarely or never sing while in South America. All are sexually dimorphic (some strikingly so), and in some species immature plumages also differ from those of adults. As they are seen more frequently here, immature and nonbreeding plumages are emphasized. Wintering warblers are most numerous and diverse in n. and w. South America and on offshore islands, with the scarcer species either straggling in from Panama or overshooting from the West Indies.

Parula pitiayumi

TROPICAL PARULA

PLATE: 8

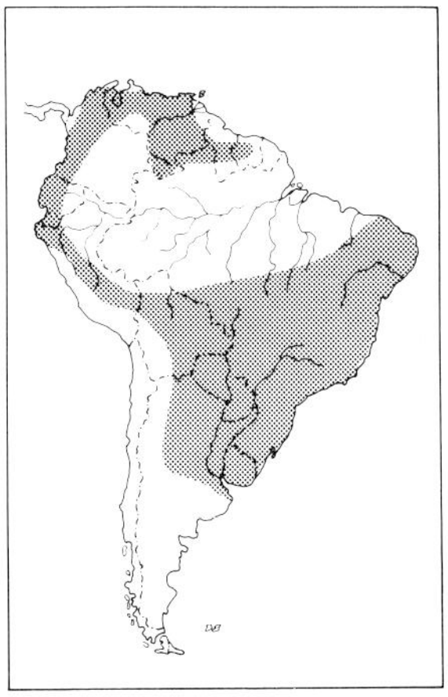

IDENTIFICATION: 11 cm (4¼"). *Mostly dull blue above* with *black foreface* (extending back only to around eyes) and olive patch on mid-back; 2 white wing-bars (varying in width). Below mostly bright yellow, with *throat and chest at least washed with ochraceous;* crissum white.
SIMILAR SPECIES: Combination of small size, bluish upperparts, and bright yellow below render this attractive warbler virtually unmistakable, but compare to various migrant species.
HABITAT AND BEHAVIOR: Fairly common to common and widespread in forest borders, deciduous and gallery woodland, and *chaco* scrub; avoids more humid lowlands (in humid regions found primarily in montane areas). Found singly or in pairs, usually foraging at considerable heights; often accompanies mixed flocks of tanagers, other warblers, etc. Males sing persistently, a buzzy trill, typically "tsip-tsip-tsip-tsip-tsip tsrrrrrrrrrip" but with many variations.
RANGE: Locally in Guyana and Suriname (Brownsberg Reserve), extreme n. Brazil (Roraima), much of Venezuela, w. Colombia (not in lowlands east of Andes), w. and e. Ecuador (in east only on Andean slopes), nw. and e. Peru (on Pacific slope south to Lambayeque and w. Cajamarca, on e. slope only in Andes), n. and e. Bolivia, Paraguay, s. and e. Brazil (north to Piauí, Maranhão, Goiás, and Mato Grosso), Uruguay, and n. Argentina (south to La Rioja, San Luis, Córdoba, and n. Buenos Aires); Trinidad and Tobago. Also Mexico to Panama. To about 2500 m.

Vermivora chrysoptera

GOLDEN-WINGED WARBLER

IDENTIFICATION: 12 cm (4¾"). *Bold facial pattern and golden wing-bars* mark this scarce warbler. Male gray above with bright yellow fore-

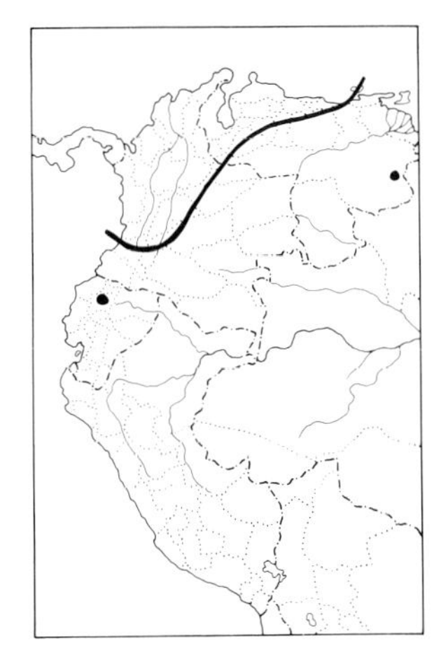

crown and *large yellow patch on wing-coverts.* Sides of head and underparts white to pale grayish, with *black patch on ear-coverts* and *black throat.* Female similar but with duller yellow forecrown and black of facial pattern *replaced by gray.* Immatures similar to respective adults but mantle may be tinged olive and underparts washed with pale yellow.

SIMILAR SPECIES: Male's striking facial pattern of black, white, and yellow make it easily recognized; female usually shows enough to be known as a Golden-wing (though pattern can be faint, in which case often the bold yellow on the wing is a better mark).

HABITAT AND BEHAVIOR: Rare to uncommon n.-winter resident (Sept.–Mar.) in forest canopy and borders and in lighter woodland. Usually seen singly. Forages at all levels but mostly quite high; regularly probes into dead leaf clusters. Often accompanies mixed flocks of insectivorous birds, especially of tanagers and other warblers. Based on studies on their breeding grounds, overall numbers of Golden-wings appear to be declining.

RANGE: Nonbreeding visitor to n. and w. Venezuela (east to Anzoátegui, with 1 sighting from e. Bolívar), w. Colombia (not recorded in lowlands east of the Andes), and nw. Ecuador (1 sighting from Pichincha). Breeds e. North America, wintering from s. Mexico south. Mostly 500–2000 m.

Vermivora peregrina

TENNESSEE WARBLER

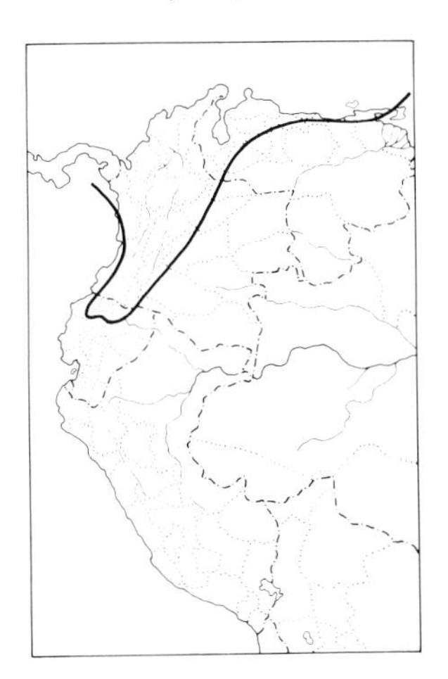

IDENTIFICATION: 12 cm (4¾″). Adult male bright olive above with gray crown and nape, *narrow white superciliary,* and dusky line through eye. Entire underparts white. Female similar but with crown and nape more olive, superciliary tinged yellow, and throat and chest washed with pale yellow. Immature uniform olive above with *narrow yellowish superciliary* and usually a single whitish wing-bar. Below more or less yellowish, brightest on breast and palest on belly and crissum (*latter usually white*). See V-33. Adult plumage assumed Jan.–Mar.

SIMILAR SPECIES: Can usually be known by its plain appearance in conjunction with the thin superciliary and very slender bill. Cf. various vireos (all of which have heavier bills and more sluggish habits), Lesser Greenlet (smaller and chunkier, with shorter tail; greenlet has white eye-ring rather than a superciliary), immature Yellow Warbler (which never has a superciliary, though it may show a single wing-bar), and female Chestnut-vented Conebill.

HABITAT AND BEHAVIOR: Locally common n.-winter resident (mostly Oct.–Apr.) in forest and woodland borders, clearings with scattered trees, and groves and gardens in agricultural areas; seems most numerous in Santa Marta Mts. Often in small groups, usually foraging actively high in trees. Searches mostly for insects but also seems very attracted to flowering *Erythrina* trees and to a lesser extent other flowering trees and shrubs.

RANGE: Nonbreeding visitor to n. and w. Venezuela (east to Sucre), w. Colombia (not recorded in lowlands east of the Andes), and n. Ecuador (recent sightings south to Pichincha, Napo, and Pastaza).

Breeds n. North America, wintering from Mexico south. Mostly 500–2000 m, less often to sea level and occasionally higher.

Dendroica petechia

YELLOW WARBLER

PLATE: 8

IDENTIFICATION: 12.5 cm (5″). Both n. migratory forms and resident forms occur (the latter being strictly coastal); males of both types are *mostly bright yellow.* Male of migrant type predominantly *bright yellow* with variable amount of *chestnut streaking* on breast and flanks; mantle more olive yellow with wings and tail duskier, latter with *large yellow patches* (obvious in flight). Female similar but duller yellow overall, with chestnut streaking faint or absent. Resident races of Caribbean coast have males with *entire head, throat, and chest rufous-chestnut,* while those of Pacific coast have *rufous-chestnut more or less restricted to crown,* with sides of head and throat at most tinged rusty. Birds from islands off Venezuelan coast are also only rufous-capped. Females of resident races are similar to migratory females but tend to be more olive or grayish above and grayish or whitish below.

SIMILAR SPECIES: The racial and individual variation is somewhat confusing, but the migrant forms can be recognized by their overall yellow appearance, with the prominent dark eye in contrast, and the yellow in the tail (latter not shown by other warblers). Residents are usually to be known by their coastal (*usually mangrove*) habitat and males by their rufous-chestnut hoods or crowns. Cf. immature Tennessee Warbler (can be quite similar but always shows a superciliary and never has the yellow in the tail) and Prothonotary Warbler (with contrasting blue-gray wings). Bicolored Conebill also occurs in mangroves along Caribbean coast (though the 2 species seem to exclude each other and rarely or never occur together), and immatures can be confused with dull Yellow Warblers, though their orange (not dark) eyes and pale legs should set them apart.

HABITAT AND BEHAVIOR: Migratory forms are fairly common (mainly Sept.–Apr.) in various semiopen habitats, especially near water, while the residents can be common but are very local and essentially restricted to mangroves along the coast. Both types are active and conspicuous birds, foraging in trees and shrubbery at all levels but often coming near to the ground or flying across open areas. Attention is drawn to them by their near-constant, excitable chipping. Migrants do not sing, but residents give a bright, lively, fast song which is variable but often seems to be composed of a series of their calls with an emphasized note at the end; it sounds rather different from the song of n. birds.

RANGE: Nonbreeding visitor south to cen. Peru, n. Bolivia, and Amaz. Brazil; resident locally along Caribbean coast of Colombia and Venezuela (east to Sucre) and on Pacific coast of Colombia, Ecuador, and nw. Peru (Tumbes); Trinidad and Tobago (as a migrant). Also breeds North America, Middle America, the West Indies, and the Galápagos Is.; N. Am. breeders wintering from Mexico south. Mostly below 1000 m, but occasionally recorded much higher (to over 2000 m), perhaps primarily as a transient.

NOTE: The 3 groups now considered to comprise this species were formerly considered full species in their own right: the *aestiva* group (Yellow Warbler) breeding in North America, the *petechia* group (Golden Warbler) of the West Indies, and the *erithachorides* group (Mangrove Warbler) of coastal Middle and South America. The latter 2 seem closer to each other than either is to *aestiva*, and they perhaps deserve to be separated as a full species from the n. Yellow Warbler.

Dendroica striata

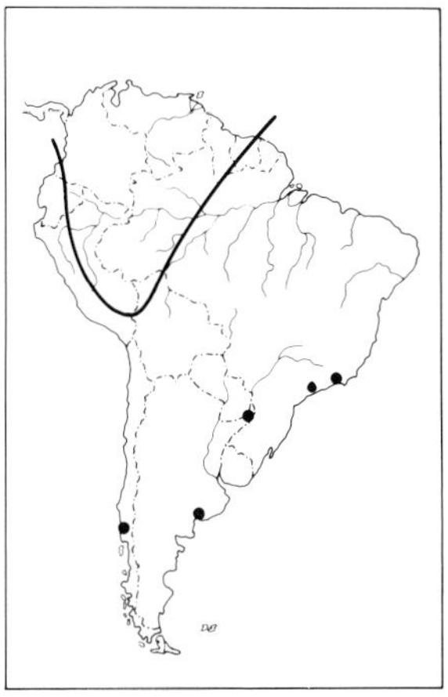

BLACKPOLL WARBLER

PLATE: 8

IDENTIFICATION: 13 cm (5"). Nonbreeding plumage: olive above streaked with blackish on crown and mantle; 2 whitish wing-bars. Below pale greenish yellow to pale yellowish, with *fine dusky olive streaking across breast and down flanks;* belly and (especially) crissum white. Extent of streaking below varies, but some almost always present (least in immature females, which also may show yellowish buff crissum). Breeding plumage male very different, with striking *black crown and white cheeks,* black and olive streaked upperparts, and white underparts with *black malar stripe and bold black streaking down sides;* this plumage is assumed by March. Breeding female similar to birds in nonbreeding plumage but more sharply streaked blackish below.

SIMILAR SPECIES: Dapper black and white males are easily identified, but in South America this plumage is seen only briefly. Nonbreeders, however, are very dull and easily confused, especially with Bay-breasted Warbler (see below). See also nonbreeding plumage Blackburnian Warbler and female Cerulean Warbler.

HABITAT AND BEHAVIOR: Uncommon to fairly common n.-winter resident (mostly Sept.–Apr., occasionally to May) in forest borders, lighter woodland, and clearings with scattered trees. Basically an arboreal bird, often feeding high in trees, though it comes lower when migrating. Considering its abundance in North America, this species seems surprisingly scarce or little noticed on its S. Am. wintering grounds. Judging from the number of collected specimens, it may be more numerous in s. Venezuela than elsewhere; in n. Venezuela, however, it appears to occur mostly as a transient (A. Altman). H. Sick (*Wilson Bull.* 83 [2]: 198–200, 1971) found them feeding regularly in tamarind trees (*Tamarindus indicus*) in residential areas.

RANGE: Nonbreeding visitor to Guianas (only casual?), Venezuela, Colombia, e. Ecuador, e. Peru, and w. Amaz. Brazil; occasionally overshoots far to the south of its normal wintering range (e.g., 1 Chile record and 2 from e. Argentina) and perhaps regular locally in se. Brazil (Rio de Janeiro and s. São Paulo); Trinidad and Tobago. Breeds n. North America, migrating through West Indies. Mostly below 1000 m, though notably higher (to 2500 m or more) on migration.

Dendroica castanea

BAY-BREASTED WARBLER

IDENTIFICATION: 13 cm (5"). Nonbreeding plumage *very* closely resembles that of Blackpoll Warbler; note that *Bay-breasted is known mainly from west of the Andes,* whereas *Blackpoll winters almost exclusively east of the Andes* (occurring west and north of them on migration). Not all individuals can be certainly identified, but the key marks are the *buff*

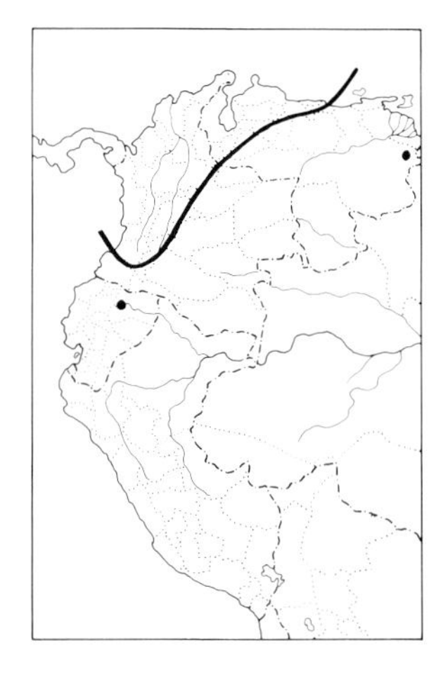

(not so yellowish or greenish) *tone to the underparts,* with *streaking absent or virtually so* (much less than is typically shown in Blackpoll); the *trace of chestnut on the flanks* (shown by virtually all adults and by some immatures; virtually all birds show at least a trace by Jan.–Feb.); and the *dingy buffy whitish crissum* (never the pure white shown by most Blackpolls). Leg color is *not* an entirely reliable criterion, for while all Bay-breasteds have dark legs, and while many Blackpolls have pale legs (pinkish or straw-colored), some Blackpoll legs can look dark. Breeding plumage male very different and unmistakable: *crown, throat, chest, and sides dark chestnut* with black forehead and face and *large, pale buff patch on sides of neck;* upperparts otherwise streaked black and buffy olive, and with whitish lower underparts. This plumage is assumed in March. Female similar but with pattern much duller.

HABITAT AND BEHAVIOR: Fairly common to common n.-winter resident (Oct.–Apr.) in forest borders, second-growth woodland, and clearings with scattered trees. Behavior similar to Blackpoll Warbler's; as with that species, 1 or a few often accompany mixed flocks of resident insectivorous birds.

RANGE: Nonbreeding visitor to n. and w. Venezuela (east mainly to Distrito Federal, with 1 sighting from e. Bolívar), w. Colombia (not recorded in lowlands east of E. Andes), and n. Ecuador (1 sighting from w. Napo and Pastaza); Trinidad (a few sightings). Breeds n. North America; winters mostly from Panama south, migrating through Central America and Greater Antilles. Mostly below 800 m, occasionally much higher (to 3100 m in Venezuela).

Dendroica cerulea

CERULEAN WARBLER

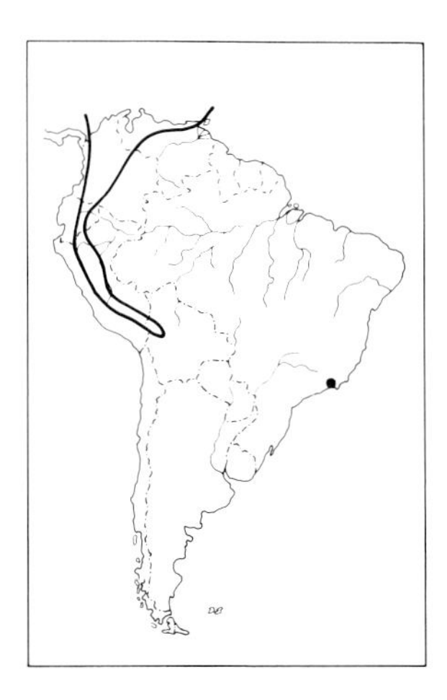

IDENTIFICATION: 12 cm (4¾"). Male *azure blue above,* brightest on crown, and streaked black on back; wings and tail duskier, former with 2 bold white wing-bars. Below white with *narrow black band across chest* (sometimes incomplete) and some black streaking on sides. Female more *bluish green above* with *narrow, pale yellowish superciliary* and no streaking on back; wings and tail as in male. Below pale yellowish to whitish, usually devoid of streaking (at most a little on sides). Immatures similar to female, though males may already show a faint chest band, and females tend to be greener (less blue) above and yellower below.

SIMILAR SPECIES: Bluish overall tone to males as well as the chest band (conspicuous from below even when dorsal color cannot be seen) are both good marks. Females and immatures are more confusing; compare especially to both Bay-breasted and Blackpoll Warblers (both of which are notably larger and streaked on back, do not have as prominent a superciliary, etc.).

HABITAT AND BEHAVIOR: Uncommon n.-winter resident (mostly Oct.–Mar.) in canopy and borders of forest and woodland. Probably to some extent overlooked due to its propensity to forage well above the ground and its quiet behavior. Usually noted singly, most often while accompanying mixed flocks of other insectivorous birds.

RANGE: Nonbreeding visitor to n. and w. Venezuela (east to Sucre), w. Colombia (east to e. slope of E. Andes and Macarena Mts. of w. Meta), and on e. slope of Andes and the adjacent lowlands in e. Ecuador, e. Peru, and nw. Bolivia (La Paz and Beni), with 2 sightings from se. Brazil (Rio de Janeiro). Breeds e. United States; winters mostly in South America, migrating through Central America and Greater Antilles. Mostly 500–2000 m.

Dendroica pensylvanica

CHESTNUT-SIDED WARBLER

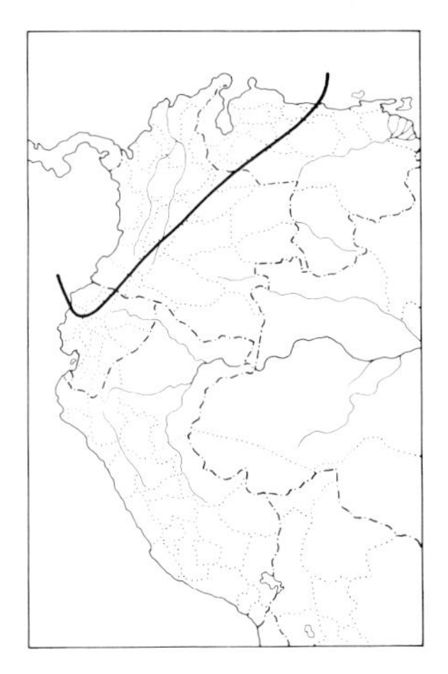

IDENTIFICATION: 12.5 cm (5″). Rare straggler to Colombia, Venezuela, Ecuador, and Netherlands Antilles. Nonbreeding plumage: *bright yellowish olive green above* with narrow white eye-ring; wings and tail duskier with 2 yellowish wing-bars. *Entire underparts grayish white,* extending up over lower cheeks; adults generally show at least some *chestnut on sides.* Breeding plumage male has *bright yellow crown,* black-streaked back, whiter cheeks and underparts, and a black stripe back from eyeline and malar streak, latter connecting to *chestnut patch on sides;* female similar but duller. Breeding plumage assumed by March.

SIMILAR SPECIES: Bright lemon green upperparts in conjunction with the whitish underparts are distinctive, even when the chestnut on the sides is not evident; by Jan. almost all birds show a trace of the chestnut. Nonbreeding plumage Bay-breasted Warbler is larger, never shows an eye-ring, tends to be buffier (not as white) below, etc.

HABITAT AND BEHAVIOR: Rare n.-winter visitant (recorded Oct.–Apr., once in May) to forest borders, secondary woodland, and clearings. Forages very actively, often with drooped wings and characteristically cocked tail.

RANGE: Nonbreeding visitor to w. and n. Colombia (Tolima, Santander, and Valle records), w. Ecuador (1 Pichincha record), and n. Venezuela (3 Aragua records and 1 from Miranda); Netherlands Antilles. Breeds e. North America, wintering mostly Guatemala to Panama.

Dendroica fusca

BLACKBURNIAN WARBLER

PLATE: 8

IDENTIFICATION: 12.5 cm (5″). *Yellow or orange outlining dark cheeks* in all plumages. Adult male unmistakable: *mostly black above* with white back striping and *large white wing-patch; fiery orange on center of crown as broad stripe encircling the black ear-coverts and* (especially bright) *on throat and chest;* underparts fade to pale yellow on breast and white on lower belly; sides and flanks streaked black. This plumage can be seen except soon after its arrival (Oct.–Nov.), when even older males are duller below and mottled with olive above. Adult females and immatures *resemble adult males in pattern but are much duller:* they are mostly grayish olive above streaked with blackish (most solidly black in immature males), have the bright orange replaced by yellow of varying intensities (dullest in immature females), and have 2 white wing-bars rather than a solid patch.

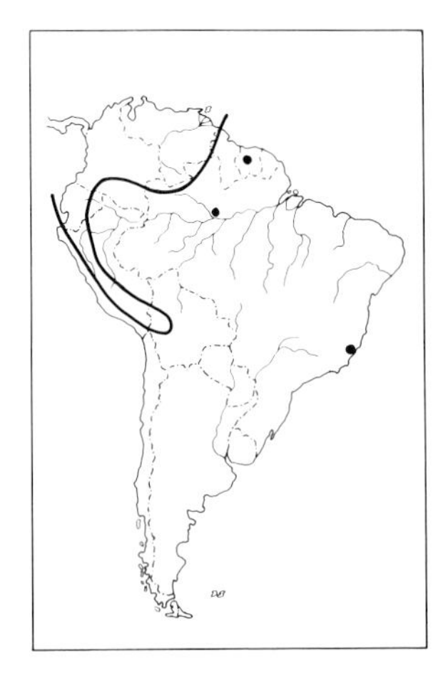

SIMILAR SPECIES: Blackburnians in immature or adult female plumage are best known by their *sharply outlined facial pattern,* yellow on the anterior underparts fading to white on belly, and side streaking. Cf. Black-throated Green and Cape May Warblers (both rare).
HABITAT AND BEHAVIOR: Common n.-winter resident (mostly Sept.–Apr., a few May records) in forest canopy and borders, lighter secondary woodland, and clearings with scattered trees. Especially numerous in montane regions (and may winter almost exclusively in such areas, occurring lower only as a transient); commonest in the Colombian Andes, where it is easily the most numerous of the n. warblers. Forages actively at all levels but most often high; small loose groups often accompany mixed foraging flocks of tanagers, flycatchers, etc.
RANGE: Nonbreeding visitor to Venezuela (including the *tepui* region of the south) and Colombia (not recorded from lowlands of the southeast) and in Andes and adjacent lower areas of Ecuador, Peru, and nw. Bolivia (a few reports from La Paz); occasionally overshoots beyond its normal wintering range (sightings from Suriname and from near Manaus and in Espírito Santo, Brazil). Breeds e. North America; winters mostly in South America, migrating through Central America and the Greater Antilles. Mostly 500–2500 m.

Dendroica tigrina

CAPE MAY WARBLER

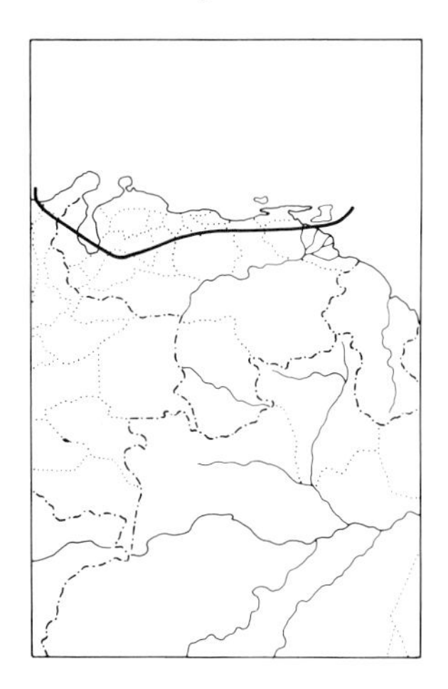

IDENTIFICATION: 12.5 cm (5″). Rare straggler to Colombia, Venezuela, and Caribbean islands. Distinctive *yellow patch on sides of neck* in all plumages (may be dull in immature females). Nonbreeding plumage (adults brighter than immatures): olive grayish above with the yellowish neck patch and *greenish yellow rump*; wings duskier with 2 whitish wing-bars. Yellowish or whitish below, variably (but often *heavily*) *streaked dusky.* Breeding plumage male notably brighter: yellowish green above streaked black and with blackish crown, *chestnut cheeks and bright yellow neck patch,* large white wing-patch, and bright yellow below heavily streaked black. Breeding female drabber but with same pattern, lacking only the chestnut cheeks. Breeding plumage gradually assumed Feb.–Mar.
SIMILAR SPECIES: Among other warblers regularly reaching South America, perhaps most resembles Blackburnian in immature plumage; Blackburnian has streaking below restricted to sides, much more prominent superciliary, etc.
HABITAT AND BEHAVIOR: Rare n.-winter visitant (recorded Oct.–Apr.) to clearings, gardens, and woodland borders; mostly near the coast. There perhaps has been a genuine increase in numbers reaching South America correlated with a general population increase on its breeding grounds. On its wintering grounds it seems very attracted to flowering trees.
RANGE: Nonbreeding visitor to n. Colombia (1 record from Magdalena), n. Venezuela (recent records from Portuguesa, Distrito Federal, and several offshore islands); Trinidad and Tobago (almost regular on the latter) and Netherlands Antilles. Breeds n. North America; winters mostly in West Indies.

Dendroica virens

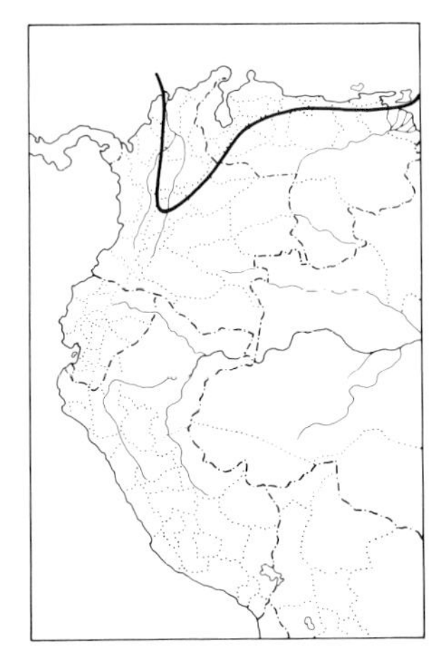

BLACK-THROATED GREEN WARBLER

IDENTIFICATION: 12.5 cm (5″). Rare straggler to Colombia, Venezuela, and Caribbean islands. *Contrasting yellow face* in all plumages. Adult male olive above with sides of head mostly bright yellow (cheeks faintly outlined olive); wings and tail blackish with 2 bold white wing-bars. *Throat and chest black,* extending down as black streaking on sides; underparts otherwise white, sometimes tinged yellow on crissum. Female similar but not as bright and with less black below; throat usually yellowish. Immatures even duller (especially females), often showing next to no black on throat or chest and with blackish streaking on sides faint or absent.

SIMILAR SPECIES: Immatures can be confused with immature Blackburnian Warbler but are more solidly yellow on face and lack back streaking.

HABITAT AND BEHAVIOR: Rare n.-winter visitant (mostly Oct.–Apr.) in forest canopy and borders and in clearings with scattered trees. Behavior much like Blackburnian Warbler's. Recorded fairly regularly from the Santa Marta Mts.

RANGE: Nonbreeding visitor to n. Colombia (numerous records from Magdalena, several also from Cundinamarca) and n. Venezuela (recorded Zulia, Aragua, and Miranda); Trinidad and Netherlands Antilles. Breeds e. North America, wintering both in Middle America and West Indies.

Mniotilta varia

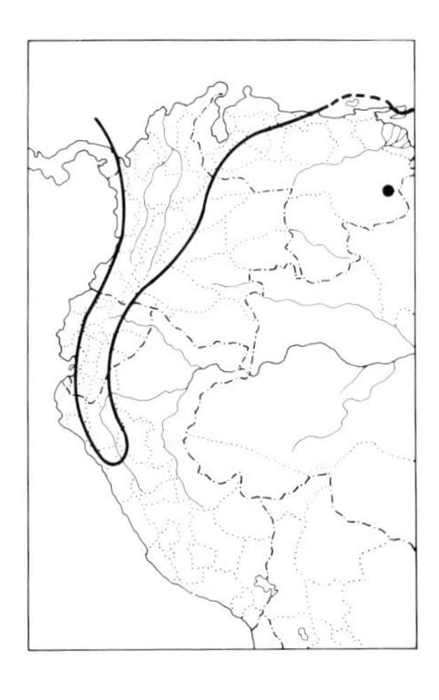

BLACK-AND-WHITE WARBLER

IDENTIFICATION: 12.5 cm (5″). *Bold black and white streaking* in all plumages. Male's streaking is more crisply delineated and it has black cheeks and throat; streaking below more extensive but center of belly white. Females and immatures have whitish cheeks and are whiter below generally, with blurry blackish streaking across breast and down flanks. See V-33.

SIMILAR SPECIES: Black and white appearance and distinctive behavior (see below) render this species virtually unmistakable.

HABITAT AND BEHAVIOR: Uncommon n.-winter resident (mostly Sept.–Mar.) in forest borders and secondary woodland. Generally found singly, often accompanying mixed flocks. *Regularly creeps over trunks and larger limbs,* gleaning bark for insects.

RANGE: Nonbreeding visitor to n. and w. Venezuela (east mainly to Miranda, with 1 sighting from e. Bolívar), w. Colombia (not recorded in lowlands east of Andes), w. Ecuador, and n. Peru (recent sightings from Lambayeque, Amazonas, and La Libertad); Trinidad. Breeds e. North America, wintering from s. United States to South America. Mostly below 2000 m.

Setophaga ruticilla

AMERICAN REDSTART PLATE: 8

IDENTIFICATION: 12.5 cm (5″). Adult male *mostly black* with *large orange patches on wings and sides and at base of tail;* belly white. See

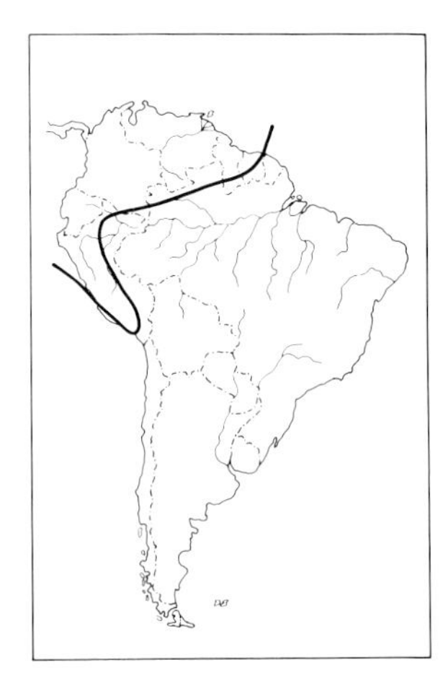

V-34. Females and immatures grayish olive above, grayest on head with narrow white spectacles; mostly whitish below; patches on wings and sides and at base of tail as in male but *yellow instead of orange.* Immature males require 2 years to acquire full adult plumage: they become blacker (first on the face and foreneck but often splotchy) and their patches gradually become more orange.

SIMILAR SPECIES: Regardless of plumage, the orange or yellow patches are unique and, together with its distinctive behavior (see below), render this species easily recognized. Does not resemble the neotropical genus *Myioborus* in color and is apparently not closely related.

HABITAT AND BEHAVIOR: Uncommon to locally fairly common n.-winter resident (mostly Sept.–Apr.) in forest borders, lighter woodland, and shrubby areas with scattered trees. Often most numerous in mangroves, elsewhere usually in small numbers. Behavior is active and animated, with tail frequently fanned and wings partially spread as if to show off its bright colors. Often sallies short distances, snapping up flying insects from the air or hovering while picking them up from foliage.

RANGE: Nonbreeding visitor to Guyana and Suriname (few records), Venezuela, extreme n. Brazil (Roraima), Colombia (not recorded from se. lowlands), Ecuador, and Peru (few records, but seen south to Apurímac and Arequipa); Trinidad and Tobago. Breeds North America, wintering from s. United States south. Mostly below 1500 m (but recorded occasionally to 3000 m).

GROUP B

Various wood-warblers of lower growth (all migratory from North America).

GENERA: *Dendroica, Protonotaria, Seiurus, Wilsonia,* and *Oporornis*

A second assemblage of N. Am. breeding warblers which pass the n. winter regularly in South America is presented here. These species tend to remain near the ground (often on the ground, especially in *Seiurus*). They are often more difficult to see than those in the previous section, for many typically forage in dense undergrowth; most, however, readily respond to persistent squeaking. One rare *Dendroica,* the Black-throated Blue, is included here, but the rare Common Yellowthroat is not, it being placed with the resident members of its genus (see below).

Dendroica caerulescens

BLACK-THROATED BLUE WARBLER

IDENTIFICATION: 13 cm (5″). Rare straggler to Colombia, Venezuela, and Caribbean islands. Dapper male unmistakable: *dark grayish blue above* (tinged greenish in immatures) with *bold small white speculum* at base of primaries. *Sides of head, throat, and chest as well as sides contrastingly black;* underparts otherwise white. Female much duller: unstreaked brownish olive above with narrow whitish superciliary and partial eye-ring and *whitish wing-patch* similar to male's (but sometimes smaller and rarely absent). Dull yellowish brown to buff below.

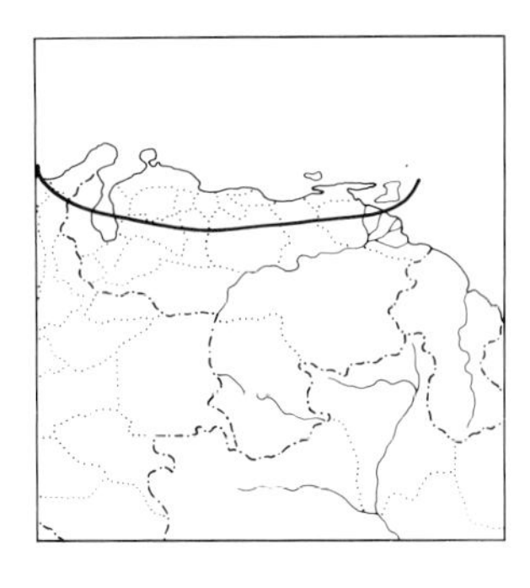

SIMILAR SPECIES: Female relatively drab but can be known by its brownish overall coloration and the speculum (small but usually prominent), a mark shared by no other similar bird.

HABITAT AND BEHAVIOR: Rare n.-winter visitant (recorded Sept.–Mar.) to lower growth of woodland and forest borders. Often relatively inconspicuous but at times easily attracted by squeaking.

RANGE: Nonbreeding visitor to n. Colombia (single records from Magdalena and Guajira) and n. Venezuela (single records from Aragua and Guárico); Trinidad and Netherlands Antilles. Breeds e. North America, wintering mostly in West Indies.

Protonotaria citrea

PROTHONOTARY WARBLER

IDENTIFICATION: 13.5 cm (5¼"). Male has *head and underparts bright orange-yellow* (brightest on head and foreneck); white on lower belly and crissum. Back bright olive; *wings, rump, and tail blue-gray,* tail with white at base (conspicuous in flight). Female similarly patterned but duller, less orange; head tinged olive.

SIMILAR SPECIES: Yellow Warbler never shows the blue-gray wings and has yellow (not white) in tail. Cf. also female Hooded Warbler.

HABITAT AND BEHAVIOR: Locally fairly common to common n.-winter resident (mostly Sept.–Mar.) in mangroves (where it is especially numerous) and woodland and scrub, usually near water. Commonest along or near the coast. Usually forages rather low in shrubbery and trees; single birds or groups are readily attracted by squeaking, but the species usually does not accompany mixed flocks.

RANGE: Nonbreeding visitor to Guyana and Suriname (few records), n. Venezuela, w. Colombia, and nw. Ecuador (Esmeraldas and Pichincha); Trinidad and Tobago. Breeds e. United States, wintering in Central and South America. Mostly below 500 m, but occurs higher when migrating (once recorded to 3300 m).

Seiurus noveboracensis

NORTHERN WATERTHRUSH

PLATE: 8

IDENTIFICATION: 14–15 cm (5½–5¾"). Olive brown above with *prominent yellowish to buff superciliary.* Pale yellowish to whitish below, *boldly streaked with dark brown* except on throat (where spotting faint) and lower belly.

SIMILAR SPECIES: Shape, behavior, and streaking below distinctive among regularly occurring birds in South America, but compare to rare Louisiana Waterthrush and Ovenbird. Cf. also Buff-rumped and River Warblers, whose behavior and water-edge habitat are reminiscent.

HABITAT AND BEHAVIOR: Locally fairly common to common n.-winter resident (mostly Sept.–Apr.) in mangroves and in woodland and shrubbery *near water;* less numerous in Ecuador, Peru, and Brazil. Usually found singly; feeds on the ground at edge of water or in damp places, walking sedately, slowly bobbing its tail and rearparts up and down. Its loud metallic call, "tchink," is distinctive and frequently draws attention to the bird as it skulks in some swampy recess; the call is also often given in flight.

RANGE: Nonbreeding visitor to Guianas, Venezuela, n. Brazil (recorded only from n. Pará, but probably more widespread), Colombia, n. Ecuador (recorded Esmeraldas, Napo, and Pastaza), and ne. Peru (mouth of Río Curaray in Loreto); Trinidad and Tobago. Breeds n. North America, wintering from s. United States south. Mostly below 2000 m.

Seiurus motacilla

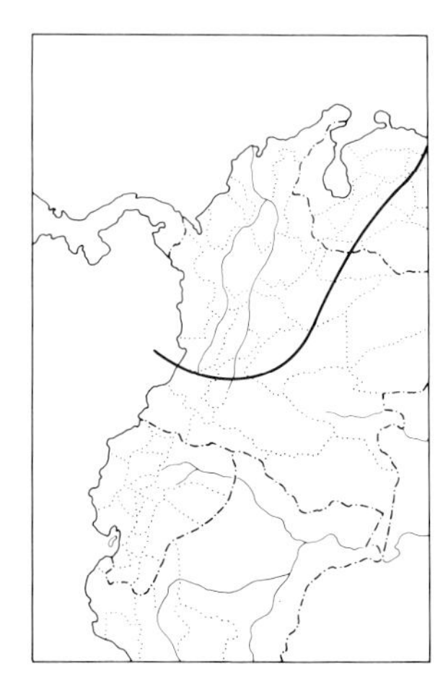

LOUISIANA WATERTHRUSH

IDENTIFICATION: 15.5 cm (6″). A rare straggler to Colombia, Venezuela, and Netherlands Antilles. Resembles much more numerous Northern Waterthrush; slightly larger, with longer bill and brighter pinkish legs. Best told by the *wider and whiter superciliary,* notably broader and whitest behind the eye (not uniformly narrow and buffyish as in Northern), and by the often quite obvious contrast between *buffy flanks* and white ground color of rest of underparts (Northern can look buffy yellowish below but always more or less uniform, never with flanks in contrast).

HABITAT AND BEHAVIOR: Rare n.-winter visitant (recorded Oct.–Feb., but doubtless occurs earlier, as in Panama it regularly arrives in Aug.). Typically associated with margins of *running* water, most often in forested or at least shady areas, but on migration may occur elsewhere (though almost always near water of some sort). Rarely or never occurs in the mangrove swamps where Northern Waterthrush is so prevalent. General behavior and metallic call very like Northern Waterthrush's.

RANGE: Nonbreeding visitor to w. and n. Colombia (south to Valle and Meta; most frequent in the Santa Marta region) and w. Venezuela (1 record from Lara and a sighting from Mérida); Netherlands Antilles. Breeds e. United States, mainly wintering in Middle America and West Indies.

Seiurus aurocapillus

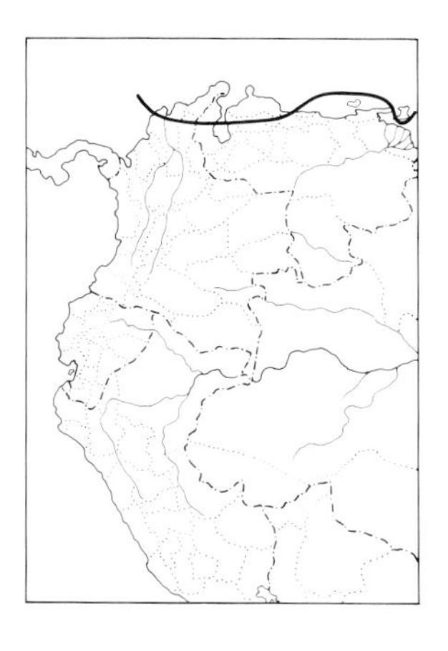

OVENBIRD

IDENTIFICATION: 14 cm (5½″). A rare straggler to Colombia, Venezuela, and Caribbean islands. Olive above with *bold white eye-ring* and *dull orange coronal stripe bordered with black.* Mostly white below, conspicuously *streaked black across breast and down flanks,* flanks also tinged olive.

SIMILAR SPECIES: Shape, posture, and overall coloration much as Northern Waterthrush's. Ovenbird, however, is not associated with water and it has an eye-ring rather than a superciliary; the waterthrush lacks all crown striping.

HABITAT AND BEHAVIOR: Rare n.-winter visitant (mostly Oct.–Apr.) to lower growth of woodland and forest. Shy and unobtrusive, Ovenbirds are probably often overlooked on their wintering grounds. They are usually seen walking on the ground, often slowly bobbing their

rearparts up and down. They are readily attracted by squeaking, when they may fly in and walk back and forth on a horizontal limb.
RANGE: Nonbreeding visitor to n. Colombia (several records from Magdalena) and n. Venezuela (a few records from Falcón, Aragua, and offshore islands); Trinidad and Tobago and Netherlands Antilles (where perhaps annual). Breeds North America, wintering mostly in Middle America and West Indies.

Wilsonia canadensis

CANADA WARBLER

PLATE: 8

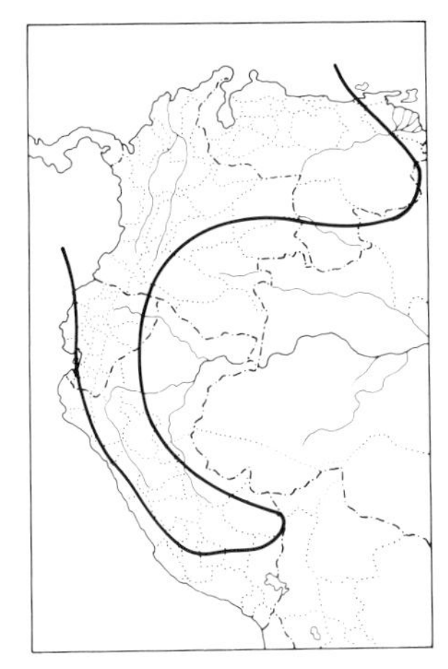

IDENTIFICATION: 13.5 cm (5¼"). Always best known by the "*necklace*" *of streaks across chest*. Male *bluish gray above* except for black on forecrown and sides of head and *prominent yellow spectacles*. Bright yellow below with *conspicuous band of black streaks across chest;* crissum white. Female similar but is somewhat paler (less bluish) dorsally, has fainter necklace and reduced black facial markings. Immatures have even fainter necklace (sometimes almost lacking) and may be tinged olive on back.
SIMILAR SPECIES: In the few Canadas in which the necklace is barely discernible or lacking, look for the spectacles and uniform gray coloration above. Cf. female Hooded and Prothonotary Warblers; among resident warblers, Canada most resembles the range-restricted Gray-and-gold (which lacks spectacled look, has a coronal stripe, etc.).
HABITAT AND BEHAVIOR: Fairly common n.-winter resident (mostly Oct.–Apr.) in lower growth of forest and secondary woodland and their shrubby borders. Most numerous in montane areas, occurring in lowlands primarily as a transient. Behavior quite active and restless, gleaning insects from foliage and occasionally making short aerial sallies. Often with mixed flocks.
RANGE: Nonbreeding visitor to Venezuela (including the *tepui* region of the south), Colombia (not recorded from se. lowlands), Ecuador (where scarce on the w. slope), e. Peru (south to Cuzco and Madre de Dios), and extreme n. Brazil (Roraima). Breeds n. North America, wintering mostly in South America, migrating through Middle America (where a few overwinter). Mostly below 2000 m.

Wilsonia citrina

HOODED WARBLER

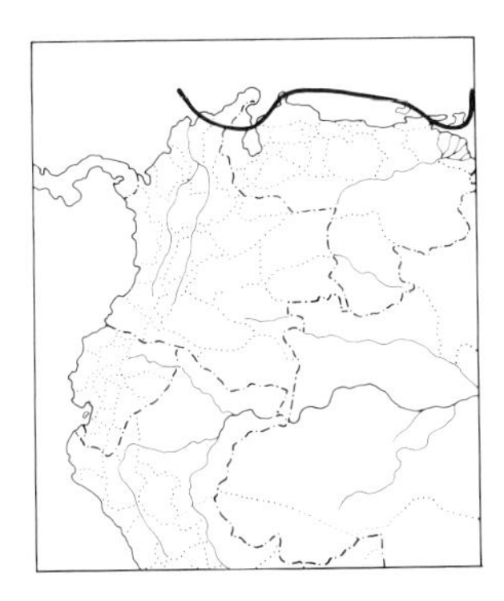

IDENTIFICATION: 14 cm (5½"). A rare straggler to Colombia, Venezuela, and Caribbean islands. Male unmistakable with *striking black hood enclosing bright yellow forehead and face*. Otherwise olive above and bright yellow below; inner web of outer tail feathers mostly white. Female lacks the black (or it shows merely as a scattering of a few black feathers) but is otherwise like male (including bright yellow underparts *extending to forehead and face* and the *white in tail*).
SIMILAR SPECIES: Female Prothonotary Warbler has all-yellow head, contrasting blue-gray wings, and white lower belly and crissum. Immature Canada Warblers lacking their necklace (diagnostic when present) are vaguely similar. Cf. also Flavescent Warbler (superficially

similar but with yellow superciliary and lacking white in tail) and female *Geothlypis* yellowthroats.

HABITAT AND BEHAVIOR: Rare n.-winter visitant (recorded Oct.–Apr.) to lower growth of woodland and borders. Forages actively, often making short aerial sallies and frequently flicking its tail open, exposing the white.

RANGE: Nonbreeding visitor to n. Colombia (several records from Magdalena) and n. Venezuela (Zulia and offshore is.); Trinidad and Netherlands Antilles. Breeds e. United States, wintering mostly in Middle America.

Oporornis philadelphia

MOURNING WARBLER

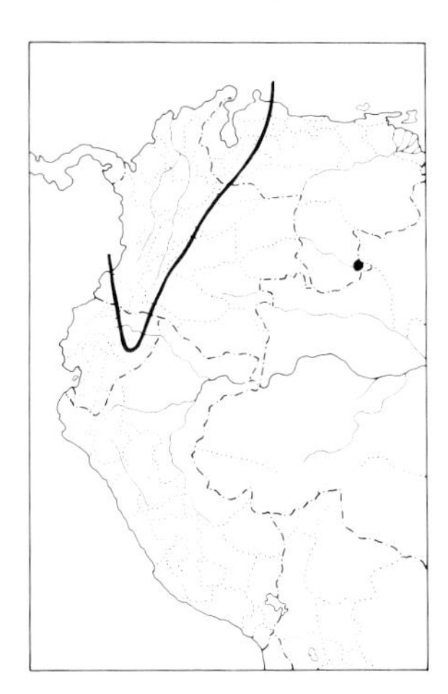

IDENTIFICATION: 13 cm (5″). Male has *gray hood which becomes black on chest and sometimes throat* (often with scaly effect); otherwise olive above, yellow below. Female similar but duller, with no black on the bib and often with throat tinged yellowish buff; may show an incomplete white eye-ring. Immatures often lack the hooded effect entirely or it is tinged brownish and most pronounced on sides of chest; an eye-ring is usually present, typically broken in front and often behind the eye but occasionally virtually complete.

SIMILAR SPECIES: Connecticut Warbler has a complete white eye-ring in all plumages; immature Mournings showing more or less complete eye-rings are therefore difficult to distinguish, but are smaller with eye-ring less prominent, have a less brownish hood, and throat is typically paler (yellowish or buff). Note, too, that they *mostly* separate out by range. Cf. also White-lored Warbler (of Santa Marta Mts.).

HABITAT AND BEHAVIOR: Uncommon to fairly common n.-winter resident (mostly Oct.–Apr.) in shrubby woodland and forest borders and in grassy clearings with scattered bushes and tangles of vegetation. A skulker, hopping about on or near the ground, most often near water. Almost invariably found singly and does not tend to associate with mixed flocks.

RANGE: Nonbreeding visitor to w. Venezuela (also an isolated record from Sierra Parima in e. Amazonas), w. Colombia (not recorded in e. lowlands at any distance from the Andes), and e. Ecuador. Breeds n. North America, migrating mostly through Middle America, wintering from Nicaragua south. Mostly below 2000 m.

Oporornis agilis

CONNECTICUT WARBLER

IDENTIFICATION: 14 cm (5½″). In all plumages shows a *conspicuous, complete white or buffy whitish eye-ring*. Male mostly olive above with *gray hood* (palest on throat, purest gray on chest). Lower underparts dull yellow. Female similar but *hood more grayish brown or even brownish buff*, palest on throat. Immatures similar to female but with duller and even browner hood and more brownish olive upperparts.

SIMILAR SPECIES: Most likely confused with Mourning Warbler. Though adult males are relatively easily distinguished, based on the

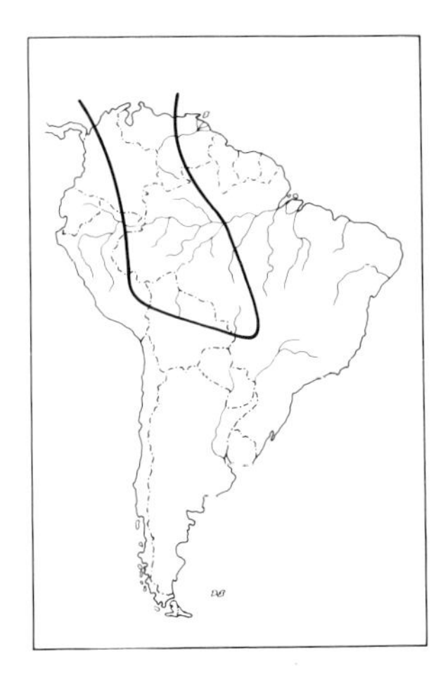

presence or absence of an eye-ring, females and immatures are often *very* tricky. In most Mournings the eye-ring is incomplete, and it usually is not as prominent as in the Connecticut. Other helpful aids are the Connecticut's longer under tail-coverts (reaching more than halfway down underside of tail and imparting a short-tailed appearance), larger size, and browner hood and duller yellow underparts.

HABITAT AND BEHAVIOR: Rare to locally uncommon n.-winter resident (mostly Oct.–Apr.) in woodland and forest borders and shrubby clearings; doubtless often overlooked, as it is an inveterate skulker. Not a well-known bird in South America, where from the scatter of records from across Amazonia it appears to overwinter widely but thinly. Sometimes seen walking on the ground, when it bobs its head and elevates its tail (Mourning Warbler hops when it is on the ground).

RANGE: Nonbreeding visitor to w. and cen. Venezuela (east to Distrito Federal), n. and e. Colombia, w. and cen. Amaz. Brazil (south to n. Mato Grosso), and se. Peru (1 record from Explorer's Inn in Madre de Dios). Breeds n. North America, migrating mostly through West Indies. Mostly below 2000 m, but recorded much higher (to 4200 m in Venezuela) as a transient.

Oporornis formosus

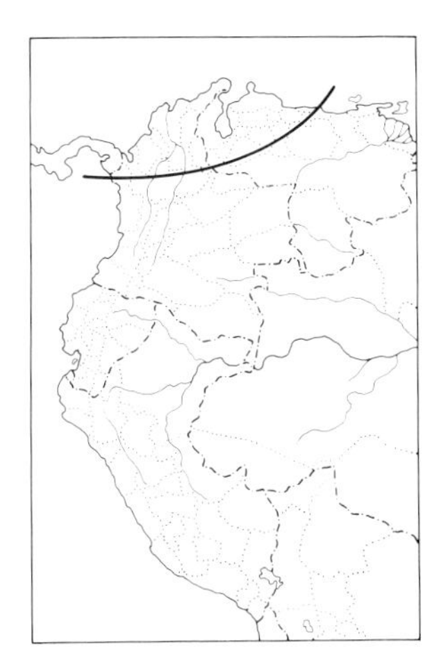

KENTUCKY WARBLER

IDENTIFICATION: 13.5 cm (5¼″). A rare straggler to Colombia, Venezuela, and Netherlands Antilles. Mostly olive above with *black forecrown and sides of head and neck* (forming "sideburns") and *prominent yellow supraloral stripe and incomplete eye-ring* separating them. Entirely bright yellow below. Female has less extensive black on face (crown often looks flecked and is grayer), while in immatures the black is sometimes entirely lacking.

SIMILAR SPECIES: Most likely to be confused with immature Canada Warbler (which may lack the diagnostic necklace), but Canada is basically gray (not olive) above; facial patterns of the species can be quite similar. Cf. also the various *Geothlypis* yellowthroats (males of which do show black on face, but none also shows yellow around eye).

HABITAT AND BEHAVIOR: Rare n.-winter visitant (mostly Oct.–Mar.) to undergrowth of forest and woodland. Very inconspicuous on its wintering grounds and doubtless often overlooked. In Panama regularly found in attendance at army ant swarms.

RANGE: Nonbreeding visitor to n. Colombia (several records from Magdalena) and n. Venezuela (recorded from Zulia, Táchira, Mérida, and Aragua); Netherlands Antilles. Breeds e. United States, wintering mostly in Middle America.

Granatellus Chats

A distinctive small genus of presumed warblers (actually the rather stout bill is more like a tanager's) represented in South America by 1 species (2 others in Mexico and Guatemala). The rosy red or pink color is unique among the S. Am. warblers.

Granatellus pelzelni

ROSE-BREASTED CHAT

PLATE: 9

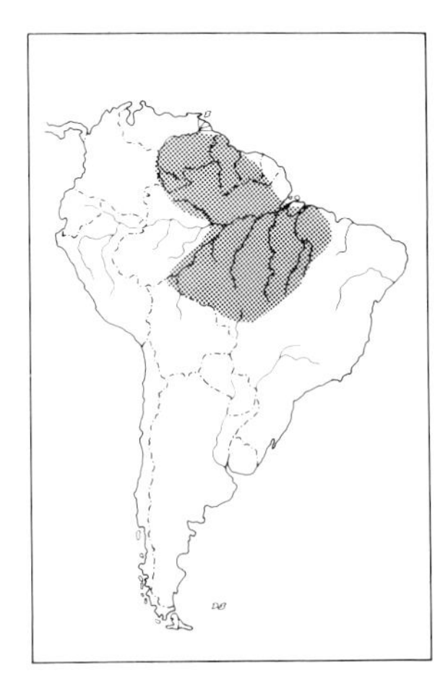

IDENTIFICATION: 12–12.5 cm (4¾–5″). Bill blackish, lower mandible basally pale gray. Male has crown and sides of head black with *prominent white postocular stripe;* otherwise *mostly blue-gray above,* with black tail. Throat white, narrowly outlined with black; *remaining underparts mostly rosy red,* with white on flanks. *Paraensis* (e. Brazil south of the Amazon east of lower Rio Tocantins) similar but with only forecrown black and no white on flanks. Female *blue-gray above,* lacking the black on head. *Forehead, postocular, face, and most of underparts cinnamon buff* (richest on face, whitest on throat and belly), but *crissum contrastingly pink.* Tail black as in male.

SIMILAR SPECIES: Pretty males are unmistakable, and basic color pattern of females (gray above, buff below) also virtually unique, but cf. Bicolored and other conebills; none of the latter shows the postocular or the pink crissum.

HABITAT AND BEHAVIOR: Uncommon in canopy and borders of woodland and deciduous forest. Typically forages well above the ground, often in viny tangles at mid-levels in openings or at edge, but sometimes comes lower, into shrubbery along streams or rivers or at forest borders. Usually in pairs and often with mixed flocks of other insectivorous birds; usually perches quite horizontally and sometimes cocks or fans its tail. Male's song is a series of 5 to 6 clear, sweet, evenly pitched notes, "t-weet, t-weet, t-weet . . ." or "sweet, sweet, tuwee-tuwee-tuwee-tuwee," while call (both sexes) is a sharp, dry "jrrt," often given in series.

RANGE: Guianas (not recorded French Guiana but surely occurs), s. Venezuela (Amazonas and Bolívar), e. and cen. Amaz. Brazil (west to Rio Negro and Rio Madeira, east to w. Maranhão, and south to n. Mato Grosso and n. Goiás), and n. Bolivia (Pando and Beni). To 850 m (in Venezuela).

Myioborus Redstarts

The *Myioborus* redstarts comprise a widespread and familiar but taxonomically difficult group of warblers perhaps most noted for their delightfully animated and trusting behavior. They are found in forest and woodland the length of the Andes and in the mountains and *tepuis* of Venezuela and nearby areas. Each is readily recognized as a redstart (we here take cognizance of those who would change the group name to "whitestart": while agreeing that that would certainly be more accurate—there is no red, only white, in the tail—we feel that the name "redstart" is simply too well entrenched to be changed at this late date), being essentially gray and yellow with conspicuous white outer tail feathers (frequently exposed by fanning). However, the various species are taxonomically confusing. Essentially what happens is that at elevations above the range of the widespread Slate-throated Redstart (which shows only minor subspecific variation) there occurs a series of different forms, both in the Andes and on the *tepuis*. These latter vary primarily in their head patterns (crown color, presence or absence of spectacles and their color, etc.) but also in the intensity of yellow below. Establishing correct species limits among these forms, all of which are allopatric, is essentially arbitrary, for certain characters seem to vary in an almost random manner. For simplicity's sake, we have mostly retained the *Myioborus* taxonomy of Meyer de Schauensee (1966, 1970).

GROUP A

Redstarts with *slaty throat* (yellow in all others).

Myioborus miniatus

SLATE-THROATED REDSTART

PLATE: 9

IDENTIFICATION: 13–13.5 cm (5–5¼"). *Slaty gray above and on throat;* small crown patch chestnut (often concealed). Remaining underparts bright yellow, with crissum white. *Outer tail feathers conspicuously white.*

SIMILAR SPECIES: No other redstart has the gray of the upperparts extending down over the throat. Slate-throated tends to occur lower than most other species of redstarts.

HABITAT AND BEHAVIOR: Common to very common in montane forest and woodland and in borders. Widespread and tolerant of considerable habitat disturbance, this is one of the most numerous and conspicuous birds at lower elevations in the Andes. Perhaps less numerous in the *tepuis*. Its song is often heard but not very memorable, a series of weak, colorless notes, sometimes faintly musical but more often rather squeaky or "chirpy."

RANGE: Coastal mts. of n. Venezuela (Sucre westward) and Andes of w. Venezuela (north to Trujillo), Colombia, Ecuador, Peru (on w. slope south only to Cajamarca), and n. and e. Bolivia (south to Chuquisaca); Colombia's Santa Marta Mts. and Perijá Mts. on Colombia-Venezuela border, and on the *tepuis* of s. Venezuela (Bolívar and Amazonas) and adjacent Guyana and extreme n. Brazil (Roraima). Also Mexico to Panama. Mostly 700–2500 m.

GROUP B

Ornatus/melanocephalus group. Andes from Santa Martas to n. Bolivia; *all allopatric.*

Myioborus ornatus

GOLDEN-FRONTED REDSTART

PLATE: 9

IDENTIFICATION: 13–13.5 cm (5–5¼"). Andes of Colombia and extreme w. Venezuela. *Foreface* (back to mid-crown and around eye) *and entire underparts bright yellow tinged orange;* rearpart of head blackish, sometimes with a small yellowish or white patch on lower ear-coverts and occasionally with a patch of rufous on center of crown. Otherwise mostly dark olivaceous gray above; tail black, with outer feathers conspicuously white. Nominate race (Colombia's E. Andes south to Bogotá and in adjacent Venezuela) has white ocular area extending narrowly to forehead and chin and its crown and underparts are more lemon yellow (without the orange tinge); see V-34, C-47.

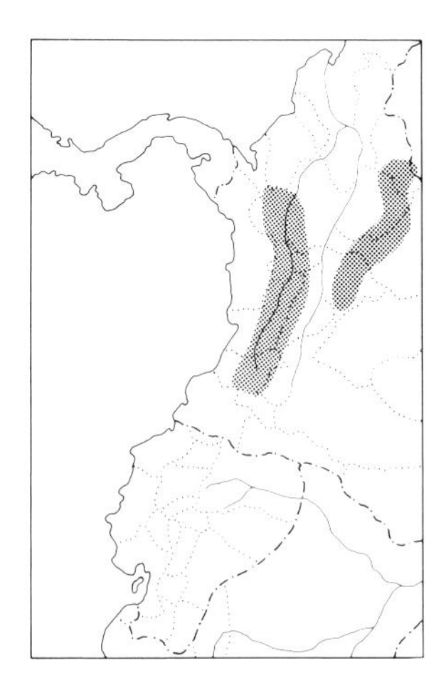

SIMILAR SPECIES: White-fronted Redstart (which replaces it to the north) has black (not yellow) crown with rufous patch in center. Spectacled Redstart (which replaces it to the south) has mostly rufous crown with less yellow around eye, but note that it is a variable form.
HABITAT AND BEHAVIOR: Common in montane forest, forest borders, stunted woodland near timberline, and adjacent scrub. Active and confiding, often one of the more numerous birds at high elevations in its range; regularly moves about in small groups of up to 4 to 6 birds, either on its own or as part of a mixed flock. Tends not to posture with drooped wings and fanned tail quite as often as Slate-throated Redstart. Its song is a series of high-pitched, rather unmusical notes which may continue for as long as 15 to 20 seconds, sometimes as a repetition of a single note or phrase, sometimes more variable. More often heard is its "tsip" contact note, frequently given as it flies from tree to tree.
RANGE: Andes of extreme w. Venezuela (w. Táchira) and Colombia (except in Nariño). Mostly 2400–3400 m.

NOTE: See comments under Spectacled Redstart.

Myioborus melanocephalus

SPECTACLED REDSTART

PLATE: 9

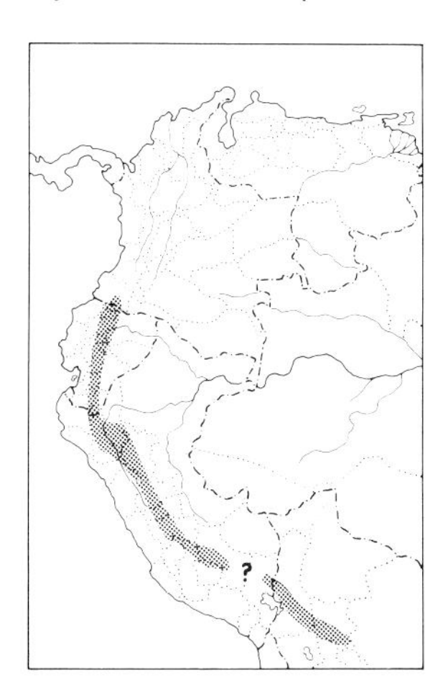

IDENTIFICATION: 13–13.5 cm (5–5¼"). *Andes of s. Colombia to Bolivia.* Head pattern varies, but always shows *prominent yellow lores and broad ocular area* (the "spectacles"). Proceeding from the south, nominate race and *bolivianus* (cen. Peru to Bolivia) have crown and sides of head mostly black, with *yellow forehead and ocular area* (the "spectacles"). Otherwise gray above; tail blacker, with outer feathers conspicuously white. Below bright yellow, except for white crissum. *Malaris* (of n. Amazonas, Peru) similar but with broader black malar stripe completely separating yellow ocular area from underparts; *griseonuchus* (W. Andes of nw. Peru) like *malaris* but with a *rufous crown patch. Ruficoronatus* (the common Ecuadorian form) has mainly black head and nape, with *rufous crown patch.* At n. edge of range (mostly in Nariño, Colombia) there occurs a "variant" of *ruficoronatus* with varying amounts of yellow on forecrown and lores (yellow feathers often are black-tipped).
SIMILAR SPECIES: The variation is confusing but bear in mind that except in Bolivia the only other redstart present throughout its range is the very different Slate-throated. In s. Colombia the Spectacled's range approaches that of the Golden-fronted, but latter does not normally show rufous on crown (an occasional aberrant individual does sometimes show a trace).
HABITAT AND BEHAVIOR: Common in montane forest, forest borders, low woodland near timberline, and scrub. Behavior much like other redstarts' (see under Golden-fronted). In Bolivia tends to be more associated with humid forest than is Brown-capped Redstart.
RANGE: Andes of s. Colombia (Nariño), Ecuador, Peru (south on w. slope to Cajamarca), and nw. Bolivia (La Paz, Cochabamba, and w. Santa Cruz). Mostly 2000–3300 m.

NOTE: J. T. Zimmer (*Am. Mus. Novitates* 1428, 1949) presents evidence for considering the n. *ruficoronatus* group as conspecific with the s. *melanocephalus* group; the link appears between *griseonuchus* (n. Peru west of the Río Marañón) and *malaris* (n. Peru east of the Río Marañón), both of which share the broad black malar stripe separating the ocular ring from the underparts. Given the nature of the variation within *M. melanocephalus* as presently defined, we suspect that *ornatus* too will eventually be found to intergrade with *melanocephalus ruficoronatus*. Indeed, the "variants" mentioned above (with increased yellow on foreface) could be viewed as evidence of some interbreeding, as could the occasional presence of some rufous in the crown of *ornatus chrysops*. Even *albifrons* of Venezuela and *flavivertex* of the Santa Martas could easily be considered part of this expanded species, the name of which would have to be *M. ornatus* (Spectacled Redstart). Field study with judicious collecting is still needed in the areas of potential contact.

Myioborus albifrons

WHITE-FRONTED REDSTART

PLATE: 9

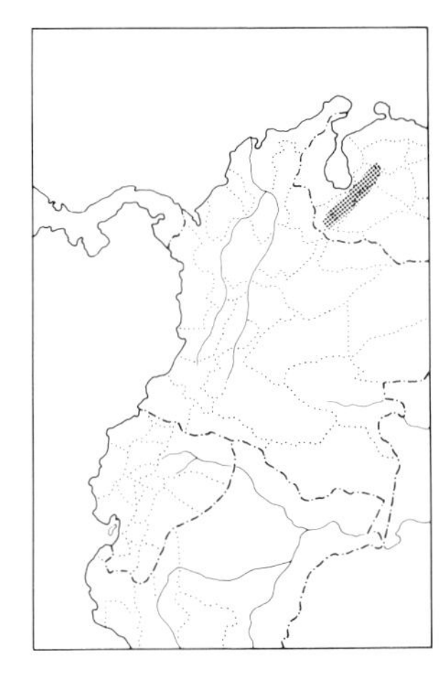

IDENTIFICATION: 13–13.5 cm (5–5¼"). *Andes of w. Venezuela.* Crown black with *median crown patch cinnamon-rufous,* feathers tipped black; *forehead, lores, and ocular area white.* Sides of head and rest of upperparts gray; tail blackish with outer feathers conspicuously white. Below bright yellow; crissum white.

SIMILAR SPECIES: Golden-fronted Redstart (which replaces this species in extreme w. Venezuela and southward) has yellow (not black) forecrown. Cf. also Slate-throated Redstart.

HABITAT AND BEHAVIOR: Common in montane forest and woodland and in shrubby borders. Behavior like other redstarts' (see under Golden-fronted). Numerous on the Queniquea road in n. Táchira and above Mérida.

RANGE: Andes of w. Venezuela (Trujillo, Mérida, and Táchira except the extreme south). 2200–3200 m.

NOTE: See comments under Spectacled Redstart.

Myioborus flavivertex

YELLOW-CROWNED REDSTART

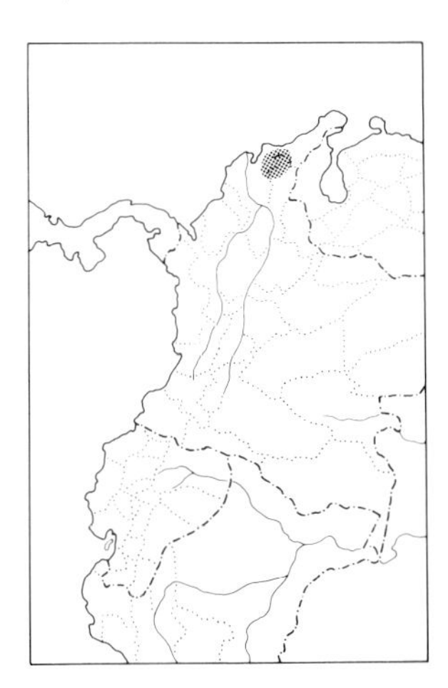

IDENTIFICATION: 13 cm (5"). *Santa Marta Mts., Colombia.* Head and nape mostly black with *conspicuous yellow median crown patch;* loral spot and upper eyelid buffyish. Above olive; tail blackish, with outer feathers conspicuously white. Underparts bright yellow. See C-47.

SIMILAR SPECIES: The only other redstart in the Santa Martas is the very different Slate-throated (gray on throat, no yellow on crown, etc.); the 2 normally segregate by altitude. Yellow-crowned Redstart is the only member of the *melanocephalus* complex with yellow in crown and an olive back.

HABITAT AND BEHAVIOR: Common in montane forest and woodland and in shrubby borders. Numerous along upper part of the San Lorenzo ridge road. Behavior much like other redstarts' (see under Golden-fronted).

RANGE: Santa Marta Mts. of n. Colombia. Mostly 2000–3000 m, occasionally somewhat lower.

NOTE: See comments under Spectacled Redstart.

GROUP C

Brunniceps group. S. Andes and Venezuela (especially *tepuis*); *all allopatric.*

Myioborus castaneocapillus

TEPUI REDSTART

PLATE: 9

Other: Brown-capped Redstart (in part)

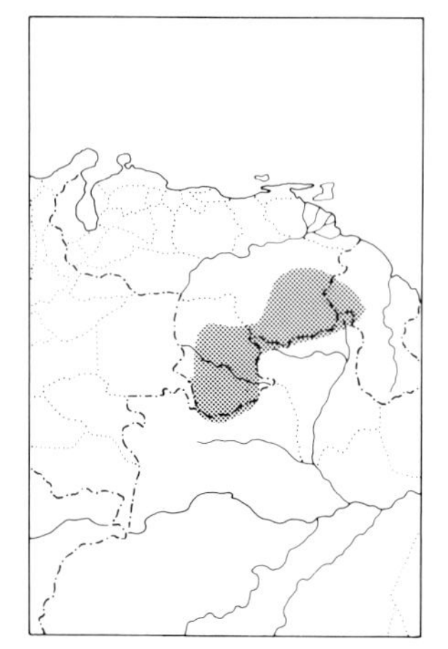

IDENTIFICATION: 13 cm (5″). *Tepuis of Venezuela and adjacent areas* (we regard the Andean form, *brunniceps,* as a distinct species; see below). *Crown rufous-chestnut,* with *short white supraloral stripe and broken eye-ring.* Above dull olive gray, tail duskier with outer feathers conspicuously white. Below bright yellow to orange-yellow; crissum white. *Duidae* (various remote *tepuis* in cen. Amazonas and sw. Bolívar) has more conspicuous white supraloral and eye-ring than *maguirei* (Cerro de la Neblina) and nominate race (Gran Sabana region).

SIMILAR SPECIES: This species is not known to occur on the same mts. as the other 2 endemic redstarts of the *tepui* region, the White-faced and Guaiquinima. May overlap with quite different Slate-throated Redstart on lower slopes of *tepuis.*

HABITAT AND BEHAVIOR: Common in forest, woodland, and borders and adjacent shrubby clearings. Behavior similar to other redstarts', as is its song, a thin unmusical chipper, starting slowly, gradually speeding up and descending in pitch (T. Davis; RSR). Easily seen along the Escalera road.

RANGE: *Tepuis* of s. Venezuela (Bolívar and cen. and s. Amazonas) and adjacent Guyana and extreme n. Brazil (Roraima). 1200–2200 m.

NOTE: Here considered a species distinct from *M. brunniceps* (true Brown-capped Redstart) of the Bolivian and Argentinian Andes. Though the 2 are undeniably similar from a plumage standpoint, we feel that the striking differences in their voices (otherwise rather uniform within the genus) together with their great range disjunction amply justify treating both as full species. This is particularly the case given that 2 other forms, *cardonai* and *albifacies,* both endemic to the *tepuis,* are presently classified as species (though they seem undeniably close to *M. castaneocapillus,* with all replacing each other allopatrically).

Myioborus brunniceps

BROWN-CAPPED REDSTART

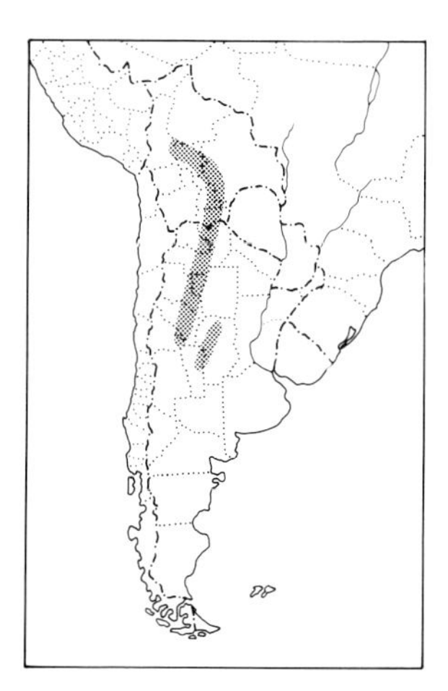

IDENTIFICATION: 13 cm (5″). *Andes of Bolivia and n. Argentina.* Resembles Tepui Redstart (the 2 have usually been considered conspecific) but is purer gray above with contrasting olive green back and with wider partial white eye-ring and supraloral stripe.

SIMILAR SPECIES: The similar Tepui Redstart is geographically very far removed. True Brown-capped may come into contact with Spectacled Redstart in Bolivia (though so far not known); it has a quite different facial pattern, lacking the rufous crown and with yellow (not white) spectacles. Cf. also Slate-throated Redstart.

HABITAT AND BEHAVIOR: Common in forest, woodland, and borders and adjacent shrubby clearings. In Bolivia found mainly in deciduous forest and alder woodland (with Spectacled taking over in more humid forests on n. slope of Andes in La Paz and Cochabamba), though farther south (where Spectacled does not occur) the Brown-capped also

occurs in more humid montane forests. Behavior much like other redstarts', but its song differs quite radically (and notably from the morphologically similar Tepui Redstart), being an even, fast, thin, sibilant trill with slight crescendo effect; N. Am. observers will be reminded of a Blackpoll Warbler.
RANGE: Andes of Bolivia (north to se. La Paz and Cochabamba) and nw. Argentina (south to La Rioja and in hills of w. Córdoba and n. San Luis). 500–3000 m.

NOTE: See comments under Tepui Redstart.

Myioborus pariae

PARIA REDSTART

Other: Yellow-faced Redstart

IDENTIFICATION: 13 cm (5"). *Paria Peninsula in Sucre, ne. Venezuela.* Similar to Tepui Redstart (nominate race) but with *yellow frontlet, supraloral streak, and eye-ring* (forming "spectacles").
SIMILAR SPECIES: The only other redstart on the Paria Peninsula is the very different Slate-throated.
HABITAT AND BEHAVIOR: Recorded from "humid cloud forest" (Meyer de Schauensee and Phelps 1978). Virtually nothing has been published on this species in life, but it probably differs little or not at all from its congeners.
RANGE: Paria Peninsula in Sucre, Venezuela. 800–1200 m.

NOTE: Part of the *M. castaneocapillus* complex and possibly (as with *M. albifacies* and *M. cardonai*) best considered conspecific with it. As this species is restricted to the Paria Peninsula and as only its spectacles are yellow (by no means its entire face), we feel a clarifying name change calling attention to its restricted range is eminently appropriate.

Myioborus cardonai

GUAIQUINIMA REDSTART

Other: Saffron-breasted Redstart

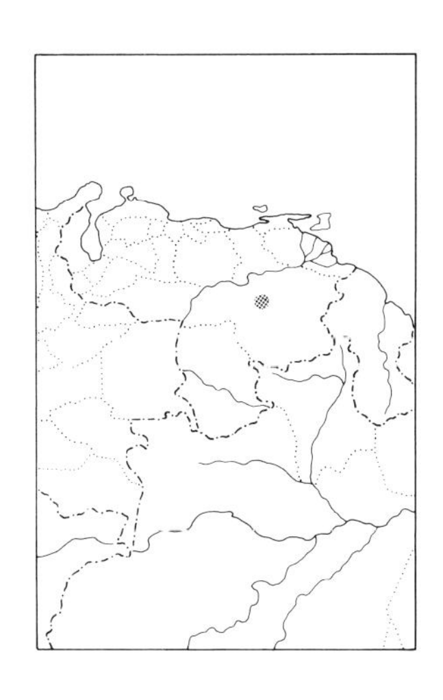

IDENTIFICATION: 13 cm (5"). *Only on remote Cerro Guaiquinima in w.-cen. Bolívar, Venezuela. Crown black,* with narrow white eye-ring and white chin; sides of head otherwise gray. Upperparts olivaceous gray, tail blacker with outer feathers conspicuously white. Below bright yellow-orange.
SIMILAR SPECIES: The only other redstart on Cerro Guaiquinima is the very different Slate-throated. Cf. also White-faced and Brown-capped Redstarts (on different *tepuis*).
HABITAT AND BEHAVIOR: Recorded from "cloud forest" (Meyer de Schauensee and Phelps 1978). Virtually nothing has been published on this species in life, but it probably differs little or not at all from the other redstarts.
RANGE: S. Venezuela (Cerro Guaiquinima in w.-cen. Bolívar). 1200–1600 m.

NOTE: See comments under White-faced Redstart. As *M. cardonai*'s breast is the *same* color as that of *albifacies* and *castaneocapillus duidae,* it seems misleading to imply that it is different by calling it "Saffron-breasted" (in any case, the color is not exactly saffron); far better, we believe, to stress its extremely limited distribution by calling it the Guaiquinima Redstart.

Myioborus albifacies

WHITE-FACED REDSTART

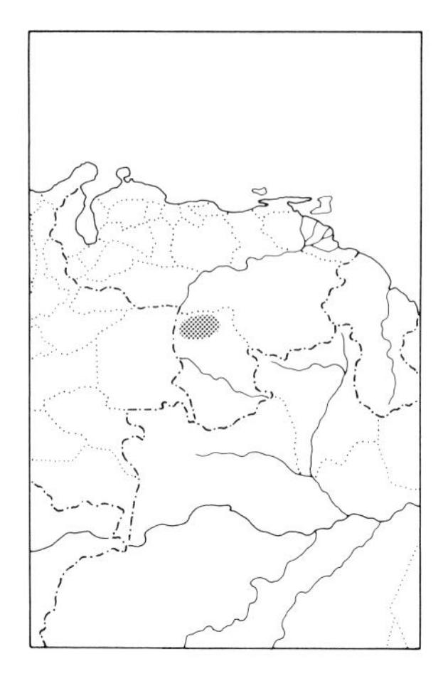

IDENTIFICATION: 13 cm (5"). *Several remote tepuis in nw. Amazonas, Venezuela. Crown black,* contrasting sharply with *pure white sides of face.* Above olivaceous gray; tail blacker with outer feathers conspicuously white. Below bright yellow-orange. See V-34.

SIMILAR SPECIES: The only other redstart on the 3 *tepuis* from which this species has been recorded is the strikingly different Slate-throated. Cf. also Guaiquinima and Tepui Redstarts (on different *tepuis*).

HABITAT AND BEHAVIOR: Recorded from "rain and cloud forest" (Meyer de Schauensee and Phelps 1978). Virtually nothing has been published on this species in life, but it probably differs little or not at all from other redstarts.

RANGE: S. Venezuela (nw. Amazonas on Cerros Guany, Yaví, and Paraque). 900–2250 m.

NOTE: We question whether *M. cardonai* and *M. albifacies* should be considered full species. Except for their black crowns, both are much like *M. castaneocapillus duidae,* to which they are clearly linked. As *M. melanocephalus* demonstrates in the Andes, crown color (and even color of the face) is quite plastic in *Myioborus* and, although striking to our eyes, seems not to act as an effective isolating mechanism. At the very least, these 2 could almost certainly be lumped together; it could be called Black-crowned Redstart, *M. cardonai.*

Geothlypis Yellowthroats

The yellowthroats are a familiar and common (though rather skulking) group of shrubbery-inhabiting warblers. Males are characterized by their black facial areas (usually appearing as a mask), while both sexes are bright yellow below. The base of the lower mandible and the legs are pinkish in all yellowthroats. Yellowthroats reach their highest diversity in Mexico, with 1 species breeding in North America (and rarely migrating to South America); only 2 are resident in South America. This genus is closely related to *Oporornis.*

Geothlypis aequinoctialis

MASKED YELLOWTHROAT

PLATE: 9

IDENTIFICATION: 13–14 cm (5–5½"). Male has *black mask extending from above bill back to lower cheeks, outlined in gray;* above otherwise olive green. Below bright yellow, tinged olive on flanks. *Auricularis* (w. Ecuador and Pacific nw. Peru) similar but notably smaller and with *black restricted to frontlet, lores, and around eye; peruviana* (of upper Río Marañón valley in nw. Peru) similar to *auricularis* in plumage but like nominate group (with *velata*) in size. Female lacking most of male's head pattern: *crown and cheeks tinged gray* (sometimes barely discernible) and with faint (but usually evident) *narrow yellow supraloral stripe and eye-ring;* otherwise olive above and bright yellow below, like male tinged olive on flanks.

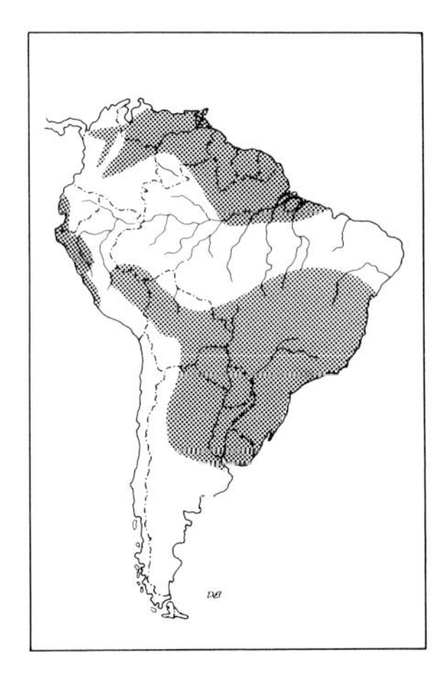

SIMILAR SPECIES: Cf. Olive-crowned Yellowthroat (overlapping with Masked only in w. Ecuador), and in Colombia/Venezuela see also Common Yellowthroat (a rare n. migrant). Female's pattern and color (aside from the touch of gray on the head) are similar to Flavescent Warbler's, though that species has a prominent yellow eyebrow and rather different shape and behavior, etc. Female also superficially like several of the *Pseudocolopteryx* doraditos (Tyrannidae).

HABITAT AND BEHAVIOR: Locally common in shrubby borders and clearings and in pastures and marshes with rank, grassy cover. Tend to skulk in thickets, where they remain inconspicuous unless excited by squeaking, or may be seen in brief tilting flight between bushes. Their numbers are more apparent when they are breeding, for males then often take to fairly exposed perches to deliver their attractive, sweet, and warbling song (overall effect rather finchlike, quite fast except for the introductory "swee, swee" notes). Its call is also distinctive, a protracted fast chattering which drops in pitch and strength.

RANGE: N. Colombia (south in Magdalena valley in Huila and in e. lowlands to Meta), Venezuela (not the extreme northwest or in s. Amazonas), Guianas, and lower Amaz. Brazil; w. Ecuador (north to Manabí and Pichincha) and w. Peru (south on Pacific coast to Ica and in upper Río Marañón valley of Cajamarca and La Libertad); se. Peru (north to Cuzco), n. and e. Bolivia, s. Brazil (north to Mato Grosso, Goiás, s. Piauí, and Bahia), Paraguay, Uruguay, and n. Argentina (south to n. San Luis, Córdoba, and n. Buenos Aires); Trinidad. Also sw. Costa Rica and w. Panama. Mostly below 1500 m.

NOTE: The w. *auricularis* group (with *peruviana*) may deserve full species status: male's facial pattern differs from nominate *aequinoctialis* group's by at least as much as do many presently recognized species of yellowthroats in Middle America. Information on voice differences, if any, is needed. The taxonomic status of *chiriquensis* of Costa Rica and Panama likewise remains uncertain.

Geothlypis semiflava

OLIVE-CROWNED YELLOWTHROAT

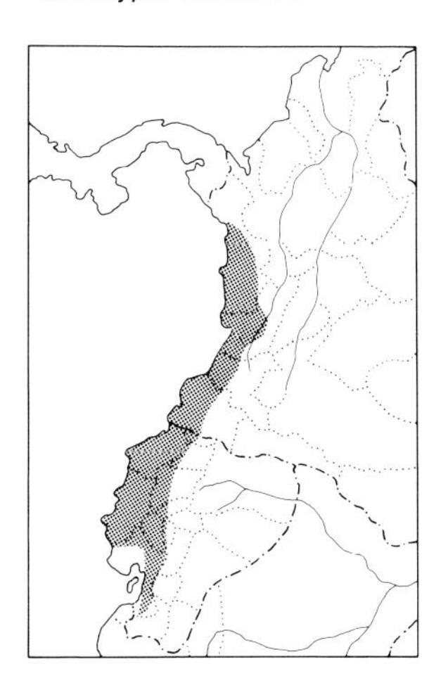

IDENTIFICATION: 13.5 cm (5¼"). *W. Colombia and w. Ecuador.* Male mostly olive green above with *broad black mask extending from forecrown back over entire cheeks to sides of neck.* Below bright yellow, with olive sides and flanks. Female lacks male's mask: forecrown and sides of head yellowish olive; otherwise olive green above. Below bright yellow, sides and flanks olive.

SIMILAR SPECIES: Overlaps with Masked Yellowthroat *only* in w. Ecuador (where the 2 occur in close proximity at certain locations). Males are easily distinguished (Olive-crowned having *much* larger mask than the sympatric race of Masked, *auricularis,* and lacking all gray on crown), but females can be difficult unless seen well (or accompanying a male). Look for Masked's yellow supraloral and eye-ring and its gray on crown and cheeks, both lacking in Olive-crowned.

HABITAT AND BEHAVIOR: Uncommon to fairly common in shrubbery and thickets along borders of forest and woodland and in grassy clearings and pastures with scattered bushes; usually but not always near water. Generally skulks in pairs in thick vegetation and not often seen

unless males are singing, when they may mount to a more exposed perch. The song is rich and musical, somewhat hesitantly phrased (pattern reminiscent of certain *Basileuterus*), starting slowly with several 2-syllabled phrases, ending with a rather jumbled twitter; much more complex than song of Masked Yellowthroat. Call is a somewhat nasal "chee-uw" or "cheh, chee-uw."
RANGE: Pacific w. Colombia (north to s. Chocó, crossing W. Andes to Cauca valley near Cali) and w. Ecuador (south to El Oro; absent from more arid areas). Also Honduras to w. Panama. Mostly below 1500 m.

Geothlypis trichas

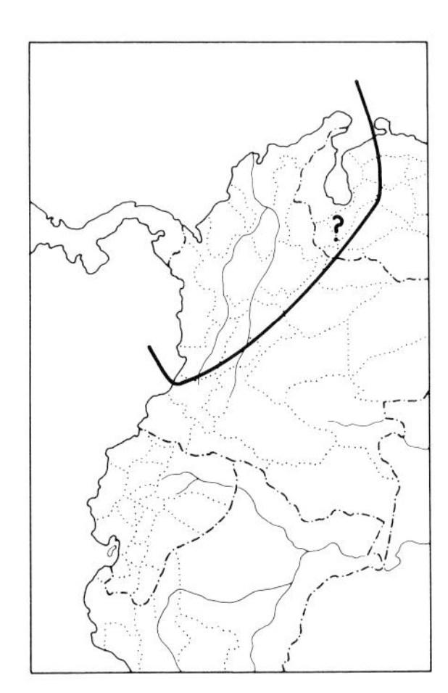

COMMON YELLOWTHROAT

IDENTIFICATION: 12.5 cm (5″). A rare straggler to Colombia and Venezuela. Resembles Masked Yellowthroat but smaller. Male differs in having *black mask distinctly bordered above by whitish or pale grayish band* and in having yellow throat and breast *fading to dingy whitish on belly,* tinged brownish on flanks. Female differs in lacking Masked's gray on forecrown and in having *yellow more or less restricted to throat* (not being entirely bright yellow below), with lower underparts dull whitish to brownish. Immature Commons are often browner (not so olive) above.
SIMILAR SPECIES: Cf. also Olive-crowned Yellowthroat (as in Masked, *all yellow below in both sexes*).
HABITAT AND BEHAVIOR: Rare n.-winter visitant (mostly Oct.–Apr.) to grassy and shrubby areas, often near water. Tends to skulk in dense vegetation but responds readily to squeaking. Its call note, a distinctive husky "tchek," often reveals its presence; call of Olive-crowned Yellowthroat similar (but that of Masked is different).
RANGE: Nonbreeding visitor to s. and n. Colombia (south to Valle) and Venezuela (1 specimen from an unknown locality); Netherlands Antilles. Breeds North America and Mexico, wintering mostly from s. United States to Panama and in West Indies.

Basileuterus Warblers

The *Basileuterus* genus comprises a large, complicated group of rather dull-plumaged warblers. Most are either striped on the crown or have a fairly prominent pale superciliary; olive, gray, or yellow predominates otherwise. The genus is widespread in South America but reaches its greatest diversity in the Andes. Many species present identification difficulties, and there are numerous taxonomic problems yet to be resolved. The genus itself is very similar to the *Hemispingus* tanagers, and the allocation of several species has shifted back and forth (perhaps telling us that the distinctions between the 2 groups, as presently understood, are themselves artificial). Some feel additionally that the "*Phaeothlypis*" group should be split off from *Basileuterus,* and though this may be a useful distinction from a field standpoint, it may have no taxonomic validity, as there are several seemingly "intermediate" species (see below). *Basileuterus* warblers forage actively through lower growth of forest and woodland, typically in pairs or small groups which often associate with mixed flocks. Except for "*Phaeothlypis,*" however, they rarely drop to the ground itself. Songs are a notable feature of many species. So far as known, nests are always dome-shaped with a side entrance; these are often situated on banks.

GROUP A

The "*citrine*" group. A small group of bright olive and yellow *Basileuterus*; *no* lateral crown striping. Strictly Andean (but note that superficially similar Flavescent Warbler of lowlands is *not* placed here, where in the past it would have been associated, but rather with the "*Phaeothlypis*" group, where we feel it belongs).

Basileuterus luteoviridis

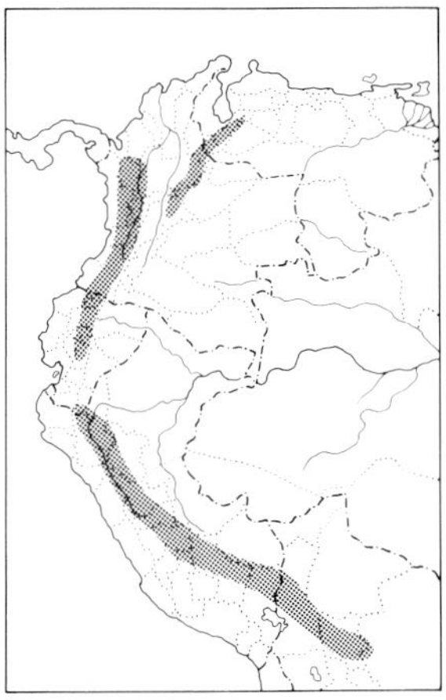

CITRINE WARBLER

PLATE: 9

IDENTIFICATION: 14 cm (5½"). Legs yellowish brown. Dull olive above with *short yellow superciliary, broadest in front* and barely extending back to over eye, and dusky lores. Below dull yellow, broadly washed olive on sides and flanks. *Richardsoni* (locally in Colombia's W. Andes) *duller and much less yellow generally,* with superciliary whitish; buffier below, with whitish throat. Moving south, *striaticeps* (e. Peru from Amazonas to Cuzco) resembles nominate race (e. slope of Andes from Ecuador north) but has *longer yellow superciliary* extending to behind eye. *Euophrys* (s. Peru in Puno and in n. Bolivia) looks, however, quite different: its yellow superciliary extends farther back than even in *striaticeps* and it has *crown, as well as lores, black or blackish* (imparting a much more contrasty effect).

SIMILAR SPECIES: Cf. the very similar Pale-legged Warbler. S. race of Citrine is remarkably like Black-crested Warbler (not found south of n. Peru). Citrine Warbler is also often confused with Superciliaried Hemispingus, though that species has a superciliary which is long and narrow (with the effect of "wrapping around" the ear-coverts) and white in all but the Venezuelan race (there yellow; thus the 2 are particularly apt to be confused here). In s. Peru cf. also Parodi's Hemispingus. Oleaginous Hemispingus typically occurs at lower elevations and never shows as prominent a superciliary as the Citrine.

HABITAT AND BEHAVIOR: Locally common in lower growth of montane forest and woodland and their borders. Most often in small groups or pairs which frequently accompany mixed flocks of other insectivorous birds, gleaning in foliage. The song in e. Peru is a long series of short, fast, sweet notes, rising and falling erratically and often becoming louder and then softer; it may go on for several minutes.

RANGE: Andes of w. Venezuela (Mérida and Táchira), Colombia (south in W. Andes to Cauca), e. Ecuador, e. Peru, and nw. Bolivia (La Paz, Cochabamba, and w. Santa Cruz). Mostly 2300–3400 m.

NOTE: See comments under Pale-legged Warbler.

Basileuterus signatus

PALE-LEGGED WARBLER

IDENTIFICATION: 13.5 cm (5¼"). *Closely* resembles Citrine Warbler, especially its *striaticeps* race (with which partially sympatric). Legs *slightly* paler, brownish yellow (but this is of marginal value in the field). Overall size slightly smaller and bill proportionately somewhat smaller. Bear in mind that in Bolivia and extreme s. Peru (Puno) comparison need only be made to the comparatively distinctive race of Citrine, *euophrys,* with black on crown, etc. Elsewhere where the 2 occur together (in Peru from Junín to Cuzco) it is probably best to go by elevation (Pale-legged tending to occur lower, though the 2 overlap broadly in Bolivia), less prominent superciliary, and voice.

HABITAT AND BEHAVIOR: Locally fairly common in lower growth of montane forest, woodland, and shrubby borders. Usually in pairs, which forage actively through undergrowth, sometimes coming to edge but more often inside; tend to forage somewhat lower than Citrine Warbler (though there is much overlap). The song in Cuzco is a distinctive series of fast, high chippers, at first rising, then dropping in pitch but becoming somewhat louder; it can often be heard over the roar of the Río Urubamba at the foot of the Machu Picchu ruins.

RANGE: E. slope of Andes in cen. and s. Peru (north to Junín), Bolivia, and extreme nw. Argentina (Salta and Jujuy); also an anomalous record from Cundinamarca, Colombia, which requires further confirmation. Mostly 2000–2800 m, occasionally somewhat higher.

NOTE: We cannot concur with J. T. Zimmer (*Am. Mus. Novitates* 1428, 1949) in his belief that *signatus* is closely related to *B. flaveolus,* so different are they in behavior, voice, and distribution. However, we remain puzzled as to the relationship between *B. signatus* and the various races of *B. luteoviridis*. In particular, it seems strange that *signatus* and *luteoviridis striaticeps* seem to segregate altitudinally in Peru, while in Bolivia *signatus* and *luteoviridis euophrys* are sympatric. *Euophrys,* though now considered a race of *B. luteoviridis,* is strange in that in its appearance and at least to some extent its behavior it harks back to *B. nigrocristatus* (of which it was formerly considered the s. race). Comparative behavioral data on all the taxa in this complex are badly needed: it is quite possible that the traditional taxonomy followed here will not prove to be the final word.

Basileuterus nigrocristatus

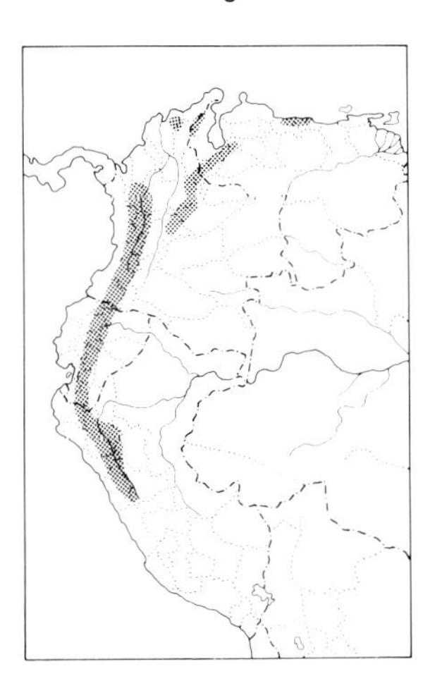

BLACK-CRESTED WARBLER

IDENTIFICATION: 13.5 cm (5¼"). *Andes of Venezuela to n. Peru.* Bright yellowish olive above with *conspicuous black center of crown,* bordered below by short yellow superciliary; *lores also black.* Below bright yellow, washed olive on sides and flanks. See V-33, C-46.

SIMILAR SPECIES: Where Citrine Warbler overlaps the range of this species it lacks any black on head and is not as bright olive or yellow. Race of Citrine found farther south (*euophrys*) is quite like Black-crested but has longer, more prominent yellow superciliary, etc.

HABITAT AND BEHAVIOR: Common in dense shrubby lower growth at borders of montane forest and woodland and in adjacent overgrown clearings or regenerating scrub. Often at edge of stands of *Chusquea* bamboo but avoids interior of forest. Usually found in pairs, tending to skulk in heavy cover, but somewhat excitable and with patience can usually be at least glimpsed. The often heard song starts slowly with a series of sharp "chit" notes, then gradually accelerates into a much more musical phrased song, ending with a crescendo of "chew" notes and a characteristic "chitty-chitty-chew."

RANGE: Coastal mts. of n. Venezuela (Aragua to Distrito Federal) and Andes of w. Venezuela (Trujillo south), Colombia, Ecuador, and n. Peru (south to Ancash and La Libertad); Colombia's Santa Marta Mts. and Perijá Mts. on Colombia-Venezuela border. Mostly 2500–3400 m, lower in n. Venezuela and w. Ecuador (to 1500–2000 m).

GROUP B

The "*gray-headed*" group. These 5 species may not all be closely related (at least their relationships are obscure), but they do all have decidedly gray heads, which sets them apart from the other *Basileuterus* rather well. All but 1 (the w. endemic, Gray-and-gold) are subtropical in distribution.

Basileuterus fraseri

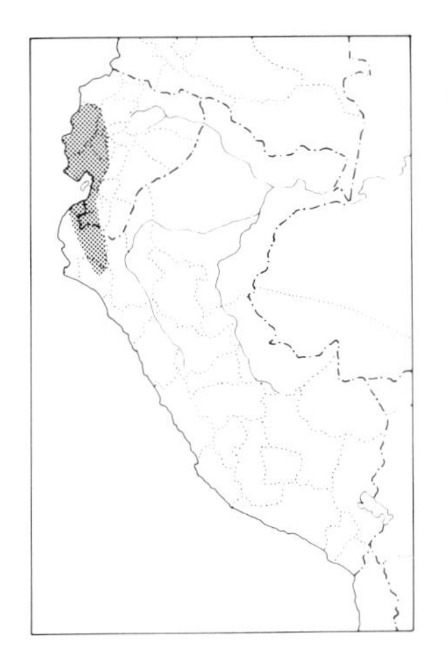

GRAY-AND-GOLD WARBLER

PLATE: 9

IDENTIFICATION: 14 cm (5½"). *W. Ecuador and nw. Peru.* Mostly *bluish gray above* with black crown and yellow coronal streak (usually at least partially hidden); *short supraloral streak white,* and center of back tinged olive. Below entirely bright yellow. *Ochraceicrista* (w. Ecuador except in El Oro and Loja) has ochraceous orange coronal streak.

SIMILAR SPECIES: The only *Basileuterus* warbler over most of its range and habitat. A brightly colored species in any case, easily recognized by the blue tone to its upperparts (not equalled in any congener). Cf. Three-banded Warbler and immature Canada Warbler.

HABITAT AND BEHAVIOR: Locally common in lower growth of deciduous woodland and scrub. Forages quite actively, but usually remains in dense cover and often hard to see clearly for long. Usually in pairs or small groups, most often not with flocks. The song is a fairly short series of somewhat burry but musical notes, "tee, tidididdeedeecheéchee," with some variation; overall effect is quite like that of the Russet-crowned Warbler. Gray-and-gold is easily seen in woodland in the Chongon Hills west of Guayaquil.

RANGE: W. Ecuador (north to Manabí and s. Pichincha; absent from more humid areas) and nw. Peru (south to Lambayeque). To about 1900 m.

Basileuterus griseiceps

GRAY-HEADED WARBLER

PLATE: 9

IDENTIFICATION: 14 cm (5½"). *Ne. Venezuela. Head mostly slaty gray* with slightly grizzled whitish effect on cheeks and *distinct white supraloral streak;* upperparts otherwise olive. Below bright yellow.

SIMILAR SPECIES: No other "gray-headed" *Basileuterus* occurs in this species' limited range. Golden-crowned Warbler has much more prominent head striping, orange-rufous median crown, etc.

HABITAT AND BEHAVIOR: Recorded from "cloud forests, second growth, clearings" (Meyer de Schauensee and Phelps 1978). Essentially unknown in life; presumably its behavior is similar to Russet-crowned Warbler's. Considered "probably endangered" due to extensive deforestation over much of its known range (C. Parrish); more information is needed.

RANGE: Coastal mts. of ne. Venezuela (Anzoátegui, Monagas, and sw. Sucre). 1200–1600 m.

Basileuterus coronatus

RUSSET-CROWNED WARBLER

PLATE: 9

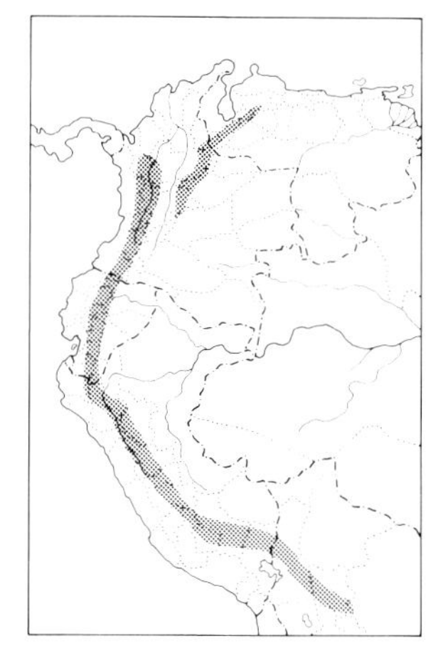

IDENTIFICATION: 14 cm (5½"). *Center of crown orange-rufous, bordered by black stripe* continuing back to nape, with another black stripe passing through eye onto cheeks; *otherwise superciliary, sides of head, and nape gray.* Above olive. *Throat and upper chest paler gray,* remaining underparts bright yellow, tinged olive on sides and flanks. A pair of races (including *castaneiceps*) from w. slope of Andes in s. Ecuador and nw. Peru has *entirely grayish white underparts,* while *orientalis* (e. Ecuador) is similar but tinged yellow on belly.

SIMILAR SPECIES: Birds with whitish underparts should be instantly recognizable, as no other *Basileuterus* in the Andes is so colored below. Yellow-bellied races can usually be known by the considerable extent of gray on head and neck and the grayish throat; smaller Golden-crowned Warbler might be confused. Gray-throated Warbler of w. Venezuela and ne. Colombia (apparently usually at *lower* elevations than Russet-crowned) is quite similar, but grayish on underparts extends lower (down over breast) and it has a smaller *yellow* coronal patch which is *not* bordered with black.

HABITAT AND BEHAVIOR: Fairly common in lower growth of montane forest and well-developed secondary woodland and borders. Usually in pairs, less often in small groups, which hop about quite actively and are not usually too hard to see; sometimes they rise up into midlevels (to about 10 m above ground), but more often they remain low. Their charming musical phrased song is frequently heard (and is similar throughout their broad range): after a few stuttering chips, it continues with a "teetu-teetu-teetu, tututeé?"; or an almost questioning "tee-tu, teetititu, tutu-teé?" is almost answered (presumably by the first bird's mate) by a "tee-tu, teetititu, tutu-tú."

RANGE: Andes of w. Venezuela (north to Trujillo), Colombia, Ecuador, Peru (south of w. slope to Cajamarca), and nw. Bolivia (La Paz and Cochabamba). Mostly 1500–2500 m, occasionally to 2800 m.

Basileuterus conspicillatus

WHITE-LORED WARBLER

IDENTIFICATION: 13.5 cm (5¼"). *Colombia's Santa Marta Mts.* Somewhat similar to Russet-crowned Warbler (with which it has been considered conspecific) but with narrower (hence much less noticeable) orange or orange-yellow coronal streak, *prominent white supraloral streak and broken eye-ring,* and with only lores blackish (no stripe behind eye). See C-46.

SIMILAR SPECIES: Golden-crowned Warbler lacks the obvious white around the eye, has throat yellow like the rest of the underparts (not contrastingly grayish white), and is grayer (not so olive) above. Santa

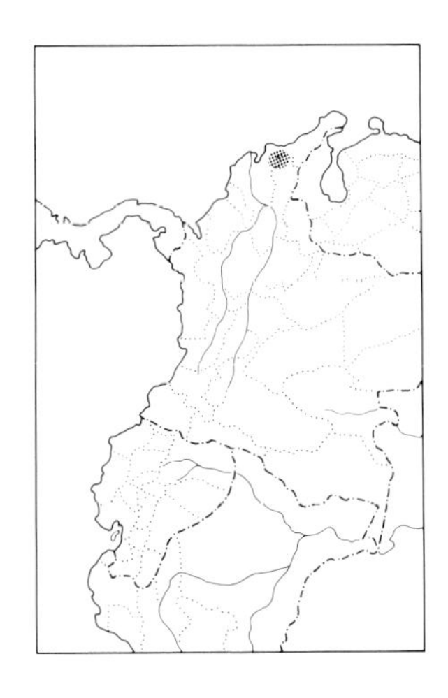

Marta Warbler (the other endemic *Basileuterus* on the Santa Martas) has very different facial pattern. Cf. also Russet-crowned and Gray-throated Warblers (neither of which is found in the Santa Martas). Female Mourning Warbler (also with broken eye-ring and grayish throat, etc.) is vaguely similar.

HABITAT AND BEHAVIOR: Common in undergrowth of montane forest and well-developed secondary woodland and borders. Behavior similar to Russet-crowned Warbler's (voice?). Easily seen along the San Lorenzo ridge road above Minca.

RANGE: Santa Marta Mts. of Colombia. 750–2200 m.

NOTE: This form has enjoyed a convoluted taxonomic history, having been described as a full species, then considered a race of *B. cinereicollis* (*Birds of the Americas,* vol. 13, part 8), then considered a race of *B. coronatus* (Meyer de Schauensee 1966, 1970; but not Meyer de Schauensee 1964, then being considered distinct), and again as a full species (*Birds of the World,* vol. 14). We too consider it sufficiently distinct from both *coronatus* and *cinereicollis* to warrant treating it as a full species; it differs as much from both of them as they do from each other, and those 2 occur in virtual sympatry.

Basileuterus cinereicollis

GRAY-THROATED WARBLER

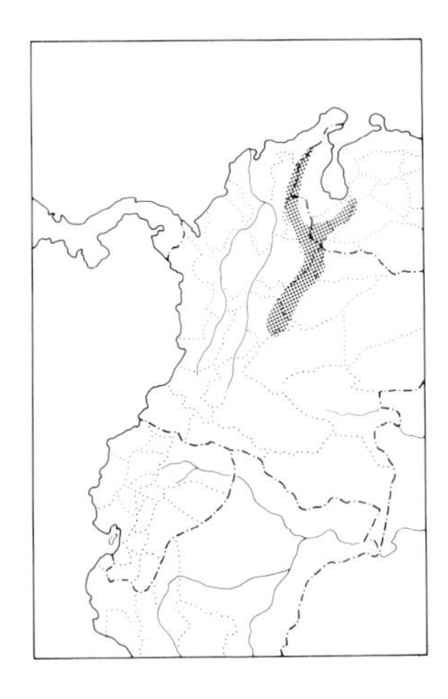

IDENTIFICATION: 14 cm (5½"). *W. Venezuela and ne. Colombia.* Resembles Russet-crowned Warbler (probably some overlap, though Gray-throated usually occurs lower), but *coronal streak narrower* (hence less conspicuous) *and yellow* (not orange-rufous) and with only vague blackish on sides of crown (unlike Russet-crown's distinct black lateral stripes) and *pale grayish extending down over breast* (not just to throat and upper chest). See V-33, C-46.

SIMILAR SPECIES: Besides the Russet-crown, might be confused with the Mourning Warbler.

HABITAT AND BEHAVIOR: Recorded from lower growth of humid montane forest and forest borders. Behavior probably much like Russet-crowned Warbler's, but very little is known about this species. It may now be becoming rare and local due to extensive deforestation at the elevations from which it has been recorded, but it should remain numerous in the still-remote Perijá.

RANGE: Andes of w. Venezuela (Mérida and Táchira) and ne. Colombia (south to Cundinamarca and w. Meta); Perijá Mts. on Colombia-Venezuela border. 800–2100 m.

GROUP C

The "*stripe-headed*" group. A complex group, but all show prominent coronal and/or lateral head striping. Otherwise usually quite dull. Most species are found on lower mountain slopes.

Basileuterus tristriatus

THREE-STRIPED WARBLER

PLATE: 9

IDENTIFICATION: 13 cm (5"). Color of underparts varies, but *facial pattern distinctive* over most of its mainly Andean range. *Coronal streak yellowish buff bordered laterally by broad black stripes;* superciliary also yellowish buff; *cheeks black,* with small white patches below eye and

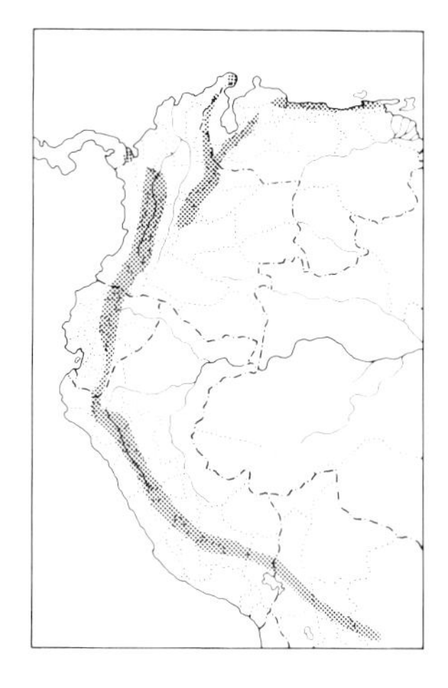

another on lower neck. Otherwise olive above. Below mostly dull yellowish buff, more whitish on throat and mottled olive across breast and on sides. Foregoing applies to *auricularis* group of Colombia and w. Ecuador; *baezae* (ne. Ecuador) similar but yellower below, with nominate race (e. Peru south to Cuzco) even brighter yellow below. *Punctipectus* group (extreme s. Peru and n. Bolivia) reverts to being dull yellowish buff below and also shows quite distinct olive mottling on breast (imparting a quite different spotted effect). Finally, the *meridanus* group (most of the Venezuelan range) and *tacarcunae* (extreme nw. Colombia) have mostly olive cheeks, with the black reduced to a thin postocular streak; see V-33.

SIMILAR SPECIES: Over most of its range the Three-striped can be known by its complex facial pattern, with black on cheeks in addition to the black stripes on either side of crown (this pattern shared among *Basileuterus* only by range-restricted Santa Marta Warbler, where Three-striped does not occur). In Venezuela, where the black cheeks are lacking, it might be confused with sympatric races of Golden-crowned Warbler, but latter is yellow (not drab) below and has orange-rufous (not dull buff) coronal streak; cf. also Venezuelan races of Oleaginous Hemispingus.

HABITAT AND BEHAVIOR: Fairly common to common in lower growth of montane forest and well-developed secondary woodland; generally inside forest, much less frequent at edge. Forages actively, most often in small groups which are usually with mixed flocks of understory birds; generally not at all difficult to see. Gives a variety of twittery, often quite squeaky calls, none particularly distinctive.

RANGE: Coastal mts. of n. Venezuela (east to the Paria Peninsula) and Andes of w. Venezuela (north to Lara), Colombia, w. (south to e. Guayas) and e. Ecuador, e. Peru, and nw. Bolivia (La Paz, Cochabamba, and w. Santa Cruz); Perijá Mts. on Venezuela-Colombia border, and Serranía Macuira on Guajira Peninsula and Cerro Tacarcuna, Colombia. Also Costa Rica and Panama. Mostly 1000–2000 m.

Basileuterus basilicus

SANTA MARTA WARBLER

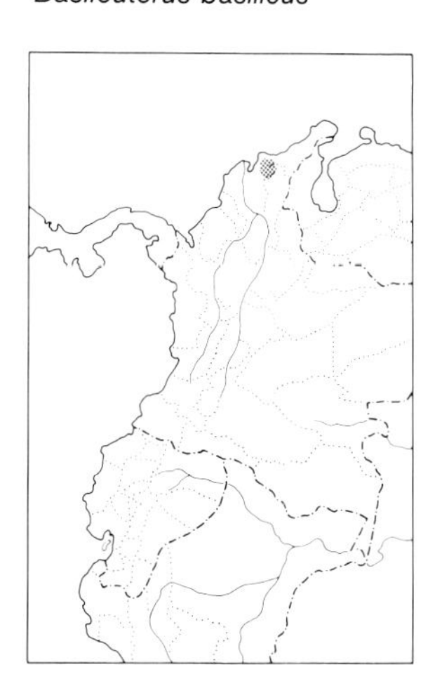

IDENTIFICATION: 14 cm (5½"). Unmistakable in its limited range *high in Santa Marta Mts. of Colombia. Head and neck mostly black, with white coronal stripe, long white superciliary arching around to join white crescent on ear-coverts, and white patch below eyes.* Otherwise olive green above. Below mostly bright yellow, tinged olive on sides; center of throat white flecked dusky. See C-46.

SIMILAR SPECIES: The ornate head pattern easily identifies this Santa Marta endemic; it recalls that of smaller Three-striped Warbler (not found on the Santa Martas).

HABITAT AND BEHAVIOR: Rare to uncommon in lower growth and borders of montane forest and secondary woodland, especially where there are dense stands of *Chusquea* bamboo. Usually in pairs or small groups, usually remaining in dense cover and often hard to see clearly. Regularly with mixed flocks. Look for it above the forestry station on

the San Lorenzo ridge, but this species is much scarcer than the White-lored Warbler and is restricted to higher elevations.
RANGE: Santa Marta Mts. of Colombia. Mostly 2300–3000 m.

NOTE: Originally described as a *Hemispingus,* which in behavior it rather resembles. Some have thought it allied to *B. tristriatus,* but it is markedly larger, with different behavior and altitudinal distribution. It seems that this and the Santa Marta Antpitta (*Grallaria bangsi*) are the 2 most divergent of the "Santa Marta endemics."

Basileuterus ignotus

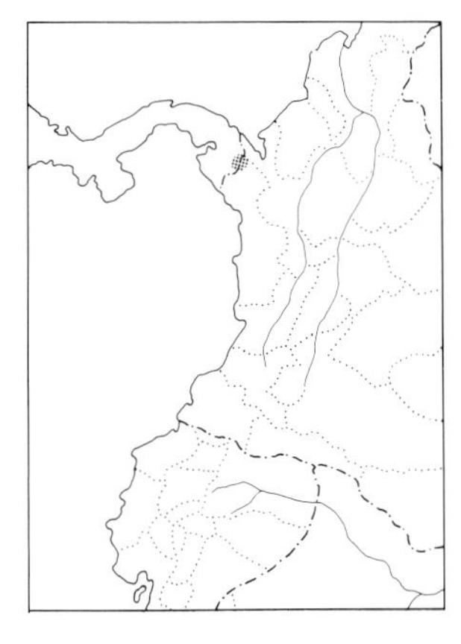

PIRRE WARBLER

IDENTIFICATION: 13 cm (5"). *Cerro Tacarcuna in nw. Colombia. Crown chestnut* margined narrowly with black, *pale greenish yellow forehead and long superciliary,* and sides of head greenish mixed black. Otherwise greenish olive above, creamy yellow below with some grayish olive on breast. See P-32.
SIMILAR SPECIES: Not likely to be confused in its small range; cf. Three-striped Warbler.
HABITAT AND BEHAVIOR: Recently recorded from lower growth of elfin forest and woodland (*fide* J. Hernandez). Very little known in Colombia. In Panama usually seen in pairs, sometimes accompanying mixed flocks; may forage up into mid-story.
RANGE: Nw. Colombia (slopes of Cerro Tacarcuna in extreme nw. Chocó). Also e. Panama. 1200–1400 m.

Basileuterus bivittatus

TWO-BANDED WARBLER

PLATE: 9

IDENTIFICATION: 13.5–14 cm (5¼–5½"). Coronal streak yellow or orange-rufous mixed with yellow (all orange-rufous in *roraimae* of the *tepuis*), bordered laterally by broad black stripe; *superciliary olive,* with *short yellow supraloral streak* and broken eye-ring; dusky stripe from lores back through eye. Above otherwise olive. Below bright yellow, clouded with olive on sides and flanks.
SIMILAR SPECIES: Cf. the *very* similar Golden-bellied Warbler (with which sympatric in se. Peru). Golden-crowned Warbler is smaller with pale grayish (not yellow and olive) superciliary, gray (not olive) effect to sides of head and neck, and less olive on sides; it and the Two-banded are marginally sympatric both in Bolivia and in the *tepui* region, but in area of overlap Two-banded apparently occurs more at higher elevations.
HABITAT AND BEHAVIOR: Locally fairly common in lower growth of humid forest and well-developed woodland; most numerous in foothill zone. In Peru very partial to bamboo (D. Stotz). Usually in pairs or small groups, often moving with mixed flocks of undergrowth birds. The song in Argentina is a fast, jumbled, warbling phrase with a rather rich musical quality reminiscent of a peppershrike's; pairs frequently duet, female following the male with a very similar phrase.
RANGE: E. slope of Andes in s. Peru (Cuzco and Puno), Bolivia, and extreme nw. Argentina (Jujuy and n. Salta); *tepuis* of s. Venezuela

(Bolívar and Amazonas) and adjacent Guyana and extreme n. Brazil (Roraima). Mostly 700–1800 m.

Basileuterus chrysogaster

GOLDEN-BELLIED WARBLER

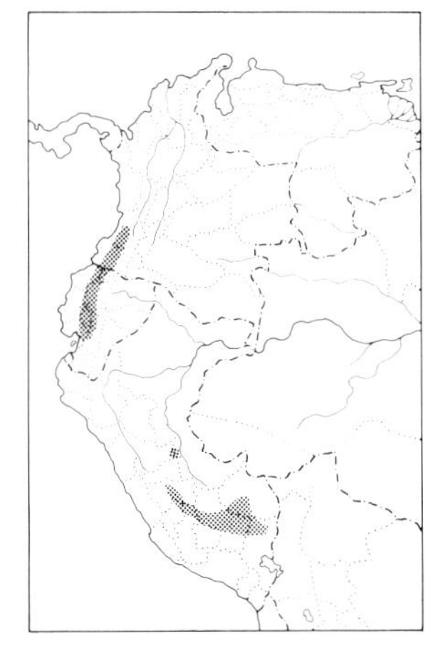

IDENTIFICATION: 13 cm (5″). *Closely* resembles Two-banded Warbler (with which nominate race overlaps in s. Peru), but *smaller.* Coronal streak orange-yellow. In area of overlap, Golden-bellied has *narrow and quite yellow superciliary* (not olive with yellow only as supraloral stripe), *lateral crown stripes not so black* (more a dark olive mixed with dusky), and little or no eye-ring. *Chlorophrys* (w. Colombia and w. Ecuador), however, has *all-olive superciliary* with at most a trace of yellow on supraloral; see C-46.

SIMILAR SPECIES: Golden-crowned Warbler has never been found sympatrically with Golden-bellied, but they approach each other in w. Colombia; Golden-crowned always shows a grayish white (not yellow or olive) superciliary, and the race in w. Colombia is much grayer-mantled.

HABITAT AND BEHAVIOR: Locally fairly common in lower growth of humid forest and better developed woodland; most numerous in the foothill zone. Usually in pairs or small groups which forage actively through thickets and dense undergrowth, regularly higher than most *Basileuterus* warblers; often with mixed flocks. The song in Ecuador is very thin and wiry, a fast buzzy "t-t-t-t-t-t-tzzzzzzz" (its quality and pattern will remind N. Am. observers of Cerulean Warbler). Numerous at various foothill localities in w. Ecuador, including the grounds of the Tinalandia Hotel.

RANGE: Sw. Colombia (north to w. Valle) and w. Ecuador (south to e. Guayas); e. slope of Andes in cen. and s. Peru (Huánuco and Junín south to Puno). Mostly 300–1200 m.

Basileuterus culicivorus

GOLDEN-CROWNED WARBLER

PLATE: 9

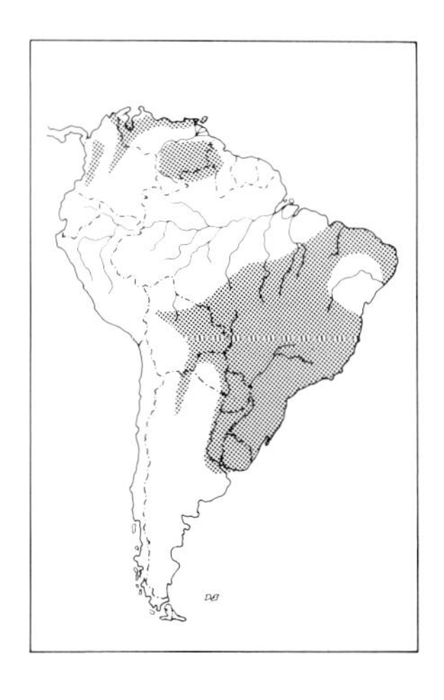

IDENTIFICATION: 12.5 cm (5″). *Coronal streak orange-rufous* bordered laterally by broad black stripes; *superciliary whitish to pale gray,* with dusky streak from lores back through eye. Upperparts otherwise grayish olive. Below uniformly bright yellow. Races of *cabinisi* group (n. Venezuela west across Colombia) have distinctly *grayer mantles* than the *auricapillus* group (of e. and s. South America) and their coronal streak tends to be yellower.

SIMILAR SPECIES: This widespread, usually common *Basileuterus* should be learned well as a basis of comparison with others which are similar. Two-banded/Golden-bellied pair is quite similar, differing most notably in their olive (not grayish white) superciliary. Cf. White-bellied Warbler (overlapping broadly with Golden-crowned across s. Brazil but with whitish underparts) and Three-banded Warbler (limited area in s. Ecuador and n. Peru, hence no overlap). Three-striped Warbler has different facial pattern, etc.

HABITAT AND BEHAVIOR: Fairly common to common in lower growth of humid forest and woodland and in shrubby borders and overgrown clearings adjacent to some woodland or forest; persists even in coffee groves and other plantations if these are shaded and have some dense natural undergrowth. Active and restless foragers, they move about almost constantly, often flicking their wings and jerking their tails; they are usually easy to see and often seem quite fearless, sometimes approaching the observer closely, all the while chitting nervously. Small groups or pairs are often with mixed flocks of understory birds; these flocks often seem to form around the warblers, which act as flock "leaders." The song in s. South America is a short series of fairly musical "twee" notes which gradually become louder, with the next to last often most accentuated.

RANGE: Coastal mts. of n. Venezuela (east to the Paria Peninsula) and locally in Andes of w. Venezuela and Colombia (E. Andes south to Meta, slopes above the Cauca valley and in upper Río Patía valley in s. Cauca, and on Pacific slope in Valle); Colombia's Santa Marta Mts. and Perijá Mts. on Colombia-Venezuela border; *tepui* region of s. Venezuela (Bolívar and Amazonas) and adjacent Guyana and extreme n. Brazil (Roraima); e. and s. Brazil (north to n. Mato Grosso, s. Maranhão, and Ceará), n. and e. Bolivia (Beni east through Santa Cruz), e. Paraguay, Uruguay, and ne. Argentina (south to n. Buenos Aires); Trinidad. Also Mexico to Panama. Mostly below 1800 m.

Basileuterus trifasciatus

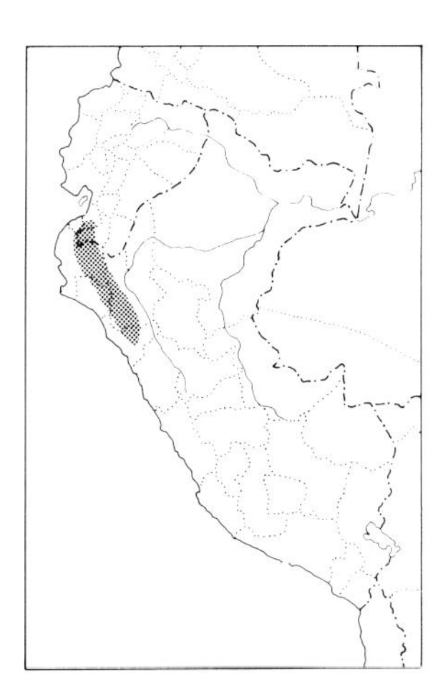

THREE-BANDED WARBLER

IDENTIFICATION: 12.5 cm (5"). *S. Ecuador and nw. Peru.* Resembles olive-backed races of Golden-crowned Warbler (no overlap), but *coronal streak olive gray* (not orange-rufous) and *throat and chest grayish white* (not all bright yellow below).

SIMILAR SPECIES: Golden-bellied Warbler (not known to overlap, though their ranges approach each other closely in w. Ecuador) has yellow coronal stripe and olive (not grayish white) superciliary and its throat and chest are yellow. Cf. also Gray-and-gold Warbler (which occurs in some of the same woodlands as does the Three-striped).

HABITAT AND BEHAVIOR: Locally fairly common in lower growth of forest, advanced second growth, and shrubby borders. Behavior much like Golden-crowned Warbler's.

RANGE: Lower Andean slopes of s. Ecuador (El Oro and Loja) and nw. Peru (south to La Libertad). Mostly 500–2000 m.

NOTE: Some have suggested that *trifasciatus* might best be considered a race of *B. culicivorus,* but we feel that it (with *nitidior*) has diverged enough to warrant full species status. In any case, Meyer de Schauensee (1966) erred in stating that *trifasciatus* most resembled the Santa Marta race of *culicivorus* (*indignus*), which it does not.

Basileuterus hypoleucus

WHITE-BELLIED WARBLER

PLATE: 9

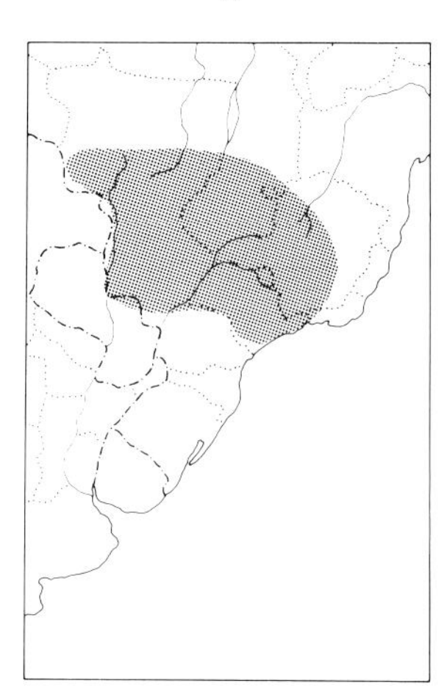

IDENTIFICATION: 12.5 cm (5″). *Interior s. Brazil. Coronal streak orange-rufous* bordered laterally by broad black stripes; superciliary white, with blackish streak from lores back through eye. Upperparts otherwise grayish olive. *Below whitish,* clouded with gray on sides, and tinged pale yellow on crissum.

SIMILAR SPECIES: Resembles Golden-crowned Warbler in overall shape, appearance, and behavior but underparts obviously white and not yellow. White-striped Warbler is notably larger with no rufous in crown, a bolder white eyestripe, etc.

HABITAT AND BEHAVIOR: Locally fairly common in lower growth of deciduous woodland, gallery forest, and shrubby overgrown borders. Behaves much like Golden-crowned Warbler; the 2 overlap broadly but may never actually occur together, with Golden-crown in more humid forest and woodland. The commonest song seems to be a fast, spritely, somewhat musical "cheetitty-chee-chee-chee-chee-cheé-chu"; many other chipping notes are given as it forages. Quite easily seen in gallery woodland at Brasilia Nat. Park (including the woods around the big public pool).

RANGE: Interior s. Brazil (Mato Grosso, s. Goiás, s. Minas Gerais, and n. São Paulo) and adjacent ne. Paraguay (Puerto Pinasco and along lower Río Apa). To about 1000 m.

Basileuterus rufifrons

RUFOUS-CAPPED WARBLER

PLATE: 9

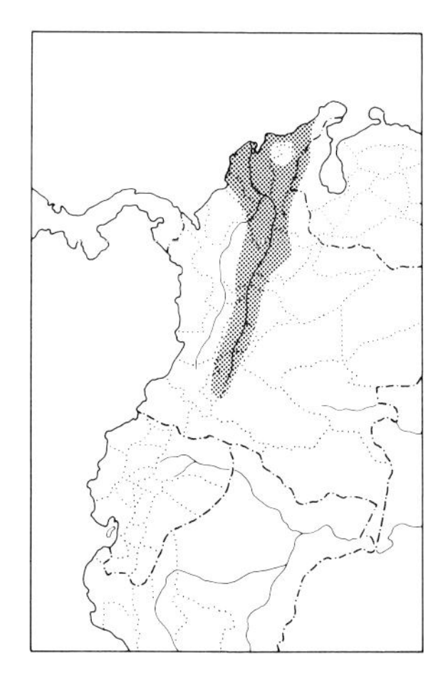

IDENTIFICATION: 13 cm (5″). *N. Colombia and adjacent Venezuela. Crown and ear-coverts brick red,* with intervening long white superciliary; lores dusky, bordered below by whitish malar area. Otherwise olive green above; entirely bright yellow below.

SIMILAR SPECIES: No other *Basileuterus* warbler has the rufous-chestnut on the ear-coverts as well as the crown.

HABITAT AND BEHAVIOR: Fairly common in shrubby clearings, thickets, and woodland borders. Most often in loosely associated pairs, generally remaining close to the ground; inquisitive and usually not difficult to see. Frequently holds its tail cocked. The song is a fast and dry phrase (e.g., "chit-cha-chup-cha-chuweépa"), somewhat variable but typically with accented ending.

RANGE: N. and w. Colombia (Santa Marta region south in Magdalena valley to Huila and on adjacent lower Andean slopes) and adjacent w. Venezuela (slopes of Perijá Mts.). Also Mexico to Panama. Mostly below 1300 m (rarely to 1900 m).

GROUP D

The "*Phaeothlypis*" subgenus. Five species of semiterrestrial warblers, all but 1 (the Flavescent) of which are associated with water to some extent. All also tend to spread their tails, often swinging them from side to side in a characteristic manner. Unlike other *Basileuterus,* these are found mostly in the lowlands. The Flavescent Warbler resembles the "citrine" group in plumage (it has even been suggested that it was conspecific with *B. signatus*) but is close to *Phaeothlypis* in most other respects (including behavior and voice), hence we place it here. The White-striped and White-rimmed Warblers bridge the gap toward the 2 classic "typical" *Phaeothlypis* warblers, Buff-rumped and River. For this reason we feel it is unwise to formally split *Phaeothlypis* from *Basileuterus*, as some recent authors have advocated (most recently, the 1983 AOU Check-list), though we do set it apart as a useful field grouping.

Basileuterus flaveolus

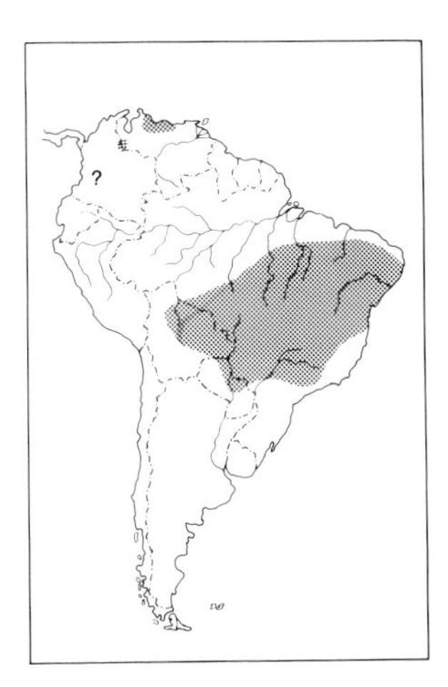

FLAVESCENT WARBLER

IDENTIFICATION: 14.5 cm (5¾"). Bill dusky; *legs orange-yellow to flesh yellow.* Bright olive above with short yellow superciliary and dusky lores. Below bright yellow, extending up over cheeks. See V-33, C-46.
SIMILAR SPECIES: Superficially similar to Citrine and Pale-legged Warblers but with completely different *lowland* distribution (not in the Andes) and different semiterrestrial behavior and vocalizations (see below). More likely to be confused with female of regularly sympatric Masked Yellowthroat but with more prominent yellow superciliary, no gray on crown, and different behavior (see below).
HABITAT AND BEHAVIOR: Locally common on or near the ground in deciduous and gallery woodland and thickets in shrubby overgrown clearings. Usually in pairs, walking or hopping on the ground, often swiveling its spread or fanned tail from side to side. The song in Brazil is loud and musical, a rather fast "titi, teetee, teetee, chéw-chéw-chéw-chéw" with characteristic accented ending; overall pattern and quality quite like those of Buff-rumped Warbler and others in the *Phaeothlypis* subgenus.
RANGE: N. Venezuela (Falcón, Lara, and Táchira east to Miranda and Guárico) and adjacent ne. Colombia (Norte de Santander); e. Bolivia (Santa Cruz), n. Paraguay, and much of interior s. Brazil (north to se. Pará, s. Maranhão, and Ceará; south to s. Mato Grosso and São Paulo; absent from coastal southeast). Mostly below 1000 m.

Basileuterus leucophrys

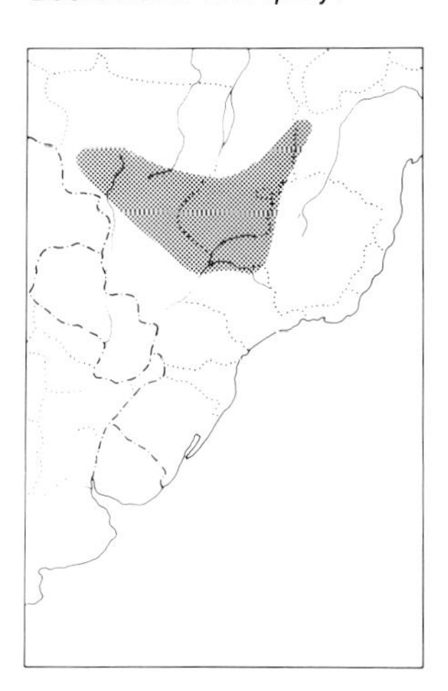

WHITE-STRIPED WARBLER

PLATE: 9

IDENTIFICATION: 14.5 cm (5¾"). *Local in s.-cen. Brazil.* Bill dusky; legs flesh yellow. *Crown slaty,* bordered below by *long, broad, white superciliary;* blackish line through eye, with lower cheeks whitish grizzled dusky. *Upperparts otherwise bronzy olive.* White below, mottled or washed with gray across breast; lower flanks and crissum pale buff.
SIMILAR SPECIES: Not known to overlap with White-rimmed Warbler, which in any case lacks this species' very bold *superciliary* (has a prominent supraloral and partial eye-ring instead), and has purer olive back and not as heavy a bill. White-striped is, however, frequently sympatric with White-bellied Warbler, a notably smaller bird with an orange-rufous coronal stripe and different behavior.
HABITAT AND BEHAVIOR: Uncommon in undergrowth of gallery forest, almost invariably near water; relatively few specimens have been taken, though it is, for example, quite numerous in Brasilia Nat. Park.

Usually in pairs which hop and walk on the ground, remaining in dense cover and often hard to see. The tail is frequently spread and slightly raised, and it regularly is swiveled from side to side. Has a beautiful loud tinkling song, the individual notes very pure and melodic (e.g., "kli-kli-kli-kleeu, klee," with numerous variations).
RANGE: S.-cen. Brazil (locally in s. Mato Grosso, s. Goiás, w. Bahia, w. Minas Gerais, and nw. São Paulo). To about 1000 m.

Basileuterus leucoblepharus

WHITE-RIMMED WARBLER

PLATE: 9

Other: White-browed Warbler

IDENTIFICATION: 14.5 cm (5¾"). *Se. Brazil and adjacent areas.* Bill dusky; legs yellowish flesh. Head mostly gray with indistinct blackish stripes on either side of crown; *bold white supraloral and partial eye-ring.* Upperparts otherwise olive. Below white, *mottled with gray across breast;* flanks tinged olive, becoming greenish yellow on crissum.
SIMILAR SPECIES: Cf. White-striped Warbler (no known range overlap), which has prominent long white superciliary, etc. White-rimmed often occurs in exactly the same habitat as nominate race of River Warbler and its facial pattern is not dissimilar (aside from having an eye-ring), but it is much grayer below (*with no buff*).
HABITAT AND BEHAVIOR: Fairly common to common in undergrowth of forest and secondary woodland; often near water, but not as tied to it as is River (or White-striped) Warbler. Usually in pairs which work through dense growth, often hopping on ground and generally not at all shy. The tail is frequently spread and slowly raised up and down or moved sideways. Its frequently heard song is a series of musical notes, initially rather high in pitch, sibilant, and slowly delivered, gradually becoming lower, louder, and more run together; the call is a very sharp penetrating "pseeyk."
RANGE: Se. Brazil (Rio de Janeiro south to Rio Grande do Sul), e. Paraguay (north to Canendiyu), ne. Argentina (west to e. Formosa, e. Chaco, and ne. Santa Fe; south to Entre Ríos), and Uruguay. To at least 1600 m.

NOTE: We believe it unlikely that this species is allied with the widely disjunct *B. griseiceps* of Venezuela, as S. Olson (*Bull. B.O.C.* 95: 101–104, 1975) has suggested. Rather, we feel that it is closer to *B. leucophrys* despite differences in bill and tail shape, and to *B. rivularis*. The previous English name of *B. leucoblepharus*, "White-browed Warbler," is very misleading, as the species shows merely a short white supraloral stripe, quite different from the true white "brow" of the White-striped Warbler. As *blepharus* means "eyelid," and as the partial white eye ring is a conspicuous feature of this bird (and diagnostic within its subgenus), we have opted to rename it the "White-rimmed Warbler."

Basileuterus fulvicauda

BUFF-RUMPED WARBLER

PLATE: 9

IDENTIFICATION: 13.5 cm (5¼"). *Obvious buff rump and basal portion of tail* easily identify this water-loving warbler. Mostly brownish olive above, crown grayer and with narrow buff superciliary and lower cheeks; *rump and basal portion of tail buff.* Below buffy whitish, whitest on throat and middle of breast.

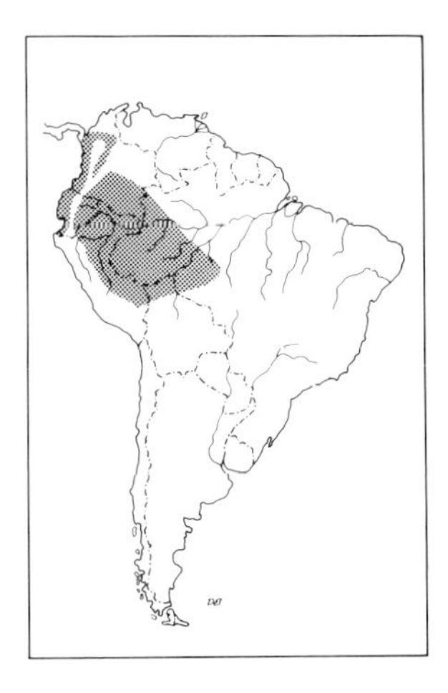

SIMILAR SPECIES: Virtually unmistakable on account of the pale rump and basal tail (which can, however, be mostly hidden when bird is at rest). Otherwise similar River Warbler lacks this. Waterthrushes are streaked below, also lack the buff on rump and tail.

HABITAT AND BEHAVIOR: Fairly common along streams and rivers in forested areas; almost never found at any distance from water. Generally associated with running water (and most numerous in hilly areas), though at least in Central America also found in swampy places and even mangroves. Mostly terrestrial, hopping (rarely if ever actually walking) on ground or at water's edge, frequently swinging its fanned tail from side to side. Often flushes repeatedly ahead of an approaching observer or passing boat, usually with a loud "chip." The song is a rising and accelerating crescendo of loud "tew" notes, ending with a series of enunciated "chéws," easily heard over the roar of turbulent streams.

RANGE: W. Colombia (absent from arid parts of Caribbean north), w. Ecuador, and extreme nw. Peru (Tumbes); se. Colombia (north to Meta and Vaupés), e. Ecuador, e. Peru, and w. Amaz. Brazil (east to the Rio Madeira drainage); probably also occurs in n. Bolivia. Also Honduras to Panama. Mostly below 1000 m.

NOTE: We follow the 1983 AOU Check-list in considering *fulvicauda* as a species distinct from *B. rivularis*. As A. H. Miller (*Proc. Biol. Soc. Wash.* 65: 13–17, 1972) has pointed out, the near-continuous signaling of its pale rump and basal tail probably serves as an effective isolating mechanism where they come into contact (as seems likely in Amaz. Brazil and near the Peru-Bolivia border, though no such zone seems yet to have been located). On the other hand, *B. f. significans* of se. Peru does seem to approach *B. r. bolivianus* in several characters (as J. T. Zimmer pointed out in his description of *significans*).

Basileuterus rivularis

RIVER WARBLER

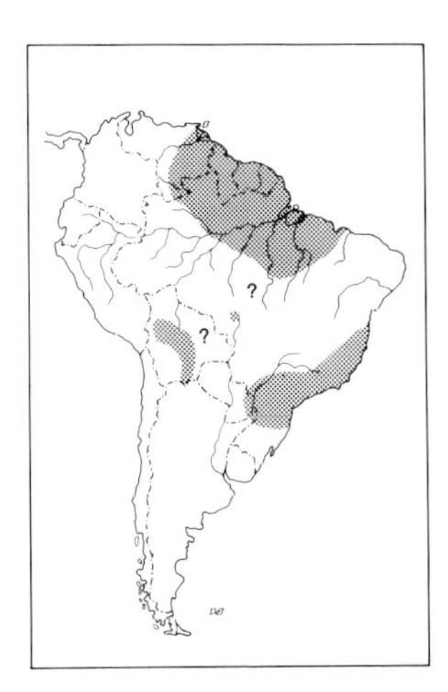

IDENTIFICATION: 13.5 cm (5¼"). Similar to Buff-rumped Warbler but *lacking* the buff rump and basal tail (these concolor with rest of upperparts, being olive). See V-33. Nominate race (se. South America) differs in having vague blackish lateral crown stripes.

SIMILAR SPECIES: Cf. Buff-rumped Warbler. White-rimmed Warbler has head pattern somewhat similar to that of nominate *rivularis* but has white eye-ring, grayish (rather than buffyish) underparts.

HABITAT AND BEHAVIOR: Fairly common along streams and rivers and in swampy areas in forested regions. Behavior and voice similar to Buff-rumped Warbler's, though more often found in swampy terrain.

RANGE: Guianas, e. and s. Venezuela (north to e. Monagas and Delta Amacuro; west to cen. Amazonas), and lower Amaz. Brazil (west to Rios Negro and Tapajós and east to n. Maranhão); s. cen. Brazil (in n. Mato Grosso at Serra dos Parecís), and se. Brazil (se. Bahia to Santa Catarina, possibly to Rio Grande do Sul), e. Paraguay, and ne. Argentina (Misiones); e. base of Andes in Bolivia (La Paz to w. Santa Cruz and in Tarija). To about 1400 m (in Bolivia).

NOTE: See comments under Buff-rumped Warbler.

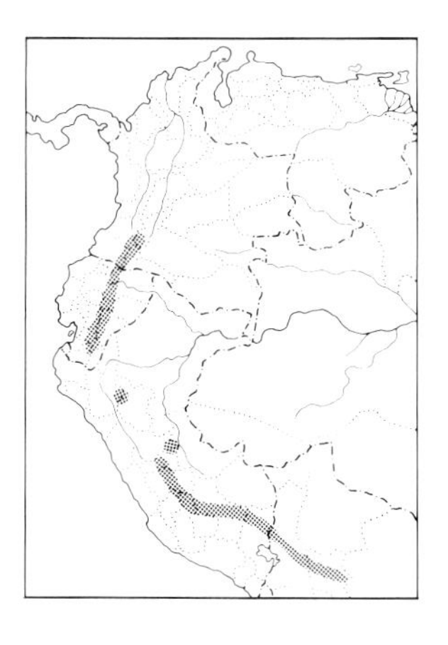

SIMILAR SPECIES: Unique among the flower-piercers is this yellow-eyed species; the eye stands out prominently against the bird's dark blue plumage.

HABITAT AND BEHAVIOR: Locally fairly common to common in humid montane forest and forest borders. Usually seen singly or in pairs, almost always with mixed flocks in which *Tangara* tanagers predominate. Generally forages rather high above ground; despite its hooked bill (obviously suitable for flower piercing), most often seen gleaning along branches and among bromeliads or in viny tangles and only infrequently at flowers.

RANGE: E. slope of Andes in sw. Colombia (north to Caquetá), e. Ecuador, e. Peru (few records from n. Peru), and nw. Bolivia (La Paz and Cochabamba). 1100–2300 m, rarely higher.

Diglossa indigotica

INDIGO FLOWER-PIERCER

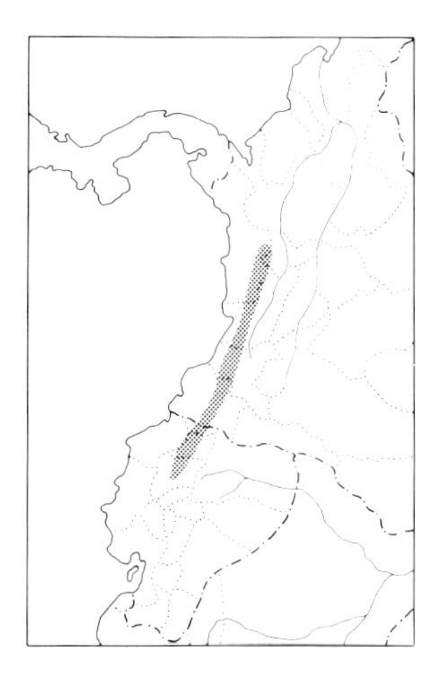

IDENTIFICATION: 11.5 cm (4½"). *W. Colombia and nw. Ecuador. Iris bright red. Brilliant deep ultramarine blue,* with flight feathers blackish edged paler blue. Lores and narrow eye-ring black (inconspicuous in the field). See C-47.

SIMILAR SPECIES: The most intensely blue of all the flower-piercers, with usually prominent fiery red eye. Mainly occurs lower than other *Diglossa,* but toward upper end of its elevational range may occasionally overlap with the much larger Masked Flower-piercer (which is duller blue and has obvious black mask).

HABITAT AND BEHAVIOR: Local, but can be fairly common to common in very humid lower montane forest (especially cloud forest). Behavior much like Deep-blue Flower-piercer's, though perhaps more frequent at flowers. Seems not to be found in many accessible localities, but is numerous in cloud forest along road above the town of Junín in Nariño, Colombia.

RANGE: Pacific slope of W. Andes in w. Colombia (north to s. Chocó) and nw. Ecuador (south to Pichincha, but few recent records). 700–2200 m, but most numerous 1000–1500 m.

Diglossa major

GREATER FLOWER-PIERCER

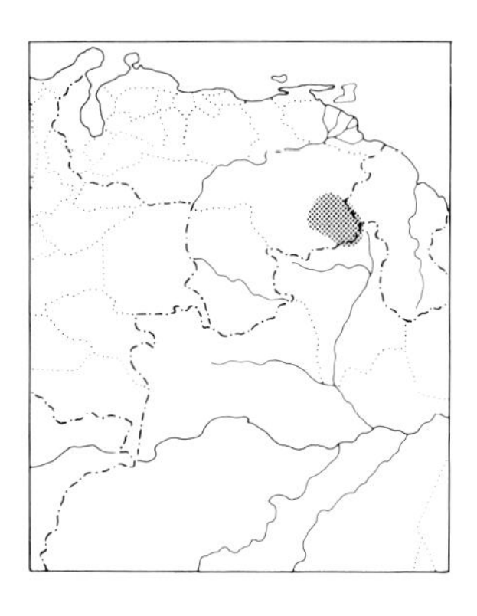

IDENTIFICATION: 16.5 cm (6½"). *Tepuis of s. Bolívar, Venezuela.* Noticeably *larger* than any of the more familiar *Diglossa.* Mostly bluish slate, darker above. *Crown and mantle with fine silvery bluish shaft streaks* and with *black mask* over forehead, sides of head, and chin; mask bordered below by whitish malar streak. Underparts with silvery shaft streaks in some races. *Crissum chestnut.* See V-34.

SIMILAR SPECIES: *No other flower-piercer in its range;* in Amazonas cf. Scaled Flower-piercer.

HABITAT AND BEHAVIOR: Fairly common to common in forest borders, shrubby openings, and low forest on *tepui* summits and slopes. Apparently usually found singly or in pairs; has not yet been seen to

actually pierce flowers. Small numbers can usually be found in the stunted forest growing on white-sand soils near Km 121 on the Escalera road, becoming one of the more numerous birds on the talus slopes of Cerro Roraima itself (C. Parrish). Its voice is a scratchy rattling, usually introduced by a series of tinkling notes, very different from the voices of other *Diglossa* (T. Meyer).
RANGE: S. Venezuela in e. Bolívar and adjacent n. Brazil in Roraima (slopes of Cerro Uei-tepui); probably also on slopes of Cerro Roraima in Guyana. 1300–2800 m.

Diglossa duidae

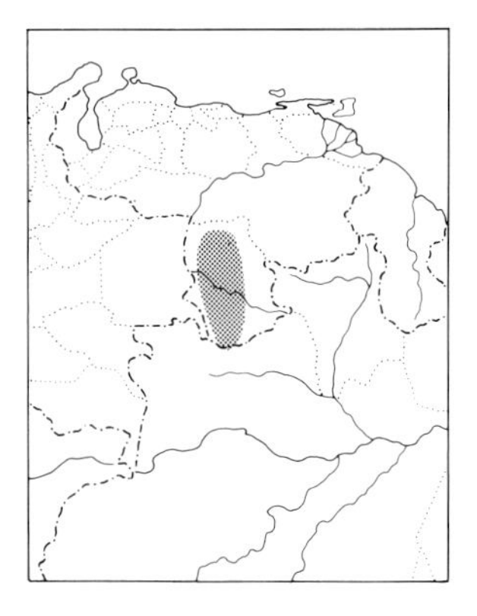

SCALED FLOWER-PIERCER

IDENTIFICATION: 14 cm (5½"). *Tepuis of Amazonas, Venezuela.* Mostly bluish slate, blacker on head (but *without a mask*). *Feathers of breast centered whitish,* resulting in scaly appearance; belly and crissum paler gray (with *no chestnut*). Often shows single (or more infrequently double) pale gray wing-bar.
SIMILAR SPECIES: *No other flower-piercer in range;* in Bolívar cf. Greater Flower-piercer (no known overlap).
HABITAT AND BEHAVIOR: Not well known, but to judge from the number of specimens in museum collections must be locally very numerous. Behavior doubtless much like that of other *Diglossa.*
RANGE: S. Venezuela in Amazonas and adjacent n. Brazil in Roraima (slopes of Cerro de la Neblina). 1400–2300 m.

GROUP B

Lafresnayii and *carbonaria* groups (first 3 in the former, then 4 in the latter; their plumages vary in a complex but essentially parallel way).

Diglossa lafresnayii

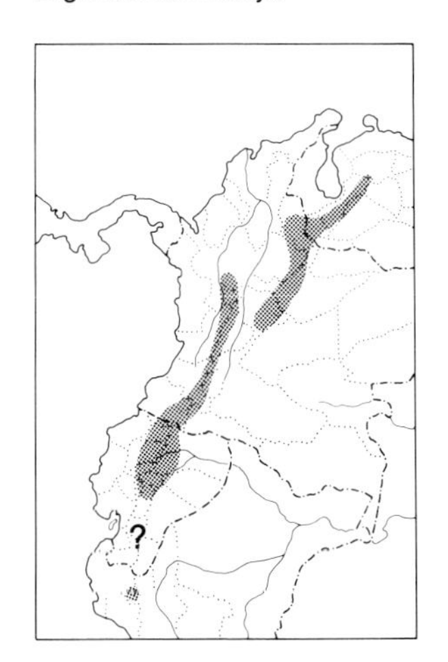

GLOSSY FLOWER-PIERCER

PLATE: 10

IDENTIFICATION: 14.5 cm (5¾"). *Glossy black* with *bluish gray shoulders.*
SIMILAR SPECIES: Closely resembles Black Flower-piercer and often found with it. For full discussion, see Black Flower-piercer.
HABITAT AND BEHAVIOR: Common in shrubby thickets and low trees near timberline and at borders of forest and in shrubby clearings somewhat lower in temperate zone. Widespread, and like Black Flower-piercer often found well away from actual forest. Usually seen singly or in pairs, sometimes accompanying mixed flocks of other tanagers and various highland passerines, but at least as often apart from them. May perch in the open on top of a bush when giving its fast, twittery song (which may go on for several minutes), but otherwise often seems shy and restless, and it may be hard to obtain a clear view of it.
RANGE: Andes of w. Venezuela (north to Trujillo), Colombia (absent from W. Andes north of Nariño), Ecuador, and n. Peru (Cajamarca). Mostly 2700–3700 m.

NOTE: We have followed F. Vuilleumier (*Am. Mus. Novitates* 2831: 1–44, 1969) in considering *gloriosissima* of w. Colombia and *mystacalis* of Peru (south of the Marañón) and Bolivia as species apart from *D. lafresnayii* of w. Colombia to n. Peru (north of the Marañón).

Diglossa gloriosissima

CHESTNUT-BELLIED FLOWER-PIERCER

Other: Glossy Flower-piercer (in part)

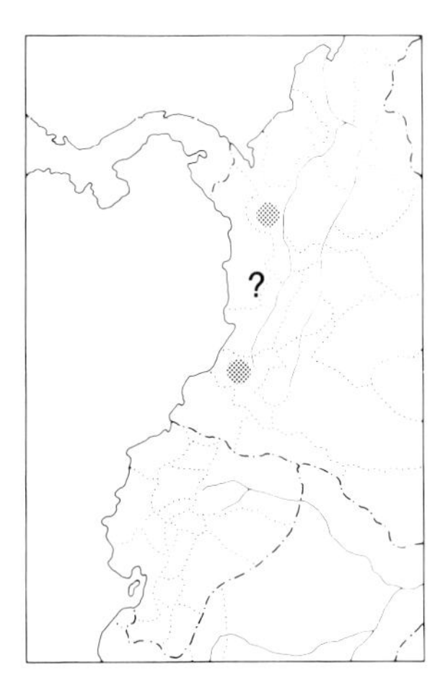

IDENTIFICATION: 14.5 cm (5¾"). *Locally in W. Andes of Colombia. Mostly glossy black,* but *breast and belly rufous-chestnut* and shoulder patch blue-gray. See C-47.

SIMILAR SPECIES: Black-throated Flower-piercer, which overlaps with this species, is mainly rufous-chestnut below (with black restricted to throat patch, not extending over chest, and outlined by moustachial streak). Merida Flower-piercer, of the *carbonaria* complex, is very similar but is found only in the Venezuelan Andes. Overall pattern recalls that of Blue-backed Conebill (but note bill differences, its blue back, etc.).

HABITAT AND BEHAVIOR: Nothing appears to be on record, but probably does not differ from Glossy Flower-piercer (which it replaces).

RANGE: W. Colombia (W. Andes in Antioquia and in Cauca on Cerro Munchique; may occur on other high peaks, as yet not explored). 3000–3700 m.

NOTE: See comments under Glossy Flower-piercer.

Diglossa mystacalis

MOUSTACHED FLOWER-PIERCER

PLATE: 10

Other: Glossy Flower-piercer (in part)

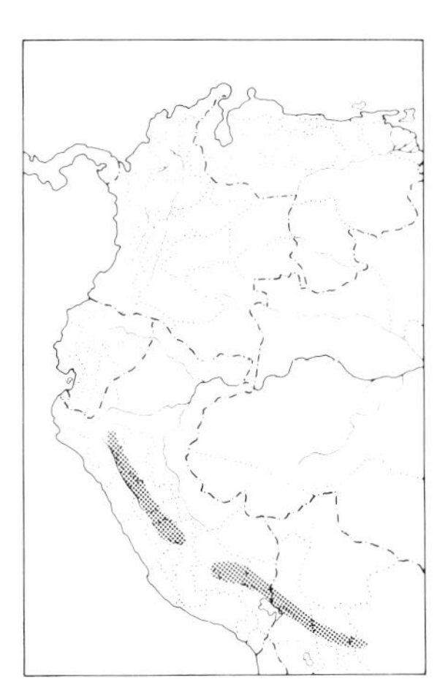

IDENTIFICATION: 14.5 cm (5¾"). *Peru and Bolivia.* The *bold moustache* (either white or rufous) is the mark of this variable *Diglossa.* All 4 races are otherwise mostly black with rufous crissum; *the 2 n. races also have a broad pectoral band* (rufous to white). From north to south: *unicincta* (Amazonas to n. Huánuco, Peru) has white moustache and rufous and buff pectoral band; *pectoralis* (s. Huánuco and Junín, Peru) similar but with rufous and white pectoral band; *albilinea* (Cuzco and Puno, Peru) has no pectoral band and a buffy white moustache; nominate race (Bolivia) is similar but has rufous moustache. Immature-plumaged birds seem particularly numerous in this species: they have whitish streaking on underparts and a pale pink basal bill.

SIMILAR SPECIES: Nothing really similar; Black-throated Flower-piercer also has rufous moustache but its underparts are mainly chestnut (not mostly black).

HABITAT AND BEHAVIOR: Uncommon in shrubby clearings and forest and woodland borders from timberline down into the upper temperate zone. Usually noted singly or in pairs, only infrequently accompanying mixed flocks. Moves rapidly inside low vegetation, tending to remain hidden, usually only briefly appearing in the open. The song is a series of short, fast, sweet notes, rising and falling in pitch in an irregular fashion. Regularly found near the tunnel on the Carpish ridge, Huánuco, Peru, and in the *yungas* of La Paz, Bolivia.

RANGE: E. slope of Andes in Peru (north to Amazonas) and nw. Bolivia (La Paz, Cochabamba, and extreme w. Santa Cruz). Mostly 2500–4000 m.

NOTE: See comments under Glossy Flower-piercer.

Diglossa humeralis

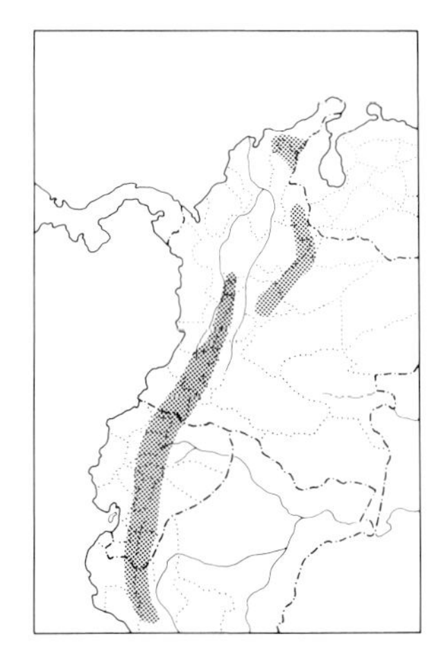

BLACK FLOWER-PIERCER

Other: Carbonated Flower-piercer (in part)

IDENTIFICATION: 13.5–14 cm (5¼–5½"). Very closely resembles widely sympatric Glossy Flower-piercer, presenting a difficult identification problem. In Santa Marta and Perijá Mts. only this species is found, making things easy; the race there (*nocticolor*) is all black with dark gray rump (hard to see in the field) and no gray on shoulders. From Colombia's Cen. Andes south, the all-black *aterrima* race is found, differing from Glossy most obviously in its *lack of blue-gray shoulders;* see C-47. It is in Colombia's E. Andes, where both Glossy and the *humeralis* race of Black (which *has* a shoulder patch) are found, that problems are greatest. Here one can try to distinguish Black on the basis of its slightly smaller size and proportionately smaller bill, the purer (not so blue) gray tone to its shoulder patch, and its usually duller black plumage; see V-34. However, the task won't be easy, and many "black flower-piercers" will never be definitely identified to species.

HABITAT AND BEHAVIOR: Fairly common in shrubby areas, forest borders, and gardens; even less of a forest-associated species than Glossy Flower-piercer and perhaps tending to occur at lower elevations. Overall numbers have probably increased due to the partial clearing of much temperate zone forest. Very active, aggressively defending its feeding area not only from other flower-piercers but also from hummingbirds, conebills, etc. Its song is a brief, rapid twittering.

RANGE: Andes of Colombia (except at n. end of both W. and Cen. Andes, where replaced by *D. brunneiventris*), Ecuador, and n. Peru (north and west of the Marañón in Piura and Cajamarca); also Colombia's Santa Marta Mts. and Perijá Mts. on Colombia-Venezuela border.

NOTE: G. Graves (*Condor* 84 [1]: 1–14, 1982) and F. Vuilleumier (Nat. Geogr. Soc. Res. Rpts., 1975 Projects) are the last to have dealt with this complex. The former considered the *D. carbonaria* group to consist of 4 species, the latter (somewhat tentatively) as 3. Venezuelan *D. gloriosa* seems isolated from *D. humeralis* by the low "Táchira Depression" in extreme w. Venezuela, while the ranges of *D. humeralis* and *D. brunneiventris* approach each other closely in n. Peru, with no evidence of hybridization. Hybridization in Bolivia between *D. brunneiventris* and *D. carbonaria* is evidently very limited and may be secondary in nature; we follow Graves in considering them as allospecies.

Diglossa gloriosa

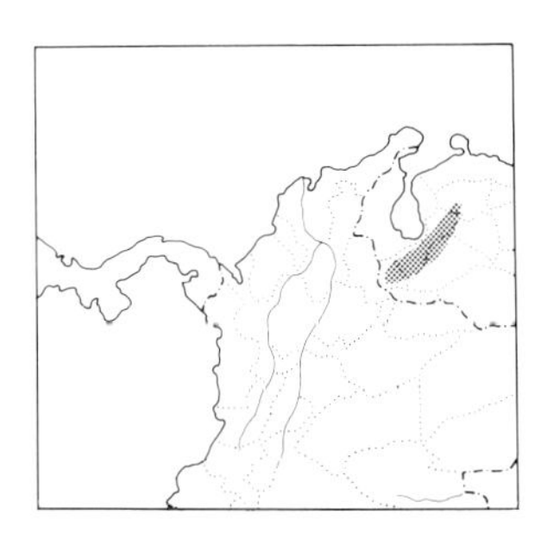

MERIDA FLOWER-PIERCER

Other: Carbonated Flower-piercer (in part), Coal-black Flower-piercer

IDENTIFICATION: 13.5 cm (5¼"). *Venezuelan Andes.* Mostly black with gray shoulders, narrow inconspicuous superciliary, and rump; *lower breast and belly contrastingly chestnut,* with dark gray flanks. See V-34.

SIMILAR SPECIES: Glossy Flower-piercer has no chestnut below. Rusty Flower-piercer male is much paler above (gray rather than black) and is uniform cinnamon below (not black-chested). No overlap with Chestnut-bellied Flower-piercer. Cf. also Blue-backed Conebill.

HABITAT AND BEHAVIOR: Uncommon and seemingly somewhat local in shrubby areas and forest borders. Behavior similar to Black Flower-piercer's and like that species not especially associated with forest. Can be seen in shrubbery near the upper Río Santo Domingo in Mérida (upstream from the town of that name).

RANGE: Andes of w. Venezuela (Trujillo to n. Táchira). Mostly 3000–4000 m.

NOTE: See comments under Black Flower-piercer. No English name seems to have been suggested for this range-restricted form. As it exhibits no easily singled-out field character, we have opted to emphasize its limited range (which centers on Mérida, Venezuela).

Diglossa brunneiventris

BLACK-THROATED FLOWER-PIERCER

PLATE: 10

Other: Carbonated Flower-piercer (in part)

IDENTIFICATION: 14 cm (5½"). *N. Colombia (local), Peru, and adjacent Bolivia.* Mainly black above with gray shoulders and dark gray rump. *Center of throat black,* with *remaining underparts mostly rufous-chestnut,* this color also extending up as moustache separating the throat and sides of head; flanks gray.

SIMILAR SPECIES: Moustached Flower-piercer is always mostly black below (though the various races do have rufous or white moustaches and breast crescents). Chestnut-bellied Flower-piercer (locally sympatric with Black-throated in Colombia) has black throat and chest, with only lower underparts chestnut. See also male Rusty Flower-piercer.

HABITAT AND BEHAVIOR: Fairly common in shrubby areas and forest borders, both in dry as well as humid areas; locally also in *Polylepis* woodlands. Behavior much like Black and Gray-bellied Flower-piercers'.

RANGE: Andes of n. Colombia (n. end of both W. and Cen. Andes, where not very well known) and in Peru (north to Amazonas and s. Cajamarca), n. Chile (1 record from Tarapacá), and nw. Bolivia (La Paz). Mostly 2500–4000 m, occasionally lower and higher.

NOTE: See comments under Black Flower-piercer. The recently described n. Colombia form (*vuilleumieri;* see G. Graves, *Bull. B.O.C.* 100 [4]: 230–232, 1980) is similar to the nominate form, despite its being highly disjunct geographically.

Diglossa carbonaria

GRAY-BELLIED FLOWER-PIERCER

PLATE: 10

Other: Carbonated Flower-piercer (in part)

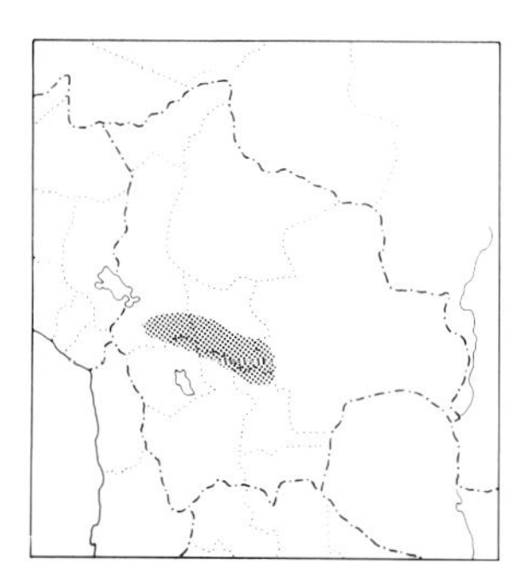

IDENTIFICATION: 14 cm (5½"). *Bolivia.* Mostly black with gray shoulders and rump; *lower breast and belly gray* (merging with slight scaly effect into black chest); crissum rufous.

SIMILAR SPECIES: The only flower-piercer in either the *lafresnayii* or the *carbonaria* complexes with a gray belly; sympatric race of Moustached Flower-piercer is all black below with a rufous moustache.

HABITAT AND BEHAVIOR: Fairly common at borders of humid forest, both near and below timberline, and in shrubby terrain and lighter (e.g., *Alnus*) woodland; ranges down locally on arid intermontane valley slopes. Usually found singly or in pairs, working nervously through foliage, pausing to probe flowers but usually not lingering long. Quite readily found at various localities around Cochabamba (e.g., near pass on road to Villa Tunari).

RANGE: W. Bolivia (s. La Paz, Cochabamba, w. Santa Cruz, and n. Chuquisaca). Mostly 2500–4000 m.

NOTE: See comments under Black Flower-piercer. As this is the only member of either the *carbonaria* complex or *lafresnayii* complex to have a gray belly, it seems best to highlight this in the form's English name.

GROUP C *Albilatera* group; sexes *differ*.

Diglossa albilatera

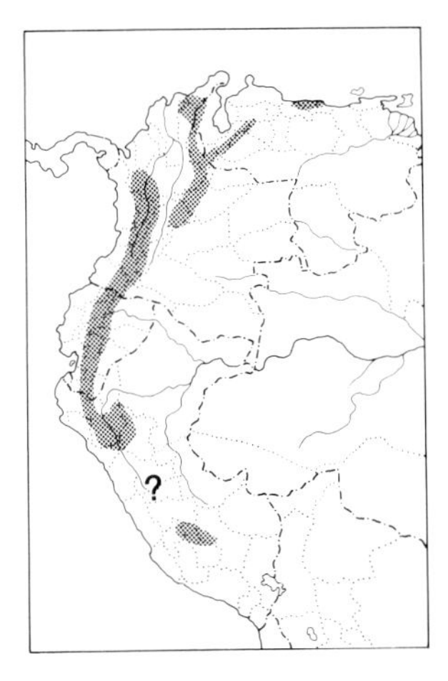

WHITE-SIDED FLOWER-PIERCER

PLATE: 10

IDENTIFICATION: 13 cm (5″). Male almost entirely dark slaty, with *partially concealed white tuft on sides and under wing-linings*. Female olive brown above, sometimes with a faint buff wing-bar. *Paler buffy brown below,* richest on breast; *white tuft on sides and under wings* similar to male's but somewhat smaller; see V-34. Immature like female but with blurry olive brown streaking on breast.

SIMILAR SPECIES: This nervous little *Diglossa* is almost always exposing at least some white along and under the wing, a character shared by no other species (except the range-restricted Venezuelan Flower-piercer). On the rare occasions when it is not, the blacker Black Flower-piercer might be confused with it (especially its race with no gray on shoulders).

HABITAT AND BEHAVIOR: Fairly common to common at borders of humid montane forest and in adjacent shrubby clearings and even gardens. Most often in pairs which forage actively, often flicking their wings, at lower and middle levels; these often move independently, though at times they also join mixed flocks. The song is a distinctive short, unmusical trill. Most numerous in Colombia and Venezuela, gradually less so southward in its range.

RANGE: N. coastal mts. of Venezuela (Aragua to Miranda) and Andes of w. Venezuela (north to Trujillo), Colombia, Ecuador, and nw. (south to Cajamarca at Chaupe) and e. Peru (south to Cuzco on Cordillera Vilcabamba); Colombia's Santa Marta Mts. and Perijá Mts. on Venezuela-Colombia border. Mostly 1800–2800 m.

Diglossa venezuelensis

VENEZUELAN FLOWER-PIERCER

IDENTIFICATION: 13.5 cm (5¼″). *Ne. Venezuela.* Resembles White-sided Flower-piercer (no overlap), but *larger* and with white tuft on sides smaller (but should still be evident in the field). Male is notably *blacker* (not slaty gray). Female has *head yellowish olive* (brightest on foreface), but this can be duplicated by some juvenal White-sideds.

SIMILAR SPECIES: Only sympatric *Diglossa* is the Rusty: males very different, but females are somewhat similar (Rusty is notably smaller and lacks white at sides).

HABITAT AND BEHAVIOR: Not well known; apparently uncommon in humid forest, second-growth woodland, and borders. Has recently been found in patches of *Clusia*-dominated forest on Cerro Negro (by P. Lau) and near Melenas on beginning of trail up Cerro Humo (by M. Lentino and M. L. Goodwin, both *fide* C. Parrish). May, like the Gray-headed Warbler, which shares its limited range, deserve threatened status due to destruction of much of its potential forest habitat.
RANGE: Coastal mts. of ne. Venezuela in limited area in s. Sucre and nw. Monagas. 1600–2500 m.

Diglossa sittoides

RUSTY FLOWER-PIERCER

PLATE: 10

Other: Slaty Flower-piercer

IDENTIFICATION: 12 cm (4¾"). *Above bluish gray,* with face somewhat blackish in some races. *Below entirely cinnamon.* Female very dull, brownish olive above and dingy yellowish buff below, *obscurely streaked dusky on throat and breast.*
SIMILAR SPECIES: Distinctly bicolored males are unlike any other *Diglossa* and are actually more likely (despite bill shape) to be confused with one of the montane conebills (especially Rufous-browed); general coloration not too dissimilar from that of female Tit-like Dacnis. Female rather resembles female and (especially) immature White-sided Flower-piercer but is more olivaceous (less rich buffy) overall and lacks that species' white patch on sides.
HABITAT AND BEHAVIOR: Uncommon to fairly common but rather local in shrubby clearings and gardens and at borders of woodland and forest. Occurs in both humid and fairly arid areas, but not found in extensively forested regions; thus has probably increased in some regions due to deforestation. Generally found singly or in pairs, most often independent of mixed flocks.
RANGE: N. coastal mts. of Venezuela (Yaracuy east to Miranda and in s. Sucre), Andes of w. Venezuela (north to s. Lara), Colombia, Ecuador, Peru (south on w. slope to Lima), w. Bolivia, and nw. Argentina (south to Tucumán); Colombia's Santa Marta Mts. and Perijá Mts. on Colombia-Venezuela border. Mostly 1500–3000 m, occasionally higher (up to over 4000 m) or lower (perhaps as seasonal visitant to 600–800 m).

NOTE: We follow Vuilleumier (1969) and the 1983 AOU Check-list in considering the S. Am. *sittoides* group as a species distinct from Middle American *D. baritula* and *D. plumbea.*

Oreomanes Conebills

A large conebill with a long, pointed bill; mostly gray and rufous. Distinctive behavior, creeping on limbs of *Polylepis* trees in the high Andes. The genus is apparently close to *Conirostrum* (nominate subgenus), and 1 apparent intergeneric hybrid has been collected.

Oreomanes fraseri

GIANT CONEBILL

PLATE: 10

IDENTIFICATION: 16.5 cm (6½"). *Rather long, sharply pointed bill. Mostly gray above* with short chestnut eyeline surmounted by whitish frosting on forecrown; *white cheeks. Entire underparts chestnut,* with thighs gray. Nominate race (Ecuador and s. Colombia) has all-gray crown; see C-48.

SIMILAR SPECIES: Nothing really similar; besides plumage, note its distinctive *Polylepis* habitat, behavior.

HABITAT AND BEHAVIOR: Rare to locally uncommon in groves of *Polylepis,* principally or entirely above true timberline. Quite unobtrusive and never really numerous even in seemingly ideal habitat. Found singly or in well-separated pairs, and often located after hearing the bird scaling off pieces of the *Polylepis* tree's flaking bark. Hitches along trunks and horizontal limbs much like a *Sitta* nuthatch (though never hops down headfirst); often remains in the same tree for some time but then may fly a considerable distance (even to another grove of trees). Occasionally pauses from its feeding to give its fairly musical but repetitious song, "cheet, cheeveét-cheeveét." In many parts of its range may now be considerably more local and less numerous than formerly, a decline attributable to felling of *Polylepis* for firewood by ever-expanding human populations (this seems especially the case in Bolivia). Can be found along the Papallacta road east of Quito, Ecuador, and along the road between Huánuco and Cerro de Pasco, Peru.

RANGE: High Andes of sw. Colombia (Nariño; no recent records), Ecuador (south to Azuay and Loja), Peru (locally from Ancash to Puno and Tacna), and w. Bolivia (few records, but recorded locally south to s. Potosí). 3300–4500 m.

Nephelornis Parduscos

An obscure, plain brown bird of uncertain affinities (but presumably closest to the tanagers), only recently discovered in the Andes of Peru.

Nephelornis oneillei

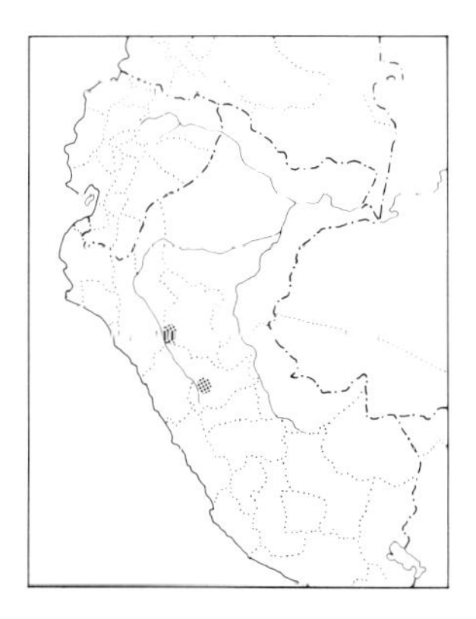

PARDUSCO

PLATE: 10

IDENTIFICATION: 13 cm (5"). *Local near timberline on e. slope of Andes in cen. Peru.* Dull olive brown above, outer primaries narrowly edged ochraceous. Ochraceous buff below, palest on throat.

SIMILAR SPECIES: This plain tanager or tanager relative is devoid of really distinctive marks, but in its limited range should easily be known from its *uniform brownish appearance* (paler below) and *slender, pointed bill.* In coloration recalls certain female or immature *Diglossa* (but note very different bill shape).

HABITAT AND BEHAVIOR: Locally common in patches of low elfin woodland near timberline, occurring primarily at their edges. Found in groups of up to 5 to 15 individuals, moving rather rapidly, often as an independent unit but sometimes joining mixed flocks of other tan-

agers, etc. Not at all shy. Forages by deliberately gleaning from leaves and stems. Best known from the Carpish ridge region of Huánuco, but only in areas well above the highway (accessible only by trail).
RANGE: Andes of cen. Peru in e. La Libertad and Huánuco. 3000–3500 m.

NOTE: A newly described species: G. Lowery and D. Tallman (*Auk* 93: 415–428, 1976). Its affinities remain obscure, but it may be most closely allied to *Urothraupis*.

Xenodacnis Dacnises

Small, stubby-billed bird of high *Polylepis* woodlands and adjacent scrub in Andes of Peru and s. Ecuador. Male blue, female with blue on crown.

Xenodacnis parina

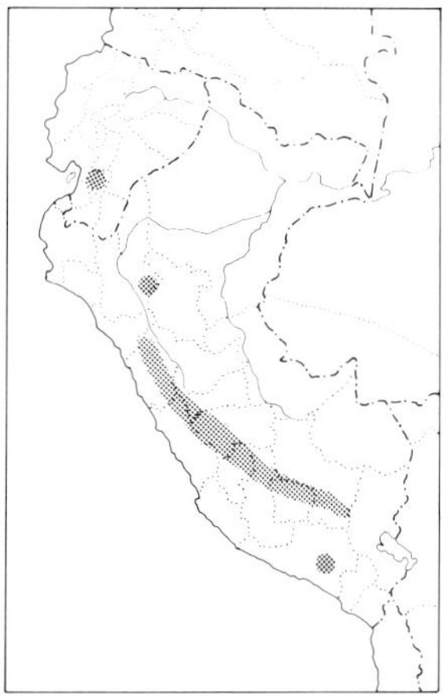

TIT-LIKE DACNIS

PLATE: 10

IDENTIFICATION: 13–14 cm (5–5½"). Short, pointed bill. *Entirely deep blue* with narrow glistening paler blue streaking (the streaking not too conspicuous at a distance). Female grayish brown above with *blue forecrown and ocular area,* wing- and tail-edging, and upper tail-coverts. *Mostly cinnamon buff below,* paler on median belly. Nominate race (north to Junín in Peru) is rather different: smaller, with streaky effect at best very indistinct and somewhat paler, more grayish blue ground color; female has *entire crown and nape blue* and is slightly brighter cinnamon buff below.
SIMILAR SPECIES: The all-blue male is virtually unmistakable in its range given decent light, and even in poor light (when it can look black) bill shape will distinguish it from any *Diglossa*. Female's contrasting blue on head serves to make her easily recognized.
HABITAT AND BEHAVIOR: Uncommon to locally fairly common in shrubbery, low woodland (frequently at the edge of *Polylepis* groves), and forest borders from just below timberline to (locally) well above it. Forages almost entirely by gleaning from the underside of leaves of low *Gynoxys* shrubs and *Polylepis* trees; usually in pairs or small groups of up to 3 to 5 birds, occasionally with mixed flocks. Readily seen at various localities in Cuzco, Peru (e.g., on the Abra Malaga road) and also at Las Cajas Recreation Area above Cuenca, Ecuador.
RANGE: Andes of sw. Ecuador (Azuay) and Peru (Amazonas south to Cuzco and Arequipa). 3000–4000 m.

Conirostrum Conebills

Small warblerlike tanagers with sharply pointed bills. Plumage patterns vary, but blue or gray predominates in both. Found in Andean forests and adjacent shrubbery, mostly at high elevations (Cinereous also regular along Pacific coast). Note that the lowland group of *Conirostrum* conebills is dealt with separately (pp. 222–224).

GROUP A

"Typical" conebills; sexes similar.

Conirostrum cinereum

CINEREOUS CONEBILL

PLATE: 10

IDENTIFICATION: 12–12.5 cm (4¾–5"). The common Andean conebill with *distinctive L-shaped wing-patch.* S. nominate race (Bolivia and e. Peru north to Huánuco) is much grayer overall than the 2 w. and n. races: leaden gray above with darker crown and *bold white superciliary;* wings blackish with L-shaped white patch (formed by wing-bar and speculum at base of primaries). Below paler gray, becoming buffier on belly, deepest on crissum. *Littorale* (w. Peru and n. Chile) is more olivaceous gray above with crown concolor and is notably buffier below (lacking any gray); while *fraseri* (Ecuador and Colombia) is slightly larger, olivaceous brown above with darker crown and buffy whitish superciliary, and is entirely ochraceous-buff below (see C-48).
SIMILAR SPECIES: Tonalities vary from gray to brown or buffy, but this species' pattern does not vary, and the combination of the superciliary and the wing-patch is characteristic.
HABITAT AND BEHAVIOR: Generally common (though less numerous at n. end of its range) in shrubby areas, woodland borders, gardens, and semiopen areas with scattered trees and bushes; occurs in both humid and arid regions and ranges up to near timberline in places, but avoids extensively forested areas. Forages actively in foliage, usually in pairs or small groups; may accompany flocks, but also often moves about on its own.
RANGE: Andes of sw. Colombia (Cauca and Nariño), Ecuador, Peru, n. Chile (Tarapacá), and nw. Bolivia (La Paz and Cochabamba); absent from e. slope of Andes. Mostly 2500–4000 m, though *littorale* of w. Peru and Chile regularly ranges down to sea level.

Conirostrum tamarugense

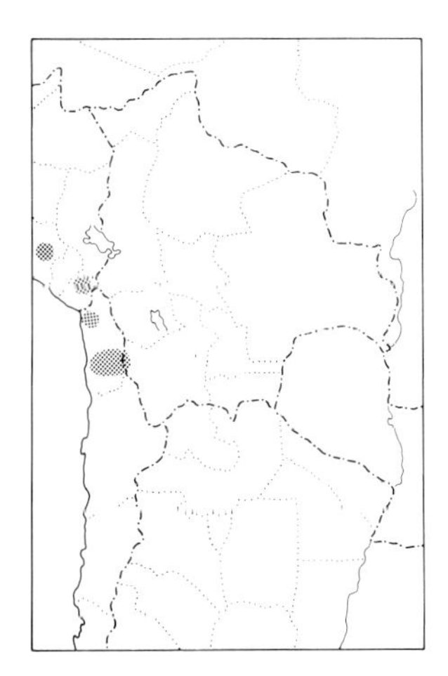

TAMARUGO CONEBILL

IDENTIFICATION: 12.5 cm (5"). *Sw. Peru and n. Chile.* Dark gray above with *cinnamon-rufous superciliary* and white L-shaped patch on wings. *Throat and center of upper chest also cinnamon-rufous;* remaining underparts whitish, tinged buffy medially; crissum cinnamon. Sexes similar (bird originally described as female being actually the immature). Immature similar but more brownish gray above and all whitish below (lacking the throat patch but retaining the superciliary).
SIMILAR SPECIES: Liable to be confused (and occurs sympatrically) only with Cinereous Conebill, with which it shares the wing-marking, but has rufous (not white) superciliary, and the male has cinnamon-rufous on throat and crissum.
HABITAT AND BEHAVIOR: Locally fairly common in groves of "tamarugos" (a small tree, *Prosopis tamarugo*) in n. Chile and in low scrubby woodland patches with some *Gynoxys* shrubs or *Polylepis* trees in s. Peru. Forages in pairs or small groups, gleaning in foliage; may occur in mixed flocks with Cinereous Conebill (in Chile) or in loose association with various Furnariids, etc. (in Peru). Has been found

in the sparse woodland patches above Chiguata along the road from Arequipa to Puno, Peru (but may be only seasonal here, seemingly not present at least Dec.–Mar.) and has also been recorded at least occasionally in the outskirts of Arica, Chile.

RANGE: Sw. Peru (Arequipa and Tacna) and n. Chile (Tarapacá). To about 4050 m (in Peru), but altitudinal distribution details remain to be worked out.

NOTE: A newly described species: A. W. Johnson and W. R. Millie, in Johnson (1972: 3–8). The correct spelling of the species name is, however, *tamarugense* (E. Mayr and F. Vuilleumier, *J.f Orn.* 124 [3]: 227, 1983).

Conirostrum ferrugineiventre

WHITE-BROWED CONEBILL

PLATE: 10

IDENTIFICATION: 12 cm (4¾″). *E. slope of Andes in Peru and Bolivia.* Gray above with *black crown* and *long white superciliary. Bright rufous below,* throat whitish, and with inconspicuous dusky submalar streak.

SIMILAR SPECIES: Sympatric races of larger Black-eared Hemispingus all differ in their prominent black chins, but overall pattern is otherwise fairly similar (and 1 race agrees in having white superciliary); they occur mostly or entirely at *lower* elevations. Male Rusty Flower-piercer lacks the eyestripe and has a hooked bill. Cf. also Three-striped Hemispingus and female Slaty Tanager.

HABITAT AND BEHAVIOR: Uncommon to fairly common at borders of humid montane forest and in shrubby areas; ranges up to timberline. Usually in pairs or singly and most often with mixed flocks, gleaning rather deliberately among foliage and on smaller stems. Regularly found in high elevation temperate forest on the Abra Malaga road in Cuzco, Peru, and at the "Siberia" forest above Comarapa along the Cochabamba–Santa Cruz highway in Bolivia.

RANGE: Andes of e. Peru (north to San Martín) and nw. Bolivia (La Paz, Cochabamba, w. Santa Cruz). 3000–3600 m.

Conirostrum rufum

RUFOUS-BROWED CONEBILL

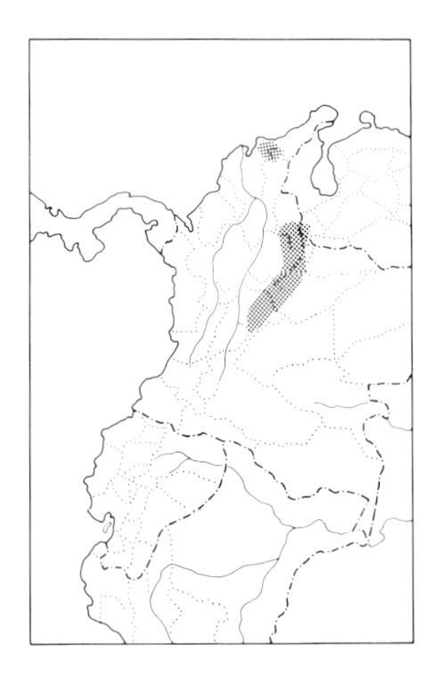

IDENTIFICATION: 12.5 cm (5″). *Mainly Andes of ne. Colombia.* Similar to White-browed Conebill but slightly larger (with longer tail) and with *rufous superciliary, sides of head, chin, and forehead.* See C-48.

SIMILAR SPECIES: Male Rusty Flower-piercer has all gray and blackish head (no rufous superciliary, etc.), is paler below, and has hooked bill. Cf. also Rufous-crested Tanager (at most limited overlap).

HABITAT AND BEHAVIOR: Uncommon to fairly common in shrubby areas and in low, stunted woodland and borders (not in continuous true montane forest), ranging up to timberline. Behavior differs little or not at all from White-browed Conebill's. Can be seen, though is not numerous, in Andes east of Bogotá (occasionally even in the city itself).

RANGE: Andes of extreme w. Venezuela (sw. Táchira) and ne. Colombia (E. Andes from Norte de Santander to Cundinamarca); Santa Marta Mts. of Colombia. 2650–3300 m.

Conirostrum sitticolor

BLUE-BACKED CONEBILL

PLATE: 10

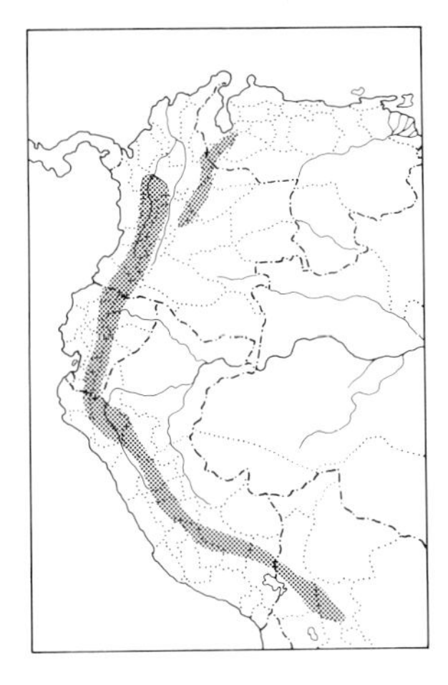

IDENTIFICATION: 13 cm (5″). The most colorful conebill. Head, neck, and chest black, with long blue postocular stripe extending back to join *blue mantle;* wings and tail mainly blackish. *Lower underparts rich rufous, sharply demarcated from black chest.* Nominate race (Colombia, Ecuador, n. Peru, and extreme w. Venezuela in Perijá Mts. and sw. Táchira) lacks the blue postocular seen in *intermedium* (w. Venezuela in Mérida and Táchira), though it begins to reappear in n. Peru. *Cyaneum* (Peru north at least to Huánuco and in Bolivia) regains the blue eyestripe (and it extends farther forward, to above eye); its *throat and chest are dusky blue* (not black).

SIMILAR SPECIES: Overall pattern similar to that of certain range-restricted *Diglossa* (but these all lack the blue above, have hooked bills, etc.).

HABITAT AND BEHAVIOR: Fairly common to common in humid montane forest and forest borders (particularly the latter), especially at higher elevations up to near timberline. In pairs or small groups which glean and probe actively in foliage, often briefly hanging from their perches and reaching out. Almost always with mixed flocks of tanagers and other montane birds. Seems considerably more numerous in Colombia and Ecuador than it is in Bolivia and most of Peru.

RANGE: Andes of w. Venezuela (north to Mérida), Colombia (though seems unrecorded in W. Andes north of Cauca), Ecuador, nw. and e. Peru (on w. slope south to Cajamarca), and nw. Bolivia (La Paz and Cochabamba). Mostly 2500–3500 m.

GROUP B

All dark with blue or white *crown* (female olive with bluish cap).

Conirostrum albifrons

CAPPED CONEBILL

PLATE: 10

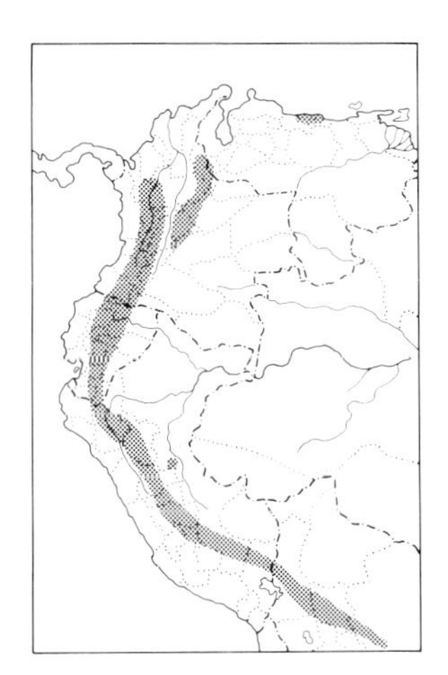

IDENTIFICATION: 13–13.5 cm (5–5¼″). Mainly black, with *dark glossy blue crown* and dark blue suffusion across shoulders, scapulars, lower back, rump, and even crissum (but the blue is often hard to discern and thus *often looks all black;* furthermore, in s. races, ranging north to cen. Peru, blue suffusion is reduced). Birds from w. Venezuela and Colombia's E. and Cen. Andes (nominate and *centralandinum*) have *striking snowy white crown* but are otherwise similar to *atrocyaneum* (Colombia's W. Andes south through Ecuador to n. Peru); see V-34. *Cyanotum* (n. Venezuela) reverts to having blue crown with inconspicuous silvery white streaks on forehead. Female has *crown grayish blue to blue* (brightest in s. birds) with gray nape; otherwise bright olive above, brightest on rump. Throat and breast bluish gray, remaining underparts yellowish green.

SIMILAR SPECIES: White-crowned males unmistakable, but blue on crown of other races is dark and can be hard to see; *near-constant tail wagging* and overall shape with slender, very pointed bill can be helpful under poor lighting conditions. Female perhaps most likely to be confused with Gray-hooded Bush-Tanager (they often are in the same

flock and both tail-wag), but note bush-tanager's larger size, thicker pink bill (n. part of range), much yellower lower underparts, etc.
HABITAT AND BEHAVIOR: Locally common in canopy and borders of humid montane forest; much more of a tall forest bird than other *Conirostrum*. Forages actively by gleaning and probing among foliage and on twigs and stems. Generally found in pairs or small groups, accompanying mixed flocks of tanagers and other insectivores.
RANGE: N. coastal mts. of Venezuela (Aragua and Distrito Federal) and Andes of sw. Venezuela (Táchira), Colombia, Ecuador (south on w. slope to e. Guayas), e. Peru, and nw. Bolivia (La Paz, Cochabamba, and w. Santa Cruz). Mostly 2000–2800 m.

Dacnis Dacnises

Small warblerlike tanagers with distinctive short, pointed bills. Iris often contrasting (yellow, red, or white). Strongly dimorphic in plumage, males being much brighter (often predominantly blue), females green or brown. Active and arboreal in humid forest and borders, but a surprising proportion (all but the Blue, Black-faced, and Yellow-bellied) are either rare or have very small ranges. A mainly S. Am. genus, though several species extend to s. Middle America.

GROUP A

"Standard" dacnises; typically black and turquoise.

Dacnis venusta

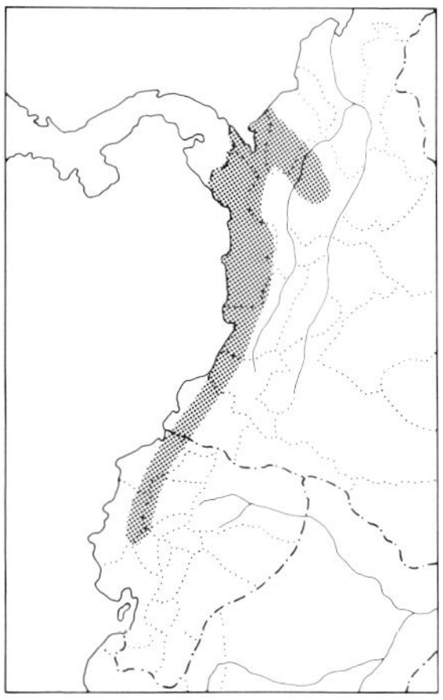

SCARLET-THIGHED DACNIS

PLATE: II

IDENTIFICATION: 12 cm (4¾"). *W. Colombia and nw. Ecuador.* Male's iris bright red, duller red in female. Male *mainly bright blue above and black below.* Crown and nape, sides of head and neck, center of back, rump, and scapulars turquoise blue. Forehead, lores, sides of back, wings, tail, center of throat, and entire underparts black. Thighs are scarlet, but these are usually hidden (and thus not a useful field character). Female *dull greenish blue above,* brightest on cheeks, scapulars, and rump, duskier on back, wings, and tail. *Below dingy buffy grayish,* buffiest on lower belly and crissum. See C-47, P-25.
SIMILAR SPECIES: Distinctive male might be mistaken for larger Blue-necked Tanager (which has blue continuing down over entire throat, solid black back, opalescent wing-coverts, etc.). Female is dull, but nothing in its range is really similar; Blue Dacnis female is green below with blue head, while female Viridian and Black-faced Dacnises lack any blue tone above, are more olive below, and have yellow irides.
HABITAT AND BEHAVIOR: Uncommon to locally fairly common in humid forest canopy and borders and adjacent clearings with scattered trees. Most numerous northward on Pacific slope and in foothill areas. Found in pairs or small groups, often feeding independently of flocks, though does at times accompany them. Active, and generally remains well above the ground except when coming lower to feed (e.g., at low *Miconia* trees).

RANGE: N. and w. Colombia (Pacific lowlands and lower w. slopes of W. Andes, and in humid lowlands north of Andes east to middle Magdalena valley in e. Antioquia) and nw. Ecuador (south to n. Guayas). Also Costa Rica and Panama. Mostly 200–700 m, but occasionally to sea level and up to 1000 m.

Dacnis cayana

BLUE DACNIS

PLATE: 11

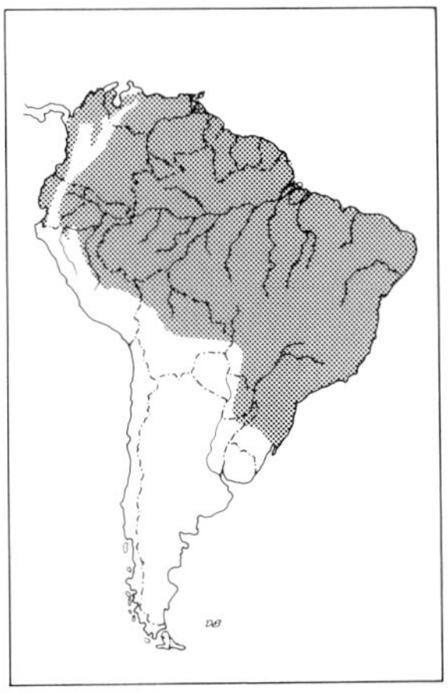

IDENTIFICATION: 12.5 cm (5"). Bill longer and more pointed than that of other *Dacnis,* with pinkish flesh base (especially gape). Legs flesh to pink (both sexes). Male *mostly turquoise blue,* with black lores, mantle, wings, tail, and throat; wings edged turquoise blue. Female *mainly bright green,* paler below, with *bluish head* and grayish throat. *Baudoana* and *coerebicolor* (w. Colombia and w. Ecuador) are *notably darker,* more ultramarine blue in male (closer to the blue of male *Cyanerpes,* though not so purplish), while female's head is purer blue; see C-48.

SIMILAR SPECIES: Almost everywhere the commonest and most widespread dacnis, so it is useful to learn to recognize both sexes quickly as a basis for comparison. In Colombia cf. Viridian and Turquoise Dacnises, and in se. Brazil cf. Black-legged Dacnis; all of these species are rare. Female's blue head and conical bill normally clinch identification, but if seen poorly she can be confused with various all-green immature *Tangara* tanagers (some of which are barely any bigger than the dacnis).

HABITAT AND BEHAVIOR: Common in forest borders, second-growth woodland, clearings with scattered trees, and (at least in interior Brazil) deciduous and gallery woodland. Usually in pairs, sometimes in small groups (but never the large aggregations *Cyanerpes* honeycreepers are sometimes found in), which actively glean in foliage and also frequently come to fruiting trees and shrubs. Regularly accompanies mixed flocks.

RANGE: Widespread in more humid lowlands south to sw. Ecuador (El Oro), n. and e. Bolivia (south to La Paz, Cochabamba, and Santa Cruz), e. Paraguay, ne. Argentina (Misiones), and s. Brazil (south to Rio Grande do Sul); Trinidad. Also Honduras to Panama. To about 1000 m.

Dacnis nigripes

BLACK-LEGGED DACNIS

IDENTIFICATION: 11 cm (4¼"). *A rare dacnis endemic to e. Brazil.* Male resembles much commoner male Blue Dacnis. Black throat patch slightly smaller, black on mantle slightly less extensive, and wing blacker (no blue edging on flight feathers). Best distinguished by *blackish or dusky* (not reddish) *legs* and *notably shorter tail.* Female very different from female Blue Dacnis, actually recalling female of geographically far-removed Scarlet-thighed Dacnis; its legs are also dusky. Brownish olive above, tinged greenish blue on forecrown, cheeks, scapulars, and rump; *uniform dull pale buffyish below.*

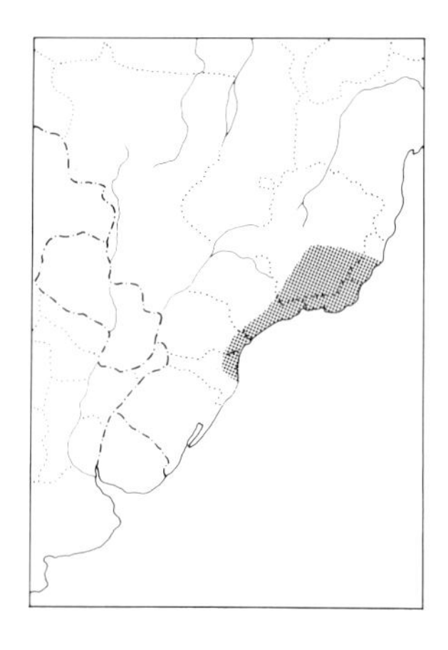

SIMILAR SPECIES: Male must be distinguished from male Blue Dacnis with care, though female should be readily recognizable if seen well (ideally as a pair).

HABITAT AND BEHAVIOR: Apparently rare in forest borders and woodland. Not well known, and seemingly erratic or perhaps migratory. Recent records from Rio de Janeiro at Magé have been from austral winter months, when groups have been seen feeding in flowering trees with other birds, including Blue Dacnises (*fide* L. P. Gonzaga). The species may be threatened by deforestation.

RANGE: Locally in lowlands of se. Brazil (Espírito Santo and e. Minas Gerais south to ne. Santa Catarina). To 850 m.

NOTE: For more details on this rare species, see L. P. Gonzaga (*Iheringia*, Ser. Zool., Porto Alegre 63: 45–58, 1983).

Dacnis viguieri

VIRIDIAN DACNIS

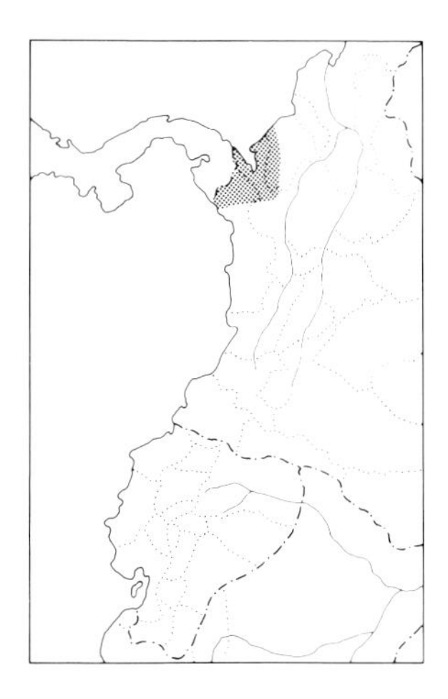

IDENTIFICATION: 11.5 cm (4½"). *Nw. Colombia. Iris yellow* (both sexes). Male *mainly bluish green,* rump somewhat more turquoise. Lores, upper back, and tail black; *wings mostly green with contrasting black primaries.* Female mostly dull green above with dusky lores; pale greenish yellow below, shaded with olive on sides and flanks; wings dull green with contrasting blackish primaries; tail black. See C-48, P-32.

SIMILAR SPECIES: Male Blue Dacnis is bluer (no green) with black throat patch and dark eye; Viridian's bicolored effect on wing (green above, black below) also quite different. Female much duller than female Blue Dacnis and without its blue on head; it more closely resembles female of w. race of Black-faced Dacnis (and both have yellow irides), but latter is notably brighter yellow on median lower underparts. Cf. also female Green Honeycreeper.

HABITAT AND BEHAVIOR: Little known, but from large number of specimens taken at certain localities within its limited range (e.g., Río Juradó in Chocó) may be locally not uncommon. Perhaps most numerous in hilly country rather than flat coastal lowlands (though has also been recorded from latter). Forages in canopy of humid forest and forest borders, with behavior similar to that of other *Dacnis.*

RANGE: Nw. Colombia (n. Chocó, nw. Antioquia, and sw. Bolívar). Also e. Panama. To about 600 m.

Dacnis lineata

BLACK-FACED DACNIS

PLATE: II

IDENTIFICATION: 11.5 cm (4½"). *Iris yellow* (both sexes). Male *mostly bright turquoise blue,* with *black lores, mask across sides of head, mantle, wings, and tail.* Scapulars and edging to tertials also blue. Center of belly and crissum, as well as under wing-coverts (sometimes protruding as tuft in front of wings) white. Female much more drab: dull brownish olive above, somewhat grayer on head; dull pale grayish below, whitest on center of belly and more olivaceous on flanks. See V-34. Races west of Andes (including *aequatorialis*) have male with blue of a greener turquoise hue, *center of belly and crissum bright yellow,*

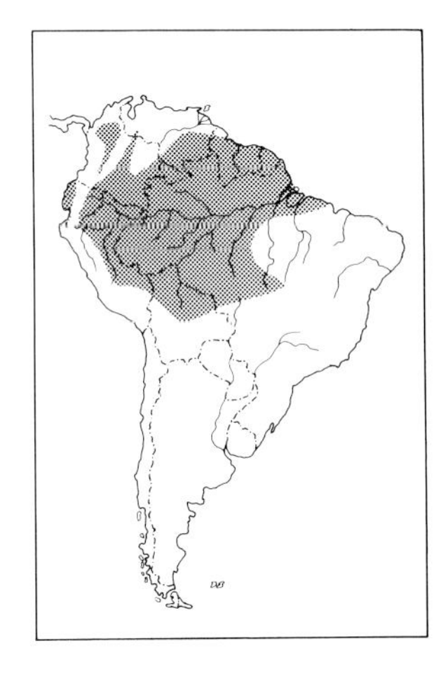

with this same color also showing as a prominent tuft in front of the wing; female more olivaceous generally, and also with *bright yellow center of belly* as well as the yellow tuft in front of the wing.

SIMILAR SPECIES: Male nearly unmistakable, but in w. Colombia cf. the rare Turquoise Dacnis. Nondescript females often best known by the company they keep; especially east of the Andes best mark often the yellow iris (shared by female White-bellied Dacnis, but red in female Yellow-bellied).

HABITAT AND BEHAVIOR: Fairly common to locally common in canopy and borders of humid forest (both *várzea* and *terra firme*) and in secondary woodland and clearings with scattered trees (away from forest especially when attracted by fruiting trees). Found in pairs or small groups which glean actively in foliage but also eat a great deal of fruit; most often encountered accompanying mixed flocks.

RANGE: Guianas, s. and sw. Venezuela (Bolívar and Amazonas, and in sw. Táchira), e. Colombia (north to Meta and Guainía), e. Ecuador, e. Peru, n. Bolivia (south to La Paz, Cochabamba, and w. Santa Cruz), and most of Amaz. Brazil (south of the Amazon south to n. Mato Grosso and e. Pará in the Belém region); n. and w. Colombia (humid lowlands along n. base of Andes and south in Magdalena valley to s. Tolima); w. Ecuador (Esmeraldas south to El Oro). To about 1300 m.

NOTE: The yellow-bellied trans-Andean group seems very distinct, though no one seems to have remarked on this situation recently, and it may well be reasonable to consider it a full species, *D. egregia*. However, for the present we continue to maintain the *egregia* group as conspecific with *D. lineata*. Yellow-tufted Dacnis would seem an appropriate English name should it be split.

Dacnis hartlaubi

TURQUOISE DACNIS

Other: Turquoise Dacnis-Tanager

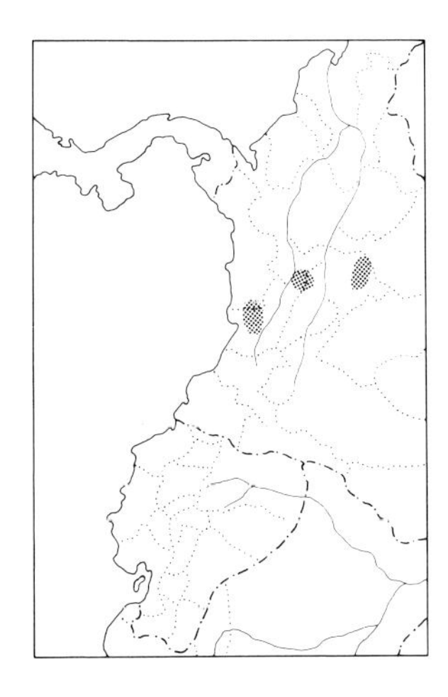

IDENTIFICATION: 11 cm (4¼"). *A rare, local endemic of the Colombian Andes.* Bill stubbier, less pointed than that of other *Dacnis*. *Iris yellow* (both sexes). Male closely resembles male Black-faced Dacnis but is *all turquoise blue below* (no white or yellow on median belly), except for *black throat,* and has *narrow blue eye-ring*. See C-48. Female also like Black-faced, *dull brown above,* wing feathers edged pale tawny-buff; grayish buff below, becoming yellowish white on median breast and belly.

SIMILAR SPECIES: Other than Black-faced Dacnis, male most likely to be confused with various mainly blue *Tangara* tanagers (e.g., Golden-naped), though all of these have *dark* irides. Nondescript female would be difficult to distinguish from female Black-faced Dacnis but is browner (not olive) above with wing-edging, buffier below. Female *Hemithraupis* tanagers are yellower, especially below.

HABITAT AND BEHAVIOR: Little known, with few recent observations. Rather sluggish behavior of one recalled a *Tangara* (S. Hilty). Recorded from humid forest borders and clearings with scattered trees. Now very rare and local due to the great extent of deforestation on the lower slopes of the Andes; may deserve endangered status. Has been seen recently at Bosque Yotoco west of Buga in Valle and in remnant patches of montane forest west of Bogotá.

RANGE: Very locally in w. Colombia (both slopes of W. Andes in Valle, w. slope of Cen. Andes in Quindío, and w. slope of E. Andes in Cundinamarca). Mostly 1300–2200 m, once down to 300 m.

NOTE: We follow *Birds of the World* (vol. 13, pp. 387–388) in merging the monotypic genus *Pseudodacnis* in *Dacnis;* its inclusion in *Dacnis* only marginally increases the variance of bill size in that genus, and its overall color and pattern clearly fall with *Dacnis* (as does its yellow iris). We abbreviate the species' English name to reflect this generic allocation.

GROUP B

Miscellaneous other dacnises (all distinctive).

Dacnis berlepschi

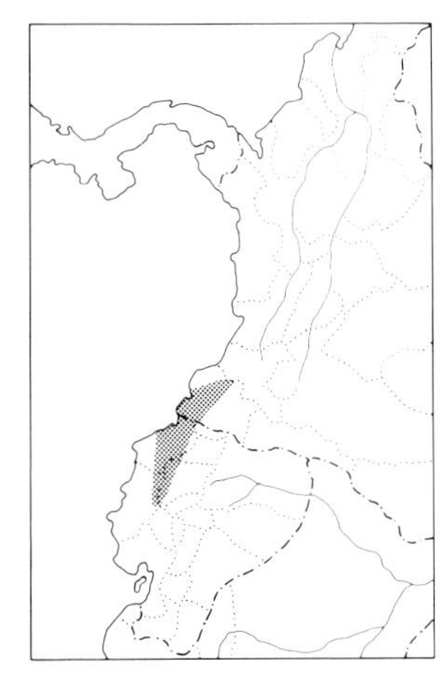

SCARLET-BREASTED DACNIS

PLATE: II

IDENTIFICATION: 12 cm (4¾"). *A rare endemic of sw. Colombia and nw. Ecuador.* Iris yellow (both sexes). *Upperparts, throat, and chest shining dark blue,* with *conspicuous light silver blue streaking on mantle* and solid rump of that color. *Breast flame scarlet,* fading to orange-tinged buff on belly. Wings and tail mainly black. Female entirely brown above (no olivaceous), paler buffier brown below, palest on center of belly, but with *narrow band of flame scarlet across breast.*

SIMILAR SPECIES: Male essentially unmistakable (though pattern somewhat recalls montane Blue-backed Conebill's), as is female due to its unique and quite conspicuous breast band.

HABITAT AND BEHAVIOR: Little known. Recorded from wet forest borders and advanced secondary growth in foothills. Behavior probably does not differ from that of other dacnises; appears never to be very numerous. Has been recently seen above Junín, along the road to Tumaco in w. Nariño, Colombia.

RANGE: Extreme sw. Colombia (w. Nariño) and nw. Ecuador (south to Pichincha). 200–1200 m.

Dacnis albiventris

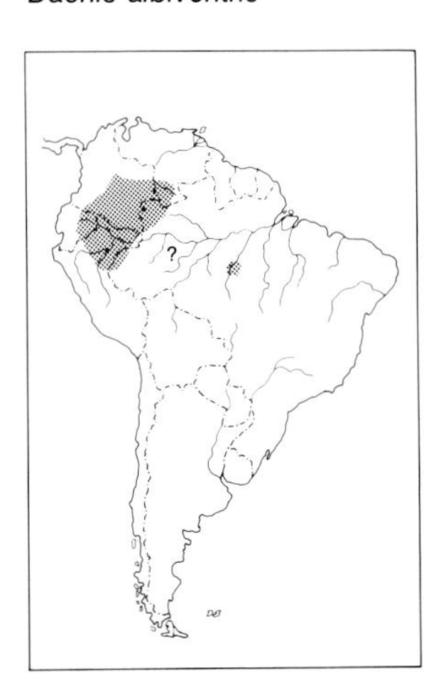

WHITE-BELLIED DACNIS

PLATE: II

IDENTIFICATION: 11.5 cm (4½"). *A rare dacnis restricted to Amazonia. Iris yellow* (male definitely; probably also female). Male *mostly bright cobalt blue; lores and mask over ear-coverts black.* Wings and tail black, wing-coverts broadly edged cobalt blue. *Center of lower breast and belly and crissum contrastingly white.* Female *green above,* brightest on rump; throat grayish white, with *remaining underparts greenish yellow,* yellowest on center of belly.

SIMILAR SPECIES: Male is a darker blue than other dacnises (compare especially to Black-faced, which also has yellow iris and generally similar pattern). Male's cobalt blue color is similar to that found in several sympatric *Tangara* tanagers (especially Turquoise), but their patterns differ in various ways. The obscure female is perhaps unlikely to be identified away from accompanying male; it most resembles Black-faced Dacnis female but is much greener above and yellower below

(compared to the drab race found in upper Amazonia). Check also various other small female tanagers (e.g., Guira, Orange-headed, etc.).
HABITAT AND BEHAVIOR: Little known; apparently rare over its ample range. Recorded from canopy and borders of humid forest (both *terra firme* and *várzea*). Has been seen accompanying mixed flocks made up primarily of various *Tangara* tanagers and other dacnises.
RANGE: S. Venezuela (s. Amazonas), se. Colombia (definitely recorded only from w. Putumayo, but presumably more widespread), e. Ecuador (very few records), ne. Peru (Loreto), and cen. Amaz. Brazil (upper Rio Cururú in se. Pará, and near Manaus; presumably also westward). To about 400 m.

Dacnis flaviventer

YELLOW-BELLIED DACNIS

PLATE: 11

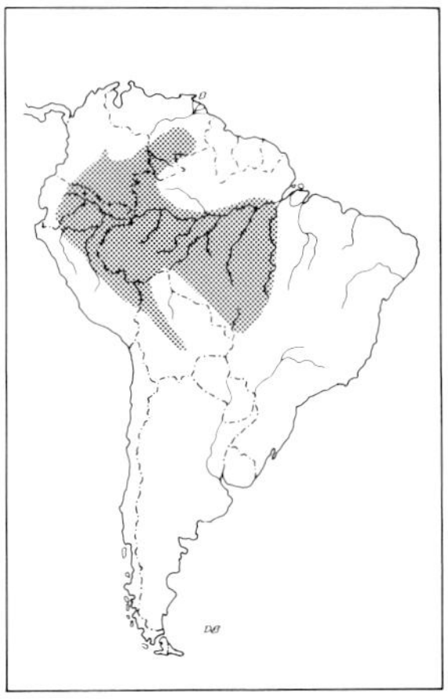

IDENTIFICATION: 12.5 cm (5"). *Iris red* (both sexes). Male very different from other dacnises, being *mostly yellow and black.* Crown and nape dull green; otherwise mostly black above with *yellow rump and scapulars. Center of throat black; otherwise* (including malar stripe on sides of throat) *yellow below,* somewhat mottled black on breast. Female olive brown above, somewhat darker on wings and tail. Below pale dull brownish buff, *obscurely mottled on throat and breast,* becoming unmarked pale buff on center of belly.
SIMILAR SPECIES: Male virtually unmistakable. Female resembles female Black-faced Dacnis but is slightly larger and more mottled on breast and has red (not yellow) iris.
HABITAT AND BEHAVIOR: Fairly common to common in canopy and borders of humid forest and in adjacent clearings with scattered trees and secondary growth. Especially numerous in *várzea* forest. Most often in small groups which usually remain high in trees but sometimes come lower along streambanks or at edge. Behavior much like Blue Dacnis's, and like that species both gleans for insects and eats fruit; also regularly seen probing for nectar at flowering trees.
RANGE: S. Venezuela (w. Bolívar and Amazonas), se. Colombia (north to w. Meta and Vaupés), e. Ecuador, e. Peru, n. Bolivia (south to La Paz, Cochabamba, and w. Santa Cruz), and w. and cen. Amaz. Brazil (north of the Amazon east to the Rio Negro, south of it east to drainage of Rio Xingú). Mostly below 500 m (but occasionally to over 1400 m, at least in Bolivia)

Cyanerpes Honeycreepers

Slender decurved bills and brightly colored legs in both sexes. Males mostly bright blue, females green and streaky below. Often in groups, foraging for fruit and at flowers, usually in canopy of humid forest. Widespread in tropical lowlands, with several species extending north to Mexico (Red-legged also on Cuba).

Cyanerpes caeruleus

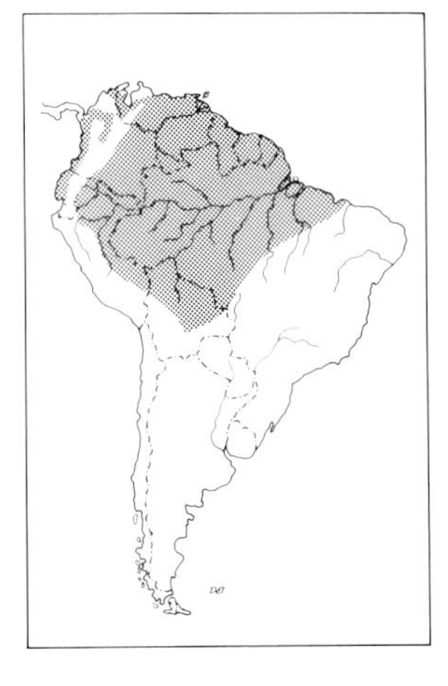

PURPLE HONEYCREEPER

PLATE: 11

IDENTIFICATION: 11 cm (4¼"). Bill long (especially so in *longirostris* of Trinidad), slender, and decurved; *legs bright yellow* (male) or olive yellow to greenish gray (female). Male *mostly purplish blue*, slightly paler on head; lores, *throat*, wings, and tail *black*. Female mostly green above, but with *conspicuous buff lores* and narrow white streaking on sides of head. Throat buff, separated from lores by narrow blue malar streak; remaining underparts broadly streaked green and yellowish, becoming clear pale yellowish on center of belly.

SIMILAR SPECIES: In nw. Colombia cf. the very similar Shining Honeycreeper. Short-billed Honeycreeper has, as its name implies, a short bill, but otherwise both sexes rather resemble this species aside from leg color differences (pinkish in both sexes of Short-billed). Compare also to the more distinct Red-legged Honeycreeper.

HABITAT AND BEHAVIOR: Fairly common to locally common in canopy and borders of humid forest and adjacent secondary growth and clearings with scattered trees. More confined to essentially forested areas than is Red-legged Honeycreeper, though there is some overlap and often the 2 are found in the same flock. Usually in groups high in trees, sometimes in concentrations of 10 or more birds (especially around rich food sources); seems quite omnivorous (as are other *Cyanerpes* honeycreepers), not only gleaning for insects in foliage and eating fruit but also regularly taking nectar from flowers.

RANGE: Widespread in forested lowlands south to w. Ecuador (Guayas), n. Bolivia (south to La Paz, Cochabamba, and n. Santa Cruz), and Amaz. Brazil (south to n. Mato Grosso, se. Pará, and n. Maranhão); Trinidad. Also e. Panama. Mostly below 1200 m, rarely to about 1600 m.

Cyanerpes lucidus

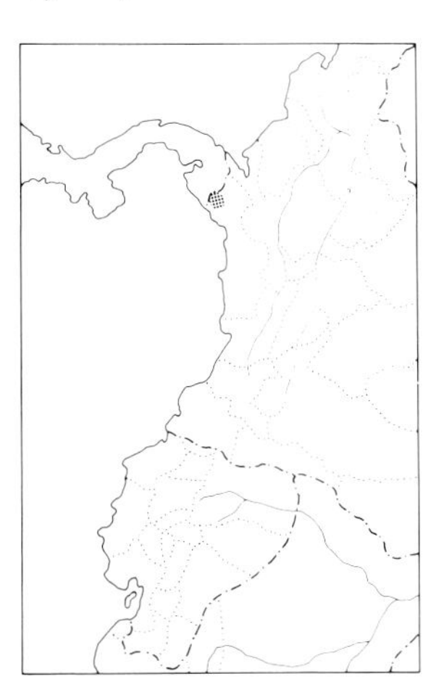

SHINING HONEYCREEPER

IDENTIFICATION: 11 cm (4¼"). *Nw. Colombia.* Male *very* similar to male Purple Honeycreeper (limited overlap), but throat patch slightly more extensive and rounded at bottom, bill slightly shorter than sympatric race of Purple (*chocoanus*), and blue of a slightly paler hue. Female easier but still basically similar; best distinguished by *dusky* (not obviously buff) *lores,* bluish gray crown (contrasting with green back), paler buff throat, and pronounced *blue streaking on breast* (instead of blurrier green to bluish green streaking).

HABITAT AND BEHAVIOR: Little known in South America (where its range is very limited). In Panama locally common in canopy and borders of humid forest, particularly in foothills. Behavior as in Purple Honeycreeper (of which it may be only a race, sympatric breeding not having yet been demonstrated).

RANGE: Extreme nw. Colombia (nw. Chocó in Río Juradó drainage). Also Mexico to Panama. In Panama up to about 800 m.

Cyanerpes nitidus

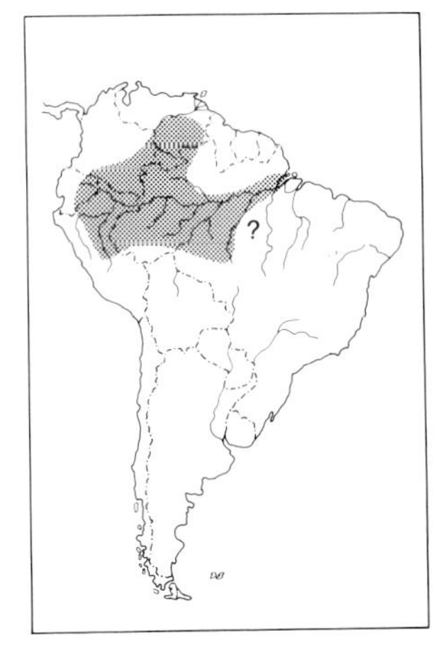

SHORT-BILLED HONEYCREEPER

PLATE: II

IDENTIFICATION: 9.5 cm (3¾"). *Short black bill; legs pale pink to orangy flesh. Mostly bright deep blue,* with black lores, *throat patch extending down over chest,* wings, and tail. Female green above with *dusky lores.* Throat pale buff, separated from lores by narrow blue malar streak; remaining underparts broadly streaked green and buffy whitish, becoming clear pale buffyish on center of belly.

SIMILAR SPECIES: Both sexes resemble respective sexes of slightly larger and definitely longer-billed Purple Honeycreeper but have pink to flesh-colored legs (not yellow as in male or dusky as in female Purple). Male additionally differs by its larger black bib, female by its dusky (not buff) lores and lack of streaking on sides of head. Female also resembles female Red-legged Honeycreeper (both have dull pinkish legs), but latter is larger with longer bill, is more olive (not grass green) above with whitish superciliary and facial streaking, and lacks blue malar streak.

HABITAT AND BEHAVIOR: Uncommon in canopy and borders of humid forest (both *terra firme* and *várzea*) and in secondary growth and clearings with scattered trees. Behavior similar to that of other *Cyanerpes* honeycreepers and often with them in the same mixed flock; usually the least numerous of its genus, but in some areas (e.g., the Iquitos, Peru, region) can be commoner.

RANGE: S. Venezuela (Bolívar and Amazonas), se. Colombia (north to Caquetá, Vaupés, and Guainía), e. Ecuador (few records), e. Peru (apparently not south of Junín and Ucayali), and w. Amaz. Brazil (north of the Amazon east to lower Rio Negro, south of it east to nw. Mato Grosso). To about 400 m.

Cyanerpes cyaneus

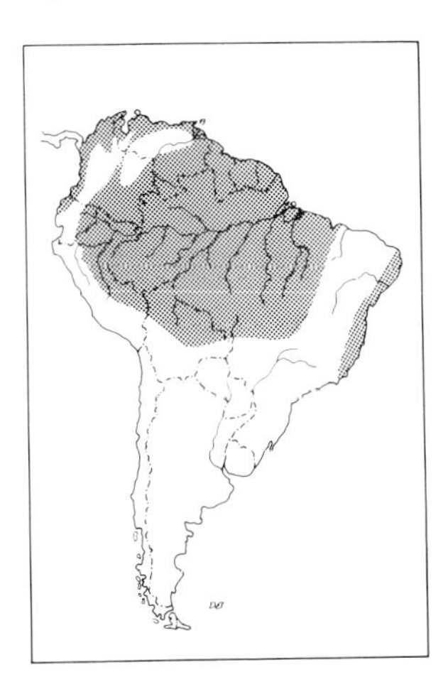

RED-LEGGED HONEYCREEPER

IDENTIFICATION: 11.5–12 cm (4½–4¾"). Bill long, slender, decurved; *legs intense red* (male) *or reddish* (female). Male *mostly bright purplish blue* with *contrasting pale turquoise crown.* Lores, mantle, wings, and tail black; *under wing-coverts bright yellow.* Female dull green above with indistinct superciliary and facial streaking whitish. Below pale yellowish, *blurrily streaked greenish* (especially on throat and breast). *Under wing-coverts yellow.* Male has a postbreeding "eclipse" plumage (1 of the few neotropical birds which has) in which it resembles female but with black back, wings, and tail and brighter red legs. See V-34, C-47.

SIMILAR SPECIES: Purple Honeycreeper is smaller with shorter tail; male has bright yellow (not red) legs and black throat patch, while female has dusky or greenish (never reddish) legs, buff throat and lores, and blue malar streak, etc.

HABITAT AND BEHAVIOR: Locally common in forest borders, secondary woodland, and clearings with scattered trees. Ranges into drier areas than Purple Honeycreeper, though the 2 are sympatric in many areas and often range in the same flock. An active, restless bird, often

perching high on exposed branches, then flying long distances while giving constant "tsip" notes; also gives an ascending "zhree" call. Like other *Cyanerpes* rather an omnivore, eating both fruit (readily coming to feeding tables where these are in use), insects, and nectar. Usually in groups, sometimes quite large (up to 15 to 20 or more birds).
RANGE: Widespread in lowlands (though much less numerous or absent across portions of w. Amazonia) south to w. Ecuador (Manabí and Pichincha), n. Bolivia (La Paz and nw. Santa Cruz), and cen. Brazil (locally east and south to n. Mato Grosso, Goiás, and Maranhão); coastal e. Brazil (Pernambuco south to Espírito Santo); Trinidad and Tobago, and Margarita Is. off Venezuela and Gorgana Is. off w. Colombia. Also Mexico to Panama, and Cuba. Mostly below 1000 m, locally or seasonally higher in small numbers.

Chlorophanes Honeycreepers

Larger than *Cyanerpes,* with stouter yellow bill. Usually only in pairs, but behavior otherwise similar (and often together). Female not streaky.

Chlorophanes spiza

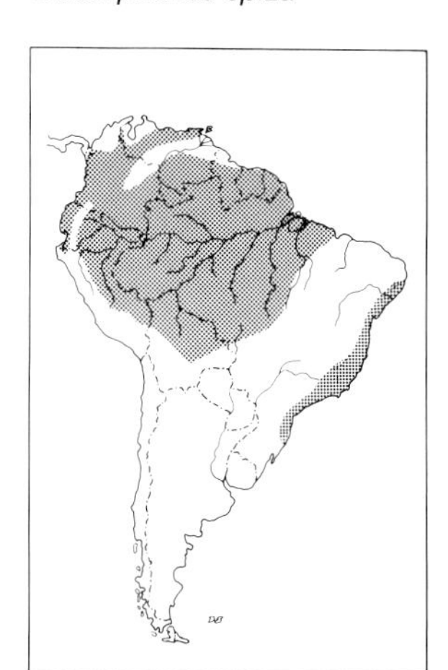

GREEN HONEYCREEPER

PLATE: II

IDENTIFICATION: 14 cm (5½"). *Bill yellow with black culmen* (duller in female) stouter and less decurved than in *Cyanerpes* honeycreepers. Iris dark red (male), brown (female). Male *mostly glistening green to* (in some races) *bluish green,* with *contrasting black head.* Wings and tail duskier. Female *mostly bright green,* paler below, with yellower throat, ocular area, and center of belly.
SIMILAR SPECIES: Patternless female can be confused with a number of species, but *note bill shape;* compare especially to certain immature *Tangara* tanagers (e.g., Bay-headed), while female Blue Dacnis has shorter straight bill and bluish head.
HABITAT AND BEHAVIOR: Fairly common in canopy and borders of humid forest (in Amazonia, both *terra firme* and *várzea*) and secondary woodland, ranging in smaller numbers up into more montane areas. Comes out into clearings rather less often than most *Cyanerpes* honeycreepers and *Dacnis.* Usually singly or in pairs (only rarely small groups), generally remaining high in trees; an inveterate flock follower.
RANGE: Widespread in forested lowlands south to nw. Peru (Tumbes), n. Bolivia (south to La Paz, Cochabamba, and w. Santa Cruz), and Amaz. Brazil (south to n. Mato Grosso, se. Pará, and n. Maranhão); coastal e. Brazil (Pernambuco south to e. Santa Catarina); Trinidad. Absent from Santa Marta region of n. Colombia and from nw. Venezuela and much of the llanos region of Venezuela and ne. Colombia. Also Mexico to Panama. Mostly below 1500 m, but higher (to 2300 m) in Colombian Andes.

Coerebinae

BANANAQUIT

THE SOLE remaining member of what formerly was considered the Honeycreeper family (Coerebidae) is the Bananaquit, all its other genera now generally considered to fall in the Thraupinae. Even the Bananaquit is sometimes considered to be a tanager (or a warbler), but on account of its distinctive nest and several morphological attributes we leave it in its own subfamily, placing it here for convenience.

Coereba Bananaquit

A familiar small bird, widespread in lowlands but most numerous coastally (and omnipresent on most of the West Indies), with a short decurved bill and usually a distinctive long white eyestripe. Builds a large untidy globular nest (very unlike cup nests of dacnises, honeycreepers, and most tanagers).

Coereba flaveola

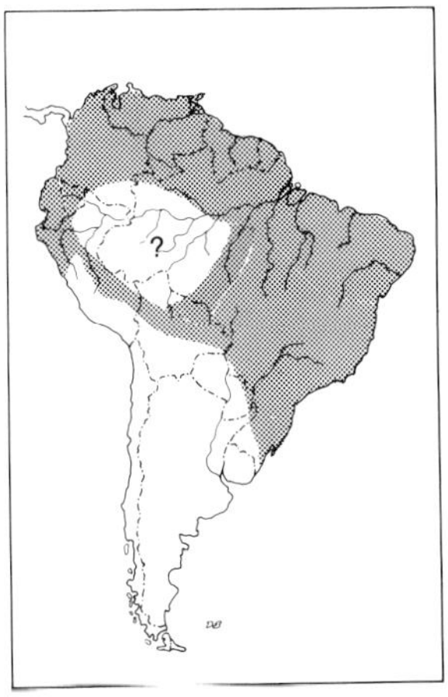

BANANAQUIT

PLATE: II

IDENTIFICATION: 10.5–11 cm (4–4¼"). *Rather short decurved bill* (distinctly longer in *magnirostris* of upper Marañón valley, Peru). Above dusky olive to sooty gray, blackest on crown and sides of head and with *striking long white superciliary;* rump dull greenish to rather bright yellow, and usually a white wing-speculum (obscure or absent in some races). *Throat gray,* contrasting with *bright yellow remaining underparts.* S. birds tend to be palest above, while *luteola* (n. Colombia, most of Venezuela, and some offshore Caribbean islands, including Trinidad and Tobago) is black above. Races on Netherlands Antilles are also black above, and this extends to *black throat.* Two races from small island groups off Venezuela (Los Roques and Los Testigos) are all sooty or blackish, the latter tinged olivaceous below.

SIMILAR SPECIES: On S. Am. mainland, readily recognized on basis of prominent eyebrow, short decurved bill, and gray and yellow underparts. Potential for confusion greater on Caribbean islands, with the bird's tendency toward melanism, but on these its overall shape and actions normally preclude problems.

HABITAT AND BEHAVIOR: Abundance varies greatly, with numbers greatest along n. coast of South America, and though widespread in tropical lowlands elsewhere, Bananaquits can be scarce or even absent from extensively forested regions. Found in a wide variety of semiopen to open habitats in both humid and fairly arid regions. An active, seemingly nervous bird which may be seen foraging at any tree level; primarily nectarivorous, but also eats fruit. May become very tame

around habitations, on some islands even to the point of hopping onto dining tables in order to obtain sugar. Frequently gives a twittering, somewhat shrill and buzzy song.
RANGE: Widespread in lowlands south to nw. Peru (La Libertad and Ancash), n. and e. Bolivia (La Paz, Cochabamba, and Santa Cruz), e. Paraguay (Alto Paraná), ne. Argentina (Misiones), and s. Brazil (Rio Grande do Sul); Trinidad, Tobago, and other Caribbean offshore islands. Apparently absent from much of w. and cen. Amazonia. Also Mexico to Panama and most of West Indies. Mostly below 1500 m, locally higher in small numbers.

Conirostrum Conebills

Small, warblerlike birds; mostly bluish gray, paler below. Found in edge or semiopen habitats across lowlands. These 4 species represent the *Ateleodacnis* subgenus, quite different from the montane true *Conirostrum* group (and perhaps even deserving generic recognition). The latter group has been dealt with separately (pp. 208–212). The genus was formerly in the Coerebidae (Honeycreepers), and some would now place it in the Parulinae; we consider it to be Thraupine.

Conirostrum bicolor

BICOLORED CONEBILL

PLATE: 11

IDENTIFICATION: 11.5 cm (4½″). *Locally in coastal areas and along Amazon. Iris reddish orange; legs dusky pink. Nondescript* aside from soft-part colors. Above light grayish blue; below dingy grayish buff, usually lighter on throat, median belly, and crissum. Sexes alike, but immature more olivaceous above and *usually tinged yellow below* (especially on throat and breast; more pronounced in some birds than others); see V-34.
SIMILAR SPECIES: In coastal areas immatures may be confused with females and (especially) immatures of Yellow Warbler, but note pale legs and orangy iris; be aware that breeding in this plumage has been recorded. Along the Amazon, cf. the very similar Pearly-breasted Conebill. Female Hooded Tanager is slightly larger but rather resembles this species except for its white lores and obviously yellow eyes and legs. Female of Chestnut-vented Conebill (which also occurs on Amazon islands and shorelines) is definitely olive on back, with gray only on head and nape.
HABITAT AND BEHAVIOR: Locally common in mangroves along coast and in open woodland and shrubby regenerating areas on islands and shores of larger rivers in Amazon basin (principally the Amazon itself). Occurs in active pairs or small groups, mostly in upper and middle levels, but can easily be attracted lower by pishing (one of the few resident S. Am. birds to respond so readily). Forages primarily by gleaning in foliage and on small branches.

RANGE: Locally in coastal n. Colombia (Magdalena River mouth and Guajira Peninsula), Venezuela, Guianas, and e. Brazil (south to São Paulo); also up the Amazon in Brazil to lower Rio Madeira, and in ne. Peru (in Loreto along the Amazon and up the Río Napo) and e. Ecuador (1 recent sighting from island in Río Napo near Limoncocha); Trinidad. To about 200 m.

Conirostrum margaritae

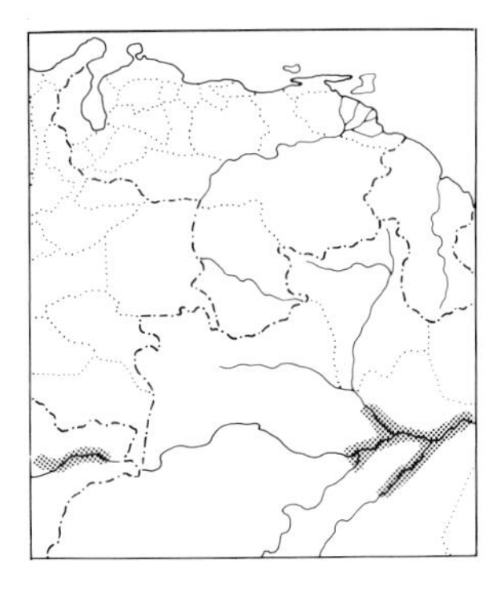

PEARLY-BREASTED CONEBILL

IDENTIFICATION: 11.5 cm (4½"). *Very locally along Amazon.* Iris pale tan; legs pinkish to pale brownish flesh. Closely resembles better-known Bicolored Conebill (with which locally sympatric); upperparts somewhat paler bluish gray and *underparts* (including facial area) *uniform pale pearly gray with no buff tinge.*

HABITAT AND BEHAVIOR: Poorly known, but apparently similar to Bicolored Conebill's (though not in mangroves). Birds seen on Isla Ronsoco, ne. Peru, were in open *Cecropia* woodland (G. Rosenberg).

RANGE: Very locally along the middle Amazon in Brazil (from near mouth of Rio Jamunda up to near mouth of Rio Negro) and along the upper Amazon in ne. Peru (near mouth of Río Napo). To about 100 m.

Conirostrum speciosum

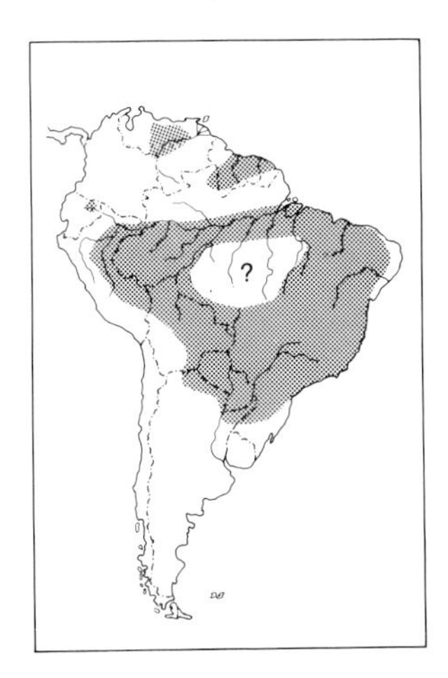

CHESTNUT-VENTED CONEBILL

PLATE: 11

IDENTIFICATION: 11 cm (4¼"). *Grayish blue above,* wings and tail duskier with white wing-speculum. Notably paler bluish gray below, lightest on belly; *crissum chestnut.* N. birds (Amazonia, Guianas, and Venezuela) are *markedly darker and more uniform bluish gray,* though still somewhat paler below than above; wing-speculum smaller or absent; see V-34, C-47. Female has *crown and nape bluish gray, contrasting with fairly bright olive green upperparts.* Below grayish white tinged with buff (more pronounced on some birds than others); crissum pale buff.

SIMILAR SPECIES: Male is the only small gray bird with a chestnut vent. Female more nondescript and easily confused, though almost invariably a male will be present to clarify the situation. Her overall pattern is strikingly like that of male Tennessee Warbler (though lacking latter's superciliary); the 2 would rarely or never be found together.

HABITAT AND BEHAVIOR: Fairly common in lighter woodland and scrub, gallery forest, and (in Amazonia) *várzea* forest on river islands and shorelines. Arboreal and warblerlike, gleaning actively in foliage, usually fairly high in trees. Most often in pairs or small groups and frequently with mixed flocks.

RANGE: Llanos of cen. Venezuela (Apure and Cojedes east to Anzoátegui); Guianas and adjacent n. Brazil (Roraima); Amaz. Brazil west to n. Bolivia, e. Peru, and e. Ecuador (where apparently only 1 old record); widely in e. and s. Brazil (south to n. Rio Grande do Sul, but seemingly absent from ne. coastal region), e. Bolivia (Santa Cruz south), Paraguay, and n. Argentina (south to Salta, Chaco, and Corrientes). To about 1000 m.

Conirostrum leucogenys

WHITE-EARED CONEBILL

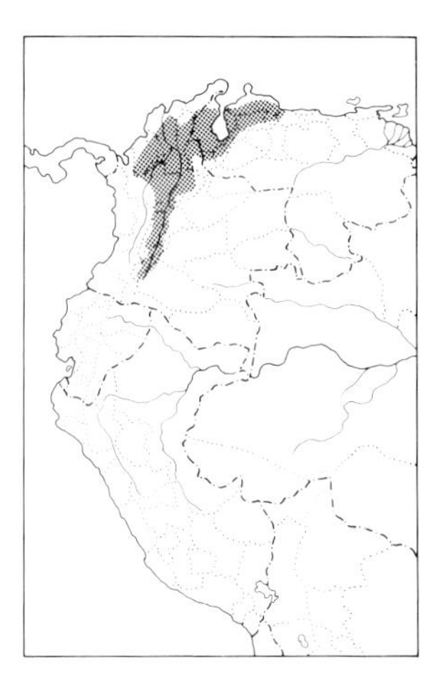

IDENTIFICATION: 9.5 cm (3¾"). *W. Colombia and nw. Venezuela.* Male mostly dark bluish gray above, with *black crown and nape* and *conspicuous white ear-patch* and *white rump.* Wings and tail blackish, feathers edged gray, whiter on lower belly, with crissum chestnut. Female bluish gray above with *whitish rump. Superciliary, sides of head, and entire underparts pale buff.* See V-34, P-32.
SIMILAR SPECIES: The smallest conebill; even when high in trees and in poor light can often be recognized by its size and *short tail.* Male distinctive, female (usually with male) less so, but latter can be known by *uniform* color of its underparts. Bicolored Conebill (different habitat) somewhat similar, but larger with pale iris and legs.
HABITAT AND BEHAVIOR: Uncommon to locally fairly common in canopy and borders of deciduous woodland and forest, secondary growth with scattered tall trees, and gallery forest. Usually in pairs, less often small groups, which glean actively and probe into flowers; frequently with mixed flocks. Generally remain very high in trees.
RANGE: N. Colombia (upper Río Sinú valley in Bolívar east across humid Caribbean lowlands to Santa Marta area and Perijá Mts., south in Magdalena valley to Huila) and nw. Venezuela (Maracaibo basin east across n. lowlands to Aragua). Also Panama. Mostly below 600 m, but recorded to 1300 m in Venezuela.

Chrysothlypis Tanagers

Only 1 species in South America (another in s. Middle America), the male being one of the most spectacular tanagers. Overall form and behavior much like the more widespread *Hemithraupis* tanagers'.

Chrysothlypis salmoni

SCARLET-AND-WHITE TANAGER PLATE: 11

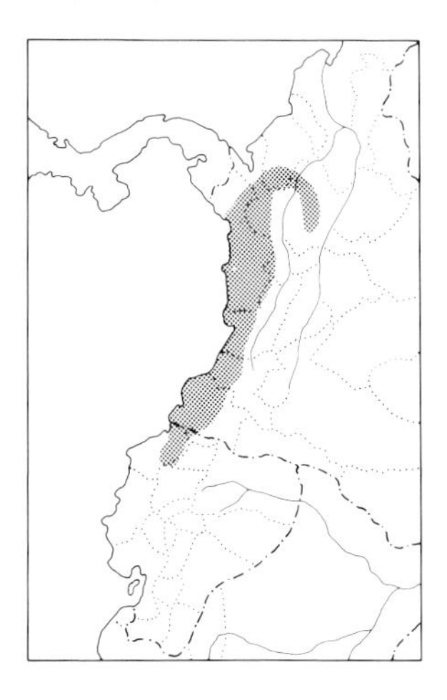

IDENTIFICATION: 13 cm (5"). *W. Colombia and nw. Ecuador.* Male unmistakable, *mostly a vibrant scarlet above, on throat and chest, and on median stripe down underparts* to crissum; breast and belly otherwise white. Wings and tail browner. Female olive above with somewhat bronzy sheen; dull yellowish buff below, broadly whiter on sides and flanks (in pattern reminiscent of male); see C-48.
SIMILAR SPECIES: The intense red of the males renders them unmistakable at virtually any range; much duller females can be confusing when they are alone (though typically they move in groups with at least 1 male), but all similar species have at least some yellow below.
HABITAT AND BEHAVIOR: Locally fairly common in wet and humid foothill forest borders. Usually in small groups which glean actively in foliage, generally remaining in or near canopy. May come lower to fruiting trees at edge or along roads. Frequently accompanies mixed flocks of other tanagers, but also may move about independently. Readily found in lower foothills along the old Buenaventura road in Valle, Colombia.

RANGE: W. Colombia (w. slope of W. Andes north to s. Chocó and west along n. base of W. and Cen. Andes to middle Magdalena valley in e. Antioquia) and nw. Ecuador (Esmeraldas). To about 1100 m, but mostly 200–800 m.

NOTE: We follow *Birds of the World* (vol. 13, pp. 275–276) in merging the monotypic genus *Erythrothlypis,* in which this species was formerly placed, into *Chrysothlypis.* We would venture further that *Chrysothlypis* could easily be merged into *Hemithraupis,* for the 2 are very similar in morphology and behavior. The Black-and-yellow Tanager (*Chrysothlypis chrysomelas*) of Costa Rica and Panama occurs virtually on the border with Colombia (on Cerro Tacarcuna and Cerro Quía), but not yet definitely within Colombian territory. Male mostly bright yellow with black ocular area, back, wings, and tail; female olive above, wings duskier and with yellowish edging, bright yellow below.

Hemithraupis Tanagers

Small tanagers with pointed bills; males are strongly patterned, with yellow and rufous (or black) predominating; females much duller. Lowland forests and woodlands.

Hemithraupis guira

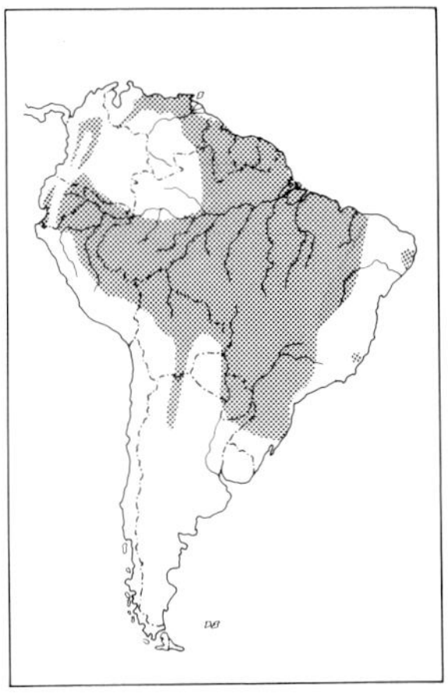

GUIRA TANAGER

PLATE: 11

IDENTIFICATION: 13 cm (5"). Bill yellowish with black culmen. *Lores, sides of head, and throat blackish to dark brown,* bordered above by *yellow superciliary* and behind by *large yellow patch on sides of neck.* Otherwise bright olive above with orange-rufous rump. *Breast also orange-rufous;* lower underparts pale yellowish, tinged gray on flanks and brighter yellow on crissum. *Guirina* (w. Colombia to nw. Peru) has more ochraceous superciliary. Female bright olive above with *short yellow superciliary and eye-ring,* yellowish rump in some races. Wings duskier edged yellow or bright olive. *Pale yellow below,* washed olive on breast and tinged gray on flanks.

SIMILAR SPECIES: Boldly patterned males are not likely to be confused (though in se. Brazil cf. Rufous-headed Tanager). Females aren't so easy, however. Yellow-backed Tanager female is very close but tends to be darker olive above and brighter yellow below and lacks the yellow brow and eye-ring. Juvenal Orange-headed Tanager also similar but likewise lacks the yellow on the face. Various female and immature wood-warblers might also be confused.

HABITAT AND BEHAVIOR: Fairly common to locally common in borders of humid forest, deciduous forest and woodland, and gallery woodland. In areas of extensive tall humid forest found primarily near openings or in secondary habitats, e.g., on river islands. Basically arboreal, foraging mostly by gleaning actively in foliage; less often comes to fruiting trees. Usually in pairs or small groups, frequently accompanying mixed flocks.

RANGE: Locally in w. Colombia (mostly in Magdalena and Cauca River valleys) and w. Ecuador (Pichincha to El Oro); Guianas, locally in n. and e. Venezuela (absent from the northwest, most of the llanos region, and Amazonas), locally in e. Colombia (mainly in southeast in Putumayo and Amazonas, also in Norte de Santander), e. Ecuador, e. Peru, n. and e. Bolivia, most of Amaz. and interior Brazil (south to Rio Grande do Sul; absent from nw. Amazonia, and only a few isolated records from e. coastal region), e. Paraguay, and ne. (Corrientes and Misiones) and nw. (south to Tucumán) Argentina. Mostly below 1500 m, locally to 2000 m in smaller numbers.

Hemithraupis ruficapilla

RUFOUS-HEADED TANAGER

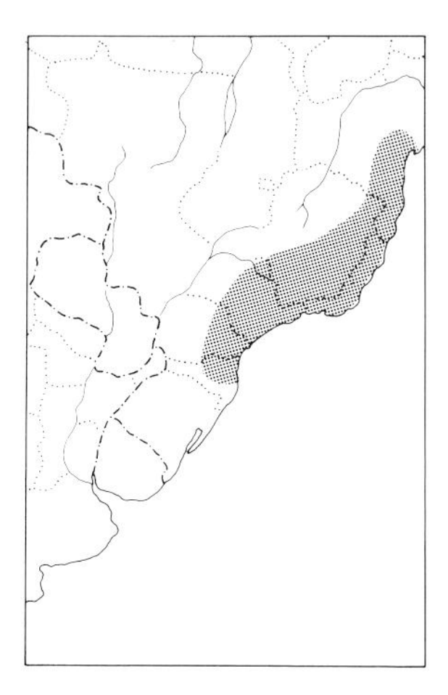

IDENTIFICATION: 13 cm (5"). Overall shape and pattern recall better-known Guira Tanager, which it replaces in *se. Brazil* (possibly with slight overlap). *Head deep rufous,* slightly paler on throat; no superciliary, but with *large yellow patch on sides of neck.* Otherwise bright olive above with orange-rufous rump. *Breast also orange-rufous;* lower underparts pale gray, tinged pale yellow in center. Female very like female Guira.

SIMILAR SPECIES: Male Guira Tanager has dark mask and throat boldly outlined in yellow (not an all-rufous head) and thus is easily distinguished from Rufous-headed; females of the 2 species probably cannot be separated in the field. Yellow-backed Tanager female is slightly darker olive above and not so pale yellow below and lacks yellow eye-ring and supraloral.

HABITAT AND BEHAVIOR: Locally fairly common in forest borders, light woodland, and secondary growth. Behavior similar to Guira Tanager's. Frequent in lower parts of Itatiaia Nat. Park, and can also be seen in Rio's Botanical Garden and Tijuca Nat. Park.

RANGE: Se. Brazil (s. Bahia and e. Minas Gerais south to e. Santa Catarina). To about 1100 m.

NOTE: This species and *H. guira* are known to hybridize at least occasionally along their zone of contact in interior se. Brazil (H. Sick); they may prove to be conspecific, but until more is known of the situation we prefer to maintain both as full species.

Hemithraupis flavicollis

YELLOW-BACKED TANAGER

PLATE: 11

IDENTIFICATION: 13–13.5 cm (5–5¼"). Bill blackish above, pinkish to yellow below. Male *mostly black above* with *bright yellow lower back and rump;* small white wing-speculum usually present. *Throat bright yellow;* remaining underparts mostly white, mottled or scaled irregularly with black, especially on chest and sides; crissum also bright yellow. *Albigularis* (n. Colombia except nw. Chocó) has *yellow of throat restricted to sides* (center of throat whitish) and small yellow supraloral streak; see C-52. *Peruana* (se. Colombia, e. Ecuador, ne. Peru) has yellow patch on lesser wing-coverts. Female *dark olive above;* wings duskier, *broadly edged yellowish olive. Mostly yellow below,* brightest on throat and breast and grayer on sides and flanks. Females west of Andes have *breast and belly whitish* (contrasting with yellow chest and crissum).

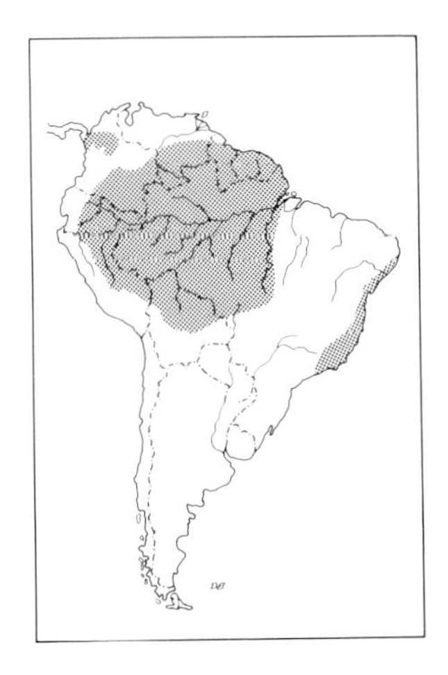

SIMILAR SPECIES: Distinctively patterned black, yellow, and white males should not be confused, whatever the race. Females readily confused with female Guira Tanager but are darker above with more strongly marked wings and deeper yellow below, and they lack yellow eye-ring and short superciliary; *habitat also helps,* for Yellow-backed is largely restricted to canopy of tall forest, with Guira more often at edge and in lower, secondary habitats.

HABITAT AND BEHAVIOR: Fairly common in canopy and to a lesser extent borders of humid forest (both *terra firme* and *várzea,* but mostly the former); occasionally out into clearings, when it may come lower, but basically a forest bird. Almost invariably found with mixed flocks of various tanagers (e.g., *Tangara* and *Tachyphonus*) and other birds, usually remaining very high and as a result often hard to see clearly. Gleans rather actively in foliage; also eats small fruits.

RANGE: Guianas, s. Venezuela (Bolívar and Amazonas), se. Colombia (north to Meta and Guainía), e. Ecuador, e. Peru, n. Bolivia (south to La Paz and Cochabamba), and Amaz. Brazil (north of the Amazon east to Amapá, south of it east to ne. Mato Grosso and the lower Rio Tapajós); coastal e. Brazil (Pernambuco south to Rio de Janeiro); n. Colombia (Chocó east along humid n. base of the Andes to the middle Magdalena valley in w. Santander). Also e. Panama. To about 800 m.

Nemosia Tanagers

A very distinct pair of basically gray and white tanagers (males with black heads). One species is widespread in lighter woodland, while the other may be extinct in se. Brazil.

Nemosia pileata

HOODED TANAGER

PLATE: 11

IDENTIFICATION: 13–13.5 cm (5–5¼"). *Iris and legs yellow;* bill of female pale flesh below. *Head and sides of neck black,* extending to sides of chest; *lores white. Otherwise grayish blue above* and white below, flanks faintly tinged gray (but immaculate white below in *hypoglauca* of n. Venezuela and Colombia). Female similar but lacking the black cap and tinged buff below (especially on sides of throat and breast).

SIMILAR SPECIES: The snappy males are readily known, but the pattern is reminiscent of the differently shaped Black-capped Warbling-Finch (with white in tail, etc.). Females, though lacking the black, can be known by their basically blue-gray and whitish coloration; note the yellow soft-parts and white lores. Females most closely resemble the Bicolored Conebill (with which sympatric along Amazon), but conebill is smaller, lacks the loral patch, has a dark bill, and is more uniform dingy gray-buff below (not as white).

HABITAT AND BEHAVIOR: Fairly common to common in lighter deciduous woodland, gallery forest, shrubby clearings with scattered trees, and plantations. In Amazonia essentially restricted to lighter *várzea* woodland and forest borders, especially on islands and along riverbanks. Usually in small groups of up to 6 or so birds which glean and perch mostly in semiopen parts of trees; often alone, but also regularly with other tanagers and various small insectivores.

RANGE: Caribbean n. Colombia (east to n. Córdoba and south in Magdalena valley to Bolívar), n. Venezuela, Guianas, much of Brazil (south to Mato Grosso, São Paulo, Minas Gerais, and Espírito Santo), Paraguay (mostly east of Paraguay River), n. Argentina (in northwest in Jujuy and Salta, in northeast only in Corrientes), n. and e. Bolivia, e. Peru (more local in the southeast), and extreme se. Colombia (along Amazon River). To about 600 m.

Nemosia rourei

CHERRY-THROATED TANAGER

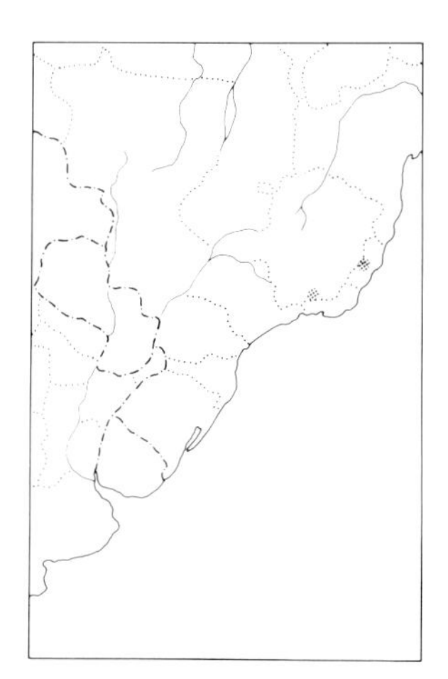

IDENTIFICATION: 14 cm (5½"). *Se. Brazil; very rare* (known from only 1 specimen). Resembles Hooded Tanager but larger, with *bright red throat and chest,* no white loral spot, and purer gray crown and upperparts (with black on head restricted to forehead and across sides of head). It is not known whether sexes differ.

HABITAT AND BEHAVIOR: Virtually unknown. A group of 8 were reported seen in forest canopy in 1941, for the only record since the collection of the type in the 19th century; these were independent of a flock (H. Sick). May be severely endangered (or possibly even already extinct) due to loss of forest habitat over vast majority of its presumed range.

RANGE: Se. Brazil (se. Minas Gerais at Muriahié on n. bank of Rio Paraiba do Sul, the type locality; and in Espírito Santo near Jatiboca, at 900 m).

Thlypopsis Tanagers

Small warblerlike tanagers, all with contrasting orange or rufous on crown or head. Mostly found in Andean woodlands and shrubby areas, with only 1 in lowlands east of the Andes. The genus seems closely related to *Hemithraupis.*

GROUP A

Grayish mantle; no yellow below.

Thlypopsis sordida

ORANGE-HEADED TANAGER

PLATE: 12

IDENTIFICATION: 13.5 cm (5¼"). *Crown and sides of head rufous-orange,* becoming bright yellow on lores, around eye, and on throat. Above otherwise gray tinged olivaceous. Below dingy buffy grayish, becoming whitish on median breast and belly. Nominate race (e. and s. parts of range) has underparts more strongly tinged with buff below than do more n. *chrysopis* and *orinocensis.*

SIMILAR SPECIES: The only *Thlypopsis* tanager over most of its *lowland range;* here readily recognized by its contrasting rufous-orange head.
HABITAT AND BEHAVIOR: Fairly common to locally common in early successional vegetation along rivers and on islands in the Amazon basin (and locally along lower Orinoco) and southward in lighter and gallery woodland and *cerrado*. The only lowland *Thlypopsis,* the other species occurring in montane or foothill regions. Gleans actively, warblerlike, at all levels but usually not very close to the ground and rarely or never inside continuous forest or woodland. Frequently found with mixed flocks of insectivorous birds.
RANGE: Extreme se. Colombia (along Amazon and in e. Nariño), e. Ecuador, e. Peru, n. and e. Bolivia, interior s. and e. Brazil (Acre and Mato Grosso east through s. Pará to Maranhão, Piauí, Ceará, and Pernambuco and south to n. São Paulo), Paraguay (local), and n. Argentina (south in west along base of Andes to Tucumán and in northeast south along Paraguay River to ne. Santa Fe and in Corrientes); also along lower Orinoco River in e. Venezuela (n. Bolívar and se. Anzoátegui). To about 1000 m.

Thlypopsis inornata

BUFF-BELLIED TANAGER

PLATE: 12

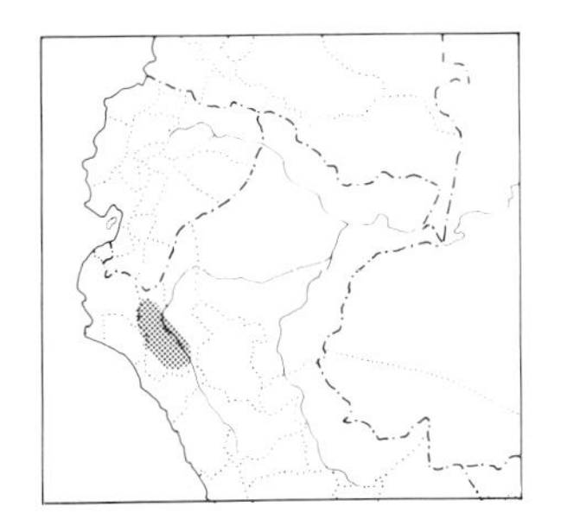

IDENTIFICATION: 12.5 cm (5"). *Nw. Peru.* Crown mostly orange-rufous, but with *narrow frontal band, ocular area, cheeks, and entire underparts quite uniform buff.* Otherwise gray above.
SIMILAR SPECIES: Much paler on face and breast than Rufous-chested Tanager and with contrasting deep rufous cap. Cf. also Brown-flanked Tanager (no known overlap).
HABITAT AND BEHAVIOR: Nothing appears to be on record.
RANGE: Nw. Peru (Amazonas and Cajamarca in drainage of upper Río Marañón). Mostly 600–2000 m.

Thlypopsis ornata

RUFOUS-CHESTED TANAGER

PLATE: 12

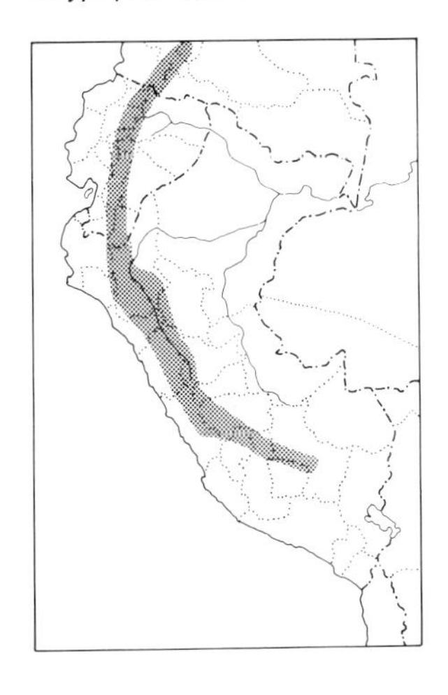

IDENTIFICATION: 12.5 cm (5"). *Entire head and underparts orange-rufous* (richest on crown), with *contrasting white median belly.* Upperparts otherwise olivaceous gray.
SIMILAR SPECIES: Cf. both Brown-flanked and Buff-bellied Tanagers: Brown-flanked is very similar but has grayish brown (not rufous) flanks, while Buff-bellied is uniform pale buff below (without the contrasting white median belly).
HABITAT AND BEHAVIOR: Locally fairly common at forest borders and in woodland and shrubby areas; mostly in humid regions, but also found in forest patches in more arid areas. Active behavior like that of other members of genus; usually in pairs or small groups, often with mixed flocks.
RANGE: Andes of sw. Colombia (only w. slope of Cen. Andes in Cauca), w. Ecuador, and Peru (south on w. slope to Lima, on e. slope to Cuzco). Mostly 1800–3200 m.

Thlypopsis fulviceps

FULVOUS-HEADED TANAGER

IDENTIFICATION: 12.5 cm (5″). *Venezuela and adjacent ne. Colombia. Entire hood rufous,* slightly paler on throat. Otherwise *dark gray above and pure light gray below,* becoming whitish on median belly. Female similar but throat even paler, a dull light buff. See V-36, C-53.

SIMILAR SPECIES: Easily recognized in its range by the rufous hood in sharp contrast to the gray of the body; the grayest *Thlypopsis.*

HABITAT AND BEHAVIOR: Uncommon to locally fairly common in forest borders, woodland, and at least in some areas (e.g., around Caracas) in scrub and suburban gardens. Found singly or in pairs (occasionally more together), regularly accompanying mixed flocks. Gleans actively like other members of its genus, but unlike them also found quite regularly associated with tall humid forest (though most often at edge).

RANGE: N. Venezuela (coastal mts. from Carabobo east to the Paria Peninsula and in the Andes of Mérida), Perijá Mts. on Venezuela-Colombia border, and extreme ne. Colombia (Cesar and Norte de Santander). Mostly 1000–2000 m.

Thlypopsis pectoralis

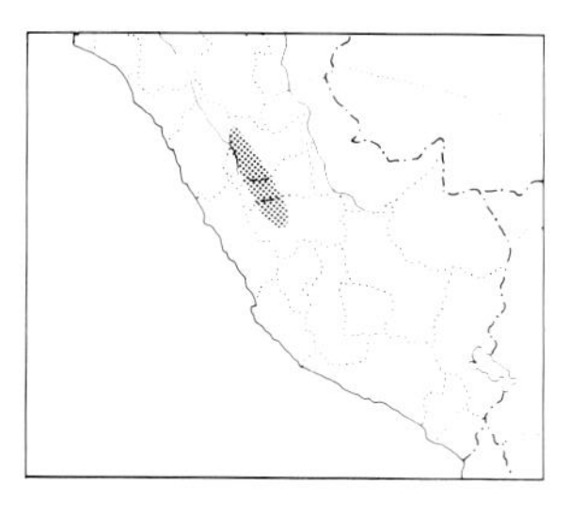

BROWN-FLANKED TANAGER

IDENTIFICATION: 12.5 cm (5″). *Cen. Peru.* Resembles Rufous-chested Tanager but *flanks pale grayish brown* (not orange-rufous). Back somewhat purer gray and crown slightly deeper rufous.

HABITAT AND BEHAVIOR: Reported to inhabit bushy growth with low trees bordering cultivation or along small streams on mountain slopes in semiarid areas (J. Zimmer). Apparently not uncommon at least locally.

RANGE: Andes of cen. Peru in Huánuco and Junín. 1800–3200 m.

GROUP B *Olive mantle; all-yellow underparts.*

Thlypopsis ruficeps

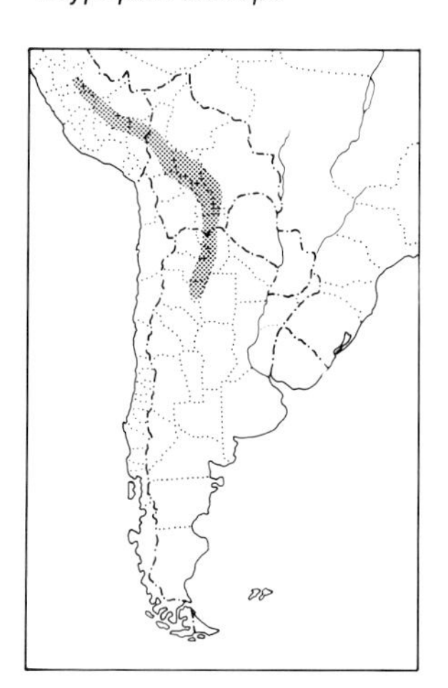

RUST-AND-YELLOW TANAGER

PLATE: 12

IDENTIFICATION: 12.5 cm (5″). *Head orange-rufous,* contrasting with otherwise bright olive upperparts. *Below bright yellow,* olivaceous on sides and flanks.

SIMILAR SPECIES: With its entirely bright yellow underparts imparts a very different impression from other *Thlypopsis* tanagers. Cf. Fulvous-headed Brush-Finch (with similar color pattern but very different shape and behavior).

HABITAT AND BEHAVIOR: Fairly common in shrubby Andean woodlands and their borders, especially where alders are frequent. Active arboreal behavior like that of other *Thlypopsis;* often seen in pairs, frequently as members of mixed flocks with warbling-finches, warblers, etc.

RANGE: Andes of e. Peru (north to Huánuco), w. Bolivia, and nw. Argentina (south to Tucumán). Mostly 1200–3200 m, lower southward.

Chlorochrysa Tanagers

Small, brilliant, iridescent tanagers of humid Andean forests. Females are duller (unlike *Tangara,* in which sexes usually are similar), immatures even more so.

Chlorochrysa nitidissima

MULTICOLORED TANAGER

PLATE: 12

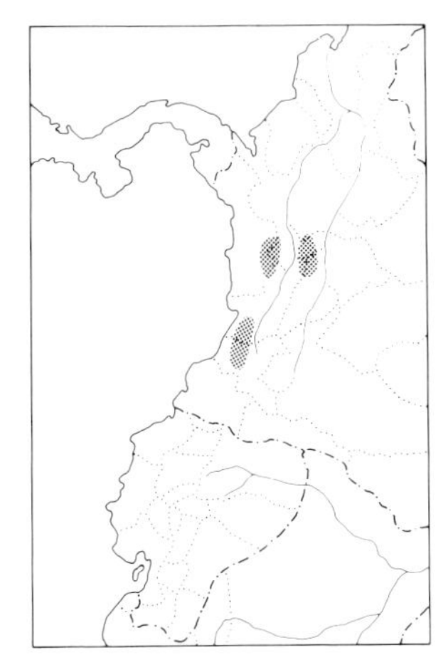

IDENTIFICATION: 12.5 cm (5″). *W. Colombia.* Showy male has *vivid colors and pattern. Foreface and throat bright yellow,* becoming glistening golden green on hindcrown and nape. *Saddle across back lemon yellow,* rump glistening greenish blue; wings and tail mostly emerald green. *Patch on sides of neck black, bordered below by chestnut.* Below mostly glistening blue, with black down center of breast and belly. Female similar but colors and pattern much duller; *retains highly contrasting yellow face and black spot on sides of neck,* but no yellow back or black below.

SIMILAR SPECIES: The incredibly colorful male is obviously unmistakable given a good view, but in dull light or fog might carelessly be confused with Saffron-crowned Tanager. Females, unless very young (when they are all green), are sufficiently similar to male to be readily recognized; usually the yellow foreface stands out the most.

HABITAT AND BEHAVIOR: Fairly common locally in humid montane forest and forest borders, to a lesser extent out into adjacent second growth and clearings with a few large trees left standing. Almost invariably found in association with a flock dominated by other tanagers; they move actively in pairs or small groups, rarely remaining long in 1 tree. Like other *Chlorochrysa* tanagers, it gleans foliage in a warbler- or vireolike manner and tends to take relatively little fruit. Found fairly readily in patches of forest remaining above Cali, Colombia (e.g., at Pichindé and near the "West Crest" on the highway to Buenaventura).

RANGE: W. Colombia (mostly on e. slope of W. Andes from sw. Antioquia to Cauca, also sparsely on adjacent w. slope; at n. end of Cen. Andes in Quindío and Caldas). Mostly 1400–2000 m, locally lower on Pacific slope.

Chlorochrysa calliparaea

ORANGE-EARED TANAGER

PLATE: 12

IDENTIFICATION: 12.5 cm (5″). *Above mostly bright shining emerald green,* with small golden orange spot on center of crown and *orange band across rump.* Throat black with *burnt orange patch on sides of neck.* Chest emerald green; *breast and belly mostly deep cobalt blue,* returning to emerald green on flanks and crissum. *Bourcieri* (south to Huánuco, Peru) is similar but lacks well-defined blue area on underparts (same area merely tinged bluish). See C-48. *Fulgentissima* (north to Puno, Peru) is also similar but has yellow crown patch, bright vermilion neck patch, and *deep violet purple throat and median lower underparts.* Female much duller (especially *bourcieri*), *mostly green* without the pattern but

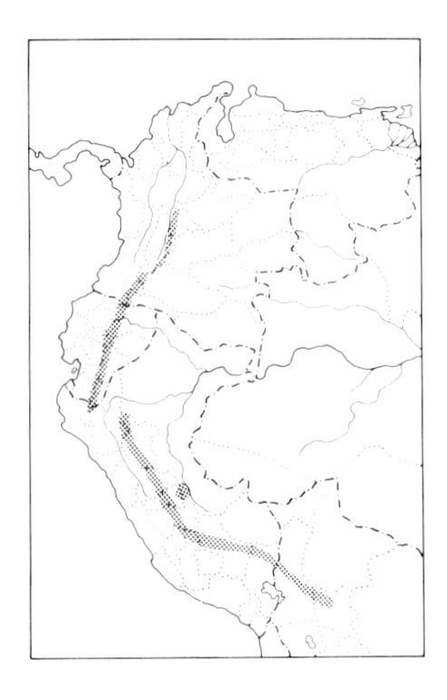

retaining the orange rump-band; throat brownish gray. Nominate race shows an echo of male's blue pattern below, while in *fulgentissima* comparatively more of male's color and pattern is retained (including neck patch).

SIMILAR SPECIES: Dazzling males are unlikely to be confused given reasonable light (contrasting black throat is a help even in poor light). Much duller females can be trickier; for them the orange rump-band (conspicuous in flight) is often the best clue apart from their consort (rarely is a lone female observed).

HABITAT AND BEHAVIOR: Fairly common in humid montane forest and forest borders, also regularly visiting trees in adjacent clearings. Usually seen in pairs or small groups which are almost always with mixed (but tanager-dominated) flocks. Forages actively, mostly by gleaning in foliage, employing a variety of acrobatic techniques; does not eat as much fruit as, for instance, many *Tangara* tanagers do.

RANGE: Andes of Colombia (on e. slope north to w. Caquetá; also in Magdalena valley on w. slope of E. Andes from Cundinamarca south to Huila), e. Ecuador, e. Peru, and Bolivia (La Paz and Cochabamba). Mostly 1100–1700 m.

Chlorochrysa phoenicotis

GLISTENING-GREEN TANAGER

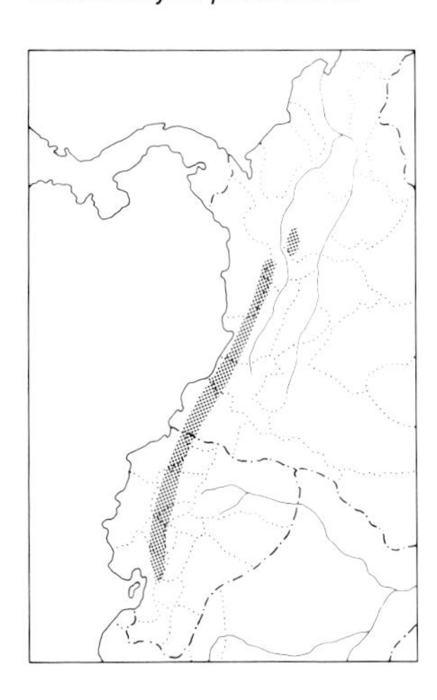

IDENTIFICATION: 12.5 cm (5"). *Sw. Colombia and w. Ecuador. Almost entirely bright glistening emerald green.* Small inconspicuous patches below and behind eye glistening gray, latter bordered behind by tiny tuft of orange. Shoulders also glistening gray (but often hidden). See C-48. Female slightly duller.

SIMILAR SPECIES: Orange-eared Tanager is almost the same green but is much more patterned and in any case does not overlap at all with this species. Female Green Honeycreeper is about the same size and shape but is nowhere near as vivid green and has a slightly decurved yellowish bill. Immature Multicolored, Rufous-winged, and Bay-headed Tanagers are mostly *dull* green (not the bright, vibrant green of this species).

HABITAT AND BEHAVIOR: Fairly common but very local in very wet mossy forest borders, less often out into adjacent clearings and second growth. Behavior similar to Orange-eared Tanager's. Quite readily seen at the Tokio forest above Queremal in Valle, Colombia (and also very common in Nariño above Junín), but seems less numerous in the Ecuadorian portion of its range (perhaps only a reflection of the difficult access to good cloud forest there).

RANGE: Andes of w. Colombia (Pacific slope north to s. Chocó and at n. end of Cen. Andes in Antioquia) and w. Ecuador (mostly in northwest south to Pichincha, but also several recent sightings from Bolívar and e. Guayas). Mostly 700–1500 m.

Montane *Tangara* Tanagers

Tangara tanagers comprise a widespread, complex, and numerous group—no other S. Am. genus has as many species. We have divided them approximately in half, here treating the montane (mostly Andean) species; be aware that there is an overlap zone on lower slopes where both lowland and montane species may occur. The *Tangara* tanagers are legendary for their colorful and often complicated plumage patterns, so much so that they are difficult to categorize in a useful way. The sexes are usually alike. Up to half a dozen or more *Tangara* may be found in the same flock in prime subtropical forest; they range mainly in the canopy, coming lower at the edge, foraging both for fruit and insects. The various species forage for fruit in much the same way, posing the interesting ecological question: how do so many apparently similar species manage to coexist in the same area? The answer, as demonstrated by the studies of S. Hilty and others, appears related to their insect-foraging strategies, some species searching mainly mossy limbs, others mainly small branches, others mainly foliage, etc.

GROUP A

Striking blue head, mostly black body.

Tangara cyanicollis

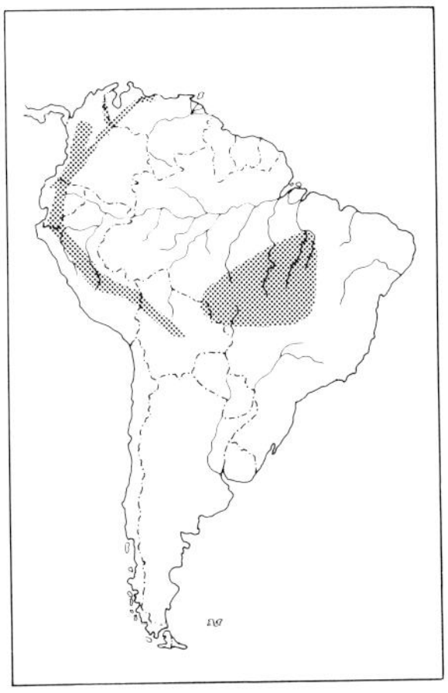

BLUE-NECKED TANAGER

PLATE: 12

IDENTIFICATION: 13 cm (5″). *Head mostly bright turquoise blue, contrasting with black mantle and breast.* Throat usually more purple and lores black. Shoulders and rump glistening straw to yellowish green; flight feathers edged yellowish green. Belly mostly violet blue, mixed with black toward center. *Hannahiae* (Venezuela and ne. Colombia) and *melanogaster* (Brazil) have all-black belly (with no blue), while *cyanopygia* (w. Ecuador) has rump turquoise blue.

SIMILAR SPECIES: One of the most stunning members of a spectacular genus. Not likely to be confused, but pattern superficially similar to Masked and Golden-hooded Tanagers' (both with white bellies, blue shoulders, differently colored hoods, etc.). Cf. also male Scarlet-thighed Dacnis.

HABITAT AND BEHAVIOR: Fairly common to common in a variety of semiopen montane habitats ranging from low secondary woodland to trees and shrubbery in gardens and partially cultivated areas. A nonforest tanager, but found mostly in humid regions; doubtless increasing due to deforestation over large parts of its range. Usually forages independent of mixed tanager flocks, generally in pairs or small groups; mostly frugivorous.

RANGE: Mts. of n. Venezuela (east to s. Carabobo and n. Guárico) and Andes of w. Venezuela, Colombia, Ecuador (south on w. slope to El Oro and Loja), e. Peru, and nw. Bolivia (La Paz and Cochabamba); Perijá Mts. on Venezuela-Colombia border; and in cen. Brazil (scattered localities in n. Mato Grosso, s. Pará, and s. Goiás). Mostly 500–2000 m, locally to near sea level in w. Ecuador.

GROUP B

Montane *Tangara* with *crown typically brighter.*

Tangara arthus

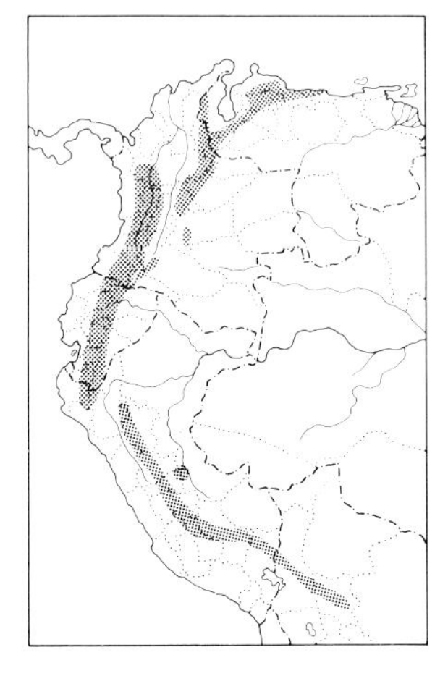

GOLDEN TANAGER

PLATE: 12

IDENTIFICATION: 13.5 cm (5¼"). *Predominantly bright golden to ochraceous yellow* with *contrasting large black patch on ear-coverts.* Back streaked black, and wings and tail mostly black. Underparts vary, with throat and chest rufous-chestnut in s. races (north to Amazonas, Peru), only a wash of rufous in *aequatorialis* (e. slope of Andes in Ecuador and adjacent Colombia and Peru), and more or less uniform in others except in nominate race of Venezuela. The latter is quite divergent, with yellow throat and median belly but *pectoral band and sides and flanks broadly chestnut;* see V-35.
SIMILAR SPECIES: Nothing likely to be confused. Pattern same as the largely green Emerald Tanager's, while yellow overall color recalls Silver-throated Tanager; both of these found only on Pacific slope.
HABITAT AND BEHAVIOR: Fairly common to common in humid montane forest and forest borders. Usually found in pairs or small groups, almost invariably in association with mixed flocks of other species of tanagers. Eats considerable fruit, but also forages a good deal for insects, typically by moving along mossy limbs and peering underneath.
RANGE: Coastal mts. of n. Venezuela (Falcón east to Miranda) and Andes of w. Venezuela (w. Lara and Yaracuy south), Colombia, Ecuador (south on w. slope to El Oro), e. Peru, and nw. Bolivia (La Paz and Cochabamba); Perijá Mts. on Venezuela-Colombia border and Macarena Mts. of e. Colombia. Mostly 1000–2500 m.

Tangara xanthocephala

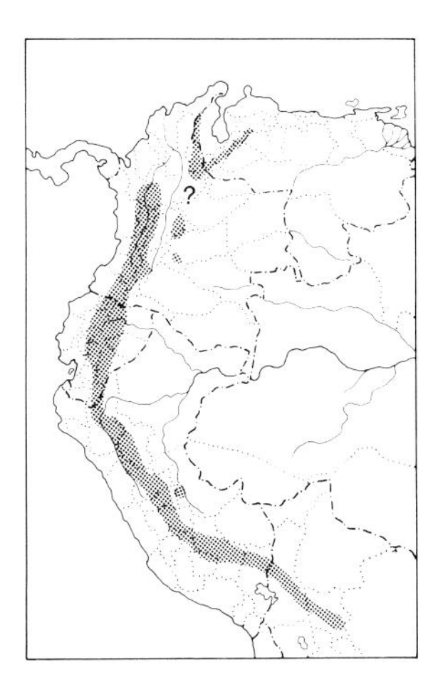

SAFFRON-CROWNED TANAGER

PLATE: 12

IDENTIFICATION: 13.5 cm (5¼"). *Crown saffron yellow to orange* (most orange in s. *lamprotis*); *most of rest of head yellow,* with black lores, ocular area, throat, and nuchal collar. Otherwise mostly opalescent bluish green; back streaked black, with wings and tail black edged bluish green. Center of belly and crissum cinnamon buff. N. *venusta* (south to n. Peru) has saffron yellow crown (concolor with rest of head; see V-35), while nominate race (cen. Peru) has crown intermediate (golden yellow) toward *lamprotis* (Bolivia and adjacent Peru).
SIMILAR SPECIES: W. race of Flame-faced Tanager (which is only orange, not scarlet, on the face) resembles Saffron-crowned, differing most strikingly in solid black back (no streaky effect) and prominent opalescent patch on shoulders (lacking in Saffron-crowned). Golden-eared Tanager has black hindcrown and malar stripe, solidly opalescent green throat (no black). Cf. also female of smaller Multicolored Tanager.

HABITAT AND BEHAVIOR: Fairly common to common in humid montane forest and forest borders, in smaller numbers out into adjacent advanced secondary growth and clearings with scattered large trees. Most often in small groups (up to 8 to 10 birds), regularly accompanying mixed flocks of other tanagers. Usually conspicuous, foraging for insects in the semiopen among smaller branches and limbs; also eats considerable fruit. Most numerous on e. slope of Andes.

RANGE: Andes of w. Venezuela (s. Lara south), Colombia, Ecuador (south on w. slope to Chimborazo), e. Peru, and nw. Bolivia (La Paz, Cochabamba, and w. Santa Cruz); Perijá Mts. on Venezuela-Colombia border, and Macarena Mts. of e. Colombia. 1300–2400 m.

Tangara parzudakii

FLAME-FACED TANAGER

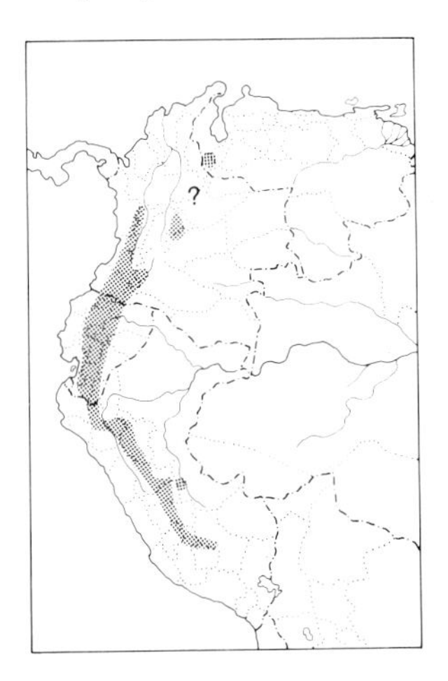

IDENTIFICATION: 14–14.5 cm (5½–5¾"). *Forecrown and cheeks scarlet,* becoming *bright golden yellow on hindcrown, nape, and sides of neck.* Lores, ocular area, upper throat, and bar extending back under cheeks black. Otherwise mostly black above with *large patch on shoulders and rump silvery green opalescent.* Below also silvery green opalescent, becoming buff on median belly and crissum. See V-35, C-49. *Lunigera* (Pacific Colombia and Ecuador) is slightly smaller and lacks fiery red on face; its forecrown and cheeks are orange.

SIMILAR SPECIES: A strikingly colored and patterned tanager which should not be confused, but cf. Saffron-crowned Tanager (particularly in area of overlap with *lunigera* race of Flame-faced).

HABITAT AND BEHAVIOR: Fairly common in humid montane forest and forest borders. Usually in pairs or small groups, and almost always with mixed flocks of tanagers, etc. Forages, much like Golden Tanager, by sidling along mossy limbs and frequently peering underneath; also consumes much fruit (e.g., *Cecropia* catkins).

RANGE: Andes of extreme sw. Venezuela (sw. Táchira), Colombia (local in E. Andes and not in Cen. Andes apart from Huila), Ecuador (south on w. slope to El Oro), and e. Peru (south to Cuzco). Mostly 1500–2500 m, lower (to 1000 m or even lower) on Pacific slope.

Tangara nigroviridis

BERYL-SPANGLED TANAGER

PLATE: 12

IDENTIFICATION: 13.5 cm (5¼"). *Boldly spangled with opalescent green or blue. Crown, nape, and rump opalescent green,* black bases of feathers showing through in places; small mask and back black. Wings and tail black edged greenish blue. *Below black heavily spangled with opalescent green,* paler on center of belly. *Cyanescens* (Venezuela, most of w. Colombia, and w. Ecuador) has bluer spangles below than nominate group (e. slope of Andes from Colombia south).

SIMILAR SPECIES: Male Black-capped Tanager lacks the spangled effect and has black (not opalescent) crown. Cf. also Metallic-green Tanager (also without spangly effect).

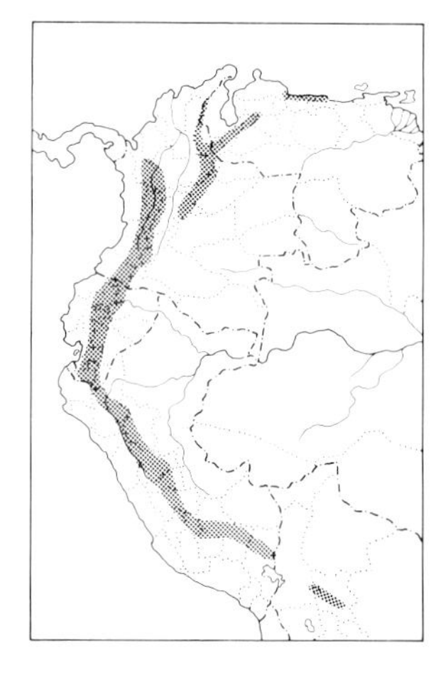

HABITAT AND BEHAVIOR: Fairly common to common in humid montane forest and especially forest borders and in adjacent secondary growth and clearings with scattered large trees. Usually in pairs or small groups which frequently associate with mixed flocks of other tanagers. Eats much fruit, but also frequently forages for insects among smaller branches and in foliage.

RANGE: Coastal mts. of n. Venezuela (Carabobo to Miranda) and Andes of w. Venezuela (north to s. Lara), Colombia, Ecuador (south on w. slope to El Oro and Loja), e. Peru, and nw. Bolivia (La Paz and Cochabamba); Perijá Mts. on Venezuela-Colombia border and Macarena Mts. in e. Colombia. Mostly 1500–2500 m.

NOTE: Green-naped Tanager (*Tangara fucosa*) is known from the Panama side of Cerro Tacarcuna and may occur in extreme nw. Colombia as well. It is mostly black above and on throat and chest, with green patch on nape, opalescent spangling on sides of neck, and blue spangling on chest; rump opalescent blue; breast and belly cinnamon buff.

Tangara vassorii

BLUE-AND-BLACK TANAGER

PLATE: 12

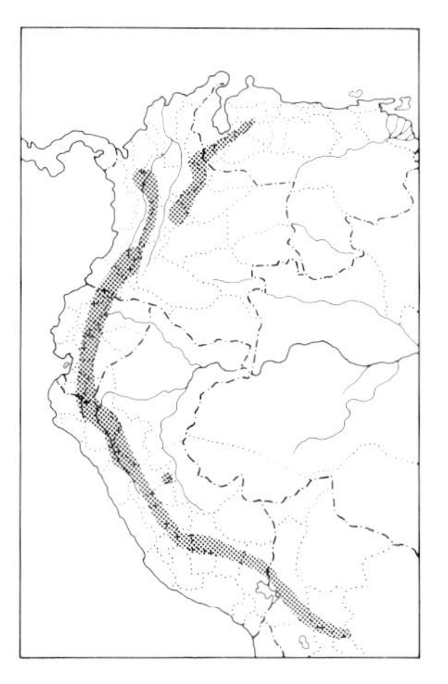

IDENTIFICATION: 13 cm (5″). *Mostly shining deep cobalt blue. Lores, wings, and tail contrastingly black;* shoulders and a single broad wing-bar the same blue. *Branickii* (ne. Peru from Amazonas to Huánuco) resembles nominate race (n. part of range south to nw. Peru) but has *head and neck paler and more opalescent bluish green,* contrasting with the bright blue back. *Atrocaerulea* (Bolivia and Peru north to Huánuco) loses that effect but does have head and neck more greenish blue (slight contrast with rest of body) with *opalescent straw-colored nape patch,* and has *solid black back* and *conspicuous black spotting below.* Females duller than males of their respective races.

SIMILAR SPECIES: S. race quite strongly resembles sympatric race of Golden-naped Tanager (see below), though that occurs at lower elevations. Pure blue coloration also may recall Masked Flower-piercer (often together), which, however, lacks the black on wings and tail and has a red eye.

HABITAT AND BEHAVIOR: Fairly common to common in humid montane forest, forest borders, and shrubbery (at times right up to treeline). Occurs higher than other *Tangara* tanagers (almost exclusively in temperate zone) and thus tends to flock more with various mountain-tanagers (*Anisognathus, Iridosornis*), hemispinguses, conebills, and brush-finches. Regularly occurs in quite large groups of up to a dozen or more birds; forages actively, mostly in foliage, and usually seems restless, rarely remaining for long in 1 tree. May engage in some seasonal altitudinal movements.

RANGE: Andes of w. Venezuela (north to Trujillo), Colombia, Ecuador, Peru (south on w. slope to Cajamarca), and nw. Bolivia (La Paz, Cochabamba, and w. Santa Cruz). Mostly 2000–3400 m northward, somewhat lower southward.

GROUP C Montane *Tangara* with *crown typically black.*

Tangara chrysotis

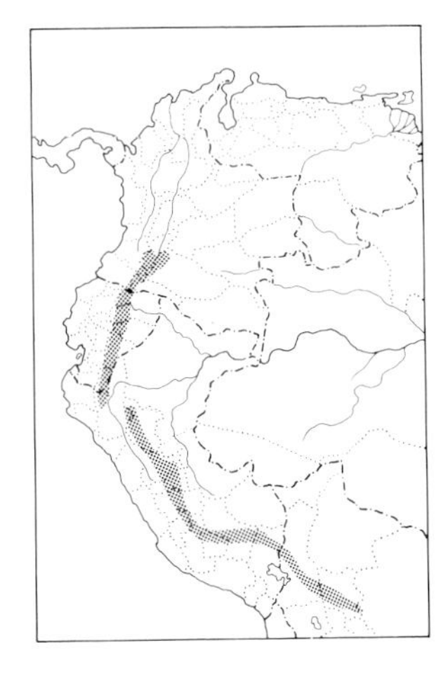

GOLDEN-EARED TANAGER

PLATE: 12

IDENTIFICATION: 14 cm (5½″). *Crown and bold malar streak black, separated by coppery gold ear-coverts* and *long opalescent green superciliary extending forward.* Mantle black streaked opalescent green, rump and upper tail-coverts entirely opalescent green. Wings and tail black edged opalescent green. Below mostly opalescent emerald green, but *center of breast, belly, and crissum cinnamon-rufous.*

SIMILAR SPECIES: A lovely tanager, superficially like the much duller Metallic-green (which lacks streaking above and the burnished gold on the face, is not as deeply rufous on median underparts, etc.). Saffron-crowned Tanager has yellow (not black) crown and nape, is paler buff on median lower underparts, etc.

HABITAT AND BEHAVIOR: Uncommon to locally fairly common (seemingly more numerous northward) in humid montane forest and forest borders. Usually seen in pairs, almost always with mixed flocks of other tanagers. Forages for insects primarily by hopping along mossy horizontal branches, pausing to probe clumps. Also eats much fruit. Regularly found in substantial numbers along the road to Mocoa in w. Putumayo, Colombia, and near San Rafael Falls along the Lago Agrio road in Napo, Ecuador.

RANGE: Andes of sw. Colombia (upper Magdalena valley in Huila and on e. slope of E. Andes from w. Caquetá south), e. Ecuador, e. Peru, and nw. Bolivia (La Paz and Cochabamba). Mostly 1100–2300 m, but apparently only above 1600 m in Colombia.

Tangara labradorides

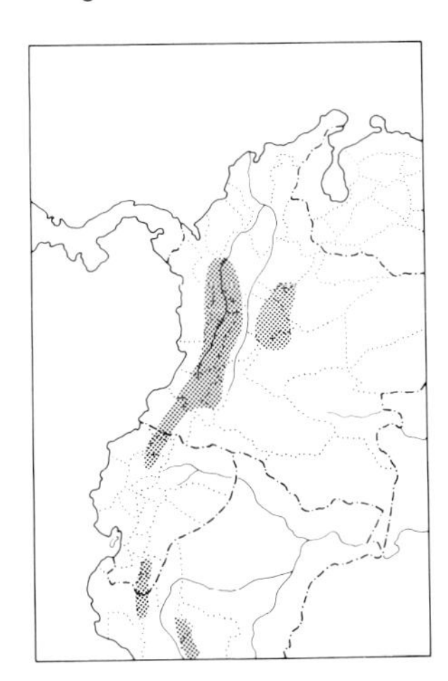

METALLIC-GREEN TANAGER

IDENTIFICATION: 13 cm (5″). Compared to its boldly patterned and brightly colored congeners, *relatively dull. Mostly opalescent bluish green,* with *center of crown and nape, small mask, and scapulars black;* wings and tail black edged bluish green. Center of belly and crissum dull buff. See C-48.

SIMILAR SPECIES: Dull and with no really characteristic field marks; no other tanager is so basically opalescent green (can look quite blue in some lights) and black without bright colors. Cf. Black-capped Tanager (male of which has much more gray and solid black crown, female of which is basically greenish without the black) and also Blue-browed and Golden-naped Tanagers.

HABITAT AND BEHAVIOR: Locally common (especially in Colombia; less numerous in remainder of its range) in humid montane forest borders, secondary woodland, and shrubby clearings or clearings with scattered tall trees. Less of a true forest bird than many of its congeners. Usually in pairs or small groups which usually accompany mixed flocks of other tanagers. Besides eating fruit, also forages for

insects by working along smaller branches and searching twigs and foliage. Frequently seen in various areas around Cali, Colombia.
RANGE: Andes of W. Colombia (in E. Andes north only to Cundinamarca), Ecuador in northwest (south to Pichincha) and southeast (Zamora-Chinchipe, above the town of Zamora), and ne. Peru (south to San Martín). Mostly 1500–2400 m.

Tangara cyanotis

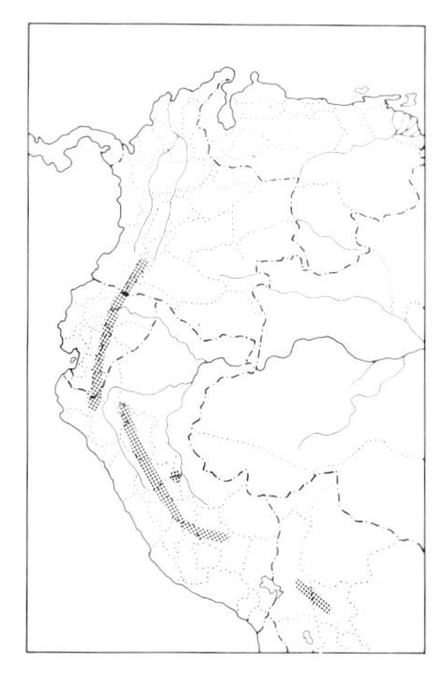

BLUE-BROWED TANAGER

IDENTIFICATION: 12 cm (4¾"). *Mostly black above* (including sides of head) with *long broad superciliary extending back from above eye turquoise blue.* Shoulders and rump also turquoise blue; wings and tail black, former edged turquoise blue. Mostly turquoise blue below, becoming buff on median underparts from lower breast to crissum. See C-48. Nominate race (Bolivia) differs from *lutleyi* (remainder of range) in having *forepart of cheeks* blue (not all black) and *mantle dusky greenish blue* (not black).
SIMILAR SPECIES: Should not be confused: no other blue and black tanager or dacnis shares the eyebrow. Cf. Metallic-green Tanager (much duller and with only limited overlap).
HABITAT AND BEHAVIOR: Uncommon in humid montane forest and forest borders. Behavior much like many of its congeners', and like them most often found in mixed flocks. When foraging for insects tends to search among smaller branches and terminal leafy areas. Seen regularly along the Cordillera Azul road east of Tingo María, Peru, and near San Rafael Falls on the Lago Agria road in Ecuador.
HABITAT: Andes of sw. Colombia (upper Magdalena valley in Huila, and on e. slope of E. Andes from w. Putumayo south), e. Ecuador, e. Peru, and nw. Bolivia (La Paz and Cochabamba). 1400–2200 m.

Tangara ruficervix

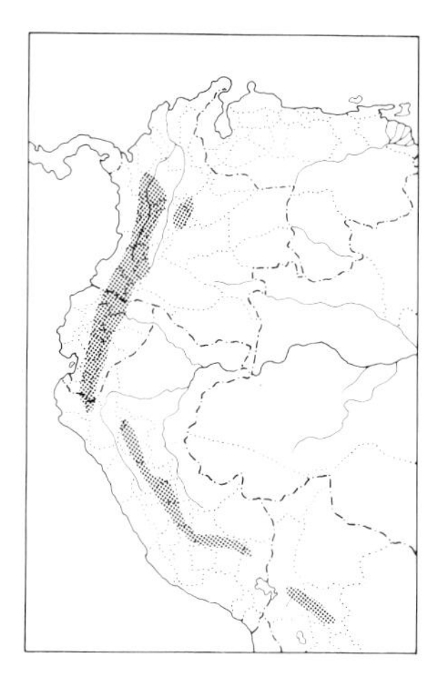

GOLDEN-NAPED TANAGER

PLATE: 12

IDENTIFICATION: 13 cm (5"). *Mostly turquoise blue,* duskier bases of feathers showing through in places. *Band across hindcrown golden buff,* bordered anteriorly and posteriorly by black and then violet blue. Forehead, ocular area, and chin black. Wings and tail black edged turquoise blue. Median belly and crissum buffy whitish to buff. *Fulcivervix* group (north to cen. Peru) gives quite different impression: it is a notably darker, *more cobalt blue* generally, and its golden nape band is not bordered with black.
SIMILAR SPECIES: S., darker blue birds are quite likely to be confused with sympatric race of Blue-and-black Tanager (which also has an opalescent straw nape patch); Blue-and-black is, however, solid black across the back and has quite conspicuous black spotting below and less black on foreface. Otherwise no other mostly blue tanager has golden on the hindcrown (latter can be a bit hard to see at times).
HABITAT AND BEHAVIOR: Fairly common in humid montane forest and forest borders and in secondary woodland and shrubby regenerating clearings. Like its congeners usually found in pairs or small groups,

most often accompanying mixed flocks of other tanagers, etc. Takes considerable fruit, but also frequently seen foraging for insects, searching along larger limbs as well as gleaning on foliage and among twigs. RANGE: Andes of Colombia (north in E. Andes to Cundinamarca), Ecuador (south on w. slope to El Oro), e. Peru, and nw. Bolivia (La Paz and Cochabamba). Mostly 1500–2400 m, ranging somewhat lower on Pacific slope.

Tangara rufigenis

RUFOUS-CHEEKED TANAGER

IDENTIFICATION: 13 cm (5″). *N. Venezuela. Mostly opalescent bluish green,* darker on back. *Mask from lores back over cheeks and chin rufous.* Wings and tail blackish, wing-coverts edged bluish green and flight feathers edged tawny olive. Center of belly buffy whitish, becoming buffy fawn on crissum. See V-35.

SIMILAR SPECIES: Rather dull, but in its limited range known by its general opalescent green coloration and the rufous cheeks (latter often not very prominent, however). Perhaps most likely to be confused with female Black-capped Tanager, which has duskier crown and streaky effect below (besides lacking the rufous cheeks).

HABITAT AND BEHAVIOR: Uncommon in humid forest and forest borders. Behavior similar to many other forest-associated *Tangara* tanagers'. Regularly seen near the pass on the Rancho Grande road in Henri Pittier Nat. Park, Aragua.

RANGE: Coastal mts. of n. Venezuela (s. Lara east to Distrito Federal). 900–2050 m.

Tangara viridicollis

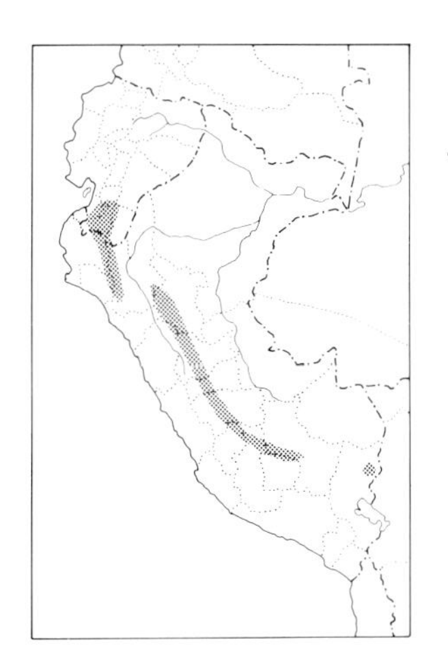

SILVER-BACKED TANAGER

PLATE: 12

Other: Silvery Tanager

IDENTIFICATION: 13 cm (5″). *Ecuador and Peru.* Crown, nape, and *most of underparts black. Throat shining greenish coppery. Mantle shining gunmetal gray;* wings and tail black edged dark bluish gray. Flanks and crissum also gunmetal gray. Female has crown dusky brown but upperparts otherwise shining green. *Throat and sides of head dull greenish coppery* with streaky effect, becoming greener on chest and mostly gray on breast and belly, with flanks tinged green. *Fulvigula* (Ecuador and nw. Peru) has mantle more opalescent than nominate race's (e. Peru) and its throat is somewhat darker and purer coppery. Female *fulvigula* has more golden brown crown and brighter coppery throat.

SIMILAR SPECIES: Overall aspect, with black underparts and pale shiny mantle, quite distinctive among *Tangara* tanagers except for the similar Straw-backed (see below). Black-capped Tanager has similar gray on mantle and lower underparts, but its throat is green. See also the very localized Sira Tanager.

HABITAT AND BEHAVIOR: Uncommon to fairly common but rather local in humid forest and forest borders and adjacent shrubby regenerating clearings. Often found in areas where forest is patchy. Active, flocking behavior much like other *Tangara* tanagers'. Readily seen along the Río Urubamba below the Machu Picchu ruins in Cuzco, Peru.

RANGE: Andes of s. Ecuador (El Oro and Loja) and Peru (south on w. slope to Lambayeque and s. Cajamarca, on e. slope from Amazonas to Puno). Mostly 1000–2200 m.

NOTE: As this bird is not predominantly silvery, we have opted to slightly modify its English name to "Silver-backed"; dorsal color also represents the simplest way to separate *T. viridicollis* from the closely related *T. argyrofenges.*

Tangara argyrofenges

STRAW-BACKED TANAGER

PLATE: 12

Other: Green-throated Tanager

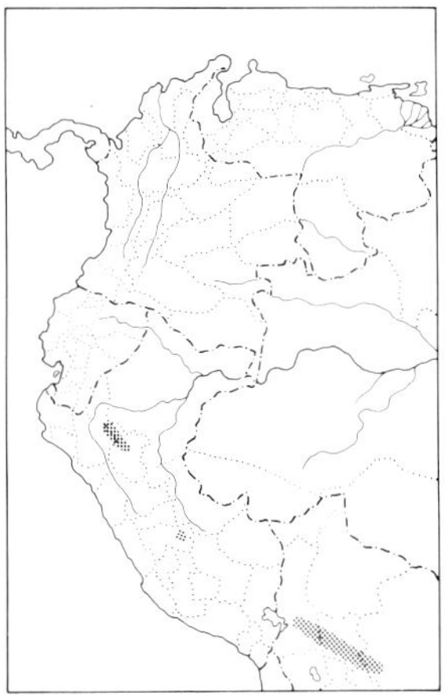

IDENTIFICATION: 13 cm (5"). *Peru and Bolivia.* Pattern resembles that of better-known Silver-backed Tanager. *Throat opalescent green* (not coppery). *Mantle shining opalescent straw color* (not gray), as are rump and flanks (but crissum is black). Female has crown dusky with feathers edged dull green. *Mantle shining yellowish green;* wings and tail black, feathers edged green. *Throat, sides of neck, and chest shining silvery green* with streaky effect. Lower underparts gray, with sides and flanks yellowish green.

SIMILAR SPECIES: Compare to Silver-backed Tanager (though they mostly do not occur together). Females would be quite tricky to tell apart, but female Straw-backed has considerably yellower tone to the green mantle and its throat is more silvery green (not coppery).

HABITAT AND BEHAVIOR: Apparently rare and local in humid forest and forest borders. Poorly known, but behavior probably differs little from that of other members of its species group (including Silver-backed, Black-capped, and Sira Tanagers).

RANGE: Andes of ne. Peru (very locally in s. Amazonas, w. San Martín, and Junín) and nw. Bolivia (*yungas* of La Paz, Cochabamba, and w. Santa Cruz). Mostly 1300–1700 m.

NOTE: This species' most striking field character (and the one which most readily distinguishes it from closely related *T. viridicollis*) is its shining straw-colored mantle. It does have a green throat, but so too do *T. heinei* and *T. phillipsi.* We thus have selected a new, distinctive English name for it.

Tangara heinei

BLACK-CAPPED TANAGER

IDENTIFICATION: 13 cm (5"). *Crown and nape black.* Upperparts otherwise shining silvery bluish gray. *Throat, sides of neck, and chest opalescent green with streaky effect.* Breast and belly silvery bluish gray. Female virtually identical to female Straw-backed Tanager but the shining green of its mantle lacks any yellow tone. Over most of its range female known by its *dusky crown* (feathers green-edged), contrasting with bright shining green back, and *streaky opalescent green effect on throat, sides of neck, and chest. Belly gray,* with flanks green. See V-35, C-49.

SIMILAR SPECIES: Cf. Silver-backed and Straw-backed Tanagers. In its range male readily identified by combination of black crown and

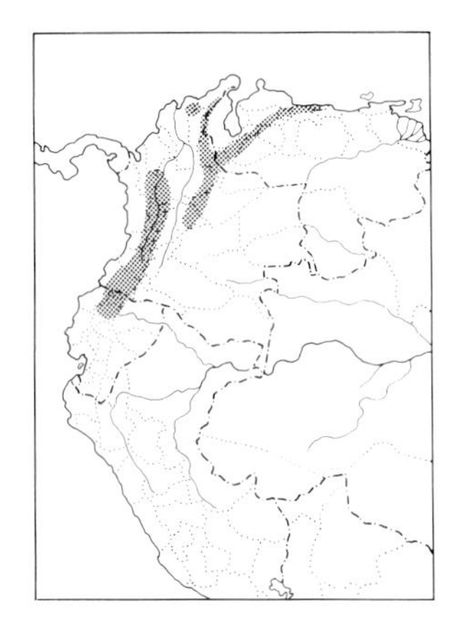

green throat. Female resembles female Black-headed Tanager (and the 2 occur sympatrically), but latter is paler and grayer on crown, throat, and chest.

HABITAT AND BEHAVIOR: Usually uncommon (locally somewhat more numerous) in humid montane forest borders and shrubby clearings with scattered trees. Not a true forest tanager. Behavior similar to its many congeners', eating much fruit and searching for insects mostly by inspecting small branches.

RANGE: Coastal mts. of n. Venezuela (Distrito Federal and Guárico west to Yaracuy) and Andes of w. Venezuela (north to s. Lara), Colombia, and n. Ecuador (south on w. slope to Pichincha, on e. slope only in Napo); Colombia's Santa Marta Mts. and Perijá Mts. on Colombia-Venezuela border. Mostly 1300–2200 m.

Tangara phillipsi

SIRA TANAGER

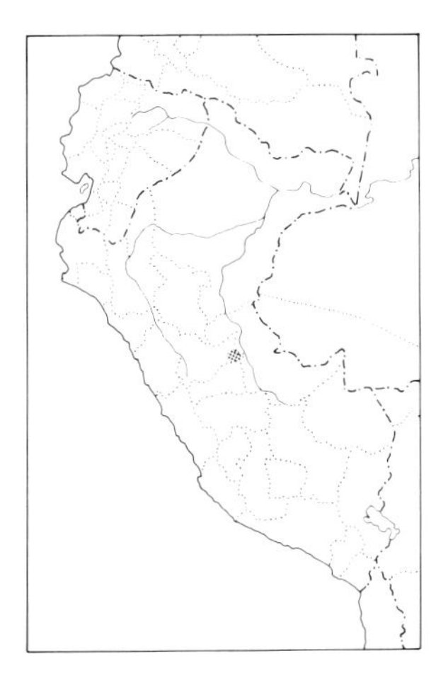

IDENTIFICATION: 13 cm (5"). Found only on the remote *Cerros del Sira* in e. Peru. Male resembles male Black-capped Tanager but has *breast, median belly, and sides of neck* (connecting to crown) *black,* and green on throat is more restricted (not extending down over chest) and with less of a streaky effect. Female resembles female Black-capped but has darker gray breast and mid-belly and less of a streaky effect on throat.

SIMILAR SPECIES: Both sexes of Silver-backed Tanager (which has been recorded with this species) have a coppery-colored throat.

HABITAT AND BEHAVIOR: Fairly common in canopy and borders of humid montane forest and cloud forest. Frequently with flocks of other tanagers.

RANGE: E. Peru (Cerros del Sira in Huánuco). 1300–1570 m.

NOTE: A newly described species: G. Graves and J. Weske (*Wilson Bull.* 99 [1]: 1–6, 1987).

GROUP D

Montane *Tangara* with *contrasting black head* and *mostly opalescent body;* females mostly olive.

Tangara cyanoptera

BLACK-HEADED TANAGER

IDENTIFICATION: 13.5 cm (5¼"). *Venezuela and adjacent regions.* Male's pattern recalls Golden-collared Honeycreeper (no overlap), but stout shape of black bill typical of *Tangara* tanagers. *Hood black,* contrasting sharply with *shining opalescent straw remainder of body plumage.* Wings and tail also contrastingly black, flight feathers rather broadly edged ultramarine blue. Female very different and much duller. *Crown pale grayish;* upperparts otherwise pale green, becoming yellower on rump. Wings and tail blackish, feathers broadly edged pale green. *Throat and breast grayish white* with streaky effect; *belly pale yellow,* tinged green on flanks. See V-35 and C-49. *Whitelyi* of *tepuis* is notably duller with less opalescence generally and a rather strong mottled effect below, resulting in a much less neat appearance; it also lacks the blue wing-edging.

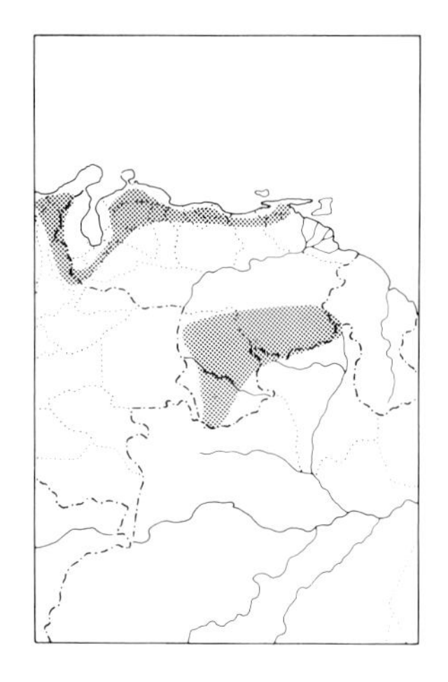

Female *whitelyi* very dingy, with pronounced flammulated effect below.
SIMILAR SPECIES: Male unmistakable in its range. Female easily confused with female Black-capped Tanager but is paler generally on the foreparts (lacking the dusky-capped effect) and belly is pale yellow (rather than green and gray).
HABITAT AND BEHAVIOR: Fairly common to common in humid forest borders, secondary woodland and plantations, and clearings with scattered trees. Behavior similar to that of other *Tangara;* usually forages for insects by searching among leaves and outer branches.
RANGE: Coastal mts. of n. Venezuela (Sucre west to Falcón), Andes of w. Venezuela (north to s. Lara) and extreme ne. Colombia (Norte de Santander), Perijá Mts. on Venezuela-Colombia border, Santa Marta Mts. of Colombia, and *tepui* region of s. Venezuela (Bolívar and Amazonas) and adjacent n. Brazil (n. Roraima) and w. Guyana. Mostly 800–1800 m.

NOTE: *Whitelyi* may well represent a distinct species.

Iridophanes Honeycreepers

Plumage pattern remarkably like that of Black-headed Tanager, but bill shape quite different, as is color of both bill and iris. We thus remain reluctant to sink the genus in *Tangara,* though this has been advocated by several recent authorities (e.g., *Birds of the World,* vol. 13). Furthermore, its behavior seems most reminiscent of *Chlorophanes* honeycreepers.

Iridophanes pulcherrima

GOLDEN-COLLARED HONEYCREEPER PLATE: 12

IDENTIFICATION: 12 cm (4¾″). Bill fairly long and slender (quite different from a *Tangara*'s), blackish with mandible mostly yellow; iris dark red. *Hood, upper back, and scapulars black,* the head and back separated by a *narrow orange-yellow nuchal collar. Remainder of back and rump opalescent greenish straw. Wings mostly ultramarine blue;* tail black, feathers edged blue. Below pale opalescent greenish yellow, whiter on center of belly. *Aureinucha* (Pacific slope) has longer bill and sootier head (in male). Female dull olive above; *wing-coverts and flight feathers edged bluish green.* Usually shows *echo of male's golden nuchal collar.* Below dull yellowish buff, breast and sides tinged olive.
SIMILAR SPECIES: Cf. male of remarkably similar Black-headed Tanager (not known to overlap); otherwise male is unmistakable. Female potentially much more confusing (but usually will be accompanying a male, thus solving the problem); note bill shape (rather like a short Green Honeycreeper's), faint nuchal collar, and blue-green wing-edging.
HABITAT AND BEHAVIOR: Uncommon to locally fairly common at borders of humid forest and in adjacent clearings with scattered trees; much more numerous on e. slopes of Andes, and seemingly rare throughout its Pacific slope range. Usually seen in pairs or small

groups, often with mixed tanager flocks; more active and restless than most *Tangara* tanagers. Frequently seen feeding on *Cecropia* catkins. Readily seen near San Rafael Falls on the Lago Agrio road in e. Ecuador.

RANGE: Locally in Andes of sw. Colombia (Pacific slope in Valle and Nariño, head of Magdalena valley in Huila, and e. slope of E. Andes in w. Caquetá and w. Putumayo), Ecuador (in northwest recorded south to Pichincha, with few or no recent reports; more numerous and widespread on e. slope), and e. Peru (south to Cuzco). 900–1900 m.

Lowland *Tangara* Tanagers

The montane members of this very large genus having been dealt with previously, we turn here to its many lowland representatives. Like their montane relatives, most are brightly colored and boldly patterned (reaching their extreme in the gaudy Paradise Tanager), though some are notably duller. The sexes are usually alike, the major exception being in the "Scrub Tanager complex." Behavior of *Tangara* in the lowlands is basically similar to that found in montane areas, though usually diversity is not as great (so there are not as many species in a single flock).

GROUP E

Miscellaneous *n. Colombian and Pacific slope Tangara.*

Tangara palmeri

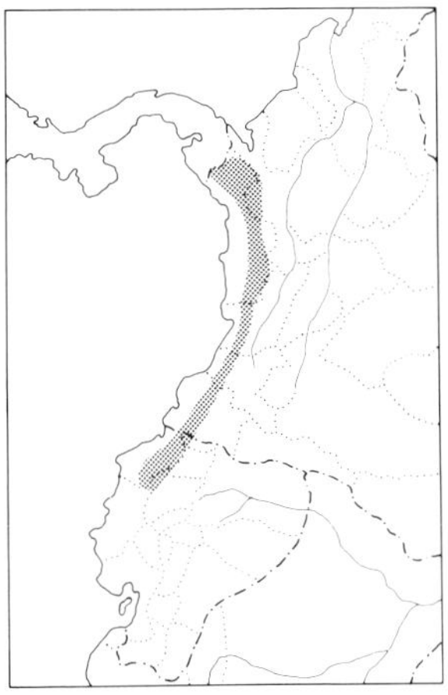

GRAY-AND-GOLD TANAGER

PLATE: 13

IDENTIFICATION: 14.5 cm (5¾"). *A rather large, mostly gray and black tanager.* Mostly pale gray above and whitish below. Small black mask and opalescent green back. Wings and tail black broadly edged gray. *Sprinkling of black spots* down sides of neck and across chest forms a "necklaced" effect; some opalescent gold also on chest.

SIMILAR SPECIES: A distinctively patterned *Tangara*. The opalescent colors are often hard to discern in the field. Cf. Golden-hooded Tanager.

HABITAT AND BEHAVIOR: Fairly common in canopy and edge of humid forest, primarily in foothill areas. Occurs in rather noisy pairs or small groups which usually are with mixed flocks of other tanagers, etc. Often perch conspicuously in treetops for fairly long periods but may then fly a long distance, leaving the flock to straggle after them. Mostly frugivorous (e.g., at *Cecropia*), but rarely drop to melastomaceous shrubs like so many other *Tangara*. Regularly seen along lower Buenaventura road in Valle, Colombia.

RANGE: Pacific slope of w. Colombia and nw. Ecuador (south to Pichincha). Also e. Panama. Mostly 300–1000 m.

Tangara rufigula

RUFOUS-THROATED TANAGER

PLATE: 13

IDENTIFICATION: 12 cm (4¾"). *Blackish and speckly. Head and neck black* with *rufous throat*. Back black, feathers edged golden green, giving scaly look; rump mostly opalescent green. Wings and tail black

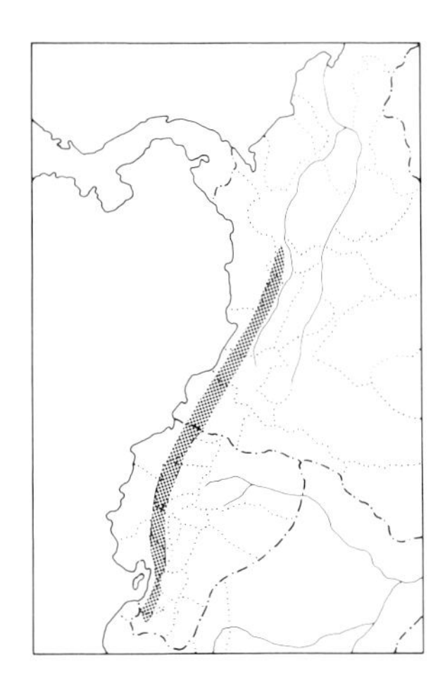

edged opalescent green. Feathers of breast and flanks opalescent green centered black, giving *conspicuously spotted effect.* Center of belly whitish, crissum buff.

SIMILAR SPECIES: Though the rufous throat can be hard to make out, Rufous-throated is readily recognized by combination of black head and scaly or spotted appearance of back and breast. Cf. Black-capped and Beryl-spangled Tanagers (both at higher elevations).

HABITAT AND BEHAVIOR: Fairly common to locally common in mossy cloud forest, especially at shrubby forest borders. Pairs or small groups often accompany mixed flocks of other tanagers, etc., gleaning in foliage as well as taking small fruits. Though numerous locally, most of the areas where this species is especially frequent are difficult of access; 1 exception is above Junín, in Nariño, Colombia, where it and the Moss-backed Tanager are the most numerous tanagers.

RANGE: Pacific w. Colombia (north to s. Chocó) and w. Ecuador (south to El Oro). Mostly 800–1800 m.

Tangara inornata

PLAIN-COLORED TANAGER

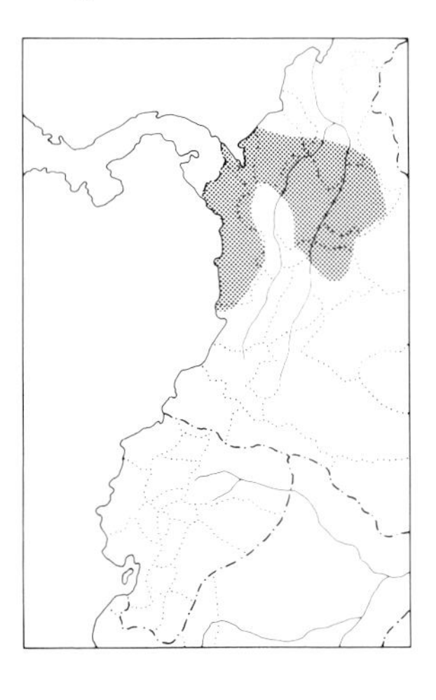

IDENTIFICATION: 12 cm (4¾"). *N. Colombia.* Unusual among the colorful and highly patterned *Tangara* tanagers is this dull-plumaged and uniform species. *Mostly leaden gray,* paler below, becoming white on center of breast and belly. Lores and ocular area dusky. Wings and tail blackish; lesser wing-coverts blue (but usually hidden). See C-48, P-27.

SIMILAR SPECIES: Cf. Blue-gray and Palm Tanagers, both of which can look gray in poor light but are obviously larger.

HABITAT AND BEHAVIOR: Uncommon in borders of humid forest, clearings with scattered trees, and secondary woodland. Usually found in small groups of 3 to 6 birds, sometimes with mixed flocks but more often moving independently. Searches for insects along limbs, but also takes much fruit (e.g., often at *Cecropia* trees).

RANGE: N. Colombia (south on Pacific slope to Valle and in humid Caribbean lowlands along n. base of Andes to middle Magdalena valley in Cundinamarca and Caldas). Also Costa Rica and Panama. To 1200 m (perhaps occasionally higher).

Tangara icterocephala

SILVER-THROATED TANAGER

PLATE: 13

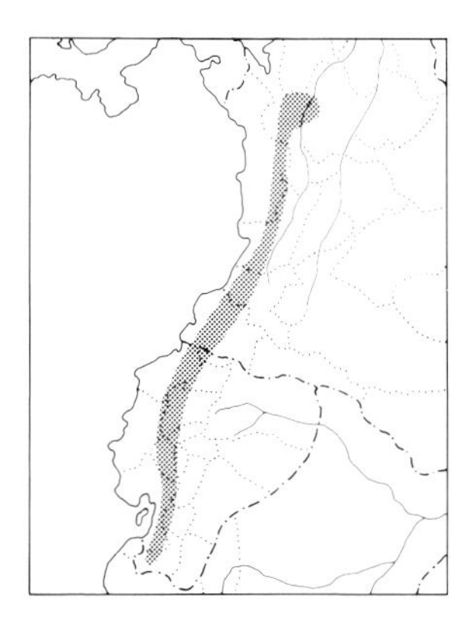

IDENTIFICATION: 13 cm (5"). *Mostly bright golden yellow.* Mantle streaked with black; wings and tail black edged bright green. *Large throat patch silvery greenish white,* bordered at sides by long narrow black malar streak. Female not as bright, with duller, streakier crown.

SIMILAR SPECIES: Virtually unmistakable, but might carelessly be confused with Golden Tanager.

HABITAT AND BEHAVIOR: Fairly common to common in humid forest, forest borders, and adjacent clearings with scattered trees. Often in quite large groups, up to a dozen or more birds, which frequently accompany mixed flocks of other tanagers, etc. Eats mostly fruit, foraging for insects mostly by searching slender moss-covered branches. Its call is a characteristic buzzy "bzeet," harsher than its congeners'.

RANGE: W. Colombia (Pacific slope of W. Andes and at n. end of Cen. Andes) and w. Ecuador (south to El Oro and w. Loja). Also Costa Rica and Panama. Mostly 500–1300 m.

GROUP F

"Scrub Tanager complex." *Opalescent* greenish blue to buff, with *rufous crown* (usually) and *black mask on face.*

Tangara vitriolina

SCRUB TANAGER

PLATE: 13

IDENTIFICATION: 14 cm (5½"). *Colombia and n. Ecuador.* Sexes similar (female slightly duller). *Crown rufous; sides of head black,* forming a mask. Otherwise mostly dull silvery greenish above, paler silvery grayish below, becoming buffy whitish on median belly and crissum. Wings and tail mainly bluish green.
SIMILAR SPECIES: A rather dull tanager, readily recognized in its semiopen, often arid habitat and small overall range. At most minimal overlap with Burnished-buff Tanager (which is basically found east of the Andes).
HABITAT AND BEHAVIOR: Common in grassy scrub and brushy areas, agricultural regions, and trees around habitations. Occurs primarily in arid regions (e.g., in rain-shadow valleys), but has also spread into deforested regions in more humid areas. Found mostly in pairs which usually forage independently of other birds, though at times they may gather with other tanagers (e.g., Blue-gray and Flame-rumped) at fruiting trees.
RANGE: W. Colombia (from w. slope of E. Andes westward; mostly in Magdalena and Cauca River drainages, but also in Dagua and Patía valleys on Pacific slope, and spreading locally into cleared areas in more humid parts of Pacific slope; ranges north to Antioquia and Santander) and nw. Ecuador (south to Pichincha, in cen. valleys north of Quito). Mostly 500–2200 m.

Tangara meyerdeschauenseei

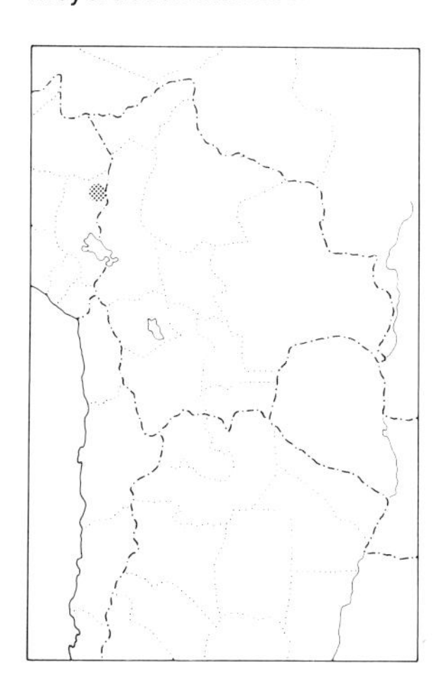

GREEN-CAPPED TANAGER

IDENTIFICATION: 14 cm (5½"). *Very local in s. Peru.* Resembles Scrub Tanager, differing in its greenish straw (not rufous) crown, greenish turquoise (not black) ear-coverts, greener (less blue) back and scapulars, and more uniform grayish buff underparts (lacking whitish on lower belly). Sexes alike, female slightly duller.
SIMILAR SPECIES: Note that Scrub Tanager's range lies far to the north of this species' (in Colombia and n. Ecuador). Burnished-buff Tanager occurs in *lowlands* some 150 km to the north of this species' range, but it differs strikingly in its shiny straw buff (not shiny bluish green) mantle, golden rufous (not greenish straw) crown, etc.
HABITAT AND BEHAVIOR: Fairly common in dense scrub and gardens in a deep, apparently naturally quite arid intermontane valley and in partially cleared areas on nearby slopes (originally covered with montane forest). Behavior similar to that of other members of the Scrub Tanager group.

RANGE: E. slope of Andes in extreme s. Peru (Puno, in and around Sandia, in the headwaters of the Río Inambari). About 2000 m.

NOTE: A newly described species: T. S. Schulenberg and L. Binford (*Wilson Bull.* 97 [4]: 413–420, 1985).

Tangara cayana

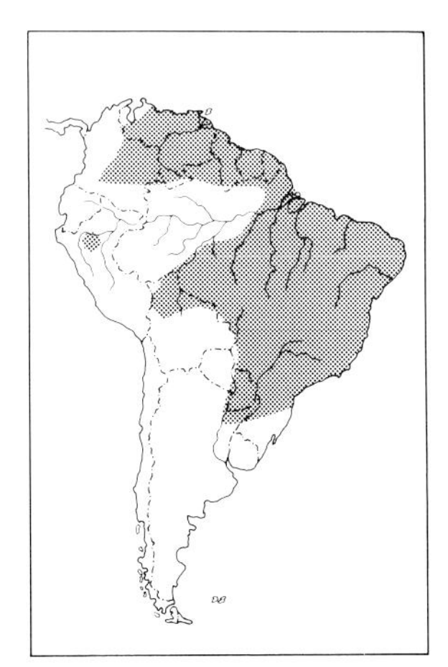

BURNISHED-BUFF TANAGER

PLATE: 13

IDENTIFICATION: 14 cm (5½″). Wide-ranging in open to semiopen areas but with 2 different "types." S. and e. *flava* group (north to mouth of Amazon and west in interior Brazil to se. Pará and sw. Mato Grosso) is *shiny ochraceous buff above,* with wings and tail mainly bluish. *Broad area from sides of head and throat to median belly black,* contrasting sharply with ochraceous sides, flanks, and crissum. Female much duller, mostly green above with crown somewhat more rufescent and *dusky mask;* wings and tail mainly greenish. Throat whitish, remaining underparts buffy whitish tinged greenish, becoming buff on crissum. N. *cayana* group (south locally to n. Bolivia) has similar pattern but lacks the black except on mask; it has a *golden rufous crown* but is otherwise shiny straw buff above. Below mostly pale buffyish with *shiny violet suffusion over throat and median underparts* (varies in extent and deepness). Female similar but facial pattern and underparts duller (latter without the violet) and with mantle and rump dull greenish. See V-35, C-49. *Huberi* (of Marajó Is. in mouth of Amazon) is apparently intermediate, with male resembling *flava* group but with black median underparts duller and less well defined, paler mantle.

SIMILAR SPECIES: Both types are readily recognized over most of their ranges, females in part by their dullness. In ne. Colombia may have limited overlap with Scrub Tanager (with obvious rufous crown and generally duller and greener plumage in both sexes). Cf. female Chestnut-backed and Black-backed Tanagers in s. Brazil.

HABITAT AND BEHAVIOR: Usually fairly common to common (may be less numerous and more local where suitable habitat is limited in extent) in savannas and *cerrado* with scattered trees, scrub, gallery woodland, and forest borders and secondary growth and gardens in more humid areas (usually where at least partially deforested). Occurs in pairs or small groups, foraging at all levels in trees and shrubbery; usually active and conspicuous, often perching in the open.

RANGE: Ne. Colombia (east of Andes south to Meta and Vaupés, also recorded locally on w. slope of E. Andes in Norte de Santander) and locally throughout Venezuela (absent from forested and coastal desert regions) east locally through extreme n. Brazil (Roraima) and the Guianas to e. Amaz. Brazil (south to Rio Grande do Sul), e. Paraguay, and ne. Argentina (Corrientes and Misiones), west over s. Amaz. Brazil to n. Bolivia (Beni) and very locally in e. Peru (arid intermontane valleys, e.g., in San Martín, and in lowlands in Pampas de Heath of Madre de Dios). Mostly below 1200 m (locally higher in Venezuela).

Tangara preciosa

CHESTNUT-BACKED TANAGER

PLATE: 13

IDENTIFICATION: 14.5 cm (5¾"). *Crown, sides of head, nape, and back shining coppery rufous;* lores black. Rump and wing-coverts shiny ochraceous; greater coverts at least tinged lavender. Wings and tail black broadly edged blue. *Below shining bluish green,* bluest on median underparts; lower flanks and crissum rufous. Female has *head and nape dull coppery rufous.* Otherwise green above, wings and tail blackish edged green. Below pale silvery greenish.
SIMILAR SPECIES: Bright and attractive males are readily recognized (but see Black-backed Tanager); duller females resemble female Burnished-buff Tanager, but only lores dusky (lacking the full black mask).
HABITAT AND BEHAVIOR: Fairly common in forest and forest borders (often in regions where Araucarias are frequent) and adjacent clearings and second growth. Usually in pairs or small groups, foraging at all levels in trees and often with mixed flocks of insectivorous birds. Perhaps most numerous in Rio Grande do Sul, Brazil.
RANGE: Se. Brazil (north to São Paulo), se. Paraguay, Uruguay, and locally in ne. Argentina (south to Entre Ríos and at least formerly to n. Buenos Aires). Apparently somewhat migratory. To about 1000 m.

NOTE: Some believe *T. preciosa* and *T. peruviana* represent partially localized morphs of the same species, with black-backed types occurring northward (situation somewhat confused by partial migration of s. chestnut-backed types). For now we prefer to maintain the 2 as full species, pending thorough study of the situation; polymorphism is not known in any other tanager. If merged, the name *T. peruviana* has priority (Polymorphic Tanager).

Tangara peruviana

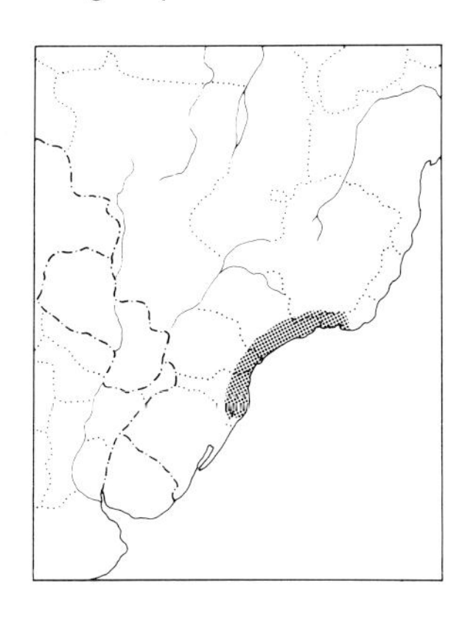

BLACK-BACKED TANAGER

IDENTIFICATION: 14.5 cm (5¾"). *Se. Brazil.* Resembles Chestnut-backed Tanager. Male differs only in having *shining black back,* in sharp contrast to the coppery rufous nape and shiny opalescent wing-coverts and rump. Females of the 2 are apparently indistinguishable.
HABITAT AND BEHAVIOR: Little known, but presumably very similar to Chestnut-backed Tanager. Recorded from thick shrubbery and woodland borders, both near the coast and in hilly areas (locally in the Serra do Mar). Known from near the Boraceia Forest Reserve in ne. São Paulo. Possibly at risk due to deforestation, but more information needed.
RANGE: Se. Brazil (Rio de Janeiro south to e. Santa Catarina). To about 700 m.

GROUP G

Predominantly "*golden green*" *Tangara.*

Tangara gyrola

BAY-HEADED TANAGER

PLATE: 13

IDENTIFICATION: 13.5–14 cm (5¼–5½"). *Head brick red,* usually bordered behind by narrow golden nuchal collar and usually with small golden area on shoulders. Otherwise bright green above. *Color of un-*

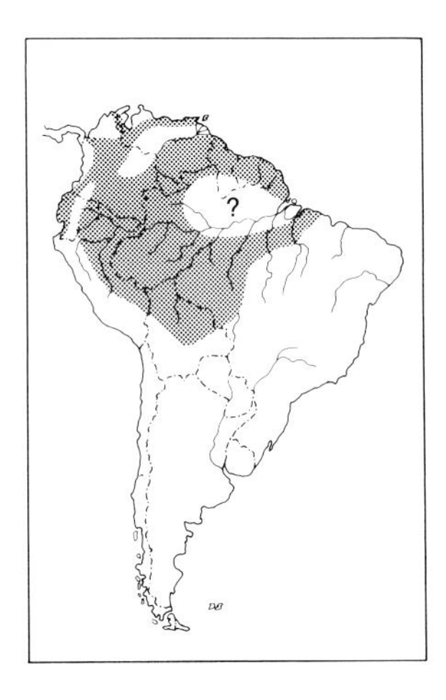

derparts varies: n. races (ne. Colombia, n. Venezuela, and Trinidad) are *bright green below,* while various w. and Amaz. races are *turquoise blue below* with green confined to crissum; nominate (Guianas, s. Venezuela, n. Brazil) is intermediate. *Albertinae* (Amaz. Brazil south of Amazon) has broader golden nuchal band merging into golden green back and brick red (not golden) shoulder. Females of all races are duller (but still basically similar to respective males); immatures are even more nondescript, and in some the head color is only faintly indicated.

SIMILAR SPECIES: The contrasting reddish chestnut head combined with the green and blue body plumage easily distinguish this lovely tanager. On Pacific slope of Colombia and Ecuador, cf. Rufous-winged Tanager.

HABITAT AND BEHAVIOR: Generally fairly common to common (more local and often less numerous in Amazonia, especially distant from the Andes) in canopy and borders of humid forest and adjacent clearings and gardens with large trees. Often in pairs or small groups and frequently with mixed flocks of other tanagers. Feeds mostly on fruit, also by gleaning for insects along horizontal limbs (regularly peering underneath).

RANGE: Widespread in forested lowlands and lower mountain slopes (up to about 1500 m in most areas, locally a little higher) south on Pacific slope to sw. Ecuador (El Oro) and east of Andes south to n. Bolivia (La Paz, Cochabamba, and w. Santa Cruz) and s. Amaz. Brazil (w. and n. Mato Grosso and e. Pará); Trinidad. Absent from Pacific coastal lowlands (below 500 m) of w. Colombia, from nw. Venezuela and the llanos region of Venezuela and ne. Colombia, and apparently from at least part of lower Amaz. Brazil north of the Amazon River. Also Costa Rica and Panama.

Tangara lavinia

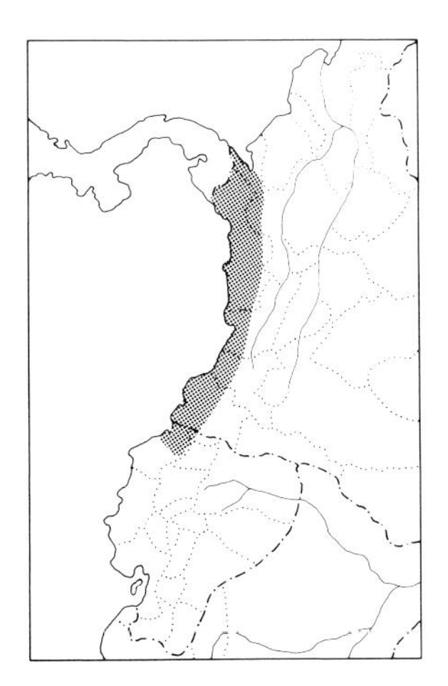

RUFOUS-WINGED TANAGER

IDENTIFICATION: 13 cm (5″). *Mainly w. Colombia.* Overall pattern reminiscent of Bay-headed Tanager, but male brighter. *Head and most of wing bright brick red.* Otherwise mostly bright grass green, with *shining golden yellow back* and turquoise blue on center of throat and median lower underparts. See C-49. Female and immature duller (especially the latter), with no yellow on back and only a little rufous on wings and often not much on head (*can look mostly green with rufous wing-patch*).

SIMILAR SPECIES: A striking and well-named tanager, for the reddish to rufous on wings is normally present regardless of age stage. Superficially similar Bay-headed Tanager never shows this, has yellow above restricted to a narrow collar, and *in area of overlap is mainly blue below.*

HABITAT AND BEHAVIOR: Fairly common in canopy and borders of humid forest and secondary woodland. Behavior similar to Bay-headed Tanager's and sometimes is with it. Readily found in lowlands around Buenaventura, Colombia.

RANGE: W. Colombia and extreme nw. Ecuador (Esmeraldas); Gorgana Is. (off Colombian coast). Also Guatemala to Panama. Mostly below 500 m, occasionally to 1000 m.

Tangara schrankii

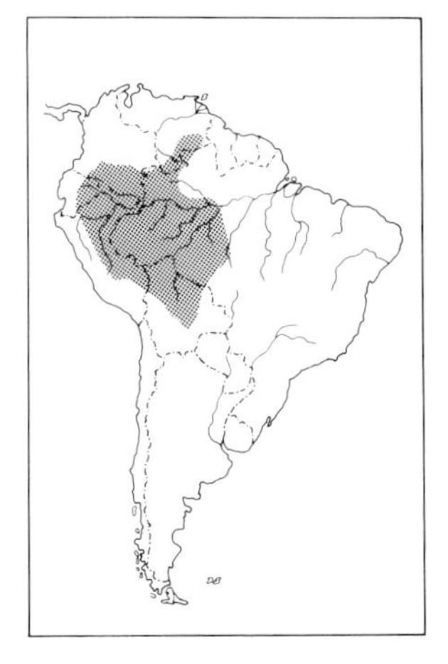

GREEN-AND-GOLD TANAGER

PLATE: 13

IDENTIFICATION: 13.5–14 cm (5¼–5½"). *Mostly bright green and yellow with conspicuous black forehead and mask.* Patch on crown and rump yellow; above otherwise bright green streaked black. Wings and tail black, feathers edged greenish blue. Median underparts broadly yellow, but throat, sides, and flanks green. Female similar but not quite as bright overall, with green crown and less yellow on rump.

SIMILAR SPECIES: This w. Amaz. *Tangara* is most likely to be confused with smaller Yellow-bellied Tanager, which, however, lacks the mask and is boldly spotted with black. Cf. similar Emerald Tanager (found only west of the Andes; no overlap with Green-and-gold).

HABITAT AND BEHAVIOR: Fairly common to common in canopy and borders of humid lowland forest (both *terra firme* and *várzea*) and to a lesser extent also out into adjacent clearings. Generally in pairs or small groups which regularly accompany mixed flocks of other tanagers (especially Paradise). Forages mostly in upper strata, gleaning insects from leaves and eating fruit. May come lower to fruiting trees at edge or in clearings (and also regularly accompanies insectivorous flocks in forest understory).

RANGE: S. Venezuela (s. Bolívar and s. Amazonas), se. Colombia (north to Caquetá and Vaupés), e. Ecuador, e. Peru, n. Bolivia (south to La Paz, Cochabamba, and nw. Santa Cruz), and w. Amaz. Brazil (north of the Amazon east to the lower Rio Negro, south of it to nw. Mato Grosso). To about 900 m (along base of Andes and on slopes of *tepuis*).

Tangara florida

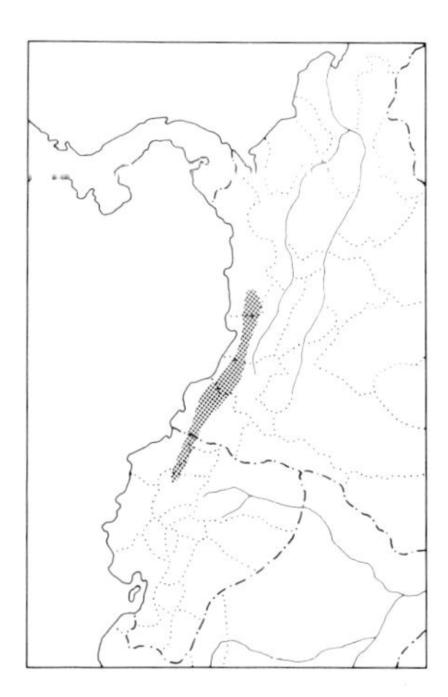

EMERALD TANAGER

IDENTIFICATION: 13 cm (5"). *Mainly w. Colombia. Mostly bright emerald green. Crown bright yellow,* with small black patch encircling bill and *bold rectangular patch on ear-coverts.* Back streaked black; wings and tail black edged green. Rump and center of belly also yellow. See C-48. Female similar but somewhat duller generally and with no yellow on crown.

SIMILAR SPECIES: Nothing really similar in range; cf. Blue-whiskered Tanager (no yellow on crown, with black throat and blue whisker, etc.). Pattern (but of course not overall color) reminiscent of Golden Tanager. Green-and-gold Tanager is found only east of the Andes.

HABITAT AND BEHAVIOR: Uncommon in canopy and borders of humid mossy forest. Usually in pairs, most often seen accompanying mixed flocks of other tanagers. Searches for insects mostly by inspecting mossy branches, primarily their undersides. Can be found in small numbers along lower Buenaventura Road, Colombia.

RANGE: Pacific slope of W. Andes in w. Colombia (north to s. Chocó) and nw. Ecuador (Esmeraldas and Pichincha). Also Costa Rica and Panama. Mostly 500–1200 m, locally slightly higher or lower.

Tangara johannae

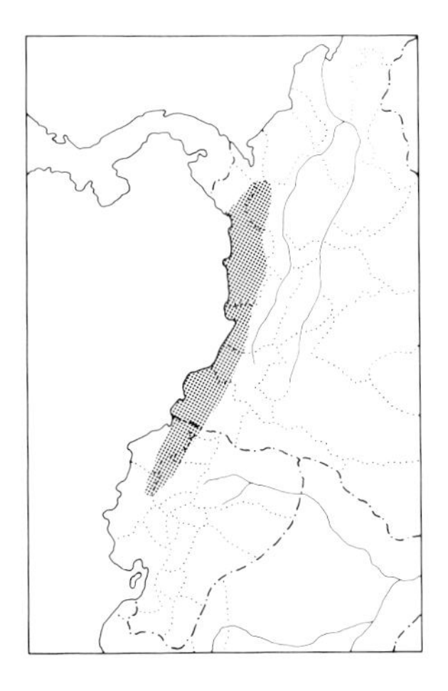

BLUE-WHISKERED TANAGER

IDENTIFICATION: 13.5 cm (5¼"). *Mainly w. Colombia. Throat and foreface black* bordered narrowly with blue; *conspicuous broad blue malar streak* (the "whisker"). Otherwise mainly bright golden green, with black-streaked back, yellow rump, and grayish center of belly and crissum. Wings and tail black conspicuously edged blue. See C-48.
SIMILAR SPECIES: Emerald Tanager has yellow crown and black ear-patch, lacks the black throat.
HABITAT AND BEHAVIOR: Rare to uncommon at borders of wet forest and in secondary woodland. Found singly or in pairs, usually while accompanying a large mixed flock of other tanagers, etc. Behavior much like that of other *Tangara*. Can be seen around Buenaventura, Colombia (but numbers always small).
RANGE: Pacific lowlands of w. Colombia (north to n. Antioquia) and w. Ecuador (south to n. Los Ríos). Below 700 m.

Tangara punctata

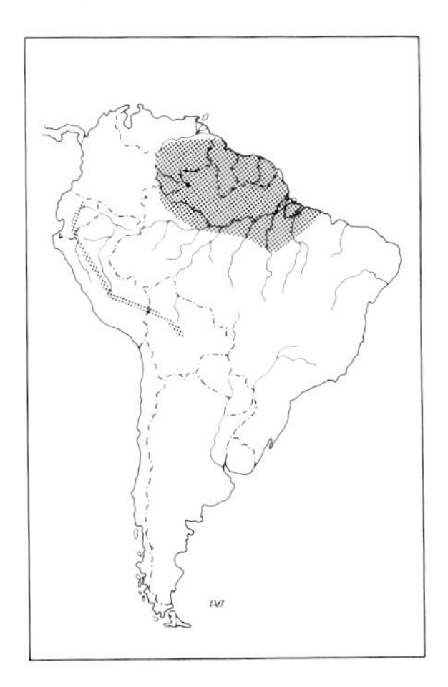

SPOTTED TANAGER

PLATE: 13

IDENTIFICATION: 13 cm (5"). *Very spotted or scaly overall.* Mostly bright green above, feathers centered black, giving scaly effect. *Face, throat, and breast bluish white* becoming white on belly, feathers throughout centered black, giving spotted effect. Lower flanks and crissum tinged yellowish green. Wings and tail black *edged yellowish green.*
SIMILAR SPECIES: Speckled Tanager is very similar but has yellower face (especially forehead, brow, and cheeks) and greenish turquoise (not yellowish green) wing-edging. The 2 are known to overlap only on the slopes of the s. Venezuelan *tepuis.* Yellow-bellied Tanager also similar but is basically green below (not bluish and white) with median belly bright yellow.
HABITAT AND BEHAVIOR: Uncommon to fairly common in humid forest canopy and borders and to a lesser extent also adjacent clearings. Favors foothill areas and apparently is most numerous in such regions (e.g., along the base of the Andes). Found in pairs or small groups, usually accompanying mixed flocks of tanagers, honeycreepers, etc. Usually remains high, searching for insects in outer foliage, coming lower mainly to obtain fruit.
RANGE: Lower slopes of Andes on e. side in e. Ecuador (north to Napo), e. Peru, and nw. Bolivia (La Paz and Cochabamba); s. Venezuela (Bolívar and Amazonas), the Guianas, and n. and e. Amaz. Brazil (south to the lower Rio Negro and e. Pará south of the Amazon). Mostly 500–1500 m (to sea level in e. Amazonia).

Tangara guttata

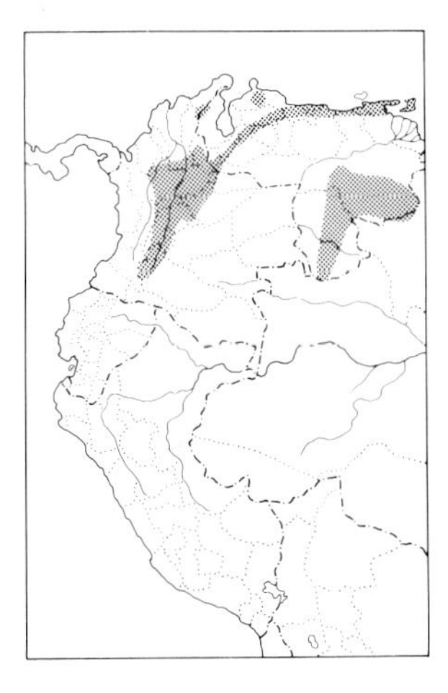

SPECKLED TANAGER

IDENTIFICATION: 13.5 cm (5¼"). Closely resembles slightly smaller Spotted Tanager. *Forehead and face* (especially the superciliary and orbital area) *tinged yellow* (not bluish), brightest in *trinitatis* of Trinidad but clearly evident in all races. *Wing feathers edged greenish turquoise* (not bright yellowish green). Scaly effect usually not as crisply defined above, but equally as or more boldly spotted below. See V-35, C-49.

HABITAT AND BEHAVIOR: Uncommon to fairly common in canopy and borders of humid forest and also in secondary woodland and adjacent clearings with large trees. Usually in pairs or small groups, most often with flocks of other tanagers. Eats mostly fruit, obtained at varying heights above ground; insects are gleaned mainly by searching underside of leaves.

RANGE: Locally in w. Colombia (both slopes of N.-Cen. Andes, south on e. slope to Tolima, e. slope of E. Andes south to Caquetá, and in Macarena Mts.) and more widely in mts. of Venezuela (Andes from Táchira to s. Lara, coastal mts. from Falcón east to Paria Peninsula, and on slopes of *tepuis* in Amazonas and Bolívar); also adjacent n. Brazil (Roraima), and a recent report from Suriname (Donahue and Pierson 1982); Trinidad. Also Costa Rica and Panama. Mostly 700–1700 m, but occurs lower in Trinidad.

Tangara xanthogastra

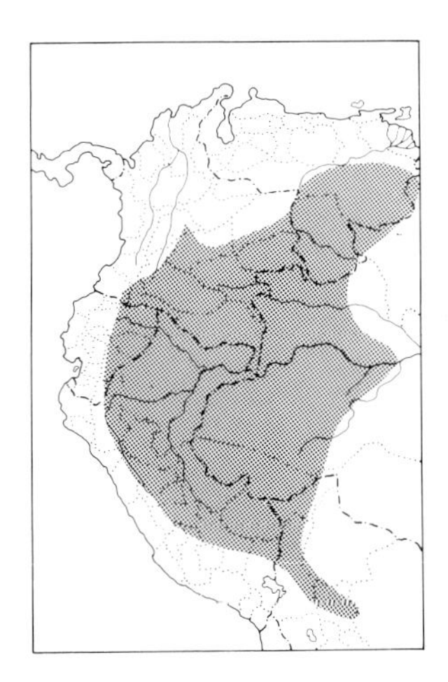

YELLOW-BELLIED TANAGER

PLATE: 13

IDENTIFICATION: 12 cm (4¾"). *Mostly bright emerald green* (slightly duller and darker above). Center of feathers from crown to back and on throat and breast centered with black, *imparting spotted or scaly appearance*. Belly unspotted, *bright yellow* with bright green on flanks. Rump green, unspotted. Wings and tail black, edged bluish green.

SIMILAR SPECIES: Spotted Tanager lacks any yellow below (its belly is mostly white) and has pale bluish ground color on face, throat, and breast (not bright green). Cf. also Speckled Tanager (only limited sympatry, on slopes of *tepuis* in s. Venezuela), which looks yellow (not green) around face but lacks yellow on belly. From below, Green-and-gold Tanager, with its yellow median underparts and green sides, can look deceptively like this species (and the 2 are often in the same flock); however, Green-and-gold lacks spotting on throat and breast and has broad black mask and yellow rump.

HABITAT AND BEHAVIOR: Fairly common in canopy and borders of humid forest (both *terra firme* and *várzea*) and in adjacent secondary woodland. Usually in pairs or small groups accompanying mixed foraging flocks of other tanagers, etc. Behavior similar to that of many other lowland forest-based *Tangara*, but seems less inclined to venture out into clearings and usually remains quite high in trees. Forages for insects by gleaning from leaves, mostly in outer foliage. Perhaps commonest on the slopes of the *tepuis* of s. Venezuela, where often one of the numerically dominant tanagers.

RANGE: Se. Colombia (north to Meta and Vaupés), e. Ecuador, e. Peru, and n. Bolivia (south to La Paz and Cochabamba) east into w. Amaz. Brazil (east to lower Rio Purús and in upper Rio Negro drainage) and s. Venezuela (Amazonas and Bolívar). To about 1500 m on *tepuis* of Venezuela, but only rarely even to 1000 m along base of Andes.

Tangara varia

DOTTED TANAGER

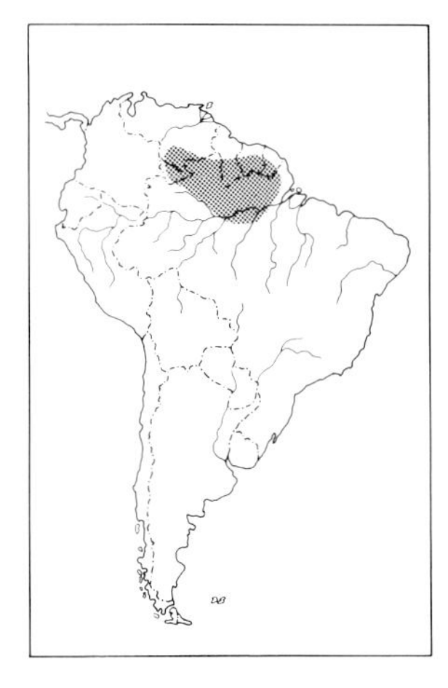

IDENTIFICATION: 11.5 cm (4½″). *Ne. South America. A small, essentially uniform bright grass green tanager.* Black bases to feathers of crown and underparts occasionally show through, resulting in vaguely speckled or dotted effect. Back tinged bluish. Wings and tail black broadly edged greenish blue (*looking uniform bluish in the field*). See V-35. Female similar but lacks male's slight dotted effect and its wing-edging is green (so effect is uniformity with rest of upperparts).

SIMILAR SPECIES: *Great caution is urged* as this scarce species is readily confused with immatures of various other small *Tangara,* some of which (e.g., Bay-headed) can also look basically uniform green. This species' small size, lack of yellow below, and in male the bluish wings and slight dotted effect are all helpful. Dotted Tanager is actually smaller than Blue Dacnis, so female can also be confused with female of that species (which has blue head).

HABITAT AND BEHAVIOR: Rare to locally uncommon in canopy and borders of humid forest and second-growth woodland; near Manaus, Brazil, especially at edge and in second-growth woodland (D. Stotz). Near Manaus noted as seen singly or in pairs, always with mixed tanager flocks.

RANGE: S. Venezuela (Amazonas and sw. Bolívar), Suriname (no confirmed recent reports) and French Guiana (presumably also in Guyana), and ne. Brazil (few records: known only from Manaus area and on lower Rio Tapajós south of the Amazon, but probably more widespread). To 300 m.

GROUP H

Predominantly blue and black Tangara (mainly Amaz.).

Tangara nigrocincta

MASKED TANAGER

PLATE: 13

Other: Black-banded Tanager

IDENTIFICATION: 13 cm (5″). *Head mostly pale lavender blue* with small black mask and pale green foreface. Back and *broad pectoral band black.* Shoulders and rump bright blue. Wings black broadly edged green; tail mainly black. Center of lower breast and belly white, flanks blue. Female similar but duller, with breast band dusky instead of black.

SIMILAR SPECIES: Replaced west of Andes by similar Golden-hooded Tanager (with mostly golden hood). Blue-necked Tanager has similar overall pattern but has head much more intense blue, golden opales-

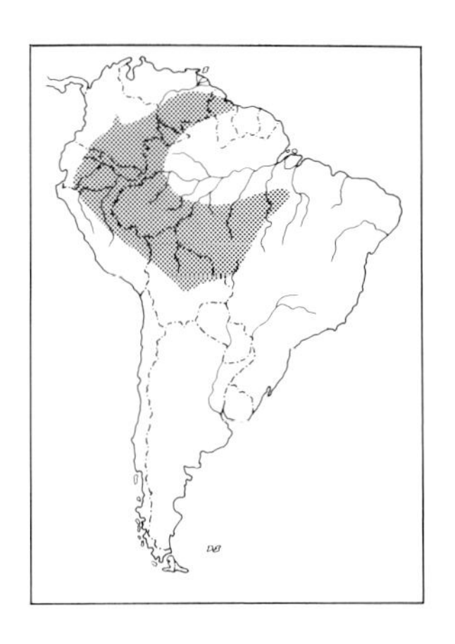

cent (not blue) shoulders, and no white below; it is mainly found at higher elevations (some overlap along e. base of Andes). In its Amaz. range might perhaps be confused with Turquoise Tanager.

HABITAT AND BEHAVIOR: Uncommon and seemingly somewhat local in borders and to a lesser extent canopy of humid forest (especially *várzea*) and in shrubby clearings. Usually in pairs, often with flocks of various tanagers, etc., but almost always greatly outnumbered by other species (e.g., Paradise and Green-and-gold). Forages mostly well above ground.

RANGE: Guyana, s. Venezuela (Bolívar and Amazonas), se. Colombia (north to Meta and e. Vichada), e. Ecuador, e. Peru, n. Bolivia (south to La Paz and nw. Santa Cruz), and w. and s. Amaz. Brazil (east to se. Pará in the Serra dos Carajás). To about 900 m.

Tangara larvata

GOLDEN-HOODED TANAGER

Other: Golden-masked Tanager

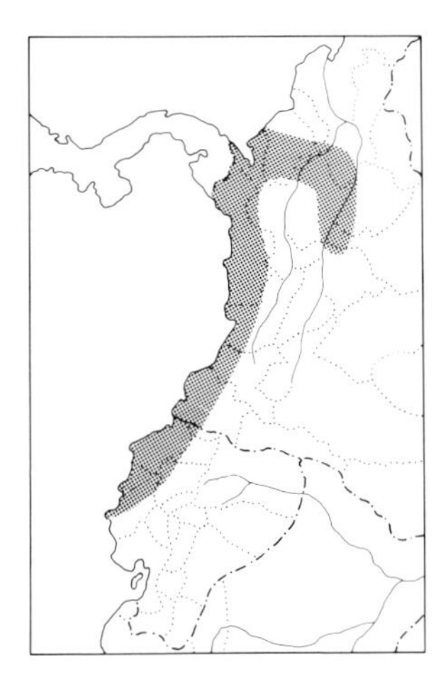

IDENTIFICATION: 13 cm (5"). *W. Colombia and w. Ecuador.* Pattern much like Masked Tanager's (of east of the Andes), but no sexual dimorphism. *Hood mostly golden,* with small black mask and violet blue foreface. Rump and wing-coverts paler, more turquoise blue. Only narrow edging on black wings and tail. See C-48, P-27.

SIMILAR SPECIES: Overlaps with somewhat similarly patterned Blue-necked Tanager in w. Ecuador; note, however, Blue-neck's deep blue hood, opalescent wing-coverts, and lack of white on lower underparts. Cf. Masked Tanager (no overlap).

HABITAT AND BEHAVIOR: Common in shrubby clearings and gardens, secondary woodland, and forest borders. Usually found in small groups, sometimes with flocks of other tanagers, etc. Forages at all levels, but principally on fruit (gleaning less than many of its congeners); often in semiopen situations, even in mostly deforested areas.

RANGE: N. and w. Colombia (entire Pacific slope and in humid lowlands north of the Andes south in Magdalena valley to n. Tolima) and nw. Ecuador (south to Manabí). Also Mexico to Panama. Mostly below 1000 m.

NOTE: We follow E. Eisenmann (*Condor* 59 [4]: 257–258, 1957), *Birds of the World* (vol. 13), and the 1983 AOU Check-list in considering trans-Andean and Middle American *T. larvata* as a species distinct from Amaz. *T. nigrocincta.* The English name "Golden-hooded," first suggested by Ridgely (1976), seems preferable to the oft-employed "Golden-masked": the hood is golden-colored, while the mask is black.

Tangara mexicana

TURQUOISE TANAGER

IDENTIFICATION: 14 cm (5½"). *Above mostly black* with cobalt blue forecrown, face, shoulders, and rump. *Below mostly cobalt blue* (black bases to feathers showing through irregularly as spotting), *but center of breast, belly, and crissum contrastingly bright yellow.* N. races (Guianas

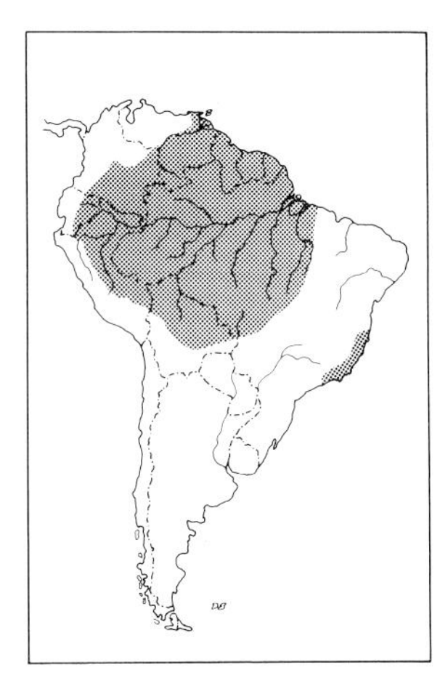

and Brazil north of the Amazon east through Venezuela to extreme e. Colombia) have median lower underparts paler, creamier yellow and bright turquoise shoulders. See V-35, C-49. *Brasiliensis* (se. Brazil) has the blue of a *paler, more silvery* hue (including shoulders) and *white median lower underparts.*

SIMILAR SPECIES: In the field often looks very dark except for the yellow or white on the belly. Cf. Opal-rumped and Opal-crowned Tanagers (both of which also often appear very dark, especially from below, when their opalescent rumps cannot be seen). Masked Tanager is smaller and has paler lavender blue hood, broad pectoral band, etc.

HABITAT AND BEHAVIOR: Fairly common to common in forest borders (especially *várzea* in Amazonia), secondary woodland, clearings with scattered trees, and plantations and gardens; basically a nonforest tanager. Usually in small groups of up to 6 or so birds which generally move independently of other birds (unlike so many other *Tangara* spp.). Feeds mostly on fruit, usually at rather high levels, and forages comparatively infrequently for insects.

RANGE: E. and s. Venezuela (north to Delta Amacuro and Sucre), Guianas and Amaz. Brazil (south to n. Goiás and n. Mato Grosso), east to se. Colombia (north to Vichada and Meta), e. Ecuador, e. Peru, and n. Bolivia (south to La Paz, Cochabamba, and Santa Cruz); coastal se. Brazil (s. Bahia south to Rio de Janeiro); Trinidad. To about 1000 m.

Tangara callophrys

OPAL-CROWNED TANAGER

PLATE: 13

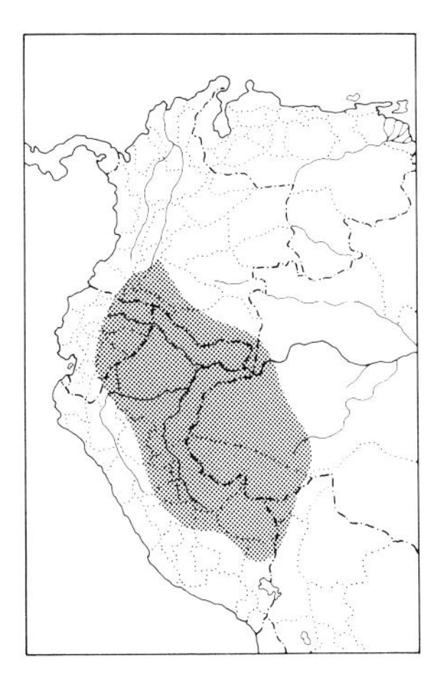

IDENTIFICATION: 14.5 cm (5¾"). *W. Amazonia.* Mostly black above with *opalescent straw forecrown and broad superciliary* as well as lower back and rump; upper tail-coverts blue. *Underparts mostly shining deep blue,* with center of lower belly and crissum black; shoulders and wing-edging the same deep blue.

SIMILAR SPECIES: Better known Opal-rumped Tanager is generally similar but races sympatric with this species lack any opalescent area on head and have chestnut (not black) median lower belly and crissum.

HABITAT AND BEHAVIOR: Uncommon in canopy and borders of humid forest (primarily *terra firme,* but also in *várzea*). Usually in pairs or small groups, often accompanying mixed flocks of other tanagers (including Opal-rumped), etc. Generally remains high, either searching branches for insects (often peering under the limbs) or coming to fruiting trees (when it may drop a little lower).

RANGE: Se. Colombia (Caquetá and Putumayo), e. Ecuador, e. Peru (south to Puno), n. Bolivia (Pando), and w. Amaz. Brazil (east to upper Rios Purús and Juruá). To about 1000 m.

Tangara velia

OPAL-RUMPED TANAGER

PLATE: 13

IDENTIFICATION: 14 cm (5½"). Mostly black above with *opalescent straw lower back and rump;* upper tail-coverts shining deep blue. Fore-

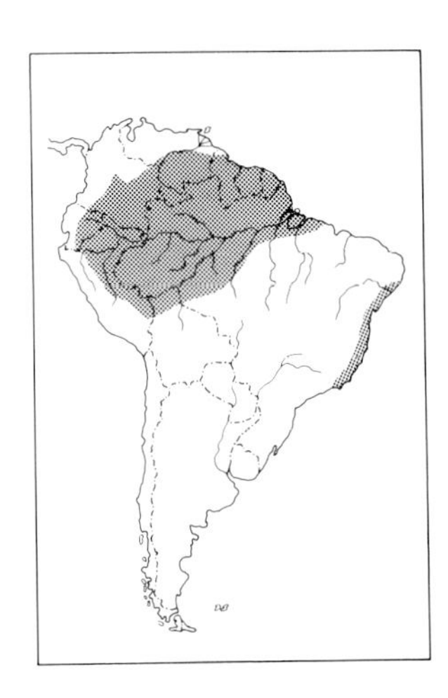

head, sides of head, and throat deep blue; *remaining underparts mostly shining purplish blue,* with *chestnut center of belly and crissum.* Shoulders shining purplish blue; otherwise wings and tail black, feathers edged blue. See C-49, V-35. *Signata* (lower Amaz. Brazil south of the Amazon) like *iridina* (most of range) but with narrow opalescent band between blue forehead and black crown, while nominate *velia* (Guianas and lower Amaz. Brazil north of the Amazon) is likewise similar to *iridina* but with paler greenish blue forehead, sides of head, wing-coverts, and wing-edging. *Cyanomelaena* (coastal e. Brazil) is, however, strikingly different, with *silvery grayish blue breast, sides, and flanks,* more pronounced black-spotted effect below (especially across chest), and fairly broad opalescent band across forehead.

SIMILAR SPECIES: Opal-crowned Tanager (with which this species overlaps in w. Amazonia) has prominent opalescent crown and eyebrow and black (not chestnut) median lower belly and crissum. Pattern of se. Brazilian *cyanomela* recalls that of sympatric race of Turquoise Tanager, but latter has silvery blue (not opalescent) rump and large white area on median lower underparts.

HABITAT AND BEHAVIOR: Uncommon to fairly common in canopy and borders of humid forest (both *terra firme* and *várzea*) and to a lesser extent in adjacent clearings with scattered tall trees and in more advanced secondary growth. Usually in pairs or small groups, frequently with mixed flocks of other tanagers, etc. Like the Opal-crowned, basically a forest tanager, rarely coming low or straying far from forest cover, foraging for insects primarily by searching along larger horizontal limbs.

RANGE: Se. Colombia (north to Meta and e. Vichada), s. Venezuela (Amazonas and Bolívar), and Guianas south to e. Ecuador, e. Peru, n. Bolivia (Beni), and Amaz. Brazil (south to nw. Mato Grosso and e. Pará in the Belém area); coastal e. Brazil (Pernambuco south to Rio de Janeiro). Mostly below 500 m (locally recorded somewhat higher, as in Venezuela).

NOTE: *Cyanomelaena* may well represent a distinct species.

GROUP I

Multicolored Tangara; all but 1 species *mainly e. Brazil.*

Tangara chilensis

PARADISE TANAGER

PLATE: 13

IDENTIFICATION: 14 cm (5½"). Unmistakable with multicolored (but to some excessively gaudy) pattern. *Head brilliant yellowish apple green* (feathers somewhat stiffened and scalelike), contrasting with black eye-ring. Otherwise mostly black above with *bright scarlet lower back and rump.* Throat violet; *remaining underparts deep turquoise blue,* with black center of lower belly and crissum. Shoulders also turquoise blue; wing otherwise mainly black. Ne. races (west to Meta, Colombia, and the Rio Negro drainage in nw. Brazil) and *chlorocorys* (of the upper Huallaga valley in e. Peru) have *upper back flame red but rump bright yellow.* See V-35, C-49.

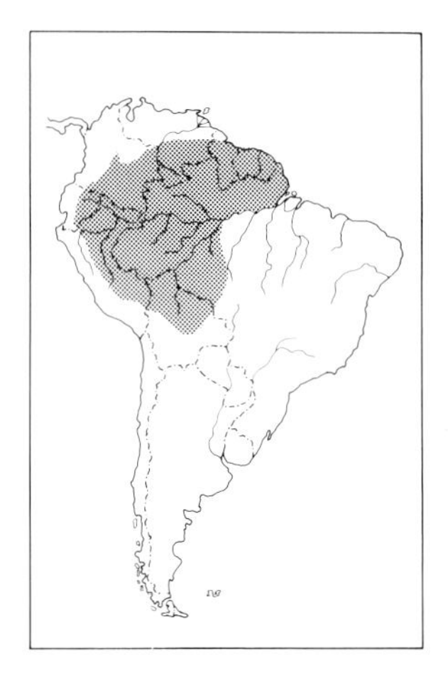

HABITAT AND BEHAVIOR: Common in canopy and borders of humid forest (both *terra firme* and *várzea*) and in trees and shrubs of adjacent clearings. Generally moves about in rather large groups (up to a dozen or more birds) which often appear to be one of the nuclear species around which canopy mixed flocks form, but also wanders about independent of other birds. Though essentially a forest species, where usually remains in upper strata, Paradise Tanagers also fairly often strike out across fairly large clearings, perhaps pausing partway in an isolated tree or shrub, sometimes even remaining for some time if a productive fruiting tree is located.
RANGE: Guianas, e. Venezuela (Bolívar and Amazonas), and se. Colombia (north to Guainía and Meta), e. Ecuador, e. Peru, n. Bolivia (south to La Paz, Cochabamba, and w. Santa Cruz), and w. Amaz. Brazil (east to the Manaus area and nw. Mato Grosso). To about 1500 m.

Tangara seledon

GREEN-HEADED TANAGER

PLATE: 13

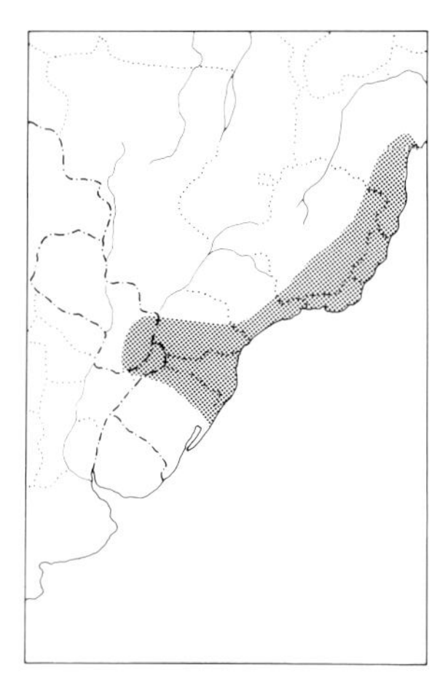

IDENTIFICATION: 13.5 cm (5¼"). *Se. South America. Head and chin bluish green to turquoise,* becoming shining yellowish green on upper back and sides of neck. *Scapulars and middle of back black;* rump bright yellow-orange; upper tail-coverts bright green. Wing-coverts purplish blue; flight feathers black, broadly edged bright green. *Lower throat black; breast and belly turquoise blue,* becoming bright green on flanks and crissum. Female more dully colored, but pattern same as male's.
SIMILAR SPECIES: This lovely tanager's complex and colorful pattern renders it almost unmistakable; no other tanager in its range shares the blue-green head, black mid-back and throat, and orange rump.
HABITAT AND BEHAVIOR: Common in humid forest and forest borders, also coming out into secondary woodland, shrubby clearings, and gardens, but absent from the large expanses of recently deforested land over much of its former range. Thus while it is still numerous locally (and seems to persist well in partially deforested areas), overall numbers have surely declined substantially. Behavior similar to Paradise Tanager's: frequently in quite large groups (up to 10 or more birds) which are often nuclear in a mixed flock, foraging at all levels but mostly high (coming lower only at edge and in clearings).
RANGE: Se. Brazil (s. Bahia south to n. Rio Grande do Sul), ne. Argentina (Misiones), and e. Paraguay (Canendiyu, Alto Paraná, and Itapúa). To about 1300 m.

Tangara fastuosa

SEVEN-COLORED TANAGER

IDENTIFICATION: 13.5 cm (5¼"). *Ne. Brazil.* Pattern resembles Green-headed Tanager's. *Lower underparts dark ultramarine blue* (not turquoise and green). Lacks yellowish green on upper back and sides of neck. Wing-coverts pale turquoise blue (not violet blue). Bill quite notably heavier.
SIMILAR SPECIES: Nothing really similar in its limited range.

HABITAT AND BEHAVIOR: Little known, though it was illustrated as early as 1648 by Marcgrave. Locally fairly common in canopy and borders of humid lowland and foothill forest. Behavior much like Green-headed Tanager's, typically foraging in pairs or small groups, frequently with mixed flocks. Because of very extensive deforestation over most of its small range, the Seven-colored Tanager (actually there seem to be only 5 or 6) is considered to be officially "vulnerable," but no detailed information on its actual status is available. The species is avidly sought as a cage bird, and this practice is also thought to have depleted its numbers. Substantial numbers remain in the still quite extensive forests of the Serra Branca near Murici, Alagoas (as of late 1987).

RANGE: Coastal ne. Brazil (e. Pernambuco and Alagoas). To at least 550 m.

Tangara cyanocephala

RED-NECKED TANAGER

PLATE: 13

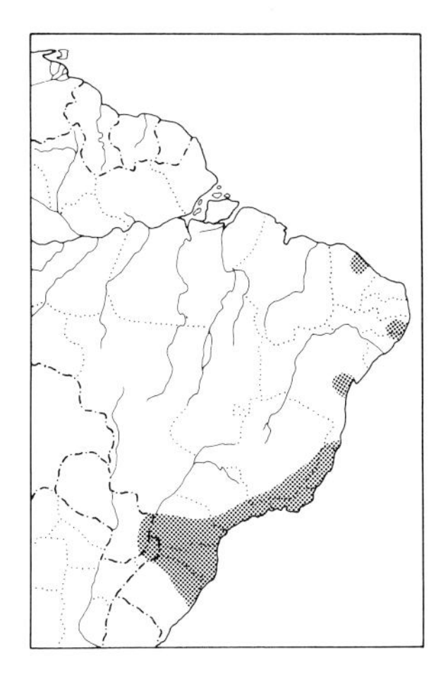

IDENTIFICATION: 13 cm (5"). *Mainly e. Brazil.* Forehead black and *crown violet blue,* with narrow pale blue band separating them and around eye. *Cheeks and broad nuchal collar scarlet.* Scapulars and back black, lower back and rump bright green. Wings and tail black broadly edged green; lesser wing-coverts tipped orange-yellow, forming narrow band. Throat violet blue, remaining underparts bright green. Female duller, but pattern much the same as male's; back green mottled dusky.

SIMILAR SPECIES: This species' red neck band (prominent even in females and immatures) is unique among the *Tangara* tanagers.

HABITAT AND BEHAVIOR: Locally fairly common in humid montane forest and forest borders. Usually in pairs or small groups, and often with mixed flocks of other tanagers, etc. Forages actively, mostly remaining quite high (especially when gleaning for insects among smaller branches), at times coming lower to fruiting trees. This stunning tanager is readily seen at the Nôva Lombardia Reserve above Santa Teresa in Espírito Santo, Brazil; smaller numbers occur in the Tijuca forests above Rio itself.

RANGE: E. Brazil (n. Ceará, Pernambuco, e. Bahia, and Espírito Santo south to n. Rio Grande do Sul), ne. Argentina (few or no recent records), and e. Paraguay (Canendiyu, and probably also Alto Paraná). Mostly 400–1000 m.

Tangara desmaresti

BRASSY-BREASTED TANAGER

PLATE: 13

IDENTIFICATION: 14 cm (5½"). *Se. Brazil. Above shining emerald green,* streaked black except on sides of head and rump. Forehead black; *forecrown and eye-ring turquoise blue.* Small patch on center of throat black; *breast brassy ochre.* Belly mostly bluish green, center of lower belly yellow. Shoulder brassy ochre; wings and tail otherwise black, feathers broadly edged bright green.

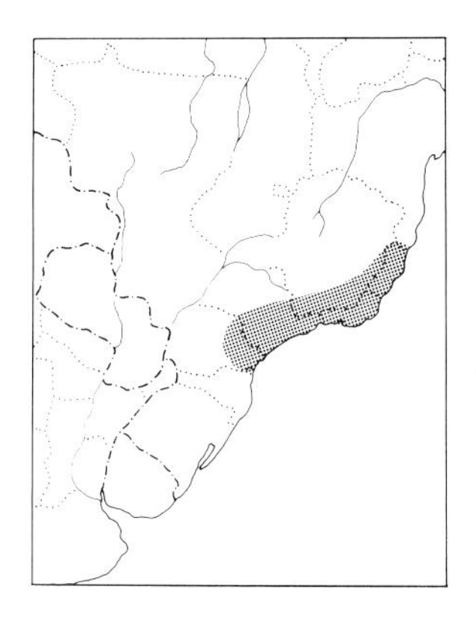

SIMILAR SPECIES: Gilt-edged Tanager is similarly patterned but is golden above (not green) and mostly blue below (including the breast).

HABITAT AND BEHAVIOR: Locally common in humid montane forest and borders, to a lesser extent also adjacent secondary woodland. Behavior much like that of many other forest-based *Tangara* tanagers. As this tanager favors montane forests, it probably has not decreased as much as have many of the se. Brazil endemics found in the lowlands. Common at Itatiaia Nat. Park, where it occurs at elevations above those of Gilt-edged.

RANGE: Se. Brazil (Espírito Santo south to e. Paraná). Mostly 500–1800 m.

Tangara cyanoventris

GILT-EDGED TANAGER

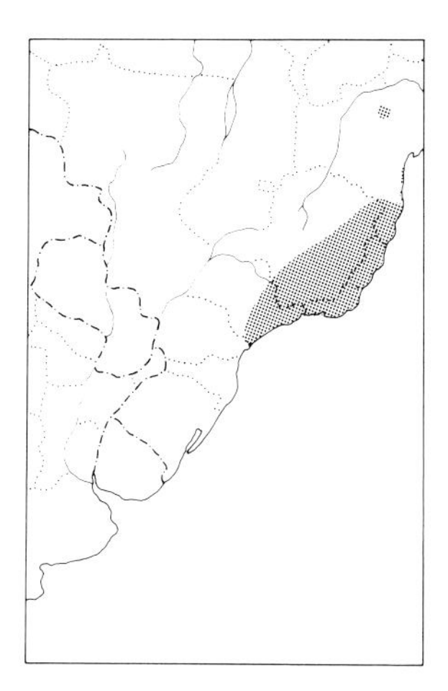

IDENTIFICATION: 13.5 cm (5¼"). *Se. Brazil.* Pattern similar to Brassy-breasted's, but *above mostly golden yellow, brightest on head and especially spectacles,* streaked with black (most broadly on back, least on face). Forehead and throat black. *Remaining underparts mostly shining turquoise blue,* center of lower belly green, and crissum yellow. Wings and tail black, feathers broadly edged bright green.

SIMILAR SPECIES: Brassy-breasted Tanager is mostly green above (not golden) and has its "brassy" (not blue) breast.

HABITAT AND BEHAVIOR: Locally common in humid montane forest and forest borders; in smaller numbers in secondary woodland and clearings with scattered trees. Behavior much like that of many other forest-based *Tangara* tanagers. Common at the Nôva Lombardia Reserve above Santa Teresa, Espírito Santo; present in smaller numbers at lower elevations of Itatiaia Nat. Park.

RANGE: Se. Brazil (s. Bahia and se. Minas Gerais south to e. São Paulo). Mostly 400–1000 m.

Euphonia Euphonias

Euphonias are small, short-tailed tanagers with stubby bills. There is strong sexual dimorphism, with males typically being steely blue above (often with yellow forecrown) and yellow below (with or without a dark throat); females are duller, typically olive and yellow. A few species diverge from this "standard" pattern and are enumerated separately below. Identification points are often subtle, both in males and females. Euphonias are arboreal, with most species being found in humid forest, a few out into clearings and savannas; they tend to be quite conspicuous, moving about in pairs or small groups which often perch in the open. Their various calls tend not to be distinctive enough for identification but are far-carrying and do often attract one's attention. Primarily frugivorous, euphonias are unusual in that they consume large quantities of mistletoe berries. Their nests are domed with side entrances, a construction unique among the tanagers except for the chlorophonias.

GROUP A

"Typical" euphonias with *throat entirely or mostly yellow.*

Euphonia laniirostris

THICK-BILLED EUPHONIA

PLATE: 14

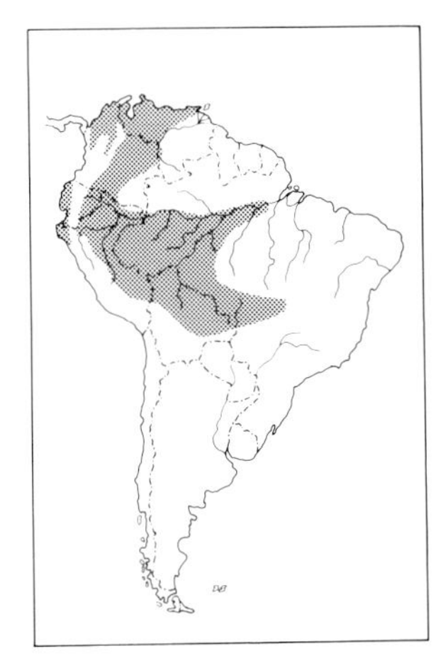

IDENTIFICATION: 11.5 cm (4½"). Bill slightly thicker than other similar euphonias', but this is of limited value in the field. Above mostly glossy steel blue; forecrown (to just behind eye) yellow. *Entire underparts (including throat) bright yellow to ochraceous yellow. Melanura* (w. Amazonia) has all-black tail, lacking the white tail spots found in all other races. *Hypoxantha* (w. Ecuador and nw. Peru) has more extensive yellow crown patch. Female olive above, yellow below, clouded olive across breast and on sides. Also frequently seen are immature males, which resemble females but may have yellow forecrown and blue-black face.

SIMILAR SPECIES: This and the Violaceous Euphonia are the *only male euphonias with all-yellow underparts.* Female resembles female Orange-crowned and Velvet-fronted Euphonias but is larger with thicker bill than either. Female closely resembles female Violaceous (but the 2 are largely allopatric).

HABITAT AND BEHAVIOR: Common and widespread in forest borders, secondary woodland and scrub, clearings and gardens, and even mostly agricultural areas with scattered trees; occurs in both humid and arid areas. Usually in pairs or small groups and often with mixed flocks of other tanagers (including other euphonias). Forages at all levels, but most often rather high. Both sexes (but especially males) frequently mimic vocalizations of other birds (especially their contact and alarm notes), with these being interspersed with their own varied repertoire of musical and chattering notes. Both sexes also often give a musical "chweet" or "peem-peem."

RANGE: Colombia (Caribbean lowlands and Cauca and Magdalena valleys, and widely east of the Andes, but absent from Pacific lowlands south of n. Chocó), n. and w. Venezuela (east to Sucre and south about to the Orinoco River), w. and e. Ecuador (on Pacific slope from Esmeraldas south), nw. and e. Peru (south on Pacific slope to sw. Cajamarca), n. Bolivia (south to La Paz, Cochabamba, and Santa Cruz), and w. and cen. Amaz. Brazil (east along Amazon to lower Rio Tapajós and around Santarém, and in w. Mato Grosso south to the upper Paraguay River drainage). Also Costa Rica and Panama. Mostly below 1200 m, occasionally higher.

Euphonia violacea

VIOLACEOUS EUPHONIA

IDENTIFICATION: 11.5 cm (4½"). Resembles Thick-billed Euphonia, which it *replaces* (with limited overlap) *in e. South America.* Male richer and more ochraceous below (especially on throat and breast). Best distinction is *smaller extent of yellow crown patch* (reaching only to di-

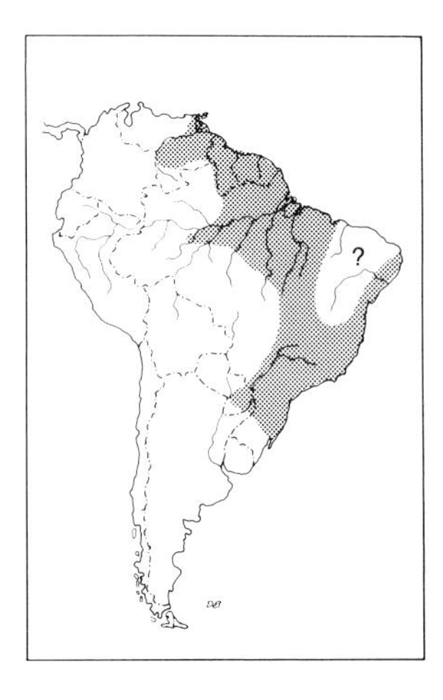

rectly above eye). White tail spots always present, unlike *melanura* race of Thick-billed, which helps to differentiate the 2 along the middle Amazon. See V-36. Female probably indistinguishable in the field from female Thick-billed; go by range.

SIMILAR SPECIES: Male Green-chinned Euphonia is rather similar but has upper throat greenish blue and underparts purer yellow. In most of range female is *only euphonia with yellow and olive underparts;* in ne. South America easily confused with female Finsch's, though latter is smaller with notably slimmer bill.

HABITAT AND BEHAVIOR: Fairly common to common and widespread in forest borders, gallery woodland, and clearings and gardens; seems to avoid (or is less numerous in) more arid open areas. General behavior similar to Thick-billed Euphonia's, but male is an even better and more frequent mimic, with a repertoire that extends to an amazingly wide variety of species (even hawks, parrots, toucans, and jays).

RANGE: E. and s. Venezuela (Sucre south to Bolívar and n. Amazonas), Guianas, lower Amaz. and most of e. Brazil (in Amazonia west to lower Rio Madeira and Mato Grosso; apparently absent from *caatinga* zone of northeast; south to n. Rio Grande do Sul), e. Paraguay, and ne. Argentina (Misiones); Trinidad and Tobago. To about 1000 m.

Euphonia chalybea

GREEN-CHINNED EUPHONIA

PLATE: 14

Other: Green-throated Euphonia

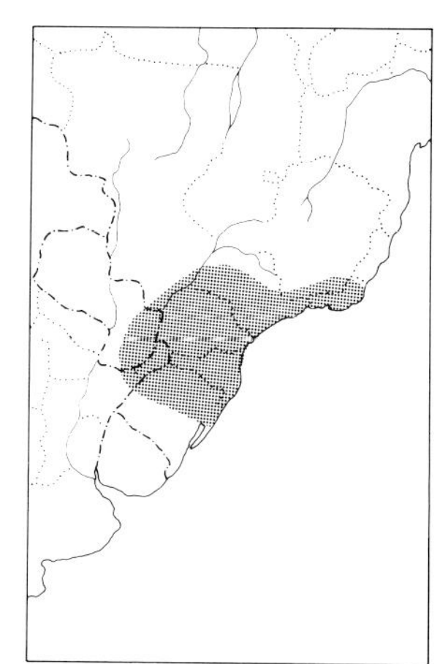

IDENTIFICATION: 11.5 cm (4½"). *Se. South America.* Bill very thick (more so than Thick-billed's). *Upperparts steely greenish blue, extending to sides of head and chin.* Small band on forehead yellow, as are *entire underparts* (except for chin). No white in tail. Female olive above and *mostly gray below,* extending up to sides of neck. *Chin, sides, flanks, and crissum olive yellowish.*

SIMILAR SPECIES: Male most likely confused with male Violaceous Euphonia (and often occurs with it). Male differs in the dark chin, greener tone to upperparts, and lighter yellow underparts. Female Violaceous is olive and yellow below (no gray); female Green-chinned also easily taken for female Chestnut-bellied Euphonia, but latter has chestnut crissum.

HABITAT AND BEHAVIOR: Uncommon at forest borders and in clearings with large trees. Usually in pairs, and often with forest-based flocks of other tanagers. Not well known, but can be seen at Iguaçu Falls, particularly on the Brazilian side. Has doubtless declined substantially due to forest destruction over much of its range.

RANGE: Se. Brazil (Rio de Janeiro south to Rio Grande do Sul), e. Paraguay, and ne. Argentina (Misiones and Corrientes). To about 500 m.

NOTE: Previously called the "Green-throated Euphonia." From that name one would assume that its pattern would be much like that of the many other dark-throated euphonias and that its entire throat is green. In fact, only the chin and sides of the head are dark (most of the throat is actually yellow), hence our clarifying name change.

GROUP B "Typical" euphonias with *dark throat.*

Euphonia minuta

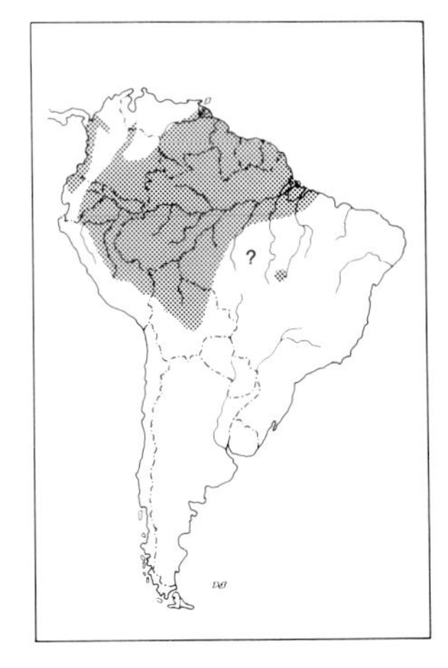

WHITE-VENTED EUPHONIA

PLATE: 14

IDENTIFICATION: 9.5 cm (3¾"). Above steely blue-black with *only forehead yellow* (extends to just before eye). Throat black glossed purplish; remaining underparts bright yellow, with *extreme lower belly and crissum white.* Female olive above. Throat grayish white; *broad breast band, sides, and flanks olive yellow; center of belly and crissum whitish.*

SIMILAR SPECIES: Both sexes unique among euphonias (except male of very different Tawny-capped) in having white on crissum. Female additionally known by her yellowish pectoral band; in other female euphonias with which she might be confused underparts are mostly gray with yellow or olive confined to sides (not extending across breast).

HABITAT AND BEHAVIOR: Uncommon to fairly common in forest canopy and borders (in Amazonia both *terra firme* and *várzea*) and to a lesser extent also secondary woodland. Infrequent or absent in cleared or semiopen areas, and in general a relatively unfamiliar euphonia due to its usually remaining high up. Behavior otherwise much like many other euphonias'. Usual call is a single "peem," most often *not* doubled.

RANGE: N. and w. Colombia (Pacific lowlands and forested lowlands along n. base of Andes south in Magdalena valley to e. Antioquia) and nw. Ecuador (south to Pichincha); Guianas, s. Venezuela (mostly south of the Orinoco in Amazonas and Bolívar; also s. Táchira along e. base of Andes), e. Colombia (except the llanos region of the northeast), e. Ecuador, e. Peru, n. Bolivia (south to La Paz, Cochabamba, and nw. Santa Cruz), and Amaz. Brazil (south to nw. Mato Grosso and e. Pará in the Belém region; 1 record also from ne. Mato Grosso). Also Mexico to Panama. To about 1000 m.

Euphonia chlorotica

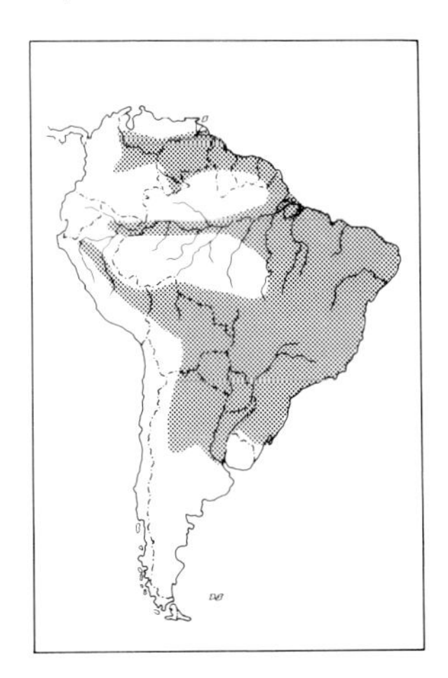

PURPLE-THROATED EUPHONIA

PLATE: 14

IDENTIFICATION: 10 cm (4"). Above steely blue-black, purpler on head; forecrown (back to just behind eye) yellow. Throat black glossed purple, remaining underparts bright yellow. Female olive above with yellower forecrown. *Mostly grayish white below;* sides, flanks, and crissum greenish yellow. Immatures, and apparently all adult females of n. race *cyanophora* (ne. Colombia, s. Venezuela, extreme n. Brazil), are *entirely greenish yellow below.*

SIMILAR SPECIES: Cf. confusingly similar Trinidad Euphonia. Also closely resembles Orange-bellied Euphonia. Orange-bellied male is more ochraceous below and on crown patch, while female Orange-bellies have deeper yellow forehead, gray nape, and strong buff tinge below. White-vented Euphonia male has smaller yellow forecrown patch and white on lower underparts, while female is much more broadly olive yellow below. Female Thick-billed and Violaceous Euphonias are similar to female *cyanophora* but are larger with heavier bills.

HABITAT AND BEHAVIOR: Fairly common and widespread in forest borders and adjacent clearings, gallery woodland, and *caatinga* and *chaco* woodland and scrub. Usually outnumbers Violaceous Euphonia in drier areas, with the converse in more humid regions, but both species are widely syntopic. Usually in pairs or small groups, sometimes with mixed flocks, but as often away from them. Frequently utters a clear, far-carrying "beem-beem."
RANGE: Ne. Colombia (south to Meta and Vichada), s. Venezuela (s. Táchira and Apure east across Amazonas and Bolívar), Guianas (few or no recent records except from Guyana), Amaz. and most of e. Brazil (up the Amazon to the Iquitos, Peru, area), e. Peru (north to Cajamarca, mostly along base of Andes), n. and e. Bolivia, Paraguay, and n. Argentina (south to La Rioja, Córdoba, and Entre Ríos). Locally to over 2000 m, but mostly below 1200 m.

Euphonia trinitatis

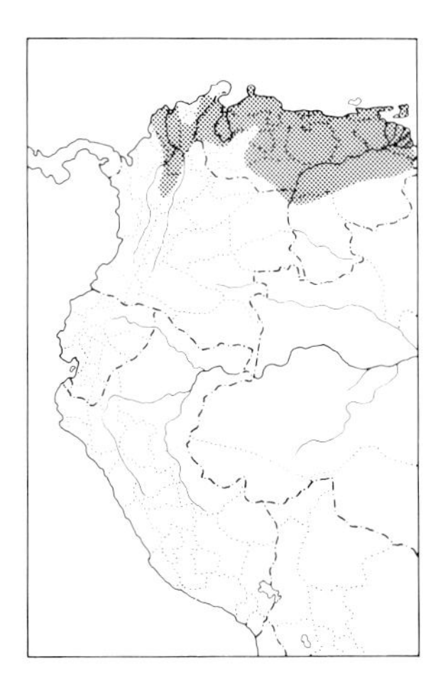

TRINIDAD EUPHONIA

IDENTIFICATION: 9.5 cm (3¾"). *Venezuela and n. Colombia.* Closely resembles slightly larger Purple-throated Euphonia (and perhaps conspecific, though the 2 seem to overlap in s. Venezuela). See V-36, C-53. Male virtually identical, but with slightly more extensive crown patch (extending well back over eye) and bluer gloss on throat. Females easier, because comparison need only be made with *cyanophora* race of Purple-throated, in which underparts are entirely greenish yellow. Thus the mostly pale gray underparts of this species (yellowish olive confined to sides and crissum) are normally distinctive, but one must beware immature Trinidads (which also are all greenish yellow below). In short, best to go by range.
HABITAT AND BEHAVIOR: Uncommon to fairly common (most numerous in arid areas) in deciduous and gallery woodland, clearings with secondary growth and scattered trees, gardens, and arid scrub; in more humid regions also forest borders. Usually in pairs or small groups, with behavior much like other euphonias'. Typical "beem-beem" song is like Purple-throated's.
RANGE: Caribbean n. Colombia (Atlántico east to s. base of Guajira Peninsula and south to middle Magdalena valley in e. Antioquia) and n. Venezuela (south to Apure, n. Amazonas, and n. Bolívar); Trinidad. Mostly below 500 m, occasionally higher.

NOTE: It is difficult to justify maintaining this as a species distinct from *T. chlorotica.* Hellmayr considered it a subspecies (*Birds of the Americas,* vol. 13, part 9), but then J. T. Zimmer (*Am. Mus. Novitates* 1225: 9–12, 1943), with reservations, regarded it as a full species because *trinitatis* and *chlorotica cyanophora* seemed to have been collected at the same localities in Venezuela's lower Orinoco valley. He added, however, that "I have confidence that . . . *trinitatis* is not specifically distinct from *chlorotica.*" All subsequent treatments have listed the 2 as full species. We are not aware of any new evidence on this point and thus reluctantly follow recent authors, but like Zimmer we suspect they may eventually be shown to be conspecific.

Euphonia concinna

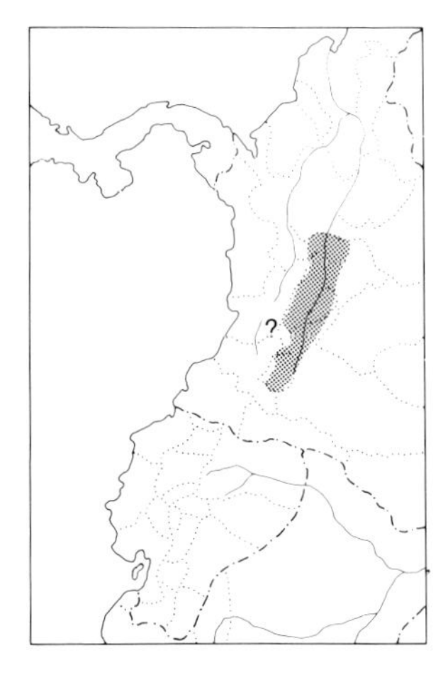

VELVET-FRONTED EUPHONIA

IDENTIFICATION: 10 cm (4″). *Mainly upper Magdalena valley, Colombia.* Male *closely* resembles male Purple-throated and Trinidad Euphonias; *black frontlet is very narrow* and virtually invisible under field conditions. *Forecrown to just above eye yellow.* Underparts yellow washed with ochraceous (not pure yellow). *No white on tail.* See C-53. Female dull olive above, grayer on nape, and with *yellowish forecrown; dull yellow below.*

SIMILAR SPECIES: Best to go by range with this scarce, local species. Does not occur with either Purple-throated or Trinidad Euphonia. Male also closely resembles male Orange-bellied Euphonia (potentially sympatric races of latter have yellow forecrown only faintly tinged ochraceous). Note, further, male Velvet-front's absence of white in tail, but probably easier to go by accompanying females (which have very different underparts, yellow in Velvet-fronted, buffy grayish in Orange-bellied); typical habitats also differ (this more in open country, Orange-bellied forest-based). Female very similar to female Thick-billed Euphonia but is smaller and seems to consistently show yellowish forecrown (lacking in Thick-bill).

HABITAT AND BEHAVIOR: Uncommon in woodland and agricultural areas with scattered trees, primarily in arid semiopen parts of Magdalena valley. Behavior resembles that of other nonforest euphonias.

RANGE: W. Colombia in upper Magdalena valley of Tolima, w. Cundinamarca, and Huila (also 1 report from upper Cauca valley in Valle). Mostly 200–1000 m, locally higher.

Euphonia xanthogaster

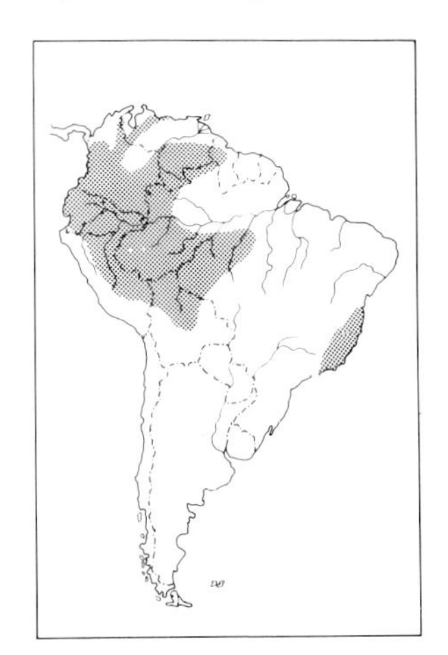

ORANGE-BELLIED EUPHONIA

PLATE: 14

IDENTIFICATION: 11 cm (4¼″). Above mostly steel blue, with blacker sides of head and nape; forecrown (to behind eye) yellow tinged ochraceous. Throat black tinged purple; remaining *underparts yellow tinged ochraceous. Brunneifrons* (se. Peru) has rufous crown patch and stronger ochraceous tinge to underparts, while in *ruficeps* (Bolivia) and *badissima* and *exsul* (ne. Colombia and n. Venezuela) *crown patch chestnut* and median underparts rich ochraceous. Female mostly olive above, with forehead tinged tawny-yellowish and *nape gray. Mostly buffy grayish below,* buffiest on belly, with sides and flanks yellowish olive. Females of the races named above have foreheads dull rufous to chestnut.

SIMILAR SPECIES: Male very similar to male Purple-throated Euphonia, but all races are more ochraceous below (some deeply so; none are bright pure yellow) and on crown; females of the 2 are notably different, Orange-bellied being buffier and dingier below and with characteristic gray nape and tawnier forehead. Orange-crowned Euphonia male is also similar but has notably larger crown patch and deeper and more ochraceous underparts than sympatric Orange-bellieds. Female closest to female Rufous-bellied Euphonia but with more ochraceous forehead and dingier underparts with buffy yellowish (not tawny) crissum; often best to check attending males.

HABITAT AND BEHAVIOR: Common and widespread in forest, forest borders, and secondary woodland, only occasionally far out into clearings. *The commonest euphonia at subtropical elevations,* and also locally numerous in many humid lowland areas. Usually in pairs or small groups, foraging at all levels but usually not very high, and also not infrequently inside forest at lower levels (where normally is the only euphonia present). Frequent calls are a "dee-dee-deét" and a nasal "cheea-cheea-cheea."

RANGE: Humid lowlands and subtropical zone in n. Guyana (locally), Venezuela (Andes and n. coastal mts. east to Distrito Federal; also slopes of *tepuis* in Amazonas and s. Bolívar), Colombia (widespread in forested lowlands east of the Andes, on Andean slopes, and entire Pacific slope; not Santa Marta Mts.), Ecuador, nw. (Tumbes) and e. Peru, nw. Bolivia (La Paz and Cochabamba), and locally in w. Amaz. and se. (s. Bahia to Rio de Janeiro) Brazil. Mostly below 2000 m.

Euphonia anneae

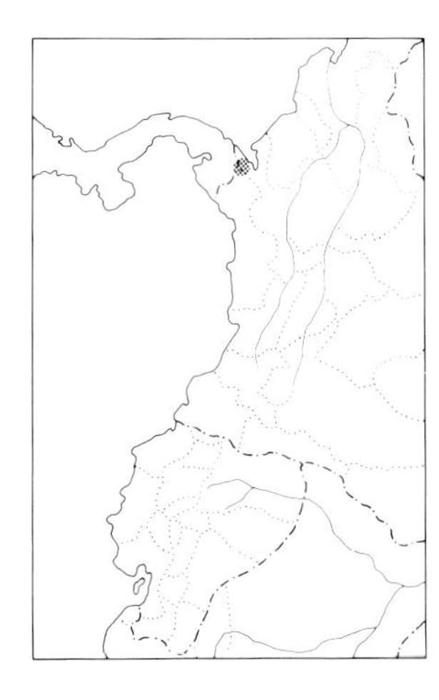

TAWNY-CAPPED EUPHONIA

IDENTIFICATION: 11 cm (4¼"). *Nw. Colombia.* Resembles Orange-bellied Euphonia (no known overlap), especially races of that species with chestnut forecrowns (e.g., *ruficeps*). However, *entire crown is rufous* (not just forecrown) and crissum is white (not yellow). See P-27. Female is *notably dark and dingy,* rather grayish below, with *dull brownish to rufous forecrown.*

SIMILAR SPECIES: Male is the only euphonia with *fully rufous crown.* Both sexes of Fulvous-vented Euphonia have rufous on lower underparts (especially crissum).

HABITAT AND BEHAVIOR: Not well known in the S. Am. portion of its range. In Middle America, where in appropriate humid forest and forest border habitat it is fairly common, its behavior is similar to that of Orange-bellied Euphonia. Like that species it forages primarily in forest understory, often accompanying mixed flocks, coming to edge primarily to feed at fruiting shrubs and low trees.

RANGE: Nw. Colombia (nw. Chocó on e. slopes of Cerro Tacarcuna on Panama border). Also Costa Rica and Panama. In Middle America mostly 600–1300 m.

Euphonia saturata

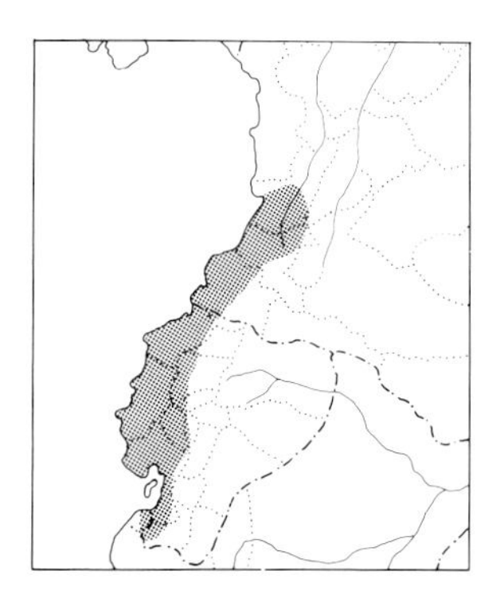

ORANGE-CROWNED EUPHONIA

PLATE: 14

IDENTIFICATION: 10 cm (4"). *Mainly sw. Colombia and w. Ecuador. Entire crown deep yellow-orange.* Otherwise glossy dark violaceous above and on throat. *Remaining underparts deep orange-ochraceous. No white in tail.* Female very plain, olive above and yellow below, strongly shaded olive on sides, flanks, and crissum.

SIMILAR SPECIES: In its range handsome, richly colored male likely confused only with sympatric races of commoner Orange-bellied Euphonia, which have smaller and yellower crown patch, are not nearly as deeply ochraceous below (only a tinge), and have white spots in tail. Female very close to female Thick-billed; note its smaller size and thinner bill.

HABITAT AND BEHAVIOR: Uncommon to rare and seemingly local, but recorded from a variety of humid to fairly dry habitats, ranging from humid forest borders to deciduous woodland, clearings, and gardens. Usually in pairs, often with mixed flocks and regularly with other euphonias (but almost always outnumbered by them); usually forages well above ground. Call is a simple "peem-peem," usually doubled like that of so many other euphonias.
RANGE: Sw. Colombia (both slopes of W. Andes from Valle south to Nariño; also locally in upper Cauca valley), w. Ecuador (Esmeraldas to El Oro), and extreme nw. Peru (Tumbes). Mostly below 1000 m.

Euphonia finschi

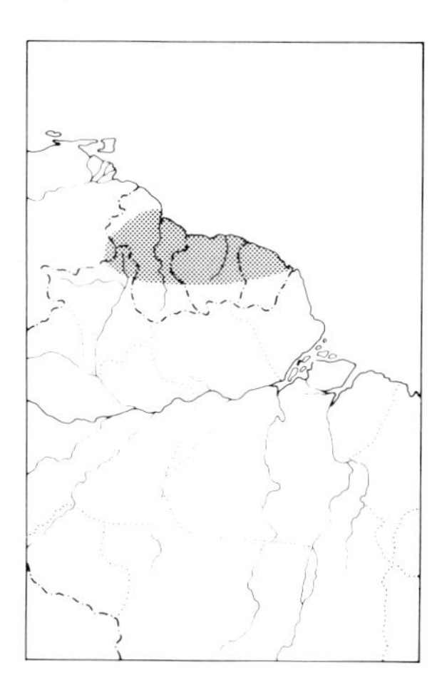

FINSCH'S EUPHONIA

IDENTIFICATION: 10 cm (4"). *Ne. South America.* Forecrown (to just above eyes) yellow; otherwise steel violet above and on throat. Chest yellow, merging into *deep, rich, ochraceous lower underparts.* No white in tail. Female olive above with more yellowish forecrown; olive yellow below, clearest yellow on center of belly.
SIMILAR SPECIES: Pattern of male suggests Purple-throated Euphonia, but much more richly colored below (not pure yellow) and lacking white in tail. Female *very* similar to female Violaceous Euphonia but slightly smaller with slenderer bill. That it frequently occurs in pairs will thus be a help.
HABITAT AND BEHAVIOR: Uncommon in forest and forest borders; in Suriname seems to occur primarily in coastal swampy forest and in gallery forests in savanna areas. Not well known. "Peem-peem" call is much like Violaceous Euphonia's (D. Stotz).
RANGE: Guianas, extreme se. Venezuela (s. Bolívar), and extreme n. Brazil (n. Roraima). To about 1200 m.

Euphonia fulvicrissa

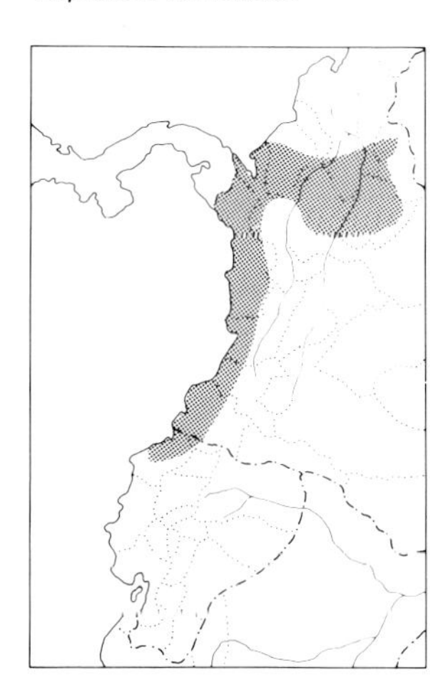

FULVOUS-VENTED EUPHONIA

IDENTIFICATION: 10 cm (4"). *W. Colombia and nw. Ecuador.* Forecrown (to just above eyes) yellow; otherwise steel blue to violet blue above and on throat. Remaining underparts mostly yellow, with *center of belly and crissum tawny-fulvous. Purpurascens* (Nariño, Colombia, and Ecuador) lacks white tail spots, while in *omissa* (rest of Colombian range except nw. Chocó) these are very small. Female mostly olive above, but often tinged grayish blue on nape and with *rufous forecrown.* Yellowish olive below, with *center of belly and crissum tawny.* See C-53, P-27.
SIMILAR SPECIES: In its range both sexes readily recognized by the fulvous stripe on lower underparts and crissum.
HABITAT AND BEHAVIOR: Fairly common in humid forest and borders, secondary woodland, and shrubby regenerating clearings. Usually in pairs or small groups, often keeping fairly low (especially at edge but also inside forest, where sometimes accompanies *Myrmotherula* antwren flocks). Usual call is a gravelly chattering "treah-treah," frequently heard.

RANGE: N. and w. Colombia (Pacific slope and forested lowlands along n. base of Andes south in Magdalena valley to e. Antioquia and w. Santander) and extreme nw. Ecuador (Esmeraldas). Also Panama. Mostly below 800 m.

GROUP C

"Atypical" euphonias.

Euphonia cyanocephala

GOLDEN-RUMPED EUPHONIA

PLATE: 14

Other: Blue-hooded Euphonia (in part)

IDENTIFICATION: 11 cm (4¼"). *Crown and nape bright turquoise blue* with narrow black frontlet. Otherwise glossy purplish black above with *bright yellow rump*. Sides of head and throat black; remaining underparts orange-yellow. *Insignis* (s. Ecuador) has frontlet ochraceous yellow backed by black line. Female olive above with blue crown and nape and tawny frontlet. Below olive yellow, yellowest on center of belly.

SIMILAR SPECIES: Male virtually unmistakable. Female's pattern somewhat reminiscent of larger female Chestnut-breasted Chlorophonia, but much duller overall (olive rather than bright green) and lacking chlorophonia's chestnut line under the blue crown and its green-yellow demarcation on underparts.

HABITAT AND BEHAVIOR: Uncommon and rather local in clearings with scattered trees and forest or woodland borders. Not particularly associated with forest (and absent from continuous forest) and sometimes found even in gardens or parks of suburban or urban areas. Found in pairs or small groups, usually perching rather high in trees and generally not associating with mixed flocks.

RANGE: Coastal mts. of n. Venezuela (Paria Peninsula west) and Andes of w. Venezuela, Colombia (in W. Andes north only to Valle), Ecuador, Peru, Bolivia, and nw. Argentina (south to Tucumán); se. Venezuela (s. Bolívar), Guyana, and Suriname; se. Brazil (north to s. Bahia, Minas Gerais, and s. Mato Grosso), e. Paraguay (few or no recent records), and ne. Argentina (south to n. Santa Fe and Entre Ríos); Trinidad. Mostly 500–2800 m, occasionally also in lowlands adjacent to montane areas.

NOTE: We follow the 1983 AOU Check-list in considering the S. Am. *cyanocephala* group as a species distinct from the other 2 "Blue-hooded Euphonias," true *E. musica* of the West Indies (Antillean Euphonia) and *E. elegantissima* of Middle America (Blue-hooded Euphonia).

Euphonia rufiventris

RUFOUS-BELLIED EUPHONIA

PLATE: 14

IDENTIFICATION: 11.5 cm (4½"). *Amazonia*. Entirely glossy steel blue above (*no crown patch*) and on throat and chest. *Lower underparts mostly rich tawny*, with golden pectoral patch often showing along sides of wing. Female olive above, tinged gray on nape and with more yellowish forehead. Below mostly gray with olive yellow chin, incomplete breast band, and flanks; *crissum tawny*.

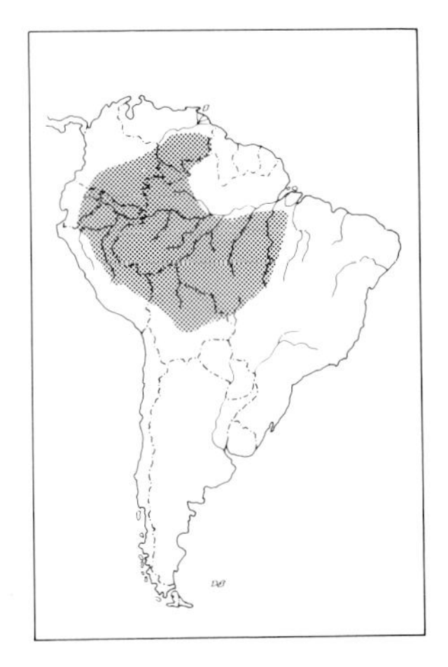

SIMILAR SPECIES: Except for the following 3 species, the male is the only euphonia with dark upperparts lacking a crown patch (beware the fact that in some races of Orange-bellied crown patch is dark and sometimes hard to discern). Female closely resembles female Golden-sided Euphonia (limited overlap) but has tawny (not grayish) crissum. Female White-lored Euphonia is all gray below (lacking olive on sides and chin) except for having yellowish olive (not tawny) crissum. Female Orange-bellied Euphonia also very similar but has more ochraceous (not olive yellow) forehead and is dingier below. Cf. also Plumbeous Euphonia.

HABITAT AND BEHAVIOR: Uncommon to locally fairly common in canopy and borders of humid forest (both *terra firme* and *várzea*). Usually in pairs, generally remaining well up in trees and often frustratingly difficult to see well as they forage in dense epiphytic vegetation, infrequently perching in the open. Regularly with mixed flocks in the canopy. Often heard is its fast, gravelly call "dr-dr-dr-drt" (in quality reminiscent of Fulvous-vented Euphonia's, but shorter and usually in sequences of 4 to 5 notes).

RANGE: S. Venezuela (Bolívar and Amazonas), se. Colombia (north to Meta and Guainía), e. Ecuador, e. Peru, n. Bolivia (south to La Paz, Cochabamba, and w. Santa Cruz), and w. and cen. Amaz. Brazil (north of the Amazon east to the lower Rio Negro region, south of the Amazon east to nw. Mato Grosso and the lower Rio Xingú). Mostly below 500 m.

Euphonia pectoralis

CHESTNUT-BELLIED EUPHONIA

PLATE: 14

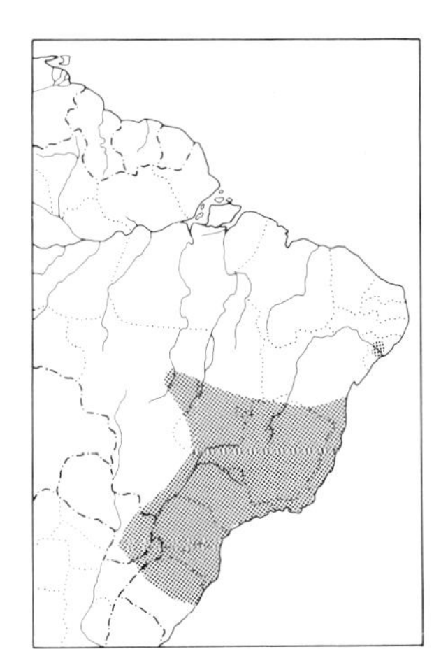

IDENTIFICATION: 11.5 cm (4½"). *Se. South America. Mostly glossy steel blue.* Lower underparts deep chestnut; *yellow pectoral patch* near bend of wing. Female olive above, grayer on nape. Below mostly gray, but broadly olive on sides and flanks and with *rufous crissum.*

SIMILAR SPECIES: In its range nothing really resembles the very dark male (can look all black with yellow patch on sides of chest). Female resembles female of geographically far-removed Rufous-bellied Euphonia, though crissum rufous (not tawny) and with more olive on sides.

HABITAT AND BEHAVIOR: Fairly common in forest canopy and borders. Found in pairs or small groups, often associating with mixed tanager flocks. Regularly perches in open, often in *Cecropia;* in general a much more conspicuous euphonia than the Rufous-bellied. Call similar to that species'. Though widespread and generally numerous in most humid forest areas remaining in its range (perhaps especially common in e. Paraguay and at Iguaçu Falls), this species' overall numbers have doubtless declined substantially as a result of massive deforestation over much of its original range.

RANGE: E. Brazil (e. Alagoas and from s. Bahia, Minas Gerais, s. Goiás, and w. Mato Grosso south to n. Rio Grande do Sul), e. Paraguay, and ne. Argentina (Misiones). To about 1300 m.

Euphonia cayennensis

GOLDEN-SIDED EUPHONIA

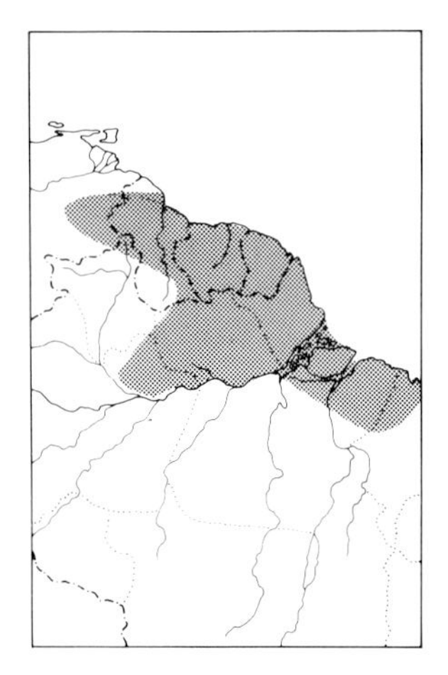

IDENTIFICATION: 11.5 cm (4½"). *Ne. South America.* Resembles Chestnut-bellied Euphonia (no overlap). Male similar (same golden pectoral patch) but *underparts all* glossy steel blue (no chestnut). See V-36. Female most resembles female Rufous-bellied Euphonia but has *olive grayish* (not tawny) *crissum.*

SIMILAR SPECIES: Female very similar to female White-lored Euphonia, but that species lacks Golden-sided's olive chin and sides (all pale gray below, with yellowish olive restricted to lower flanks and crissum) and *has whitish lores.*

HABITAT AND BEHAVIOR: Fairly common in canopy and borders of humid forest; occasionally forages lower (both at edge and inside forest). General behavior and voice similar to Rufous-bellied Euphonia's, but (like Chestnut-bellied Euphonia) seems somewhat more conspicuous, regularly perching in semiopen at forest edge.

RANGE: Se. Venezuela (Bolívar), Guianas, and ne. Brazil (north of the Amazon from Amapá west to the lower Rio Negro, south of it in e. Pará in the Belém area, and in nw. Maranhão). Mostly below 600 m.

Euphonia plumbea

PLUMBEOUS EUPHONIA

PLATE: 14

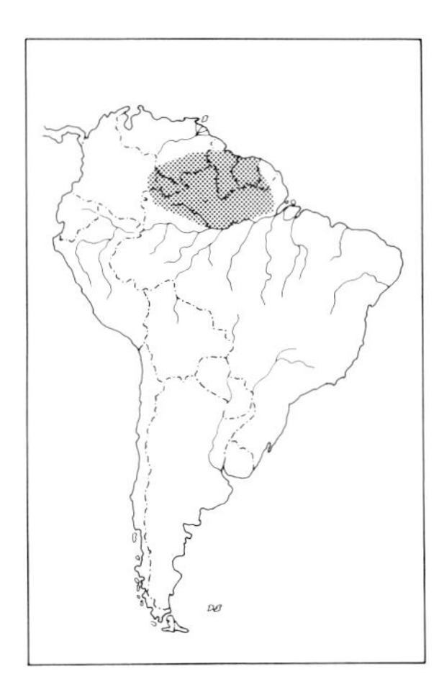

IDENTIFICATION: 9.5 cm (3¾"). *Ne. South America.* Male *mostly glossy steel gray* (no yellow on head). Sharply contrasting rich yellow breast and belly; flanks flecked with gray. Female echoes male's pattern but lacks the gloss. *Crown and nape gray;* otherwise dark olive above. *Throat and chest pale gray;* lower underparts greenish yellow, brightest on center of belly.

SIMILAR SPECIES: Pattern of gray and yellow male slightly resembles that of male Rufous-bellied Euphonia. Cf. also White-lored Euphonia.

HABITAT AND BEHAVIOR: Rather rare (locally more numerous) in forest borders and shrubby clearings and in shrubbery and low woodland at edge of savanna areas. Usually in pairs, less often in small groups; sometimes with mixed flocks but perhaps as often away from them. Its call is a high clear "peee" or "weee, peee-peee," often repeated several times, while the song is a long series of jumbled twittery notes into which its call note is often incorporated (T. Meyer).

RANGE: Guyana and Suriname, s. Venezuela (Bolívar and Amazonas), extreme e. Colombia (sightings from e. Guainía), and locally in n. Brazil (north of the Amazon from upper Rio Negro drainage east to w. Amapá). Mostly below 300 m (recorded higher in Venezuela).

Euphonia chrysopasta

WHITE-LORED EUPHONIA

PLATE: 14

Other: Golden-bellied Euphonia

IDENTIFICATION: 11–11.5 cm (4¼–4½"). *Amazonia to the Guianas.* Male relatively dull compared to other male euphonias, but has *conspicuous white loral area and chin.* Mostly olive above with slight glossy sheen; hindcrown and nape gray. *Mostly golden yellow below with vague olive mottling* across breast and flanks. Female similar to male above

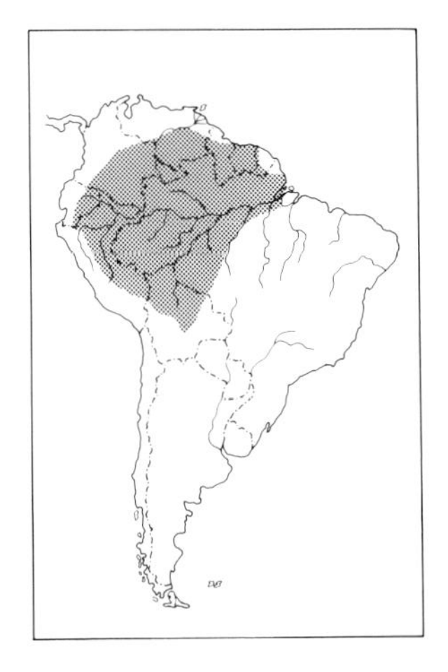

(including *white lores and area around base of bill*) but mostly pale gray below; lower flanks and crissum olive yellow, yellower on latter.
SIMILAR SPECIES: Female resembles female Golden-sided and Rufous-bellied Euphonias but told by white lores, absence of olive on chin and sides, and olive crissum (not tawny as in Rufous-bellied). No other euphonia shows the white lores, though habit of generally remaining high up may make these difficult to discern.
HABITAT AND BEHAVIOR: Uncommon in canopy and borders of humid forest (both *terra firme* and *várzea*) and, less often, out into trees and shrubbery in adjacent clearings (primarily to visit fruiting trees). Usually in pairs and often with mixed tanager flocks in the canopy; generally remain well above ground, but sometimes come lower at edge or in clearings. Usually not very conspicuous, rarely perching in the open. Male's song is a buzzy, explosive "pitz-week," sometimes lengthened into a series of sputtering notes.
RANGE: Suriname (no definite records from Guyana or French Guiana, though should occur in both), s. Venezuela (Bolívar and Amazonas), e. Colombia (north to Meta and e. Vichada), e. Ecuador, e. Peru, n. Bolivia (south to La Paz, Cochabamba, and w. Santa Cruz), and much of Amaz. Brazil (north of Amazon east to Amapá, south of it to w. Mato Grosso and lower Rio Tapajós). To about 900 m.

NOTE: A large number of euphonias have golden bellies, so this species' old name, "Golden-bellied Euphonia," while not exactly inaccurate is misleading in implying that it differs in this regard. On the other hand, *E. chrysopasta*'s white lores are unique in the genus and are a striking field character.

Euphonia mesochrysa

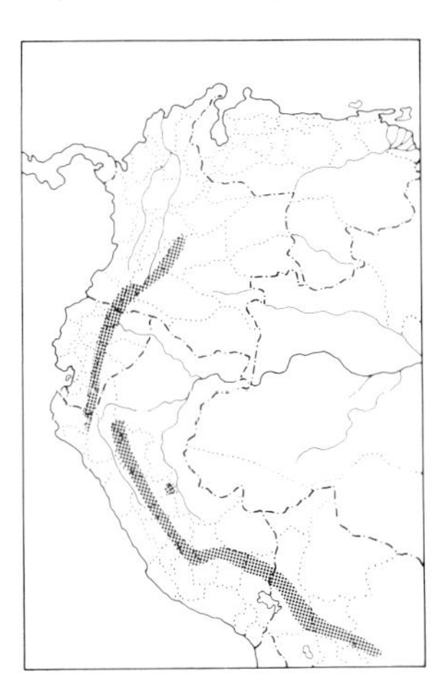

BRONZE-GREEN EUPHONIA

IDENTIFICATION: 10 cm (4"). *E. slope of Andes.* Like White-lored, male relatively dull compared to other male euphonias but has *prominent yellow forehead.* Mostly olive above with slight glossy sheen; hindcrown and nape gray. Throat, chest, and sides yellowish olive, with *remaining underparts rich ochraceous yellow* (brightest in s. birds). See C-53. Female like male above but without yellow forehead. Mostly yellowish olive below, with center of breast and belly gray.
SIMILAR SPECIES: Male most resembles male White-lored Euphonia but lacks white on lores and has bright yellow forehead as well as more ochraceous tone to underparts. Beware immature males of other euphonias (most frequently Thick-billed), which may get yellow forecrown while plumage still basically olive. Female recalls various female euphonias with gray and olive underparts, though White-lored has yellowish olive only on extreme lower underparts (none on sides, etc.), while Rufous-bellied has tawny (not yellowish olive) crissum. Female also rather like female Orange-bellied Euphonia, but note absence of that species' drab buffy effect below and its yellowish to dull rufous forehead. *Bronze-green occurs at higher elevations than other euphonias except the Orange-bellied.*
HABITAT AND BEHAVIOR: Rare to uncommon (seems slightly more numerous southward) in canopy and borders of humid montane forest. Behavior differs little from that of other forest-based euphonias.

Call is a gravelly "treeuh, treeuh," most often doubled (reminiscent of Fulvous-vented Euphonia's).
RANGE: Andes of Colombia (e. slope of E. Andes north to Meta; also at head of Magdalena valley on e. slope of Cen. Andes in Huila), e. Ecuador, e. Peru, and nw. Bolivia (south to Cochabamba and w. Santa Cruz). Mostly 1000–2000 m, occasionally higher or lower.

Chlorophonia Chlorophonias

Larger and plumper relatives of *Euphonia*, the genus *Chlorophonia* differs in its mostly bright emerald green plumage (females somewhat duller). Full adult males are relatively infrequent in 2 species (Yellow-collared and Blue-naped). Their arboreal behavior is similar to the euphonias', though they tend not to be as conspicuous, often remaining hidden in dense foliage. Like euphonias, they eat large quantities of mistletoe berries and build globular domed nests. Found primarily in montane forests (Blue-naped also in lowlands of se. South America).

Chlorophonia pyrrhophrys

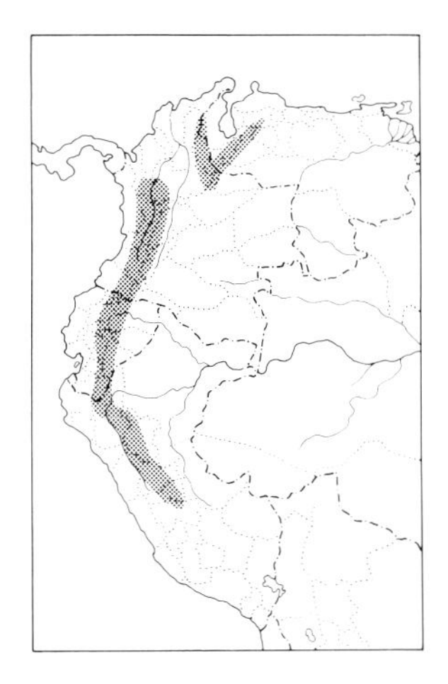

CHESTNUT-BREASTED CHLOROPHONIA PLATE: 14

IDENTIFICATION: 11.5 cm (4½"). *Andes of Venezuela to Peru. Multicolored and strongly patterned. Crown deep cobalt blue* narrowly outlined with black. Above mostly deep green, rump bright yellow. Throat, sides of head and neck, and chest bright emerald green, separated from bright yellow underparts by narrow black band; *median lower underparts and crissum chestnut.* Female echoes male's pattern but its *border to blue crown is chestnut,* yellow rump is lacking, and lower underparts are dull greenish yellow. See V-35, C-49.
SIMILAR SPECIES: The only chlorophonia with a wholly blue crown and (in male) chestnut on underparts; in Blue-naped, blue is confined to the nape. Female Golden-rumped Euphonia has somewhat similar pattern and colors.
HABITAT AND BEHAVIOR: Uncommon and rather local in humid montane forest and forest borders, sometimes coming out into adjacent clearings with scattered tall trees. Usually in pairs, almost always remaining high in trees; may move with mixed flocks, but at least as often found independently of them. Perhaps most numerous in Andes of Venezuela.
RANGE: Andes of w. Venezuela (north to Trujillo), Colombia (where local, mostly in W. and Cen. Andes), n. Ecuador (on w. slope in Pichincha, on e. slope in Napo), and e. Peru (recorded only from Piura, Amazonas, La Libertad, Huánuco, and Pasco); also Perijá Mts. on Venezuela-Colombia border. Mostly 1500–2500 m.

Chlorophonia cyanea

BLUE-NAPED CHLOROPHONIA PLATE: 14

IDENTIFICATION: 11.5 cm (4½"). *Entire head, throat, and chest bright grass green;* ocular area pale blue. *Nuchal collar, center of back, and rump bright blue;* wings and tail green. *Breast and belly bright yellow.* Races

found in n. Venezuela and on Colombia's Santa Marta Mts. have yellow frontlet and green back; *roraimae* of the *tepuis* also has the yellow frontlet but retains blue back of widespread group. Females duller, but all races are more or less alike: mostly green above with *blue restricted to ocular area and the nuchal band* (also the rump in *psittacina* of the Santa Martas). Throat and chest green, lower underparts dull greenish yellow.
SIMILAR SPECIES: Overall pattern, particularly the sharp break between green chest and yellow belly, recalls other chlorophonias. Lacks obvious blue crown of both sexes of Chestnut-breasted, while smaller Yellow-collared has distinctively colored soft-parts (and male is very gaudy otherwise).
HABITAT AND BEHAVIOR: Uncommon and somewhat local in canopy and borders of humid forest and forest borders and in adjacent clearings with scattered large trees. Often in small groups, usually remaining well above the ground; sometimes with mixed flocks, but more often moves about independently. Surprisingly inconspicuous under most circumstances, often remaining motionless, hidden in dense foliage, for long periods. Presence often made known by its plaintive call, "peeng," similar to other chlorophonias'. Seems most numerous in Santa Marta Mts.
RANGE: Coastal mts. of n. Venezuela (Paria Peninsula west) and Andes of w. Venezuela (north to Lara), Colombia (on Pacific slope of W. Andes south only to Valle), e. Ecuador (few records, some from lowlands adjacent to Andes), e. Peru, and nw. Bolivia (south to Cochabamba and w. Santa Cruz); Santa Marta Mts. of Colombia and Perijá Mts. on Venezuela-Colombia border; *tepui* slopes of s. Venezuela (Bolívar and Amazonas), adjacent n. Brazil (Roraima), and n. Guyana (where also recorded in lowlands); se. Brazil (s. Bahia and e. Minas Gerais south to n. Rio Grande do Sul), e. Paraguay, and ne. Argentina (Misiones). Mostly 500–2000 m.

Chlorophonia flavirostris

YELLOW-COLLARED CHLOROPHONIA

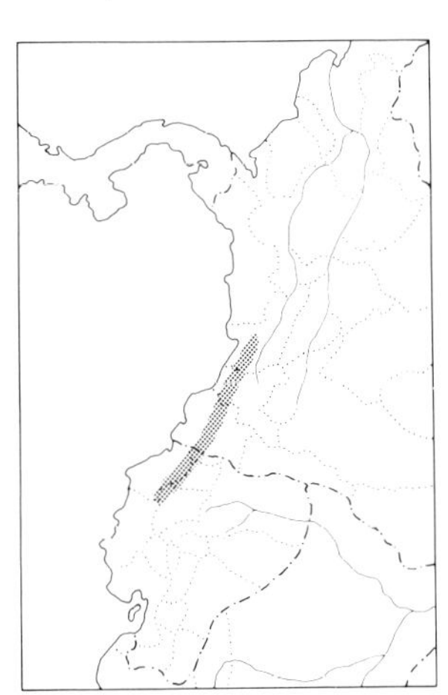

IDENTIFICATION: 10 cm (4"). *Sw. Colombia and nw. Ecuador. Bill and legs salmon orange; iris white,* with *yellow eye-ring.* Mostly bright green above and on throat and chest; *broad nuchal collar and rump bright yellow.* Breast and belly bright yellow, separated from green sides and flanks. Female *predominantly bright green* (lacking the yellow above except around eye); center of breast and belly bright yellow. See C-48.
SIMILAR SPECIES: Stunning full males are unmistakable, while the mostly green females and immatures (usually far more frequent than adult males) can be readily recognized by their unusual soft-part colors. Cf. the larger Blue-naped Chlorophonia.
HABITAT AND BEHAVIOR: Rare to uncommon and very local in canopy and borders of very humid forest on lower slopes. Now seen fairly regularly in its limited range, though until very recently it was an almost unknown bird. Usually in small groups (though flocks of up to 80 birds have been seen, all but a few in non–adult male plumage), remaining well up in trees and most often moving independently of mixed flocks. Attention is often drawn to these small birds by their

distinctive plaintive "peeeeeeee" call, often given in flight. Look for it along the lower Buenaventura road in Valle, Colombia; numbers seen seem to fluctuate (perhaps engages in seasonal movements of some sort?), with largest numbers especially July–Aug., though some can usually be found at other seasons as well.
RANGE: Pacific slope of W. Andes in sw. Colombia (recorded Valle and Nariño) and nw. Ecuador (Esmeraldas and Pichincha). Also e. Panama (once in Darién). Mostly 400–1500 m.

Tersina Tanagers

Formerly placed in a monotypic family, the Tersinidae, the Swallow Tanager is now considered only a tribe in the tanager subfamily. It differs from all tanagers in its broad flat bill and most particularly in its hole-nesting behavior.

Tersina viridis

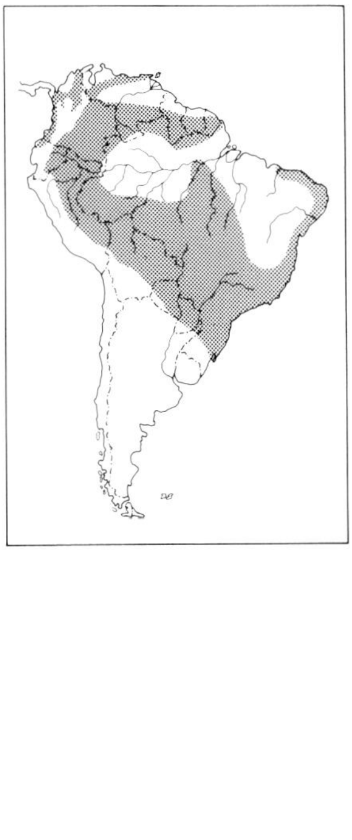

SWALLOW TANAGER

PLATE: 14

IDENTIFICATION: 14.5–15 cm (5¾–6″). Bill broad and flat. Male *mostly shiny turquoise blue* (somewhat variable depending on light); *frontlet, foreface, and throat black.* Center of belly white; flanks barred black. Female *mostly bright green,* mottled grayish on frontlet, foreface, and throat. Belly pale yellow with *conspicuous green barring on flanks.* Immature males, resembling female but with scattered areas of blue coming in, are frequently seen.
SIMILAR SPECIES: Color of males and their habit of perching on high exposed branches bring to mind various cotinga males. Females somewhat resemble various other green tanagers, but green flank-barring unique and easy to see. Habit of moving about in *monospecific flocks* also often an aid.
HABITAT AND BEHAVIOR: *Very erratic* (can be locally or seasonally common, but also absent across wide areas or for most of year in some regions, then reappearing) at humid forest borders, clearings with scattered trees, and gallery woodland. Most numerous in hilly areas (and may always breed in such regions) and infrequent or very seasonal in arid areas. Notably gregarious, occurring in flocks of up to scores of birds, especially when not breeding; tend to separate out when nesting, though loose aggregations may form where suitable nest sites (holes in steep banks, bridges, walls, or even buildings) are frequent. Often perch conspicuously on open branches, from there often sallying out after passing insects; also consume much fruit. Quite noisy, a loud explosive unmusical "tzeep" call being given most often, both in flight and when perched.
RANGE: Widespread in more humid lowlands (mostly) south to nw. Ecuador (Pichincha), n. and e. Bolivia, e. Paraguay (also recorded as seasonal visitant to *chaco* of w. Paraguay), ne. Argentina (Misiones), and s. Brazil (south to Rio Grande do Sul); Trinidad; absent from llanos and lower Orinoco valley in Venezuela and the coastal Guianas, cen. and e. Amaz. Brazil, and the *caatinga* region of interior ne. Brazil. Also e. Panama. To about 1500 m, occasionally higher.

Catamblyrhynchinae

PLUSHCAP

FORMERLY considered the sole member of a separate family, the Plushcap is of uncertain affinity but now generally is given subfamily rank within the expanded Emberizidae, probably closest to the tanagers. It ranges in the Andean forest.

Catamblyrhynchus Plushcap

The Plushcap is a distinctive Andean bird of uncertain affinities; it was formerly called a "finch," which it likely is not.

Catamblyrhynchus diadema

PLUSHCAP

PLATE: 15

Other: Plush-capped Finch

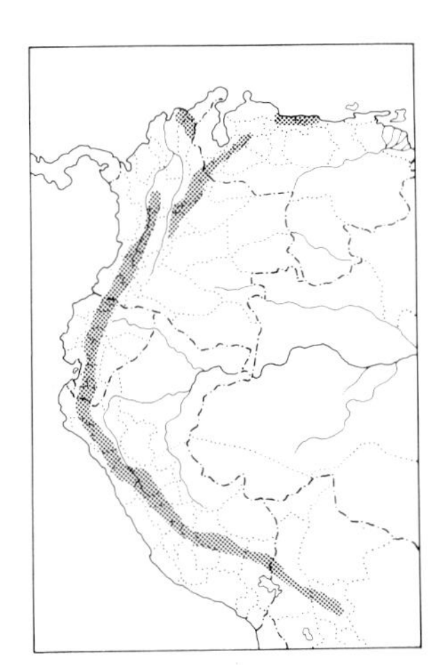

IDENTIFICATION: 14 cm (5½"). *Thick stubby black bill. Forecrown bright yellow,* feathers short and plushlike; hindcrown and nape black. Otherwise dark gray above. *Cheeks, sides of neck, and entire underparts rich chestnut.* Underparts of s. *citrinifrons* slightly less dark, more rufous-chestnut. Immature much duller, brownish olive above, paler olivaceous buff below; somewhat older birds begin to acquire yellow on forecrown. SIMILAR SPECIES: Golden-crowned Tanager has similar overall pattern (in particular sharing the yellow on crown) but is basically blue below (not chestnut). Superficially also resembles Black-eared Hemispingus, with which it regularly is found (even in same flock on occasion), but the hemispingus lacks the yellow crown patch, etc. Dull immature can be confusing, but note its distinctive bill.

HABITAT AND BEHAVIOR: Uncommon in lower growth of humid montane forest and forest borders, especially where there are extensive stands of *Chusquea* bamboo. Usually rather inconspicuous; 1 or 2 birds often follow mixed flocks moving through understory and middle levels of forest. Frequently feeds on bamboo stems, clinging to them like a chickadee (*Parus* sp.), pushing its bill into stems and leaf nodes and prying them open. Song (not frequently heard) is a long series of almost random chips and twitters (overall quality rather hummingbirdlike).

RANGE: Coastal mts. of n. Venezuela (Aragua and Distrito Federal) and Andes of w. Venezuela (north to Trujillo), Colombia, Ecuador, Peru (south on w. slope to Lambayeque), nw. Bolivia (La Paz, Cochabamba, and w. Santa Cruz; probably also southward along e. slope), and nw. Argentina (Jujuy); Colombia's Santa Marta Mts. and Perijá Mts. on Colombia-Venezuela border. Mostly 2300–3500 m, lower in n. mts. of Venezuela.

NOTE: Calling *Catamblyrhynchus* a "finch" seems inappropriate given its uncertain affinities, and we prefer to simply name it the "Plushcap."

Urothraupis Bush-Tanagers

A distinct, monotypic genus; we consider it closest to *Chlorospingus*, but others have considered it to be allied to *Atlapetes* (and its plumage pattern is indeed reminiscent). A predominantly black and gray bird of timberline areas in the n. Andes.

Urothraupis stolzmanni

BLACK-BACKED BUSH-TANAGER

PLATE: 15

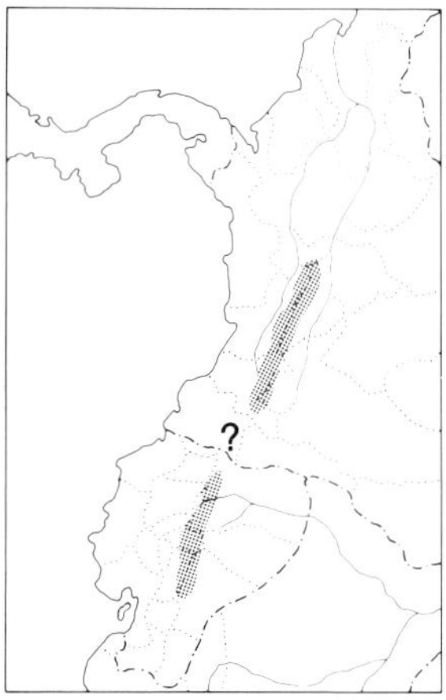

IDENTIFICATION: 15 cm (6″). *S. Colombia and Ecuador. Entirely black above.* Throat white, chest white vaguely *flecked* with gray; breast and belly gray *mottled* with white; flanks and crissum dark gray.

SIMILAR SPECIES: This black, gray, and white tanager is perhaps more likely to be mistaken for an *Atlapetes* brush-finch, though none has the same mottled or flecked effect. Slaty Brush-Finch is most similar (and occurs with this species), but it has a chestnut crown, prominent white wing-speculum, etc.

HABITAT AND BEHAVIOR: Locally fairly common in low woodland and dense shrubby areas, especially near and just below timberline. Most often found in small groups of up to about 6 birds which forage actively in foliage, gleaning and probing for insects. Frequently associate with mixed flocks, but also sometimes found on their own. Quite readily seen in shrubbery at Puracé Nat. Park in Cauca, Colombia, and in *Polylepis* groves on the Papallacta pass east of Quito, Ecuador.

RANGE: Andes of sw. Colombia (recorded only from Cen. Andes from Caldas-Tolima border south to Cauca; probably also in e. Nariño) and e. Ecuador (south to Morona-Santiago on the Altar/Sangay massif). Mostly 3000–4000 m.

Chlorospingus Bush-Tanagers

A common and conspicuous group of small, stocky tanagers found in the Andes. Olive and gray predominate, with facial patterns (e.g., a postocular spot or stripe) varying between or even within species; many have a pale iris (colors varying from white to pale brown or gray). They often range about in sizable flocks, their noisy contact vocalizations often drawing attention (though their true songs are not often heard).

GROUP A

"Capped" appearance (cap usually gray or brown), with yellow pectoral band.

Chlorospingus ophthalmicus

COMMON BUSH-TANAGER

PLATE: 15

IDENTIFICATION: 14–14.5 cm (5½–5¾″). *Very variable.* Best marks for most races are *gray to brown head contrasting with pale iris or white postocular spot* (*both* in most or all of Venezuela, ne. Colombia, and

nw. Bolivia), *white throat variably speckled dusky* (tendency toward more speckling northward), and *greenish yellow to yellow pectoral band*. Otherwise olive above, grayish white to pale gray below, with yellowish olive sides, flanks, and crissum. Only obvious deviations from this are enumerated below. *Flavopectus* of ne. Colombia is only race with both a dark iris and no white postocular, while *nigriceps* of the Cen. Andes and upper Magdalena valley has blackest head, sharply contrasting with creamy white iris (but no postocular). *Phaeocephalus* of Ecuador and northernmost Peru is the dingiest race, with pale yellow-orange to pinkish iris and no postocular. *Cinereocephalus* of cen. Peru (Junín to Cuzco) is the most divergent subspecies, as it lacks the yellow pectoral band altogether, it being replaced by a buffy wash across the breast; its iris is pale gray. Birds from Argentina and s. Bolivia (north to Santa Cruz) have dark iris and strong white postocular (*argentinus*), while in Cochabamba, Bolivia, there is also a pronounced buff tinge on throat (*fulvigularis*), and in La Paz the buff is lost and the iris becomes pale again.

SIMILAR SPECIES: The geographic variation can be confusing. Problems most likely with Ashy-throated Bush-Tanager, but that *always* has a dark iris and a whiter, unspeckled throat contrasting with a sharply defined yellow pectoral band.

HABITAT AND BEHAVIOR: Usually common in humid montane forest and forest borders and to a lesser extent also secondary woodland and clearings with scattered trees and bushes; in Bolivia and Argentina also in somewhat drier situations. In many areas one of the most abundant and frequently seen birds, but in others (e.g., Ecuador) unaccountably scarce and local. Typically forage in groups (sometimes up to 10 to 20 birds) which usually are with mixed flocks of other tanagers, etc.; often such flocks seem to coalesce around the bush-tanagers. Search leaves and branches for insects at all levels and also consume much fruit. Rather noisy, with a variety of undistinctive "chep" call notes and also a song which consists of a series of "chips" followed by a trill, though with some geographic variation.

RANGE: Coastal mts. of n. Venezuela (Miranda west to Falcón) and Andes in w. Venezuela (north to Lara), Colombia (largely absent from W. Andes except at n. end), Ecuador (very local on e. slope; on w. slope only in El Oro), e. Peru, w. Bolivia, and nw. Argentina (south to Tucumán); Perijá Mts. on Venezuela Colombia border. Also Mexico to Panama. 800–2600 m.

Chlorospingus canigularis

ASHY-THROATED BUSH-TANAGER

PLATE: 15

IDENTIFICATION: 14 cm (5½"). *Mostly Colombia to Peru. Head gray* with darker gray ear-coverts surmounted by *narrow white postocular stripe;* upperparts otherwise clear olive, contrasting with gray head. Below grayish white, with *chest crossed by bright yellow pectoral band;* sides and crissum olive yellow. Only *signatus* (of e. Ecuador and Peru) has the postocular; in other races head merely gray with darker auricular region; see C-54.

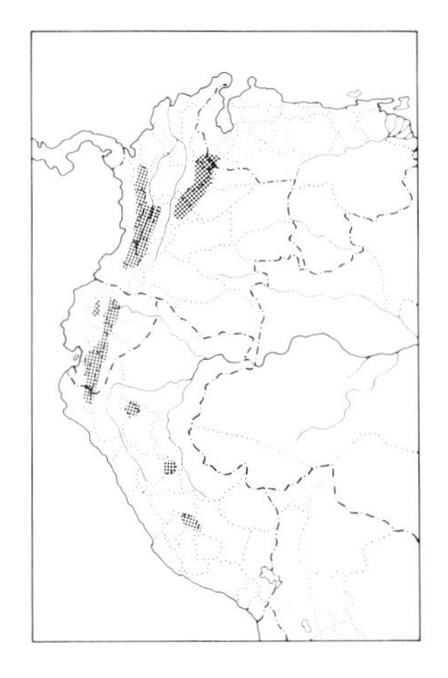

SIMILAR SPECIES: The distinct pectoral band distinguishes the Ashy-throated from all bush-tanagers except the Common. Best told from that species by its invariably dark iris and unspotted throat; the shape of *signatus*'s postocular streak is markedly different from the spot shown by many races of Common. North of Bogotá, Colombia, nominate race of Ashy-throated may overlap with very similar *flavopectus* race of Common, which is perhaps only distinguished by its faintly speckled throat.

HABITAT AND BEHAVIOR: Fairly common in humid montane forest and forest borders. Usually in small groups which often associate with mixed flocks of other tanagers, etc. Tends to forage high, mainly in canopy and subcanopy, gleaning for insects in foliage and on branches, but also takes considerable fruit.

RANGE: Andes of extreme sw. Venezuela (s. Táchira), Colombia, Ecuador (on w. slope in Pichincha and from e. Guayas to El Oro), and e. Peru (recorded locally south to Cuzco; FM specimens). Also Costa Rica and w. Panama. Mostly 900–2000 m.

GROUP B

Nondescript species, either olive or grayish below.

Chlorospingus semifuscus

DUSKY BUSH-TANAGER

PLATE: 15

Other: Dusky-bellied Bush-Tanager

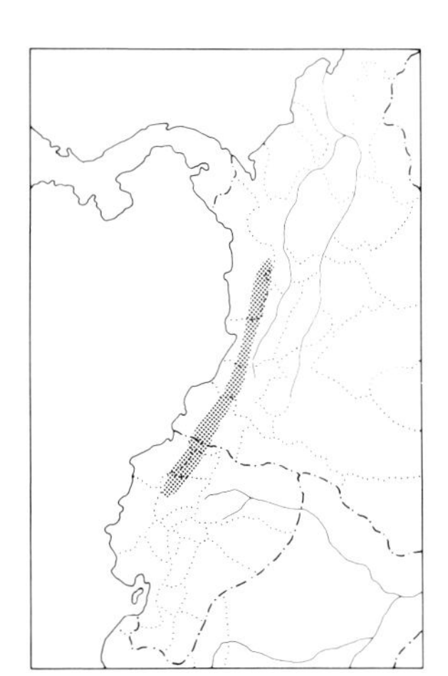

IDENTIFICATION: 14.5 cm (5¾"). *W. Colombia and nw. Ecuador. Iris yellowish white* in n. *livingstoni,* but usually *reddish brown* in nominate ssp. (north to Nariño, Colombia). *Head and neck dark gray* contrasting with olive upperparts; *below brownish gray,* paler on median belly, and yellowish olive on flanks and crissum. *Livingstoni* (south to Cauca, Colombia) is purer gray below.

SIMILAR SPECIES: *Very dark and dull.* Yellow-throated Bush-Tanager is not as dark overall and has obvious yellow on throat. Yellow-green Bush-Tanager is olive (not gray) below. Cf. Ochre-breasted Tanager (also very dull, and sympatric, but larger with heavier bill and obviously buffy below).

HABITAT AND BEHAVIOR: Common in humid montane forest and forest borders; especially numerous in Pichincha, Ecuador. Usually occurs in groups (sometimes quite large, up to 10 to 20 birds) which forage actively and noisily at all levels, but principally in lower and middle strata. Frequently with mixed flocks of other tanagers, etc.

RANGE: Pacific slope of W. Andes of Colombia (north to s. Chocó) and nw. Ecuador (south to Pichincha). Mostly 700–2000 m.

NOTE: We have shortened the English name to reflect the fact that *most* of the bird is drab and dusky and because its belly is actually partly olive (*not* as dusky as on head, breast, etc.).

Chlorospingus tacarcunae

TACARCUNA BUSH-TANAGER

IDENTIFICATION: 14 cm (5½"). *Extreme nw. Colombia. Iris pale* (yellowish white to grayish white with fine reddish streaks). Mostly olive

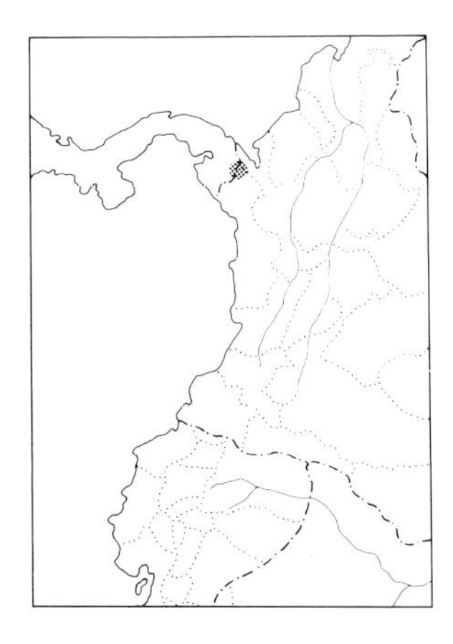

green above, slightly browner on head, with no white postocular spot. *Throat and chest yellow,* with lower underparts more olive yellowish. See P-29, C-52.

SIMILAR SPECIES: Rather dull overall, but can be known by its pale eye and nearly uniform upperparts. Does not overlap with any other bush-tanager. Yellow-throated Bush-Tanager is mostly gray below.

HABITAT AND BEHAVIOR: Little known in the very restricted Colombian portion of its range. In Panama locally fairly common in forest and forest borders, perhaps especially in rather low elfin cloud forest on exposed ridges and peaks. Usually in small groups which tend to forage rather low, often in association with other tanagers.

RANGE: Extreme nw. Colombia (e. slope of Cerro Tacarcuna in nw. Chocó). Also e. Panama. In Panama 900–1400 m.

Chlorospingus flavovirens

YELLOW-GREEN BUSH-TANAGER

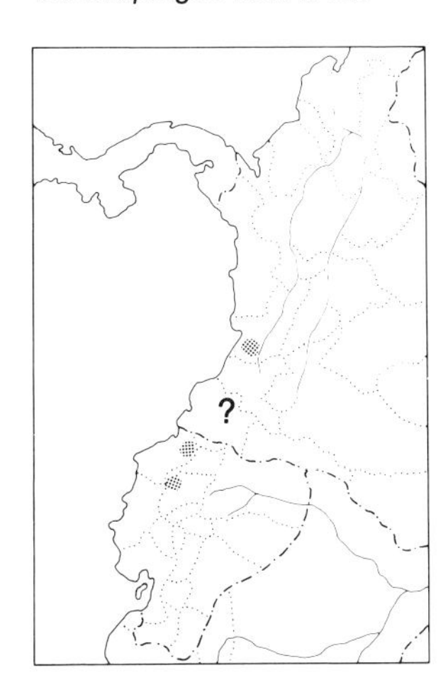

IDENTIFICATION: 14.5 cm (5¾"). *Very local in w. Colombia and nw. Ecuador.* Iris brown. *Dull olive above; olive yellow below,* yellowest on throat, median belly, and crissum. See C-52.

SIMILAR SPECIES: A drab, very uniform-looking species. Occurs sympatrically with only 1 other bush-tanager, the Yellow-throated (which has yellow throat contrasting with gray underparts and a pale iris). Perhaps most resembles Olive Tanager (though not known to occur with it); Olive is notably larger and has a proportionately heavier bill.

HABITAT AND BEHAVIOR: Local and uncommon in wet mossy forest and forest borders. Occurs in pairs or groups of 3 to 4 birds which hop on mossy epiphyte-laden trunks and branches, usually in middle strata or subcanopy (thus higher than is the norm for Yellow-throated), and often associate with mixed flocks (S. Hilty).

RANGE: Pacific slope of W. Andes in Colombia (known from only 1 ridge, in upper Anchicayá valley in Valle) and nw. Ecuador (recorded very locally from Ibarra and Pichincha). 900–1100 m.

GROUP C

Grayish below with *contrasting bright yellow on throat.*

Chlorospingus parvirostris

YELLOW-WHISKERED BUSH-TANAGER PLATE: 15

Other: Short-billed Bush-Tanager

IDENTIFICATION: 14.5 cm (5¾"). *Iris whitish to pale gray.* Olive above. *Sides of throat mustard to golden yellow* (the effect being flaring yellow whiskers), with remaining underparts dingy brownish gray. Nominate race (se. Peru and Bolivia) has *bright canary yellow whiskers,* purer gray underparts.

SIMILAR SPECIES: Resembles Yellow-throated Bush-Tanager rather closely, but usually occurs at higher elevations; the yellow-whiskered effect of this species is prominent (in area of overlap Yellow-throated has yellow of throat cutting more or less evenly across lower throat) and its lores are olive (not gray). At close range iris color also appears to differ consistently (*parvirostris* not showing brown).

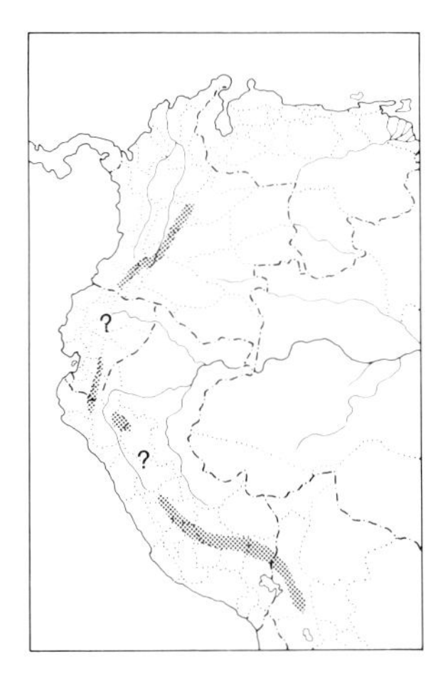

HABITAT AND BEHAVIOR: Fairly common but often somewhat local in humid montane forest and forest borders. Usually occurs in pairs or small groups, both alone and as a member of mixed foraging flocks (where at times appears to be one of the nuclear species around which such flocks form). Forages quite actively at all levels, but principally in lower and middle strata.

RANGE: E. slope of Andes in s. Colombia (north to Cundinamarca, and the head of the Magdalena valley in Huila), Ecuador (Zamora-Chinchipe), Peru (locally), and nw. Bolivia (La Paz). Mostly 1400–2200 m.

NOTE: While this species' bill is very slightly shorter than that of its sibling species, *C. flavigularis,* this is hardly a useful field character. Far more helpful in the area of sympatry is the configuration of the yellow on the throat, which we describe in the bird's new English name.

Chlorospingus flavigularis

YELLOW-THROATED BUSH-TANAGER

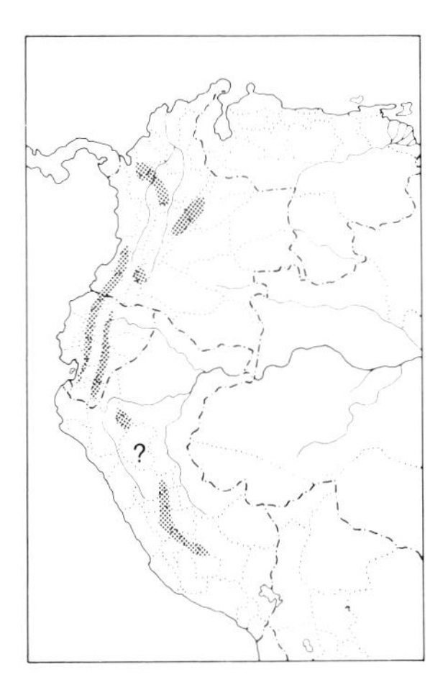

IDENTIFICATION: 15 cm (6″). Confusingly similar to Yellow-whiskered Bush-Tanager. *Iris hazel* (thus differing from Yellow-whiskered). Olive above, but with *distinctly gray lores* (latter lacking in Yellow-whiskered). Throat bright yellow (*without "flaring whisker" effect*); otherwise light gray below, tinged olive on lower flanks and crissum. Foregoing applies to nominate race found on e. slope of Andes (where overlap is possible with Yellow-whiskered); see C-54. *Marginatus* of Pacific slope differs in being dingier gray below (thus resembling Yellow-whiskered, but there is no overlap); see C-52.

HABITAT AND BEHAVIOR: Fairly common to common in humid foothill and lower subtropical forest. Generally replaces Yellow-whiskered Bush-Tanager at somewhat lower elevations. Behavior similar, but may tend to forage somewhat lower.

RANGE: Lower slopes of Andes in Colombia (from upper Sinú and lower Cauca valleys south to head of Magdalena valley in Huila, and on e. slope from Boyacá south), e. Ecuador, and e. Peru (south locally at least to Cuzco on the Cordillera Vilcabamba; possibly farther); on Pacific slope in sw. Colombia (north to Valle) and w. Ecuador (south to El Oro). Also Panama (*hypophaeus*; same species?). Mostly 700–1600 m.

Cnemoscopus Bush-Tanagers

This monotypic genus is close to *Hemispingus* but differs in its slightly more robust overall shape and heavier bill (distinctively pink south to n. Peru). Temperate zone forests in the Andes.

Cnemoscopus rubrirostris

GRAY-HOODED BUSH-TANAGER PLATE: 15

IDENTIFICATION: 15 cm (6″). *Mainly Colombia to Peru. Bill and legs pinkish* in more n. nominate race but dark grayish in *chrysogaster* (Peru

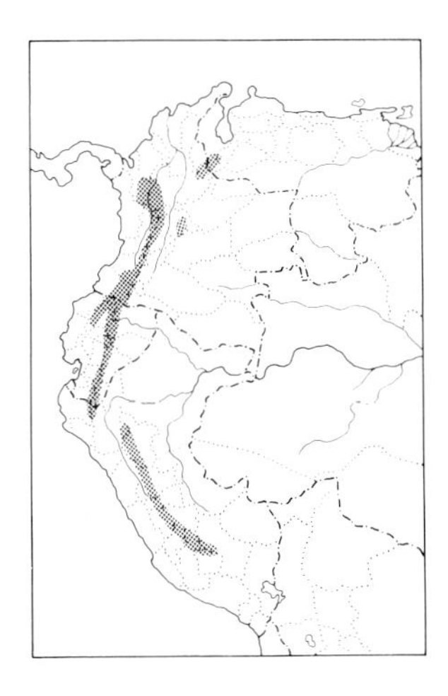

south of the Marañón). *Entire head, throat, and chest gray,* throat and chest somewhat paler. Otherwise olive above, yellow below. *Chrysogaster* has throat and chest paler gray than nominate, brighter yellow belly.

SIMILAR SPECIES: Gray-hooded effect in conjunction with the *near-constant tail wagging* is distinctive. In poor light female Capped Conebill, which also wags tail a great deal, can be confused. *Chlorospingus* bush-tanagers are more heavily built (not as slender as this species), and while some have gray on head, none has such a bright yellow belly. Shape and overall color pattern of Superciliaried Hemispingus similar, but it always shows an eyestripe and lacks gray on throat and chest.

HABITAT AND BEHAVIOR: Fairly common to locally common in humid montane forest and forest borders. Seems especially numerous in areas where native alders (*Alnus* spp.) are prevalent. Arboreal, foraging both along limbs and in foliage, mostly in middle and upper strata. Regularly with mixed flocks of other tanagers, etc. Tail wagging at times seems to involve entire rear half of body, which teeters like a waterthrush's does.

RANGE: Andes of w. Venezuela (s. Táchira), Colombia, nw. and e. Ecuador (south on w. slope to Pichincha), and e. Peru (south to Cuzco on Cordillera Vilcabamba). Mostly 2100–3000 m.

Hemispingus Hemispinguses

A confusing, difficult group of rather dull-plumaged, mostly olive tanagers; their head patterns provide their best identification clues. Taxonomy within the genus (which should perhaps be subdivided) is still not fully understood. Several species remain poorly known (2 were only discovered in the 1970s), while others have often been confused with *Basileuterus* warblers—with good reason, so similar are the 2 supposedly distantly related genera. The various hemispinguses are restricted to the Andes, where they are found mostly in temperate zone forests, foraging mostly in lower growth and at edge.

GROUP A

Underparts yellow to olive.

Hemispingus atropileus

BLACK-CAPPED HEMISPINGUS

PLATE: 15

IDENTIFICATION: 16–18 cm (6¼–7"). *Mainly Colombia to Peru. Head black* with *long narrow white superciliary.* Above otherwise rather bright olive. Throat and breast ochraceous yellow, becoming paler yellow on center of belly and tinged olive on flanks. Foregoing applies to Peruvian *auricularis* (north to Amazonas south of the Río Marañón); nominate race (south through Ecuador) is larger, sootier on head with more buffy whitish superciliary and less broad ear-patch, darker olive above, and much more olive below (with ochraceous tone confined to throat). See C-53.

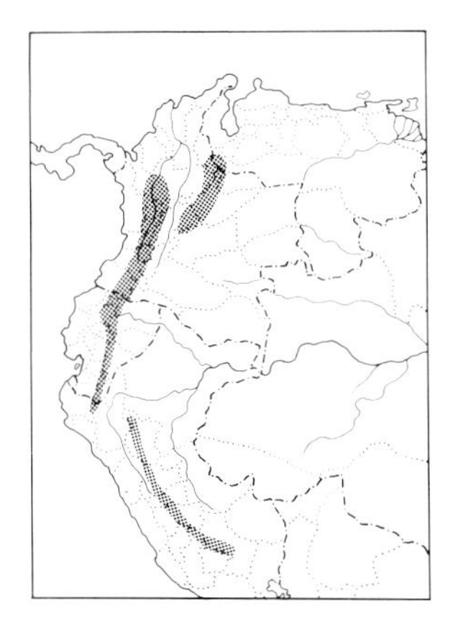

SIMILAR SPECIES: In Colombia and Ecuador most likely to be confused with sympatric Superciliaried Hemispingus, which, however, is smaller and has less black on head and is yellower below. In Peru, where Superciliaried is white below, cf. also the very local Parodi's Hemispingus (with yellow, not white, superciliary).

HABITAT AND BEHAVIOR: Fairly common to common in humid temperate zone forest, especially in dense, shrubby borders where there is a dense growth of *Chusquea* bamboo. Forages in pairs and small groups (up to 6 or so birds), often with mixed flocks of other insectivorous tanagers, brush-finches, etc. Usually conspicuous and rather tame.

RANGE: Andes of sw. Venezuela (s. Táchira), Colombia, Ecuador (south on w. slope only to Pichincha), and e. Peru (south to Cuzco). Mostly 2600–3400 m.

Hemispingus calophrys

ORANGE-BROWED HEMISPINGUS

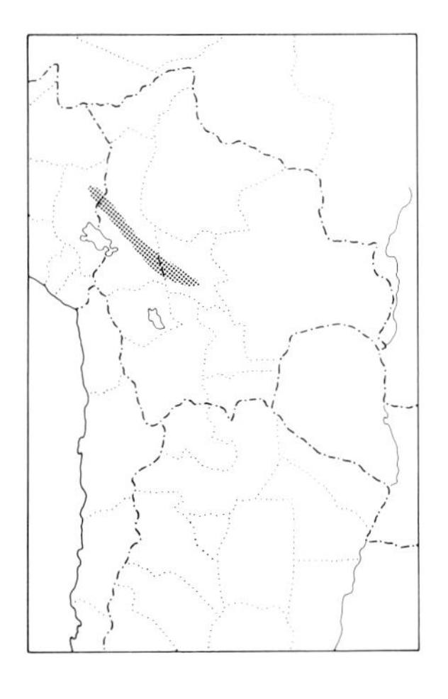

IDENTIFICATION: 16 cm (6¼"). *Mainly Bolivia.* Overall pattern and colors recall Black-capped Hemispingus (especially *auricularis*). *Superciliary notably wider and orange-ochraceous* (not white) and with small orange spot on ear-coverts. Crown sooty and cheeks dusky (not the deep black of Black-capped Hemispingus).

HABITAT AND BEHAVIOR: Not well known. Reported to be locally fairly common in humid temperate zone forest. Forages, principally in bamboo, in small groups either independently or with mixed flocks of insectivorous birds; rather inconspicuous (V. Remsen). Can be found near Chuspipata in the *yungas* of La Paz, Bolivia.

RANGE: Andes of se. Peru (Puno) and nw. Bolivia (La Paz and Cochabamba). 3000–3500 m.

NOTE: Following the arguments presented by J. Weske and J. Terborgh (*Wilson Bull.* 86 [2]: 97–103, 1974), we consider *H. calophrys* as a distinct species and not a race of *H. atropileus*.

Hemispingus parodii

PARODI'S HEMISPINGUS

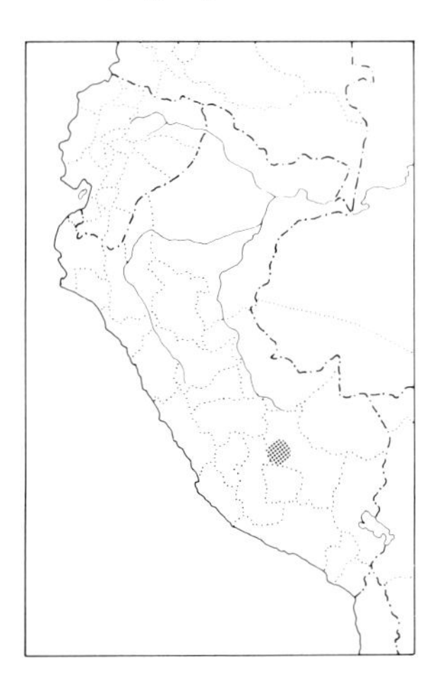

IDENTIFICATION: 16 cm (6¼"). *Local in s. Peru.* Overall pattern recalls Black-capped Hemispingus (especially *auricularis*). *Long superciliary golden yellow* (not white). Crown sooty blackish, but cheeks dusky olive (thus not as black as in Black-capped Hemispingus, and cheek patch not as neatly demarcated below and to rear as in that species). Throat yellow (with little or no ochraceous tinge).

SIMILAR SPECIES: Other than Black-capped Hemispingus (which replaces Parodi's at slightly lower elevations on certain mountains in se. Peru), also liable to be confused with Citrine Warbler; warbler, however, has an olive (not blackish) crown and shorter superciliary (extending back to only just past the eye, not around ear-coverts as in the hemispingus).

HABITAT AND BEHAVIOR: Only recently described. Uncommon in elfin forest slightly below timberline and in temperate zone forest, especially where there is much *Chusquea* bamboo. Behavior similar to

other hemispinguses', usually in pairs or small groups which forage actively through lower growth and shrubbery at edge; often with mixed flocks. Has recently been found along the road through high montane forest at Abra Malaga.
RANGE: Andes of se. Peru (Cuzco). 3000–3500 m.

NOTE: A newly described species: J. Weske and J. Terborgh (*Wilson Bull.* 86 [2]: 97–103, 1974).

Hemispingus superciliaris

SUPERCILIARIED HEMISPINGUS PLATE: 15

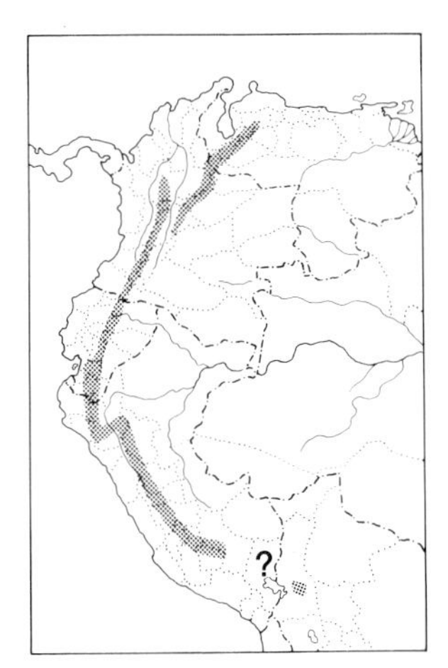

IDENTIFICATION: 14 cm (5½"). Two very different-appearing "types," 1 found only in n. Peru (with white underparts, etc.). In most of range, mostly olive above and bright yellow below, with *forecrown and cheeks gray to blackish* and *white superciliary. Chrysophrys* (w. Venezuela) similar but has olive crown and long yellow superciliary; see V-36. In n. and cen. Peru (Amazonas to Junín) *leucogaster* group is entirely *gray above* with white superciliary; *whitish below,* tinged gray on breast and flecked with dusky on malar area and throat, and tinged buff on lower belly. Birds of s. Peru and Bolivia revert to being yellow-bellied.
SIMILAR SPECIES: The variation is confusing. Yellow-bellied races easily confused with Citrine Warbler, which, however, has an entirely olive crown and a broader but short superciliary that is yellow (not white). Cf. also Black-capped Hemispingus (with much more black on head and long white superciliary) and in Venezuela the Oleaginous Hemispingus. White-bellied birds are most apt to be confused with Drab Hemispingus (lacking superciliary and with pale eye) and perhaps also with Cinereous Conebill (which has prominent white wing-markings).
HABITAT AND BEHAVIOR: Uncommon to fairly common in canopy and edge of humid temperate zone forest and in adjacent shrubby second growth. Usually in pairs or small groups which regularly accompany mixed-species flocks of other insectivorous birds. More arboreal than the Citrine Warbler (considerable overlap in preferred strata, however).
RANGE: Andes of w. Venezuela (north to Trujillo), Colombia (except in W. Andes), sw. and e. Ecuador (on w. slope only in Azuay and El Oro), nw. and e. Peru (on w. slope south only to Cajamarca), and nw. Bolivia (La Paz). Mostly 2200–3200 m.

NOTE: We here follow the treatment first proposed by J. T. Zimmer (*Am. Mus. Novitates* 1367, 1947), in which he considered both *chrysophrys* (of Venezuela; Yellow-browed Hemispingus) and the *leucogaster* group (of Peru; White-bellied Hemispingus) conspecific with *superciliaris.* Prior to then both had been considered full species, and further research could again raise them to that rank. A study of the contact zones (if any) between *chrysophrys* and nominate *superciliaris,* and between *leucogaster* and *urubambae,* would be of great interest.

Hemispingus frontalis

OLEAGINOUS HEMISPINGUS PLATE: 15

IDENTIFICATION: 14–15 cm (5½–6"). *Dull olive above* with *indistinct yellowish superciliary* most pronounced in front of eye. *Below dingy yel-*

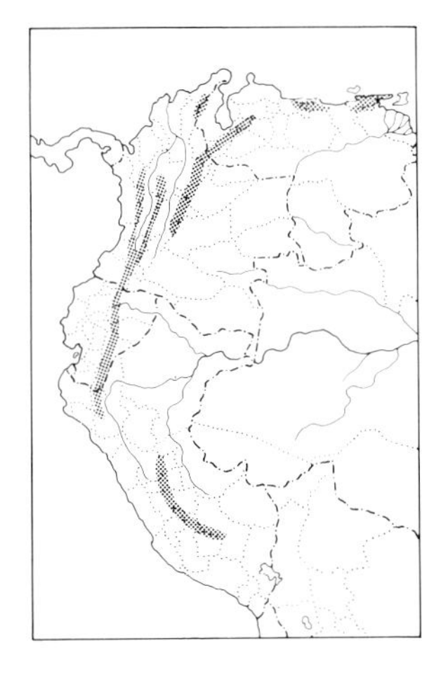

lowish, tinged olive especially on sides. Venezuelan races are more ochraceous below (progressively so eastward) and superciliary is more conspicuous and more or less strongly tinged ochraceous; see V-36.

SIMILAR SPECIES: A dingy and uniform-looking bird, *best known by the absence of distinctive field marks.* Most likely to be confused with yellow-bellied forms of Superciliaried Hemispingus. These are always yellower below, and in most of zone of sympatry they have more prominent white superciliary (in Venezuelan Andes more similar, but Oleaginous has ochraceous tinge to superciliary and underparts); note also habitat and elevation differences. Citrine Warbler also somewhat similar but smaller with much broader superciliary. Cf. also the rare Yellow-green Bush-Tanager.

HABITAT AND BEHAVIOR: Fairly common but inconspicuous in lower growth inside humid montane forest, only occasionally emerging to borders. Usually seen singly or in pairs, gleaning in foliage; often accompanying mixed flocks of undergrowth birds (e.g., various *Basileuterus* warblers).

RANGE: Coastal mts. of n. Venezuela (east to Paria Peninsula) and Andes of w. Venezuela (north to s. Lara), Colombia (on w. slope south only to Cauca), e. Ecuador, and e. Peru (south to Cuzco); Perijá Mts. on Colombia-Venezuela border. Mostly 1500–2500 m.

NOTE: The various Venezuelan races have such different coloration and even pattern (broader superciliary) that it almost seems justifiable to regard them as a full species, *H. ignobilis* (Ochraceous Hemispingus). We opt, however, not to do so, mainly because of the seemingly clinal nature of the variation (i.e., birds farthest from typical Andean populations are the most different).

Hemispingus reyi

GRAY-CAPPED HEMISPINGUS

IDENTIFICATION: 14 cm (5½″). *W. Venezuela. Crown gray,* contrasting with otherwise olive upperparts (*no superciliary*). Below yellow, tinged olive on sides and flanks. See V-36.

SIMILAR SPECIES: Venezuelan form of Superciliaried Hemispingus has yellow superciliary and olive (not gray) crown. Oleaginous Hemispingus there has ochraceous yellow superciliary, lacks gray crown.

HABITAT AND BEHAVIOR: Locally fairly common in montane forest and (especially) shrubby forest borders. Usually forages fairly close to the ground, often in small groups of 6 or so birds which are often accompanying mixed flocks of small insectivorous birds. Easily seen along the Queniquea road in Táchira.

RANGE: Andes of w. Venezuela (Trujillo to Táchira). Mostly 2200–3000 m.

GROUP B

Underparts pale grayish; iris pale.

Hemispingus verticalis

BLACK-HEADED HEMISPINGUS

PLATE: 15

IDENTIFICATION: 15 cm (6″). *Mainly Colombia and Ecuador. Iris straw-colored to whitish. Mostly gray,* darker above and paler below, with center

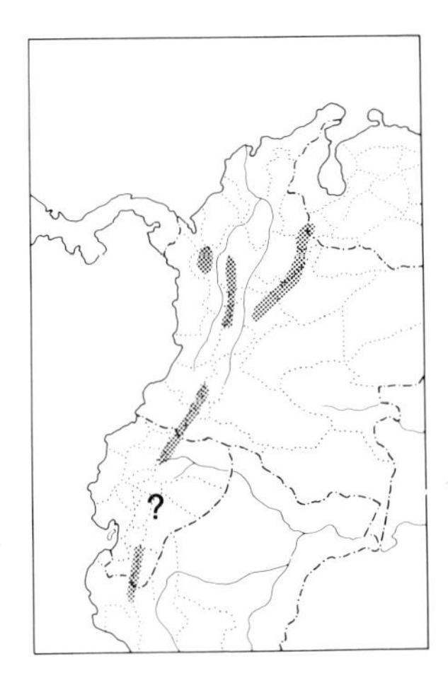

of belly whitish. *Head mostly black,* with *long, pale, clay-colored median crown stripe* extending to nape. Immature lacks black on throat.
SIMILAR SPECIES: A neat, sleek, small tanager, perhaps most likely to be mistaken for a brush-finch (e.g., Slaty), though much smaller and with pale eye, etc. Cf. also Black-backed Bush-Tanager (with which it sometimes occurs in the same flock).
HABITAT AND BEHAVIOR: Uncommon and somewhat local in dense wet temperate zone forest and shrubby edges and in elfin woodland, mostly at or just below timberline. Usually seen in pairs or small groups, regularly accompanying mixed flocks of other insectivorous birds. Characteristically perches on the *top* of foliage.
RANGE: Andes of sw. Venezuela (s. Táchira), Colombia (in W. Andes known only from Páramo Frontino at n. end), e. Ecuador (locally), and extreme n. Peru (Cerro Chinguela on Piura-Cajamarca border). Mostly 2700–3600 m.

Hemispingus xanthophthalmus

DRAB HEMISPINGUS

PLATE: 15

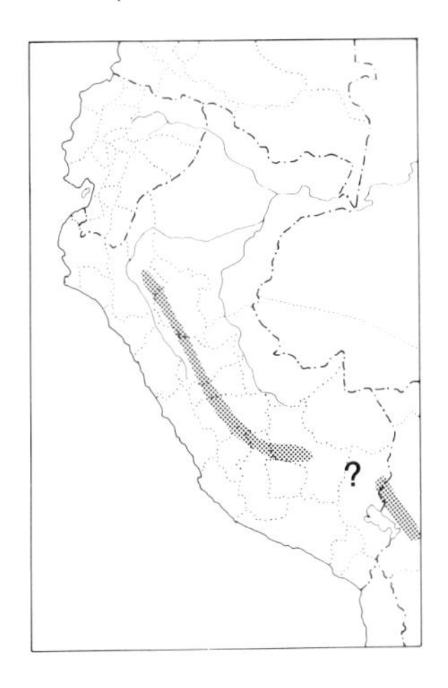

IDENTIFICATION: 14.5 cm (5¾"). *E. Peru and adjacent Bolivia. Iris straw-colored to whitish.* Uniform brownish gray above, slightly browner on head. Dingy grayish white below, whitest on throat and belly and tinged buff on lower flanks and crissum.
SIMILAR SPECIES: As the name indicates, a *notably drab* bird whose only really distinctive mark is the pale eye. In its range and habitat most likely to be confused with the white-bellied forms of Superciliaried Hemispingus, which, however, have prominent white superciliary and dark iris.
HABITAT AND BEHAVIOR: Uncommon to locally fairly common in temperate zone forest and shrubby borders and in elfin woodland; often found near timberline northward, but to south usually found lower. Forages in pairs or small groups which often accompany mixed flocks of insectivorous birds in the canopy or shrubby edges. Characteristically hops on top of foliage.
RANGE: Andes of Peru (north to cen. Amazonas, south of the Marañón River) and nw. Bolivia (La Paz). 2500–3200 m.

GROUP C

Underparts ochraceous to rufous.

Hemispingus trifasciatus

THREE-STRIPED HEMISPINGUS

PLATE: 15

IDENTIFICATION: 14 cm (5½"). *S. Peru and nw. Bolivia. Above* (including center of crown) *brownish olive.* Sides of crown and face blackish, with *long narrow superciliary buffy whitish. Below tawny-ochraceous,* paler and yellower on center of belly.
SIMILAR SPECIES: Because of its head pattern perhaps most likely to be confused with various *Basileuterus* warblers, none of which, however, is so ochraceous below (all are yellower or more olive). Cf. Black-

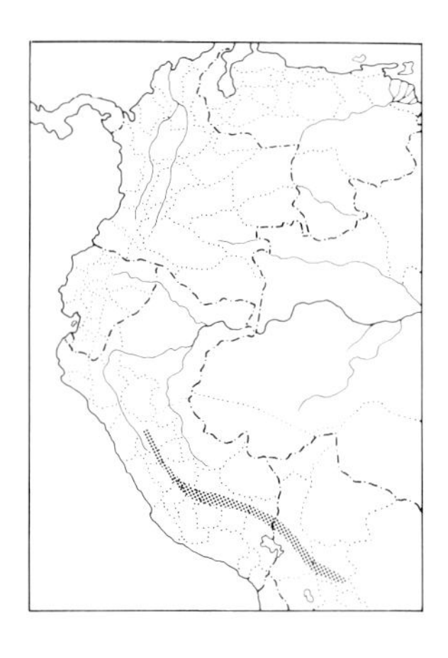

eared Hemispingus, which is typically found at subtropical elevations (lower than this species), and which is grayer above (including crown) and has much less prominent superciliary (if any is present at all) in races sympatric with Three-striped.

HABITAT AND BEHAVIOR: Fairly common in humid temperate zone forest and forest borders. Usually observed in small groups of up to 4 to 6 birds, often accompanying mixed foraging flocks of other small insectivorous birds. Feeds actively, gleaning in foliage. Seems to replace Superciliaried Hemispingus above about 3000 m. Quite readily found along the road through high montane forest at Abra Malaga in Cuzco, Peru.

RANGE: Andes of cen. and s. Peru (north to Huánuco) and n. Bolivia (La Paz and Cochabamba). Mostly 2900–3600 m.

Hemispingus melanotis

BLACK-EARED HEMISPINGUS

PLATE: 15

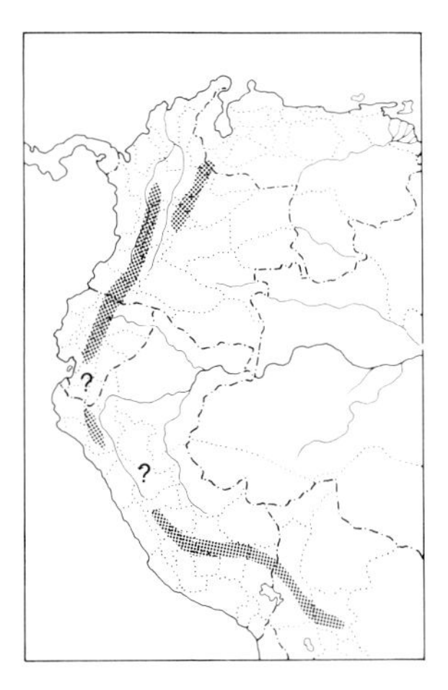

IDENTIFICATION: 14–15 cm ($5\frac{1}{2}$–6"). Highly variable. Nominate *melanotis* (w. Venezuela, most of Colombia, and e. Ecuador) has *crown and nape gray,* becoming grayish olive brown on mantle. *Cheeks broadly black* surmounted by narrow indistinct pale gray superciliary, whiter above lores, and with white tear-shaped spot on lower eyelid. *Below cinnamon buff,* whitish on center of belly. *Ochraceus* (w. Nariño, Colombia, and w. Ecuador) much duller, lacking superciliary and has dusky cheeks. *Berlepschi* (Peru from south of Marañón to Cuzco) is similar to nominate but has black chin and somewhat richer ochraceous breast and is clearly intermediate toward *castaneicollis* (se. Peru and Bolivia), which has obvious *black upper throat, rufous chest,* and *fairly prominent long white superciliary.* Most divergent is the *piurae* group (n. Peru on w. slope); these have *black crown* with *wider, bold white superciliary* and *almost uniform cinnamon-rufous underparts* (apart from black throat).

SIMILAR SPECIES: Variation confusing, but all races known by combination of black cheeks, superciliary (be it white or pale gray), and cinnamon to rufous underparts. Might be confused with Rufous-crested Tanager (also basically gray and cinnamon-rufous, but lacking black and white on head, etc.) or Fawn-breasted Tanager (with blue on crown and rump); neither of these is a forest interior species. Cf. also Slaty-backed Hemispingus.

HABITAT AND BEHAVIOR: Uncommon to fairly common but local in lower growth of humid montane forest, especially on slopes or in ravines with a dense growth of *Chusquea* bamboo. Infrequent at edge and generally rather inconspicuous. Usually found singly or in pairs, gleaning actively in foliage, and often with mixed flocks of insectivorous birds.

RANGE: Andes of extreme sw. Venezuela (s. Táchira), locally in Colombia (not in W. Andes except in Nariño), Ecuador (on w. slope seemingly only in Chimborazo), Peru (on w. slope south to s. Cajamarca; locally on e. slope), and nw. Bolivia (La Paz, Cochabamba, and w. Santa Cruz). Mostly 1700–2500 m.

NOTE: The striking plumage variation found among the various forms now considered as races of *H. melanotis* leads us to suspect that more than 1 species is involved; more study is needed.

Hemispingus goeringi

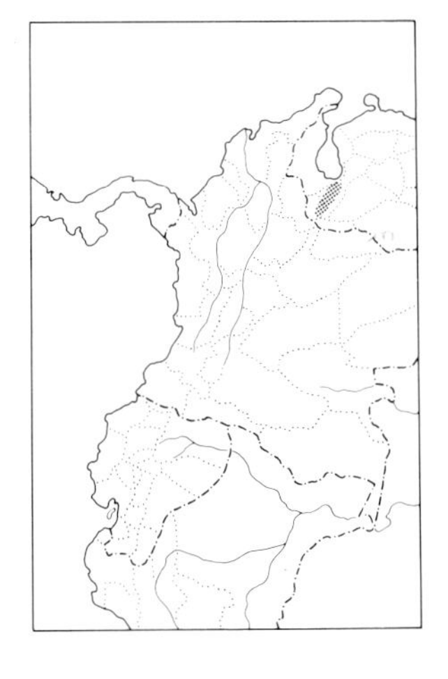

SLATY-BACKED HEMISPINGUS

IDENTIFICATION: 14.5 cm (5¾"). *W. Venezuela.* Resembles Black-eared Hemispingus. Similar to that species' *piurae* group, it differs only in slaty back, lacking the black throat, and in having the superciliary not quite reaching the base of the bill. See V-36. Compared to nominate *melanotis* (the most nearly sympatric race of that species), Slaty-backed has black (not gray) crown with conspicuous white superciliary, is much darker and more uniform gray above (no olivaceous), and has much richer cinnamon-rufous underparts.
HABITAT AND BEHAVIOR: Very poorly known. Apparently rare in lower growth of montane forest, perhaps especially where bamboo is present. C. Parrish (pers. comm.) has twice observed this species in the Páramo Zumbador area of Táchira; on both occasions single birds were moving within 1 m of the ground while with a mixed flock.
RANGE: Andes of w. Venezuela in Mérida and n. Táchira. 2400–3200 m.

Hemispingus rufosuperciliaris

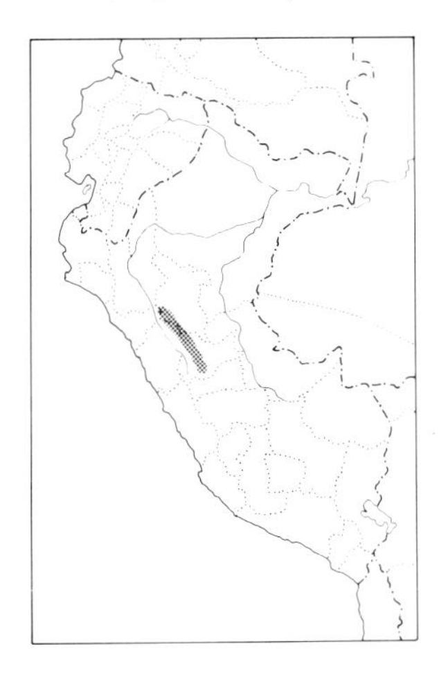

RUFOUS-BROWED HEMISPINGUS

PLATE: 15

IDENTIFICATION: 15 cm (6"). *Local near timberline on e. slope of Andes in Peru.* Crown and sides of head deep black, separated by *broad cinnamon-rufous superciliary.* Otherwise deep slaty above. *Mostly cinnamon-rufous below,* with flanks brownish gray.
SIMILAR SPECIES: Virtually unmistakable in its restricted range and habitat. Cf. Orange-browed Hemispingus.
HABITAT AND BEHAVIOR: Uncommon and local in undergrowth of humid montane forest and woodland, especially near and just below timberline. Inconspicuous, typically foraging in pairs on or near the ground, often in stands of *Chusquea* bamboo. Sometimes found with mixed flocks, but perhaps more often moves independently of them. A contact note is an ascending "tseee" (M. Robbins). Best known from the Carpish ridge region of Huánuco (but only very rarely found as low as the road itself).
RANGE: Andes of e. Peru in e. La Libertad and Huánuco. 2600–3100 m.

NOTE: A newly described species: E. R. Blake and P. Hocking (*Wilson Bull.* 86 [4]: 321–323, 1974).

Pipraeidea Tanagers

A monotypic genus of uncertain affinities: possibly closest to *Euphonia,* which it resembles in its small size and simple overall pattern, but has also been considered related to *Delothraupis.*

Pipraeidea melanonota

FAWN-BREASTED TANAGER

PLATE: 16

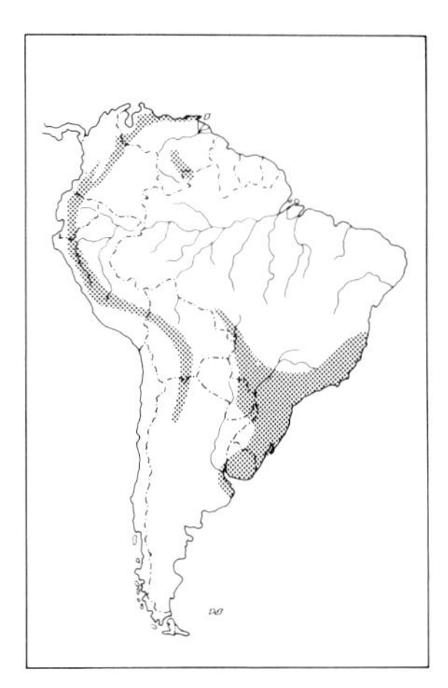

IDENTIFICATION: 14 cm (5½"). Iris bright red in Andes and Venezuela (*venezuelensis*), browner in se. South America. *Crown and nape bright pale blue,* with *broad black mask* through eyes into ear-coverts. Otherwise dusky blue above, but with pale blue rump (same color as crown). *Below uniform fawn buff.* Female with same pattern but much duller.
SIMILAR SPECIES: Superficially resembles male of smaller Golden-rumped Euphonia, but that species has black throat and yellow rump. In Peru and Bolivia cf. also Chestnut-bellied Mountain-Tanager.
HABITAT AND BEHAVIOR: Uncommon to locally fairly common in lighter woodland, forest borders, and clearings with scattered bushes and trees; rare or absent from mostly forested regions and possibly increasing in overall numbers due to deforestation. Usually seen singly or in pairs, most often in the semiopen, but may perch at virtually any height. Typically they do not accompany mixed flocks.
RANGE: N. coastal mts. of Venezuela (Sucre west) and in Andes of w. Venezuela (w. Lara and Mérida), Colombia (recorded from Pacific slope of W. Andes only in Valle and Nariño), Ecuador, Peru (south on w. slope to Lima), w. Bolivia, and nw. Argentina (Salta, Jujuy, and Tucumán); se. Brazil (s. Bahia to Rio Grande do Sul), e. Paraguay (north to Canendiyu and west to Paraguarí), ne. Argentina (Misiones), and Uruguay; *tepui* region of s. Venezuela (s. Bolívar and Amazonas) and extreme n. Brazil. Mostly 1500–2500 m, but regularly lower on Pacific slope (to 700 m or lower), and in se. South America from sea level to about 1200 m.

Delothraupis Mountain-Tanagers

An attractive, mainly blue and rufous tanager of the Peruvian and Bolivian Andes. It is here considered to be a monotypic genus separate from *Dubusia,* following J. T. Zimmer (*Am. Mus. Novitates* 1262, 1944) and *Birds of the World* (vol. 13), but we remain uncertain whether this is the best course (Meyer de Schauensee 1966, 1970 considered them congeneric).

Delothraupis castaneoventris

CHESTNUT-BELLIED MOUNTAIN-TANAGER

PLATE: 16

IDENTIFICATION: 17 cm (6¾"). *Above mostly dark blue,* paler and more silvery blue on crown with streaky or frosty effect, contrasting with black sides of head and neck. *Below mostly rufous, deepening on belly,* with whitish chin and conspicuous *black submalar streak.*
SIMILAR SPECIES: In size and overall pattern most resembles Fawn-breasted Tanager but with brown (not red) eye, deeper rufous (not buff) underparts, rump concolor (not pale blue) with remaining upperparts, and black whisker. Cf. also Rufous-crested Tanager (limited overlap in cen. Peru) and Black-eared Hemispingus.

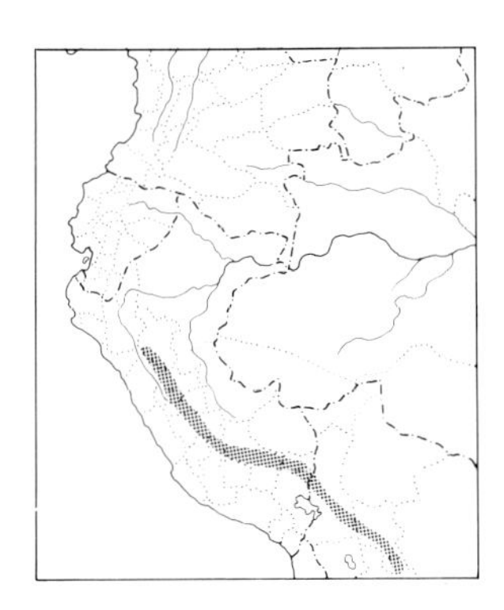

HABITAT AND BEHAVIOR: Uncommon in humid montane forest and forest borders, especially near and just below timberline. Found singly or in pairs, frequently with mixed foraging flocks. Forages for insects rather deliberately, typically working along mossy horizontal branches, usually in middle and upper strata; also eats much fruit. Its song is a simple, somewhat musical "tsee, tee-u-ee" (not too dissimilar from Buff-breasted Mountain-Tanager's, though not nearly as loud or forceful).
RANGE: Andes of cen. and s. Peru (north to e. La Libertad) and nw. Bolivia (La Paz, Cochabamba, and w. Santa Cruz). Mostly 2400–3400 m.

Creurgops Tanagers

A pair of rather heavy-billed tanagers with simple, mostly gray or gray and rufous color patterns.

Creurgops dentata

SLATY TANAGER PLATE: 16

IDENTIFICATION: 15.5 cm (6″). *Se. Peru and nw. Bolivia.* Male *mostly slaty gray,* with *wide chestnut crown* very narrowly bordered with black. Female mostly slaty gray above with blackish crown and nape and narrow white superciliary. *Sides of neck, entire breast, and flanks rufous* (brightest on breast); median throat and belly white.
SIMILAR SPECIES: Dapper gray male with his prominent chestnut crown is patterned unlike any other tanager. Females look so different that they might not even be recognized as belonging to the same species (and in fact at first they were not). Their basically gray and rufous coloration is reminiscent of numerous other tanagers (e.g., the much less arboreal Black-eared Hemispingus).
HABITAT AND BEHAVIOR: Uncommon in humid montane forest and forest borders. Behavior very similar to Rufous-crested Tanager's, which it replaces southward. Can be found with fair regularity at appropriate elevations in the Chapare along the road between Cochabamba city and Villa Tunari in Cochabamba, Bolivia.
RANGE: Andes of se. Peru (north to Cuzco at Aputinye and on the Cordillera Vilcabamba) and nw. Bolivia (La Paz and Cochabamba). Mostly 1600–2200 m.

Creurgops verticalis

RUFOUS-CRESTED TANAGER PLATE: 16

IDENTIFICATION: 16 cm (6¼″). *Mainly Colombia to Peru. Gray above and cinnamon-rufous below.* Coronal patch also cinnamon-rufous and outlined with blackish, but the patch is usually not conspicuous in the field. Female similar but lacks rufous in crown and is slightly paler below.
SIMILAR SPECIES: Black-eared Hemispingus always has at least some black on head. Male Rusty Flower-piercer has similar color pattern (no

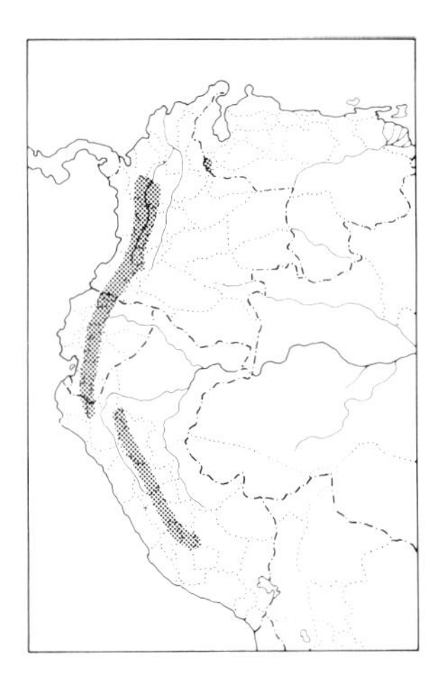

crown patch) but is much smaller with different bill, behavior, etc. Cf. also Chestnut-bellied Mountain-Tanager (much bluer above with silvery on crown, a black submalar streak, etc.).

HABITAT AND BEHAVIOR: Fairly common but somewhat local in humid montane forest and forest borders. Usually found singly or in pairs, often accompanying mixed flocks of other tanagers, etc. Mostly insectivorous, foraging rather high, primarily by slowly working along limbs, carefully inspecting foliage and probing moss; also occasionally sallies after flying insects. Numerous at Finca Merenberg in Huila, Colombia.

RANGE: Andes of extreme w. Venezuela (s. Táchira), Colombia (where local and not numerous except around head of upper Magdalena valley), nw. (recently collected in Carchi; N. Krabbe) and e. Ecuador, and e. Peru (south to Ayacucho). Mostly 1400–2700 m.

Iridosornis Tanagers

A small group of richly colored, lovely tanagers of high elevations in the Andes. All 5 species are predominantly blue with sharply contrasting yellow (either on the head or the throat). They are quiet and, despite their bright colors, often inconspicuous tanagers of forest lower growth.

GROUP A *Extensive yellow throat.*

Iridosornis analis

YELLOW-THROATED TANAGER PLATE: 16

IDENTIFICATION: 16 cm (6¼"). *Mainly e. Ecuador and e. Peru.* Mostly dull blue above, more purplish on crown and nape, more greenish on back and rump; lores and sides of head black (not very contrasty, but imparting weak masked effect). *Large throat patch bright yellow, contrasting with dull buffy ochraceous underparts;* crissum dull chestnut.

SIMILAR SPECIES: See Purplish-mantled Tanager (only Pacific Colombia and Ecuador; no overlap with this species).

HABITAT AND BEHAVIOR: Uncommon and perhaps somewhat local in lower growth of humid forest, less often at forest borders. Usually found in pairs, and regularly accompanies mixed flocks, but seemingly not very active and thus often not noticed. Generally silent.

RANGE: Andes of se. Colombia (sighting from w. Putumayo; S. Hilty), e. Ecuador, and e. Peru (south to Puno). Mostly 1300–2100 m.

Iridosornis porphyrocephala

PURPLISH-MANTLED TANAGER

IDENTIFICATION: 16 cm (6¼"). *W. Colombia and nw. Ecuador.* In pattern a duplicate of Yellow-throated Tanager (no overlap), but much bluer above (deepening to purplish blue on crown and nape) and *mostly blue below* (most intense in a narrow band just below the yellow throat). Center of lower belly buff, and crissum dull chestnut. See C-50.

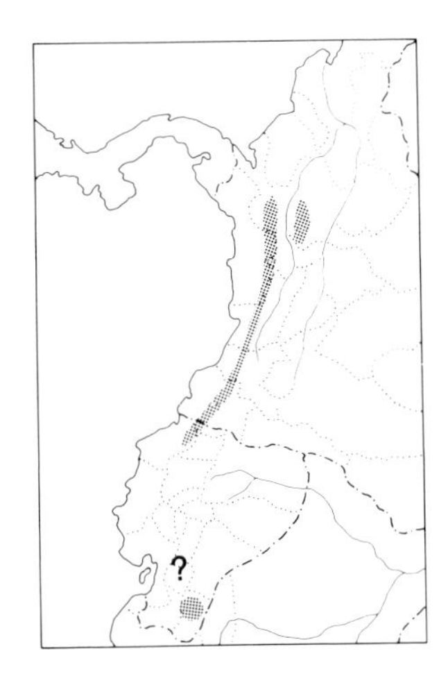

SIMILAR SPECIES: Yellow-throated Tanager (only on e. slope of Andes) has buff underparts and is duller blue above. Cf. also Golden-chested Tanager (all dark with yellow chest, not a throat patch).
HABITAT AND BEHAVIOR: Uncommon to locally fairly common in lower growth of humid forest (especially "mossy" forest), less often at forest borders. Behavior like Yellow-throated Tanager's; like that species rather sluggish and thus quite inconspicuous. Can be encountered with fair regularity at several localities in Valle and Cauca, Colombia (e.g., in the Tokio forest above Queremal on the w. slope in Valle).
RANGE: W. Andes of Colombia (mostly Pacific slope of W. Andes north to s. Chocó, but also locally on adjacent e. slope and recorded also from both slopes of n. end of Cen. Andes in Antioquia) and w. Ecuador (recorded only from Carchi and Imbabura, with an uncertain report from Loja). Mostly 1500–2200 m.

NOTE: Though the sympatry mentioned by Meyer de Schauensee (1966: 475) has in fact not been confirmed, we nonetheless continue to take the conservative course and maintain this and *analis* as full species. Nonetheless we must note that their patterns are essentially identical, and we would thus be more inclined to regard these 2 as conspecific than we would the *I. rufivertex/reinhardti* pair, despite the fact that in *Birds of the World* (vol. 13) the opposite was done.

GROUP B

Yellow on head.

Iridosornis jelskii

GOLDEN-COLLARED TANAGER

PLATE: 16

IDENTIFICATION: 15 cm (6"). *E. Peru and nw. Bolivia. Yellow collar extending up from sides of neck and onto crown* (where mixed with black), outlining black face. Otherwise blue above, duller on rump; wings and tail dusky, edged greenish blue. Throat black, *remaining underparts dull chestnut,* shaded with dull blue on sides of breast.
SIMILAR SPECIES: Yellow-scarfed Tanager is blue (not chestnut) below and has an all-black crown. Pattern also somewhat reminiscent of Saffron-crowned Tanager's.
HABITAT AND BEHAVIOR: Rare to locally uncommon in woodland and forest borders, especially near timberline. General behavior similar to Golden-crowned and Yellow-scarfed Tanagers', but usually forages more out in the open, regularly inspecting mossy limbs and trunks. Most often occurs at higher elevations than Yellow-scarfed, though there may be local overlap.
RANGE: Andes of cen. and s. Peru (north to La Libertad and San Martín) and nw. Bolivia (La Paz). Mostly 2500–3500 m.

Iridosornis rufivertex

GOLDEN-CROWNED TANAGER

PLATE: 16

IDENTIFICATION: 17 cm (6¾"). *Mainly Colombia and Ecuador. A largely bright deep blue tanager with a rich golden crown patch.* Head, throat, and neck black, with a patch of golden yellow on top of crown (like a skullcap). Otherwise mostly deep purplish blue, but wings and tail black broadly edged greenish blue. Lower belly and crissum chest-

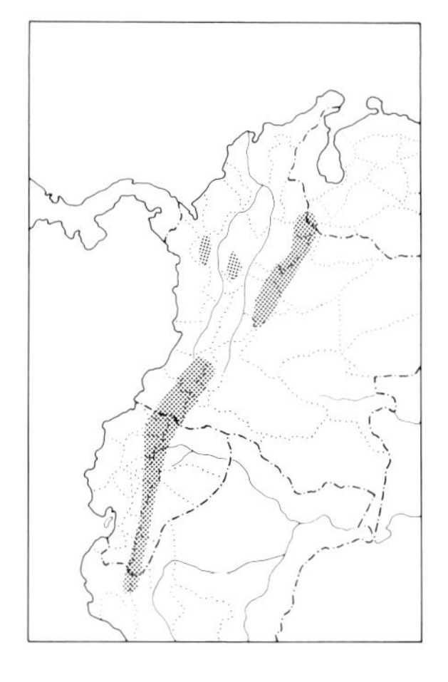

nut (but purplish blue in *caeruleoventris* of n. end of Colombia's Cen. and W. Andes).

SIMILAR SPECIES: This is one of the loveliest tanagers, but its colors (particularly the vivid blue) can look subdued in poor light. Yellow-scarfed Tanager (found to the south of this species' range) has yellow extending around nape (and none on top of crown). Cf. also Plushcap, with rather similar pattern but rufous (not mainly blue) below.

HABITAT AND BEHAVIOR: Fairly common in high-elevation shrubbery and the edge of temperate forest and woodland. Typically occurs in pairs which regularly accompany mixed flocks of other tanagers, brush-finches, tyrannids, etc. Usually forages low, and often skulks, but at times takes an exposed perch on top of a bush prior to diving back into cover.

RANGE: Andes of extreme w. Venezuela (s. Táchira), Colombia (somewhat local northward), Ecuador (on w. slope not known south of Pichincha), and extreme n. Peru (Cerro Chinguela on Piura-Cajamarca border). Mostly 2500–3500 m.

Iridosornis reinhardti

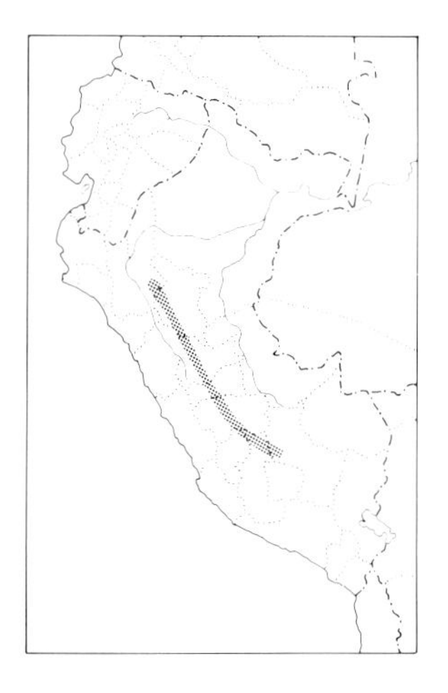

YELLOW-SCARFED TANAGER

IDENTIFICATION: 16.5 cm (6½"). Resembles Golden-crowned Tanager, replacing that species in the Peruvian Andes south of the Marañón River. Instead of having a golden crown patch, it has a *band of golden yellow extending from sides of neck up over rear of crown* (looking like earmuffs). *Lower underparts dull blue* (with no chestnut).

SIMILAR SPECIES: Golden-collared Tanager has dull chestnut (not blue) underparts and its yellow collar extends up onto crown.

HABITAT AND BEHAVIOR: Uncommon in temperate zone woodland and forest borders and in shrubbery near timberline. Behavior much like Golden-crowned Tanager's. This beautiful tanager is quite readily seen along the Carpish Pass road between Huánuco and Tingo María in Huánuco, Peru.

RANGE: Andes of Peru from s. Amazonas (Leymebamba) and San Martín south to Cuzco (Cordillera Vilcabamba). Mostly 2000–3400 m.

NOTE: This species is so distinctly different from its n. allospecies, *I. rufivertex* (from which it is separated by the Río Marañón barrier), that we see no reason for regarding the 2 as conspecific, as was done by Hellmayr and in *Birds of the World* (vol. 13). The head pattern of none of the races of *rufivertex* shows any approach to *reinhardti*.

Anisognathus Mountain-Tanagers

Another small group of boldly patterned Andean tanagers, the *Anisognathus* mountain-tanagers are larger than *Iridosornis* and colored very differently, with either all-yellow or all-red underparts. They are arboreal and tend to be common and conspicuous birds. The first 2 species discussed were formerly placed in the genus *Compsocoma* and have more conical bills; the "true" *Anisognathus* have stubbier bills.

GROUP A *Yellow median crown stripe,* more conical bill ("*Compsocoma*" subgenus).

Anisognathus somptuosus
(incorrectly called *A. flavinucha* on Plate 16)

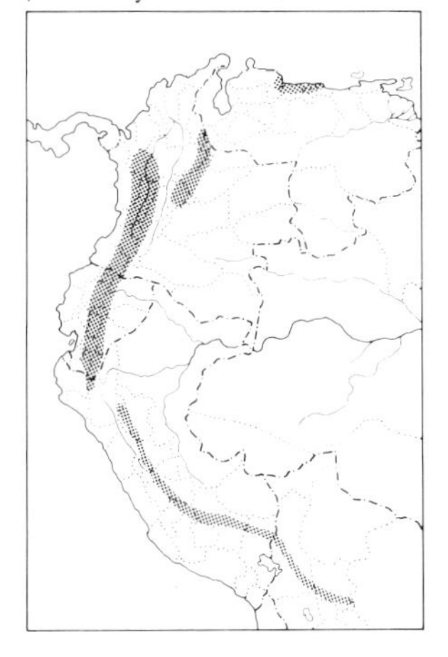

BLUE-WINGED MOUNTAIN-TANAGER PLATE: 16

IDENTIFICATION: 18 cm (7″). Most races (including *somptuosus*) mainly black above with *broad yellow stripe extending from mid-crown to nape,* inconspicuous blackish olive rump. Shoulders cobalt blue and *flight feathers broadly edged bright turquoise blue;* outer tail feathers also edged blue. Entirely bright yellow below. Races of e. Colombia (*victorini*) and ne. Ecuador (*baezae*) have back moss green (see C-50), while *cyanoptera* (of sw. Colombia and most of w. Ecuador) has flight feathers edged with same cobalt blue as the shoulders. Nominate *flavinucha* (of Bolivia and se. Peru) differs strikingly from *somptuosa* group in having a contrasting *large bright cobalt blue rump patch.*

SIMILAR SPECIES: No other mountain-tanager has such contrasty blue on the flight feathers. In w. Colombia and w. Ecuador cf. the superficially similar Black-chinned Mountain-Tanager.

HABITAT AND BEHAVIOR: Common and widespread in humid forest and forest borders and in lesser numbers even in secondary woodland and more or less isolated woodlots. Rather active and conspicuous, foraging mostly in the canopy or at edge, frequently fully in the open. Usually in groups of up to 8 to 10 birds which often are with mixed flocks (frequently behaving as if one of the nuclear species around which such flocks form), but may also move about independently.

RANGE: N. coastal mts. of Venezuela (Yaracuy to Miranda) and Andes of extreme w. Venezuela (s. Táchira), Colombia, Ecuador (south on w. slope to El Oro), e. Peru, and nw. Bolivia (La Paz and Cochabamba). Mostly 1500–2500 m, but regularly down to at least 1200 m on Pacific slope.

Anisognathus notabilis

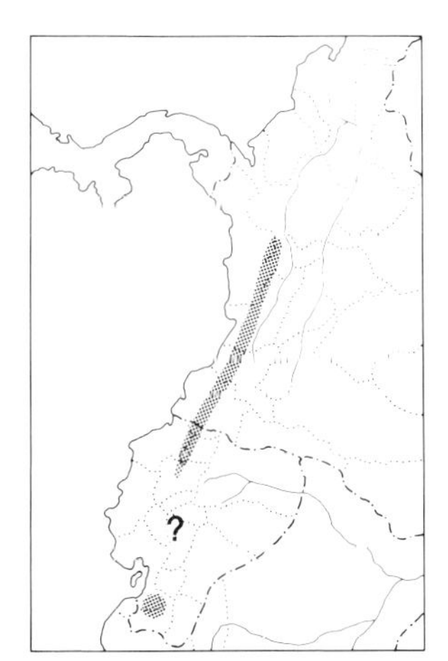

BLACK-CHINNED MOUNTAIN-TANAGER

IDENTIFICATION: 18 cm (7″). In general form recalls Blue-winged Mountain-Tanager; note *sw. Colombia and w. Ecuador range. Head and nape mostly black,* the black extending down over face to include chin; small yellow stripe down center of hindcrown. *Back and rump rather shiny yellow-olive,* contrasting with black head and mostly black wings and tail; greater coverts edged blue, but *little or no blue on flight feathers. Below entirely rich orange-yellow.* See C-50.

SIMILAR SPECIES. Overlapping race of Blue-winged Mountain-Tanager (*cyanopterus*) is basically black above (not obviously olive), more lemony (not so orange) yellow below, and has different head pattern: a wider yellow coronal stripe and entirely yellow throat, resulting in a black-masked rather than a black headed look.

HABITAT AND BEHAVIOR: Uncommon and rather local in humid (sometimes mossy) forest and forest borders. Behavior resembles Blue-winged Mountain-Tanager's, but usually in smaller groups (only up to 4 to 6 birds); like that species frequently with mixed flocks, especially of *Tangara* tanagers.

RANGE: Pacific slope of w. Colombia (north to s. Chocó in headwaters of Río San Juan) and w. Ecuador (south to Pichincha and again in El Oro and w. Loja). Mostly 800–2000 m.

GROUP B

Stubbier bill ("true" *Anisognathus*).

Anisognathus lacrymosus

LACRIMOSE MOUNTAIN-TANAGER

PLATE: 16

IDENTIFICATION: 18 cm (7″). *Venezuela to Peru.* Mostly dark bluish slate above, duskier on sides of head and neck, and dark blue on rump. *Small spot below eye* (the "teardrop") *and larger spot behind ear-coverts yellow.* Wings and tail inconspicuously edged blue. *Below deep ochraceous yellow.* Nominate race (Peru south of Marañón River) lacks the postauricular spot (retains the teardrop) and is bluer above. *Pallidorsalis* of Perijá Mts. looks quite different, with paler bluish gray upperparts and sides of head and neck yellowish olive.

SIMILAR SPECIES: The spots of yellow on the sides of the head are quite prominent, and in conjunction with the basically bicolored (slaty and yellow) pattern readily identify it.

HABITAT AND BEHAVIOR: Common in humid forest and forest borders and in more stunted (near timberline) or disturbed woodland. Usually in small groups which forage at all levels (though rarely inside forest), frequently with mixed flocks of other tanagers, etc. One of the more often seen tanagers of temperate forests in its range.

RANGE: Andes of w. Venezuela (north to Trujillo), Colombia, Ecuador (apparently absent from actual Pacific slope, though recorded locally from Azuay and El Oro), and e. Peru (south to Cordillera Vilcabamba in n. Cuzco); Perijá Mts. on Venezuela-Colombia border. Mostly 2500–3600 m.

Anisognathus melanogenys

SANTA MARTA MOUNTAIN-TANAGER

Other: Black-cheeked Mountain-Tanager

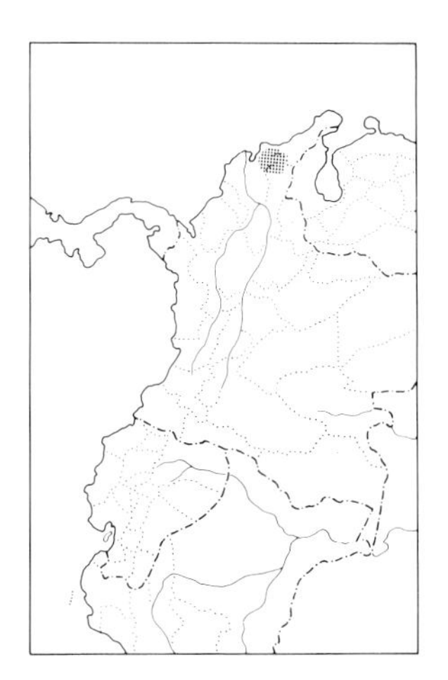

IDENTIFICATION: 18 cm (7″). *Colombia's Santa Marta Mts.* Resembles Lacrimose Mountain-Tanager (no overlap), but *crown and nape blue* (in contrast with black sides of head) and has only the yellow teardrop below eye (no postocular spot). Otherwise dull grayish blue, golden yellow below, but with thighs black. See C-50.

SIMILAR SPECIES: The only mountain-tanager of its type in the Santa Martas and as such easily identified. Cf. Santa Marta Brush-Finch.

HABITAT AND BEHAVIOR: Fairly common in forest borders and secondary woodland, less often out into regenerating cleared areas and actually inside forest. Found in pairs or small groups which forage at all levels though less often in canopy; frequently accompanies mixed flocks.

RANGE: Santa Marta Mts. of Colombia. Mostly 2000–3000 m but shifts lower during peak of rainy season (June–Sept.).

NOTE: As so many other mountain-tanagers have black cheeks, it seems far preferable to emphasize the restricted distribution of this tanager by calling it the "Santa Marta Mountain-Tanager."

Anisognathus igniventris

SCARLET-BELLIED MOUNTAIN-TANAGER PLATE: 16

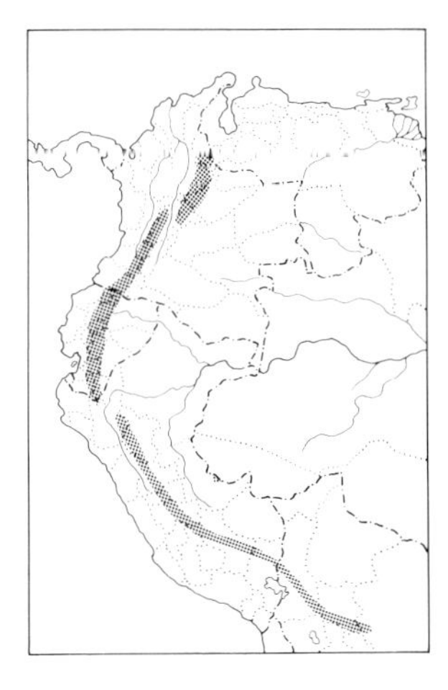

IDENTIFICATION: 18.5 cm (7¼"). A spectacular tanager, *unmistakable in its black, vivid red, and blue garb.* Mostly black above with triangular red patch on ear-coverts and bright *blue shoulders and rump* (latter visible mostly in flight). Throat and chest also black, but *lower underparts scarlet;* crissum mixed black and red, and with black thighs. Nominate race (se. Peru and Bolivia) is a duller, more bluish slate above, and its lower underparts are a more orangy red, and this color extends to the crissum. Sexes similar, but throughout range duller, more orangy birds (presumably immatures) are often seen.

HABITAT AND BEHAVIOR: Common and widespread in woodland and shrubbery near timberline and lower down in regenerating clearings and pastures with scattered trees, even in dense hedgerows. Moves about in pairs or small groups (occasionally up to a dozen or so together) which, though they usually are found in semiopen terrain, are often frustratingly difficult to see except when they fly, for they tend to perch inside heavy cover. Sometimes joins mixed flocks of other tanagers, finches, etc. but at least as often occurs alone. Not only is this one of the showiest tanagers, it also has a surprisingly lovely song, a distinctive jumbled series of bell-like tinkling notes which sound remarkably untanagerlike, usually given from concealment.

RANGE: Andes of extreme sw. Venezuela (s. Táchira), Colombia (not in W. Andes, and somewhat local northward in Cen. and E. Andes), Ecuador (both slopes), e. Peru, and nw. Bolivia (south to Cochabamba and w. Santa Cruz). Mostly 2500–3500 m, locally to 2200 m.

Bangsia and *Wetmorethraupis* Tanagers

Bangsia tanagers are chunky, rather short-tailed tanagers found in the wet forests of w. Colombia and adjacent Ecuador. Plumage patterns vary, bold and simple in some, more subdued in others, but all 4 species tend to be rather inactive birds of forest understory and mid-levels. It has been suggested that the genus *Bangsia* be merged into *Buthraupis,* but we favor maintaining it. *Wetmorethraupis* remains virtually unknown in life, but in proportions and overall color scheme seems close to *Bangsia;* it is known only from a limited area in n. Peru.

GROUP A

Mainly dark green with *contrasting color on face;* mandible flesh color.

Bangsia edwardsi

MOSS-BACKED TANAGER PLATE: 16

IDENTIFICATION: 16 cm (6¼"). *W. Colombia and nw. Ecuador.* Bill light dusky above, yellowish flesh below (looks pale in the field). Foreface, top of crown, and center of nape black, enclosing *large, round, dull blue area on sides of head and neck.* Otherwise *mossy green above,* wings and tail dusky, edged greenish blue. Throat black, with *large yellow patch on center of chest.* Remaining underparts dull green.

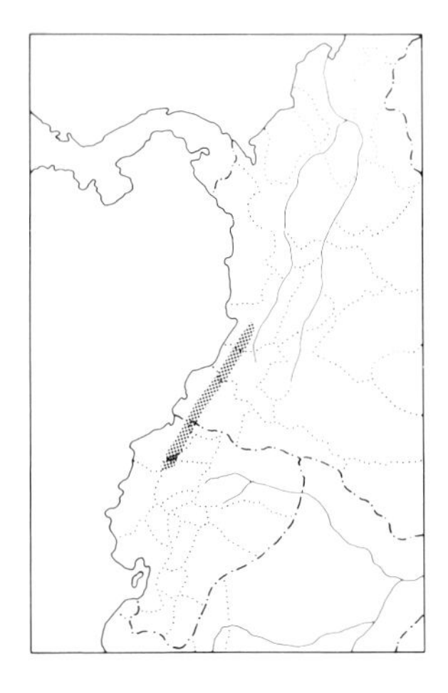

SIMILAR SPECIES: No other tanager combines the mostly blue head and prominent yellow chest patch.

HABITAT AND BEHAVIOR: Locally fairly common to common in wet, mossy forest and at forest edge, occasionally out into clearings. Usually found singly or in pairs, often accompanying mixed flocks. Rather sluggish behavior; may perch almost motionless on horizontal limb or mostly in the open, stolidly peering about. Rather quiet, but when breeding birds may mount to top of tree to give their untanagerlike song, a simple unmusical series of chippering notes, recalling certain foliage-gleaners. One of the most numerous birds at El Placer in Esmeraldas, Ecuador.

RANGE: Pacific slope of W. Andes in sw. Colombia (north to the Río Dagua valley in Valle) and nw. Ecuador (south to Pichincha). Mostly 700–1800 m.

Bangsia aureocincta

GOLD-RINGED TANAGER

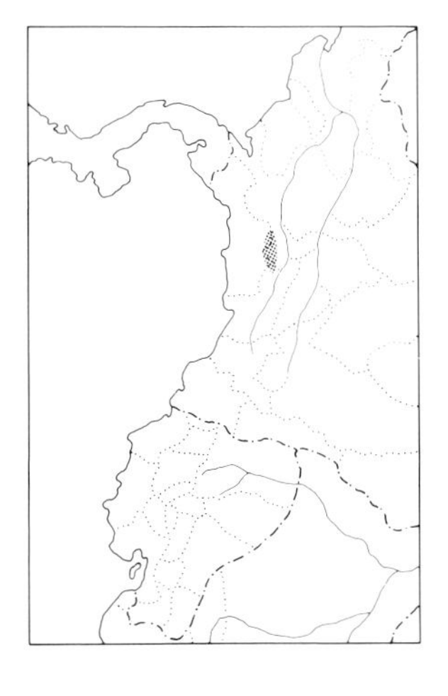

IDENTIFICATION: 16 cm (6¼"). *W. Colombia;* rare. Bill blackish above, yellowish below. Crown and nape black; *conspicuous bright yellow ring starting above eye and encircling dark ear-coverts, ending at base of bill.* In other respects very like Moss-backed Tanager. See C-50. Female has black of head replaced by dark olive.

SIMILAR SPECIES: No other tanager has a facial pattern resembling this one's. It does, however, slightly recall Slaty-capped Shrike-Vireo, a similarly sized bird which occurs with it.

HABITAT AND BEHAVIOR: Very little known. Apparently occurs in wet, mossy forest and replaces Moss-backed Tanager northward. Behavior probably is like that species'.

RANGE: Pacific slope of W. Andes in w. Colombia in s. Chocó and extreme w. Caldas (headwaters of Río San Juan) and in n. Valle (Novita Trail). 2000–2200 m.

GROUP B. *Dark blue or black with contrasting yellow.*

Bangsia rothschildi

GOLDEN-CHESTED TANAGER PLATE: 16

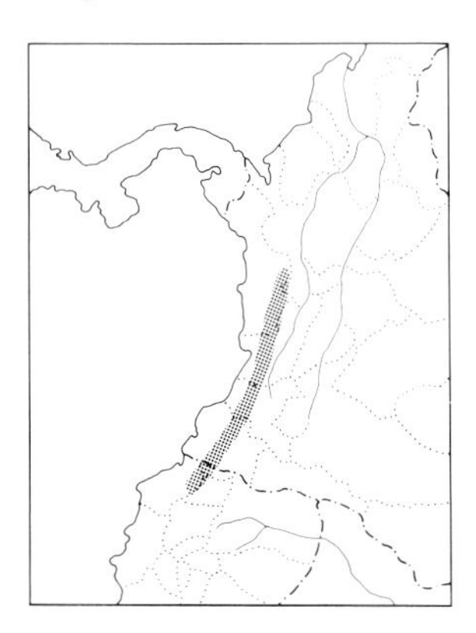

IDENTIFICATION: 16 cm (6¼"). Bill black. *Mostly dark navy blue,* with *large bright golden yellow patch on chest* and yellow crissum.

SIMILAR SPECIES: See Black-and-gold Tanager (with much more yellow on underparts, etc.). Moss-backed Tanager, which occurs in same general areas as this species (though usually higher), has green upperparts and mostly blue face. Purplish-mantled Tanager is also mostly blue and yellow but has yellow throat (rather than a big chest patch).

HABITAT AND BEHAVIOR: Fairly common in wet forest and forest borders of foothills and lower slopes. Forages singly or in pairs, regularly accompanying mixed flocks. General behavior rather sluggish and stolid, perching quietly on a horizontal limb, then hopping farther on or briefly fluttering while snatching fruit and then moving. Most often

silent, but does have an infrequently uttered wiry song as well as typical tanager contact notes. Easily found in foothills (300–500 m) along the old Buenaventura road in Valle, Colombia.
RANGE: Pacific slope of W. Andes in w. Colombia (north to s. Chocó in the upper Río Atrato drainage) and nw. Ecuador (Esmeraldas). 200–1000 m.

Bangsia melanochlamys

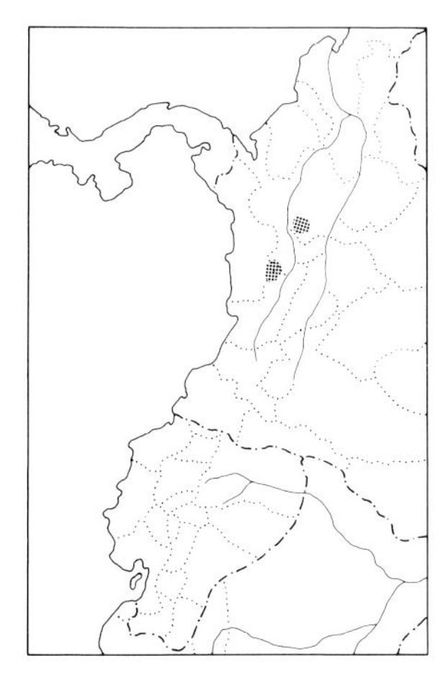

BLACK-AND-GOLD TANAGER

IDENTIFICATION: 16 cm (6¼″). *W. Colombia.* Bill black. Mostly black above with blue upper wing-coverts and upper tail-coverts. Throat black, with *large ochraceous yellow patch on center of chest, extending down broadly over median underparts* (where more golden yellow); sides and flanks black. See C-50.
SIMILAR SPECIES: Has much more yellow below than any other *Bangsia;* Golden-chested Tanager (found to the south of this species' range) has yellow only on chest patch and crissum.
HABITAT AND BEHAVIOR: Very little known. Apparently not uncommon in at least some localities in its limited range (at least formerly; some are now mostly deforested), e.g., at La Selva in Caldas (where von Sneidern collected 11 specimens in 1945) and above Valdivia, Caldas (where Carriker took an additional 11 in 1948). At the former locality recorded with (but much commoner than) Gold-ringed Tanager, another little-known Colombian endemic. Behavior unknown, but may be similar to Golden-chested Tanager's.
RANGE: W. Colombia on w. slope of N.-Cen. Andes in Caldas and Pacific slope of W. Andes in s. Chocó and w. Caldas (upper Río San Juan drainage). Mostly 1500–2200 m.

Wetmorethraupis sterrhopteron

ORANGE-THROATED TANAGER

PLATE: 16

IDENTIFICATION: 18 cm (7″). *A gaudy, unmistakable tanager of extreme n. Peru.* Black above with intense blue wing-coverts and inner flight feathers. *Throat and chest bright orange,* remaining underparts yellowish buff, with a little black on sides and flanks.
SIMILAR SPECIES: Not likely to be confused in its limited range; the orange throat is unique.
HABITAT AND BEHAVIOR: Not well known, and only recently (1964) described. Recorded from heavily forested foothill localities along e. base of Andes. In general form and plumage pattern seems reminiscent of *Bangsia,* but behavior poorly documented. Seen mostly high in forest canopy with mixed flocks of tanagers. Call reported to be a repeated, deliberate "in-chee-tooch" (J. O'Neill).
RANGE: N. Peru in Amazonas (drainage of Marañón River; recorded from headwaters of Río Cenepa and hills ne. of Bagua); may perhaps be found in adjacent Ecuador. About 600 m.

Dubusia Mountain-Tanagers

Monotypic genus of large tanagers, mainly blue and yellow with buff band across chest. Favors lower growth of temperate forest and borders in the Andes.

Dubusia taeniata

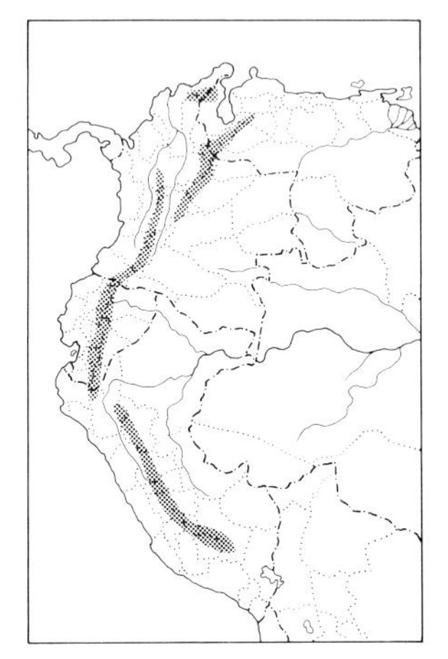

BUFF-BREASTED MOUNTAIN-TANAGER PLATE: 17

IDENTIFICATION: 18.5–19.5 cm (7¼–7¾"). Head and throat black, with *long frosty blue superciliary* extending back from forehead and onto sides of neck. Above otherwise dark blue, feathers of wings and tail edged brighter blue. *Band below black throat buff,* lower underparts bright yellow; crissum also buff. *Carrikeri* of Colombia's Santa Marta Mts. has light blue spangling on crown and *buff of breast extends up as point over center of throat. Stictocephala* of cen. and s. Peru has *brighter and more extensive blue streaking on upperparts,* extending to crown and brightest on nape, so effect of superciliary is lost.

SIMILAR SPECIES: The basic blue and yellow color pattern is reminiscent of numerous other mountain-tanagers, but none gives the effect of the frosty eyeline or more general streaking above. Often the buff chest is not conspicuous.

HABITAT AND BEHAVIOR: Usually rather uncommon (locally somewhat more numerous) at borders of humid montane forest, in dense shrubby areas with low trees, and also often inside forest in understory or middle levels; in at least some areas its presence seems associated with bamboo. Usually in pairs, less often alone, generally foraging apart from mixed flocks; generally not very conspicuous, and often seems sluggish. Much more apt to be recorded once its distinctive voice is recognized. Its song is a loud but quite sweet and musical, simple 2-note whistle, "feeeeu-bay," sometimes with 2 introductory notes, "feeeeu-feeeeu-bay."

RANGE: Andes of w. Venezuela (Trujillo to Táchira), Colombia (not in W. Andes north of Cauca), Ecuador (south on w. slope to Azuay, on e. slope to Loja), and e. Peru (Amazonas to Cuzco); Perijá Mts. on Venezuela-Colombia border, and Colombia's Santa Marta Mts. Mostly 2500–3500 m.

Buthraupis Mountain-Tanagers

Spectacular, large tanagers found at high elevations in the Andes (2 right at timberline). Patterns vary but are always bold, and all 4 species have yellow underparts. One species (the Hooded) is numerous and conspicuous, while the other 3 are much less often encountered.

Buthraupis montana

HOODED MOUNTAIN-TANAGER PLATE: 17

IDENTIFICATION: 22–23 cm (8½–9"). *A large, red-eyed mountain-tanager of temperate zone forests. Head and throat black, remaining upper-*

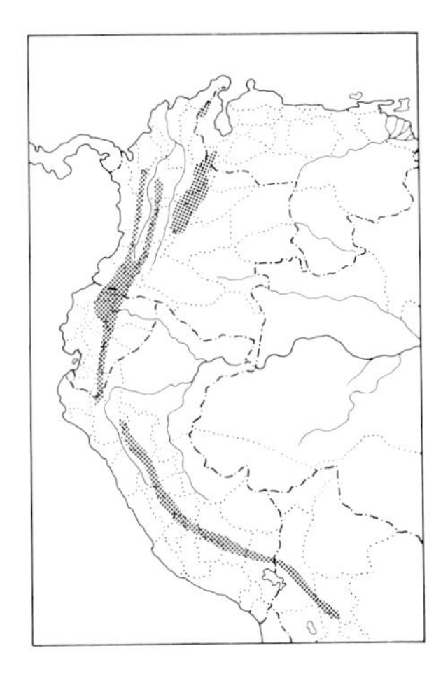

parts deep, vibrant blue (somewhat duller in *gigas* of Venezuela and ne. Colombia). Wings and tail black, broadly edged with same blue as on back. Below bright yellow, with *narrow band on lower flanks blue and thighs black* (both prominent in field). Nominate race of Bolivia is similar but has the most intense blue upperparts and also has *conspicuous pale azure-blue band on nape.*

SIMILAR SPECIES: Unlikely to be confused if seen reasonably well; even when clouds shroud the montane forest, obscuring virtually all colors, the dark band from flanks to thighs is distinctive. Cf. Black-chested Mountain-Tanager.

HABITAT AND BEHAVIOR: Usually fairly common in humid montane forest and forest borders; may stray into lower regenerating woodland or to isolated tall trees in clearings, but rarely far from mature forest. Generally occurs in conspicuous and noisy groups of 4 to 8 (sometimes more) birds which troop about the canopy, regularly taking exposed perches and often flying for long distances in the open (e.g., down a forested slope or across a canyon). May accompany mixed flocks, but at least as often away from them. Calls are frequent and loud, the most characteristic being a far-carrying "tee-tee-tee-tee," usually given in flight.

RANGE: Andes of extreme w. Venezuela (s. Táchira), Colombia, Ecuador (south on w. slope only to Pichincha), e. Peru, and n. Bolivia (south to Cochabamba); Perijá Mts. on Venezuela-Colombia border. Mostly 2000–3200 m, but lower on Pacific slope in Colombia and Ecuador (regularly to 1500 m).

Buthraupis eximia

BLACK-CHESTED MOUNTAIN-TANAGER PLATE: 17

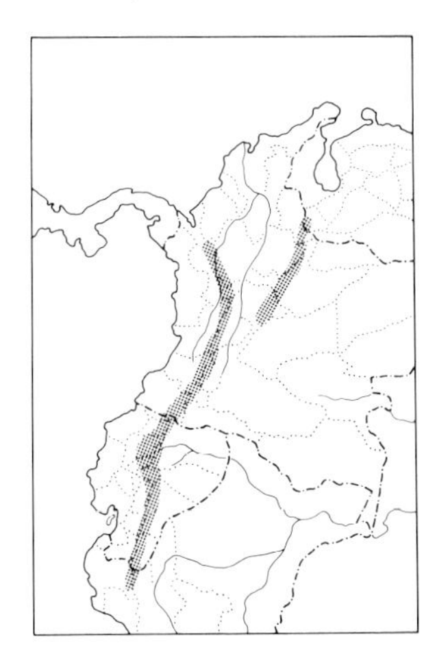

IDENTIFICATION: 20.5–21.5 cm (8–8½"). Mainly Colombia and Ecuador. *Crown and nape dark blue,* with *back and rump moss green.* Wings and tail mostly black, with blue lesser wing-coverts and inner flight feathers edged green. *Sides of head, throat, and chest black;* remaining underparts bright yellow. Nominate race of w. Venezuela and ne. Colombia has rump blue (same color as crown), but back is still moss green; otherwise like *chloronota* (rest of range).

SIMILAR SPECIES: Hooded Mountain-Tanager has entirely black head (no blue crown) and is mostly bright blue above (with no green); its iris is conspicuously red (not dark as in this species). Cf. also the very rare Masked Mountain-Tanager.

HABITAT AND BEHAVIOR: Uncommon and rather local in humid, often quite mossy montane forest and forest borders. Usually in pairs, less often alone or in small groups, most often occurring independently of mixed flocks. Nowhere near as conspicuous as Hooded Mountain-Tanager, rarely perching prominently and almost always silent. Can be found in Colombia's Puracé Nat. Park, especially at the s. 2 entrances.

RANGE: Andes of extreme sw. Venezuela (s. Táchira), Colombia, Ecuador (south on w. slope to Pichincha), and extreme n. Peru (Cerro Chinguela on Piura-Cajamarca border). Mostly 2800–3500 m.

Buthraupis aureodorsalis

GOLDEN-BACKED MOUNTAIN-TANAGER PLATE: 17

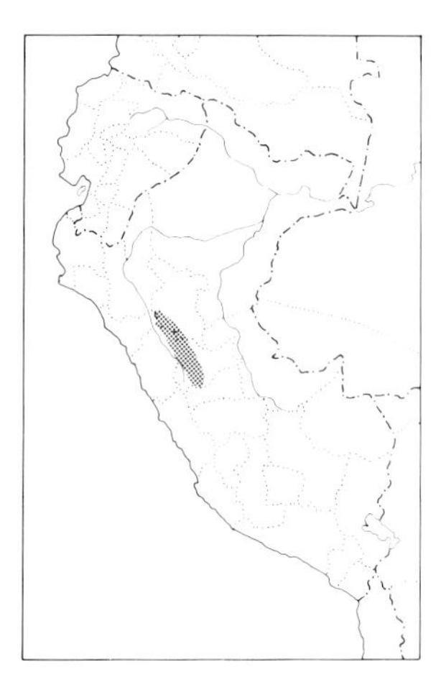

IDENTIFICATION: 22 cm (8½"). *Local near timberline on e. slope of Andes in cen. Peru.* Crown and nape deep blue, with *sides of head and neck, upper back, and throat and chest black. Back, rump, and scapulars bright orange-yellow;* wings and tail mostly black, but lesser wing-coverts deep blue. Breast and belly bright orange-yellow, with *chestnut streaking on breast;* crissum and thighs chestnut.

SIMILAR SPECIES: This spectacular, colorful tanager is unlikely to be confused in its limited known range. Overall pattern somewhat recalls Black-chested Mountain-Tanager, but differs strikingly in its golden (not green) back.

HABITAT AND BEHAVIOR: Uncommon in patches of low elfin woodland near timberline; apparently restricted to wet and frequently cloud-enshrouded ridges. Usually found in pairs, which are quiet and often rather inconspicuous. Best known from the Carpish Ridge region of Huánuco, but only in areas well above the highway (accessible only by trail).

RANGE: Andes of cen. Peru in e. La Libertad and Huánuco. 3000–3400 m.

NOTE: A newly described species: E. R. Blake and P. Hocking (*Wilson Bull.* 86 [4]: 323–324, 1974).

Buthraupis wetmorei

MASKED MOUNTAIN-TANAGER

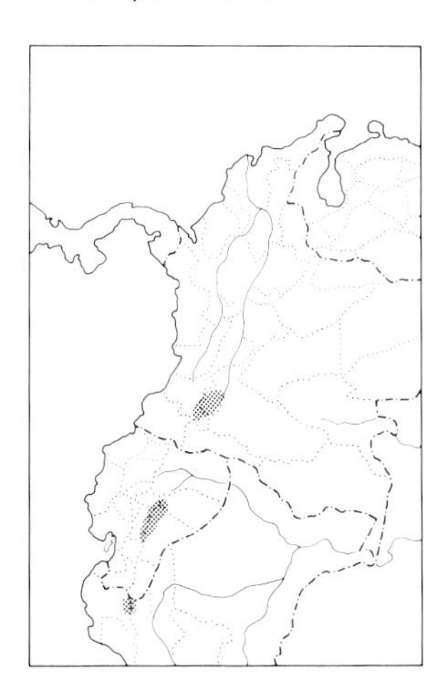

IDENTIFICATION: 20.5 cm (8"). *Local near timberline from s. Colombia to n. Peru.* Crown and nape yellowish olive, becoming pure olive on back, but *contrastingly yellow on rump.* Sides of the head and neck and chin black (forming the mask), *outlined by narrow band of yellow* (contrasting with olive crown). Remaining underparts bright yellow, sides and flanks (sometimes belly) marked with dusky. Wings and tail black, with shoulders and a single wing-bar blue. See C-50.

SIMILAR SPECIES: Does not really resemble the other *Buthraupis* tanagers: a much yellower olive above, with less blue. Black-chested Mountain-Tanager has blue crown and nape and all-green upperparts otherwise. Cf. also Black-chinned Mountain-Tanager (of lower Pacific slope only).

HABITAT AND BEHAVIOR: Rare and local in low stunted woodland near timberline. Usually found singly or in pairs, sometimes accompanying mixed flocks. Rather quiet and generally lethargic, but as it is found in semiopen conditions, often not difficult to see. To be looked for, though it is rare, in Colombia's Puracé Nat. Park, especially past the Coconuco entrance.

RANGE: Andes of sw. Colombia (w. slope of Cen. Andes in Cauca), e. Ecuador (e. slope of Sangay/Altar massif in Morona-Santiago, and also recently found on Gualaceo-Limón road by N. Krabbe), and extreme n. Peru (Cerro Chinguela on Piura-Cajamarca border). 3000–3600 m.

Chlorornis Tanagers

Unique appearance, but shape and behavior reminiscent of *Buthraupis;* legs and heavy bill red.

Chlorornis riefferii

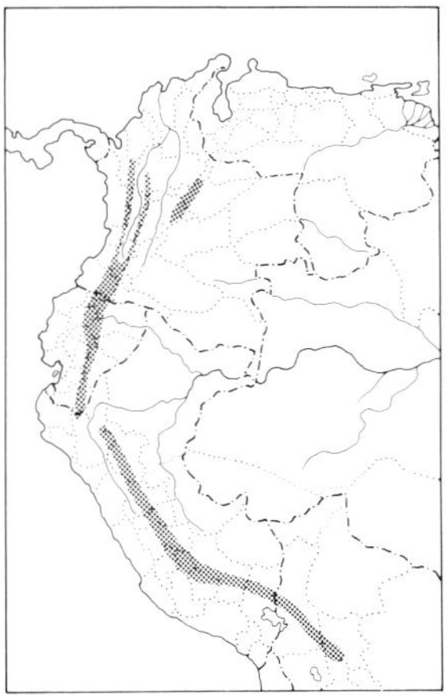

GRASS-GREEN TANAGER PLATE: 17

IDENTIFICATION: 20.5 cm (8"). *An unmistakable, almost garish bright green tanager with red and chestnut accenting. Bill and legs salmon red.* Mostly bright green, with chestnut red mask, throat, median lower belly, and crissum.

HABITAT AND BEHAVIOR: Usually common in humid montane forest and forest borders; somewhat less numerous where forest cover has been partially removed. Usually in groups of about 3 to 8 birds which regularly accompany mixed flocks of other tanagers, etc. Generally quite conspicuous, at times almost seeming to ignore one's presence; forages at all levels, but only infrequently is it found actually inside forest.

RANGE: Andes of Colombia (in E. Andes north to Boyacá), Ecuador (on w. slope south only to Pichincha), e. Peru, and nw. Bolivia (south to Cochabamba). Mostly 2200–2800 m, but also somewhat lower (to about 1800 m) on Pacific slope and southward.

Sericossypha Tanagers

A spectacular, unmistakable large black tanager with stunning snowy white crown and dark red throat. Ranges in canopy of Andean forests. The affinities of this wonderful bird have been debated (it having been suggested that it might possibly be a cotinga), but recently published anatomical evidence points to its being a valid member of the 9-primaried assemblage, with no particular reason for considering it other than thraupine (J. Morony, 1985, *Neotropical Ornithology,* AOU Monograph no. 36, pp. 382–389).

Sericossypha albocristata

WHITE-CAPPED TANAGER PLATE: 17

IDENTIFICATION: 23–24 cm (9–9½"). *A spectacular, large black tanager with stunning* (and very conspicuous) *snowy crown.* Mostly black with the white crown; *throat and chest crimson* (but color often not too apparent). Female similar but has bib dark purple (quite hard to see under normal field conditions). Immature has black bib and thus is an all-black bird with snowy white crown.

HABITAT AND BEHAVIOR: Quite rare and seemingly local in humid montane forest and forest borders. Occurs in groups of up to 10 or rarely 20 individuals which appear to have large home ranges through which they wander freely; they do not consort with mixed tanager flocks but are sometimes with caciques, jays, or other larger birds. Nor-

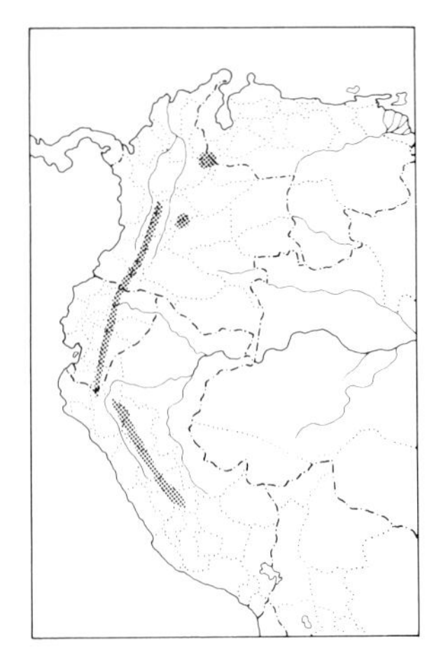

mally conspicuous, perching and hopping about in the open on treetops, often flying long distances before again alighting. Noisy (and apt to be heard long before they are seen), and in this and general comportment they remind one more of a blackbird or jay than they do a typical tanager. Most frequently uttered is a loud, arresting, jaylike "cheeeyáp," often followed by one or several shrieking "cheeeyp, cheeeyp, cheeeyp" notes given by another individual. Perhaps seen most readily in Colombia on the Mocoa road in w. Putumayo and above Florencia, Caquetá.

RANGE: Andes in extreme w. Venezuela (s. Táchira), Colombia (mostly e. slope of E. Andes, but also around head of Magdalena valley in Huila, and on w. slope of Cen. Andes in Caldas, Valle, and Cauca in Puracé Nat. Park area), e. Ecuador (where local), and e. Peru (south to Junín). Mostly 1800–3000 m.

Thraupis Tanagers

"Typical" tanagers—the genus for which the family was named—and very well known, as most species are frequent around houses and towns, with some even in parks in the center of the largest cities. Colors tend to be subdued, with a bold pattern in only 1 sex of 1 species (male Blue-and-yellow); grayish blue or olive predominates in the others. Active, conspicuous, and noisy: one often sees more *Thraupis* tanagers than any other group of birds. They seem to be mostly frugivorous.

GROUP A

Montane or temperate zone species.

Thraupis bonariensis

BLUE-AND-YELLOW TANAGER

PLATE: 17

IDENTIFICATION: 16.5–18 cm (6½–7"). *Andes and s. South America. Head and throat blue,* with lores and area around eye black. *Back and sides of chest black, rump bright orange-yellow;* wings and tail black, feathers broadly edged blue. *Breast bright orange-yellow,* becoming paler yellow on belly. Female much duller, *mostly grayish brown above* with yellowish buff rump and *dusky blue tinge on head* (especially superciliary and sides of neck); *dingy buff below,* somewhat grayer on throat and chest. Nominate race of s. lowlands has less blue on head than female *composita* (of e. Bolivia). Male of smaller *darwinii* (Andes of Ecuador to n. Chile and Bolivia) markedly different, with *olive* (not black) *back and sides of chest;* female has considerably stronger blue on head.

SIMILAR SPECIES: Gaudy males of both types are essentially unmistakable in the arid, semiopen terrain they favor. Potentially more confusing are the much more nondescript females, usually best recognized by their consorts or by their blue on head. Female Hepatic Tanager is larger and more yellow-olive; in Andes see also females of the "hooded" sierra-finches (rather similar overall coloration, but smaller with differently shaped bill, etc.).

HABITAT AND BEHAVIOR: Fairly common in lighter woodland and scrub and in trees and shrubbery in gardens and agricultural areas. Favors semiarid areas, especially in n. part of its Andean range. Usually in pairs or small family groups and where not excessively persecuted often rather tame, feeding unconcernedly on fruits or less often leaves very close to the observer. The song of the nominate race is a series of 4 to 6 rather sweet doubled notes, "sweé-sur, sweé-sur, sweé-sur . . . ," rather different from its congeners' (lacking the squeaky quality).
RANGE: Andean slopes and arid intermontane valleys (also lowlands in s. South America) in Ecuador (north to Imbabura), Peru (where also ranges down into Pacific lowlands), n. Chile (Arica), w. Bolivia (La Paz, Cochabamba, and w. Santa Cruz south through Potosí and Tarija), extreme w. Paraguay, n. and cen. Argentina (south to Mendoza, La Pampa, and n. Río Negro), Uruguay, and s. Brazil (Paraná and Rio Grande do Sul). Locally to at least 3300 m in the Andes.

Thraupis cyanocephala

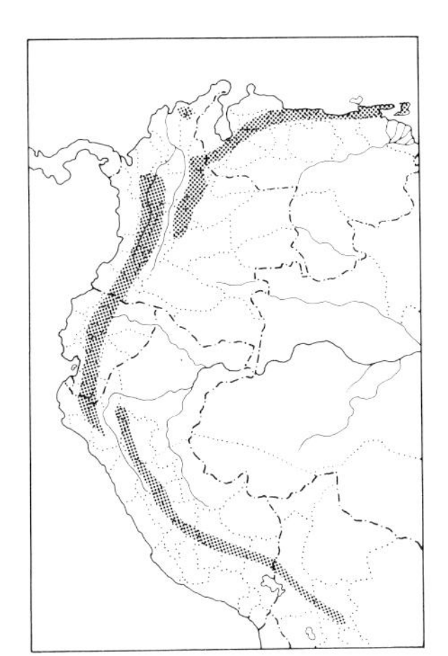

BLUE-CAPPED TANAGER

PLATE: 17

IDENTIFICATION: 18 cm (7"). *Mts. from Venezuela to Bolivia. Crown and nape shiny bright blue* with black lores and mask. Upperparts otherwise bright olive to yellowish olive. Below mostly gray to bluish gray, with *lower flanks, crissum, and thighs olive yellow* (brightest yellow in *auricrissa* of ne. Colombia and w. Venezuela; see V-37). Nominate race (Ecuador south) and races of ne. Venezuela and Trinidad have most prominent black masks; in latter races mask bordered below by speckled whitish malar streak and blue of crown and nape has streaky effect. Most strikingly different race is *olivicyanea* of coastal mts. of n. Venezuela (Aragua and Miranda): it has *entire underparts* (except for yellow thighs and crissum) *the same blue as crown and nape;* see V-37. *Hypophaea* (Lara, Venezuela) is reportedly intermediate between the latter and *auricrissa* of Venezuelan Andes.
SIMILAR SPECIES: The contrasting blue crown and yellow on lower underparts are distinctive. Cf. various mountain-tanagers.
HABITAT AND BEHAVIOR: Fairly common to common and widespread in shrubby forest borders and openings, patches of secondary woodland, and regenerating areas along roadsides. Usually quite conspicuous, often perching in the open on tops of shrubs or trees. Forages at all levels, but rarely or never inside continuous forest; generally found in pairs or small groups of up to 6 or so birds, sometimes with mixed flocks but as often away from them.
RANGE: Coastal mts. of n. Venezuela (Paria Peninsula west to Aragua) and Andes of w. Venezuela (north to Lara and Trujillo), Colombia, Ecuador, nw. and e. Peru (south on w. slope to w. Cajamarca), and nw. Bolivia (La Paz and Cochabamba); Colombia's Santa Marta Mts. and Perijá Mts. on Colombia-Venezuela border; Trinidad. Mostly 1800–3000 m, but occurs much lower in Venezuela (to 800 m) and even lower on Trinidad (to 600 m).

GROUP B *Lowland species.*

Thraupis palmarum

PALM TANAGER

PLATE: 17

IDENTIFICATION: 17–18 cm (6¾–7"). *Mostly grayish olive,* darker and suffused with glossy violaceous on back and underparts. Forecrown olive yellowish; wing-coverts chalky olive, *contrasting with black terminal half of closed wing;* tail blackish. Females tend to be more olive below (less of a violaceous suffusion). *Violivata* (sw. Colombia and w. Ecuador) shows less contrast on wings and is particularly highly glossed.
SIMILAR SPECIES: One's impression of this bird's overall color can change markedly depending on the light; frequently the best mark is its *obviously bicolored wing.* Blue-gray and Sayaca Tanagers have similar shape and behavior and often occur with it; in poor light they can look very dull and gray, but normally some bluish can be discerned. In se. Brazil cf. also Golden-chevroned Tanager.
HABITAT AND BEHAVIOR: Common in gardens and in populated or agricultural regions (especially where palms are numerous) and in shrubby clearings and forest and woodland borders (and to a lesser extent in canopy). Generally widespread (occurring in both humid and drier areas, wherever there exists appropriate habitat) and conspicuous. Appropriately named, it does seem to have a predilection for palms of various species, often feeding on or at the base of the fronds and even regularly hanging upside down from their tips. Forages for insects more than its congeners. Song and calls are rather similar to Blue-gray Tanager's.
RANGE: Widespread in lowlands south to n. and e. Bolivia (south to La Paz, Cochabamba, and Santa Cruz), e. Paraguay (no recent records), and s. Brazil (south to Rio Grande do Sul); west of Andes south to sw. Ecuador (El Oro); Trinidad. Also Honduras to Panama. To about 2000 m (rarely higher), but mostly below 1200 m.

Thraupis ornata

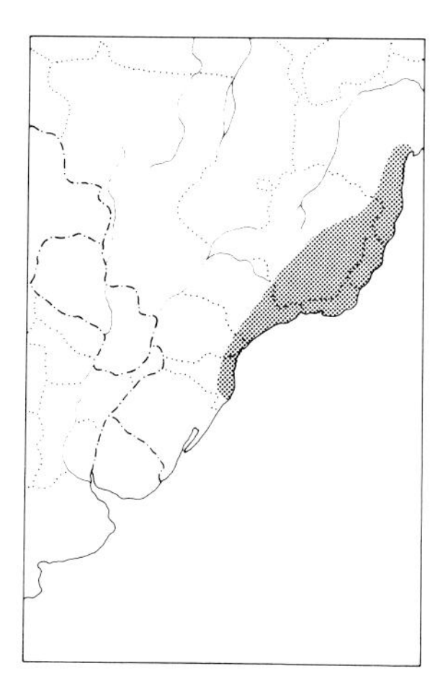

GOLDEN-CHEVRONED TANAGER

PLATE: 17

IDENTIFICATION: 17–18 cm (6¾–7"). *Se. Brazil. Head and underparts dark shiny violet blue,* brighter and bluer on crown. Mantle dull blackish blue, with *bright brassy yellow patch on shoulders,* and wing-coverts and flight feathers edged greenish. Female duller, but with *wing similar to male's;* back and underparts grayer. Immature even duller, with very little blue.
SIMILAR SPECIES: Most similar to Blue-capped Tanager of the Andes (no overlap). Sayaca and Azure-shouldered Tanagers lack yellow on wing (normally quite prominent in Golden-chevroned, even in dull female). Palm Tanager more grayish olive with differently patterned bicolored wing.
HABITAT AND BEHAVIOR: Locally fairly common to common in canopy and borders of humid forest and in adjacent shrubby clearings and gardens with large trees. Behavior similar to that of other members of genus: usually conspicuous, foraging mostly for fruit at all heights

above the ground. Occurs mostly in hilly or montane areas (e.g., numerous at Itatiaia Nat. Park), but locally also in lowlands (as in Espírito Santo and s. Bahia).
RANGE: Se. Brazil from s. Bahia and e. Minas Gerais south to e. Santa Catarina. To about 1300 m.

Thraupis episcopus

BLUE-GRAY TANAGER

PLATE: 17

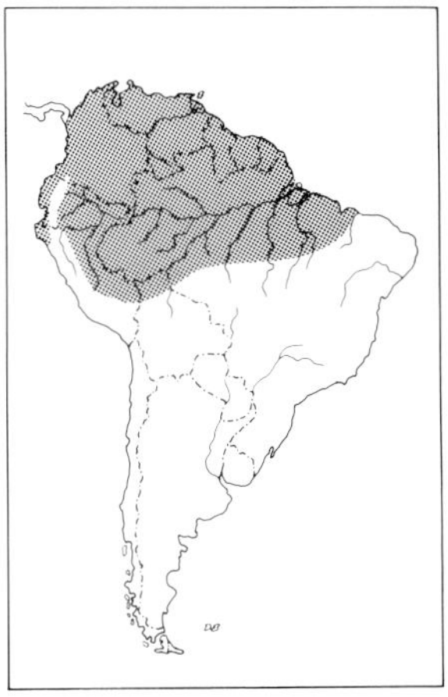

IDENTIFICATION: 16.5–17 cm (6½–6¾"). One of South America's most familiar birds: *mostly pale grayish blue,* darkest on back (in some races resulting in contrast between pale head and dark back). *Flight feathers edged bright blue,* but color of shoulders and upper wing-coverts varies. In races west of the Andes and in n. and ne. South America (south to e. Amaz. Brazil; includes nominate) these *vary from bright cobalt to paler milky blue,* whereas in races of w. and cen. Amazonia and in the upper Orinoco valley shoulders are *conspicuously white to bluish white* and there often is another broad wing-bar of the same color. Tail feathers also edged blue.
SIMILAR SPECIES: Quite similar to Sayaca Tanager; for full discussion see under that species. In general Sayaca most resembles w. and n. races of Blue-gray (i.e., *without white on wing*) and it always has greenish blue edging to wing and tail feathers, not deeper cobalt blue. In arid Caribbean north cf. also Glaucous Tanager.
HABITAT AND BEHAVIOR: Common and widespread in a variety of habitats (both humid and dry), varying from forest borders and patchy secondary woodland to gardens in settled areas and trees and shrubbery in agricultural regions. The nervously active (even ebullient) Blue-gray Tanager is one of the best-known birds over much of its range, as much for its often tame behavior as for its propensity to occur around houses. Basically arboreal, it may forage virtually anywhere; much of its diet is fruit, though it also gleans and even occasionally sallies forth for insects. The song is a jumbled series of squeaky, fast notes, with calls of a similar quality.
RANGE: Widespread in lowlands south to extreme n. Bolivia (Beni) and cen. Brazil (south to nw. Mato Grosso, se. Pará, n. Goias, and cen. Maranhão); west of Andes south to nw. Peru (Piura); Trinidad. Also Mexico to Panama. Mostly below 1500 m, less often to 2000 m and rarely even higher.

Thraupis sayaca

SAYACA TANAGER

PLATE: 17

IDENTIFICATION: 16.5 cm (6½"). Resembles Blue-gray Tanager, *replacing it south and eastward.* In zones of actual or potential sympatry, Blue-gray has white shoulders, whereas *Sayaca is plain-winged. Mostly dull bluish gray* (not as blue overall as Blue-gray), with crown and back the same color (crown contrastingly paler in Blue-gray). *Edging on wing and tail greenish turquoise blue* (not deeper cobalt blue). *Boliviana* (of n. Bolivia lowlands) seems somewhat intermediate (and indeed it was originally described as a race of *episcopus*) in that it has variably

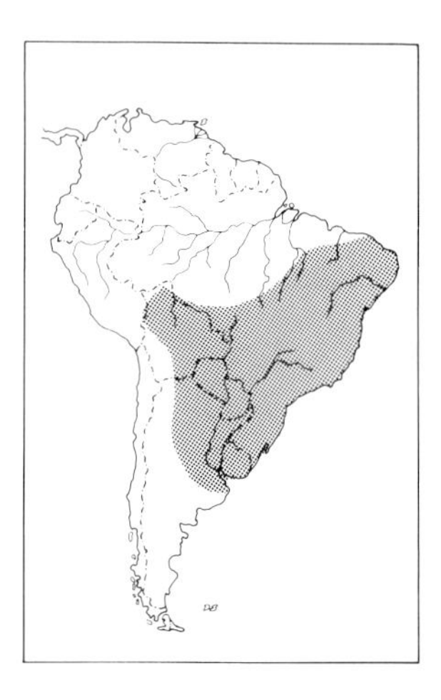

bluish white to pale blue shoulders and pale crown (recalling Blue-gray) and greenish blue edging to flight and tail feathers (typical of Sayaca).

SIMILAR SPECIES: Cf. Blue-gray Tanager and (in se. Brazil) also the larger Azure-shouldered Tanager.

HABITAT AND BEHAVIOR: Common and widespread in various habitats in both humid and dry regions, the prime requirement in the latter case being a relatively lush growth of trees and shrubbery such as is often found along watercourses or near habitations. In every way the s. replacement of Blue-gray Tanager and very similar in most aspects of its behavior, though its song is somewhat more melodic and less squeaky.

RANGE: N. and e. Bolivia (north to Beni lowlands and from Cochabamba and Santa Cruz south), s. and e. Brazil (north to nw. Mato Grosso, n. Goiás, and Maranhão), Paraguay, n. Uruguay, and n. Argentina (south to La Rioja, Córdoba, and n. Buenos Aires); also extends north in arid intermontane valleys to se. Peru (ANSP specimens from Huiro in Cuzco, and recently seen in Madre de Dios). To about 2000 m.

NOTE: It is possible that *T. sayaca* and *T. episcopus* will ultimately be best merged into 1 species unit, as first suggested by J. T. Zimmer (*Am. Mus. Novitates* 1262, 1944). The "link" between them seems to be *T. s. boliviana* of the n. Bolivia lowlands. Though originally described as a race of *episcopus* it is, as N. Gyldenstolpe (1945b: 273–275) pointed out, clearly closer to *sayaca*, though intermediate in a few respects (especially color of shoulders). *T. s. boliviana* and *T. e. mediana* are recorded as having been taken at the same localities in extreme n. Bolivia, and this reported sympatry has led systematists to continue to regard them as full species. However, whether actual sympatric populations of both forms are involved (or whether perhaps there was a mislabeling problem or a single wandering bird or migration) remains to be determined, not only in n. Bolivia but also elsewhere along their presumed zone of contact across cen. Brazil. Their primary vocalizations do differ, and this could perhaps represent a barrier to their interbreeding. Note further that *T. glaucocolpa* (which seems closely related to *T. sayaca*) and *T. episcopus* do occur syntopically in parts of Venezuela.

Thraupis glaucocolpa

GLAUCOUS TANAGER

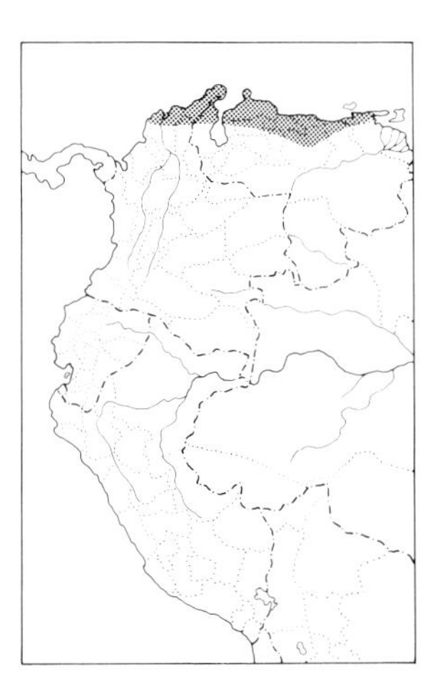

IDENTIFICATION: 16.5 cm (6½"). *Arid Caribbean lowlands.* Resembles geographically far-removed Sayaca Tanager (and until recently considered conspecific). Most obvious difference is the *large patch of dark blue on primary coverts,* contrasting sharply with the bluish green–edged flight feathers. Breast and flanks shining pale turquoise blue, contrasting with grayish throat and whitish median belly (instead of uniform bluish gray below). See C-51, V-37.

SIMILAR SPECIES: Also similar to Blue-gray Tanager (often sympatric) but has grayish head concolor with back (not contrastingly paler) and paler turquoise green wing-edging (not deep bright blue).

HABITAT AND BEHAVIOR: Generally uncommon (locally more numerous) in gallery and deciduous woodland, gardens, and scrub and trees in agricultural regions; *commonest in arid regions.* Behavior like Blue-gray and Sayaca Tanagers'.

RANGE: Locally in dry parts of n. Colombia (Bolívar east through Guajira Peninsula) and n. Venezuela (Guajira Peninsula and n. Zulia east to Sucre, inland to n. Lara and cen. Guárico); Margarita Is. (off Venezuelan coast). Mostly below 500 m.

NOTE: We follow Meyer de Schauensee and Phelps (1978) in considering *T. glaucocolpa* as a species distinct from *T. sayaca* of s. and e. South America; it is quite different from all races of *T. sayaca*. Treating *glaucocolpa* as a full species had earlier been suggested as a footnote in *Birds of the World* (vol. 13, p. 322).

Thraupis cyanoptera

AZURE-SHOULDERED TANAGER

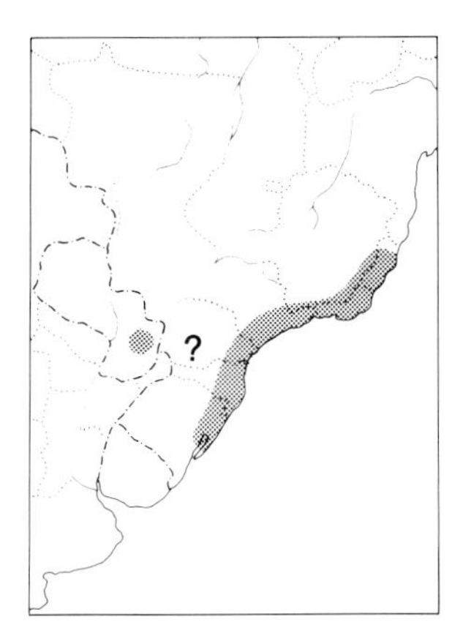

IDENTIFICATION: 18 cm (7″). *Se. Brazil.* Resembles Sayaca Tanager (with which frequently sympatric) but larger with *notably heavier bill.* Bluer (less gray) above generally with *bright deep cobalt blue shoulders* (sometimes hard to see) and bluer (less green) edging to flight feathers. Underparts similar, but with greenish tinge on flanks (not concolor as in Sayaca).
HABITAT AND BEHAVIOR: Uncommon to fairly common but local in humid montane forest canopy and borders. Found mostly on seaward slopes of the Serra do Mar. Much more of a forest bird than Sayaca Tanager, but behavior otherwise similar.
RANGE: Se. Brazil (Espírito Santo and e. Minas Gerais south to n. Rio Grande do Sul); also an old (questionable?) record from extreme e. Paraguay. Mostly 300–1600 m.

Piranga Tanagers

Typical tanagers with most males mainly red (all have at least some), females mainly olive and yellow. Fairly heavy bill. All are arboreal, in forest or woodland canopy or edge; generally in pairs (not flocks). Behavior rather sluggish, and some are more often heard than seen. The only tanager genus to have extended to North America, with 2 of these occurring in South America as migrants.

GROUP A

Bold white wing-bars (both sexes); small size.

Piranga leucoptera

WHITE-WINGED TANAGER

PLATE: 18

IDENTIFICATION: 14 cm (5½″). *Scarlet,* with back mottled blackish and *black lores, scapulars, wings, and tail; 2 broad white wing-bars.* Female has some pattern but is yellowish olive above, brightest on face; wings and tail blackish with *same 2 white wing-bars.* Below bright yellow, more orange-yellow on throat and breast in some (older?) birds.
SIMILAR SPECIES: Not easy to confuse as it is the only S. Am. tanager with such bold wing-bars; male Scarlet and Vermilion Tanagers have all-black wings.

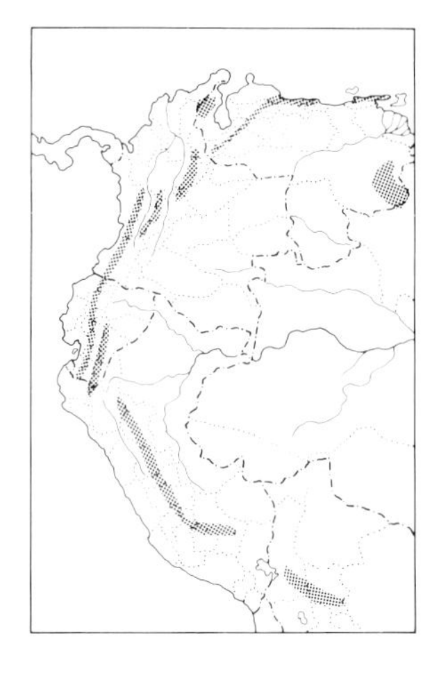

HABITAT AND BEHAVIOR: Uncommon and local in canopy and borders of humid montane forest. Often in pairs and frequently with mixed flocks of other tanagers, etc. Usually quite lethargic, foraging well above ground (as illustrated by the fact that the species is very rarely or never captured in ground-level mist-nets, unlike many other "canopy" tanagers), gleaning in foliage and on outer branches. Its call is a distinctive "tsupéet" or "wheet, tsupéet," with some variation, especially of the ending.
RANGE: N. coastal mts. of Venezuela (Paria Peninsula west) and Andes of w. Venezuela, Colombia (very local), Ecuador (south on w. slope to El Oro and on e. slope from Tungurahua south), e. Peru, and nw. Bolivia (La Paz, Cochabamba, and w. Santa Cruz); Perijá Mts. on Venezuela-Colombia border; *tepui* slopes of s. Venezuela (Bolívar) and adjacent n. Brazil (Roraima). Also Mexico to Panama. Mostly 1200–2000 m.

GROUP B

Predominantly rosy red to scarlet (males) or *olive and yellow* (females).

Piranga flava

HEPATIC TANAGER

PLATE: 18

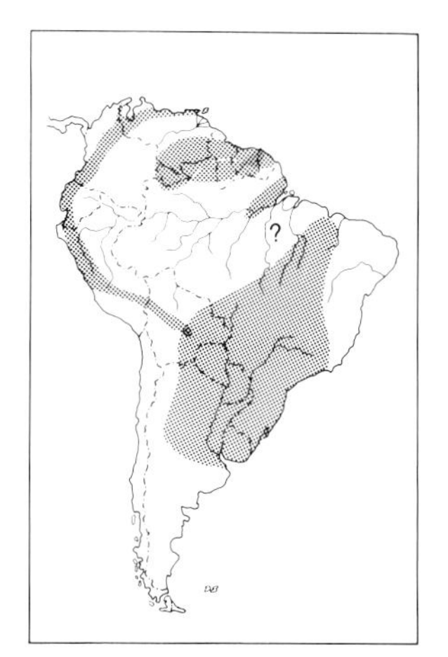

IDENTIFICATION: 18–19 cm (7–7½"). Bill dark horn above, leaden blue to yellowish horn below. Male *mostly orange-red,* paler and brighter below and on forecrown; *lores dusky.* Female yellowish olive above, also with *dusky lores; rather bright yellow below and on forecrown.* Foregoing applies to *flava* group of s. and e. South America. *Lutea* group (of montane n. and w. South America) similar but somewhat smaller; male darker and more carmine red generally, with forecrown uniform with rest of head; female darker olive above, also with uniform forecrown, and more tinged with olive below. Note that neither type shows the dusky ear-coverts so prominent in birds from sw. United States to Nicaragua (*hepatica* group).
SIMILAR SPECIES: Both sexes of Summer Tanager are confusingly similar. In color, male Summers more nearly resemble the paler *flava* group of Hepatic, but overwintering Summers are much more likely to be found in the range of darker, more brick red *lutea* group of Hepatic. Further differences involve the bill: Summer's bill is generally paler (though can be quite dusky) and always more uniform (not so bicolored), and it lacks the distinct notch found in Hepatic's upper mandible. Hepatic's dusky lores are diagnostic if they can be seen.
HABITAT AND BEHAVIOR: Uncommon to fairly common in deciduous and gallery woodland (also edge of *Araucaria* forest in s. Brazil), *cerrado,* and other types of campo and savanna vegetation (the *flava* group); or in a variety of habitats ranging from borders of humid forest and woodland to clearings and plantations and patches of light woodland in arid areas (the *lutea* group). Usually found singly or in pairs, *flava* group much more often as members of mixed flocks than *lutea* group (which usually forage alone). Basically arboreal, they move about slowly in all levels of trees, but mainly rather high up. Song of

lutea group is a fast series of rich burry phrases, while its call is a distinctive "chúhtiti" (P. Slud) or "chúp-chitup" (RSR), sometimes shortened to just the first "chúh." Call of nominate *flava* is a soft "chef," sometimes doubled and often given in flight (its song not known).
RANGE: *Flava* group in isolated savannas of s. Guyana and Suriname (Sipaliwini) and ne. Brazil north of the Amazon (scattered localities in Amapá, along the Amazon in n. Pará, and in Roraima); interior and s. Brazil (s. Maranhão and Ceará south and west to s. Mato Grosso and Rio Grande do Sul), e. Bolivia (Santa Cruz and e. Cochabamba south through Tarija), Paraguay, Uruguay, and n. Argentina (south to La Rioja, Córdoba, and n. Buenos Aires); up to about 1500 m. *Lutea* group in n. coastal mts. of Venezuela (Paria Peninsula west) and Andes of w. Venezuela, Colombia (locally; not recorded from E. Andes or slopes above Magdalena valley except in Norte de Santander, nor from W. Andes north of Valle), w. Ecuador (semiarid intermontane valleys and drier w. lowlands), w. Peru (also in coastal lowlands from La Libertad to Lima), and nw. Bolivia (La Paz, Cochabamba, and w. Santa Cruz); Santa Marta Mts. of Colombia and Perijá Mts. on Venezuela-Colombia border; slopes of *tepuis* and other low isolated mts. in s. Venezuela (Bolívar and Amazonas), adjacent n. Brazil (Roraima), Guyana, and Suriname (Brownsberg); Trinidad. Also sw. United States to Panama. Up to 2000 m (in Andes).

NOTE: It seems likely that 2 or more probably 3 species are involved: *P. flava* of s. and e. South America, mainly in the lowlands; *P. lutea* of montane n. and w. South America (extending to Costa Rica and Panama); and *P. hepatica* of sw. United States to Nicaragua (this and *lutea* could be combined, with the name *hepatica* having priority). Plumage differences between *flava* and *lutea* are often most pronounced in zones of apparent or actual sympatry (such as in Guyana with *P. l. haemalea* and *P. f. saira*), and their preferred habitats also differ strikingly. For now, however, we continue to follow the traditional single-species treatment. Should they ultimately be split, we would suggest that it would be best to continue to recognize their close relationship in their English names, calling *flava* the Lowland Hepatic-Tanager, *lutea* the Highland Hepatic-Tanager, and *hepatica* the Northern Hepatic-Tanager.

Piranga rubra

SUMMER TANAGER

IDENTIFICATION: 18–18.5 cm (7–7¼″). *Bill pale to dusky horn.* Male *all rosy red,* paler below, with dusky wings and tail. Female yellowish olive above, duskier on wings and tail; deep yellow below. Immatures resemble adult females, with males gradually assuming full adult plumage over the course of 2 years (washed or blotched with pale red in their first spring, paler and duller than full adult in their second spring; these intermediate plumages are often seen in South America).
SIMILAR SPECIES: Both sexes resemble respective sexes of Hepatic Tanager, so much so that they are often not easy to distinguish; be aware that the bills of wintering Summers are not as pale as they are in the breeding season, and that *bills of Hepatics in South America are not all dark.* Hepatic usually shows prominent dusky lores (absent or obscure in Summer); male Hepatics of *lutea* group are a darker, more carmine red generally. Cf. also Scarlet Tanager.

HABITAT AND BEHAVIOR: Locally fairly common n.-winter resident (mostly Oct.–Apr.) in second-growth woodland, shrubby clearings and gardens, and forest borders; primarily montane on its S. Am. wintering grounds. Usually seen singly, peering about sluggishly, generally well up in trees, though may perch for extended periods in the open. Its staccato call, a distinctive "pitichuck" or "pi-tuck," is often heard, but it sings rarely or not at all in South America.

RANGE: Nonbreeding visitor to w. Colombia, Ecuador, and n. Peru (mostly west of the Andes, but also in e. lowlands near their base), with more scattered records south to s. Peru and nw. Bolivia (La Paz) and in Venezuela, w. Amaz. Brazil (east to lower Rio Negro and lower Rio Tapajós), and Guyana and Suriname; Trinidad, and 2 records from Netherlands Antilles; accidental in n. Chile (Antofogasta). Breeds North America, wintering from Mexico south. To about 2000 m, occasionally higher.

Piranga olivacea

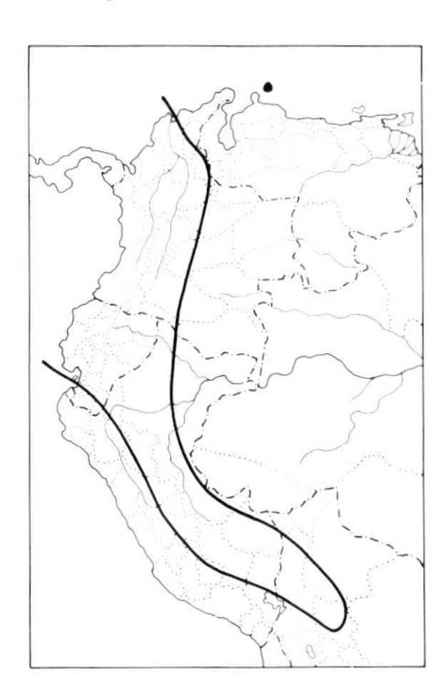

SCARLET TANAGER

IDENTIFICATION: 16.5–17.5 cm (6½–7"). Nonbreeding male olive above with *blackish wings and tail;* greenish yellow below. Males molt into breeding plumage while still on their wintering grounds (gradually assumed between Jan. and Apr.); they then are unmistakable, *vivid scarlet with jet black wings and tail.* First-year males are duller and paler orangy red. Female resembles nonbreeding male but wings and tail duskier (not as black). Immatures may show 1 or 2 pale wing-bars (usually faint).

SIMILAR SPECIES: Female Summer Tanager does not have as dark wings and tail and is deeper and usually brighter yellow below. Cf. also White-winged Tanager (much smaller, and with 2 white wing-bars).

HABITAT AND BEHAVIOR: Uncommon n.-winter resident (mostly Oct.–Apr.) in canopy and borders of humid forest and secondary woodland, less often in clearings with scattered trees and gardens (perhaps especially on migration). In general this common N. Am. breeder seems unaccountably scarce in South America, in which virtually the entire population winters; they must simply be very inconspicuous at this season. Basically arboreal, foraging deliberately in foliage, typically well up in trees; usually found singly. Fairly often heard is its distinctive call, "chip-burr"; however, it rarely or never seems to sing in South America.

RANGE: Nonbreeding transient and visitor through w. Colombia (east to lowlands along e. base of Andes), Ecuador, e. Peru, and nw. Bolivia (La Paz and Cochabamba); apparently passes n. winter primarily in upper Amazonia (and thus should also be found in westernmost Brazil); fairly regular as passage migrant on Netherlands Antilles, so should occur in w. Venezuela. Breeds North America, wintering almost exclusively in South America (a few in Panama); migrates through Middle America and in smaller numbers in West Indies. Mostly below 1500 m (higher, to 2500 m or even more, as transient).

GROUP C

Unmistakable *scarlet hood;* sexes similar.

Piranga rubriceps

RED-HOODED TANAGER

PLATE: 18

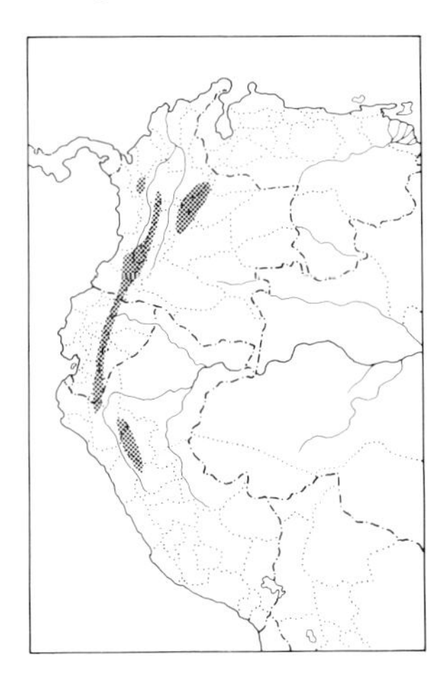

IDENTIFICATION: 18–18.5 cm (7–7¼"). *A striking, predominantly yellow and red tanager of e. slope of Andes. Entire head, throat, and breast bright scarlet,* contrasting with yellowish olive upperparts (rump yellower) and bright yellow lower underparts. Wings black with yellow shoulders; tail dusky olive. Female similar but with scarlet extending only to throat.
SIMILAR SPECIES: Molting immature male Summer Tanagers may show red first on head and breast, thus to some extent duplicating this species.
HABITAT AND BEHAVIOR: Rare to uncommon and seemingly local in canopy and borders of humid montane forest. Pairs or small groups (up to 6 or so) will often perch conspicuously in a semiopen tree and remain there for considerable periods, hopping about lethargically, but in general, a seldom-encountered bird. Usually not with mixed flocks.
RANGE: Locally in Andes of Colombia (north in E. Andes to Cundinamarca and mostly absent from w. slope of W. Andes, entirely so from w. Nariño), e. Ecuador, and ne. Peru (south to La Libertad). 1700–3000 m.

Ramphocelus Tanagers

A genus well characterized by its conspicuous pale silvery on mandible, with swollen mandible especially in male. These tanagers are, with 1 notable exception, mainly velvety maroon to bright red, with females being duller and often blacker; the exception is mostly black but can be instantly known by its puffy, bright-colored rump. All species are common and conspicuous in shrubby and edge habitats.

GROUP A

Both sexes with *bright contrasting rump* (vermilion to yellow); male otherwise black.

Ramphocelus flammigerus

FLAME-RUMPED TANAGER

PLATE: 18

Other: Yellow-rumped Tanager

IDENTIFICATION: 18.5 cm (7¼"). *W. Colombia and w. Ecuador. Bill silvery bluish with black tip.* Mostly velvety black with *bright orange-red lower back and rump.* Female mostly blackish brown above with *orange-red lower back and rump.* Below mostly yellow, with diffused but still prominent *band of orange across breast. Icteronotus* (all of range except Colombia's Cauca valley, where nominate *flammigerus* is found) has *bright lemon yellow lower back and rump* in male, while females are less dark, more olive or grayish brown above, and have *all pale yellow underparts* (no orange across breast) as well as lower back and rump; see C-51, P-28. *Hybrids between the 2 are frequent* (males with rumps of

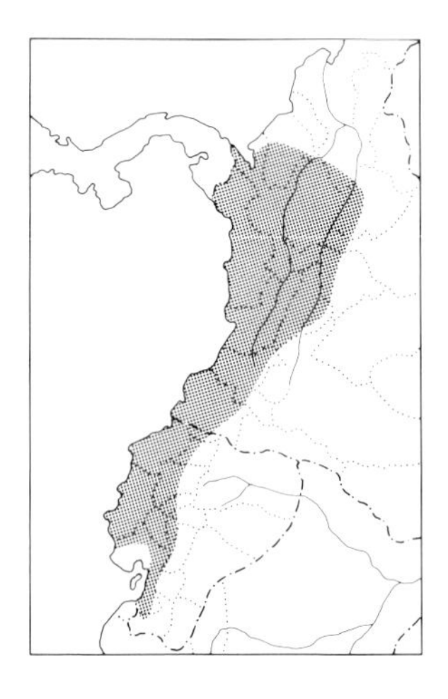

various shades of orange, females the same and with less obvious breast bands).

SIMILAR SPECIES: Males of either type should be unmistakable, and females likewise are not apt to be confused (though because of all the hybrids it was often not easy to distinguish between the 2 forms when they were considered full species). Female *Piranga* tanagers lack the bright rump; cf. female Crimson-backed Tanager.

HABITAT AND BEHAVIOR: Fairly common to common (especially numerous in Pacific lowlands) in shrubby clearings, forest borders, and plantations and gardens. Noisy and conspicuous, trooping about in the semiopen in groups of up to 6 or more birds (birds in female or immature plumage always in preponderance). Frequently seen flying across roads, when their bright rumps render them unmistakable at almost any distance. Males often perch on top of a bush or small tree with wings held drooped and their puffed-out rumps exposed.

RANGE: W. Colombia (Pacific lowlands, east along humid n. base of Andes, and up the Cauca valley to Cauca and the Magdalena valley to n. Tolima) and w. Ecuador (south in humid areas to El Oro). To about 2000 m (though *icteronotus* is infrequent above 1400 m).

NOTE: We follow *Birds of the World* (vol. 13) and the 1983 AOU Check-list in considering *R. icteronotus* conspecific with *R. flammigerus*. This was done primarily on the basis of extensive hybridization between the 2 in sw. Colombia, but we believe it is possible that this contact is secondary in nature (resulting from massive deforestation in the region) and that in due course the 2 may stabilize around the 2 parental types. Voice and behavior of the 2 are, however, similar. Lumping the 2 forms presents a rather difficult situation as far as English names are concerned, for *R. icteronotus*'s yellow rump is not really "flame"-colored; we follow the lead of the 1983 AOU Check-list in calling the enlarged species "Flame-rumped," having no better solution to offer.

GROUP B

Blackish maroon to crimson with black wings and tail.

Ramphocelus carbo

SILVER-BEAKED TANAGER

PLATE: 18

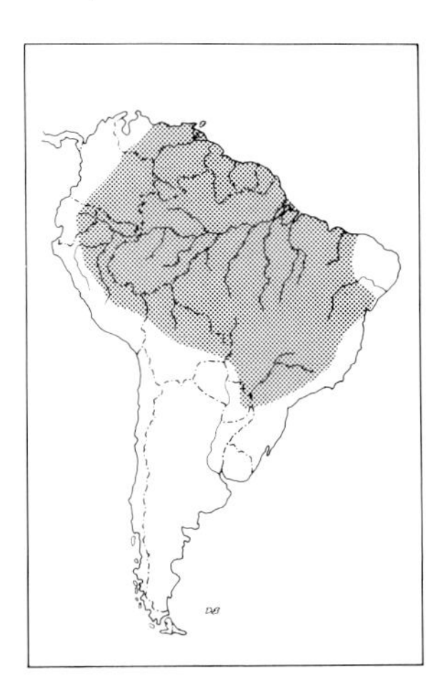

IDENTIFICATION: 18 cm (7"). *Mandible swollen, gleaming silvery white* (male), but less striking (grayish to dull silvery below) in female; bill notably larger in *magnirostris* of Trinidad. *Velvety black above with dark maroon overtones.* Throat and chest dark crimson, becoming blackish maroon on lower underparts. S. races (se. Peru to cen. Brazil and south) have lower breast and belly much blacker (contrasting more with throat and chest) and blacker mantles. Female dark reddish brown above (reddest on rump and upper tail-coverts); mostly paler duller reddish below, *darkest on throat* (which is essentially concolor with upperparts). *Atroceriseus* female (n. and e. Bolivia) is very different, *being essentially all blackish* relieved only by some rufous on median lower belly.

SIMILAR SPECIES: In favorable light males of this dark tanager are stunning and instantly recognized; even in poor light the silvery on the bill can normally be seen. Females might be confused with male Red-crowned Ant-Tanager, but throat darker (rather than paler) and behavior very different, etc. For distinctions from other members of its genus, of which *carbo* is by far the best known and most wide-ranging, see under those species.

HABITAT AND BEHAVIOR: Common to very common in shrubby clearings, forest borders, and gardens; particularly numerous near water. Invariably conspicuous, and one of the most frequently encountered birds in many areas (especially along watercourses in Amazonia). Forages in noisy groups which usually move rapidly through lower growth and shrubbery, pausing to glean in foliage or eat fruit; though usually low, they may range higher, and at times are even seen in forest canopy (perhaps especially in *várzea*). Often other birds may join in all the activity, but many times Silver-beaks are seen moving about on their own. Females and young males still in female plumage outnumber adult males. Most frequently heard call is a loud metallic "chink," often given in flight.

RANGE: Widespread in lowlands east of the Andes south to n. and e. Bolivia (La Paz, Cochabamba, and Santa Cruz), e. Paraguay (only 1 old, perhaps questionable, record), and s. Brazil (south to Paraná); absent from coastal strip of e. Brazil; Trinidad. To about 1000 m.

Ramphocelus melanogaster

HUALLAGA TANAGER

Other: Black-bellied Tanager

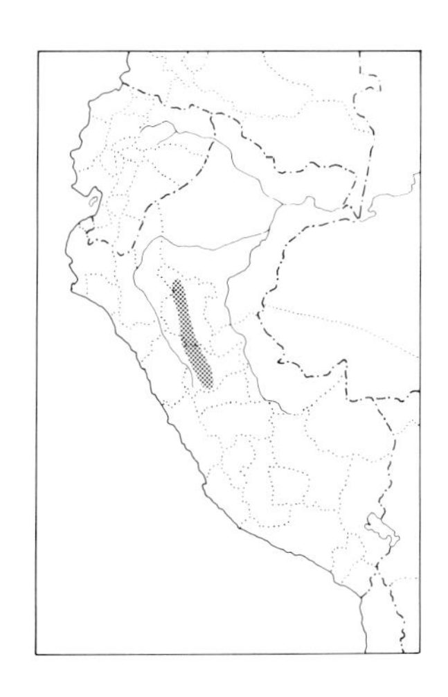

IDENTIFICATION: 18 cm (7"). *Río Huallaga valley of e. Peru. Bill as in Silver-beaked Tanager.* Dark velvety maroon above with *bright crimson lower back and rump;* wings and tail black. Throat and chest somewhat paler velvety maroon, becoming *bright crimson on lower underparts,* with black stripe down center of lower breast and belly. Female dark reddish brown above, redder on lower back and rump, and *paler rosier forehead, ocular area, and chin.* Lower throat and chest dark reddish brown, becoming paler and redder on remaining underparts.

SIMILAR SPECIES: Likely only confused with Silver-beaked Tanager, which it replaces in the Huallaga valley. Male Silver-beak is much darker overall, lacking the crimson rump and lower underparts, while female Huallaga is best known by her paler foreface (stands out quite well). Cf. also Crimson-backed Tanager.

HABITAT AND BEHAVIOR: Locally common in shrubby clearings, forest borders, and gardens. Behavior essentially identical to Silver-beaked Tanager's. Easily seen around Tingo María in Huánuco, though just over the Cordillera Divisoria to the east, in the Boquerón Canyon and beyond, the Silver-beaked Tanager is equally numerous.

RANGE: Río Huallaga valley of ne. Peru in San Martín and Huánuco. 500–1000 m.

NOTE: The specific status of this Peruvian endemic seems a bit uncertain. Males are in plumage virtually identical to geographically distant *R. dimidiatus*. So far there is no evidence to indicate that *R. melanogaster* and *R. carbo* intergrade in the lower Huallaga valley, though their ranges come very close. We feel we must maintain all 3 as full species; nonetheless we harbor the suspicion that they may eventually be shown to be conspecific. As *R. melanogaster* is no more "Black-bellied" than is *R.*

dimidiatus (and is much less so than many races of *R. carbo*), we find that formerly used name to be decidedly misleading. As its range is limited to the Huallaga River valley, that name provides a good modifier.

Ramphocelus dimidiatus

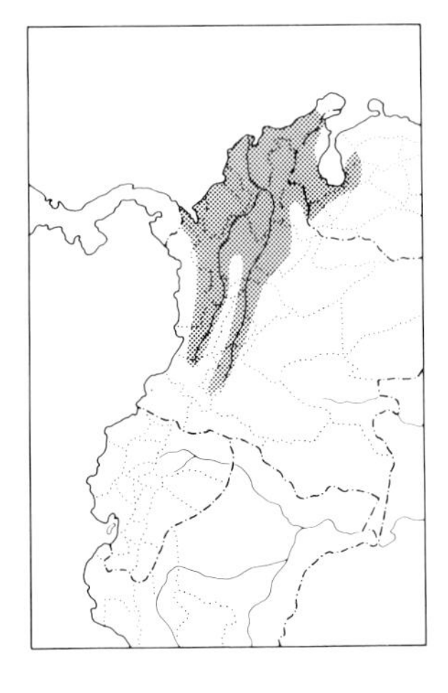

CRIMSON-BACKED TANAGER

IDENTIFICATION: 18 cm (7"). *W. Colombia and nw. Venezuela.* Male virtually identical to male Huallaga Tanager (of Peru), but back is slightly redder (less black). Female also very similar but lacks the paler forehead, ocular area, and chin. See V-37, P-28.

SIMILAR SPECIES: Does not overlap with any other member of its superspecies; *found only west of the Andes.* Female might be confused with male of either ant-tanager, but they lack the contrasting reddish rump and belly.

HABITAT AND BEHAVIOR: Fairly common to common in shrubby areas, clearings, and woodland and forest borders. Behavior much like Silver-beaked Tanager's.

RANGE: N. and w. Colombia (more humid lowlands west of E. Andes, including Cauca and Magdalena valleys, but local and much less numerous on w. slope of W. Andes, where absent south of Valle) and nw. Venezuela (Maracaibo basin). Also Panama. Mostly below 1200 m.

Ramphocelus bresilius

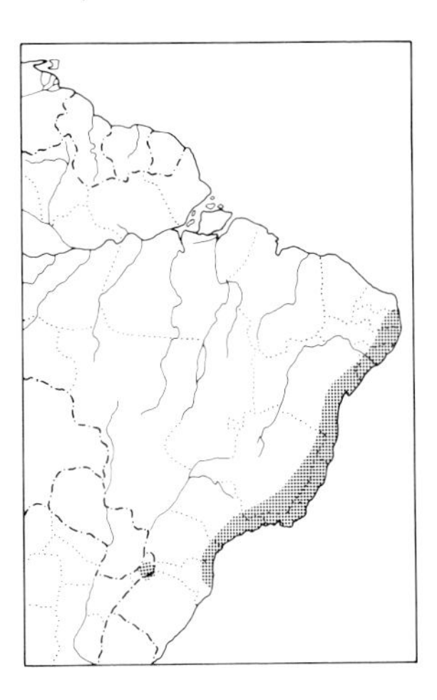

BRAZILIAN TANAGER

PLATE: 18

IDENTIFICATION: 18.5 cm (7¼"). *Coastal e. Brazil. Bill as in Silver-beaked Tanager. Mostly intense bright crimson,* slightly darker on mantle (especially in *dorsalis,* from Espírito Santo south); wings and tail black. Female grayish brown above, grayest on crown; rump more reddish brown. Throat brownish gray, remaining underparts pale reddish brown.

SIMILAR SPECIES: Stunning, brilliant males are virtually unmistakable (cf. breeding male Scarlet Tanager—no known overlap, very different bill, behavior, etc.). Female recalls female Silver-beaked Tanager but is paler and grayer generally, especially on head and throat. Cf. also male Red-crowned Ant-Tanager.

HABITAT AND BEHAVIOR: Uncommon in shrubby clearings and forest borders, especially near water; sometimes also in city parks and plazas. Behavior much like Silver-beaked Tanager's, but seen more often in pairs than in larger groups, and never seems very numerous. Relatively frequent in coastal lowlands of s. Rio de Janeiro and ne. São Paulo (less numerous in Espírito Santo and Bahia, perhaps because of trapping).

RANGE: E. Brazil (Paraíba and Pernambuco south to Santa Catarina) and ne. Argentina (an old specimen, and 1 recent record, from Misiones). Mostly below 400 m.

NOTE: Hybrids between this species and *R. carbo* have been reported from the Rio Dôce area of Minas Gerais, Brazil (H. Sick), but for now we continue to maintain *R. bresilius* as a full species.

Ramphocelus nigrogularis

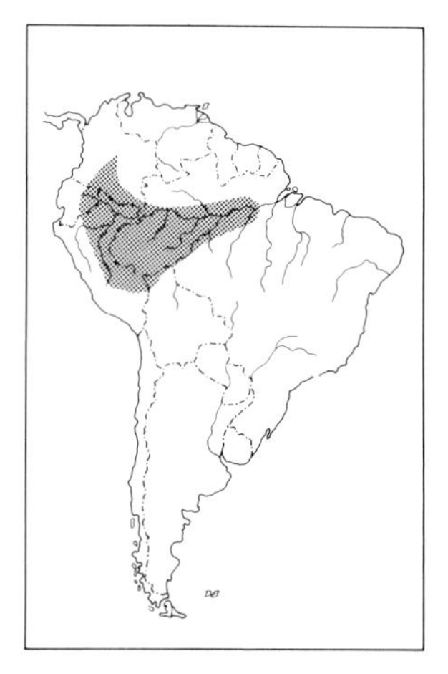

MASKED CRIMSON TANAGER

PLATE: 18

IDENTIFICATION: 18.5 cm (7¼"). *A stunning, bright, crimson and black tanager found near water in Amazonia. Bill as in Silver-beaked Tanager. Forehead back to area around eyes and throat black;* head and neck otherwise intense crimson, extending down as broad band across breast and continuing on flanks. Mantle, wings, tail, and mid-belly black; rump the same intense crimson. Female slightly duller (usually looks similar in field); immature can look much duller (brick red), but pattern already the same.

SIMILAR SPECIES: Virtually unmistakable in its Amaz. range; the bold red and black pattern differs markedly from the much darker, more maroon-red male Silver-beaked Tanager (with which Masked Crimson often occurs). Superficially resembles Vermilion Tanager, but they do not occur together (Vermilion being montane).

HABITAT AND BEHAVIOR: Fairly common to locally common in shrubbery and trees along or near water; primarily a *várzea* forest bird, though may wander to clearings on nearby high ground. Usually in groups of up to a dozen or even more birds which troop about noisily (though no one call is particularly distinctive) at all levels but usually not too high. They often forage alone, but at other times are associated with other tanagers (most often Silver-beaks). Normally anything but shy, these spectacular tanagers are frequently to be seen while boating along smaller rivers and creeks or around oxbow lakes.

RANGE: Se. Colombia (north to s. Meta and Amazonas), e. Ecuador, e. Peru, n. Bolivia (Pando), and w. Amaz. Brazil (east along Amazon itself to around mouth of Rio Tapajós). To about 500 m.

Calochaetes Tanagers

Not dissimilar from *Piranga,* but sexes alike, and with different bill.

Calochaetes concinneus

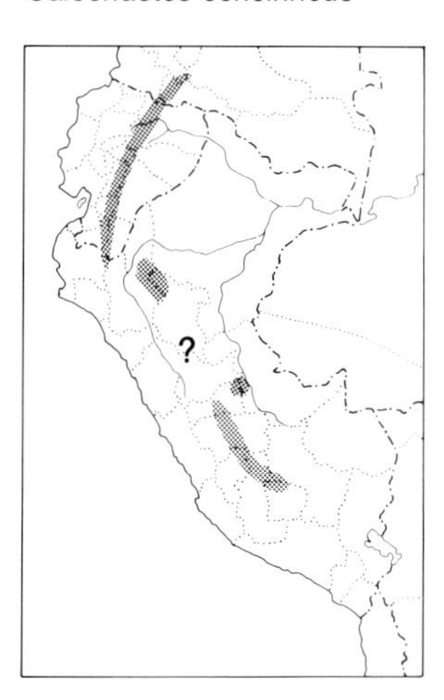

VERMILION TANAGER

PLATE: 18

IDENTIFICATION: 18 cm (7"). *E. slope of Andes.* Bill black with base of mandible silvery gray. *Mostly shiny scarlet,* with black wings and tail and black lores and ocular area extending down over throat and upper chest.

SIMILAR SPECIES: This spectacular tanager is unlikely to be confused in its limited montane range. Superficially resembles Masked Crimson Tanager, but they do not occur together (Masked Crimson being found in Amaz. lowlands). Breeding male Scarlet Tanager (might briefly occur with Vermilion during Scarlet's northward passage) lacks the black bib, as does male White-winged Tanager (also with white wing-bars, etc.).

HABITAT AND BEHAVIOR: Rare to locally fairly common in canopy and borders of humid montane forest. Generally found only singly or in pairs. Almost always encountered with mixed-species flocks of tan-

agers, foraging well above the ground, often hopping along horizontal branches, inspecting moss and epiphytes. It apparently has no true song. Vermilion Tanagers can be seen regularly in forests of Cordillera Divisoria east of Tingo María in Peru and at appropriate elevations along the road to Zamora in Ecuador.
RANGE: E. slope of Andes in se. Colombia (north to w. Caquetá), e. Ecuador, and e. Peru (south to Cuzco on Cordillera Vilcabamba). Mostly 1100–1900 m.

Habia Ant-Tanagers

A group of rather dull, infrequently seen tanagers of forest interior, usually foraging low. Their plumage is rather dull, even the red of the males being quite dusky, but 2 Colombian endemics are prominently crested. Only the Red-crowned Ant-Tanager is widespread in South America, the other 3 species being found only in Colombia. Due to the detailed studies of E. O. Willis, the behavior of the ant-tanagers is better documented than that of any other genus of tanagers.

GROUP A

Lack obvious crests; both sexes with *paler throats.*

Habia rubica

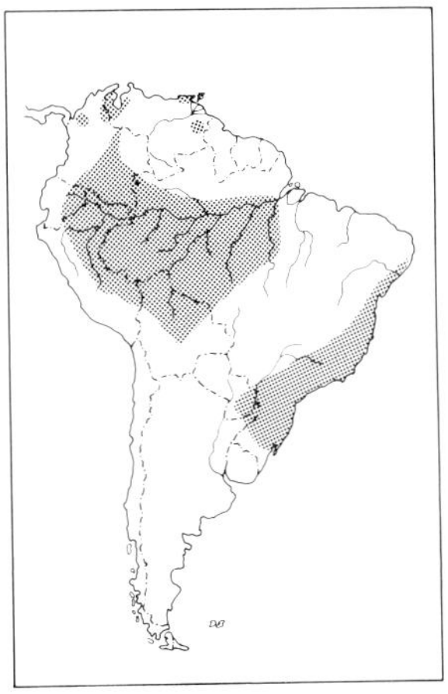

RED-CROWNED ANT-TANAGER

PLATE: 18

IDENTIFICATION: 16.5–19 cm (6½–7½"). *The only ant-tanager across its extensive S. Am. range* (not overlapping with other *Habia* species, which are found only in n. and w. Colombia). *Mostly dull brownish red,* paler and redder (often rosier) below with gray on flanks, and often contrastingly brighter on throat. *Crown patch scarlet,* bordered narrowly by thin black line (usually not apparent, and often neither is red in crown, though latter can be flared). Female *olive brown above* with *tawny-yellow crown patch* (but bordering dusky line inconspicuous at best). Dingy pale buff to dull ochraceous below, buffiest on throat. Females of Amaz. races (except for *peruviana* of e. Peru south of the Marañón and Amazon rivers, n. Bolivia, and westernmost Amaz. Brazil) are rather whitish below, especially on throat. S. races (nominate and *bahiae*) are notably larger and both sexes are relatively dark and dingy with barely contrasting throats; in females tawny crown patch often barely indicated.
SIMILAR SPECIES: Cf. Red-throated Ant-Tanager of n. Colombia (no recorded overlap, though they come close). Male Red-crowned Ant-Tanager can be confused with females of both Silver-beaked and Brazilian Tanagers, but note crown patch (when not hidden), usually paler throat, lack of redder rump, and absence of pale area on mandible. Female Fulvous-crested Tanager is more olive above with grayer head and yellow eye-ring. Female Ruby-crowned Tanager is more rufescent above and deeper ochraceous below with mottling on breast.

HABITAT AND BEHAVIOR: Uncommon to common (perhaps most numerous on Trinidad) in lower growth of humid forest and forest borders. Usually rather unobtrusive, normally found in pairs or small family groups which are often with flocks of various insectivorous antbirds (*Thamnomanes* antshrikes, *Myrmotherula* antwrens, etc.). Rarely or never at army ant swarms in South America except apparently on Trinidad. Its repeated chattering "chit" or "chiup" (often given in rapid series) scolding call is regularly heard, but male's thrushlike phrased song (which may even include imitations of other species, at least in nominate race) is given much less often, principally soon after dawn.
RANGE: Locally in humid parts of n. Colombia and nw. Venezuela (w. slope of E. Andes in Magdalena, slopes of Perijá Mts., and w. slope of Andes in Mérida and Lara; perhaps also in upper Río Sinú valley in Bolívar, Colombia); ne. Venezuela in Sucre and Anzoátegui, locally also in e. Bolívar; se. Colombia (north along base of Andes to Venezuelan frontier), e. Ecuador, e. Peru, n. Bolivia (south to La Paz, Cochabamba, and w. Santa Cruz), and Amaz. Brazil (mostly south of the Amazon, east to nw. Mato Grosso and lower Rio Xingú in Pará); e. Brazil (Pernambuco south to Rio Grande do Sul), e. Paraguay, and ne. Argentina (Misiones); Trinidad. Also Mexico to Panama. Mostly below 900 m.

Habia fuscicauda

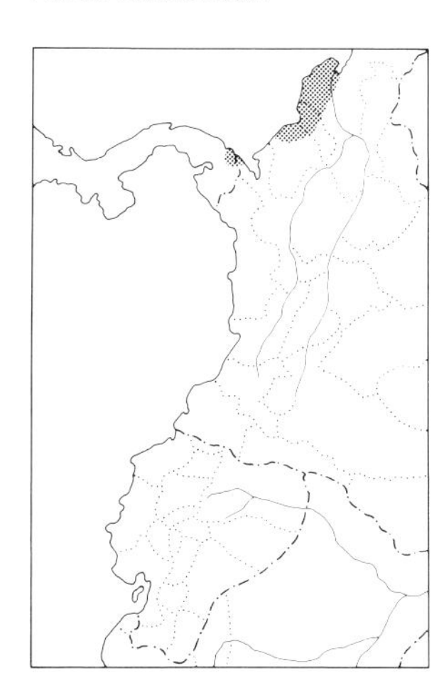

RED-THROATED ANT-TANAGER

IDENTIFICATION: 19 cm (7½"). *N. Colombia.* Closely resembles slightly smaller Red-crowned Ant-Tanager (no recorded overlap), but compared to its closest races male is darker and more carmine red (less rosy). Furthermore, it lacks any trace of the dark line bordering its red crown patch (barely apparent even in Red-crowned, however) and it has a *notably brighter, paler red throat* contrasting with remaining underparts. Female is easier; more olivaceous (not as brown) above with *no crown patch,* and with *contrasting pale yellow or yellowish ochre throat* and duskier belly. See P-28, C-51.
SIMILAR SPECIES: Sooty Ant-Tanager (no known overlap) is much grayer, has crest, etc.
HABITAT AND BEHAVIOR: Not well known in the S. Am. portion of its range. Recorded as "not rare" in the Serranía de San Jacinto in Bolívar and Sucre (J. Haffer). In Panama frequents lower growth of humid secondary woodland and forest borders, where it travels about in small active groups of its own species, typically appearing briefly to scold the observer, then quickly dropping back out of sight. They sometimes forage over swarms of army ants. Though tolerant of second growth, its continued existence in Colombia may be threatened by the massive deforestation which has taken place across much of its range.
RANGE: Locally in Caribbean n. Colombia west of mouth of Magdalena River (Córdoba to Atlántico and n. Bolívar). Also Mexico to Panama. To about 200 m.

GROUP B *Conspicuous scarlet crest;* sexes alike.

Habia gutturalis

SOOTY ANT-TANAGER

PLATE: 18

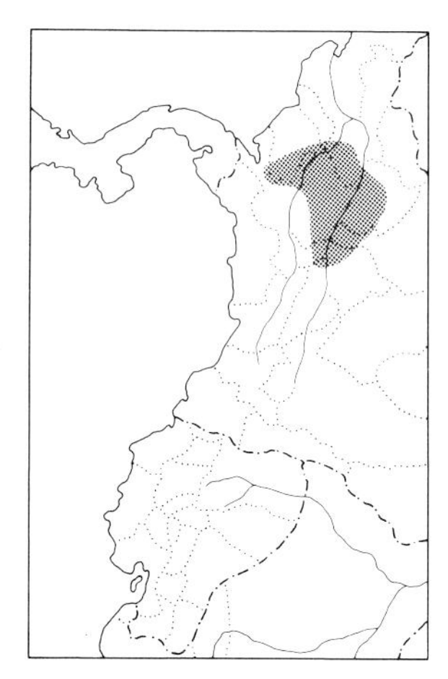

IDENTIFICATION: 19 cm (7½"). *N. Colombia. Mostly dark sooty gray,* almost blackish on sides of head and neck, relieved only by *scarlet crown patch and crest* and by *rosy red throat patch.* Female similar but duller, with paler and less bright throat patch.
SIMILAR SPECIES: Combination of dark gray plumage and the red crest should preclude confusion; note its restricted range.
HABITAT AND BEHAVIOR: Locally fairly common in lower growth of forest borders and openings and in secondary woodland. Pairs or small family groups either join mixed foraging flocks or may follow swarms of army ants. Rather noisy, with a fast, chattering call similar to Red-crowned Ant-Tanager's (E. O. Willis).
RANGE: N. Colombia in hilly country and Andean foothills from north of W. Andes (upper Río Sinú valley) east to middle Magdalena valley (where ranges south to n. Tolima). 100–1100 m.

Habia cristata

CRESTED ANT-TANAGER

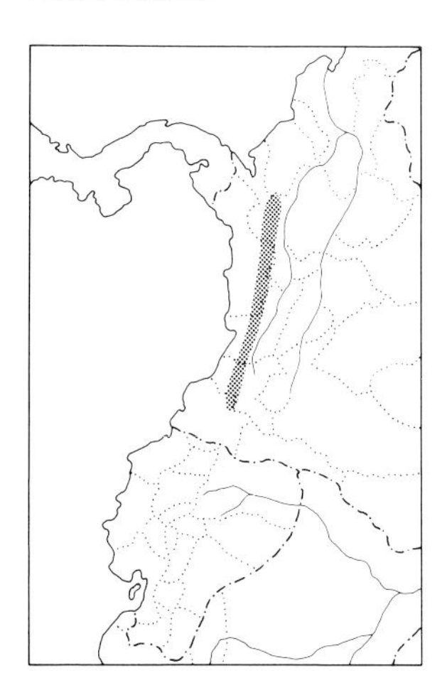

IDENTIFICATION: 18.5 cm (7¼"). *W. Colombia.* Sexes nearly alike. *Long prominent scarlet crest* imparts look of a cardinal (*Cardinalis* sp.), but usually laid nearly flat, projecting behind (raised in excitement and alarm). Dark crimson above, brightest on head and neck. *Throat and chest scarlet,* becoming grayish tinged crimson on lower underparts. See C-51.
SIMILAR SPECIES: Color pattern somewhat similar to male Red-crowned Ant-Tanager's (no overlap), but brighter and redder on head and throat and of course with the unmistakable expressive crest.
HABITAT AND BEHAVIOR: Uncommon and local in lower growth of forest and forest borders, especially in dense, low vegetation on steep banks along rushing mountain streams and in regrowth following landslides on steep slopes. Usually in small groups of up to 10 or so birds which forage like other ant-tanagers and also at least occasionally follow army ant swarms. Alarm call is a loud (though sometimes hard to discern over roar of water) "chi-veék," usually repeated several times, birds often flaring and raising crest up and down as they call. Most of its habitat is difficult of access but can be seen along the Río Pichindé above Cali (you must go down to near the river itself).
RANGE: W. Colombia on both slopes of W. Andes (Antioquia south to Cauca). 700–800 m.

Rhodinocichla Thrush-Tanagers

An odd, distinctive, mostly terrestrial tanager. Though some have speculated that it belongs in the Mimidae, anatomical evidence is to the contrary; it does have a superficially mimidlike bill, and overall shape and behavior are somewhat alike.

Rhodinocichla rosea

ROSY THRUSH-TANAGER

PLATE: 18

Other: Rose-breasted Thrush-Tanager

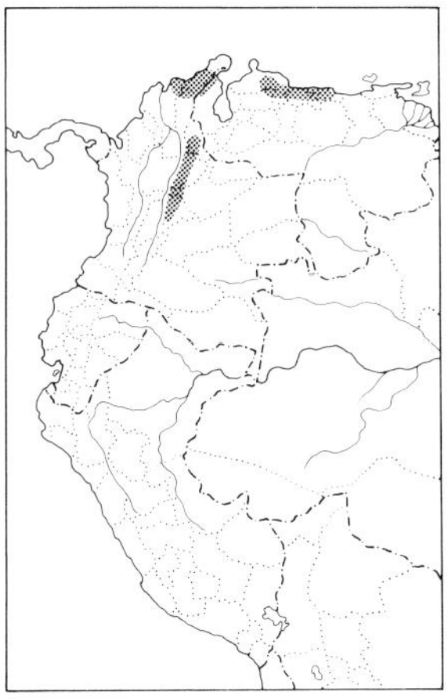

IDENTIFICATION: 20 cm (7¾"). *N. Colombia and Venezuela.* Unmistakable. Note *mimidlike bill.* Dark brownish slaty above with *long narrow superciliary rosy magenta in front of eye, whitish behind it;* bend of wing also rosy red magenta. *Below mostly rosy magenta,* with flanks broadly dusky gray. Female has identical pattern, but *rose magenta replaced by rich ochraceous. Beebei* (of Perijá Mts.) has barely indicated postocular stripe.

HABITAT AND BEHAVIOR: Uncommon and rather local in thickets and dense tangled undergrowth of deciduous woodland, second growth, and scrub, mostly in foothills and on lower slopes. A lovely bird (male's rosy pink or magenta is nearly unique among S. Am. birds) but shy and furtive, and almost always frustratingly difficult to even glimpse (though it does respond to tape playback). Largely terrestrial, single birds or pairs hop on the ground or low in undergrowth, flicking leaves and litter with their bills. Usually not with flocks. Has an excellent rich song, in quality reminiscent of certain *Thryothorus* wrens (e.g., Black-bellied, Whiskered)—a loud ringing "cho-oh, chowee" or "wheeo, chee-oh, chweeoh" with variations. Both members of the pair sing, and this is the only tanager known to sing antiphonally.

RANGE: Locally in n. Colombia (lower slopes of Santa Marta Mts. and n. end of Perijá Mts., w. slope of E. Andes in Norte de Santander, Cundinamarca, and n. Tolima) and n. Venezuela (Perijá Mts. in Zulia, Falcón and Lara east to Distrito Federal and Miranda). Also Mexico, Costa Rica, and Panama. Mostly 500–1500 m.

NOTE: We have slightly shortened the English name of this species, following Ridgely (1976) and the 1983 AOU Check-list.

Chlorothraupis Tanagers

Drab, heavy-set tanagers with large bills. Olive predominates. Found in lower growth of humid forest, less often at edge; usually in noisy groups, but often independent of mixed flocks.

Chlorothraupis carmioli

OLIVE TANAGER

PLATE: 19

Other: Carmiol's Tanager

IDENTIFICATION: 17 cm (6¾"). *A robust, quite uniform olive tanager of Andean foothill forests* (locally). Bill black, heavy. Uniform olive green above; below slightly paler olive, yellowest on throat. Males show streaky effect on throat and upper chest, obsolete in females (in which throat is uniform pale yellow); females also are paler yellowish olive below and have yellowish lores (olive in male).

SIMILAR SPECIES: A dull bird almost devoid of real field marks, but can be known by its uniform olive plumage and distinctive behavior (see below). Lemon-spectacled Tanager (no known overlap) very similar but for its yellow spectacles.

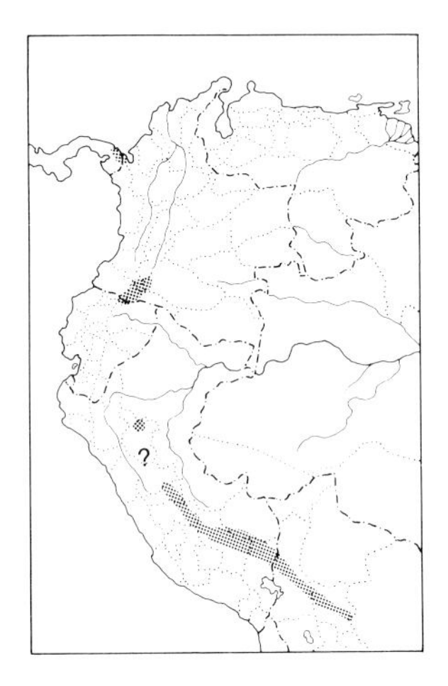

HABITAT AND BEHAVIOR: Not well known from S. Am. portion of its range, but judging from number of museum specimens available must be locally not uncommon (at least in s. Peru and Bolivia). In s. Peru seems to be found mostly on low outlying ridges east of the main Andes (J. Fitzpatrick). In Panama, common in lower growth of humid forest and forest borders, particularly in foothill areas. Here it troops about in noisy conspecific flocks of up to several dozen individuals, sometimes being joined by other species (perhaps most often the Tawny-crested Tanager). Panama birds incessantly give harsh, chattering calls, and these often herald the approach of a group; they vary, but a typical one is a repeated thrushlike "zhwek-zhwek-zhwek." Bolivia birds have *not* been heard to voice these calls, but they do occasionally give a loud blackbirdlike song of repeated phrases (up to 6 to 8 repetitions of each phrase before going on to the next), e.g., "cho-cho-cho; chee-ker, chee-ker, chee-ker, chee-ker, chee-ker; wheeo, wheeo, wheeo, wheeo . . . ," often becoming very agitated (R. A. Rowlett).
RANGE: Nw. Colombia (nw. Chocó on slopes of Cerro Tacarcuna); foothills along e. base of Andes in s. Colombia (w. Caquetá to e. Nariño) and disjunctly on e. slope of Andes from s. Peru (north to Huánuco) to nw. Bolivia (La Paz and Cochabamba). Also Nicaragua to Panama. Mostly 500–1200 m.

NOTE: The distribution of the race *frenata* makes us suspect a distinct species is involved. Information on its behavior and vocalizations remains scanty, but what exists (see above) is rather supportive. However, given its plumage similarity and in the absence of conclusive evidence, we maintain it as a subspecies of *carmioli*. We follow the 1983 AOU Check-list in calling *C. carmioli* the "Olive" rather than "Carmiol's" Tanager, though we are rather at a loss to explain why the more or less established "Carmiol's" was replaced (there being so many "olive" tanagers).

Chlorothraupis olivacea

LEMON-SPECTACLED TANAGER

Other: Lemon-browed Tanager

IDENTIFICATION: 17 cm (6¾"). *W. Colombia and nw. Ecuador.* Iris reddish brown. Similar to Olive Tanager (no known overlap) but can at once be recognized by its *bright yellow lores and prominent eye-ring* (forming the "spectacles"). Male darker olive generally, especially on most of underparts (contrasting with yellow on throat); female less dark (about same olive tone as Olive Tanager). See P-28, C-52.
SIMILAR SPECIES: Ochre-breasted Tanager is much more ochraceous (not olive and yellow) below; it replaces Lemon-spectacled at higher elevations.
HABITAT AND BEHAVIOR: Fairly common in lower growth of humid and wet forest and forest borders. Behavior much like Olive Tanager's and like it noisy, with loud, excited-sounding calls usually given in a fast series (e.g., "treu-treu-treu-treu"). Usually, however, moves in smaller groups (2 to 4 birds being the norm) and more often seems to be a member of true multispecies foraging flocks (at times seeming to act as a flock leader around which other species gather and then follow). Regularly seen in small numbers along lower Buenaventura road in Colombia.

RANGE: W. Colombia (entire Pacific slope and east along humid n. base of Andes to middle Magdalena valley in Antioquia) and nw. Ecuador (Esmeraldas). Also e. Panama. To about 800 m, rarely higher; most numerous below 400 m.

NOTE: This species does *not* have a yellow "brow," hence its formerly employed name is inaccurate. It does, however, have conspicuous yellow spectacles, hence our modification of its English name.

Chlorothraupis stolzmanni

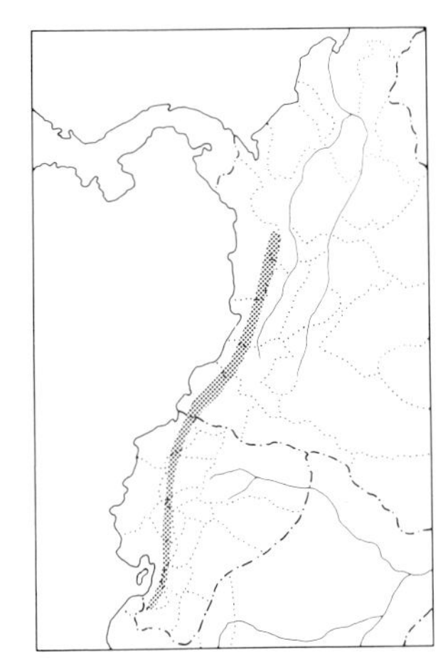

OCHRE-BREASTED TANAGER PLATE: 19

IDENTIFICATION: 18 cm (7"). *W. Colombia and w. Ecuador. Notably drab. Iris bluish gray;* often shows pinkish gape mark. Mostly dull olive above (head tinged gray in *dugandi* of Colombia). *Below mostly dull ochraceous buff,* palest on throat; breast, sides, and flanks tinged olive.
SIMILAR SPECIES: A husky, dull-colored tanager of lower slopes of W. Andes. Perhaps most easily known by the near absence of distinctive marks, though the pale eye is normally quite easy to discern; none of its congeners has the ochraceous tone below (they are mainly olive).
HABITAT AND BEHAVIOR: Uncommon to fairly common in lower growth of humid and wet montane forest and forest borders. Behavior much like Olive Tanager's, like that species foraging in noisy groups of up to 12 to 15 birds (usually fewer), moving rapidly through understory and less often middle strata but normally remaining inside forest (infrequent at edge). Sometimes these groups are joined by other birds, but at least as often they seem to move alone. Rough, chattering calls are given almost continually. Song, most often given soon after dawn and sometimes by a small group "in concert," is a series of loud harsh notes (the overall effect being cacophonous!); during the often long (up to a half hour) performance the birds are often perched out fully in the open.
RANGE: W. slope of W. Andes in w. Colombia (southernmost Chocó south) and w. Ecuador (south to El Oro). Mostly 400–1500 m.

Mitrospingus Tanagers

We suspect that the 2 species presently comprising this genus do not form a natural group as their behavior is so different. Nonetheless, morphologically they are similar, both being predominantly olive and gray with pale irides.

Mitrospingus oleagineus

OLIVE-BACKED TANAGER PLATE: 19

IDENTIFICATION: 18.5 cm (7¼"). *Tepui slopes of Bolívar, s. Venezuela, and adjacent areas.* Iris gray to (young?) reddish brown. Mostly dull olive above, including most of crown; mainly olive yellow below, with *gray throat extending to face and over forehead.* Wings and tail dusky. *Obscureipectus* (most of range other than the Roraima/Escalera area) is slightly darker and more olive below.

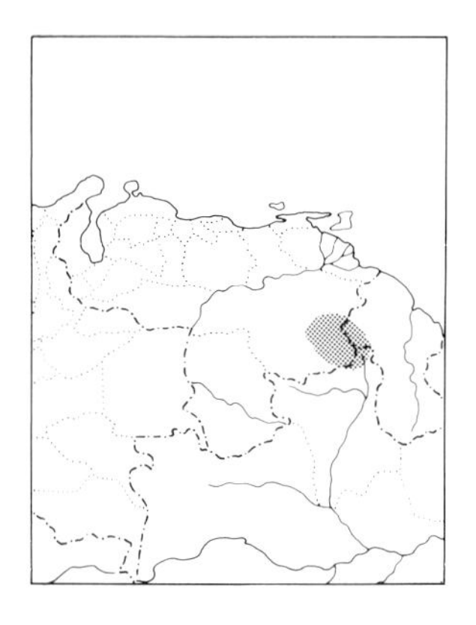

SIMILAR SPECIES: On the *tepuis* there really is nothing resembling it. Superficially much like Dusky-faced Tanager, but note far-separated ranges, different behavior, etc.
HABITAT AND BEHAVIOR: Locally common in humid forest on slopes of *tepuis*. Mostly arboreal, moving about in groups which can be quite large (up to 10 to 20 birds) and often with mixed flocks. Searches for insects by moving slowly along branches, methodically inspecting foliage. Calls frequently, a loud, harsh "shhweee," often rapidly repeated. Regularly found along the Escalera road in s. Bolívar.
RANGE: *Tepuis* of s. Venezuela (se. Bolívar only) and adjacent Guyana (Mt. Twek-quay) and n. Brazil (Cerro Uei-tepui in Roraima). 900–1800 m.

Mitrospingus cassinii

DUSKY-FACED TANAGER

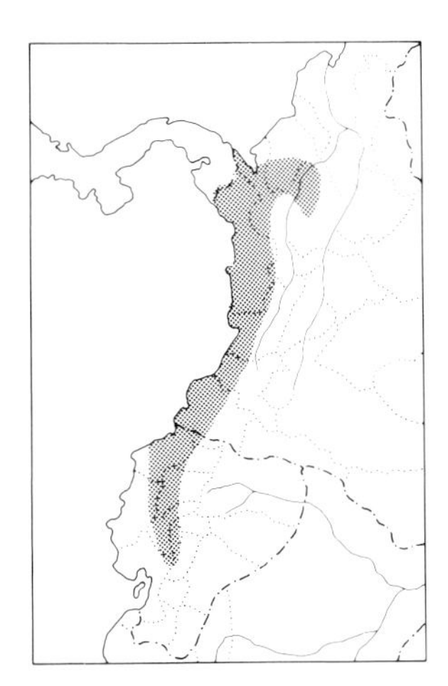

IDENTIFICATION: 18 cm (7"). *W. Colombia and w. Ecuador. Iris grayish white.* In general form resembles Olive-backed Tanager (no overlap), but smaller and with mostly slaty gray upperparts and *blackish mask extending back from forehead over sides of head* (outlining the yellow-olive crown); gray of throat more restricted; darker, oilier yellowish olive below. Dusky-face's active, nervous behavior also very different. See P-29, C-52.
SIMILAR SPECIES: The pale eye stands out conspicuously against the blackish mask and makes this dark tanager easy to recognize. Cf. Gray-headed Tanager (limited overlap).
HABITAT AND BEHAVIOR: Locally common in dense thickets and shrubbery at forest and woodland borders; especially numerous along streams but by no means restricted to such situations. Usually found in noisy groups of from 4 to 12 birds, almost always moving independently of other species. Movements typically are very rapid, and as the birds are shy, tending to remain in the undergrowth and rarely perching for long in the open, this can be a hard bird to get a good look at. Most frequently heard call is an incessant gravelly "cht-cht-cht . . ."
RANGE: W. Colombia (entire Pacific slope and extending east along humid n. base of Andes to middle Magdalena valley in e. Antioquia) and w. Ecuador (south to e. Guayas). Also Costa Rica and Panama. Mostly below 800 m.

Eucometis Tanagers

An engaging, bright olive and yellow tanager with a gray head; usually looks at least somewhat crested. Found in lower growth of woodland and forest in tropical lowlands.

Eucometis penicillata

GRAY-HEADED TANAGER PLATE: 19

IDENTIFICATION: 17–18 cm (6¾–7"). Bill black to grayish brown (most of range) or at least partly pink (*albicollis* of Bolivia and s.-cen. Brazil). Widespread nominate race has *head gray with short bushy crest*

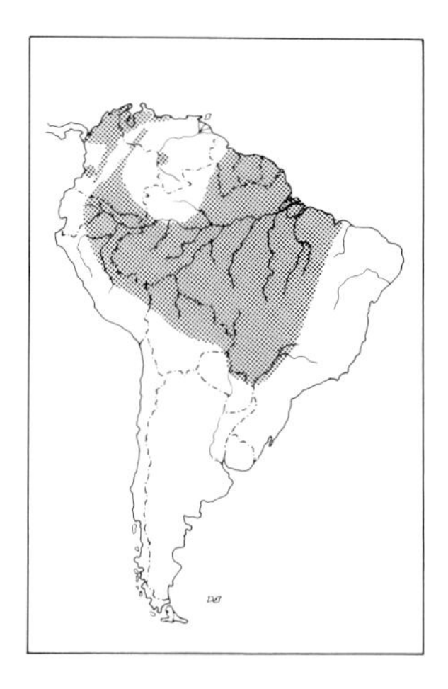

(usually laid back more or less flat, but can be flared up when excited, white bases to feathers then exposed). Above otherwise bright olive. *Throat grayish white,* with *remaining underparts rich yellow.* Races west of Andes are slightly smaller and have weaker crest with no white and darker gray head. *Albicollis,* apart from its pink bill, differs in being paler overall with buffy whitish throat and in having (particularly in male) a longer bushy crest.

SIMILAR SPECIES: Female White-shouldered Tanager has similar pattern but is much smaller and lacks any trace of a crest; it is arboreal rather than being a bird of lower growth. Cf. also Gray-hooded Bush-Tanager (strictly higher Andean slopes).

HABITAT AND BEHAVIOR: Uncommon to locally fairly common in lower growth of *várzea* forest, gallery forest, and second-growth woodland and forest borders; primarily associated with vicinity of water east of the Andes, more widespread west of them. Most often seen in pairs, less frequently in small groups. North and west of the Andes frequently noted following army ant swarms, but east of them this is rarely reported, and the species occurs more often with mixed flocks. Active and rather excitable, and may come in to investigate a squeaking or pishing sound, chattering and with crest raised. Its song is a jumbled series of sputtery high notes.

RANGE: Widespread in lowlands south to n. and e. Bolivia (south to Cochabamba and Santa Cruz), ne. Paraguay (Concepción), and s.-cen. Brazil (south to nw. São Paulo and east to Goiás and Maranhão); west of Andes only in w. Colombia (south into Cauca and Magdalena valleys and not on Pacific slope) and nw. Venezuela (east to Miranda); mostly or entirely absent from Orinoco and upper Río Negro drainages. Also Mexico to Panama. Mostly below 600 m.

Heterospingus Tanagers

Ornately patterned tanager of humid forest canopy in w. Colombia and nw. Ecuador; a congener is found in s. Middle America. Behavior much as in arboreal *Tachyphonus* group.

Heterospingus xanthopygius

SCARLET-BROWED TANAGER

PLATE: 19

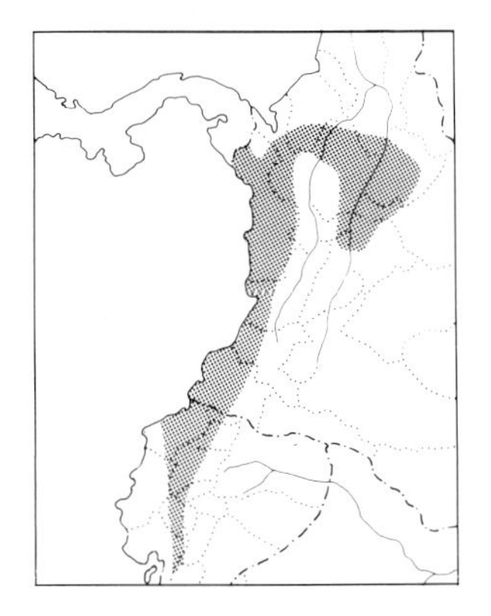

IDENTIFICATION: 18 cm (7"). *W. Colombia and w. Ecuador.* Heavy bill. Mostly black, with narrow white line above eye, *broadening into conspicuous scarlet postocular tuft* (often seeming to protrude outward); small patch on shoulders and *rump bright yellow; pectoral tuft white* (usually protrudes from under wing). Female not as gaudy and *grayer* (slaty above, ashy gray below); lacks head markings but *retains yellow rump and the white pectoral tufts;* see C-52.

SIMILAR SPECIES: Male unmistakable and female virtually so: no other mostly gray tanager shows the white patch on sides of breast or the yellow rump. Cf. Plain-colored Tanager (much smaller and lacks the yellow and white) and male Flame-rumped Tanager (yellow-rumped race).

HABITAT AND BEHAVIOR: Uncommon to locally fairly common in canopy and borders of humid and wet forest, less often out into secondary woodland or clearings with scattered trees. Usually in pairs, foraging well above ground and almost always with canopy flocks of other tanagers. Sometimes sluggish and inactive, at times perching nearly motionless for considerable periods on an exposed branch. Often given is a loud "dzeet" or "dzip."

RANGE: W. Colombia (entire Pacific coast and east along humid n. base of Andes to middle Magdalena valley, south to e. Antioquia and w. Cundinamarca) and w. Ecuador (recorded south to e. Guayas, but in recent years found only to Pichincha). Also e. Panama. Mostly below 800 m.

Lanio Shrike-Tanagers

Boldly patterned pair of tanagers, reminiscent in many ways of the arboreal *Tachyphonus* tanagers. Despite their name, the shrike-tanagers do not have an appreciably heavier or more hooked bill than the latter genus. Found in canopy of Amaz. forests, 1 species on either side of the Amazon (2 other species occur in Middle America).

Lanio versicolor

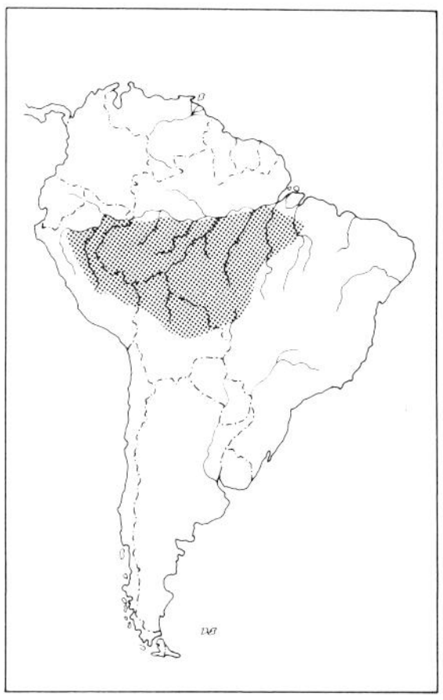

WHITE-WINGED SHRIKE-TANAGER PLATE: 19

IDENTIFICATION: 16 cm (6¼"). *Amazon basin south of Amazon River.* Head blackish with yellowish olive forecrown; back and rump ochraceous, brighter and yellower on rump. *Large area on wing-coverts white;* wings and tail otherwise blackish. Throat dark olive; remaining underparts bright yellow, tinged ochraceous on breast and flanks. Female very uniform looking, *ochraceous brown above* (brownest on wings and tail); *warm ochraceous below, yellower on center of belly.*

SIMILAR SPECIES: Does not overlap with Fulvous Shrike-Tanager (found north of the Amazon). Strongly patterned male is not likely to be confused, though female (if alone—in fact, not often the case) can cause problems: it most resembles female Flame-crested Tanager (and often with it), but latter is browner generally (without olivaceous tones), with whitish throat and uniform ochraceous buff lower underparts (*no yellow*).

HABITAT AND BEHAVIOR: Fairly common in canopy and subcanopy of humid *terra firme* forest, less often at forest borders. Usually in pairs, less often small groups, which almost invariably are found with mixed-species flocks foraging well above the ground (as a result often hard to see clearly). Often acts as a nuclear species in such flocks, with other species cuing in on its loud and arresting "tchew!" and other forceful calls.

RANGE: E. Peru (north to the Río Marañón and the s. bank of the Amazon), n. Bolivia (south to La Paz and Cochabamba), and Amaz. Brazil south of the Amazon east to nw. Mato Grosso and the lower Rio Tocantins in e. Pará. To about 900 m.

Lanio fulvus

FULVOUS SHRIKE-TANAGER

IDENTIFICATION: 17 cm (6¾"). *Amazon basin north of Amazon River.* Resembles White-winged Shrike-Tanager but is larger, *lacking visible white on the wing,* with solid black hood *extending down over throat; patch of chestnut on center of chest,* along lower margin of black throat, and with rump fulvous and lower underparts more ochraceous yellow. See V-37, C-51. Female olivaceous brown above, most olive across mantle, most *rufescent on rump;* crown duskier. Throat pale grayish buff; remaining underparts ochraceous olive, becoming *rufescent on flanks and crissum.*

SIMILAR SPECIES: Does not overlap with White-winged Shrike-Tanager (found south of the Amazon). Female resembles female Flame-crested Tanager but is larger and more rufescent on rump (this hard to see as both species are usually above you) and on lowermost underparts (belly of female Flame-crest a paler, more uniform ochraceous buff, not really rufous).

HABITAT AND BEHAVIOR: Fairly common in canopy and subcanopy of humid *terra firme* forest. Behavior as in White-winged Shrike-Tanager.

RANGE: Guianas, s. Venezuela (Bolívar and Amazonas), Amaz. Brazil north of the Amazon (rather few records; perhaps overlooked), ne. Peru (n. Loreto, Amazonas, and San Martín north of the Amazon and Río Marañón), e. Ecuador, se. Colombia (north along e. base of Andes to Arauca), and extreme sw. Venezuela (Táchira). Mostly below 900 m (but recorded higher in s. Venezuela).

Cyanicterus Tanagers

Unmistakable blue and yellow tanager found in humid forest canopy. Bill very heavy. Behavior reminiscent of the arboreal *Tachyphonus* group.

Cyanicterus cyanicterus

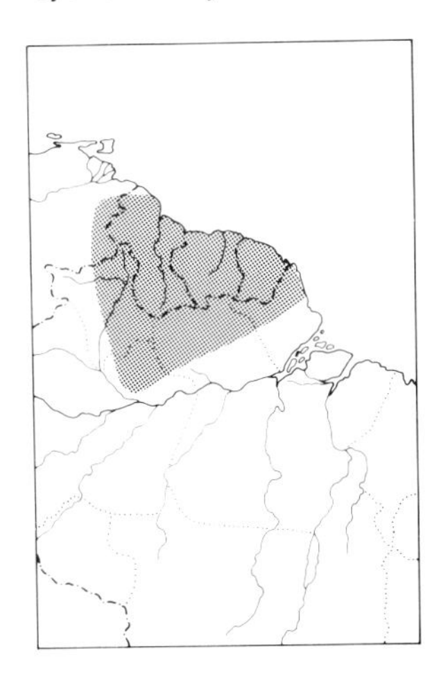

BLUE-BACKED TANAGER

PLATE: 19

IDENTIFICATION: 17 cm (6¾"). *Ne. South America.* Very heavy long bill, blackish (male), paler below (female). Iris orange to red. Male *bright cobalt blue above and on throat and upper chest;* remaining underparts bright yellow, with dark blue thighs. See V-37. Female paler overall but still a very striking bird: *mostly cerulean blue above,* with *more cobalt blue wings and tail. Lores, face, throat, and chest bright ochraceous yellow,* becoming pure yellow on lower underparts.

SIMILAR SPECIES: This superb, scarce tanager shouldn't be confused in its range (color scheme reminiscent of certain Andean mountain-tanagers).

HABITAT AND BEHAVIOR: Uncommon (but perhaps somewhat overlooked and previously under-collected) in canopy of humid lowland forest. Usually in pairs, foraging high in trees, and often with mixed-species flocks of other tanagers, etc. Its call is a very loud, arresting, high-pitched "tseee-tsew" or "tseee-tsew-tsew" (T. Meyer). Seen with some regularity at Brownsberg Reserve in Suriname, but may be most numerous in the Imataca Forest south of El Dorado in Bolívar, Venezuela.

RANGE: Se. Venezuela (e. Bolívar), Guianas, and ne. Brazil (recorded only from Manaus area on lower Rio Negro, but presumably more widespread). To about 500 m.

Tachyphonus Tanagers

A widespread group of mainly black tanagers (males), usually relieved by contrasting crest, patch on wing, rump, or flanks; females much duller and easily confused. Young males with patchy black and brown plumage are quite often seen in all species. Either the mandible or the base of the bill is usually pale bluish. Most range in humid forest, usually in the canopy. The last 3 species are rather different, being birds of shrubby semiopen country, with males looking all black. Note that our 3 groups are based on males.

GROUP A

Small, with *prominent white shoulder.*

Tachyphonus luctuosus

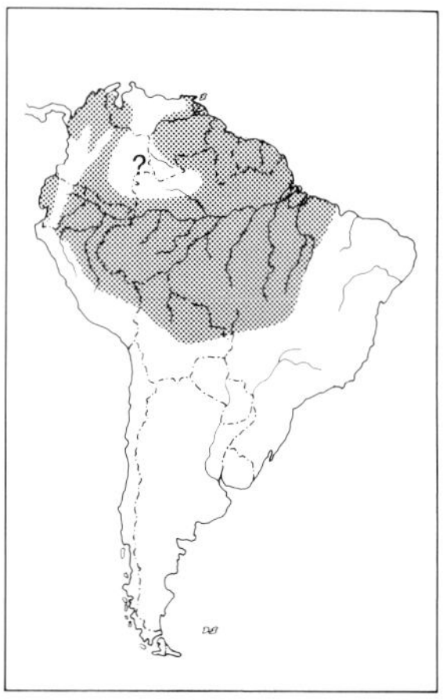

WHITE-SHOULDERED TANAGER

PLATE: 19

IDENTIFICATION: 13–13.5 cm (5–5¼″). *Glossy black* with *conspicuous white area on wing-coverts* and white under wing-linings. Female olive above with *rather contrasting gray head* (though crown may be tinged olive). Throat grayish white, with remaining underparts bright yellow, clouded olive on breast and sides.

SIMILAR SPECIES: Male White-lined Tanager is superficially similar but shows little or no white on closed wing (sometimes a little shows on bend of wing only). Female recalls a miniature Gray-headed Tanager but lacks bushy-crested effect and has very different behavior. Female Fulvous-crested Tanager has yellowish eye-ring and mostly buffy underparts, while female Yellow-crested is larger and more ochraceous below.

HABITAT AND BEHAVIOR: Fairly common to common in middle and upper strata of humid forest (both *várzea* and *terra firme*), forest borders and secondary woodland and clearings with scattered trees. Forages actively in foliage, often as one of the more numerous members of mixed flocks of insectivorous birds; also takes some fruit. Usually remains rather high, but may come lower in clearings.

RANGE: Widespread in humid-forested lowlands south to n. Bolivia (south to La Paz, Cochabamba, and Santa Cruz) and cen. Brazil (south to n. Mato Grosso, n. Goiás, and n. Maranhão); west of Andes south to w. Ecuador (to El Oro); Trinidad. Perhaps absent from much of ne. Colombia, sw. Venezuela, and nw. Brazil. Also Honduras to Panama. Mostly below 800 m.

GROUP B *Bright* (usually prominent) *coronal crest.*

Tachyphonus cristatus

FLAME-CRESTED TANAGER PLATE: 19

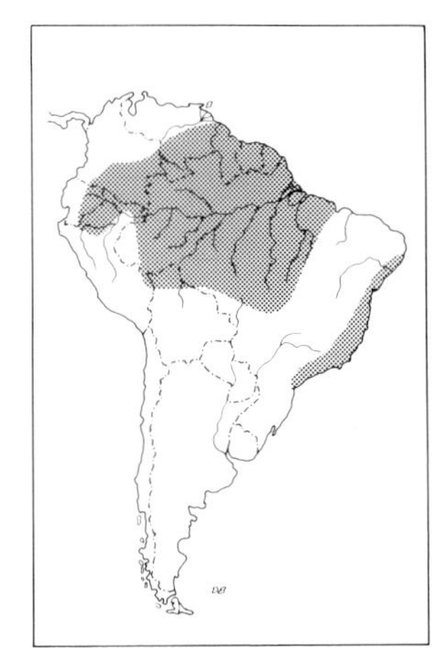

IDENTIFICATION: 16–16.5 cm (6¼–6½"). Mostly glossy black with *broad flat crest flame orange to scarlet. Rump golden buff; center of throat ochraceous buff* (often not very conspicuous); inner shoulders white (sometimes hidden), as are under wing-linings. Crest pure orange in *fallax* (e. Ecuador, ne. Peru, and adjacent Colombia) and *intercedens* (se. Venezuela, Guyana, and Suriname). *Nattereri* (w.-cen. Brazil) is slightly smaller and also has orange crest; it lacks the pale color on throat and has smaller rump patch. Female olivaceous to rufescent brown above, most rufous on rump; throat whitish, rest of *underparts rather rich ochraceous,* often deepest on crissum.

SIMILAR SPECIES: Color of male's crest may vary, but it is *always bright and conspicuous.* Overall pattern of male Fulvous-crested Tanager similar, but its crest is narrower and less prominent and it lacks any throat patch at all; furthermore, note its white pectoral patch and the chestnut on flanks. Female Flame-crest can be difficult, especially as it is usually seen only dimly, high in the canopy. It is most likely to be confused with females of either shrike-tanager, especially White-winged, as it is closer to size of that species; note shrike-tanager's more yellowish tone below, especially on belly, and its lack of whitish throat. Cf. also female Fulvous-crested Tanager and females of several *Thamnomanes* antshrikes.

HABITAT AND BEHAVIOR: Fairly common to common in middle and upper strata of humid forest (both *terra firme* and *várzea*) and to a lesser extent also forest borders and adjacent small clearings. Found in pairs or small groups, foraging well above the ground (much higher than Fulvous-crested Tanagers normally do), typically with flocks of other birds, particularly other tanagers. Gleans actively in foliage but also eats some fruit.

RANGE: Guianas, s. Venezuela (Bolívar and Amazonas), se. Colombia (north to Meta and Guainía), e. Ecuador, ne. and extreme se. (e. Madre de Dios) Peru, n. Bolivia (Beni), and Amaz. Brazil (south to n. Mato Grosso, n. Goiás, and n. Maranhão); coastal e. Brazil (Pernambuco south to e. São Paulo). Mostly below 800 m (locally higher in s. Venezuela).

NOTE: We follow J. T. Zimmer (*Am. Mus. Novitates* 1304, 1945) and *Birds of the World* (vol. 13) in considering *nattereri* of w.-cen. Brazil as a race of this species and not as a distinct species (as did Meyer de Schauensee 1966, 1970). Morphologically it is not very divergent from various races of *T. cristatus,* and it remains unknown in life (and is still definitely known only from the type specimen).

Tachyphonus rufiventer

YELLOW-CRESTED TANAGER PLATE: 19

IDENTIFICATION: 15 cm (6"). *E. Peru and adjacent lowlands.* Mostly glossy black above with yellow crown patch and yellowish buff rump; small area on wing-coverts white (sometimes hidden). *Below mainly*

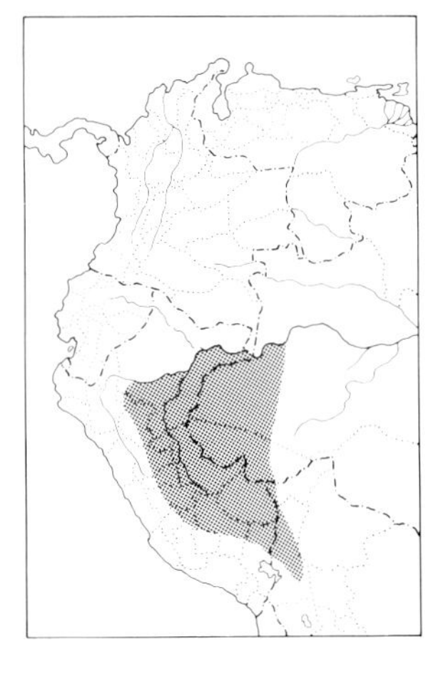

tawny, paler and buffier on throat and flanks, with *narrow black pectoral band* (occasionally almost obsolete). Female olive above with rear crown tinged gray and upper tail-coverts tinged ochraceous. Throat whitish; *remaining underparts rather bright ochraceous yellow, deepest on crissum.*

SIMILAR SPECIES: Male's pattern on underparts unlike that of any other *Tachyphonus* tanager; from below most resembles male White-winged Shrike-Tanager (and the 2 often forage in the same flock), but usually effect of black pectoral band outlining the small pale throat patch can be discerned. Crest shape more like that of Fulvous-crested than of Flame-crested Tanager (and hence often is partially hidden). Female much yellower and more olive than female Flame-crested; female White-shouldered Tanager is smaller and less ochraceous below and has more obviously gray head, while Fulvous-crested has yellowish eye-ring and is buffier below.

HABITAT AND BEHAVIOR: Fairly common in middle and upper strata of humid forest (both *terra firme* and *várzea*), less often at forest borders. Behavior very similar to Flame-crested Tanager's, which it appears to replace in its rather limited range, though they are recorded sympatrically from at least one locality in westernmost Brazil. Not uncommon in the Tingo María/Santa Elena area in Huánuco, Peru.

RANGE: E. Peru (south of the Amazon and Río Marañón), nw. Bolivia (recorded only La Paz), and w. Amaz. Brazil (east only to Rio Juruá drainage). To about 1200 m.

Tachyphonus surinamus

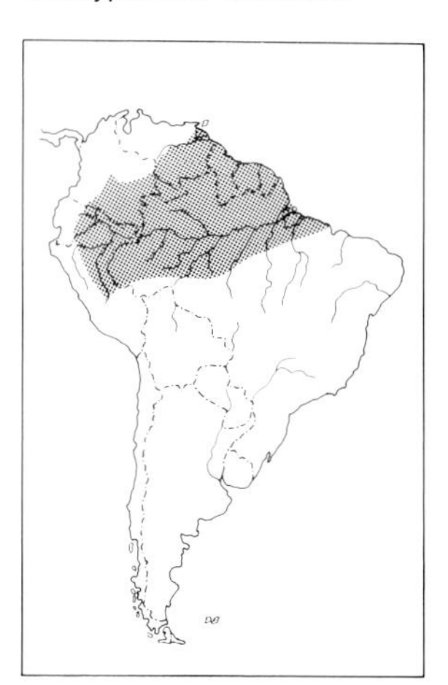

FULVOUS-CRESTED TANAGER

IDENTIFICATION: 16.5 cm (6½"). *Amazonia to the Guianas.* Male resembles Flame-crested Tanager, but tawny-buff *crown patch smaller* (with less of a crested effect), *no buff throat patch,* a fairly prominent *white pectoral tuft* (usually protrudes from under bend of wing), and *irregular chestnut patch on lower flanks.* See V-37, C-51. *Insignis* (Amaz. Brazil south of the Amazon west to lower Rio Madeira) has buff pectoral tuft, while in *napensis* (e. Peru and Amaz. Brazil south of the Amazon east to lower Rio Juruá) both crown and rump patch are rufous. Female mainly olive above with *gray head and broken but conspicuous yellowish eye-ring.* Below buffy whitish to yellowish buff, deepest on belly.

SIMILAR SPECIES: Compare to Flame-crested Tanager male. Female differs from female of smaller White-shouldered Tanager in its eye-ring and the buff (not bright yellow) tone to underparts. Cf. also female Flame-crested (much browner generally, without olive upperparts, etc.) and Yellow-crested (lacking this species' eye-ring, more ochraceous below, etc.).

HABITAT AND BEHAVIOR: Uncommon to fairly common in lower growth of humid forest (both *terra firme* and *várzea*) and forest borders. More numerous eastward; in w. Amazonia found primarily in the few areas with white sand soils. Pairs forage mostly in lower and middle strata inside forest, where they often accompany mixed flocks of antwrens, foliage-gleaners, etc. Rarely or never in canopy as is Flame-crested Tanager.

RANGE: Guianas, e. and s. Venezuela (e. Sucre and e. Monagas south through Bolívar and Amazonas), se. Colombia (north to Meta and Guainía), e. Ecuador, e. Peru (south to Junín), and Amaz. Brazil (south to n. Mato Grosso and e. Pará in the Belém area). Mostly below 500 m, but recorded higher in Venezuela.

Tachyphonus delatrii

TAWNY-CRESTED TANAGER

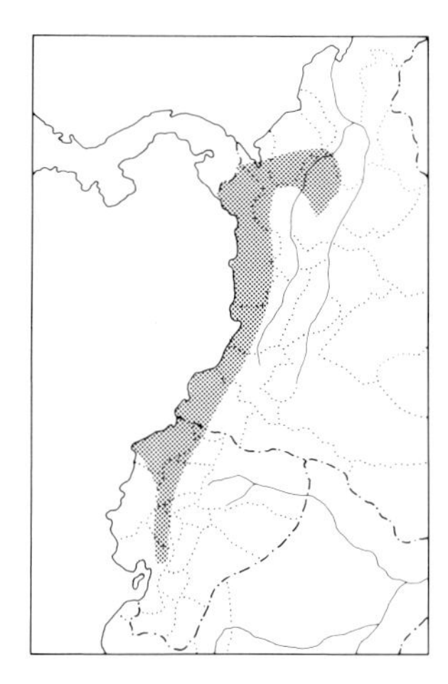

IDENTIFICATION: 14.5 cm (5¾"). *W. Colombia and w. Ecuador. All black* with *prominent golden tawny crest.* Female *uniform dull dark olive brown,* somewhat paler below. See C-52. Immature males may show the crown patch before they become black.

SIMILAR SPECIES: Male unmistakable in its range; female is darker brown than any other sympatric tanager (and almost invariably occurs with male).

HABITAT AND BEHAVIOR: Fairly common to common in lower growth of humid and wet forest and forest borders and in advanced secondary woodland. Very active and conspicuous, sweeping through understory and forest edge in fast-moving flocks. They often move so rapidly and pause so briefly that it is hard to get a good look at them. Group size is often fairly large (up to 10 to 20 or more birds); these usually travel about independently of other species, but occasionally they may be joined by other tanagers, etc. Noisy, though none of their loud "chit" calls is particularly distinctive. Especially dominant along the lower part of the old Buenaventura road in Colombia.

RANGE: W. Colombia (Pacific lowlands and east along humid n. base of Andes to middle Magdalena valley in e. Antioquia) and w. Ecuador (recorded south to Chimborazo, but in recent years found only to Pichincha); Gorgona Is. off sw. Colombia. Also Honduras to Panama. Mostly below 800 m.

GROUP C

Mainly *all black,* with *no obvious crest.*

Tachyphonus phoenicius

RED-SHOULDERED TANAGER

PLATE: 19

IDENTIFICATION: 16 cm (6¼"). *Very locally in savannas from Guianas to Amazonia.* Almost entirely glossy black; *very small area on inner shoulder red and white* (the red difficult to see under normal field conditions); under wing-linings also white. See V-37, C-51. Female *brownish gray above,* grayest on head and neck with duskier mask. *Mostly grayish white below,* with throat and median belly more whitish. Some females (perhaps younger birds) are somewhat flammulated on breast.

SIMILAR SPECIES: Male very closely resembles male White-lined Tanager (especially as the red on shoulders is so hard to see and as both species have the same white under the wing) but is markedly smaller and often looks distinctly blacker on head; sometimes best to go by accompanying females (which are very different). Over most of its

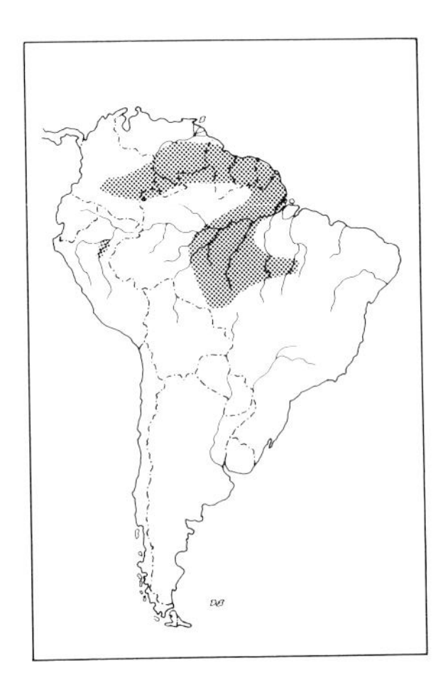

range does not occur with White-lined (main area of sympatry being the Guianas). *Dull but gray-looking female* looks very different from other female *Tachyphonus* tanagers (no buff or yellow) but might be confused with various other drab sympatric species (e.g., female Burnished-buff Tanager, which occurs in same habitat).

HABITAT AND BEHAVIOR: Locally fairly common to common in savannas and edges of woodland patches or gallery woodland; very local in w. and cen. Amazonia, where more or less restricted to isolated areas of semiopen, bushy campina vegetation. Found especially in association with areas having poor soil (e.g., sandy areas or around granite outcroppings). Usually remains low, foraging as pairs, gleaning in foliage and often perching fully in the open. Rather nervous, often flicking its wings, males sometimes exposing the red.

RANGE: Disjunct and often very locally in Amazonia (distribution being governed by existence of suitable habitat): Guianas, s. Venezuela (Bolívar and Amazonas), and e. Colombia (recorded e. Vichada, e. Guainía, Vaupés, and Meta); Amaz. Brazil (Amapá, along lower Amazon up to Rio Negro and Rio Madeira, and south to n. Mato Grosso and extreme n. Goiás); ne. Peru (very locally in sw. Loreto, n. San Martín, and n. Ucayali). To 1900 m (in Venezuela; lower elsewhere, primarily below 400 m).

Tachyphonus coronatus

RUBY-CROWNED TANAGER

PLATE: 19

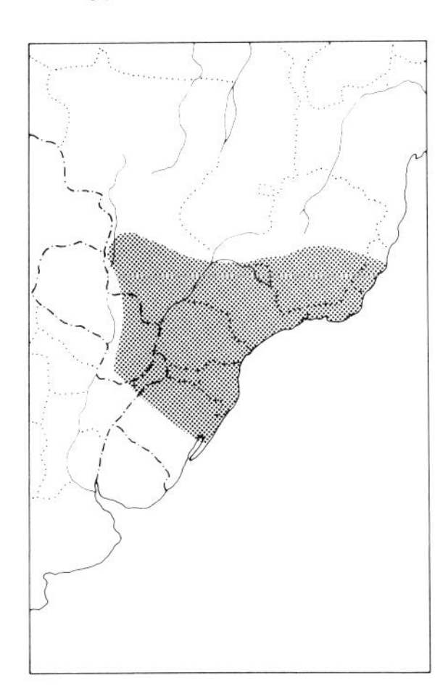

IDENTIFICATION: 18 cm (7"). *S. Brazil and adjacent areas. Entirely lustrous blue-black,* with *narrow scarlet streak on rear crown* (usually hidden); under wing-linings white. Female mostly rufous-brown above, most rufescent on rump and tail, grayer on head. *Below mostly ochraceous,* deepest on crissum and palest on throat (nearly white on chin; *considerable dusky flammulation across breast.*

SIMILAR SPECIES: Male superficially much like male Shiny and Screaming Cowbirds in plumage, but note pale bluish mandible, white under wing (usually not evident until bird flies), red in crown when visible, very different behavior, etc. Cf. also very similar White-lined Tanager. Female closest to female of sympatric race of Red-crowned Ant-Tanager, but latter much more olivaceous generally, lacks the breast flammulation, has tawny crown stripe, etc. Cf. also female Flame-crested Tanager (different arboreal behavior, etc.).

HABITAT AND BEHAVIOR: Common in lower growth and shrubby borders of forest and secondary woodland; much less a bird of clearings than the White-lined Tanager. Usually in pairs, foraging actively and conspicuously, only rarely with mixed-species flocks. Males often take prominent perches to give their leisurely but rather monotonous and repetitious song, a series of unmusical single notes, each repeated several times with a *ch* often interspersed, thus "ch-chweek . . . ch-chweek . . . ch-chweek . . ."

RANGE: S. Brazil (Espírito Santo, Minas Gerais, and s. Mato Grosso south), e. Paraguay, and ne. Argentina (n. Corrientes and Misiones). To about 1200 m.

Tachyphonus rufus

WHITE-LINED TANAGER

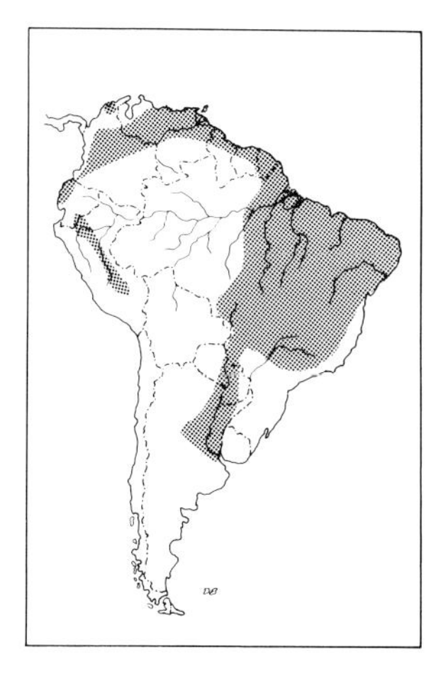

IDENTIFICATION: 18–19 cm (7–7½"). Most closely resembles male Ruby-crowned Tanager but lacks red in crown; bill slightly longer. Female *uniform rufous,* paler below. See V-37, P-29.
SIMILAR SPECIES: Overlaps with very similar (in male) Ruby-crowned Tanager only in portions of interior s. Brazil: there probably best identified by the different females (female Ruby-crowned not nearly as uniform below and has marked breast flammulation and *grayish head*). Male of smaller Red-shouldered Tanager likewise perhaps best distinguished by its attendant, very different female (fortunately all 3 of these species are usually found in pairs). Cf. also Shiny Cowbird male. Male White-shouldered Tanager has much more conspicuous white wing-coverts (in male White-lined the white generally does not show except as small tuft at bend of wing until the bird flies). Female White-lined is colored much the same as Rufous Mourner and various becards, but note bluish mandible and differently shaped bills, less arboreal behavior, etc.
HABITAT AND BEHAVIOR: Locally common in clearings and shrubby forest borders, often near water and mainly in humid areas; less numerous and much more local in some parts of range (e.g., Ecuador and Peru). On Trinidad and Tobago also regularly inside forest. Almost invariably in pairs and only infrequently with mixed flocks. Usually remains quite low and in general is not very conspicuous, as it often seems shy and rarely perches for too long in the open (but rather more in evidence on Trinidad and Tobago). Usually quiet except for contact "chek" notes; song is reported to consist of many deliberate repetitions of a single phrase, "cheép-chooi . . . cheép-chooi . . ."
RANGE: Guianas (mainly coastal region), most of n. Venezuela (south to n. Bolívar, but absent from northwest and Maracaibo basin), locally in w. and n. Colombia (mainly Cauca and Magdalena valleys and nearby Andean slopes; ne. llanos region and around Santa Marta Mts.), very locally in w. and se. Ecuador and in e. Peru (south to Cuzco); e. Brazil (extreme lower Amazon area in Amapá and around Belém, and from Maranhão and Pernambuco south to Mato Grosso, n. São Paulo, Minas Gerais, and Bahia), cen. Paraguay (not humid forested east or *chaco* of west), and n. Argentina (s. Misiones, Corrientes, and e. Formosa south to n. Córdoba and extreme n. Buenos Aires); Trinidad and Tobago. Also Costa Rica and Panama. Mostly below 1500 m (but locally follows clearings higher).

Endemic *forest* tanagers found mainly in *se. Brazil and adjacent areas.*

The following 5 species belong to monotypic genera, all of which are found only or mainly in se. Brazil and adjacent areas. They have little in common other than their distribution and habitat preferences and are placed together here partly in order to highlight the distinctive nature of the se. Brazilian avifauna. Only the Black-goggled Tanager ranges outside this region, it also occurring in the Andes of Bolivia and Peru.

Orchesticus abeillei

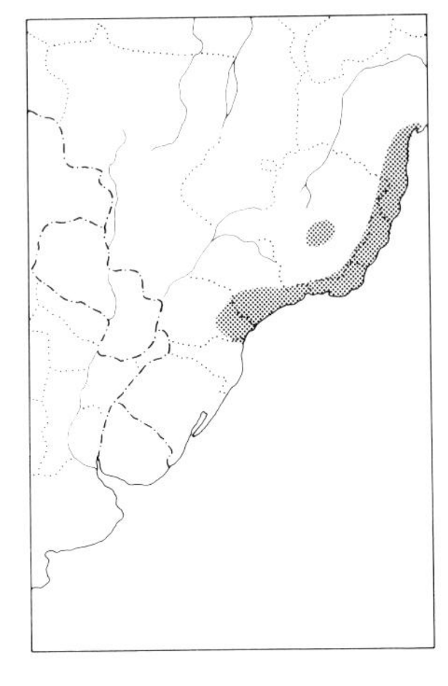

BROWN TANAGER

PLATE: 20

IDENTIFICATION: 18 cm (7"). *A strikingly foliage-gleaner-like tanager of se. Brazilian forests. Bill very stout,* blackish. Mostly dull brown above, duskier on crown and on narrow stripe through eye; *cinnamon forehead and broad superciliary.* Wings and tail more rufous. Below uniform dull cinnamon buff.

SIMILAR SPECIES: This unusual tanager looks totally unlike any other member of the Thraupinae; in fact, in both overall appearance and behavior it rather closely resembles the sympatric Buff-fronted Foliage-gleaner (the 2 are even sometimes in the same flock!). The foliage-gleaner differs most obviously in its very different bill (longer and slenderer, with hooked tip) and typical foliage-gleaner tail. Cf. also Chestnut-crowned Becard (which also is regularly with the tanager).

HABITAT AND BEHAVIOR: Uncommon to fairly common in middle and upper strata of humid forest and forest borders. Usually in pairs or small groups of up to 4 to 5 birds which regularly accompany mixed-species flocks. Forages actively, primarily by clambering along limbs, inspecting epiphytes, probing clusters of dead leaves, and gleaning leaves; often perches quite upright. Rather numerous at Itatiaia Nat. Park.

RANGE: Se. Brazil (s. Bahia and s. Minas Gerais south to e. Paraná). To about 1600 m.

Stephanophorus diadematus

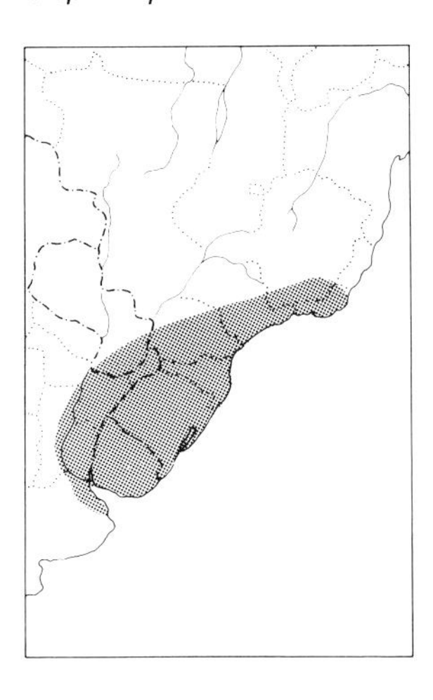

DIADEMED TANAGER

PLATE: 20

IDENTIFICATION: 19 cm (7½"). *Se. Brazil and adjacent areas.* Bill stubby. *Mostly shining dark purplish blue.* Forecrown, lores, and chin patch black (feathers of forecrown plushlike), backed by *snowy white center of crown with small red patch on top.* Wings and tail largely black, feathers edged blue. Sexes alike, but immatures duller and duskier with crown pattern obscure or absent.

SIMILAR SPECIES: Can look very dark in the field, when best known by flat-crested profile, in which usually at least the white shows; often the red projects up like a small pointed topknot.

HABITAT AND BEHAVIOR: Locally common at borders of montane forest, in *monte* woodland, and in gardens and patches of shrubbery in otherwise mostly open country. Found in pairs or small groups, often taking prominent perches on top of bushes and lower trees but then diving into cover when alarmed. Regularly with flocks of other tanagers, warbling-finches, etc., foraging at all levels, though typically not too high. Has a rich, loud, whistled song, rather grosbeaklike in character and often characterized by its distinctively upslurred ending. Common and widespread in Rio Grande do Sul, Brazil, and in Uruguay, as well as in the upper reaches of Brazil's Itatiaia Nat. Park (where it occurs right up to treeline).

RANGE: Se. Brazil (Rio de Janeiro and s. Minas Gerais south to Rio Grande do Sul; possibly also in Espírito Santo), se. Paraguay (Alto Paraná), Uruguay, and ne. Argentina (Misiones, Corrientes, ne. Santa Fe, Entre Ríos, and n. Buenos Aires). To at least 2100 m (northward occurring mostly in montane areas, dropping somewhat during austral winter).

Orthogonys chloricterus

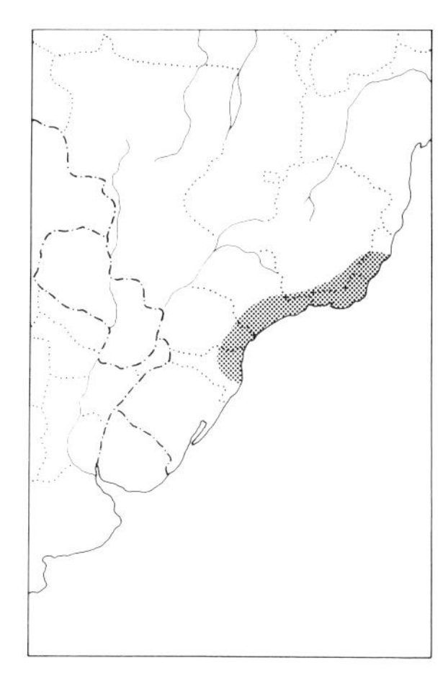

OLIVE-GREEN TANAGER

PLATE: 20

IDENTIFICATION: 20 cm (7¾"). *A large, drab, uniform olive tanager of se. Brazilian forests.* Bill rather long and slender, blackish. *Uniform olive green above, olive yellowish below.*

SIMILAR SPECIES: Superficially similar to geographically far-removed *Chlorothraupis* and *Mitrospingus* tanagers. Should be distinctive in its limited range, but compare to female Hepatic Tanager (true sympatry limited at best and with quite different behavior, etc.).

HABITAT AND BEHAVIOR: Uncommon to locally fairly common in middle and upper strata of humid, primarily montane forest. Troops about in rather large (up to 20 birds), noisy flocks, usually independent of other species; their loud, chattering calls often herald a flock's approach, but they tend to pass on rapidly, rarely lingering for long in 1 place. Unlike the *Chlorothraupis* tanagers, Olive-greens seem exclusively arboreal and usually remain well above the ground. Numerous at the Boraceia Forest Reserve in the Serra do Mar of ne. São Paulo and found in smaller numbers at Itatiaia Nat. Park.

RANGE: Se. Brazil (Espírito Santo south to e. Santa Catarina). To 1300 m.

Trichothraupis melanops

BLACK-GOGGLED TANAGER

PLATE: 20

IDENTIFICATION: 16.5 cm (6½"). *Forehead and ocular area black* (forming the "goggles"). Otherwise mostly dusky olive above with *contrasting black wings and tail* and golden yellow crest (usually mostly hidden). Below buff, deepest on crissum. Under wing-linings and pectoral patch white, sometimes protruding from under bend of wing; *in flight also shows conspicuous band of white* across base of primaries (not visible on closed wing). Female similar but lacks black on face and the yellow crest; wings and tail duskier (and hence less contrasty).

SIMILAR SPECIES: Even the less-patterned female is unlikely to be confused in its range and habitat; perhaps most resembles a dull female or immature Fawn-breasted Tanager. Cf. also females of Red-crowned Ant-Tanager and Flame-crested Tanager.

HABITAT AND BEHAVIOR: Common in lower growth of forest, better-developed secondary woodland, and gallery forest; mostly inside and infrequent at edge. Seems less numerous in Andean portion of its range. Usually in pairs or small groups (birds in male plumage noted less often than females and presumed immatures), generally foraging with mixed flocks of various flycatchers, furnariids, etc. Usually rather

bold and not difficult to observe; primarily insectivorous and regularly seen in attendance at army ant swarms. General behavior recalls the *Habia* ant-tanagers.
RANGE: Locally on e. slope of Andes in e. Peru (north to San Martín), Bolivia (south to Chuquisaca); s. Brazil (s. Mato Grosso, s. Goiás, and s. Bahia south to Rio Grande do Sul), e. Paraguay, and ne. Argentina (Misiones and ne. Corrientes). Mostly 1000–2400 m in Andes; mostly below 1200 m in se. South America.

Pyrrhocoma ruficeps

CHESTNUT-HEADED TANAGER

PLATE: 20

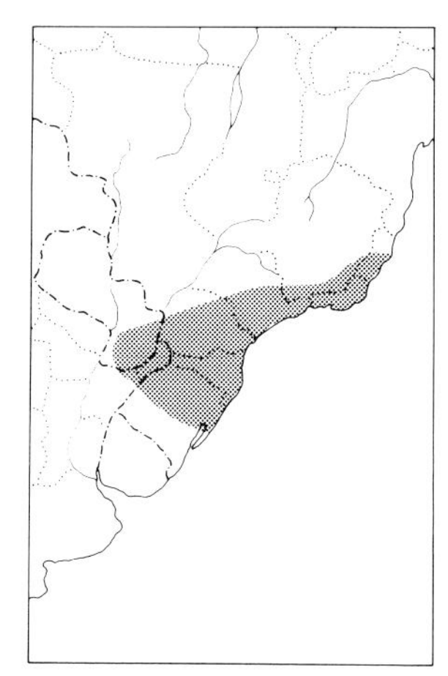

IDENTIFICATION: 14 cm (5½"). *Se. Brazil and adjacent areas. Mostly dark gray* with *contrasting chestnut hood;* forehead, lores, and chin black. Female much duller, brownish olive above with *dull cinnamon-rufous head* and buff lores and throat; dull yellowish buff below. Immature males resemble female, and they evidently require considerable time before reaching full adult plumage: birds with duller and paler gray bodies and only a little chestnut on head are frequently recorded.
SIMILAR SPECIES: Male easily recognized, and as they are usually in pairs, females will normally not be a problem. Their overall color pattern does resemble Rufous-crowned Greenlet's (note very different behavior, etc.), while immature males might conceivably be confused with Orange-headed Tanager (but are nowhere near as brightly patterned).
HABITAT AND BEHAVIOR: Uncommon in undergrowth of forest and forest borders. Usually in pairs, generally apart from flocks, foraging near the ground in dense shrubby growth or bamboo thickets; unobtrusive and often quite hard to see clearly but can be excited by squeaking. General comportment reminiscent of a brush-finch (*Atlapetes*). Its song is often a good indication of its presence: a series of 3 high sibilant notes followed by 3 (sometimes 2) similar but slightly lower ones. Can be found at Iguaçu Nat. Park, and even more commonly at Paraguay's Ybycuí Nat. Park.
RANGE: Se. Brazil (Espírito Santo south to Rio Grande do Sul), se. Paraguay, and ne. Argentina (Misiones). To about 1100 m.

Open country tanagers found mainly in *interior Brazil.*

The following 5 species are grouped together in 4 genera (*Schistochlamys* has 2 species). As with the previous set of 5 species, this group shares few morphological characteristics but tends to be associated because of distribution and habitat preferences. The White-rumped and White-banded Tanagers are classic members of Brazil's *cerrado* avian community, while the less well known Scarlet-throated Tanager has a distribution centering on the *caatinga* zone of ne. Brazil. The 2 *Schistochlamys* tanagers look quite different but are united by their preference for shrubby semiopen habitats, the Cinnamon only in e. Brazil.

Schistochlamys melanopis

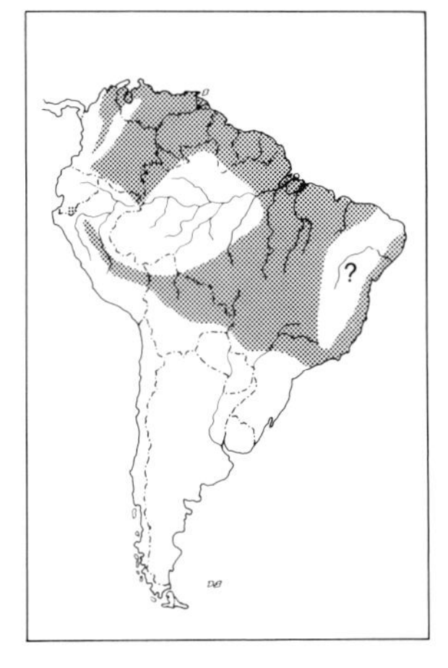

BLACK-FACED TANAGER

PLATE: 20

IDENTIFICATION: 18–18.5 cm (7–7¼"). Bill bluish gray, tipped black. *Mostly gray,* paler below, whitish on center of lower belly; ***black to brownish black forehead, face, and bib*** extending over throat and chest. Sexes alike, but immature entirely different (and may not even be recognized as the same species): olive above, more yellowish on head and with *yellow eye-ring;* paler yellowish olive below. Older immatures begin to show the black face and bib while still in olive plumage.

SIMILAR SPECIES: Gray adults with their contrasting black faces and bibs are unmistakable, but the olive immatures can be very confusing if seen apart from an adult (as sometimes will happen). It looks (and even acts) rather like a plain-looking saltator but is more uniformly olive than any; cf. also female of Hepatic and Fulvous-crested Tanagers.

HABITAT AND BEHAVIOR: Uncommon to fairly common in grassy areas with scattered bushes and trees, *cerrado,* and (in more humid areas) forest and woodland borders and shrubby clearings. Usually in pairs, often perching quite conspicuously on top of bushes and low trees or flitting low across roads; rarely much above ground. Sometimes with mixed-species flocks (especially in *cerrado*?), but at least as often independent of them. Has a rich, melodic grosbeaklike song.

RANGE: Locally in lowlands of n. and ne. Colombia (west of Andes in Magdalena and upper Cauca valleys and north to Santa Marta region; east of Andes south to w. Putumayo and Vaupés), Venezuela (locally), Guianas, lower Amaz. and interior Brazil (east to Maranhão, Goiás, Minas Gerais, and nw. São Paulo; also scattered records from e. coastal region from Pernambuco south to Rio de Janeiro), e. and n. Bolivia (Santa Cruz and Cochabamba west to La Paz), e. Peru (locally along e. base of Andes from San Martín to Cuzco), and se. Ecuador (seen in Oct. 1986 near Peruvian border south of Loja in Loja-Zamora, *fide* P. Greenfield); absent from most of middle and upper Amazon basin. Mostly below 1500 m.

Schistochlamys ruficapillus

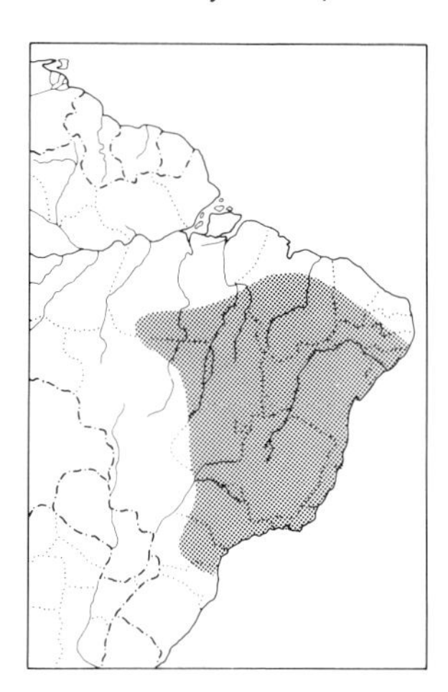

CINNAMON TANAGER

PLATE: 20

IDENTIFICATION: 18–18.5 cm (7–7¼"). *A svelte, cinnamon and gray tanager of semiopen country in e. Brazil.* Bill bluish gray tipped black. *Foreface black,* crown brown; otherwise *bluish gray above,* wings and tail duskier. *Sides of head and most of underparts cinnamon buff;* flanks pale gray, center of belly white, but crissum again cinnamon buff.

SIMILAR SPECIES: Black-faced Tanager lacks all the cinnamon buff and has much greater extension of black on face and throat. Cf. also Black-throated Saltator.

HABITAT AND BEHAVIOR: Uncommon to fairly common in semiopen grassy country with scattered bushes and low trees, at edges of woodland patches, and in gardens. Usually noted singly or in pairs, perching in low shrubbery, often quite in the open; infrequently with mixed flocks. Has a brief but musical whistled song, repeated over and over, similar to Black-faced Tanager's.

RANGE: E. Brazil (s. Pará on the Serra do Cachimbo, cen. Maranhão, Piauí, and Pernambuco south to São Paulo and e. Paraná). To about 1100 m.

Neothraupis fasciata

WHITE-BANDED TANAGER

PLATE: 20

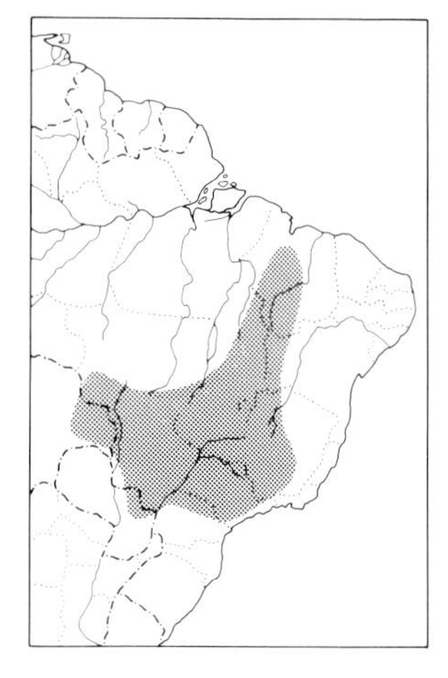

IDENTIFICATION: 16 cm (6¼"). *A distinctive, predominantly gray tanager of cerrado in interior Brazil and adjacent areas.* Above gray with *black mask from lores across ear-coverts.* Wings and tail more brownish, but wing-coverts mainly black with white band along edge of lesser coverts. Throat white; remaining underparts pale grayish, whiter on center of belly. Sexes alike, but immatures (often seen) are decidedly browner with reduced mask and are tinged yellowish below; as they age they gradually become grayer and mask becomes darker.
SIMILAR SPECIES: Virtually unmistakable in its habitat (where the total avifauna is in fact limited in variety). N. observers will at once be struck by how remarkably like a typical shrike (*Lanio*) this species is in pattern and coloration.
HABITAT AND BEHAVIOR: Uncommon to fairly common in *cerrado.* Almost always in pairs, foraging mostly by gleaning branches and leaves of *Curatella* and other *cerrado* trees, but also seems to regularly drop to the ground, presumably also in pursuit of (dislodged?) insect prey. Often a pair is present with the dispersed assemblage which passes for a *cerrado* mixed-species flock, but is never as numerous or as conspicuous as the White-rumped Tanager (with which it usually occurs sympatrically). Readily found in Brasilia Nat. Park.
RANGE: Interior e. and s. Brazil (s. Maranhão and s. Piauí south to Minas Gerais and w. São Paulo, west through s. Mato Grosso), e. Bolivia (e. Santa Cruz), and ne. Paraguay (Amambay). Mostly 500–1100 m.

Cypsnagra hirundinacea

WHITE-RUMPED TANAGER

PLATE: 20

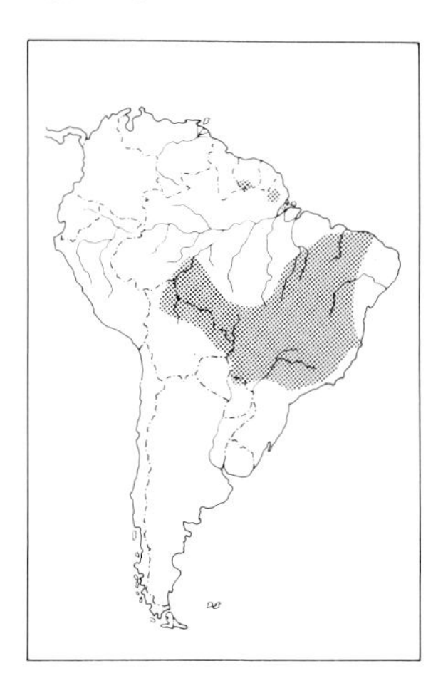

IDENTIFICATION: 16–16.5 cm (6¼–6½"). *An unmistakable black, white, and buff tanager of cerrado in interior Brazil and adjacent areas.* Mostly black above with *conspicuous white rump,* white band across wing-coverts, and white patch in primaries. *Throat rufous, paling into buff on breast;* lower underparts creamy white, tinged buff on flanks. *Pallidigula* (ne. Brazil, n. Bolivia, and adjacent sw. Brazil) similar but with throat and chest much paler, *only washed ochraceous buff.* Immatures are frequently seen: browner above but already with adult's wing and rump pattern and with rich creamy buff on underparts.
HABITAT AND BEHAVIOR: Fairly common to common in *cerrado* and in isolated campos and savannas with scattered low trees and bushes in what are otherwise mostly forested regions. Usually in small groups, typically of 4 to 6 birds, which glean in foliage and among the gnarled branches, only occasionally dropping to the ground. Pairs occasionally burst into a wonderful, loud, rollicking duet while perched in a treetop,

most often in early morning: the presumed female gives a continual low churring, while the presumed male sings simultaneously its vigorous, far-carrying "cheedoocheechoo, cheedeereeyou-chee-choo . . ." repeated a number of times; the overall effect is rather like that of the *Campylorhynchus* wrens. Numerous in Brasilia Nat. Park.

RANGE: Interior e. and s. Brazil (s. Maranhão, Piauí, and Ceará south to Minas Gerais, n. São Paulo, and s. Mato Grosso; west into Rondônia and extreme s. Amazonas in upper Rio Madeira drainage), n. and e. Bolivia (locally in Beni and Santa Cruz), and ne. Paraguay (n. Concepción); also in ne. Brazil north of the Amazon (Amapá and probably also n. Pará) and s. Suriname (Sipaliwini). To about 1100 m.

Compsothraupis loricata

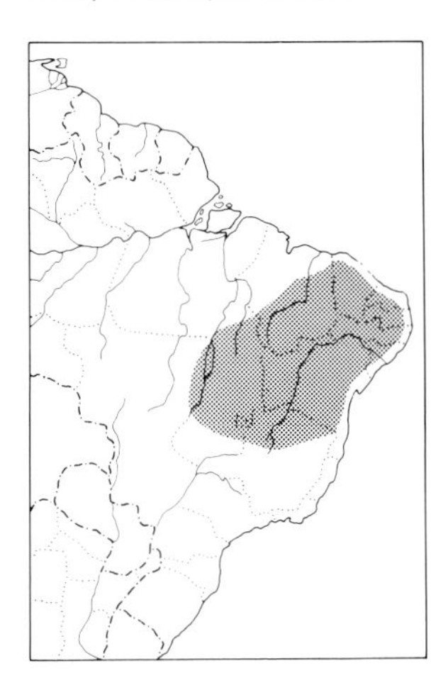

SCARLET-THROATED TANAGER

PLATE: 20

IDENTIFICATION: 21.5 cm (8½"). *Ne. Brazil. Mostly glossy blue-black* with *scarlet throat and center of chest.* Feathers of rump and flanks are basally white, but these rarely show in the field. Female similar but *lacking the scarlet bib* and the basal white of rump and flank feathers.

SIMILAR SPECIES: Males are unmistakable provided the red can be seen (and except in good light it's often inconspicuous). All-black female is much more likely to be confused, especially if seen apart from males (usually sexes are together, though adult males are often outnumbered by females); it looks rather like various blackbirds, especially the Chopi, but differs in bill shape. Cf. also female Giant Cowbird.

HABITAT AND BEHAVIOR: Uncommon to locally fairly common in *caatinga* and gallery woodland and semiopen areas around marshes and ponds or along rivers; usually not very far from water. Found in pairs or small groups of up to 6 to 8 birds; rather conspicuous, and often seems quite sluggish or even tame, perching on high exposed branches for long periods. Though currently classified as a tanager, its exact taxonomic affinities have yet to be determined, and overall its behavior in fact seems quite icteridlike. A frequently heard call is a loud, blackbirdlike "chirt," often repeated, sometimes for protracted periods. In s. Piauí seen to enter holes of dead *Mauritia* palm snags in Oct. 1977 (RSR) and may nest in such situations.

RANGE: Ne. Brazil (locally from Goiás and n. Minas Gerais north to s. and e. Maranhão, Piauí, and Ceará; mostly inland). To about 700 m.

NOTE: We have followed *Birds of the World* (vol. 13) in separating this species in the genus *Compsothraupis* rather than uniting it with *Sericossypha* as was done in Meyer de Schauensee (1966, 1970). Behavior, habitat, and range are entirely different.

Miscellaneous "*black and white*" tanagers.

A set of 4 species of boldly marked, black and white tanagers (though females of *Conothraupis* are olive, so far as known). Each species should be readily recognized; behavior of the 3 genera differs markedly, and they are not closely related.

Conothraupis speculigera

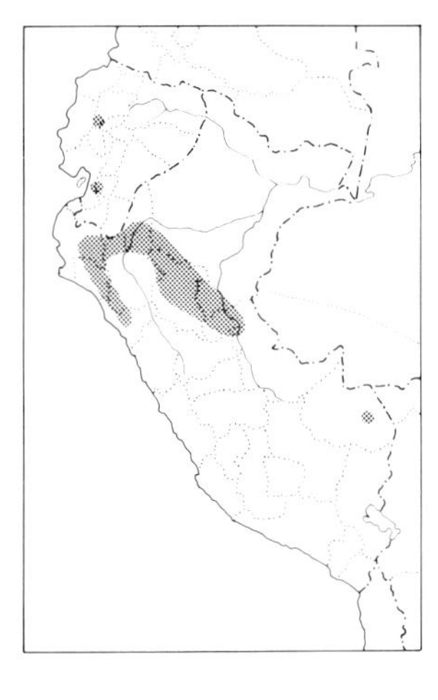

BLACK-AND-WHITE TANAGER

PLATE: 20

IDENTIFICATION: 16.5 cm (6½"). *W. Ecuador and Peru.* Iris red; bill black above, bluish below with black tip. Male *mainly glossy blue-black* with *sharply demarcated white breast and belly.* Rump dark gray, flanks washed with gray; large wing-speculum white, as are under wing-coverts. Female olive above, slightly duskier on wings and tail. *Pale yellow below, flammulated olive across breast.*

SIMILAR SPECIES: Black and white male should not be confused, but note striking plumage similarity to much smaller male Black-and-white Seedeater. Female rather dull by comparison and resembles Streaked Saltator (which is larger and has a prominent white eyebrow); female *Piranga* tanagers lack the mottled olive streaking on breast.

HABITAT AND BEHAVIOR: Generally rare and local (though judging from specimens may temporarily be more numerous at certain sites) in deciduous woodland, semiopen arid scrub, and borders of more humid forest. Exact distributional and behavioral details have yet to be worked out; possibly only a seasonal visitant to some parts of its range. Generally found singly or in pairs, often foraging rather low in shrubbery, but has also been observed perched for long periods well above ground at forest edge.

RANGE: Locally in w. Ecuador (recorded only in sw. Pichincha and Azuay) and nw. (Piura south to La Libertad) and e. Peru (Amazonas, Cajamarca, San Martín, Loreto, Ucayali, and Madre de Dios). To about 1400 m.

Conothraupis mesoleuca

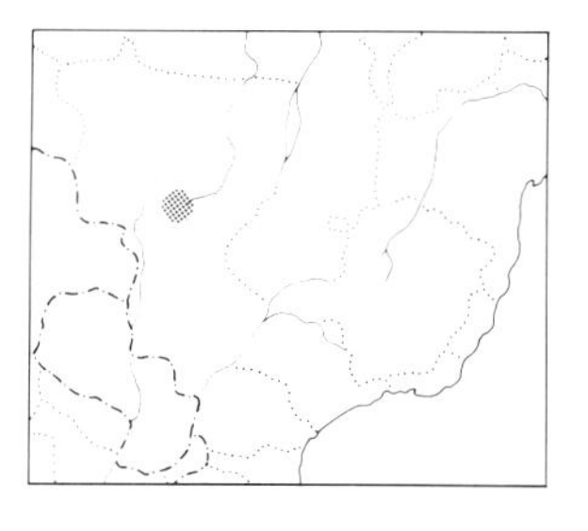

CONE-BILLED TANAGER

IDENTIFICATION: 16 cm (6¼"). Known only from a single specimen from *Mato Grosso, Brazil.* Very similar to the geographically far-removed Black-and-white Tanager, but with flanks and crissum black; wing-speculum may be smaller. Female unknown.

HABITAT AND BEHAVIOR: The 1 specimen was taken in fairly dry deciduous forest, but otherwise nothing has been recorded about it.

RANGE: Recorded only from Juruena, northeast of Cuiabá, in Mato Grosso, Brazil.

Lamprospiza melanoleuca

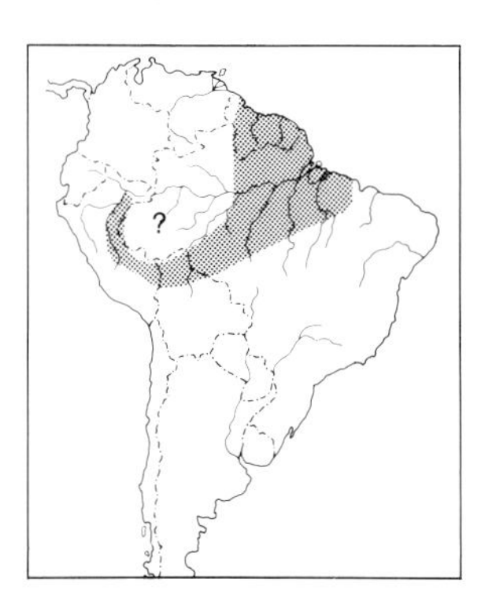

RED-BILLED PIED TANAGER

PLATE: 20

IDENTIFICATION: 17 cm (6¾"). *Bill scarlet. Glossy blue-black above and on throat and chest.* Below white, with *2 blue-black bands angling to either side from center of chest* (with almost the effect of a pair of crossed bandoliers). Female similar but with *contrasting bluish gray nape, back, and rump.*

SIMILAR SPECIES: This boldly pied tanager with its striking red bill is unlikely to be confused.

HABITAT AND BEHAVIOR: Uncommon to locally fairly common in canopy and borders of humid forest, principally in *terra firme.* Appears to be more numerous in e. Amazonia. Generally forages very high, though sometimes lower at edge of clearings, typically as small groups of 4 to 6 birds with mixed flocks of other tanagers (e.g., *Tachyphonus* spp.), etc. Apparently eats mostly fruit, but also seen to glean for insects. Rather noisy, with a frequent call being a far-carrying "purcheecheeéchur."

RANGE: Guianas, lower and s. Amaz. Brazil (south to n. Maranhão, se. Pará, and nw. Mato Grosso); n. Bolivia (Beni) and e. Peru (north to near s. bank of Amazon on Río Manití in Loreto); not recorded (but probably occurs?) from much of w. Amaz. Brazil south of the Amazon. To about 600 m.

Cissopis leveriana

MAGPIE TANAGER

PLATE: 20

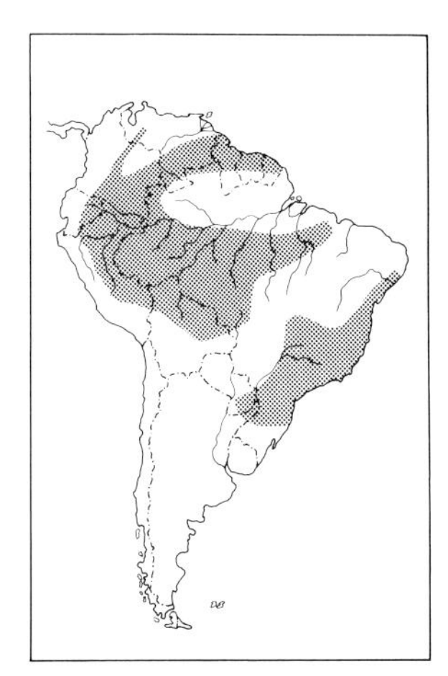

IDENTIFICATION: 25.5–29 cm (10–11¼"). *An unmistakable large, very long tailed black and white tanager with piercing yellow eyes. Glossy blue-black hood* extending over entire head to upper back and to point on middle of breast. *Scapulars, lower back and rump, and lower underparts white.* Wings and graduated tail mostly black, with tertials broadly edged white, and all of rectrices broadly tipped white (visible mainly from below). *Major* of se. Brazil and adjacent areas is considerably larger and black extends farther down back (leaving only the rump white).

HABITAT AND BEHAVIOR: Fairly common at forest borders and in shrubby clearings with scattered tall trees. Conspicuous birds, rather jaylike in appearance and mannerisms. Seen in pairs or small groups, often perching in the open. Usually not with mixed flocks. Generally remain fairly high in trees. Noisy, with a variety of arresting metallic calls, typically a loud "chenk," often given in series.

RANGE: Guianas (rare except in Guyana), s. (Bolívar and Amazonas) and w. Venezuela (Mérida, w. Barinas, Táchira, and w. Apure), e. Colombia (except in the llanos of the northeast), e. Ecuador, e. Peru, n. Bolivia (south to La Paz, Cochabamba, and Santa Cruz), and w. and s. Amaz. Brazil (east to Mato Grosso, se. Pará, and n. Maranhão); e. Brazil (Pernambuco south to n. Rio Grande do Sul and east to s. Goiás), e. Paraguay, and ne. Argentina (Misiones). To about 1400 m.

Icterinae

AMERICAN ORIOLES AND BLACKBIRDS

THE icterids are a varied group confined to the New World, where they are widespread, though they are most numerous and diverse in the tropics. Most are mid-sized to large birds, often colorful, with females usually either smaller or duller-plumaged than the males; their bills are typically conical and sharply pointed, and their feet are powerful. Many are gregarious and conspicuous birds, particularly when not breeding. A large number construct beautifully woven long pendent nests, some of these being highly colonial; others nest in less prominent situations, often on or near the ground.

Sturnella Meadowlarks and Blackbirds

A small group of gregarious and conspicuous grassland-inhabiting blackbirds. Sexually dimorphic, males being basically red and blackish, females duller and browner with little or no red; 1 species (Eastern Meadowlark) yellow below, and not sexually dimorphic. There is a notable range of bill shapes, from long and pointed to shorter and more conical; the latter represent the *Leistes* subgenus (Red-breasted and White-browed Blackbirds), until recently considered a distinct genus but subsumed by Short (1968) and in the 1983 AOU Check-list.

GROUP A

Underparts yellow with black crescent; sexes alike.

Sturnella magna

EASTERN MEADOWLARK

PLATE: 21

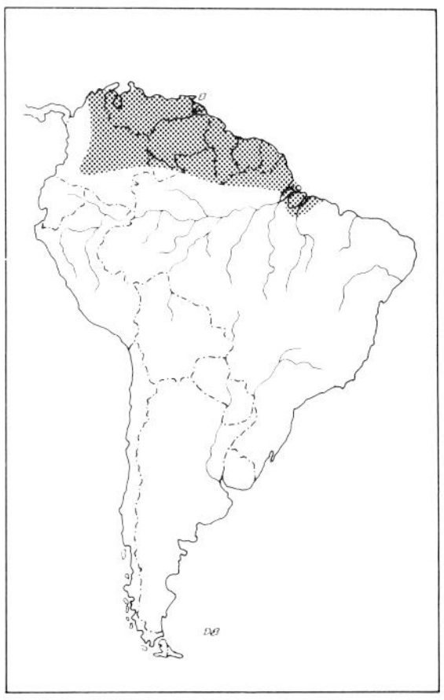

IDENTIFICATION: 21.5–23 cm (8½–9″). Chunky, with short tail. Brown above streaked buff and blackish; head broadly striped whitish and blackish brown. Lores and *entire underparts bright yellow, chest crossed by bold black crescent;* sides and flanks buffier, with sparse but bold dark brown streaking. *Outer tail feathers mostly white,* flashing conspicuously in flight. Immature duller, often with only a wash of yellow below and a trace of the chest band.

SIMILAR SPECIES: Virtually unmistakable as it is the only member of its genus which is yellow below.

HABITAT AND BEHAVIOR: Fairly common to locally common in open grasslands and pastures (in latter up locally to near treeline). Mostly terrestrial, often flicking its tail and exposing the white. Flight distinctive: a series of rapid, shallow wingbeats alternating with short glides during which wings are held stiffly downward. Song is a series of typically 3 to 5 whistled, slurred notes, often with a warbled effect at the end; overall effect intermediate between songs of "typical" Eastern and Western (*S. neglecta*) Meadowlarks of North America.

RANGE: Nonforested parts of n. and e. Colombia (Santa Marta region south in E. Andes and Magdalena valley to Huila; also entire llanos region south to Meta and east to Orinoco River), Venezuela, and Guianas, and n. Brazil (Roraima, Amapá, and ne. Pará on islands at the mouth of the Amazon and around Belém); has straggled to Bonaire. Also North and Middle America. To about 3500 m (in E. Andes of Colombia).

GROUP B

Underparts red; females duller, streakier below.

Sturnella loyca

LONG-TAILED MEADOWLARK

PLATE: 21

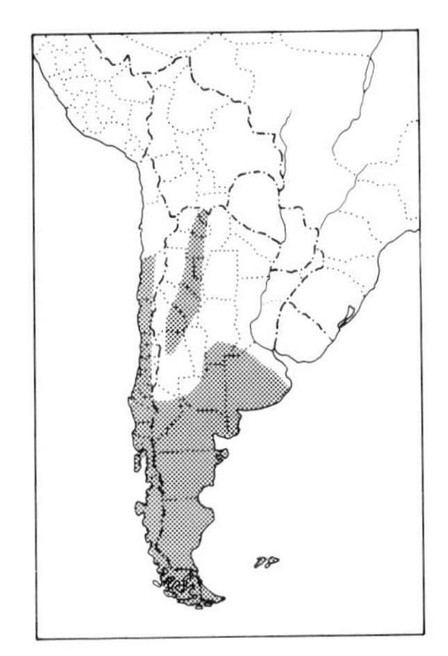

IDENTIFICATION: 24–25.5 cm (9½–10″). *Argentina and Chile. Considerably the largest and longest-tailed Sturnella; notably long bill* (especially in *falklandicus* of the Falkland Is.). Above mostly brown streaked blackish; head and sides of neck blackish, with prominent superciliary rosy red in front of eye, white behind. Under wing-coverts white. *Bend of wing and underparts predominantly rosy red,* with blackish encroaching on sides of neck; sides, flanks, and lower belly blackish with grayish streaking. Female similar to male above but paler and superciliary all white. *Throat white,* contrasting sharply with blackish cheeks and sides of neck. Remaining underparts light grayish sparsely streaked dusky, lower breast and belly tinged pinkish.

SIMILAR SPECIES: Pampas Meadowlark is notably smaller with shorter tail and has blacker belly in both sexes; furthermore, male's under wing-coverts are black (not white) and female's throat buffyish (not contrastingly white). Habitats also typically differ, Pampas mainly in pure grasslands.

HABITAT AND BEHAVIOR: Fairly common to locally common in arid grassy areas, often with scattered shrubs, and in pastures and cultivated meadows. Behavior similar to Eastern Meadowlark's, but flight stronger and more direct, and more often perches in trees and shrubs. Except when breeding usually in small flocks. Often mounts to a fence post or to the top of a bush or low tree to call or sing; the song is wheezy and loud, typically consisting of 4 short introductory notes followed by a longer one that drops off, "tshwit, tshwit, tshu-tshee? zheeeuww."

RANGE: Chile (Atacama south) and Argentina in Andean valleys of the northwest (Jujuy to n. Mendoza and Córdoba) and steppes of the south (Neuquén, La Pampa, s. Córdoba, and Buenos Aires south); Falkland Is. Apparently only an austral migrant to Córdoba. To at least 2500 m.

Sturnella bellicosa

PERUVIAN MEADOWLARK

Other: Peruvian Red-breasted Meadowlark

IDENTIFICATION: 20.5 cm (8″). *W. Ecuador to n. Chile.* Resembles Long-tailed Meadowlark (no overlap) but markedly smaller and proportionately somewhat shorter-tailed and *shorter-billed.* Male similar but note *white* (not brownish) *thighs;* red color on underparts usually

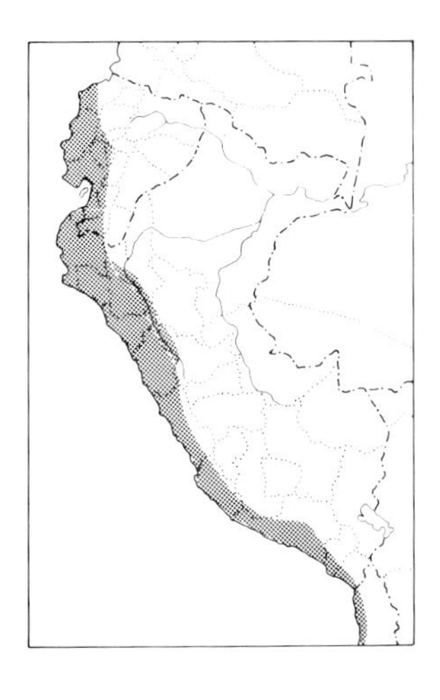

more intense, and black on sides of head and neck and lower underparts deeper (hence overall effect more contrasty). Female also blacker, with coarser streaking below; its thighs are likewise white.

SIMILAR SPECIES: The only meadowlark in its range; at its n. and s. extremities, range does approach that of Red-breasted Blackbird and Long-tailed Meadowlark, respectively.

HABITAT AND BEHAVIOR: Locally common on open pastures and in other grassy or irrigated areas. Behavior, including voice, similar to Long-tailed Meadowlark's, though during breeding season they do engage in nuptial flights during which a short version of their wheezy song is given. In sw. Ecuador often gathers in quite large flocks in the nonbreeding season (50 or more birds), and these may range into quite brushy areas (not just in grassy fields, etc.). A sharp "chack" is frequently given.

RANGE: W. Ecuador (north to coastal Esmeraldas and in intermontane valleys of Azuay and Loja), w. Peru (including various arid intermontane valleys south to Huánuco), and extreme n. Chile (Arica and Tarapacá). To about 2500 m.

NOTE: As this species is the only meadowlark occurring in coastal Peru and adjacent countries, we consider the "Red-breasted" in Meyer de Schauensee's (1966, 1970) English name superfluous and hence shorten it to simply "Peruvian Meadowlark."

Sturnella defilippii

PAMPAS MEADOWLARK

Other: Lesser Red-breasted Meadowlark

IDENTIFICATION: 20.5 cm (8"). *Mainly ne. Argentina.* Resembles Long-tailed Meadowlark and sympatric with it in parts of e. Argentina. Notably *smaller; bill and tail both shorter* proportionately. Male differs in its black (not white) under wing-coverts and in its *much blacker belly, flanks, and sides* (lacking the brownish gray). Female Pampas likewise has belly much more blackish than in female Long-tailed, and its throat is much less contrastingly white.

SIMILAR SPECIES: Fresh-plumaged birds might also be confused with White-browed Blackbird; the blackbird is smaller and has a shorter, more conical bill.

HABITAT AND BEHAVIOR: Fairly common but now local in less disturbed grasslands, its numbers and overall range having evidently been much reduced by cultivation and overgrazing. Behavior similar to Long-tailed Meadowlark's. Its song, several short notes followed by a longer wheezy one, is usually given from the ground or from a partially hidden perch in the grass, but it also (unlike Long-tailed) frequently sings during a display flight.

RANGE: Ne. Argentina (La Pampa and Buenos Aires north to Santa Fe and Corrientes; not recorded at Misiones), Uruguay (rare), and se. Brazil (also rare; 2 old records from s. Rio Grande do Sul and 1 from Paraná); perhaps only an austral migrant northward. Below 500 m.

NOTE: We consider Short's (1968: 27) English name suggestion for this species to be eminently appropriate and much shorter than Meyer de Schauensee's (1966, 1970) awkward "Lesser Red-breasted Meadowlark." The species' range is indeed essentially defined by the limits of the pampas.

GROUP C

Similar to Group B, but blacker and *bill shorter* ("*Leistes*" subgenus).

Sturnella superciliaris

WHITE-BROWED BLACKBIRD

PLATE: 21

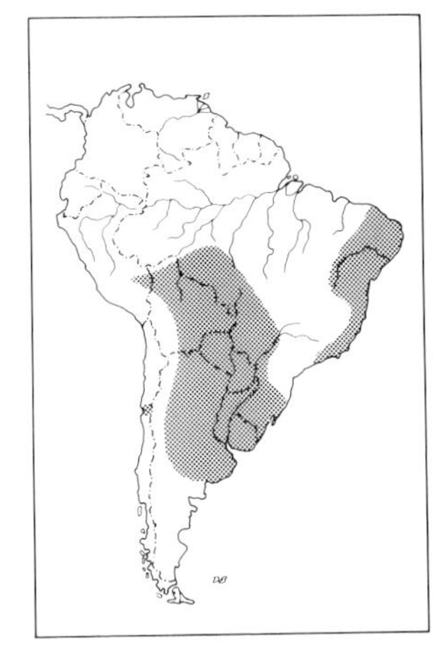

IDENTIFICATION: 18 cm (7"). Mostly black above with *prominent whitish postocular stripe. Throat and breast bright red,* belly contrastingly black. In fresh plumage feathers throughout edged pale brownish, these gradually disappearing with wear; pattern of freshly molted birds can be mostly obscured. Female broadly streaked blackish and buff above, with prominent long, pale, buff superciliary and coronal stripe. Mostly buff below, streaked dusky across chest and on belly, and variably tinged with pink on breast.

SIMILAR SPECIES: Pampas Meadowlark is similar (particularly freshly molted males) but is larger and has longer, more pointed bill; red breast of the male meadowlark is of a paler, rosier hue. Female blackbird never shows the female meadowlark's mostly black belly. Cf. also Red-breasted Blackbird and female or nonbreeding male Bobolink.

HABITAT AND BEHAVIOR: Locally common in grasslands and pastures (often quite damp); increasing and expanding its range in response to the spread of cultivation of rice and other grains, construction of airports, etc. Like the meadowlarks mostly terrestrial; also quite gregarious, especially in the nonbreeding season, but even when nesting often occurring in loose "colonies." Often perches on fence posts or wires and on top of clumps of bunchgrass. Most frequent song consists of several (most often 2) introductory notes followed by a longer buzzy trill inflected at the end, usually given during a display flight.

RANGE: Se. Peru (Madre de Dios and Puno), n. and e. Bolivia, s. Brazil (w. Mato Grosso, s. Goiás, and Rio Grande do Sul), Paraguay, Uruguay, and N. Argentina (south to Mendoza, La Pampa, and Buenos Aires); 1 record from Chile (Coquimbo); also e. Brazil from Pernambuco and Ceará south through Bahia to Rio de Janeiro and e. São Paulo, having only recently spread into the Rio/São Paulo area. Northernmost records (e.g., from Peru) may represent austral migrants or wanderers. To about 2500 m (in Bolivia).

NOTE: *S. superciliaris* may ultimately prove conspecific with *S. militaris,* but in the absence of any evidence of intergradation in the zones of potential contact, we continue to follow Short (1968) and Meyer de Schauensee (1966, 1970) in considering them full species.

Sturnella militaris

RED-BREASTED BLACKBIRD

IDENTIFICATION: 19 cm (7½"). Resembles White-browed Blackbird, *replacing it northward.* Slightly larger, with somewhat longer, more pointed bill. Male differs in *lacking the white superciliary.* Females are apparently indistinguishable in the field. See V-38, C-44.

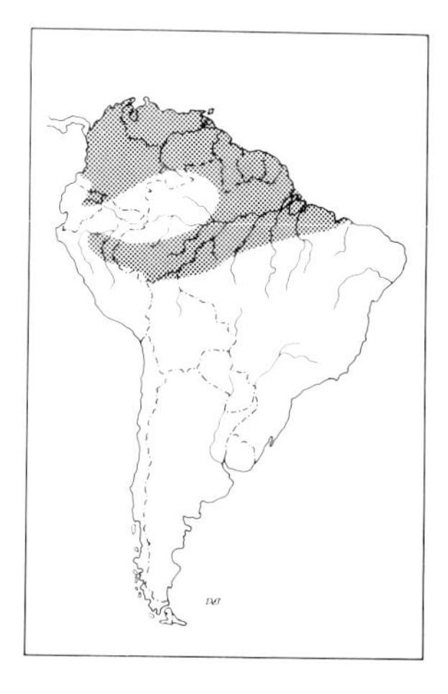

SIMILAR SPECIES: Female and nonbreeding male Bobolinks resemble female but have blunter bill and spikier tail and are yellowish buff below (never with pink tinge).
HABITAT AND BEHAVIOR: Locally common in grasslands, pastures, and ricefields. Behavior and vocalizations very similar to White-browed Blackbird's, and like that species is increasing and spreading (often colonizing newly deforested regions).
RANGE: Nonforested parts of Guianas, Venezuela, Colombia (not in extreme southeast in Amazonas, etc., but spreading into recently cleared areas in w. Caquetá and w. Putumayo; in Pacific lowlands recorded only from Guapi in sw. Cauca), e. Ecuador (cleared areas in Limoncocha/Coca region), ne. Peru (Loreto, where only 1 old record, from Jeberos on lower Río Huallaga), extreme n. Bolivia (Pando), and Amaz. Brazil (east to n. Maranhão and south to Acre and nw. Mato Grosso); Trinidad. Also Costa Rica and Panama. To about 1600 m (in Colombia).

Dolichonyx Bobolinks

A highly migratory icterid, breeding in North America, spending the n. winter mostly in n. Argentina. Usual plumage seen is basically buff with dark head striping, but on northward passage males gradually become more colorful; the Bobolink is one of the few passerine birds with 2 annual molts.

Dolichonyx oryzivorus

BOBOLINK

PLATE: 21

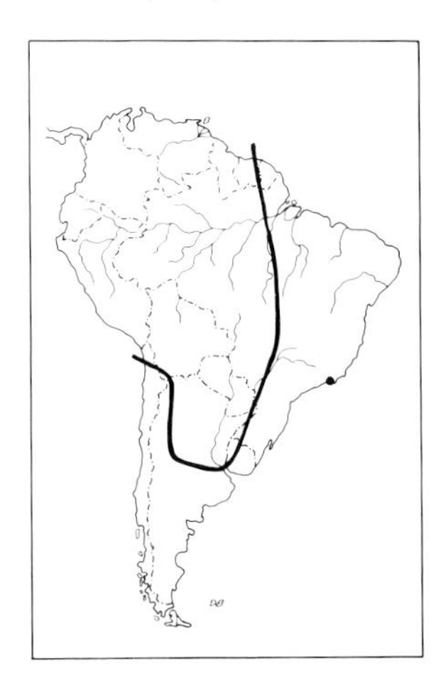

IDENTIFICATION: 17–18 cm (6¾–7"). Conical bill; *tail feathers sharply pointed.* Female and nonbreeding male buffy brown above streaked blackish and with *head broadly striped dusky and buff. Below yellowish buff* sparsely streaked dusky on sides and flanks. Males molt into their unmistakable alternate (breeding) plumage during austral midsummer, but at first it is mostly obscured by buffy tipping and edging to feathers (see illustration). These are gradually lost, so that during northward passage (Mar.–May) the pattern is increasingly evident: *mostly black* with *golden buff nuchal collar* and *white scapulars and large white rump patch.*
SIMILAR SPECIES: Female White-browed and Red-breasted Blackbirds usually show pink tinge on breast, are more coarsely streaked below, and have normally shaped and barred tails (not narrow and spiky, with tips sharply pointed). Bobolink's frequently given metallic call is unlike any vocalization of either blackbird.
HABITAT AND BEHAVIOR: Locally common transient and n. winter resident (mostly Sept.–May) in marshes, ricefields, and pastures with tall grass; resident in its n. winter quarters (mostly n. Argentina) only Jan.–Mar., occurring as transient northward across Amazonia, etc. In Argentina flocks can be very large, but on migration group size tends to be smaller. Seemingly erratic (or just inconspicuous and over-

looked?) over much of its migratory route, which is still imperfectly known. Often noted only in flight overhead, when its metallic "ink" call (diagnostic once learned) frequently draws attention; sometimes groups drop down into clearings or along rivers.

RANGE: Nonbreeding winter resident mostly in Argentina (primarily in *pantanal* area of Corrientes, Entre Ríos, ne. Santa Fe, and e. Chaco; recorded south to San Luis and Buenos Aires); widespread as a transient mainly east of Andes with smaller numbers along Pacific coast (where a few may also overwinter) south to n. Chile (single records from Arica and Antofogasta), east at least occasionally to se. Brazil and the Guianas. Breeds North America, migrating mostly through West Indies. To about 2500 m as transient, but mostly in lowlands.

Pseudoleistes Marshbirds

A pair of rather large, basically brown and yellow icterids of s. South America, bigger than the *Agelaius* blackbirds often found in the same marshes.

Pseudoleistes guirahuro

YELLOW-RUMPED MARSHBIRD — PLATE: 21

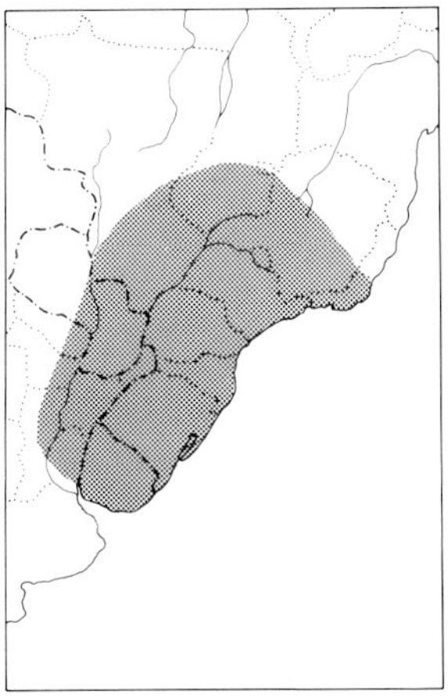

IDENTIFICATION: 23.5 cm (9¼"). *Above mostly brown,* darker on foreface and more olivaceous on back; *rump and lesser wing-coverts conspicuously yellow.* Throat and chest dark brown, contrasting sharply with *bright yellow remaining underparts.*

SIMILAR SPECIES: A distinctive, water-associated icterid of se. South America; liable to be confused only with Brown-and-yellow Marshbird, but that lacks the yellow rump and is extensively brown on the flanks. Female Saffron-cowled Blackbird is smaller, has yellow on face, etc.

HABITAT AND BEHAVIOR: Locally fairly common in marshy swales and grasslands near watercourses, typically in rolling, often hilly country. Though found most often in association with water, it also regularly forages on fields or pastures some distance away. Conspicuous and noisy birds, usually occurring in flocks of 10 to 20 or more individuals. Feeds mostly on the ground, walking about steadily and probing the earth for invertebrates, etc. Loud, rather rich and gurgling calls are frequently given and often draw attention to a group foraging in tall grass (but otherwise hidden) or to a flock flying past.

RANGE: S. Brazil (north to s. Mato Grosso, s. Goiás, and Minas Gerais; largely or entirely absent from coastal areas), e. Paraguay, ne. Argentina (south to Santa Fe and Entre Ríos and at least formerly to Buenos Aires), and Uruguay. To about 1100 m.

Pseudoleistes virescens

BROWN-AND-YELLOW MARSHBIRD

IDENTIFICATION: 23.5 cm (9¼"). Resembles Yellow-rumped Marshbird but bill slenderer, plumage *all brown above* (lacking the yellow

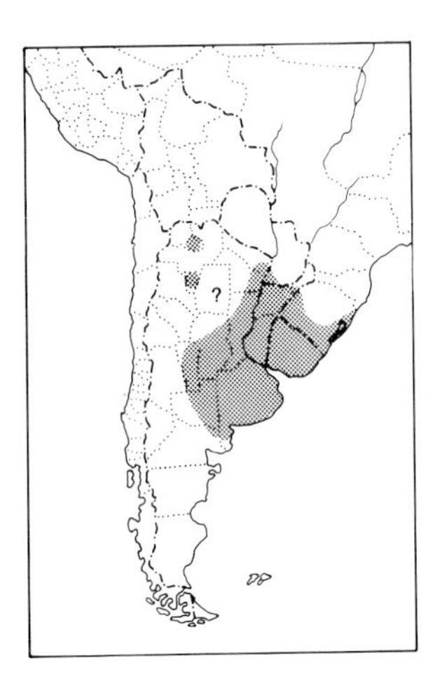

rump), less extensive yellow on wing (basically restricted to bend of wing), not as dark brown on throat and chest, and with *flanks extensively brown.*

HABITAT AND BEHAVIOR: Locally common in and around marshes and in adjacent moist grasslands and ploughed fields. Behavior similar to Yellow-rumped Marshbird's, though it is often more numerous and flock size (particularly in the nonbreeding season) tends to be larger. Though similar in quality, none of its vocalizations seem to be as loud or forceful. A characteristic and easily seen inhabitant of the "wet pampas" of e. Buenos Aires province, Argentina.

RANGE: Ne. Argentina (e. Formosa, Corrientes, and s. Misiones south to San Luis and s. Buenos Aires), Uruguay, and extreme s. Brazil (Rio Grande do Sul); in nonbreeding season occasionally wanders to nw. Argentina (north to Jujuy). Mostly below 300 m.

Gymnomystax Blackbirds

A spectacular, large, black and yellow blackbird found in semiopen areas and near water in n. South America. Conspicuous and gregarious.

Gymnomystax mexicanus

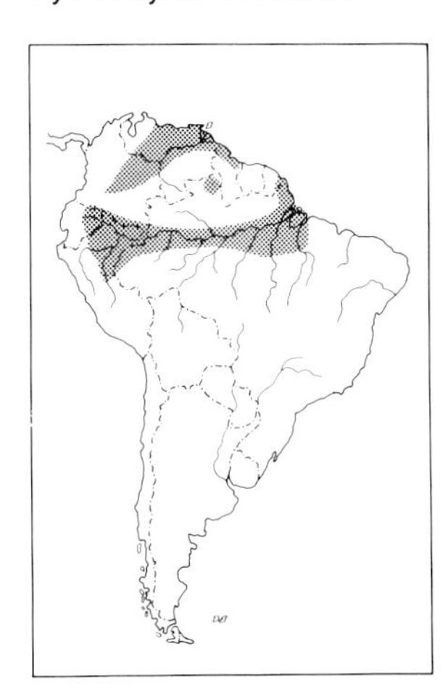

ORIOLE BLACKBIRD

PLATE: 21

IDENTIFICATION: Male 30.5 cm (12"); female 26.5 cm (10½"). *Mostly bright golden yellow,* with *contrasting black back, wings* (except for lesser coverts), *and tail.* Bare ocular area and short malar streak also black. Immature similar but yellow not quite as bright and *crown patch also black.*

SIMILAR SPECIES: With its bright yellow and black plumage could only be mistaken for an oriole or 1 of the forms of the Troupial. See also Yellow-hooded Blackbird.

HABITAT AND BEHAVIOR: Fairly common to common in open grassy (often marshy) areas with scattered trees and gallery woodland in Venezuela and ne. Colombia (especially across the llanos) and in semiopen riparian areas, marshes, and river islands in Amazonia. Usually conspicuous, occurring as pairs or scattered groups, foraging on the ground in grass or around muddy spots or near water, perching in the open on top of bushes or low trees. Generally does not consort with other birds.

RANGE: Ne. Colombia (south the Meta and Vichada), much of Venezuela (not in the northwest or south of the Orinoco River), sparsely in the Guianas (still not recorded Suriname; no recent records Guyana), Amaz. Brazil (recorded mostly along Amazon itself, also locally along lower courses of some of its s. tributaries, and recently at the Maracá Ecol. Station in Roraima), ne. Peru (Loreto, San Martín, and n. Ucayali south to the Pucallpa region), and e. Ecuador; evidently absent from blackwater region of much of n. Amaz. Brazil. To about 1000 m (in Venezuela).

Amblyramphus Blackbirds

Unmistakable large red-hooded blackbird of marshes in s. South America. Note its peculiarly shaped bill (with an almost upturned effect).

Amblyramphus holosericeus

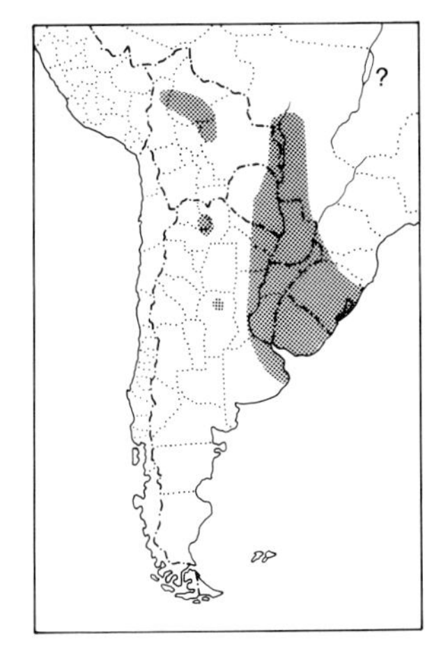

SCARLET-HEADED BLACKBIRD

PLATE: 21

IDENTIFICATION: 24 cm (9½"). Long, slender, very sharply pointed bill. Unmistakable. Mostly black, with *brilliant orange-red hood;* thighs also orange-red. Juvenal may start out all black but soon acquires some orange-red on throat and breast, and this gradually brightens and intrudes onto head and neck.

HABITAT AND BEHAVIOR: Uncommon in extensive reedbeds in marshes; necessarily rather local. Quite strictly confined to such reedbeds and only rarely found outside of them. Usually found in scattered pairs, never really flocking. Most often seen as they perch conspicuously on top of a reed or rush stem, frequently far out in the middle of the marsh. Its song is loud, clear, and melodic, a ringing "cleer-cleer-clur, clulululu," and other simpler calls have much the same quality. Readily seen in coastal s. Rio Grande do Sul, Brazil, and in e. Buenos Aires, Argentina; numbers seem smaller in the upper Paraguay River basin.

RANGE: S. Brazil (w. Mato Grosso north to the Cáceres/Poconé area and in Rio Grande do Sul; Sick records it also from n. Goiás), Paraguay (except the far northwest), ne. Argentina (e. Formosa, Chaco, and Corrientes south to Buenos Aires), and Uruguay; a seemingly isolated population in n. Bolivia (Beni and nw. Santa Cruz) and a few records (wandering birds?) from nw. Argentina (Jujuy, Salta, and Córdoba). To about 600 m (in Bolivia).

Agelaius Blackbirds

Small, mostly black icterids with straight, pointed bills; males of most species have patches of contrasting yellow or chestnut, with females browner and streakier. All are associated with marshes.

GROUP A

Bright yellow underparts or hood (male).

Agelaius flavus

SAFFRON-COWLED BLACKBIRD

PLATE: 21

IDENTIFICATION: 18–19 cm (7–7½"). Local in *extreme se. South America.* Male has *face and entire underparts rich golden yellow.* Lesser wing-coverts and rump-band also yellow. *Lores, nape, mantle, and tail glossy black.* Female olive brown above streaked dusky, with yellowish rump and lesser wing-coverts. *Superciliary and entire underparts yellow.*

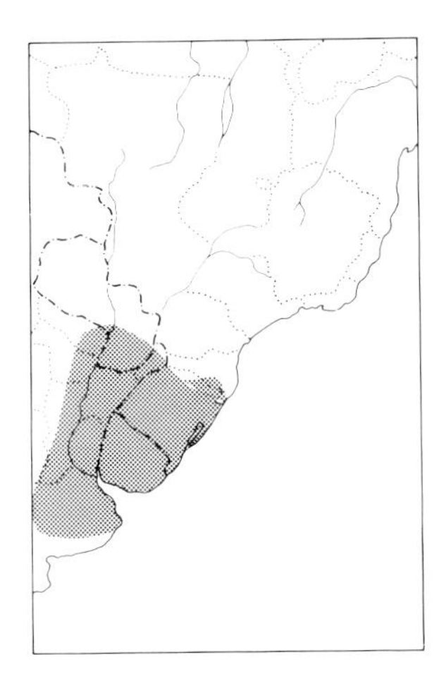

SIMILAR SPECIES: The beautiful male is unlikely to be confused. As the species usually troops about in flocks, females are normally no problem, but cf. the larger Yellow-rumped Marshbird and female Unicolored Blackbird.

HABITAT AND BEHAVIOR: Uncommon and local in open rolling terrain with small boggy swales. These swales normally have no open water and have some taller woody vegetation in them (D. Stotz). Usually in flocks, occasionally of up to 50 to 100 birds or more, which forage mostly by walking about in grasslands adjacent to the boggy depressions in which they nest and roost. A characteristic feature of Saffron-cowled flocks, at least in Rio Grande do Sul, Brazil, is the association between that species and the Black-and-white Monjita: the monjita almost seems to "lead" the blackbirds, the latter following along whenever the flycatcher changes perches. Numerous in ne. Rio Grande do Sul (in the Itaimbezinho area), but now very rare in Buenos Aires and may be declining in most parts of its range (for unknown reasons).

RANGE: Ne. Argentina (e. Formosa and Chaco south to Buenos Aires), se. Paraguay, Uruguay, and extreme s. Brazil (Rio Grande do Sul, also in adjacent s. Santa Catarina, *fide* D. Stotz). To about 1000 m.

NOTE: We follow Short (1975) in merging the monotypic genus *Xanthopsar* into *Agelaius*. As Short points out, inclusion of *Xanthopsar* in *Agelaius* "hardly extends the variation [in *Agelaius*]."

Agelaius icterocephalus

YELLOW-HOODED BLACKBIRD

IDENTIFICATION: 17–18 cm (6¾–7"). *Yellow hood extending down over chest;* otherwise all black, with lores also black. Female dull olive above obscurely streaked dusky; face (and especially superciliary) yellowish. *Throat rather bright yellow,* becoming olive on breast and brownish dusky on belly. See V-38, C-44. Female *bogotensis* of Colombia's E. Andes is darker overall, with less bright yellow throat.

SIMILAR SPECIES: Male unmistakable, while female's yellow throat, in conjunction with the marshy habitat, makes her unlikely to be confused (no known overlap with any other *Agelaius* blackbird, but cf. Pale-eyed immature). See also Oriole Blackbird.

HABITAT AND BEHAVIOR: Locally common in freshwater marshes and nearby tall wet grasslands, and (in Amazonia) in marshes and semiopen grassy and *Gynerium* cane-dominated areas on river islands and along riverbanks. Usually in small loose flocks which never seem to leave the vicinity of water; considerable numbers may congregate at roosts (on Trinidad these have been found in mangroves). Male's song is a wheezy, unmusical "gleeo, gleeeeeyr."

RANGE: N. Colombia (Caribbean lowlands from Chocó east, up Cauca valley to Valle and Magdalena valley to Huila, upper Patía valley in s. Cauca, temperate zone of E. Andes in Cundinamarca and Boyacá, and ne. llanos south to Meta and Vichada), Venezuela (except extreme south), Guianas, Amaz. Brazil (Amapá to Roraima and along Amazon itself), extreme se. Colombia (along Amazon), and ne. Peru (Loreto);

Trinidad; has straggled to Curaçao and Bonaire. A small population derived from escaped cage birds was established in marshes at Villa, south of Lima, Peru, in the 1960s and 1970s, but they may have died out of late. To about 2600 m (in Colombia's E. Andes).

GROUP B

Predominantly or entirely black (male); mostly e. or s. South America.

Agelaius ruficapillus

CHESTNUT-CAPPED BLACKBIRD

PLATE: 21

IDENTIFICATION: 18.5 cm (7¼"). Mostly glossy black; *crown and bib over throat and chest chestnut.* The chestnut is brighter and lighter, especially on crown, in *frontalis* of n. half of range. Female notably nondescript: olivaceous brown above streaked blackish on mantle; *throat and chest pale buffy fawn to tawny,* with remaining underparts dull tawny olivaceous. At some seasons many males in immature plumage are seen: these vary from being much like female except for some chestnut on bib to being much like adult male except lacking chestnut on crown.

SIMILAR SPECIES: Male unlike any other blackbird, though the chestnut can be hard to discern in poor light (especially in darker nominate race). The dull female lacks either the yellow or the streaking found in other *Agelaius* blackbirds; it most resembles female Shiny Cowbird, though the latter never shows the buff (at times almost "chamois") color on foreneck.

HABITAT AND BEHAVIOR: Uncommon to locally common in reedbeds in marshes and around lakeshores and in ricefields. Gregarious at all seasons, but especially when congregating during austral autumn when they are feeding in ricefields; semicolonial even when nesting. Rarely any distance from water. Vocalizations are variable, but many have the liquid gurgling quality typical of the genus. Particularly numerous in Rio Grande do Sul, Brazil, where it may have increased greatly due to the 20th-century spread of rice cultivation; elsewhere seems notably local and generally uncommon.

RANGE: French Guiana, e. and cen. Brazil (Amapá and islands at mouth of Amazon south to Rio Grande do Sul, west to Mato Grosso and se. Pará), e. Bolivia (north to Santa Cruz), Paraguay, Uruguay, and n. Argentina (south to Córdoba and Buenos Aires). Mostly below 500 m, occasionally to 850 m (at least in s. Brazil, *fide* D. Stotz).

Agelaius cyanopus

UNICOLORED BLACKBIRD

PLATE: 21

IDENTIFICATION: 19 cm (7½"). *Uniform glossy black.* Female has yellowish superciliary and *blackish lores and cheeks.* Mantle streaked brown and black, with *wings blackish edged rufous-chestnut. Below fairly bright oily yellow, vaguely streaked dusky;* streaking bolder and more blackish on flanks. Young birds tend to be more streaked (both above and below).

SIMILAR SPECIES: All-black male *lacks any yellow or chestnut* and hence is readily distinguished from its sympatric congeners. However, because of this uniform black plumage, male can also be confused with

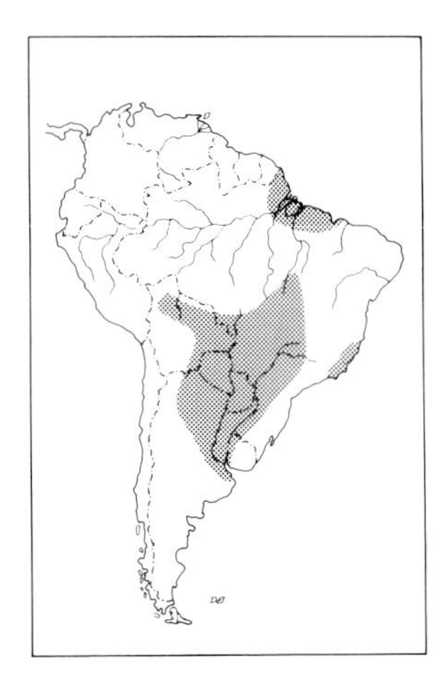

various other all-black blackbirds; cf. especially Velvet-fronted Grackle and Chopi Blackbird. Female is actually the more distinctive of the sexes, and as species usually occurs in groups, typically there is no problem.

HABITAT AND BEHAVIOR: Fairly common in reedbeds and other tall grassy vegetation in marshes and around small ponds and lagoons. Forages by walking about on floating vegetation and clambering in tall reeds and grasses, but often perches in the open on small bushes. Unlike most of its congeners seems not to emerge from marshes to feed in adjacent fields or pastures. Its song is a loud and ringing repetition of a single note, "tchew-tchew-tchew-tchew . . ." Particularly numerous and widespread in the *pantanal* along the Paraguay River from Mato Grosso to ne. Argentina.

RANGE: Ne. Brazil in lower Amazon region (Amapá to n. Maranhão); n. and e. Bolivia (northwest to Beni), sw. Brazil (s. and cen. Mato Grosso east to Goiás, w. São Paulo, Paraná, and Rio Grande do Sul), Paraguay, and n. Argentina (south to n. Buenos Aires); se. Brazil (Rio de Janeiro and seen in Espírito Santo by RSR et al.). To about 500 m.

Agelaius thilius

YELLOW-WINGED BLACKBIRD

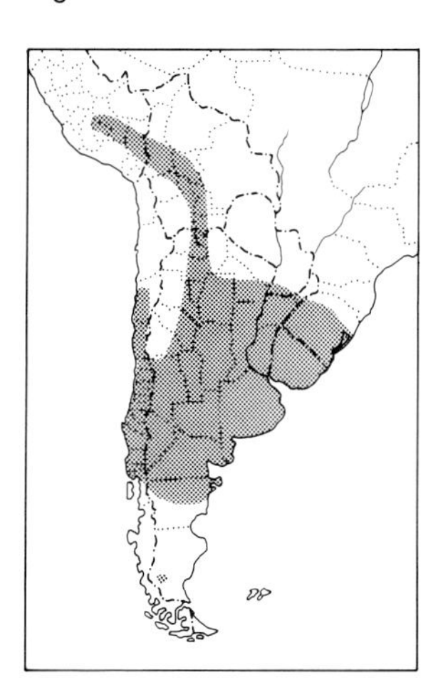

IDENTIFICATION: 18–18.5 cm (7–7¼"). *Black* with *yellow shoulders,* bend of wing, and under wing-coverts. Female streaked brown and blackish above, with superciliary whitish to pale grayish. Below pale grayish brown, *prominently streaked blackish. Yellow shoulders,* bend of wing, and under wing-coverts. Males in fresh plumage have black feathers tipped and edged brownish.

SIMILAR SPECIES: The contrasting yellow on shoulders is distinctive, though it is often smaller (and therefore less evident) in females. Female more streaked below than other female blackbirds. Cf. Chestnut-capped Blackbird, and also yellow-winged race of Epaulet Oriole (different shape and behavior, but similar plumage pattern overall).

HABITAT AND BEHAVIOR: Locally common in reedbeds of marshes and lakeshores, ranging from marshes along coast to the altiplano lakes of s. Peru and w. Bolivia. May forage in adjacent fields and meadows, but generally not far from water. Gregarious, usually occurring in small groups, and even when nesting seems colonial. Male's most frequently given song is a drawn-out, rather nasal "chree-layyy" (at times often sounding remarkably like saying "Chile," and in that country known by the onomatopoeic name of *trile*).

RANGE: Altiplano of s. Peru (Cuzco and Puno) and w. Bolivia (recorded only La Paz, Cochabamba, and Oruro); lowlands of Chile (Atacama south to Aysén), Argentina (Jujuy, e. Chaco, and Corrientes south to Chubut, with 1 recent sighting from w. Santa Cruz; N. Krabbe), se. Paraguay (Misiones), Uruguay, and extreme s. Brazil (Rio Grande do Sul). To about 4000 m (in Peru and Bolivia).

Agelaius xanthophthalmus

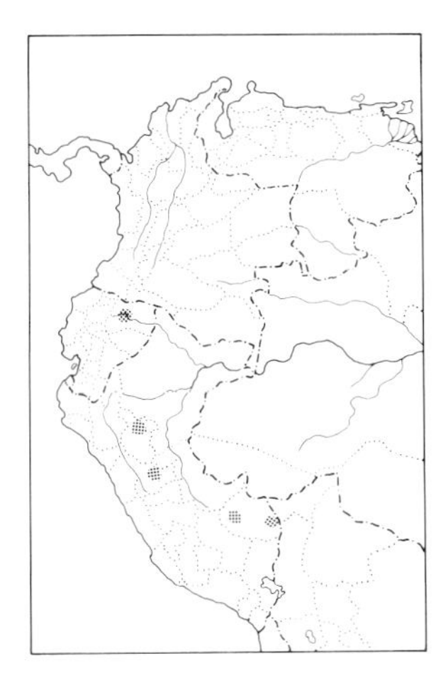

PALE-EYED BLACKBIRD

IDENTIFICATION: 20.5 cm (8"). *Local in e. Ecuador and e. Peru. Iris straw-colored to whitish. Entirely black* (both sexes). Immature more brownish black above and broadly streaked with buffyish or yellowish below.

SIMILAR SPECIES: Because of its conspicuous pale iris, this blackbird is unlikely to be confused in its limited range and restricted habitat. Other all-black blackbirds occurring with it (e.g., Velvet-fronted Grackle, Shiny Cowbird) have dark irides.

HABITAT AND BEHAVIOR: Uncommon and *very* local in floating marsh vegetation with scattered bushes along the edges of oxbow lakes. Most conspicuous in the early morning, when they tend to forage more in the open and also sing while perched atop bushes, their song a loud, ringing "tew-tew-tew-tew." Usually in pairs (unlike other *Agelaius*, they do not seem to flock). Though described only in 1969, now regularly seen both at Limoncocha, Ecuador, and at Tambopata, Peru.

RANGE: Locally in e. Ecuador (Limoncocha in Napo) and e. Peru (San Martín; near Tingo María, Huánuco; Manu Nat. Park, Hacienda Amazonia, and Tambopata Reserve, Madre de Dios). To 650 m.

Lampropsar Grackles

A slender, long-tailed, all-black icterid found locally near water across n. and cen. South America. Does not appear to be closely related to the other "grackle" genera (*Quiscalus, Macroagelaius,* and *Hypopyrrhus*).

Lampropsar tanagrinus

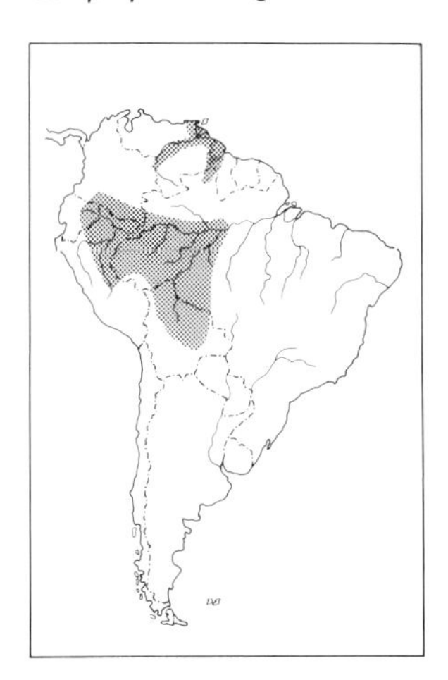

VELVET-FRONTED GRACKLE

PLATE: 22

IDENTIFICATION: Male 21.5–23.5 cm (8½–9¼"); female 19–21 cm (7½–8¼"). Rather short pointed bill. Entirely black, slightly glossed steel blue. Feathers of forecrown dense and plushlike (but this difficult to discern in the field). *Macropterus* (of upper Rio Juruá drainage in sw. Amaz. Brazil) larger than other races.

SIMILAR SPECIES: This slender, long-tailed icterid has no truly diagnostic field characters but can generally be known by its overall configuration, habitat and range, and behavior. Cf. male Shiny Cowbird and dark-shouldered races of Epaulet Oriole.

HABITAT AND BEHAVIOR: Uncommon to locally common in *várzea* forest and forest borders and in marshy vegetation around lakes and ponds. Typically in monospecific groups of up to 20 or so birds which forage in foliage at just about any level, then often fly off as a group; they sometimes accompany mixed flocks of caciques, tanagers, etc. Around lakes they may drop to floating vegetation mats and hop around, but usually they remain higher. Its call is a loud, rich, fast chuckling or gurgling (e.g., "chuk-a, chik-a, chuk" or "gluk-gluk-glí-gluk").

RANGE: Locally in e. and s. Venezuela (along Orinoco River and north to Sucre), Guyana, and extreme n. Brazil (recently reported from Maracá Ecol. Station in Roraima); w. Amaz. Brazil (along Amazon and south of it along major tributaries from Rio Madeira drainage west), se. Colombia (along Amazon near Leticia and in w. Caquetá and Putumayo), e. Ecuador, ne. Peru (south to n. Ucayali), and n. Bolivia (south to nw. Santa Cruz). To 400 m.

Gnorimopsar, Oreopsar, Curaeus, and *Dives* Blackbirds

A group of 5 all-black icterids, currently classified in 4 genera. We are confident that these blackbird genera have been "oversplit," but in the absence of comparative study adhere to the conservative position of maintaining all 4. We take note of the fact that this treatment differs from that adopted for the oropendolas (see below) but can only plead that the blackbirds remain too little known to safely make taxonomic changes. With the exception of the little-known *Curaeus forbesi,* all have allopatric distributions; likewise, so far as known, vocalizations are similar. All are conspicuous, common, gregarious birds found in open and semiopen country.

Gnorimopsar chopi

CHOPI BLACKBIRD

PLATE: 22

IDENTIFICATION: 23–24 cm (9–9½"). Glossy black; flight feathers slightly browner. *Feathers of crown and nape narrow and pointed* (often imparting a "hackled" or "streaky" effect in the field). *Diagonal groove on mandible* is often fairly conspicuous, while another on maxilla can occasionally be discerned. *Sulcirostris* (ne. Brazil, and Bolivia and nw. Argentina) somewhat larger.

SIMILAR SPECIES: This familiar and widespread blackbird should be learned well as a basis for comparison with other less numerous species; cf. especially Bolivian and Forbes' Blackbirds. Male Shiny Cowbird is glossier purple and smaller and lacks the grooves on the bill and lanceolate head feathers. Male Unicolored Blackbird differs in bill also (more slender and pointed), as well as in having very different behavior and habitat.

HABITAT AND BEHAVIOR: Common (locally abundant) and widespread in agricultural regions and around farm buildings, in savannas and pastures, and in marshy areas. Very conspicuous and often tame, usually occurring in small, noisy flocks which parade on the ground or perch in shade trees around buildings or in gallery woodland. Typically associated with palms. Their loud, whistled calls are varied and musical and are often given in flight or as a sort of agreeable medley by a flock, even in the heat of the day; often a bird leads into its rich, gurgled song by giving a series of well-spaced single "peer" or doubled "pur-peer" notes. Chopis are one of the first birds to begin singing at dawn, sometimes starting well before first light. Apparently nests in holes, both at the base of palm fronds and in cavities of dead snags; nesting is often semicolonial.

RANGE: E. and s. Brazil (extreme ne. Pará in the Belém area, Maranhão, Goiás, and Mato Grosso south), n. and e. Bolivia, extreme se. Peru (Heath Pampas in Madre de Dios), Paraguay, Uruguay, and n. Argentina (Salta, and from e. Formosa and Chaco south, rarely, to n. Buenos Aires). To about 1000 m (in Brazil).

Oreopsar bolivianus

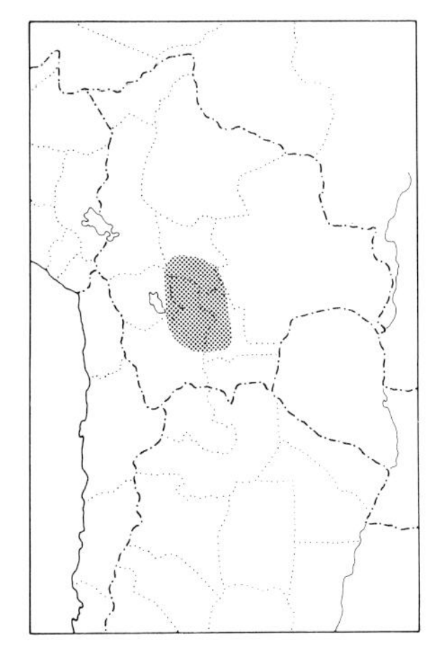

BOLIVIAN BLACKBIRD

IDENTIFICATION: 23 cm (9″). *Intermontane valleys of w. Bolivia.* Closely resembles Chopi Blackbird (no overlap). Black sootier and not as glossy, and *wings* (especially flight feathers) *browner,* more so than Chopi's. *Lacks* the "hackled" effect of Chopi's crown and nape. Bill shape similar to Chopi's, but *lacks grooves.*
HABITAT AND BEHAVIOR: Locally fairly common in arid intermontane valleys (often partly or mostly given over to agriculture) with scattered groves of trees and shrubby areas. Seems restricted to areas within a reasonable distance of cliffs, in which they nest. Found in small flocks of some 4 to 8 birds. These forage as a group on bare or grassy ground or in foliage and among *Tillandsia* moss in trees; they have also been observed to eat the fruit of cactus. Various loud calls are often given, notably a clear "chu-pee" in flight and a series of sharp "chip" and "chu-pit" notes which apparently serves as a song. Can be found in small numbers outside Cochabamba city. For more behavioral data, see G. Orians et al., *Condor* 79(2): 250–256, 1977.
RANGE: Andean valleys of sw. Bolivia (Cochabamba, Chuquisaca, and Potosí). Mostly 1500–2800 m.

Curaeus curaeus

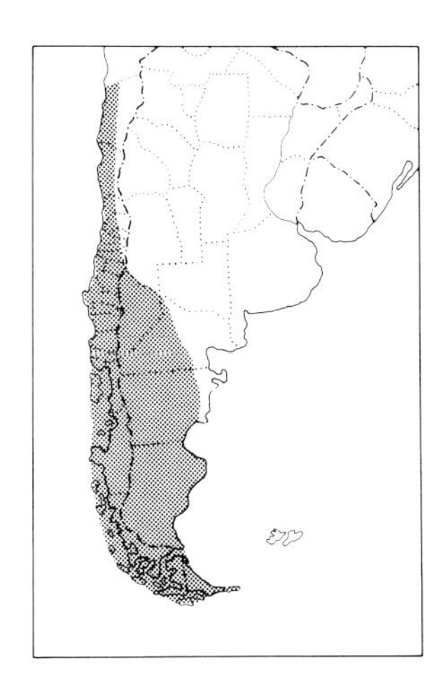

AUSTRAL BLACKBIRD

PLATE: 22

IDENTIFICATION: Male 26.5 cm (10½″); female 24 cm (9½″). *S. Argentina and Chile.* Bill quite long and conical. *Glossy black.* Feathers of head and nape narrow and pointed, with shiny shafts imparting slightly streaky look (but less so than on Chopi). Female duller and sootier, with browner wings.
SIMILAR SPECIES: The only truly black blackbird in its range (no overlap with the very similar Chopi Blackbird). Cf. male Shiny Cowbird.
HABITAT AND BEHAVIOR: Uncommon to fairly common in forest and woodland borders and in agricultural land with thick hedgerows and groves of trees. Familiar and conspicuous over much of its range (perhaps especially in Chile; scarcer southward), usually seen in small flocks except during the nesting season when pairs separate out. The song and calls are varied but on the whole musical, though interspersed are more guttural or "chack" notes; the overall effect is quite like that of the Chopi Blackbird, and like that species a group of Australs will often sing in concert, particularly at dawn or dusk from their roosting site.
RANGE: S. Argentina (north to Neuquén and s. Río Negro) and cen. and s. Chile (mostly from Coquimbo south, occasionally north to Atacama) south to Tierra del Fuego. To about 1500 m.

Curaeus forbesi

FORBES' BLACKBIRD

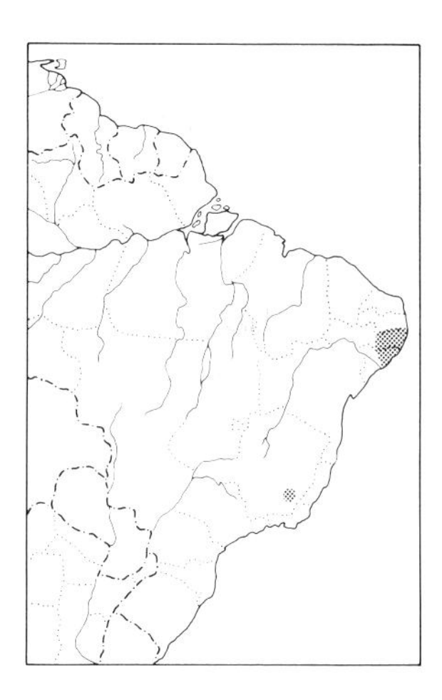

IDENTIFICATION: Male 24 cm (9½"); female 21.5 cm (8½"). Known only from a few localities in *e. Brazil.* Closely resembles both Austral and Chopi Blackbirds, but potentially sympatric only with the latter. Compared to the Chopi, Forbes' Blackbird is a sootier, less glossy black, and it would have a thinner, straighter bill (more like an Austral). It also looks like a Unicolored Blackbird, but habitat presumably differs, and there is no sexual dimorphism in plumage, etc.
HABITAT AND BEHAVIOR: Nothing certainly recorded. In 1880, when the type specimen was obtained, the collector recorded it as locally "common, flying about in large flocks . . . in the neighborhood of sugar-plantations." However, it is not known whether some or even most of the birds seen were actually Chopis. There are no subsequent published reports of field observations, though several formerly misidentified specimens have since come to light (see L. L. Short and K. E. Parkes, *Auk* 96 [1]: 179–183, 1979).
RANGE: E. Brazil in Pernambuco (Macuca), Alagoas (near Sergipe and in the Quebrangulo region), and se. Minas Gerais (Raul Soares region).

Dives warszewiczi

SCRUB BLACKBIRD

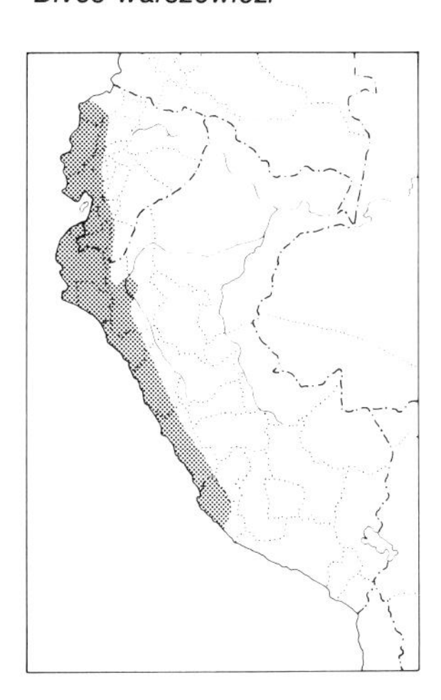

IDENTIFICATION: Male 24–28 cm (9½–11"); female 23–26.5 cm (9–10½"). Generally similar to Chopi Blackbird. Glossy black. Female slightly duller. *Kalinowskii* (w. Peru north to La Libertad) is notably larger and has proportionately longer bill; it apparently intergrades with the nominate race in La Libertad (Trujillo) and Piura (Palambla).
SIMILAR SPECIES: Does not overlap with any of the other black blackbirds, all of which are very similar. Male Shiny Cowbird could be confused, particularly with smaller n. nominate race, but cowbird is even glossier (and more purple) with shorter, more conical bill. See also male of larger Great-tailed Grackle (pale eye, long creased tail, etc.).
HABITAT AND BEHAVIOR: Common and conspicuous in agricultural land, shrubby areas and groves of trees, borders of deciduous woodland, and even verdant suburban areas. Avoids actual desert, but total range and numbers have doubtless increased due to spread of irrigation; also spreading northward into recently deforested parts of nw. Ecuador. Usually in small groups, foraging mostly on the ground but frequently perching in shrubs and trees. Its loud, ringing song and call notes often draw attention to it; seems not to sing in groups (unlike Chopi and Austral Blackbirds). Especially numerous in sw. Ecuador.
RANGE: W. Ecuador (north to Pichincha, still spreading; inland in s. intermontane valleys of Azuay, Loja, and El Oro) and w. Peru (south to Ica; inland locally into middle Río Marañón valley in Piura). Mostly below 1500 m (ranging higher locally and in smaller numbers).

NOTE: We continue to maintain *D. warszewiczi* as a species distinct from the geographically distant Mexican *D. dives,* following the 1983 AOU Check-list. In fact, it is even possible that *kalinowskii,* the much larger s. form of *warszewiczi,* may prove to be a distinct species, though voices of the 2 forms are similar. The 1983 AOU Check-list so suggests, despite the fact that intermediate specimens are known. The contact zone between the 2 forms has not been thoroughly investigated, however (see T. S. Schulenberg and T. A. Parker III, *Condor* 83 [2]: 213–214, 1981).

Molothrus Cowbirds

Rather small, short-billed blackbirds of open to semiopen country. Males of most species are black with a lustrous gloss, females duller and grayer; one, the Bay-winged, is grayer. Cowbirds are gregarious birds found in open country; they are well known for their brood parasitism.

GROUP A

Males *glossed;* females usually duller.

Molothrus rufoaxillaris

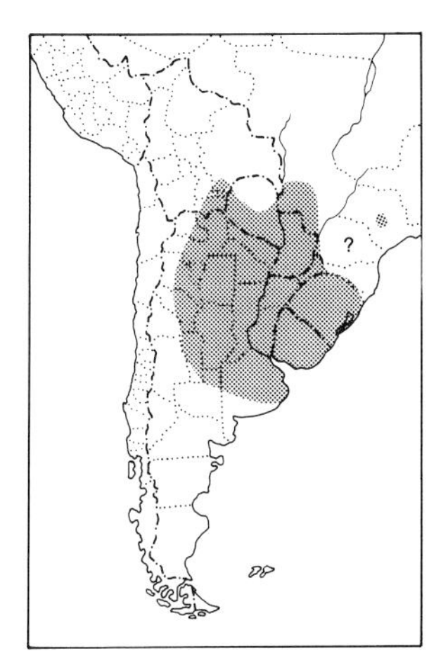

SCREAMING COWBIRD

PLATE: 22

IDENTIFICATION: 19 cm (7½"). *N. Argentina and adjacent areas.* Closely resembles male Shiny Cowbird (which in area of sympatry is somewhat larger); *sexes alike. Bill slightly shorter and stouter,* more conical and finchlike (sometimes imparting a "snub-nosed" look). Lustrous black *with only slight greenish blue sheen* (lacking the strong violet gloss of male Shiny). The rufous axillars for which this species is named are generally invisible, even in flight. Juvenals resemble adult Bay-winged Cowbird (its host) but quickly molt into adult plumage, so this usually is not a problem; rufous on wing often persists after body plumage is black.

SIMILAR SPECIES: Easily confused with Shiny Cowbird; as the 2 species seem only infrequently to flock together, often the *lack* of sexual dimorphism in a pair or groups of cowbirds is a good indication that they are Screamings. During the breeding season small numbers often associate with flocks of Bay-winged Cowbirds (D. Stotz). Voice of male Screaming diagnostic if heard (see below). Cf. also Chopi Blackbird.

HABITAT AND BEHAVIOR: Uncommon to locally common (but often passes as only uncertainly identified) in pastures, fields, and scrubby and weedy areas and around towns. Behavior similar to Shiny Cowbird's, but tends not to gather in as large groups and associates less often with cattle. Male's call is a short, loud, somewhat harsh note (but hardly a "scream"), often uttered as the bird raises its neck feathers into a ruff; a whistled song is given less often. Apparently an obligate brood parasite on the Bay-winged Cowbird in most of its range, though in Paraná, Brazil (to which it has only recently spread and where Baywings do not occur) it parasitizes the Chopi Blackbird (H. Sick).

RANGE: N. Argentina (south to San Luis, La Pampa, and Buenos Aires), Uruguay, Paraguay, s. Brazil (sw. Mato Grosso, Paraná, and Rio Grande do Sul; also a recent record from interior São Paulo and presumed wanderers in Rio de Janeiro and Pernambuco), and e. Bolivia (mainly Tarija, and 1 record from Guanacos in Santa Cruz). Evidently spreading northward in se. Brazil (elsewhere?). To about 1000 m.

Molothrus bonariensis

SHINY COWBIRD

PLATE: 22

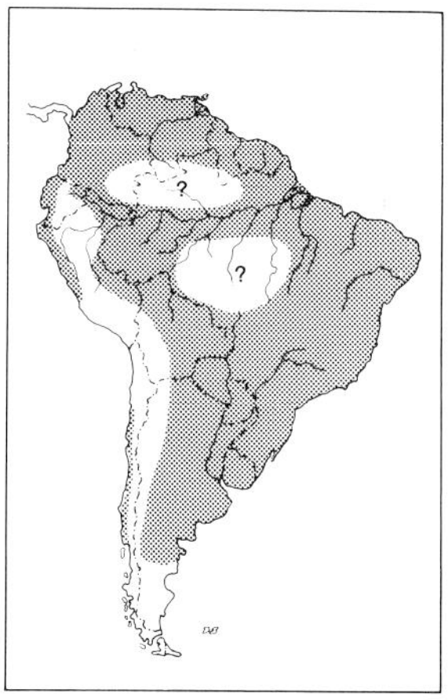

IDENTIFICATION: 18–21.5 cm (7–8½"). *Mostly glossy purplish black,* with wings (and to lesser extent tail) glossed greenish blue; gloss is strongest on head, neck, and breast. Female *more or less uniform dingy gray to brownish gray,* somewhat paler below. Most females tend to have whitish postocular streak, with this especially marked in *aequatorialis* and *occidentalis* (sw. Colombia to w. Peru). *Occidentalis* (w. Peru and adjacent Ecuador) is much the palest race overall, with underparts pale grayish white blurrily streaked dusky. Immatures more or less like respective females but tend to be more yellowish below with obscure dusky streaking. Considerable size variation, with *cabanisi* (w. Colombia) being the largest and *minimus* (Guianas, adjacent Brazil, Trinidad and Tobago) the smallest; most races are around 20.5 cm (8") long.

SIMILAR SPECIES: Often best told by the variation of plumages in a flock: many other potentially confusing species show *no* sexual dimorphism. Male very similar to Screaming Cowbird (see above), and cf. various all-black blackbirds and male White-lined Tanager. Female is your basic dull bird, and out of context or if alone (infrequent) can be difficult.

HABITAT AND BEHAVIOR: Common in a broad variety of semiopen to open habitats, but especially numerous in agricultural land with scattered groves of trees and patches of shrubbery. Colonizes recently deforested areas quite rapidly. Usually in small flocks (larger congregations at roosts) which forage on the ground, often assuming distinctive posture with tail held slightly cocked. By no means necessarily with cows, though regularly with them. Males often voice their brief, liquid, bubbling song, often accompanied by raising neck feathers. Parasitic on a variety of passerine birds.

RANGE: Widespread (though absent from extensively forested regions) south to w. Peru (south to Lima), Chile (Atacama to Aysén), and s. Argentina (Chubut); seemingly absent from portions of Amazonia (but spreading with deforestation; small numbers now present locally along major rivers); Trinidad and Tobago. Also Panama and most of West Indies (where spreading northward). Mostly below 2000 m, but locally much higher (recorded to 3500 m in w. Argentina).

Molothrus aeneus

BRONZED COWBIRD

Other: Bronze-brown Cowbird

IDENTIFICATION: 18.5–19 cm (7¼–7½"). *Locally on Caribbean coast of Colombia.* Iris red (but dark, and thus inconspicuous except in good light). *Mostly lustrous dark bronzy brown,* wings slightly glossed more purplish blue. Female slightly duller and smaller. Feathers of neck usually held puffed up, forming more or less conspicuous ruff (less so in female). See C-44.

SIMILAR SPECIES: Shiny Cowbird, which is sympatric with this species, is much less brown (male being violet-glossed black, female dull gray) and never shows the red eye.

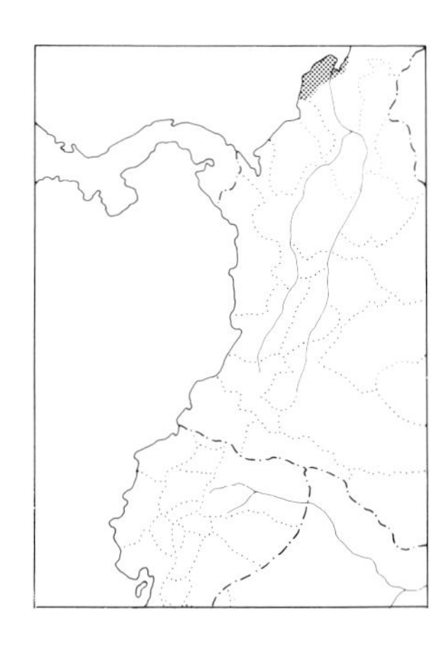

HABITAT AND BEHAVIOR: Locally uncommon to fairly common in open areas and fields with scattered bushes and low trees. Behavior similar to Shiny Cowbird's, but never seems as numerous as that species, often foraging alone. Can at times be located in agricultural land on the w. end of Isla de Salamanca, but even here numbers are usually small (and often there are more Shinies about).
RANGE: Coastal Caribbean Colombia in n. Magdalena (possibly also occurs westward in Córdoba, etc.). Does *not* occur at Leticia (that locality being in the literature because of a mix-up in a live bird shipment). Also sw. United States to Panama. Near sea level.

NOTE: We concur with A. Dugand and E. Eisenmann (*Auk* 100 [4]: 992, 1983) that the Colombian isolate *armenti* "cannot be looked upon as other than a southernmost race of the species *M. aeneus.*"

GROUP B

Both sexes "hen-plumaged" with *rufous wings.*

Molothrus badius

BAY-WINGED COWBIRD

PLATE: 22

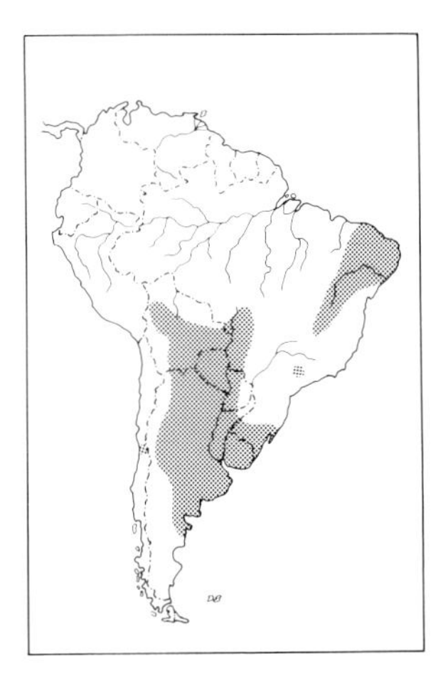

IDENTIFICATION: 18–19 cm (7–7½"). *Mostly ashy brownish gray,* somewhat browner above; *ocular area* (lores and behind eye) *dusky,* with effect of small mask. *Wings contrastingly rufous;* tail blackish. *Fringillarius* (ne. Brazil) is notably paler overall (hence dark ocular area contrasts more): browner above and much buffier below (lacking dingy grayish tone), with rufous on wings paler.
SIMILAR SPECIES: No other vaguely similar bird shows the rufous on the wings. Cf. juvenile Screaming Cowbird (a frequent brood parasite of the Bay-wing and which briefly resembles its host).
HABITAT AND BEHAVIOR: Fairly common to common in semiopen scrubby or lightly wooded terrain and in adjacent pastures or areas with tall grass. Apparently less numerous in its ne. Brazil range and much rarer (perhaps not even resident) in Chile. Most often in small flocks of up to 25 birds; usually these move about independently, but at times they are seen feeding with Shiny Cowbirds. Unlike other cowbirds, this species is not parasitic, rather building its own cup nest or (apparently less often) appropriating the domed nest of some other species (e.g., of the Firewood-Gatherer or a hornero); this has been interpreted as an early ("primitive") stage in the development of the parasitic habit. On the other hand, this species is the usual *host* of the entirely parasitic Screaming Cowbird.
RANGE: Ne. Brazil (Piauí, Ceará, and Pernambuco south to n. Minas Gerais); n. and e. Bolivia (La Paz, Cochabamba, and Santa Cruz south), s. Brazil (sw. Mato Grosso and Rio Grande do Sul, with several recent reports from São Paulo, *fide* E. O. Willis), w. and cen. Paraguay (mostly west of Paraguay River, locally just to east of it), Uruguay, and n. and cen. Argentina (south to Mendoza, Río Negro, and n. Chubut); a few records from cen. Chile. To about 2500 m (in Bolivia), occasionally higher.

Scaphidura Cowbirds

A large cowbird (only infrequently with cows), much more associated with forest than *Molothrus*. It is a brood parasite on forest-inhabiting oropendolas and caciques. Widespread in more humid lowlands.

Scaphidura oryzivora

GIANT COWBIRD

PLATE: 22

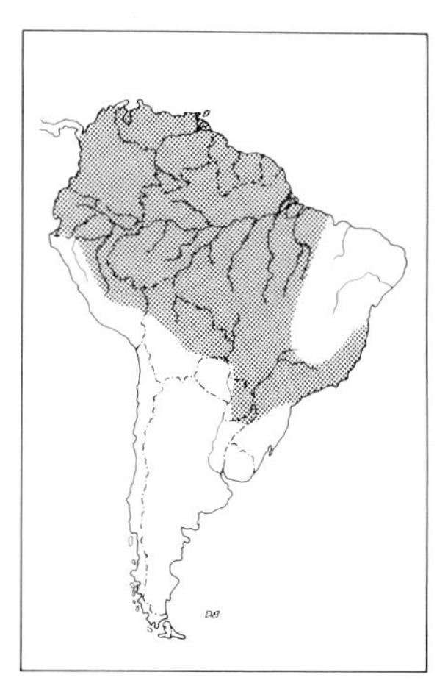

IDENTIFICATION: Male 35.5–38 cm (14–15″); female 30.5–33 cm (12–13″). Iris variable, tending to be yellow westward and southward, orange or orange-red in northeast (Guianas, etc.). Glossy purplish black with *conspicuous ruff on neck* (imparting curiously *small-headed aspect*). Female smaller, sootier black, with ruff smaller or even lacking. Immature like female but may show pale bill and iris.
SIMILAR SPECIES: Male much larger than other cowbirds, the size of similarly glossy black male Great-tailed Grackle. Even the smaller female is considerably larger than other cowbirds, but cf. other all-black blackbirds (especially the Chopi). Giant Cowbird's *silhouette and undulating flight style distinctive:* small head and large body, with rather long, pointed wings flapping rapidly a few times, then closing briefly, followed by a few more flaps.
HABITAT AND BEHAVIOR: Usually uncommon (locally more numerous) in wide variety of humid and semihumid habitats. General distribution quite closely tied to the presence of oropendola and cacique colonies (on which it is an obligate brood parasite), though when not nesting the cowbirds may disperse quite widely. Giant Cowbirds are most often seen flying, sometimes high overhead; they feed both on the ground and (less often) in trees, in the latter case at times with their hosts. In Amazonia (at least) regularly on riverbanks. Notably silent. Females linger around active oropendola and cacique colonies; at some the oropendolas try to chase them off, while in others the presence of cowbirds is tolerated. See N. G. Smith, *Nature* 219: 690–694, 1968.
RANGE: Widespread south to extreme nw. Peru (Tumbes), n. and e. Bolivia, e. Paraguay, ne. Argentina (Misiones), and s. Brazil (south to Santa Catarina); apparently absent from much of ne. Brazil; Trinidad and Tobago. Also Middle America. Mostly below 1000–1500 m, occasionally to 2000 m or higher.

Quiscalus Grackles

A pair of large, long-tailed blackbirds; males are shiny black, females browner or grayer, both sexes have pale eyes. Both S. Am. species boldly parade about on open ground, especially near n. coasts; several other species are found in Middle and North America and the West Indies. We follow the 1983 AOU Check-list and most other recent authors in considering *Cassidix* congeneric with *Quiscalus*.

Quiscalus lugubris

CARIB GRACKLE

PLATE: 22

IDENTIFICATION: Male 24.5–27.5 cm (9¾–10¾″); female 20.5–23 cm (8–9″). *Iris yellow (both sexes,* but dark in immatures); rather long slender black bill; somewhat creased, *wedge-shaped* tail. Mostly glossy purplish black, with wings more glossed greenish blue. Female blackish brown above, somewhat paler grayish brown below, lightest on throat and darkest on belly. Trinidad females are a more uniform dark chocolate brown below, with only immatures showing the paler throat and grayish tone below. Various subspecies from small islands off Venezuelan coast have bill even longer and slenderer.

SIMILAR SPECIES: No overlap with the much larger Great-tailed Grackle (see below). Male Shiny Cowbird has dark iris, normally shaped tail, etc.

HABITAT AND BEHAVIOR: Locally common in semiopen areas, both along the coast and in the llanos of interior Venezuela and Colombia; also often locally numerous in towns and around farm buildings. A very gregarious and conspicuous bird, parading about on the ground and around houses; large congregations roost in town plazas or in mangroves. Noisy birds, males especially almost continually giving voice to a variety of loud, often raucous calls, often accompanied by various strutting displays with head and neck held high.

RANGE: Ne. Colombia (llanos region south to Meta and Vichada), n. Venezuela (except in the northwest; ranges south to along the Orinoco River), Guianas, and extreme ne. Brazil (Amapá); Trinidad, Tobago, and various small islands off n. coast of Venezuela (only as a straggler to Bonaire and Aruba). Also Lesser Antilles. To about 850 m (in Venezuela).

Quiscalus mexicanus

GREAT-TAILED GRACKLE

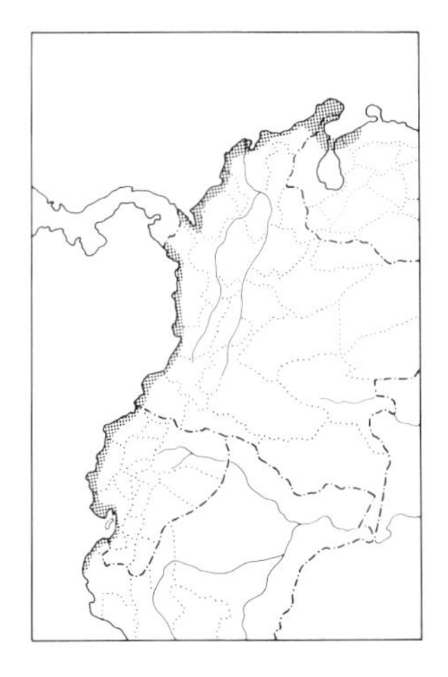

IDENTIFICATION: Male 44 cm (17″); female 33 cm (13″). *Locally on coasts of nw. South America.* Iris yellow in male, yellowish to pale brown in female; bill long and rather heavy. Male mostly glossy blue-black, with wings and tail more glossed greenish blue; *tail very long and creased,* often looking keel-shaped. Female much smaller, dusky brown above with *buffy superciliary and throat; remaining underparts pale buffy brownish;* tail shorter and less keel-shaped. Immature like female but iris dark. See P-26.

SIMILAR SPECIES: Male is so much larger and longer-tailed than otherwise similar Carib Grackle that confusion is unlikely; they are not sympatric.

HABITAT AND BEHAVIOR: Common locally in mangroves and along shorelines and on lawns and in parks in towns and cities. Behavior similar to Carib Grackle's, and like that species bold and aggressive, eating virtually anything edible. Feeds mostly on the ground but roosts in trees (sometimes even on wires), often in very large flocks.

RANGE: Coasts of nw. Venezuela (Zulia), Colombia (both Caribbean and Pacific), w. Ecuador, and extreme nw. Peru (Tumbes). Also sw. United States and Middle America. Around sea level.

Macroagelaius Mountain-Grackles

A pair of slender, long-tailed grackles with restricted distributions in n. South America. Range in small flocks through the canopy of montane forest. We have modified their English names so as to better reflect their very close relationship (they may prove to be conspecific) and their small ranges.

Macroagelaius imthurni

TEPUI MOUNTAIN-GRACKLE

PLATE: 22

Other: Golden-tufted Grackle

IDENTIFICATION: Male 28 cm (11"); female 25.5 cm (10"). *Tepuis of s. Venezuela and adjacent areas.* Long, slender bill; long tail. Glossy blue-black; *pectoral tufts golden yellow.*

SIMILAR SPECIES: Though the golden pectoral tufts are often hard to see under field conditions, this is the only all-black blackbird in its range. Compare to the geographically distant Colombian Mountain-Grackle.

HABITAT AND BEHAVIOR: Common in canopy and edge of humid forest on the slopes of the *tepuis.* Very gregarious, mostly in flocks of some 10 to 30 birds which forage in the treetops and fly rapidly in compact groups from tree to tree. Quality of vocalizations varies, with some typical "chirt-chirt" notes and also a loud, more musical song (e.g., "gleeo, glee-gleeo, shw-shwee?"—reminiscent of Chopi Blackbird); both are given while foraging, while a more screeching call is given in flight. Easily seen along the Escalera road in e. Bolívar, Venezuela.

RANGE: S. Venezuela (Bolívar and n. Amazonas), Guyana, and adjacent extreme n. Brazil (Roraima). 500–2000 m.

Macroagelaius subalaris

COLOMBIAN MOUNTAIN-GRACKLE

Other: Mountain Grackle

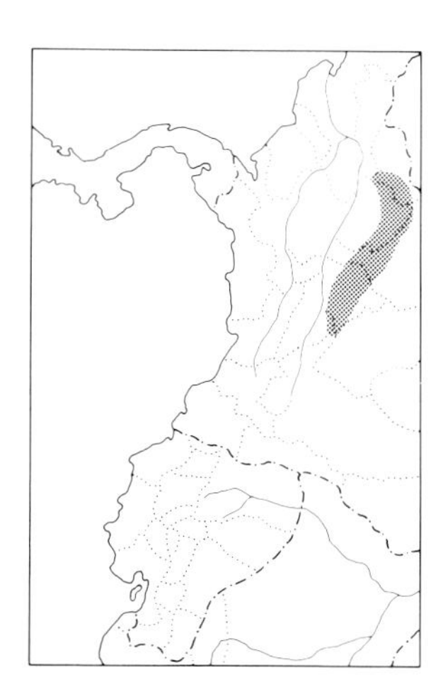

IDENTIFICATION: Male 29.5 cm (11½"); female 27 cm (10½"). *E. Andes of ne. Colombia.* Closely resembles the Tepui Mountain-Grackle (no overlap), but proportionately longer-tailed. *Axillars and under wing-coverts dark chestnut* (this hard to discern under normal field conditions). See C-45.

SIMILAR SPECIES: Female Giant Cowbird has different proportions, very different behavior, etc. See Red-bellied Grackle.

HABITAT AND BEHAVIOR: Poorly known but probably does not differ much from Tepui Mountain-Grackle; may never have been as common. Doubtless also a forest inhabitant and as a result surely much reduced in overall numbers, especially at the s. end of its limited range; likely deserves threatened status, but more information is needed.

RANGE: Ne. Colombia (E. Andes in Norte de Santander and on their w. slope from Santander south locally to w. Cundinamarca). 1950–3100 m.

Hypopyrrhus Grackles

A single species of spectacular montane grackle restricted to Colombia. Probably close to *Macroagelaius*.

Hypopyrrhus pyrohypogaster

RED-BELLIED GRACKLE

PLATE: 22

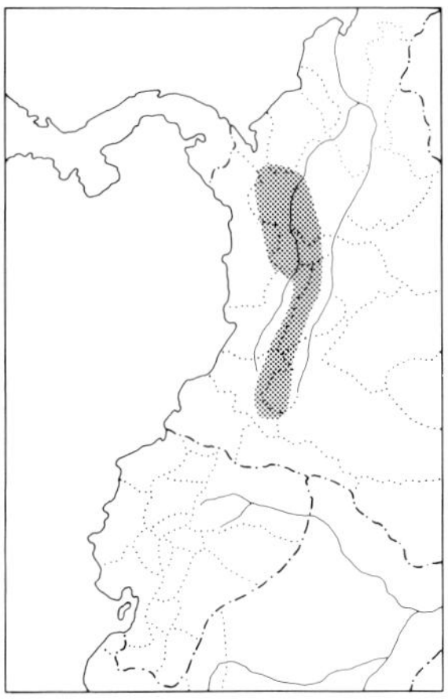

IDENTIFICATION: Male 31.5 cm (12¼″); female 27 cm (10½″). *Locally in Colombian Andes.* Unmistakable. Iris straw yellow. Glossy black with *bright red belly and crissum.* Feathers of head and throat with shiny shafts, imparting slightly streaky look.
SIMILAR SPECIES: Large size and red on underparts will immediately distinguish this spectacular but lamentably rare montane grackle. See the all-black Colombian Mountain-Grackle (no overlap).
HABITAT AND BEHAVIOR: Rare and now very local in canopy and borders of humid montane forest. Usually in small noisy groups of up to 10 or so birds, foraging actively in foliage. Often moves about with other large birds such as oropendolas, Green Jays, Blue-winged Mountain-Tanagers, etc. Emits a variety of calls, some liquid and gurgling, others more wheezy (e.g., "glok-glok, shleee-o, schleee"). Most of the localities where this species was found in some numbers during the 19th and early 20th centuries (e.g., around Medellín and on slopes of n. end of Cen. Andes) have long since been deforested. There have been a few recent sightings from the head of the Magdalena River valley (e.g., at Cueva de los Guácharos Nat. Park) and above Florencia, Caquetá, but on the whole this is a very difficult bird to find, and it surely deserves at least threatened status.
RANGE: Locally in Andes of Colombia (W. Andes south to extreme n. Valle, Cen. Andes south to n. Tolima, head of Magdalena valley in Huila, and e. slope of E. Andes in w. Caquetá). 1200–2700 m.

Icterus Orioles and Troupials

Medium-size, rather slender and long-tailed blackbirds with sharply pointed bills. Basically orange or yellow and black, with no sexual dimorphism except in the 2 n. migrants. Mainly found in n. South America (only 2 species southward), with many other species north into North America and in the West Indies. They are found in woodland and borders, none being true forest birds.

GROUP A

Mainly black.

Icterus chrysocephalus

MORICHE ORIOLE

PLATE: 22

IDENTIFICATION: 20.5–21 cm (8–8¼″). Bill rather long and slightly decurved. Black with *yellow crown* (aside from frontal band), *shoulders, rump, and thighs.* A few apparent intergrades with the following spe-

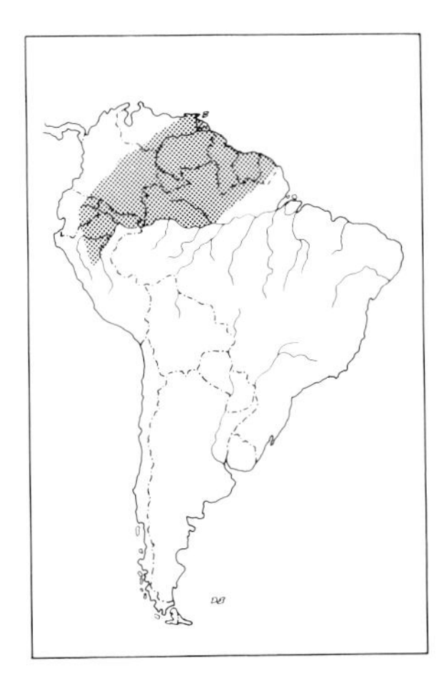

cies have been recorded from interior Suriname, French Guiana, and adjacent Brazil: these have crown variably streaked with black, etc.

SIMILAR SPECIES: Compare to Epaulet Oriole (which never shows yellow on crown, rump, etc.).

HABITAT AND BEHAVIOR: Uncommon to locally fairly common at borders of humid forest and in clearings with scattered trees and swampy areas; shows a strong association with moriche palms (*Mauritia flexuosa*). Found singly or in pairs, usually alone but occasionally following mixed flocks; searches for insects among fronds of the palms (it also usually nests in such situations) and in foliage and at flowers. Song is a slow series of sweet whistled notes, often repeated for long periods, "suweet, tweeu, suweet, tweeu, tweeu, suweet." Numbers have been depleted in some areas due to its popularity as a cage bird.

RANGE: Guianas, s. and e. Venezuela (mostly south of the Orinoco River, locally north of it in Apure, Monagas, and Sucre), e. Colombia (north to Meta and Vichada), e. Ecuador, ne. Peru (north and west of the Amazon and west of the Río Ucayali), n. Brazil (south to n. bank of the Amazon, except absent east of the Rio Negro); Trinidad. Mostly below 500 m (locally higher in Venezuela).

NOTE: We continue to regard *I. chrysocephalus* as a full species, following Short (1975) but not *Birds of the World* (vol. 14). There does appear to be some hybridization between *chrysocephalus* and nominate *cayanensis* in ne. South America, but to date its extent remains unknown, and until the situation is further clarified it seems convenient to continue to regard both as species. Both occur at the WWF sites north of Manaus, Brazil, with Moriche the more numerous (D. Stotz). It might further be pointed out that despite their plumage dissimilarities, *chrysocephalus* and nominate *cayanensis* are united in having a longer, more curved bill, while the other (more southern) races of *cayanensis* have a shorter, straighter bill.

Icterus cayanensis

EPAULET ORIOLE

PLATE: 22

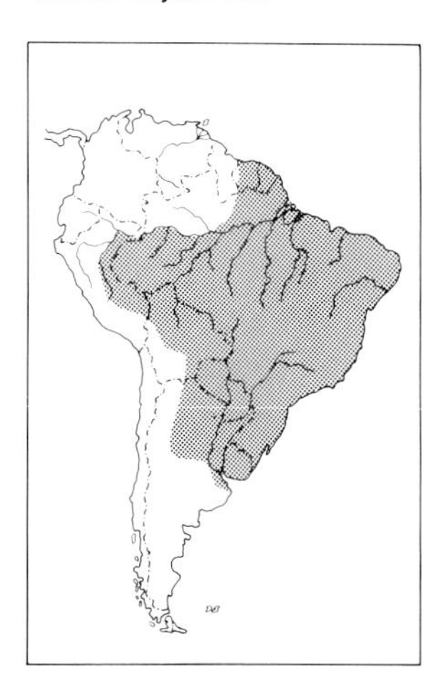

IDENTIFICATION: 20.5–21 cm (8–8¼"). Bill shorter and straighter than Moriche Oriole's (except in nominate race). All black with *shoulders chestnut to tawny* (*pyrrhopterus* and *periporphyrus* of s.-cen. South America). Nominate *cayanensis* (Amazon basin to the Guianas) similar but with *shoulders yellow to ochraceous yellow. Tibialis* (e. Brazil from Maranhão to Rio de Janeiro) has larger ochraceous yellow shoulder patch and yellow thighs and under wing-coverts.

SIMILAR SPECIES: Variable, but can always be known from closely related Moriche Oriole by *absence* of yellow on crown and rump (the 2 overlap only in part of the Guianas and adjacent Brazil; along most of the Amazon, Epaulet is on the s. bank, Moriche on the n. bank). The "epaulet" of the chestnut-shouldered races is quite dark and thus often difficult to see; it therefore can be confused with other all-black icterids (e.g., Velvet-fronted Grackle).

HABITAT AND BEHAVIOR: Uncommon to common (much more numerous southward) in woodland, forest borders, clearings, and gallery woodland in both humid and dry regions. Much less associated with moriche palms than is the Moriche Oriole, even where the palm is present. Behavior similar to that species', though southward it frequently moves in small groups, foraging actively through trees at varying heights, often twitching its tail jerkily. Voice similar to Moriche Oriole's.

RANGE: Guianas (not coastally), Brazil south of the Amazon (north of it only east of Rio Negro), e. Peru (north and west to Amazon and lower Río Ucayali), n. and e. Bolivia, Paraguay, Uruguay, and n. Argentina (south to Tucumán, Córdoba, and n. Buenos Aires). To about 900 m.

GROUP B

N. Am. migrants (only to Colombia and Venezuela); rather small, males black-headed.

Icterus galbula

NORTHERN ORIOLE

Other: Baltimore Oriole

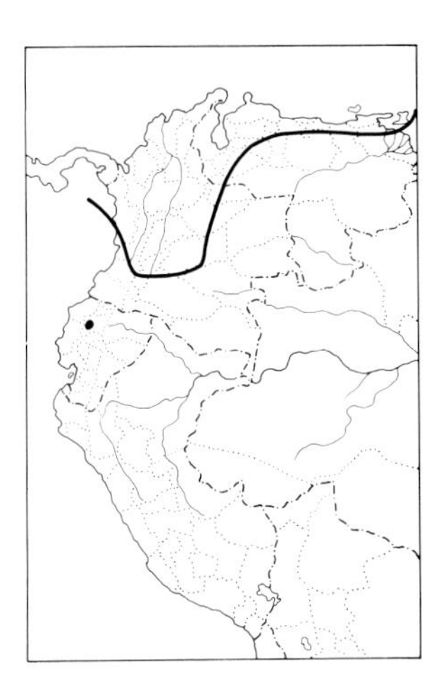

IDENTIFICATION: 19 cm (7½"). Male basically *bright orange and black: entire hood and back black:* wings mostly black with orange shoulders and a single white wing-bar and white edging; tail black with considerable orange on outer feathers; lower back and rump and underparts bright orange. Female brownish olive to brownish gray above, grayest on back; wings dusky with *2 white wing-bars. Mostly yellowish orange below, brightest on breast.* First-winter females are paler below, more whitish on throat and belly, while first-winter males resemble adult females.

SIMILAR SPECIES: Hooded pattern of male is unlike that of any other resident S. Am. oriole except for the larger Troupial (see below). Most resident S. Am. orioles are black-bibbed and none has such distinct wing-bars. Female most resembles the other n.-migrant oriole to South America, the female Orchard, but is much more orange below (not merely greenish yellow).

HABITAT AND BEHAVIOR: Uncommon to locally fairly common n.-winter resident (mostly Oct.–Apr.) in woodland and forest borders and clearings and semiopen areas with scattered trees. Feeds to a large extent in flowering trees, often in loose groups. Males occasionally give a snatch of their song. Especially frequent on n. slopes of Colombia's Santa Marta Mts. (e.g., at Minca).

RANGE: Nonbreeding visitor to n. and w. Colombia (south to Valle) and n. Venezuela (Zulia east to Miranda); single sighting from nw. Ecuador (Pichincha); a few sightings from Trinidad and Tobago. Breeds North America, wintering mostly Mexico to Panama. To about 1500 m.

Icterus spurius

ORCHARD ORIOLE

IDENTIFICATION: 17 cm (6¾"). Male *basically chestnut and black:* entire hood and back black; wings black with chestnut shoulders and a single pale wing-bar and edging; tail black; rump and underparts deep chest-

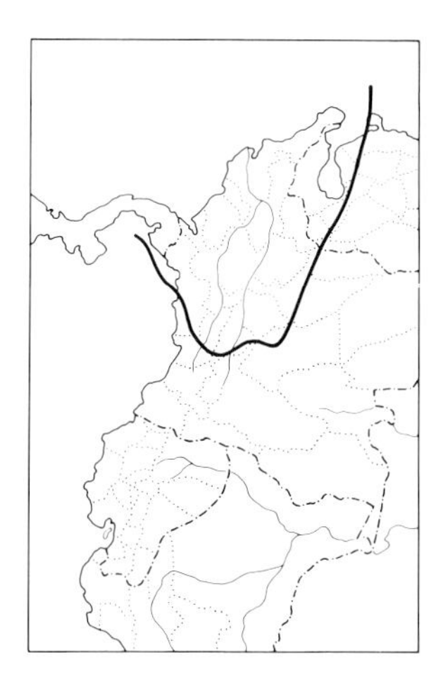

nut. Female olive above, wings duskier with 2 whitish wing-bars; *below greenish yellow*. First-year males like female but with *black throat patch* and often some chestnut patches on breast.

SIMILAR SPECIES: Adult male's dark chestnut color unique among the orioles, though its pattern resembles that of Northern Oriole. The Orchard is a notably small oriole, and this helps to distinguish the dull female; cf. especially Yellow Oriole (much yellower, with only 1 wing-bar, etc.) and Northern Oriole.

HABITAT AND BEHAVIOR: Rare to locally uncommon n.-winter resident (mostly Sept.–Mar.) in semiopen areas with scattered trees, clearings, and woodland borders. Numbers reaching South America are small; search for it especially in flowering *Erythrina* trees, to which Orchard Orioles show a strong affinity (they probe into the blossoms for nectar). Males regularly break into abbreviated versions of their rich, warbled song.

RANGE: Nonbreeding visitor to n. and w. Colombia (south to Valle) and n. Venezuela (Zulia and Aragua). Breeds e. North America, wintering mostly Mexico to Panama. To about 500 m.

GROUP C

Large and bright orange; always with "*shaggy*" *bib* and *pale iris.*

Icterus icterus

TROUPIAL

PLATE: 22

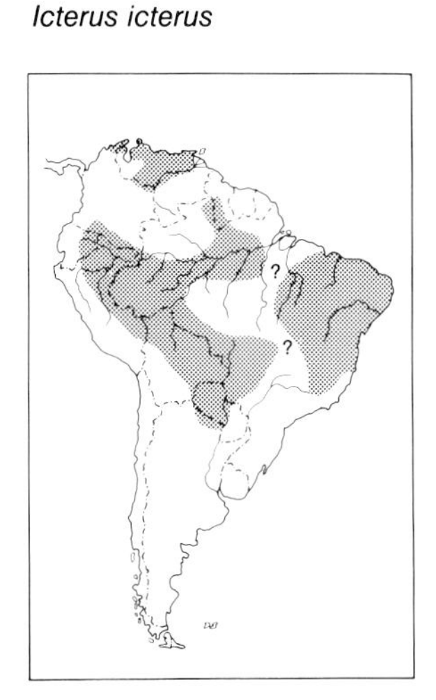

IDENTIFICATION: 23–23.5 cm (9–9¼"). Variable, but always with *yellow iris.* All forms *basically orange and black,* with feathers on black bib pointed on lower margin giving *shaggy effect.* Nominate group (including *ridgwayi*) of Venezuela and adjacent Colombia has a large bare blue ocular patch; *its entire hood is black,* separated from black back by an orange nuchal band, and it has a broad white stripe on wing from coverts to secondaries. *Metae* (sw. Venezuela and adjacent Colombia) is similar but with less black on back and orange encroaching on hindcrown. *Jamacaii* (ne. Brazil) also similar but has bare blue ocular patch greatly reduced or lacking and greater wing-covert patch orange (not white). The more divergent form, though it is geographically intermediate, is *croconotus* (with *stictifrons*), found from Amazonia south to Paraguay and sw. Brazil and north to Guyana; this has *almost completely orange back* (black reduced to the scapulars), with orange nape and crown as well; its bare ocular area is smaller than nominate group's but usually larger than *jamacaii*'s; white on wing reduced to patch on inner secondaries.

SIMILAR SPECIES: Compared to the other S. Am. orioles, the Troupial is larger, always has pale iris (dark in other orioles), and always shows shaggy effect along lower margin of bib (smoothly cut off in other orioles). Full black hood of nominate group and *jamacaii* unique among S. Am. resident orioles, though bibbed pattern of *croconotus* group similar to several others'.

HABITAT AND BEHAVIOR: Fairly common in a variety of lightly wooded, semiopen, or edge habitats; these vary with the form under consideration. Nominate group inhabits the llanos region (semiopen with gallery woodland) and arid desert scrub along the coast; *croconotus* favors the edge of *várzea* forest, margins of oxbow lagoons, and other semiopen swampy areas in Amazonia, but *stictifrons* is found in light woodland often far from water (e.g., in the Paraguayan *chaco*); *jamacaii* is found in both clearings and forest borders (mainly in more humid regions) and in deciduous *caatinga* woodland. Throughout their large range Troupials are found mainly in pairs, foraging at various levels but typically low and not accompanying mixed flocks. Songs of all forms appear to be more or less similar: a series of loud, rich, musical phrases, each often made up of 2 components (e.g., "tree-trur" or "cheer-to") and repeated slowly (often with long pauses) for up to 8 or 10 times. N. birds are known to pirate the large stick nests of Plain Thornbirds, while *croconotus* has been found doing the same to Yellow-rumped Caciques. Frequently kept as a cage bird, and this has probably reduced its numbers in some areas, especially in Venezuela and Brazil. RANGE: Venezuela north of the Orinoco River and ne. Colombia (Guajira Peninsula to se. base of Santa Marta Mts. and in n. Arauca); s. Guyana and adjacent n. Brazil; se. Colombia (Putumayo and Amazonas), e. Ecuador, e. Peru, n. and e. Bolivia, w. and cen. Amaz. Brazil (mostly south of the Amazon; east to vicinity of mouth of lower Rio Tapajós and south to w. Mato Grosso in Paraguay River drainage), w. Paraguay, and extreme n. Argentina (Formosa); ne. Brazil (Maranhão, Ceará, and Pernambuco south and west to Goiás, Minas Gerais, and Bahia); Netherlands Antilles and Margarita Is. off Venezuela. Also introduced to certain of the West Indies. Mostly below 500 m.

NOTE: We continue to recognize all the various forms most often united in *I. icterus* as conspecific, though this may be an oversimplification. There are 3 main groups: the n. nominate group, the primarily Amaz. *croconotus* group (with *stictifrons*), and *jamacaii* of e. Brazil. The first 2 almost appear to have differentiated to full species level based on plumage considerations, but *metae,* though falling clearly within the range of the nominate group, has some seemingly "intermediate" characters. *Jamacaii* returns to resembling the nominate group, though it is geographically more distant. The extent of the bare orbital skin among the 2 s. groups seems to be individually variable (though it is never as extensive as it always is in the n. nominate group). Songs of all 3 forms are substantially similar. The nature of the contact zone (if there is one) between *jamacaii* and the *croconotus* group (*stictifrons*) in s. cen. Brazil remains poorly known, and until it is better understood we consider it preferable to consider all forms as 1 species.

GROUP D

Typical "*black-bibbed*" orioles.

Icterus chrysater

YELLOW-BACKED ORIOLE

PLATE: 22

IDENTIFICATION: 21.5 cm (8½"). *Colombia and Venezuela. Mostly bright golden yellow* with sharply contrasting black facial area and bib, wings, and tail. Yellow plumage often strongly tinged ochraceous, particularly adjacent to the mask and bib; this is apparently individual variation. *Hondae* (known from only 2 specimens from the upper Magdalena valley in Colombia) is more orange-yellow and has slightly longer, slenderer bill basally horn (not bluish) and a little less black over eye.

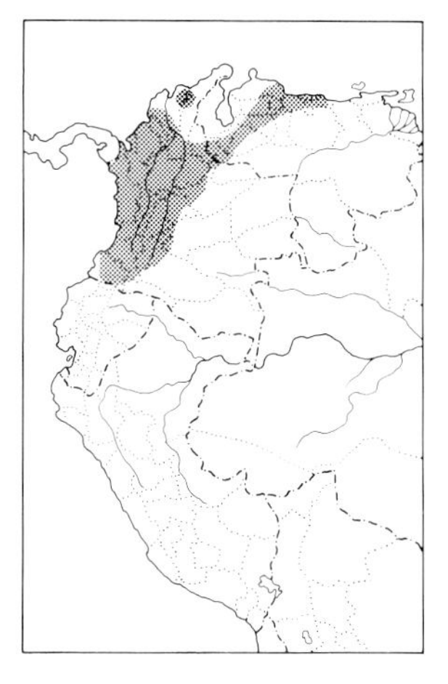

SIMILAR SPECIES: Only other oriole with all-yellow back is Yellow Oriole, which has smaller black bib and white on wing and never shows the ochraceous tone. Also note that *Yellow-backed is the only oriole normally found in the subtropical zone.*
HABITAT AND BEHAVIOR: Fairly common in forest and woodland borders and in clearings with scattered trees; especially prevalent in the foothills and highlands, though also found in deciduous woodland in the lowlands. Usually in pairs or small groups, foraging mostly in middle and upper tree levels, checking foliage for insects and probing into flowers. The loud, clear song is a deliberate series of rich, musical notes, moving up and down in pitch in a random but still pleasant way; it is often delivered from a fairly prominent high perch.
RANGE: W. Colombia (south to Nariño and east of the Andes to the Macarena Mts. in Meta; largely absent from Pacific lowlands) and n. Venezuela (east to Miranda, Portuguesa, and w. Apure). Also Mexico to Panama (except Costa Rica). To about 2500 m, occasionally higher.

Icterus nigrogularis

YELLOW ORIOLE

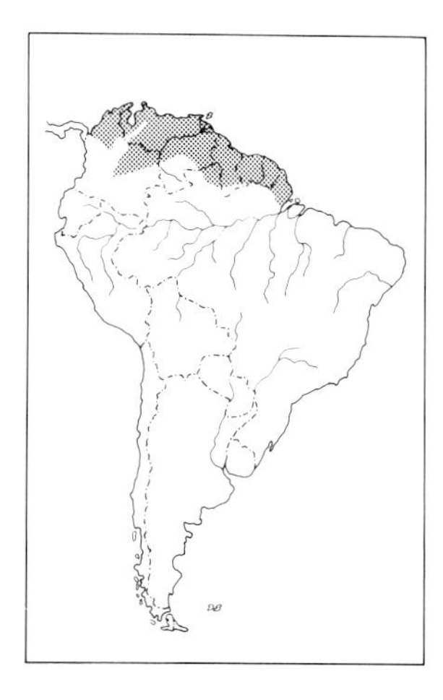

IDENTIFICATION: 20.5 cm (8″). Overall pattern reminiscent of slightly larger Yellow-backed Oriole (including the *yellow back*), but can at once be known by presence of *white on wing* (a narrow but distinct wing-bar and edging to inner flight feathers; less white on wing in *trinitatis* of ne. Venezuela and Trinidad). *Black bib less extensive,* separated from black ocular area by yellow. Overall tone *more clear yellow,* lacking any ochraceous tones. Immature more greenish yellow (particularly above, where it starts out almost olive), younger birds almost lacking the bib but already with at least some white on wing. See V-38, C-44.
SIMILAR SPECIES: Immature could be confused with female or immature male Orchard Oriole, though latter is notably smaller, etc.
HABITAT AND BEHAVIOR: Fairly common to common in arid scrub, deciduous and gallery woodland, gardens, and even the edge of mangroves; tends to avoid humid areas at least in Colombia and Venezuela (where at least to some extent replaced by Yellow-backed Oriole; the 2 are normally not sympatric). Usually in pairs or small family groups, generally conspicuous in their semiopen habitats. The song is a rich, musical phrase or series of phrases, usually with the first note separated and higher-pitched than the others.
RANGE: N. and ne. Colombia (arid Caribbean lowlands south in Río Magdalena valley to Santander and in ne. llanos region south to Meta and Vichada), Venezuela (except extreme south), Guianas, and n. Brazil (Roraima and Amapá); Trinidad and other islands off Venezuela (including Netherlands Antilles). Mostly below 500 m.

Icterus graceannae

WHITE-EDGED ORIOLE

PLATE: 22

IDENTIFICATION: 20.5 cm (8″). *W. Ecuador and nw. Peru.* Mostly golden yellow with black facial area and bib, back, wings (aside from

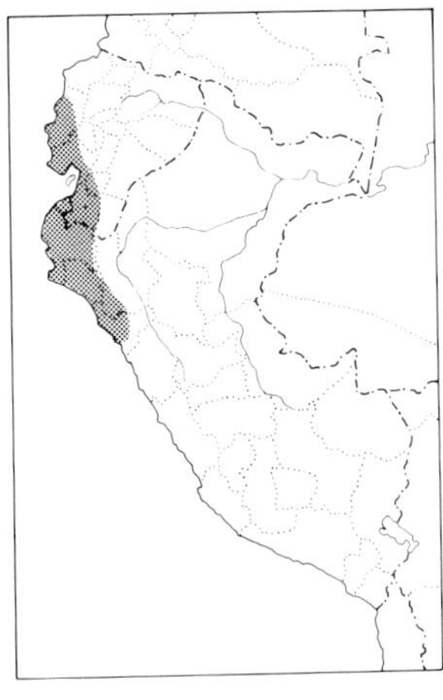

yellow lesser wing-coverts), and tail. Wings with white patch on inner flight feathers, and *outer tail feathers edged and tipped white.*

SIMILAR SPECIES: Frequently confused with slightly larger Yellow-tailed Oriole, the sympatric (or nearly so) race of which *also shows a white patch on wing;* it, however, has *outer tail feathers yellow* (with no white), resulting in different pattern.

HABITAT AND BEHAVIOR: Uncommon to common (more numerous from s. El Oro, Ecuador, south into Peru) in deciduous woodland and adjacent desert scrub and in riparian thickets. Usually in more arid areas than Yellow-tailed Oriole. Often in pairs, but general behavior not different from other orioles'. A frequently heard call is a throaty "jori-jori" or "cheerwik, cheerwik," characteristically doubled.

RANGE: W. Ecuador (north to Manabí and inland across s. Loja) and nw. Peru (south to La Libertad). Mostly below 300 m, occasionally somewhat higher.

Icterus mesomelas

YELLOW-TAILED ORIOLE

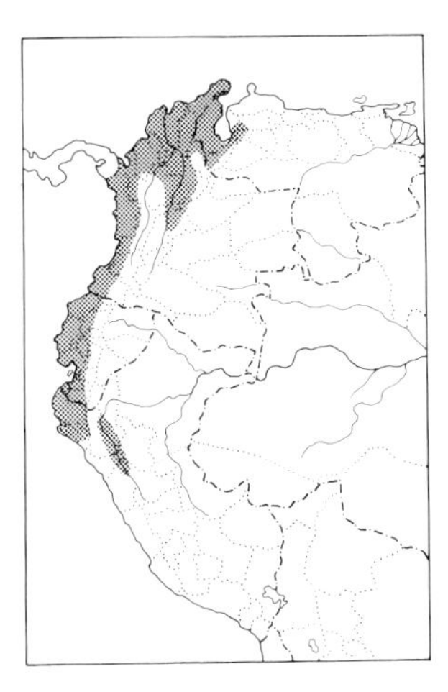

IDENTIFICATION: 21.5 cm (8½"). Resembles White-edged Oriole, but all races have *outer tail feathers mainly or entirely yellow* (with *no white*). S. race *taczanowskii* (w. Ecuador and nw. Peru) has *conspicuous white patch plus white edging on inner flight feathers,* with yellow wing-band reaching back to the white (in White-edged the yellow does not extend as far back, so the yellow and white are *not* in contact). In more n. race *carrikeri* (w. Colombia and w. Venezuela) no white is present on wing; see V-38, P-26.

SIMILAR SPECIES: Yellow in tail and obvious black back quickly distinguish Yellow-tailed from Yellow-backed Oriole. Cf. also Orange-crowned Oriole.

HABITAT AND BEHAVIOR: Fairly common in clearings, second growth, and thickets almost always near water. Most often in pairs, foraging at all heights but typically not very high. Usually conspicuous, and with loud, rich song of whistled phrases. In sw. Ecuador more numerous than White-edged Oriole (which is more restricted to arid areas), often even perching and singing from roadside wires.

RANGE: W. Venezuela (Lake Maracaibo basin in Zulia, Táchira, and Mérida), n. and w. Colombia (east to E. Andes and south in Magdalena valley to Cundinamarca), w. Ecuador, and nw. Peru (south on Pacific coast to Lambayeque and in the upper Marañón valley south to La Libertad). Also Mexico to Panama. Mostly below 500 m.

Icterus auricapillus

ORANGE-CROWNED ORIOLE

IDENTIFICATION: 20.5 cm (8"). *N. Colombia and Venezuela. Most of crown, nape, and sides of head and neck fiery orange;* forehead and broad bib black. Back, wings, and tail also black, with lesser wing-coverts yellow. Rump and entire underparts rich golden yellow. See V-38, P-26.

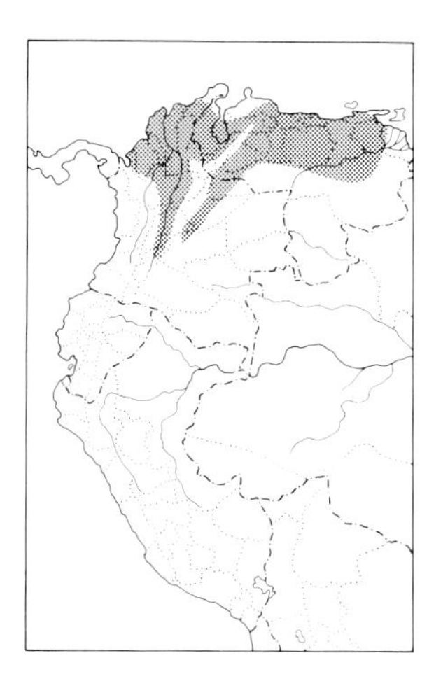

SIMILAR SPECIES: No other S. Am. oriole has the contrastingly orange crown (be aware that intensity of this orange does vary). Orange-crown's overall pattern not dissimilar from Yellow-tailed Oriole's, though lacking the latter's yellow in the tail.
HABITAT AND BEHAVIOR: Uncommon to fairly common in deciduous woodland, forest borders, and clearings with scattered trees; in general a bird of humid regions (in arid areas scarcer and always near water-courses). Usually in pairs, with behavior not differing appreciably from other orioles', though in some areas it does show a predilection for palms, from the fronds of which it may suspend its nest.
RANGE: N. Colombia (Caribbean lowlands south in Magdalena valley to Huila; east of Andes in Arauca and w. Meta) and n. Venezuela (widespread north of the Orinoco east to Sucre, south of it only in n. Bolívar). Also e. Panama. Mostly below 800 m.

Amblycercus Caciques

An all-black cacique which skulks in lower growth of woodland borders and thickets and in montane forest. We follow the 1983 AOU Check-list (but not *Birds of the World,* vol. 14) in maintaining *Amblycercus* as a distinct, monotypic genus, mostly because its nest is so different from the suspended pouches of all the *Cacicus* caciques.

Amblycercus holosericeus

YELLOW-BILLED CACIQUE

PLATE: 23

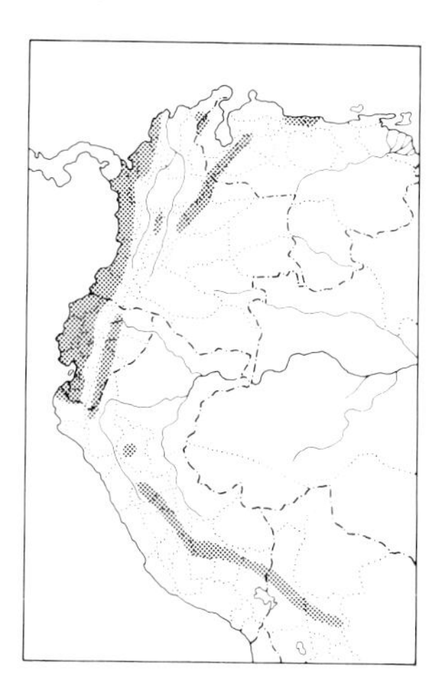

IDENTIFICATION: Male 23–23.5 cm (9–9¼"); female 22 cm (8½"). *Bill and iris pale yellow.* Entirely dull black.
SIMILAR SPECIES: General appearance and behavior very similar to Solitary Cacique's, but latter has dark (not pale) iris. So far the 2 species do not seem to have ever been found at the same locality (Solitary being mainly s. and Amaz., Yellow-billed nw. and Andean); *Yellow-billed is entirely montane except in Pacific nw. South America.*
HABITAT AND BEHAVIOR: Rare to fairly common (more numerous and widespread northward) in dense undergrowth of clearings, woodland and forest borders, and (from at least e. Ecuador south) *Chusquea* bamboo thickets of montane forest and forest borders. Usually in pairs but a great skulker, only rarely coming into the open or rising up into trees (the latter most often to visit flowers), and often seen only as it flies low across an opening or trail. In Peru sometimes joins mixed understory flocks of *Basileuterus* warblers, etc. (D. Stotz). More often recorded once its distinctive loud voice is recognized: there may be geographic variation, but nw. lowland birds (at least) most frequently give a variety of piercing clear whistles, often doubled (e.g., "whew-whew, whew-whew . . . ," often echoed by the female's single "wheee? churrr").
RANGE: Locally in lowlands and lower montane slopes (to about 2000 m) of n. and w. Colombia, w. Ecuador, and extreme nw. Peru (Tumbes); locally in montane areas (mostly 1800–3000 m) of n. and w. Venezuela

(coastal mts. in Aragua and Distrito Federal, Perijá Mts., and Andes from Mérida south), Colombia in E. Andes (Norte de Santander and Cundinamarca) and Cen. Andes (Caldas and Tolíma), se. Ecuador (north to Morona-Santiago), e. Peru (recorded at least from San Martín, Huánuco, Cuzco, and Puno), and nw. Bolivia (La Paz and Cochabamba). Also Mexico to Panama. To 3300 m (in Bolivia).

Cacicus Caciques

Rather large, mostly black icterids, many species with bright yellow or red on rump and sometimes yellow on wing. Bill pale, long, and pointed and iris pale (most often blue) in all but 1 species (the rather aberrant Solitary). Caciques are smaller than oropendolas and never show so much yellow on the tail; the 2 genera regularly consort with each other, and some caciques even nest in association with oropendola colonies (usually in adjacent trees, but sometimes in the same one). Most are conspicuous, ranging in noisy groups through the forest canopy, the main exception again being the Solitary, which skulks through undergrowth. The genus is wide-ranging in South America, most numerous in the lowlands.

GROUP A *All black.*

Cacicus solitarius

SOLITARY CACIQUE PLATE: 23

Other: Solitary Black Cacique

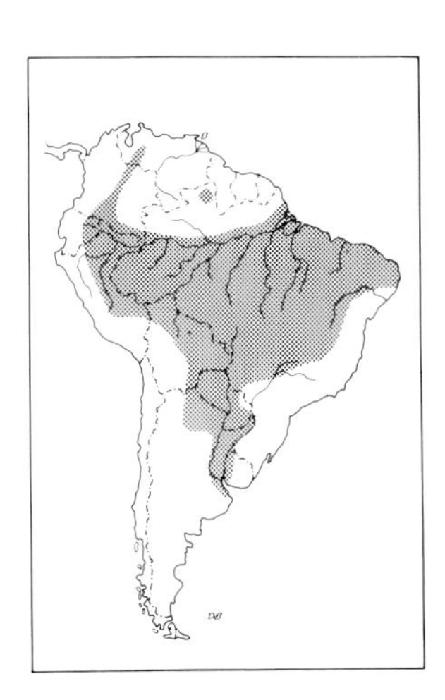

IDENTIFICATION: Male 27 cm (10½"); female 23 cm (9"). *Bill pale greenish; iris brown* (birds with pale irides are occasionally seen: juvenals?). *Entirely black.*

SIMILAR SPECIES: Similar to Yellow-billed Cacique but larger with dark (not pale) eye; their ranges do not overlap. Ecuadorian Cacique is likewise similar; its iris is pale blue, and its habits differ markedly. Red-rumped Cacique's red rump is often hidden when the bird is perched; its eye also is blue. Can even be confused with Band-tailed Oropendola, for the latter's yellow in the tail is often hidden.

HABITAT AND BEHAVIOR: Uncommon to fairly common in rank lower growth of woodland and forest borders, especially near water (e.g., in gallery woodland and *várzea* borders along rivers and lakeshores); in s. part of range also occurs in lower growth of drier woodland (e.g., in the *chaco* of w. Paraguay). Behavior similar to Yellow-billed Cacique's, being very much a skulker, though occasionally rising up into more open middle levels of trees. Often probes into leaves or under loose pieces of bark. Its voice is often the best clue to its presence: it is varied and often quite strange-sounding, with a loud "wheeeah" being most frequent (similar to a common call of the Black-capped Donacobius), but other repeated nasal or ringing notes (e.g., "kway-kway") are also given, often by both members of a pair simultaneously or in response to each other.

RANGE: Sw. Venezuela (Táchira, Barinas, and w. Apure), e. Colombia (along e. base of Andes and east into Amazonas), e. Ecuador, e. Peru, n. and e. Bolivia, most of Amaz. and interior Brazil (north of the Amazon mainly along its n. bank east into Amapá, but also reported recently from Maracá Ecol. Station in Roraima; absent from e. coastal area south of Pernambuco), Paraguay (local in east), ne. Argentina (south to Córdoba, Santa Fe, and n. Buenos Aires), and w. Uruguay. Mostly below 500 m, but to 800 m in Cuzco, Peru (*fide* D. Stotz).

NOTE: The "Black" in the English name of Meyer de Schauensee (1966, 1970) seems superfluous.

Cacicus sclateri

ECUADORIAN CACIQUE

Other: Ecuadorian Black Cacique

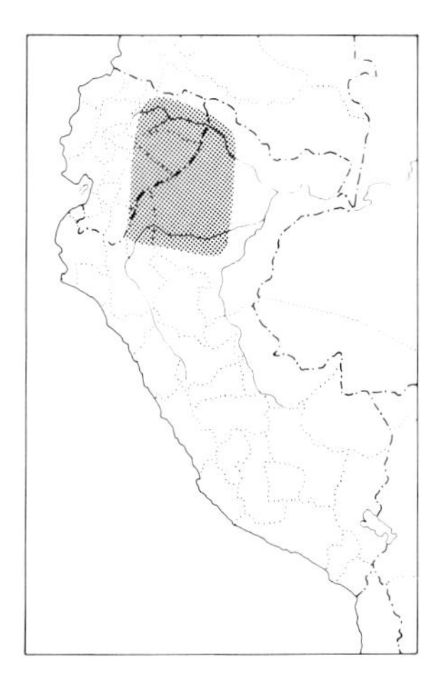

IDENTIFICATION: Male 23–23.5 cm (9–9¼"); female 19–19.5 cm (7½–8"). *E. Ecuador and adjacent Peru.* Bill pale bluish gray; *iris pale blue.* All black.

SIMILAR SPECIES: Superficially resembles Solitary Cacique (and the 2 are sympatric), but notably smaller and with light blue (not dark brown) iris; behavior also differs markedly (see below).

HABITAT AND BEHAVIOR: Not well known, but from the few specimens and sightings apparently rare (locally more numerous?) in canopy and middle levels of forest borders and woodland. Birds seen near Limoncocha in e. Ecuador foraged as pairs in a manner reminiscent of Scarlet-rumped Caciques (e.g., well up in trees, often probing into epiphytes and the bases of leaves).

RANGE: E. Ecuador (recorded only from limited area in upper Río Napo drainage, but probably more widespread) and adjacent n. Peru in Amazonas (Huampami) and n. Loreto (mouth of Río Curaray into the Napo; Río Samiria in Pacaya-Samiria Reserve, ANSP specimen). 200–600 m.

NOTE: The "Black" in the English name of Meyer de Schauensee (1966, 1970) seems superfluous.

GROUP B *Rump scarlet-red.*

Cacicus uropygialis

SCARLET-RUMPED CACIQUE

PLATE: 23

IDENTIFICATION: Male 28–30.5 cm (11–12"), female 24 cm (9½") in nominate race of Andes and Perijá Mts., but markedly smaller (male 23–24 cm [9–9½"], female 20.5–21.5 cm [8–8½"]) in *pacificus* of w. Colombia and w. Ecuador. Bill pale yellowish green or yellowish white; iris pale blue. Entirely black with *scarlet rump* (the rump usually hidden on perched birds but conspicuous in flight). *Pacificus,* in addition to its smaller size, differs also in its proportionately shorter tail and less heavy bill; see C-44, P-26.

SIMILAR SPECIES: *The only cacique with a red rump found in either subtropical Andean forests or in lowlands west of the n. Andes;* note that the 2

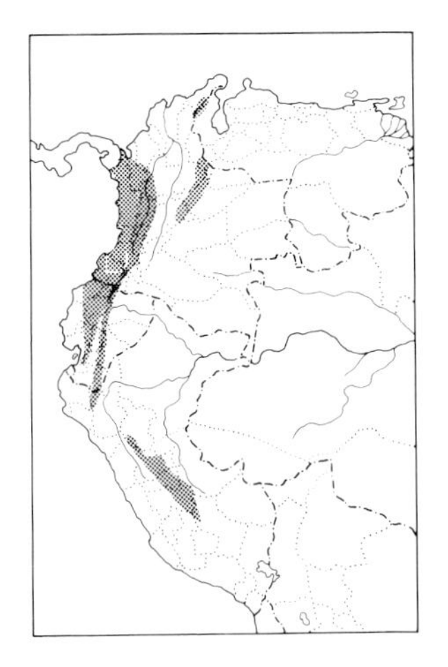

forms replace each other altitudinally and likely represent distinct species. Red-rumped Cacique's range may come locally into contact with that of Scarlet-rump on e. slope of Andes, but Red-rumped differs in having blue gloss to its black plumage and red on rump extending much farther up its back. Cf. also Yellow-billed Cacique: perched Scarlet-rumps with their red concealed can look confusingly similar, but note Yellow-bill's yellow iris and its typically different behavior and habitat.

HABITAT AND BEHAVIOR: Uncommon to locally fairly common in canopy and borders of humid subtropical (nominate race) or lowland (*pacificus*) forest and mature second-growth woodland; *pacificus* is usually more numerous than nominate, which seems particularly scarce in Peru. *Pacificus* seems commonest in foothill areas (500–700 m), but even so, at least on the w. slope of Colombia's W. Andes, there is usually an altitudinal gap between its range and that of nominate race on subtropical slopes above. Nominate race usually moves about in small groups of from 3 to 6 birds which forage through middle and upper forest levels, often accompanied by jays, tanagers, etc.; *pacificus,* on the other hand, often forages in larger groups (up to 20 or so birds) and is most often with oropendolas, fruitcrows, toucans, etc. They frequently hop along horizontal limbs, peering among foliage or probing into bromeliads; some fruit is also consumed. The commonest calls of the nominate race include a loud, ringing, jaylike "greer!" often given in series, and a liquid "wurt-wurt-wurt-wurt-wurt-wurt," sounding much like one of the calls of the donacobius. None of its calls much resemble any of the vocalizations of *pacificus,* which gives a variety of loud, musical, whistled or slashing calls, including a "treeo, trew-trew-trew-trew!"

RANGE: Andean slopes of extreme w. Venezuela (w. Táchira), Colombia (in W. Andes south only to Valle), e. Ecuador, and e. Peru (south to Cuzco); Perijá Mts. on Venezuela-Colombia border; Pacific lowlands of w. Colombia (east around n. base of W. Andes to lower Cauca River valley) and w. Ecuador (south to El Oro). Also Nicaragua to Panama. Mostly 1300–2300 m (nominate) or mostly below 800 m, occasionally somewhat higher (*pacificus*).

NOTE: We strongly suspect that the 2 S. Am. forms of this species will ultimately be demonstrated to be distinct species; morphological differences, as well as differences in their vocalizations and the gap in their altitudinal ranges, all seem to so indicate. Should this eventually be shown, we would urge that *uropygialis* (*sensu stricto*) be given the English name of Subtropical Cacique in order to draw attention to its unique range among the red-rumped caciques. This would leave Scarlet-rumped Cacique available as the name for the better-known lowland-inhabiting *C. microrhynchus* (of which *pacificus* would be considered a race; note that it has also been suggested that *pacificus* itself may merit species status as the Pacific Cacique).

Cacicus haemorrhous

RED-RUMPED CACIQUE

IDENTIFICATION: Male 27–29.5 cm (10½–11½"); female 21.5–24 cm (8½–9½"). Iris usually blue, but brown in young birds and (?) some

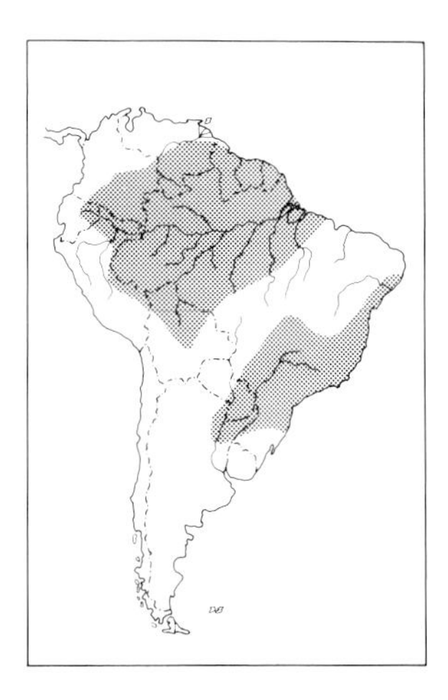

females. In general pattern resembles Scarlet-rumped Cacique. Differs in its even longer, "sharper" bill; its glossier blue-black general coloration; greater extent of its scarlet rump (extending to lower back); and its proportionately somewhat shorter and squarer tail. See C-45. *Affinis* (e. Brazil and adjacent areas) resembles nominate *haemorrhous* but is less glossy and often is quite sooty or even brownish black (particularly below and in females).

SIMILAR SPECIES: Red-rumped Cacique has at most limited overlap with Scarlet-rumped Cacique along e. base of Andes only. Though this species' red rump is more extensive than in the others, it nonetheless is often hidden at rest, causing potential confusion with Solitary Cacique (then note latter's dark, not pale blue, iris).

HABITAT AND BEHAVIOR: Varies from rare to common (tending to be scarce westward in Amazonia, but much more numerous in the Guianas and the southeast) in canopy and borders of *terra firme* forest through the Guianas and Amazonia (where rather strictly avoids *várzea*) and in forest and woodland (including deciduous forest and gallery woodland) through the southeast. Usually in flocks and often conspicuous in various forest edge habitats; for example, very numerous around Iguaçu Falls. Vocalizations include a wide variety of harsh and guttural calls, including a sharp "gwap," often repeated, frequently given in flight. Nesting is colonial, usually in slightly isolated trees at forest edge, generally (unlike Yellow-rumped Cacique) not in association with other icterids.

RANGE: Guianas, s. Venezuela (Bolívar and Amazonas), e. Colombia (north to Meta and Vaupés, probably to at least Guainía), e. Ecuador (only a few records), locally in e. Peru and in n. Bolivia (south to w. Santa Cruz), and Amaz. Brazil (south to n. Mato Grosso and east to e. Pará in the Belém area); coastal e. Brazil (Pernambuco southward and inland to s. Goiás and se. Mato Grosso), e. Paraguay, and ne. Argentina (south to ne. Santa Fe and Corrientes). To about 1000 m.

GROUP C

Rump yellow (often a wing-patch as well).

Cacicus cela

YELLOW-RUMPED CACIQUE

PLATE: 23

IDENTIFICATION: Male 27–29.5 cm (10½–11½"); female 23–25 cm (9–9¾"). Bill pale greenish yellow; iris pale blue. Mainly glossy black with large bright yellow patch on inner wing-coverts; *rump, crissum, and basal third of tail also bright yellow.* The 2 races from west of the Andes (*flavicrissus* and *vitellinus*) differ in having less yellow at base of tail, and *vitellinus* (n. Colombia) differs further in having a smaller yellow wing-patch. Females of all subspecies are a duller, sootier black.

SIMILAR SPECIES: *By far the commonest and most widespread yellow and black cacique in South America.* Learn it well as a basis of comparison with the others in this color group (only 1 of which, the very rare and localized Selva Cacique, ever occurs sympatrically with it). Unlike the

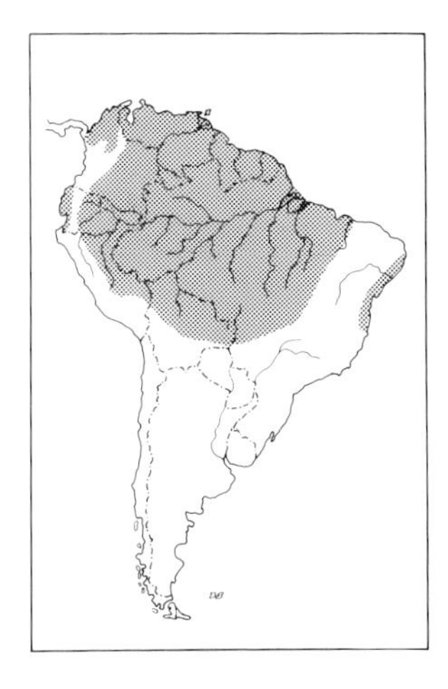

caciques with red rumps, the yellow on this species *always* shows when the bird is perched, rendering confusion with other all-black icterids unlikely. Cf. also Band-tailed Oropendola.

HABITAT AND BEHAVIOR: Usually common (especially east of the Andes) in forest borders (especially *várzea*), woodland, and various semiopen situations with scattered trees (often found even in towns and villages). Very gregarious and conspicuous (in many areas one of the most often seen birds), foraging at all levels, though infrequent low inside forest or woodland. A colonial nester, colonies sometimes situated in the same tree with oropendolas but more often in an adjacent tree or alone; groups of the pouch-shaped nests (shorter than oropendolas') are often clustered together, frequently near wasp nests. Has a variety of loud calls, some harsh, some rather liquid; interesting is the fact that nominate *cela* is an excellent and frequent mimic of other birds, though birds from west of the Andes are not known to be.

RANGE: Widespread east of Andes south to n. and e. Bolivia and cen. Brazil (Mato Grosso, Goiás, and Ceará); coastal ne. Brazil (Pernambuco to se. Bahia); west of Andes only in nw. Venezuela, n. Colombia (*not* the Pacific lowlands), and w. Ecuador (Manabí to El Oro); Trinidad. Also Panama. To about 900 m.

Cacicus chrysopterus

GOLDEN-WINGED CACIQUE

PLATE: 23

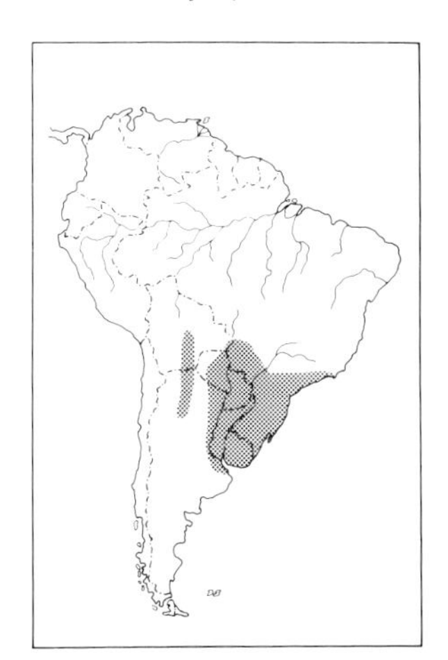

IDENTIFICATION: Male 20.5 cm (8"); female 18 cm (7"). A *small* cacique of *s. South America*. Iris bluish to yellowish white; bill pale bluish horn. Black with *yellow wing-coverts* and *yellow rump*. At times a slight shaggy-crested effect is apparent.

SIMILAR SPECIES: Only marginally sympatric with any other cacique having yellow in plumage. Yellow-rumped Cacique is larger with yellow on tail, etc.; s. race of Mountain Cacique lacks yellow on wing. Because of its small size, more likely to be mistaken for an oriole; however, sympatric races of Epaulet Oriole have chestnut (not yellow) on wing-coverts. Yellow-winged Blackbird is restricted to marshes, lacks yellow on rump, and shows less yellow on wing, etc.

HABITAT AND BEHAVIOR: Uncommon to locally common in deciduous and gallery woodland and locally also in more humid forest and forest borders. Moves about in pairs or small groups, often foraging by probing into epiphytes or under bark with its bill; also comes to flowering trees and eats some fruit. A solitary nester, the pendulous bag nest suspended from the tips of branches. Gives a variety of calls, the most frequent and characteristic being a nasal or mewing "wreyur" and a loud, ringing "gloo-gloo-gloo, gleéyu."

RANGE: E. slope of Andes in s. Bolivia (w. Santa Cruz south) and nw. Argentina (south to Tucumán); s. Brazil (north to s. Mato Grosso, São Paulo, and Rio de Janeiro), most of Paraguay (except the extreme west; more local westward into the *chaco*), Uruguay, and ne. Argentina (Formosa south to n. Buenos Aires; now rare in the latter). To about 2000 m (in Bolivia).

Cacicus koepckeae

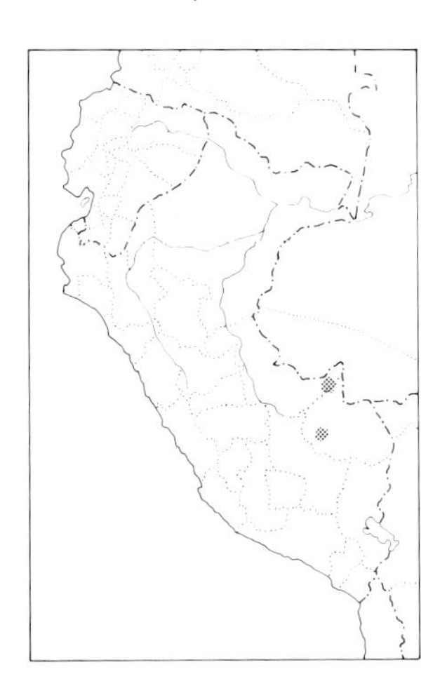

SELVA CACIQUE

IDENTIFICATION: 23 cm (9″). *Se. Peru lowlands* (where known from only 1 locality). Iris bluish white; bill bluish gray with paler tip. All black, with *only rump yellow.*

SIMILAR SPECIES: Yellow-rumped Cacique, which *vastly* outnumbers this species, has yellow on tail and wing-coverts (as well as rump).

HABITAT AND BEHAVIOR: Only described in 1965, this cacique is still known from only 2 specimens, both taken at the type locality. Virtually unknown in life. J. O'Neill's sole encounter with it consisted of a sighting of 6 individuals which were apparently bathing in a pool along a stream; when 1 was collected, the other 5 retreated into a *Heliconia* thicket. Despite this, he suspects that the Selva Cacique is an arboreal bird occurring mostly at forest borders; it may be most closely related to the Ecuadorian Cacique.

RANGE: Se. Peru (Balta on the Río Curanja in se. Ucayali; also a possible sighting from Manu Nat. Park in Madre de Dios). 300 m.

Cacicus leucoramphus

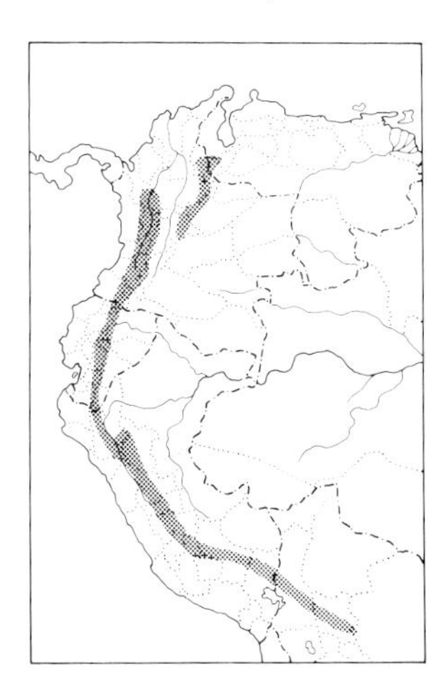

MOUNTAIN CACIQUE

IDENTIFICATION: Male 30.5 cm (12″); female 25.5 cm (10″). *Montane forests in Andes* (where it occurs higher than any other cacique except Yellow-billed). Iris pale blue; bill pale bluish horn, more yellowish or pale olive toward tip. All black, with *yellow wing-coverts and rump.* Female slightly sootier (not as glossy black); see C-45. *Chrysonotus* (north to s. Peru in Cuzco) *lacks yellow on wing,* thus appearing very different; a few individuals show some yellow fringing on feathers of wing-coverts.

SIMILAR SPECIES: Yellow-rumped Cacique, which normally occurs at much lower elevations (though they might occasionally overlap), has yellow on crissum and a shorter tail. Scarlet-rumped Cacique narrowly overlaps with s. race of this species in Cuzco, Peru (though even here they separate by altitude, Mountain replacing Scarlet-rumped at higher elevations); as this race of Mountain lacks yellow on wing, it and Scarlet-rumped resemble each other closely until their rump color (often hidden) is revealed.

HABITAT AND BEHAVIOR: Uncommon to locally fairly common in humid montane forest, primarily in temperate zone. Seems more numerous southward (e.g., the s. race can be regularly found in the *yungas* of La Paz, Bolivia). Usually forages in pairs (less often singly or in small groups) which range through the canopy and middle forest levels; they often accompany jays or larger tanagers (e.g., Hooded Mountain-Tanagers). Quite noisy, with a variety of arresting calls, typically doubled (e.g., "kay-kay").

RANGE: Andes of extreme sw. Venezuela (Táchira), Colombia (where rare and local, especially northward), e. Ecuador, e. Peru, and nw. Bolivia (La Paz and Cochabamba). Mostly 1800–3000 m.

NOTE: There are 2 distinctly different, allopatric groups within this Andean species; the n. nominate group (with *peruvianus*) and the s. *chrysonotus* of s. Peru and Bolivia. Meyer de Schauensee (1966, 1970) treated them as conspecific, but in *Birds of the World* (vol. 14, p. 147) they were considered full species. We have opted for the former course, mainly because vocalizations of the 2 forms seem similar, and because some *chrysonotus* specimens do show slight yellow tipping on the wing-coverts (as noted by Blake in *Birds of the World* and also shown in certain ANSP specimens), presumed to be an indication of intergradation with *peruvianus*. However, it should be noted that some Bolivian specimens (far from the zone of potential contact) also show this. Should they ultimately best be treated as distinct species, we would suggest that the best English names are "Northern Mountain-Cacique" for the *leucoramphus* group and "Southern Mountain-Cacique" for *chrysonotus,* so as to better indicate their close relationship.

Ocyalus Oropendolas

A still little-known oropendola of river islands and *várzea* forest along the upper Amazon and some of its major tributaries. This monotypic genus seems intermediate between the true oropendolas and the caciques, and its velvety plumage texture is unique among oropendolas, if not among icterids in general. Its nesting behavior appears to be unrecorded.

Ocyalus latirostris

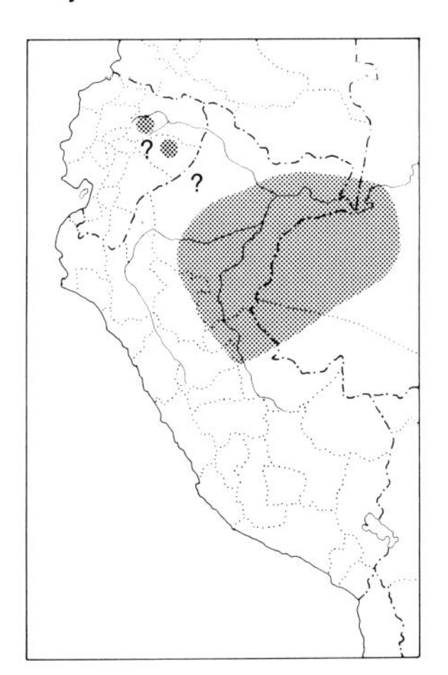

BAND-TAILED OROPENDOLA

PLATE: 23

IDENTIFICATION: Male 32–34.5 cm (12½–13½"); female 25.5 cm (10"). Limited range in *upper Amazonia.* Iris blue; small frontal shield and maxilla black tipped whitish, mandible yellowish white. *Mostly velvety black,* glossier and bluer on wings; though hard to see in the field, crown, hindneck, and upper back are dark chestnut. Tail pattern distinctive, but also often hard to see in the field (tail often looks all dark from above): *outer feathers mostly yellow, leaving black band across tip.*
SIMILAR SPECIES: Though an oropendola, this rare icterid is more likely to be mistaken for a cacique. Yellow-rumped Cacique (with which it often occurs) has yellow on wing and rump, and bill and tail pattern differ. Because Band-tail's tail pattern is often so hard to see, the birds may look all dark, and it then can be confused with Solitary Cacique (but note different soft-part colors, etc.).
HABITAT AND BEHAVIOR: Very local and seemingly uncommon to rare (possibly locally more numerous) on river islands and in *várzea* forest borders. Usually occurs in small groups (occasionally up to 20 or so birds) which frequently move about with Russet-backed Oropendolas and Yellow-rumped Caciques, seeming always to be outnumbered by them. Voice reminiscent of Solitary Cacique (V. Remsen).
RANGE: Ne. Peru (Loreto and Ucayali south to around Pucallpa), extreme se. Colombia (s. Amazonas in Leticia area), and extreme w. Brazil (upper Rio Juruá); 2 old records from e. Ecuador (Archidona and Sarayacu), but no recent reports. Mostly (entirely?) below 300 m.

Psarocolius Oropendolas

Large to very large icterids (males are among the largest of all passerine birds) whose basic color varies from bright olive through chestnut to black, but always with a mostly yellow tail. As in the caciques, bills are sharply pointed and pale, while irides are usually pale blue. Oropendolas are noisy and conspicuous birds, ranging through the forest canopy in often large flocks and sometimes gathering in immense congregations to roost. All species are colonial breeders, their long, pendulous nests hanging from an isolated tree and forming one of the characteristic features of the humid tropical zone lowlands. Taxonomy at the generic level is still disputed. We opt to follow the 1983 AOU Check-list in considering *Zarhynchus* and *Gymnostinops* as congeneric with *Psarocolius;* behavior, overall appearance, and nest shape are all basically similar. The formerly monotypic *Clypicterus* is no more divergent from *Psarocolius* than are *Zarhynchus* and *Gymnostinops,* and we thus feel that it too should be sunk into *Psarocolius.*

GROUP A

Smaller oropendolas, with swollen casque on bill.

Psarocolius oseryi

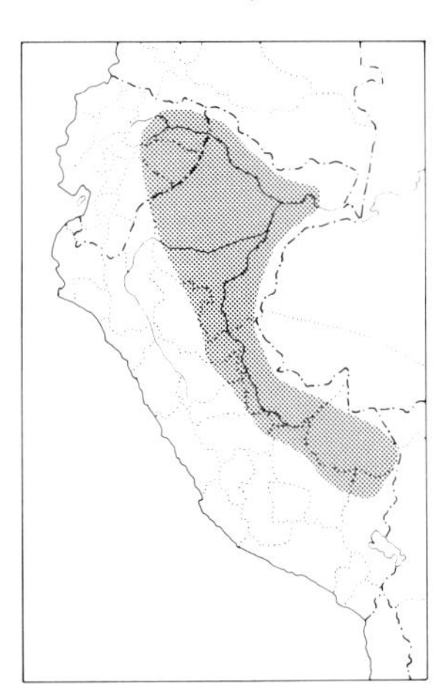

CASQUED OROPENDOLA

PLATE: 23

IDENTIFICATION: Male 36–38 cm (14–15″); female 28–29.5 cm (11–11½″). *E. Ecuador and e. Peru.* Iris blue or brown; *conspicuous frontal shield* (swollen on forehead in male) and bill yellow tipped dusky. *Mainly chestnut,* with *yellowish olive throat and breast.* Central tail feathers dusky olive, with outer feathers bright yellow.

SIMILAR SPECIES: Most likely to be confused with Olive or Green Oropendolas, both of which are larger with brightly colored soft-parts (bill and/or facial area), lack the casque, and are not as uniformly chestnut above. Cf. also Russet-backed Oropendola (especially its yellow-billed *alfredi* race), which is much more a bird of semiopen areas (not forest).

HABITAT AND BEHAVIOR: Uncommon to locally fairly common in canopy of humid *terra firme* forest, to a lesser extent also in *várzea;* rather infrequent in clearings or at edge, where usually only in flight. Usually in small groups, most often independent of other oropendolas or caciques; larger numbers congregate at nesting colonies. Flight tends to be rather undulating, in this respect rather caciquelike and differing from the steady rowing flight of most other *Psarocolius.* Males in display give a startlingly long "squaaaaaa-ooók"; at other times a variety of other typical but loud oropendola calls are uttered. Regularly seen at the Tambopata Reserve in Madre de Dios, Peru.

RANGE: E. Ecuador (where local and seemingly scarce) and e. Peru (Loreto south to Puno); must also occur in adjacent w. Brazil and n. Bolivia. Mostly below 400 m, locally to 750 m in Peru (*fide* D. Stotz).

NOTE: Formerly placed in the genus *Clypicterus.*

Psarocolius wagleri

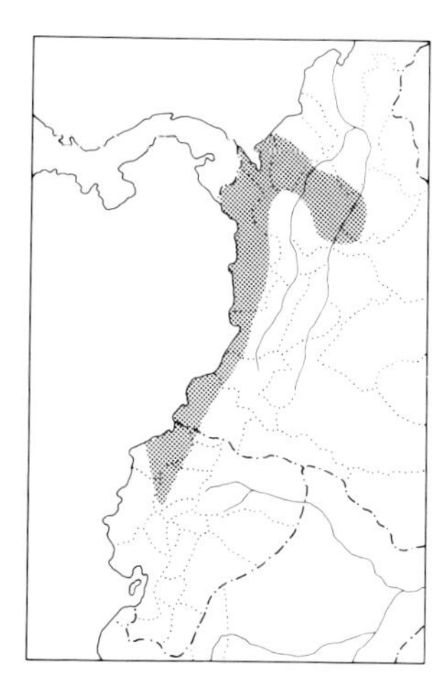

CHESTNUT-HEADED OROPENDOLA

IDENTIFICATION: Male 34–36 cm (13¼–14"); female 27–28 cm (10½–11"). *W. Colombia and nw. Ecuador.* Bill ivory to pale greenish yellow, often with dusky tip; iris pale blue; inconspicuous hairlike crest in male. Overall plumage somewhat reminiscent of Band-tailed Oropendola, while bill shape recalls that of Casqued, though frontal shield proportionately longer and not as enlarged. *Head and neck chestnut,* merging into black on most of upperparts and across breast and then into chestnut again on rump and lower belly. Tail yellow except for dusky central pair of feathers and outer web of outermost pair. See C-44, P-26.

SIMILAR SPECIES: Note its limited S. Am. range. Likely confused only with much larger Crested Oropendola, which lacks chestnut on head and neck (this color is hard to see except in good light). Cf. also Russet-backed Oropendola.

HABITAT AND BEHAVIOR: Rare to fairly common in humid forest and forest borders, taller secondary woodland, and clearings with scattered tall trees. More numerous across n. Colombia than it is farther south into Ecuador. Behavior similar to other oropendolas': gregarious, most often seen at edge (and always nesting in such a situation), but often ranging well back into upper levels of humid forest to forage, where they clamber about actively searching for a variety of prey items as well as fruit.

RANGE: Locally in n. and w. Colombia (Pacific lowlands and Caribbean lowlands east to middle Magdalena valley) and nw. Ecuador (south to Pichincha). Also Mexico to Panama. Mostly below 400 m, occasionally higher.

NOTE: Formerly placed in the genus *Zarhynchus.*

GROUP B

Typical *large* oropendolas.

Psarocolius decumanus

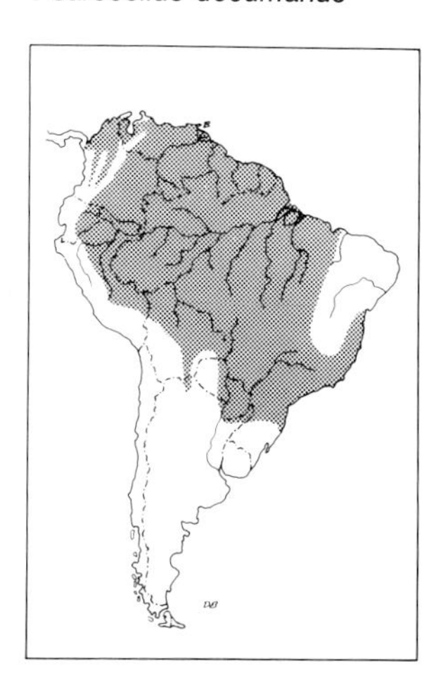

CRESTED OROPENDOLA

PLATE: 23

IDENTIFICATION: Male 46–48 cm (18–18¾"); female 36–38 cm (14–15"). *Bill ivory to pale greenish yellow:* iris blue; inconspicuous hairlike crest in male. *Mostly glossy black,* with rump and crissum dark chestnut (both hard to see in the field). *Maculosus* (Amazon south) has sparse scattering of pale yellow feathers throughout plumage. Tail bright yellow except for blackish central pair of feathers.

SIMILAR SPECIES: The only mostly black, large oropendola across most of its vast range. In n. Colombia compare to Black and Baudo Oropendolas (both with bright bare cheek patches, bicolored bill, etc.).

HABITAT AND BEHAVIOR: Uncommon to locally very common in a variety of habitats in humid to fairly dry regions, ranging from humid forest (though tending to avoid areas of extensive *terra firme* forest) to deciduous woodland to clearings or agricultural areas with scattered tall trees. Usually conspicuous and always less of a true forest oropendola than most other species. Colonies of this and other oropendolas are a characteristic sight in many tropical regions, as they are almost invariably sited in an isolated tree at forest edge or in a clearing. Displaying males are spectacular, leaning forward with fluttering wings and upraised tail so far that they look like they are going to fall from their perch, all the while uttering an almost indescribable loud gurgling and slashing call accelerating to a crescendo, accompanied by plainly audible wing flapping, etc.

RANGE: Widespread in lowlands south to n. and e. Bolivia, Paraguay (where local or absent from most of *chaco* in w. half), n. Argentina (in extreme northwest in Salta and in northeast in e. Formosa, Corrientes, and Misiones), and se. Brazil (south to Santa Catarina); west of Andes only in n. Colombia and nw. Venezuela; absent from much of ne. Brazil; Trinidad and Tobago. Also Panama. Mostly below 1200 m, but occasionally may wander higher.

Psarocolius angustifrons

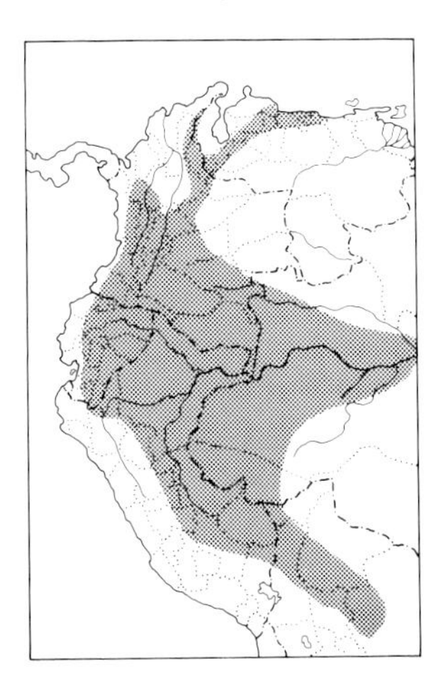

RUSSET-BACKED OROPENDOLA PLATE: 23

IDENTIFICATION: Male 44–49 cm (17–19"); female 34.5–38 cm (13½–15"). *Bill yellowish* (but typically *black* in nominate race of most of Amazonia); iris usually brown, but blue in some (old?) males. *Mostly olivaceous washed with rufous,* most rufescent on rump (or sometimes back and rump); *narrow yellow frontal band* in most races (broadest, even extending back as short superciliary, in *sincipitalis* of Colombia's Magdalena valley slopes; see C-45), but often lacking in *oleagineus* of n. Venezuela and *never* present in nominate race (see C-45). Tail mostly yellow, with central pair of feathers dusky and outer pair olive to olive tipped yellow. *Oleagineus* is the smallest and most pure olivaceous subspecies (see V-38), while *atrocastaneus* (w. Ecuador) and *salmoni* (w. Colombia) are the darkest and most rufescent races, with *salmoni*'s head being quite dusky. *Alfredi,* a yellowish-billed race found mainly on e. slope of Andes from s. Ecuador south, also occurs out into the Amaz. lowlands, at least in se. Peru.

SIMILAR SPECIES: Though variable in plumage, the montane forms of this species can usually be known by their habitat (overlapping only marginally with other oropendolas) and their yellow forehead. From cen. Peru south, cf. also the confusingly similar Dusky-green Oropendola. East of the Andes the nominate form is readily recognized by its dingy overall appearance and dark bill.

HABITAT AND BEHAVIOR: Uncommon to fairly common (in montane areas) to very common (in Amazonia) in humid lower montane forest, forest borders, and adjacent clearings and (in Amazonia) on river islands and along banks of rivers in clearings and *várzea* forest. Locally abundant in Amazonia, where great flocks can be seen just before dusk streaming in to roost on river islands. Nesting colonies are much like those of Crested Oropendola. Male's display also is similar ("falling forward" off its perch with flapping wings and raised tail), but primary vocalizations of the Amaz. form (nominate) and montane forms apparently differ: that of *salmoni* (presumably typical) is a liquid "whoop-kee-chót!" while in Amazonia it is a very different "whoo-éel-tiii-ooóp!" (S. Hilty).

RANGE: Coastal mts. of n. Venezuela (Yaracuy east to Distrito Federal and Aragua), Perijá Mts. on Venezuela-Colombia border, lower Andean slopes in w. Venezuela, Colombia, w. (south to El Oro) and e. Ecuador, e. Peru, and nw. Bolivia (south to w. Santa Cruz); Amaz. lowlands of se. Colombia (north to Meta and Vaupés), e. Ecuador, e. Peru, n. Bolivia (Beni), and w. Brazil (east to mouth of Rio Negro). To about 2000 m.

NOTE: It has been suggested that the numerous montane forms now considered part of this species are in fact a species distinct (*alfredi* has priority) from Amaz. lowland *angustifrons*. The latter differs in its black (not pale) bill, concolor forehead, and primary vocalizations and habitat. However, some supposed intermediate specimens have been reported, and until a detailed analysis of the situation has been prepared, we favor maintaining all forms as conspecific.

Psarocolius atrovirens

DUSKY-GREEN OROPENDOLA

PLATE: 23

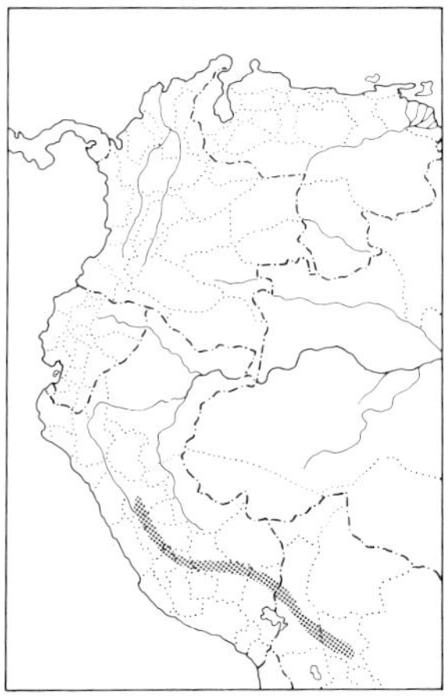

IDENTIFICATION: Male 41–43 cm (16–16¾"); female 33 cm (13"). *E. slope of Andes in s. Peru and n. Bolivia.* Bill pale greenish yellow; iris usually brown, but blue in some (old?) males. *Mostly dusky green,* becoming rufous on rump and crissum; usually shows whitish on throat and sometimes some yellow feathers on forecrown. Central and 2 outer pairs of tail feathers olive, rest yellow narrowly tipped olive.

SIMILAR SPECIES: Most likely confused with larger Russet-backed Oropendola (sympatric race of which is *alfredi*), but latter is more rufescent generally except for its generally paler head (essentially concolor in Dusky-green); it shows a well-marked yellow frontal band (usually lacking in Dusky-green, though a few specimens show it, *fide* D. Stotz). Cf. also Green Oropendola (no overlap).

HABITAT AND BEHAVIOR: Fairly common to common at borders of humid lower montane forest and adjacent clearings with scattered large trees. Usually occurs at elevations above Russet-backed Oropendola, though the 2 do overlap altitudinally to some extent; their behavior is similar. Most numerous in Bolivia, but also can be seen at the base of Machu Picchu in Cuzco, Peru.

RANGE: Se. Peru (Huánuco south to Puno) and nw. Bolivia (La Paz, Cochabamba, and w. Santa Cruz). Mostly 800–2400 m.

NOTE: Some hybridization with *P. angustifrons alfredi* may be occurring in se. Peru (D. Stotz).

Psarocolius viridis

GREEN OROPENDOLA

IDENTIFICATION: Male 46–51 cm (18–20"); female 36–38 cm (14–15"). *Bill pale greenish yellow, tipped orange-red;* iris pale blue; inconspicuous hairlike crest. *Mostly yellowish olive* (rather bright in some individuals); rump, lower belly, and crissum chestnut. Tail mostly yellow, central pair of feathers dusky olive. See V-38, C-45.

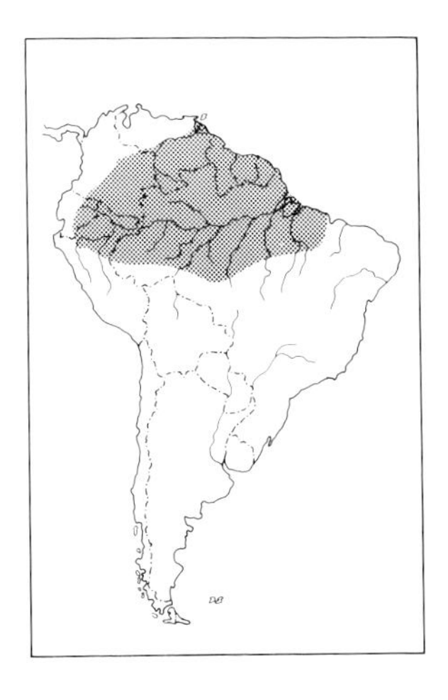

SIMILAR SPECIES: Russet-backed Oropendola of Amazonia is dingier and more rufescent generally, has all blackish bill, and is much less of a true forest bird. Olive Oropendola (of the widespread *yuracares* race) somewhat similar but brighter generally, with bright chestnut rearparts (*including wings*); its bill is black (not pale) basally, and it also shows conspicuous bare pink cheeks.
HABITAT AND BEHAVIOR: Uncommon to locally fairly common (but in general never an especially numerous oropendola, though somewhat commoner eastward) in canopy and borders of *terra firme* forest and in adjacent clearings with tall trees. General behavior similar to that of other typical *Psarocolius* oropendolas, but more of a forest-based species than, for example, Crested or Russet-backed.
RANGE: Guianas, s. and e. Venezuela (north to Monagas and s. Sucre), e. Colombia (north to Meta and Vichada), e. Ecuador, ne. Peru (south only to s. Loreto), and Amaz. Brazil (east to n. Maranhão, south to n. Mato Grosso). Mostly below 500 m.

GROUP C

Very large oropendolas, mainly black and chestnut (except 1 race of Olive); note *bare facial skin*. The "*Gymnostinops*" subgenus.

Psarocolius bifasciatus

OLIVE OROPENDOLA

PLATE: 23

Other: Para Oropendola

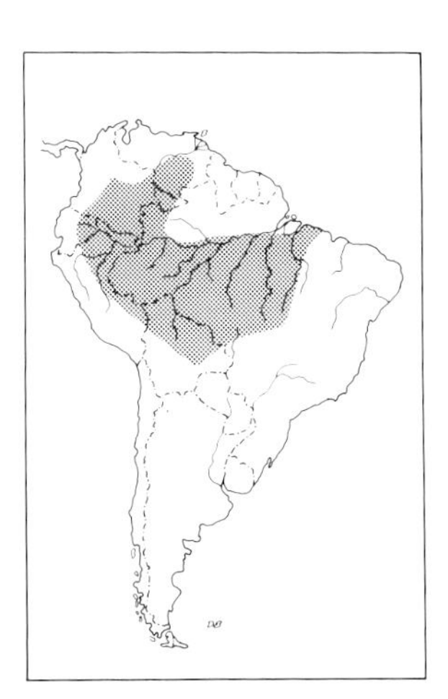

IDENTIFICATION: Male 47–53 cm (18½–20½"); female 41–43 cm (16–16¾"); nominate race consistently smaller. *Bill black tipped orange-red;* iris brown; *conspicuous bare cheek patch pink;* inconspicuous long hairlike crest. *Yuracares* (most of range) has *foreparts bright olive yellow; rearparts (including wings) bright chestnut.* Tail mostly yellow, with olive central pair of feathers. Nominate race (limited area south of Amazon in Belém area) looks very different, with *all of foreparts brownish black* (instead of olive yellow), but otherwise is similar. *Neivae* (of Rio Tocantins drainage south to Bananal Is. area on Rio Araguaia) more or less intermediate, with foreparts dusky olive (darker eastward toward range of nominate race).
SIMILAR SPECIES: *Yuracares* most likely confused with Green Oropendola, but soft-part colors and *bicolored* effect to body plumage really quite different. Nominate race recalls geographically far-distant Baudo Oropendola (see below).
HABITAT AND BEHAVIOR: Fairly common in *terra firme* forest, forest borders, and adjacent clearings with scattered tall trees; usually absent from *várzea* areas. This huge oropendola is usually seen high in canopy of forest or flying past with heavy, easily audible wingbeats. Sometimes associates with Green Oropendola when foraging or moving about, but infrequently or never nests in the same colony. Vocalizations resemble those of other large oropendolas (e.g., Crested), displaying males giving a very loud, liquid, gurgling "tek-tek-ek-ek-ek-ek-oo-guhloóp!"

RANGE: S. Venezuela (Amazonas and w. Bolívar), e. Colombia (north to Meta and Guainía), e. Ecuador, e. Peru, n. Bolivia (south to Santa Cruz), and Amaz. Brazil south of the Amazon (east to its mouth around Belém and south into n. Mato Grosso). To about 500 m.

NOTE: Following the evidence presented by Haffer (1974), we consider the widespread Amaz. form *yuracares* conspecific with *bifasciatus* (which has priority), found in a limited area south of the lower Amazon (Para Oropendola). *Neivae* is intermediate in coloration and size, and Haffer (1974:81) is probably correct in saying that "geneflow between these two forms is uninhibited." The nominate race, in fact, is very similar to *P. montezumae* of Middle America. We recognize that with the inclusion of forms in this species which are not exactly "olive" the English name is not really accurate over the species' entire range. Nonetheless, as *yuracares* (which *is* predominantly olive) is by far the most wide-ranging and well-known race of the species, we have opted to retain that familiar name. We note, further, that the name "Great Oropendola" is best reserved for the entire subgenus *Gymnostinops* if *all* its component forms are ultimately deemed conspecific.

Psarocolius guatimozinus

BLACK OROPENDOLA

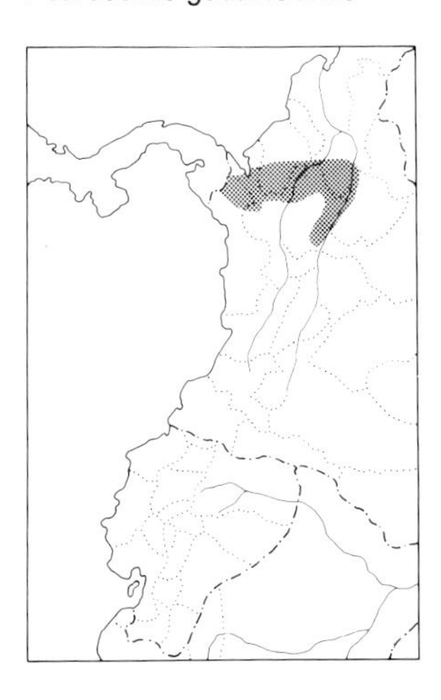

IDENTIFICATION: Male 44–48 cm (17–18¾"); female 38–40 cm (15–15½"). *N. Colombia. Bill black tipped orange-red; bare cheek patch mostly blue,* with pink strip along lower border. *Mostly black,* with back, rump, crissum, and a little on wing-coverts dark chestnut. Tail yellow, with central pair of feathers black. See C-44.

SIMILAR SPECIES: In general appearance resembles nominate form of geographically far-distant Olive Oropendola, but much blacker overall, and note difference in cheek patch colors. Cf. the very similar Baudo Oropendola. Crested Oropendola lacks cheek patch, has ivory-colored bill, etc.

HABITAT AND BEHAVIOR: Locally fairly common in humid forest and forest borders and in partially cleared areas with scattered remnant tall trees. Seems to avoid extensively forested areas and is regularly noted along rivers, where the nesting colonies are often situated. General behavior similar to other large oropendolas'.

RANGE: N. Colombia (n. Chocó east along humid n. base of W. and Cen. Andes to middle Magdalena valley in e. Antioquia). Also e. Panama. To about 800 m.

Psarocolius cassini

BAUDO OROPENDOLA

Other: Chestnut-mantled Oropendola

IDENTIFICATION: Male 44–46 cm (17–18"); female 38–40 cm (15–15½"). *Nw. Colombia.* Closely resembles nominate Olive Oropendola. Told from geographically close Black Oropendola by its *entirely pink* (not mostly blue) *bare cheek patch.* Further differs from Black in having *closed wing entirely rich chestnut,* more chestnut across back (and the chestnut is brighter, not as dark), and more chestnut on flanks.

HABITAT AND BEHAVIOR: Unknown in life. Probably very similar to Black Oropendola, whose range it approaches very closely (though not known to overlap).

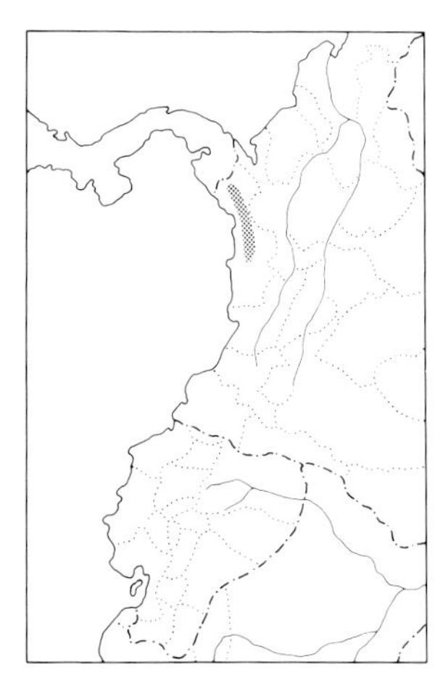

RANGE: Nw. Colombia in nw. Chocó (Río Truandó and upper Río Chocó). Recorded about 275–360 m.

NOTE: This species is undoubtedly closely related to *P. guatimozinus* and may prove to be conspecific (along with all the other members of the subgenus *Gymnostinops,* including *montezumae* of Middle America). However, the supposed intermediate *guatimozinus* from nw. Colombia on the Río Salaquí (Meyer de Schauensee 1966 : 431) appears, on reexamination, to be perfectly typical of *guatimozinus* in the amount of chestnut on its flanks, and in fact is exceeded in this regard by a certain example from Panama in the ANSP. It shows *no* approach to *cassini* in either of the latter's 2 key characters; the fully chestnut closed wing (including the secondaries) and the all-pink bare cheek patch (not blue and pink with a line of black feathers dividing them). It should be noted that *cassini's* all-pink cheek patch is identical to that found in the cis-Andean *bifasciatus* group, while the mostly blue cheek patch of *guatimozinus* is like that of Middle American *montezumae.* We suspect these strikingly different cheek patches might serve as effective isolating mechanisms, and this could explain why no actual hybrids between the 2 have been found. As *both guatimozinus* and *cassini* have chestnut mantles, calling *cassini* the "Chestnut-mantled" Oropendola seems very misleading, and we prefer to recognize the species' restricted range by giving it a geographic epithet: *cassini* is still known from only 2 localities, both of them along rivers flowing out of the isolated Serranía de Baudó.

Cardinalinae

CARDINALS, GROSBEAKS, AND ALLIES

A GROUP of about 30 S. Am. "finches" is now considered to form part of the Cardinalinae subfamily, with other members found in North America. The distinctions between this group, the Emberizine finches, and even the "tanagers" are a bit uncertain; we have allocated various genera following *Birds of the World* (vol. 13) for the most part (certain little-known "intermediate" groups pose serious difficulties). The S. Am. Cardinalines include the saltators, various forest or forest-based grosbeaks, and 3 genera of "cardinals."

Periporphyrus Grosbeaks

A monotypic genus of black-hooded grosbeak found inside humid forests of ne. South America.

Periporphyrus erythromelas

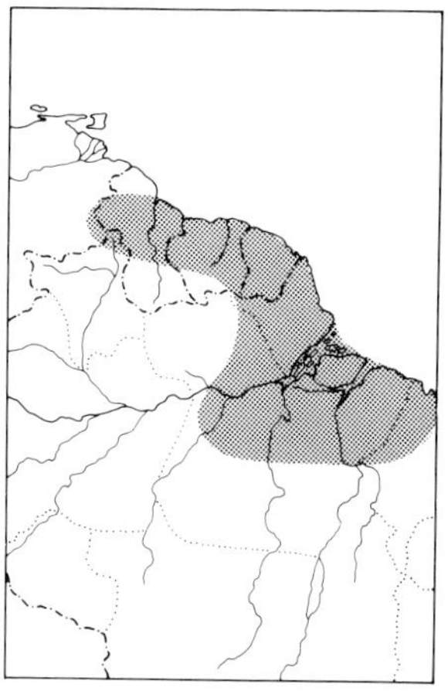

RED-AND-BLACK GROSBEAK

PLATE: 24

IDENTIFICATION: 20.5 cm (8″). *Very heavy blackish bill.* Male unmistakable, mostly *carmine above* and *rosy red below,* with contrasting *black head and throat.* Female *with pattern of male,* but upperparts olive and underparts yellow, shaded olive on sides. Immature male has olive upperparts suffused with carmine and underparts tinged orange-red.
SIMILAR SPECIES: Yellow-green Grosbeak is smaller with only foreface black; very different behavior. Certain tanagers are mostly olive or yellow, but none has the black head or massive bill.
HABITAT AND BEHAVIOR: Rare to locally uncommon in lower and middle strata inside humid forest, mostly in hilly areas. Forages singly or in well-separated pairs; rarely with mixed flocks. Male's song strongly reminiscent of Slate-colored Grosbeak's, with same quality and phraseology, but delivery notably more drawn out; "spink" call note is also similar to Slate-colored Grosbeak's (and to Northern Cardinal's). Found regularly in Suriname's Brownsberg Reserve.
RANGE: Locally in the Guianas, extreme se. Venezuela (only the Cerro Roraima area of se. Bolívar), and ne. Brazil (Amapá, Pará south of the Amazon from Rio Tapajós eastward, and nw. Maranhão). To about 500 m.

Pitylus Grosbeaks

A pair of mostly gray, red-billed grosbeaks of humid lowland forests.

Pitylus grossus

SLATE-COLORED GROSBEAK

PLATE: 24

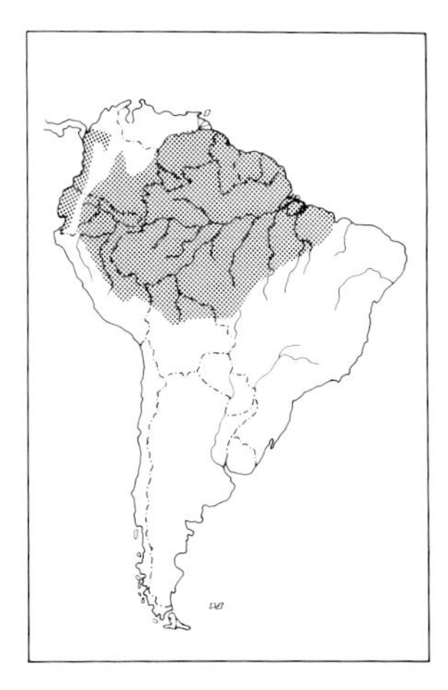

IDENTIFICATION: 20 cm (7¾"). *Heavy coral red bill.* Male *mostly dark bluish gray.* Center of throat white, with foreface and broad margin around the throat patch black. Female similar but slightly paler and more olivaceous gray and lacking black on face and around throat.
SIMILAR SPECIES: Red bill in conjunction with generally slaty plumage is distinctive. See also Black-throated Grosbeak (no overlap).
HABITAT AND BEHAVIOR: Fairly common in middle and upper strata of humid forest, less often lower, usually inside and not at edge. Generally in pairs and, though often associating with mixed flocks, usually not very conspicuous. Its fine, rich song is rather variable in its phraseology but not in its peppershrikelike quality; the frequently heard call is a cardinallike "peek."
RANGE: Widespread south to n. Bolivia (La Paz and Cochabamba) and Amaz. Brazil (south to cen. Mato Grosso, s. Pará, and n. Maranhão), Colombia, Venezuela, and the Guianas through Ecuador; on Pacific slope south to sw. Ecuador (El Oro and w. Loja). Also Honduras to Panama. To about 1200 m, locally somewhat higher.

Pitylus fuliginosus

BLACK-THROATED GROSBEAK

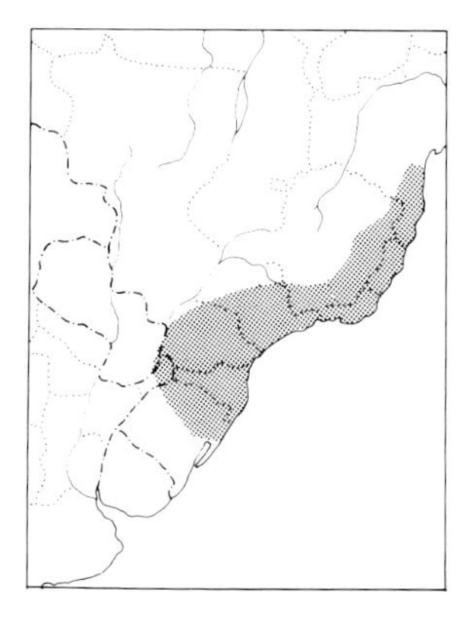

IDENTIFICATION: 22 cm (8½"). *Se. Brazil and adjacent areas.* Resembles Slate-colored Grosbeak (no overlap), but *lacks white on throat* and is somewhat darker bluish slate generally. Male has *entirely black throat and chest,* while in female these areas are dusky.
HABITAT AND BEHAVIOR: Uncommon in middle and upper strata of humid forest, less often in lower growth; usually remains inside forest. Behavior and voice much like Slate-colored Grosbeak's.
RANGE: E. Brazil (Alagoas, and se. Bahia to n. Rio Grande do Sul), se. Paraguay (recorded by Bertoni in the early 20th century in Alto Paraná, but not since), and ne. Argentina (Misiones). To about 1200 m.

NOTE: Perhaps conspecific with the geographically separated, more widespread *P. grossus.*

Caryothraustes Grosbeaks

Pair of humid lowland forest–inhabiting grosbeaks; their behavior and overall appearance differ markedly.

Caryothraustes humeralis

YELLOW-SHOULDERED GROSBEAK

PLATE: 24

IDENTIFICATION: 16 cm (6¼"). Fairly heavy blackish bill. *Crown and nape gray,* contrasting with *bright yellowish olive upperparts,* bend of wing bright yellow. *Broad facial mask black,* bordered below by *white malar stripe. Throat white conspicuously scaled black* and bordered by blackish submalar streak, remaining underparts mostly gray, with white center of belly and yellow crissum.

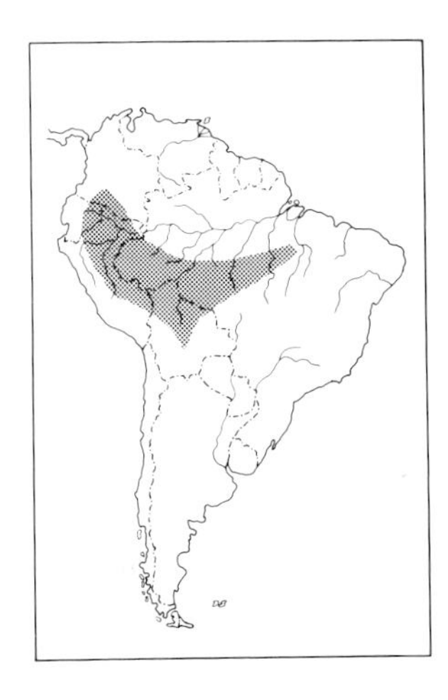

SIMILAR SPECIES: Readily recognized by its unique head pattern. Behavior also distinctive (see below).

HABITAT AND BEHAVIOR: Rare to uncommon in canopy and at edge of humid forest; the recent spate of records appears to indicate that the species is more numerous than was earlier thought. Found singly or in pairs, often with mixed foraging flocks of tanagers, honeycreepers, etc. Forages mostly by gleaning insects from foliage; also regularly perches on high, exposed branches, not moving for long periods. Though most often silent, it does give a loud, clear, inflected "suweet," quite tanagerlike (especially the Swallow Tanager).

RANGE: Colombia (known only from Bogotá trade skins, but presumably in southeast in Caquetá or Putumayo), e. Ecuador (Napo, Pastaza, and Morona-Santiago), e. Peru (Loreto, Cuzco, and Madre de Dios), n. Bolivia (La Paz and Cochabamba), and sw. Brazil (upper Rio Purús in Amazonas, Cachoeira Nazaré in Rondônia, Serra dos Carajás in se. Pará, and upper Rio Aripuana in nw. Mato Grosso). 200–1000 m.

NOTE: A very distinct species, totally unlike the other 2 species currently placed in *Caryothraustes*. May not belong in the genus, but no other name seems to be available.

Caryothraustes canadensis

YELLOW-GREEN GROSBEAK

PLATE: 24

Also: Green Grosbeak

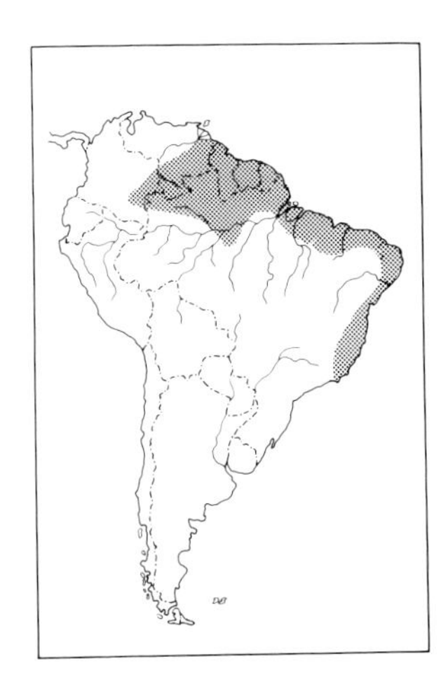

IDENTIFICATION: 17–18.5 cm (6¾–7¼"). Fairly heavy bill decidedly paler at base. *A mostly yellow-olive grosbeak with a conspicuous black face.* Above mostly yellowish olive, below mostly bright yellow. *Frontalis* of coastal ne. Brazil has black frontal band. *Brasiliensis* of coastal se. Brazil is somewhat larger with decidedly brighter yellow forecrown.

SIMILAR SPECIES: Cf. female Red-and-black Grosbeak (with all-black head and much larger bill) and Yellow-shouldered Grosbeak (with more complex facial pattern).

HABITAT AND BEHAVIOR: Fairly common to common in middle and upper strata of humid forest. Noisy and conspicuous, it troops about in pairs or flocks of up to 10 to 20 birds, often appearing to be a nuclear species around which flocks of tanagers, etc., form but at other times moving independently. The most frequent call is a loud, buzzy "dzzeet" or "dzreet," sometimes followed by a repeated "chew-chew-chew-chew," or the latter series is given alone.

RANGE: Extreme e. Colombia (Guainía and Vaupés), s. Venezuela (south of the Orinoco), the Guianas, and Amaz. Brazil (north of Amazon from upper Rio Negro east, south of Amazon from Rio Madeira east); e. Brazil (coastal Ceará to Alagoas and south to e. Minas Gerais and Rio de Janeiro). Also e. Panama (the isolated race *simulans*). To about 900 m, but mostly lower.

NOTE: Two English names for *C. canadensis* have been suggested in recent references, "Green Grosbeak" and "Yellow-green Grosbeak." After having earlier used "Green Grosbeak," Meyer de Schauensee (1970) modified it to "Yellow-green," and both Ridgely (1976) and Meyer de Schauensee and Phelps (1978) followed suit; the

1983 AOU Check-list, however, reverted to simply "Green." It really makes little difference, but as the bird's basic color is a *yellowish* olive, we opt to continue with "Yellow-green."

Saltator Saltators

Large arboreal finches found mostly at forest edge or in scrub; usually conspicuous. Bill stout, often brightly colored. Olive or gray above, typically with a black-bordered throat patch and a white superciliary (but pattern of some species is quite divergent).

GROUP A

Duller, typical saltators.

Saltator albicollis

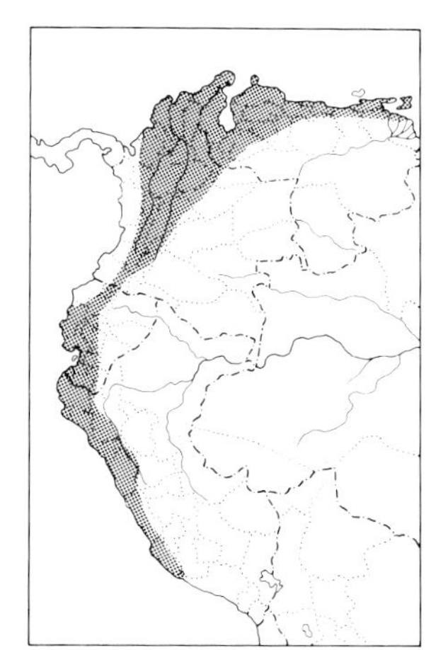

STREAKED SALTATOR

PLATE: 24

IDENTIFICATION: 19–20.5 cm (7½–8″). Bill usually (sometimes boldly) tipped yellowish or flesh. Grayish olive to olive above, grayest on head, with short white superciliary (barely extending to behind eye). Below whitish *profusely streaked olive* except on belly. Birds from sw. Ecuador and nw. Peru (s. part of range of *flavidicollis*) are *unstreaked yellowish white* with *much broader and longer superciliary;* birds believed to be immatures are streaked below and have eyebrow similar to that of typical races. Birds from w. Peru (Lambayeque south; *immaculatus*) are unstreaked whitish below and grayer above.

SIMILAR SPECIES: *The only saltator with streaking below.* Unstreaked birds look very different, but no other vaguely similar bird is uniform olive above and plain below. Cf. female of rare Black-and-white Tanager.

HABITAT AND BEHAVIOR: Common in shrubby clearings, pastures, and gardens with scattered trees and hedgerows and in arid scrub; most numerous and widespread in drier regions. Often seen in pairs (occasionally more will gather at a fruiting tree) and usually conspicuous. The song is heard frequently, a loud "tchew-tchew-tchew, tcheeér" with many variations.

RANGE: N. Colombia (Caribbean lowlands, south in Cauca and Magdalena valleys) and n. Venezuela (north of the Orinoco, east to Sucre and Monagas) and south locally on Pacific slope in more arid regions through w. Ecuador (where also in arid intermontane valleys) to w. Peru (on Pacific coast south to Ica, also in upper Marañón valley); Trinidad. Also Costa Rica to Panama and on some of the Lesser Antilles. Mostly below 1500 m, locally to 2500 m.

NOTE: The puzzling situation involving the 2 "types" of this species in w. Ecuador and nw. Peru remains to be resolved. As discussed by Hellmayr (*Birds of the Americas,* vol. 13) and Chapman (1926: 617–620), in the n. part of its range the *flavidicollis* subspecies is basically similar to typical birds, being streaked below on a yellowish-tinged ground and with a short superciliary. In the s. part, however, it is virtually unstreaked whitish below and has a much longer superciliary, and only presumed immatures are streaked below and have a short brow. Field study of the situation is needed in order to determine whether this is the correct interpretation.

Saltator coerulescens

GRAYISH SALTATOR

PLATE: 24

IDENTIFICATION: 20.5 cm (8"). *Dark to rather light gray above* (darkest in Amaz. birds, including *azarae;* palest in birds from n. Colombia and Venezuela), with short white superciliary. Throat white (tinged buff from Bolivia and cen. Brazil south) bordered by black malar stripe. Below pale grayish (darker in Amaz. birds), becoming buffy ochraceous on belly and crissum. Immatures of all races more olivaceous generally, with yellow tinge to eyestripe, throat, and belly; flight feathers rather prominently edged green.
SIMILAR SPECIES: Pattern generally resembles Buff-throated Saltator, but upperparts obviously gray (not pure olive). Immature Grayish, with green on wings, is confusingly similar to Green-winged Saltator, but latter always has white throat and much longer superciliary.
HABITAT AND BEHAVIOR: Fairly common to common in dry scrub, deciduous woodland, and pastures with scattered trees and hedgerows in drier regions; in clearings and riparian areas and other secondary growth in more humid regions (e.g., Amazonia). Usually conspicuous, often around towns and in gardens. Song varies geographically but is always musical, sometimes with more guttural notes interspersed or given as a duet; 1 common phrase is "wee-cho, chi-chi-chi, choh."
RANGE: Widespread in lowlands east of the Andes (west of them only in n. Colombia, where south into middle Magdalena valley, and nw. Venezuela) south through Amaz. and cen. Brazil (east to n. Maranhão, Piauí, w. Bahia, Goiás, and s. Mato Grosso), n. and e. Bolivia, and w. Paraguay (east to along the Paraguay River) to n. Argentina (south to La Rioja, Córdoba, and n. Buenos Aires), and extreme w. Uruguay (once in Soriano); Trinidad. Also Mexico to Costa Rica. To about 1000 m, rarely higher.

Saltator similis

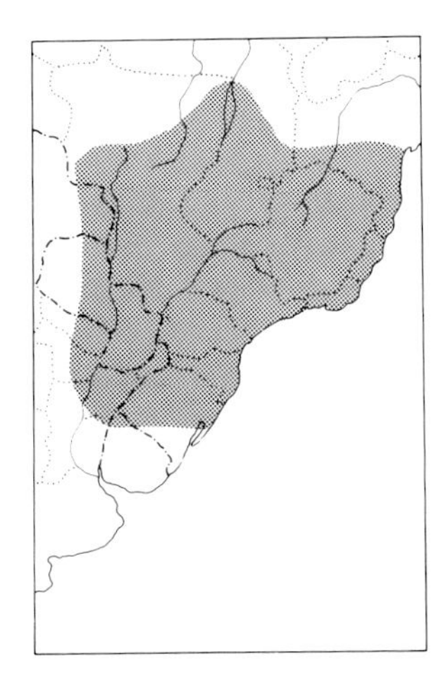

GREEN-WINGED SALTATOR

IDENTIFICATION: 20.5 cm (8"). Generally resembles Grayish Saltator (some overlap). Best distinguishing points are its *very long white superciliary* (extending to over ear-coverts; only to above eye in Grayish) and its *obviously white throat* (buff-tinged in sympatric races of Grayish), which contrasts with dull buff underparts (deeper ochraceous in *ochraceiventris* of extreme se. Brazil). Back and especially flight feathers are rather bright olive green, but remember that immature Grayish also shows green on wings.
SIMILAR SPECIES: In addition to Grayish, also resembles Thick-billed Saltator; latter shares Green-wing's long eyestripe, but throat always rather buffy (never clear white).
HABITAT AND BEHAVIOR: Usually fairly common in woodland and forest borders (including gallery woodland) and clearings. Most often found in pairs, with behavior similar to Grayish's. Song is a short series of loud, clear, whistled notes (e.g., "cheer, cheer, chwer, cheér, chir").
RANGE: Se. Brazil (north to s. Mato Grosso, Goiás, and Bahia) and extreme e. Bolivia (e. Santa Cruz) south through e. Paraguay (west to

pantanal along Paraguay River) to ne. Argentina (south to Entre Ríos and Santa Fe) and n. Uruguay. To about 1200 m.

Saltator maximus

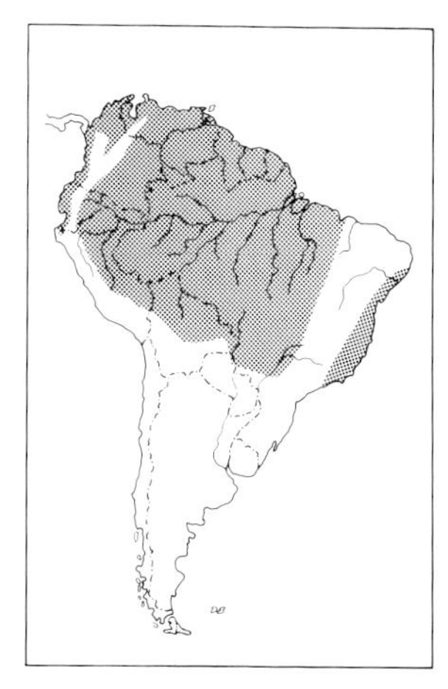

BUFF-THROATED SALTATOR

IDENTIFICATION: 20.5 cm (8″). Recalls Grayish Saltator, but *mostly rather bright olive above,* only sides of head gray, with short white superciliary (extending to just behind eye). Chin white, *lower throat buff,* both bordered by black malar stripe. Below mostly grayish; center of belly pale buff, crissum cinnamon buff. See C-55, V-39.

SIMILAR SPECIES: Widely sympatric with Grayish Saltator, which is obviously gray (not olive) above with a white (not buff) throat patch in overlap zone. Streaked Saltator also olive above but profusely streaked below.

HABITAT AND BEHAVIOR: Common in shrubby clearings with scattered trees, forest borders (and to a lesser extent the canopy, especially in *várzea*), and lighter secondary woodland; mostly in more humid regions. Usually occurs singly or in pairs, at virtually all levels (though infrequent actually inside forest). The song is frequently heard, a series (often long) of short, sweet, thrushlike phrases, individual phrases often repeated several times.

RANGE: Widespread south to nw. Peru (Tumbes), n. and e. Bolivia, extreme n. Paraguay (no recent records), and s. Brazil (south to Mato Grosso, São Paulo, and Rio de Janeiro: local in drier regions). Also Mexico to Panama. Mostly below 1200 m.

Saltator maxillosus

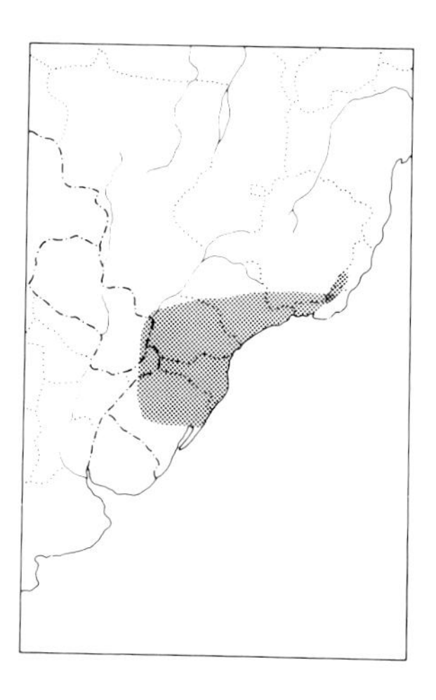

THICK-BILLED SALTATOR

PLATE: 24

IDENTIFICATION: 21 cm (8¼″). Compared to other saltators, *bill heavier and shorter, with more curved culmen* (but difference often not obvious in the field); it is black, often with orange blotch at base of mandible (orange sometimes more extensive). *Above gray* with long white superciliary. *Throat buff* bordered by black malar stripe. Remaining underparts buffy grayish, with crissum deeper buff. Female has *upperparts olive* but otherwise is like male.

SIMILAR SPECIES: The only sexually dimorphic saltator. Readily confused with Green-winged Saltator (with which often sympatric), especially the olive-backed female, but throat always buff, often rather deep (and never white).

HABITAT AND BEHAVIOR: Uncommon in canopy and borders of humid forest and woodland. Most often in pairs, foraging at all levels, and regularly with mixed-species flocks. Occurs principally at higher elevations; numerous at higher levels in Itatiaia Nat. Park, Brazil.

RANGE: Se. Brazil (Espírito Santo to n. Rio Grande do Sul), ne. Argentina (Misiones), and e. Paraguay (recorded by Bertoni in early 20th century from Alto Paraná; no recent records). To about 2200 m; rare near sea level, but does tend to occur at somewhat lower elevations during austral winter.

NOTE: Though usually associated with the *S. aurantiirostris* group, largely because of bill shape and color, this species also seems close to *S. similis* (which it resembles in its long superciliary and several other features), and in the field *S. maxillosus* would be most apt to be confused with that species. We are not aware of the details regarding reported intergradation between *S. maxillosus* and *S. aurantiirostris* in Corrientes, Argentina (see Short 1975:308) and continue to regard *maxillosus* as a full species.

GROUP B

Sides of head broadly black, bordered by (except in Black-cowled) *light eyebrow and throat.*

Saltator atripennis

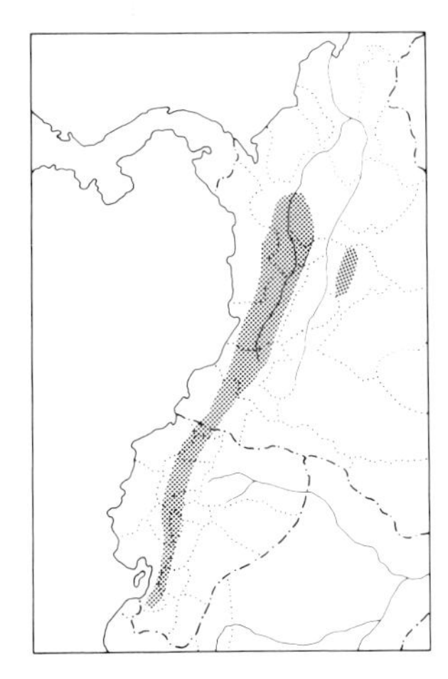

BLACK-WINGED SALTATOR

PLATE: 24

IDENTIFICATION: 20.5 cm (8″). W. Colombia and w. Ecuador. *Crown and sides of head black* with *conspicuous white superciliary and patch on ear-coverts.* Back bright olive, but *wings and tail contrastingly black.* Throat white, remaining underparts light grayish, with buffy crissum.

SIMILAR SPECIES: No other saltator has either the black wings or the large white ear-patch.

HABITAT AND BEHAVIOR: Fairly common in canopy and at edge of humid montane forest and woodland; less often out in clearings than most other saltators. Regularly accompanies mixed flocks of tanagers and other birds; usually in pairs. Its loud and lively song resembles other saltators' in quality but usually goes distinctly down-scale (e.g., "twee, twaa, toou, toweer, tweeeeear").

RANGE: W. Colombia (both slopes of W. Andes, w. slope of Cen. Andes, on e. slope only in Caldas; w. slope of E. Andes only in Cundinamarca) and w. Ecuador (south to El Oro). To about 2000 m, lowest on Pacific slope (locally almost down to sea level in Ecuador).

Saltator aurantiirostris

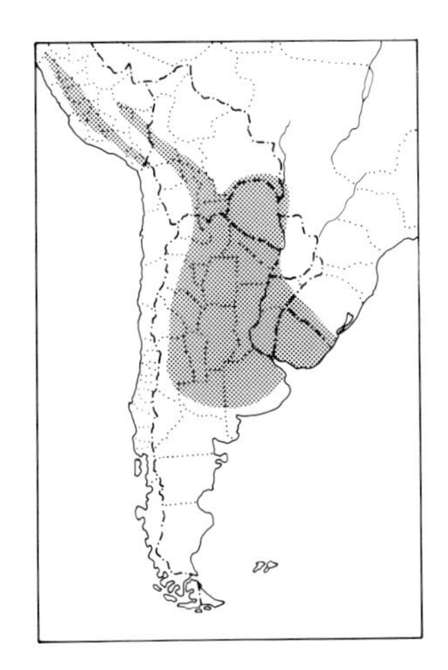

GOLDEN-BILLED SALTATOR

PLATE: 24

IDENTIFICATION: 19–20.5 cm (7½–8″). *Bill mostly or entirely orange* (apparently only adult males); blackish with *orange splotches,* especially near base (females); or mostly blackish (immatures). Above gray. *Forecrown and sides of head black,* extending down on sides of neck and across chest as *pectoral band* which encloses white to buffy white throat. *Conspicuous white postocular stripe.* Underparts buff to buffy grayish. Outer tail feathers broadly tipped white (more narrowly or not at all in Argentina). Pectoral band widest in *albociliaris* of s. Peru. Female similar but more olivaceous above with pectoral band restricted or lacking; in Andes the sexes resemble each other. Immature has yellower postocular stripe and throat patch and is grayer on sides of head and neck and olivaceous above.

SIMILAR SPECIES: The only saltator with white superciliary extending only back from eye (*not reaching to lores*). Plumage variation is somewhat confusing, but most birds show at least some orange on bill and have at least some indication of the black encircling the throat.

HABITAT AND BEHAVIOR: Fairly common to common in drier montane scrub and adjacent woodland and in *chaco* and *monte* woodlands.

Occurs mostly in pairs; sometimes rather shy, but may perch conspicuously on tops of bushes or low trees. The song (in the *chaco*) is a loud, explosive "switch-it, tchweet-a-sweéu," somewhat variable but always with the last syllable emphasized; songs in Peru are similar.
RANGE: Highlands from n. Peru (north to Cajamarca and Amazonas, south of the Río Marañón) south to extreme n. Chile (once, in Arica) and through w. and s. Bolivia, w. Paraguay, and extreme sw. Brazil (sw. Mato Grosso) to n. and cen. Argentina (south to Mendoza, La Pampa, and ne. Río Negro) and east to Uruguay and extreme s. Brazil (Rio Grande do Sul). To at least 3000 m (in the Andes).

Saltator nigriceps

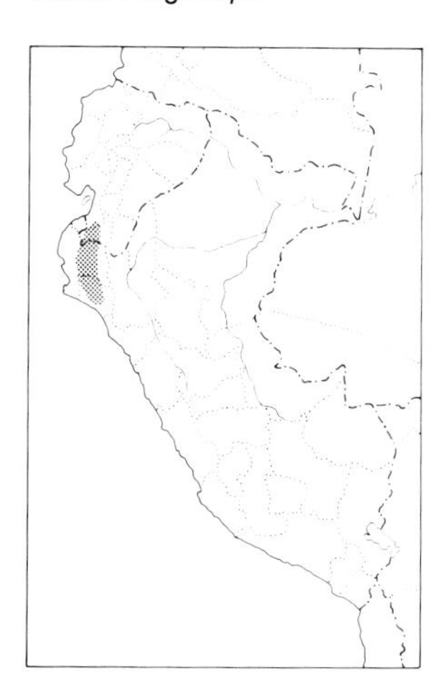

BLACK-COWLED SALTATOR

IDENTIFICATION: 22 cm (8½"). S. Ecuador and nw. Peru. Bill rather heavy, *salmon red. Mostly gray,* with *contrasting black hood* extending to chest. Belly and crissum buff. Occasionally shows a few white feathers on chin and behind eye. Outer tail feathers tipped white.
SIMILAR SPECIES: Golden-billed Saltator's bill is orange in male but never red; Golden-bill is much buffier (not so gray) below and has conspicuous whitish throat and white postocular stripe.
HABITAT AND BEHAVIOR: Uncommon in borders of humid montane forest, and in secondary scrub. Usually in pairs and often quite inconspicuous. Not well known, but probably does not differ much from Golden-billed Saltator, though perhaps more of a forest bird.
RANGE: Highlands of s. Ecuador (Loja) and nw. Peru (Piura and Lambayeque). Mostly 1000–2000 m.

NOTE: We maintain *nigriceps* as a full species while recognizing that it is close to *S. aurantiirostris* (which it replaces northward). The closest race of the latter, *iteratus,* in fact does not, however, seem to approach *nigriceps* in appearance (*albociliaris* from farther south is more similar).

Saltator orenocensis

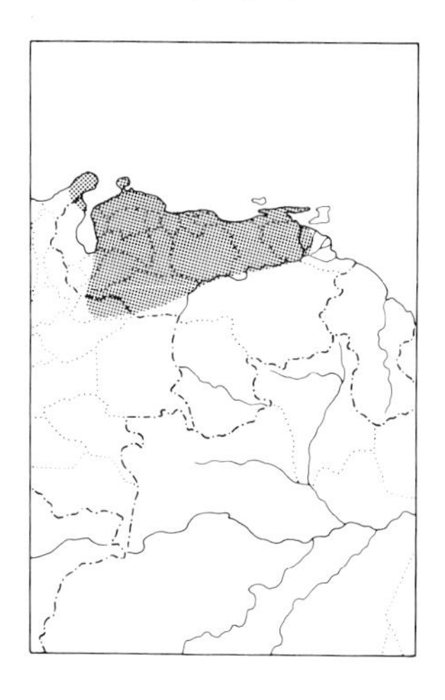

ORINOCAN SALTATOR

IDENTIFICATION: 18.5 cm (7¼"). Above mostly gray, with black sides of head and neck and *long white superciliary* extending from bill to over ear-coverts. Below mostly white, with *ochraceous buff sides and crissum. Rufescens* of nw. Venezuela and adjacent Colombia *mostly ochraceous buff below,* with only throat and center of belly white. C-55 and V-39.
SIMILAR SPECIES: This handsome saltator is unlikely to be confused in its range; no other is so bright buff below.
HABITAT AND BEHAVIOR: Fairly common in arid scrub and woodland (near Caribbean coast) and in gallery woodland and semiopen grassy areas with scattered bushes and trees (in more humid llanos). Song is a simple repetitive "cheeyir, cheeyir, cheeyir . . . ," eventually breaking into a short musical jumble, usually given from the top of a tree in the open.
RANGE: Ne. Colombia on Guajira Peninsula (west to about Riohacha) and arid nw. Venezuela (coastal Zulia, Lara, and Falcón); llanos of

Venezuela from Cojedes and Apure east to Delta Amacuro and n. Bolívar, also adjacent ne. Colombia (1 sighting from Apure). To about 400 m.

GROUP C

Three *distinctive* saltators.

Saltator atricollis

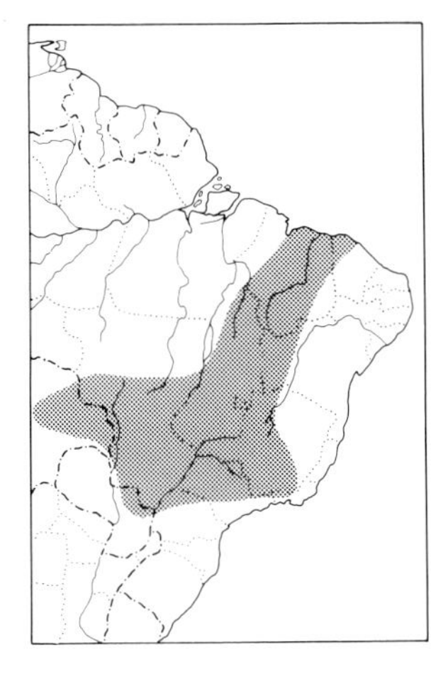

BLACK-THROATED SALTATOR

PLATE: 24

IDENTIFICATION: 20.5 cm (8″). *Bill orange,* usually with blackish ridge. *Mostly brown above,* vaguely streaked with darker brown. *Foreface and throat contrastingly black.* Underparts buffy whitish, becoming deeper cinnamon buff on flanks and crissum. Immature similar but with ashy brown foreface and throat.

SIMILAR SPECIES: Nothing really similar in its cen. S. Am. range. Golden-billed Saltator has white or buffyish throat outlined in black and a white eyestripe and is essentially gray (not brown) above.

HABITAT AND BEHAVIOR: Uncommon to locally fairly common in *cerrado* and *caatinga* scrub. Behavior quite different from other saltators', perching quite prominently on tops of bushes and low trees but sometimes dropping to ground to feed. Song is fast and jumbled but rather musical (sometimes given by several birds, more or less antiphonally), while the distinctive call is a sharply inflected "wheék . . . wheék . . ." Neither vocalization is much like any other saltator's, more resembling quality of Wedge-tailed Grass-Finch's (though song is much more complex).

RANGE: Interior ne. and cen. Brazil (from s. Maranhão and Ceará south to Minas Gerais, n. São Paulo and Mato Grosso; occasionally to Rio de Janeiro), e. Bolivia (e. Santa Cruz), and ne. Paraguay. Mostly 700–1300 m.

Saltator rufiventris

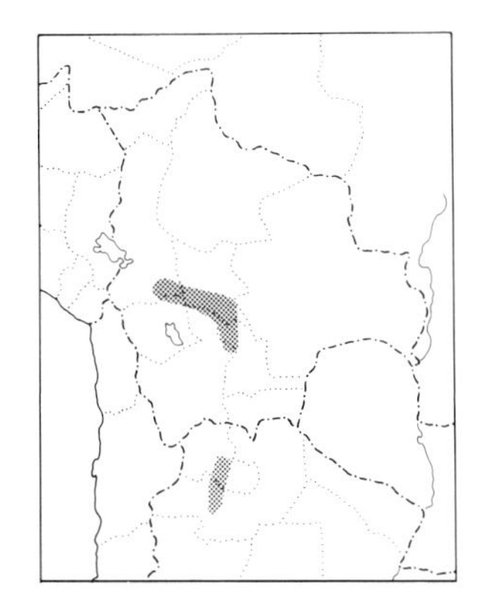

RUFOUS-BELLIED SALTATOR

PLATE: 24

IDENTIFICATION: 22 cm (8½″). A distinctive saltator of *Bolivian highlands* (mostly). Base of mandible flesh. *Mostly bluish gray; lower breast, belly, and crissum contrastingly rufous.* Long superciliary white.

HABITAT AND BEHAVIOR: Uncommon in arid montane scrub and woodland, including *Polylepis* groves, and in adjacent cultivated areas. Occurs mostly as well-separated pairs, which often are conspicuous in their fairly open habitat; though primarily arboreal, they do occasionally descend to the ground to feed. The most frequent calls are a soft "phueet-phueet" and a louder "whueet-whueet" (C. G. Schmidt).

RANGE: Highlands of w. Bolivia (La Paz, Cochabamba, and Chuquisaca; doubtless also Potosí and Tarija) and nw. Argentina (Jujuy, and seen in Nov. 1986 in Salta; RSR et al.). Mostly 3000–4000 m.

Saltator cinctus

MASKED SALTATOR

PLATE: 24

IDENTIFICATION: 21.5 cm (8½″). A rare saltator of the e. slope of the Andes. Bill black with varying amounts of red, especially on maxilla

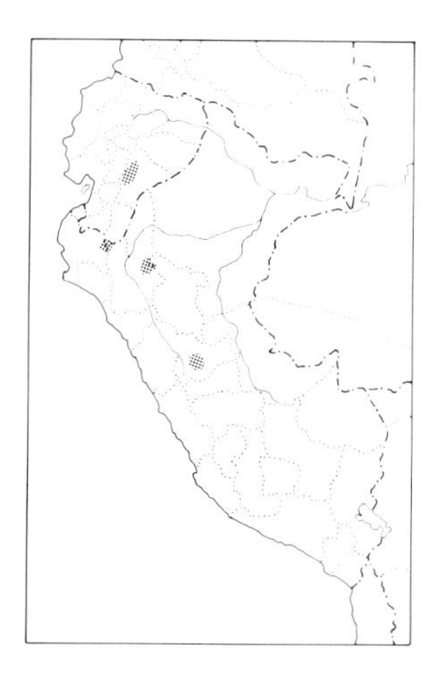

(but bill almost entirely red in some birds); iris also variable, usually yellowish orange. *Above mostly dark bluish gray. Face, sides of head, and throat black* (forming "mask"). Chest white, bordered below by broad black band. Lower underparts mostly white, flanks broadly gray; crissum gray, feathers tipped white. Tail strongly graduated, blackish, with *broad white tipping* from below (like a cuckoo's).

SIMILAR SPECIES: Does not really resemble any other saltator. Pattern vaguely reminiscent of an *Arremon* sparrow, while bluish gray color recalls Slate-colored Grosbeak.

HABITAT AND BEHAVIOR: Rare and local (probably overlooked) in lower growth of humid montane forest. Occurrence seems tied to the presence of dense stands of *Chusquea* bamboo. Behavior virtually unknown, but reported to be shy and difficult to see in its often almost impenetrable habitat.

RANGE: E. slope of Andes in s. Ecuador (Cordillera de Cutucú in Morona-Santiago; also seen recently by P. Greenfield in Podocarpus Nat. Park south of Loja) and n. and cen. Peru (recent specimens from Cerro Chinguela on the Piura-Cajamarca border, Amazonas, and Huánuco). 1679–2940 m.

NOTE: See also J. P. O'Neill and T. S. Schulenberg, *Auk* 96 (3): 610–613, 1979.

Paroaria Cardinals

Red head (crested in some) combined with gray/black upperparts and white underparts is distinctive. Shrubbery and scrub; smaller species usually near water.

GROUP A

Smaller species; *black above. Near water.*

Paroaria gularis

RED-CAPPED CARDINAL

PLATE: 25

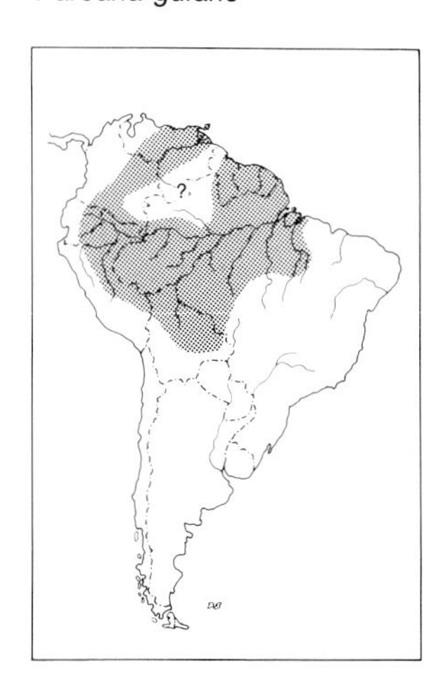

IDENTIFICATION: 16.5 cm (6½"). Bill black, base of mandible yellowish flesh; eyes orange. *Head and chin crimson-red,* with black smudge around eye. Otherwise glossy blue-black above. *Lower throat and pointed bib on chest black,* remaining underparts and partial collar white. *Nigrogenis* of Orinoco drainage has *red (not black) bib,* and black facial mask from bill onto ear-coverts. See C-55, V-39. *Cervicalis* of n. Bolivia and adjacent nw. Mato Grosso has no black on face but has black bib of nominate *gularis.* Immatures (all races) have crown and upperparts brown, with face, throat, and bib light buffy brown.

SIMILAR SPECIES: Nothing resembling it occurs in its range, though the brown immatures (without red) can be confusing. At periphery see Crimson-fronted and Yellow-billed Cardinals.

HABITAT AND BEHAVIOR: Usually common and conspicuous (less so in blackwater areas) in shrubbery and open areas along shores of lakes, ponds, and sluggish streams and rivers. Locally also in mangroves (only Trinidad?). Seen in pairs or small groups, often low over the

water; frequently perches on dead branches or stubs protruding from water. Song (not heard frequently) is a clear, sweet "suwee, chu."
RANGE: Guianas, interior and e. Venezuela, e. Colombia, e. Ecuador, e. Peru, Amaz. Brazil (east to e. Pará, south to extreme n. Goiás and n. Mato Grosso), and n. Bolivia (south to nw. Santa Cruz, Cochabamba, and La Paz); Trinidad. Mostly below 400 m.

Paroaria baeri

CRIMSON-FRONTED CARDINAL PLATE: 25

IDENTIFICATION: 16.5 cm (6½"). *Locally in cen. Brazil. Bill black.* Resembles Red-capped Cardinal but has *head shiny blue-black,* with *only forecrown and upper throat dark crimson* (looking essentially black except in strong light). Red of throat often mixed with black. *Xinguensis* of the upper Rio Xingú system is similar but has most of throat black, leaving only dark crimson malar streak.
SIMILAR SPECIES: Red-capped Cardinal has partly pale lower mandible and much more extensive brighter red on head.
HABITAT AND BEHAVIOR: Similar to Red-capped Cardinal except that at least on the Rio Araguaia it seems not as numerous as that species often is elsewhere. The nature of the zone of overlap or replacement between *P. gularis* and *P. baeri* remains to be determined; they perhaps will prove to intergrade.
RANGE: Cen. Brazil in n. Goiás and ne. Mato Grosso along middle and upper Rio Araguaia (e.g., on Bananal Is.) and in n. Mato Grosso along upper Rio Xingú and Rio Cristalino. To about 400 m.

Paroaria capitata

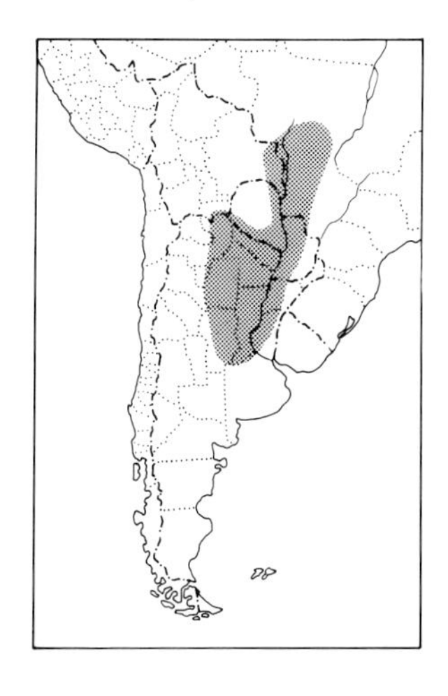

YELLOW-BILLED CARDINAL

IDENTIFICATION: 16.5 cm (6½"). *Bill pinkish yellow;* eyes orange. Resembles Red-capped Cardinal (especially *P. g. cervicalis*), with *completely red cap* and *black bib,* but has entirely pale bill and also an almost complete white nuchal collar (almost meeting on nape).
HABITAT AND BEHAVIOR: Common to locally abundant (especially in *pantanal* of Mato Grosso) around marshes, flooded grasslands, and along shores of lakes and rivers. Behavior similar to Red-capped Cardinal's, though in some areas much more numerous than that species ever is. In nonbreeding season it may gather in straggling flocks of up to hundreds of birds, with a large proportion of immatures.
RANGE: Sw. Brazil (sw. Mato Grosso in Paraguay River drainage), extreme e. (e. Santa Cruz) and s. (e. Tarija) Bolivia, w. Paraguay (east to Paraguay River), and n. Argentina (south to Tucumán, Córdoba, n. Buenos Aires, and Entre Ríos). To about 500 m.

GROUP B

Larger species; *gray above.*

Paroaria coronata

RED-CRESTED CARDINAL PLATE: 25

IDENTIFICATION: 19 cm (7½"). Bill mostly whitish. Head, crest, and long pointed bib (extending to center of breast) *bright scarlet.* Other-

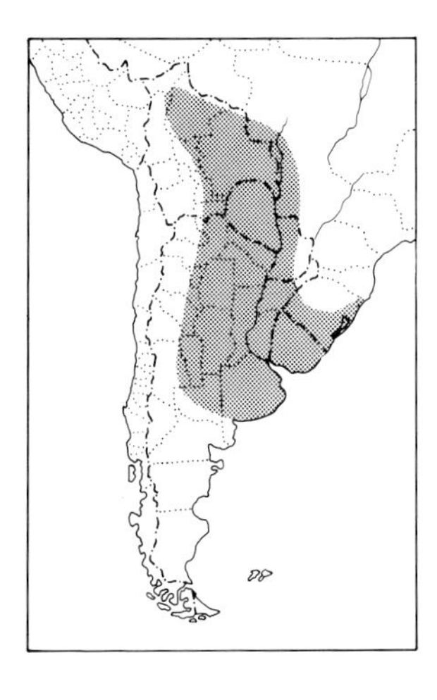

wise mostly gray above, with blacker wings and tail. Underparts and broad nuchal collar white.

SIMILAR SPECIES: See Red-cowled Cardinal. Other *Paroaria* cardinals are blacker (not gray) above and are smaller and *not crested.*

HABITAT AND BEHAVIOR: Common to locally abundant (especially in parts of *chaco*) in semiopen areas with shrubbery and scattered trees; most numerous near water, but not restricted to such situations. Usually in pairs or small groups, often feeding in the open on the ground; in some areas (especially in Argentina and Paraguay) large flocks are seen in the nonbreeding season. Has a rich melodic song, delivered deliberately and rhythmically (e.g., "weerit, churit, weer, churit," repeated several times with sequence variations). In demand as a cage bird, and as a result has declined in many populated areas.

RANGE: N. and e. Bolivia (north mainly to Santa Cruz, also recorded locally from s. Beni), extreme sw. and s. Brazil (sw. Mato Grosso and Rio Grande do Sul), w. Paraguay (east to about the Paraguay River), Uruguay, and n. Argentina (south to Mendoza, La Pampa, and Buenos Aires). To about 500 m.

Paroaria dominicana

RED-COWLED CARDINAL

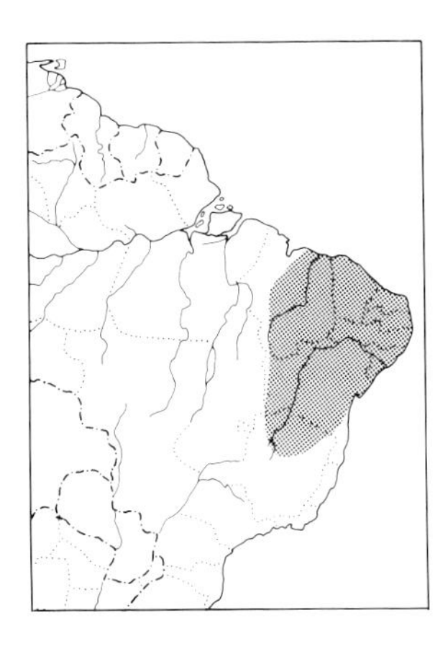

IDENTIFICATION: 18 cm (7"). *Ne. Brazil.* Lower mandible whitish. Resembles Red-crested Cardinal, though *not crested. Upper back black with white spotting,* lower back gray; overall effect mottled. *Flight feathers conspicuously edged white.*

SIMILAR SPECIES: See Red-crested Cardinal. Not known to occur with any other *Paroaria* cardinal.

HABITAT AND BEHAVIOR: Fairly common to common (especially in Bahia) in semiopen scrubby areas and lighter *caatinga* woodland. Unlike other *Paroaria* cardinals not strongly associated with water. Usually in pairs or loose groups, but may gather in large flocks when not breeding. Often kept as a cage bird.

RANGE: Interior ne. Brazil from s. Maranhão, Piauí, and Ceará south to n. Minas Gerais. Up to about 1200 m.

Cardinalis Cardinals

Long pointed crest; male basically red, female sandy buff with some red, both recalling the familiar Northern Cardinal (*C. cardinalis*) of North America. Xeric Caribbean coast.

Cardinalis phoeniceus

VERMILION CARDINAL

PLATE: 25

IDENTIFICATION: 18.5 cm (7¼"). *Arid Caribbean coast.* Stout bill pale grayish tipped blackish. *Long pointed crest.* Male *mostly bright rosy red,* wings and tail edged dusky. See C-55, V-39. Female sandy grayish above with *rosy red crest as in male;* tail tinged dark red. *Below mostly ochraceous buff.*

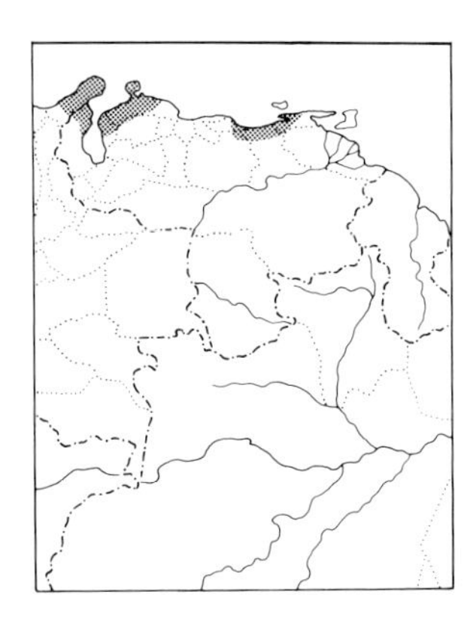

SIMILAR SPECIES: No other vaguely similar bird in range has such a long spikelike crest (usually held straight up).
HABITAT AND BEHAVIOR: Fairly common in dense thorny thickets and desert scrub along the coast. Usually in pairs and often shy, though in early morning birds (especially males) often take conspicuous perches, especially to sing. Song is a loud whistled "cheer, cheer, to-weet, to-weet, cheer, cheer," with many variations; overall it is quite reminiscent of the Northern Cardinal's (*C. cardinalis*). Regularly kept as a cage bird.
RANGE: Arid Caribbean coast in ne. Colombia (mostly on Guajira Peninsula, ranging west to ne. base of Santa Marta Mts.) and locally in n. Venezuela (east to Sucre). To about 300 m.

Gubernatrix Cardinals

Crested and boldly patterned with yellow. Locally in woodland and semiopen scrub in se. South America.

Gubernatrix cristata

YELLOW CARDINAL

PLATE: 25

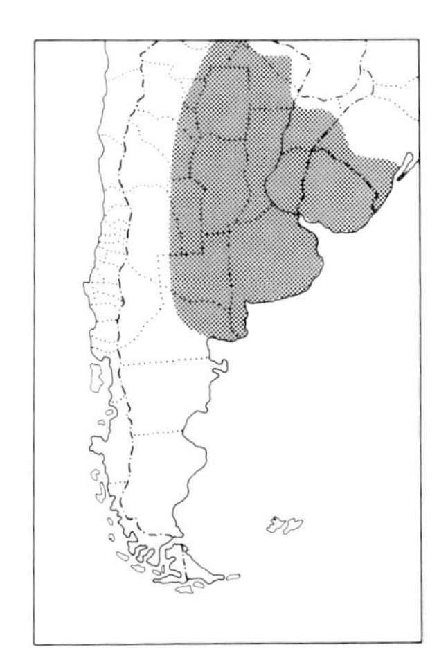

IDENTIFICATION: 20 cm (7¾"). *A strongly patterned, black-crested finch of e. Argentina and vicinity.* Male mostly olive above with *bright yellow superciliary and broad malar streak.* Outer tail feathers mostly yellow. Conspicuous *black throat patch,* otherwise olive yellow below, center of belly yellower. Female's overall pattern similar but much *grayer* (pure light gray on cheeks, breast, and sides), with *white on face* where male is yellow.
SIMILAR SPECIES: This elegant cardinal is essentially unmistakable; *nothing even vaguely similar is crested.*
HABITAT AND BEHAVIOR: Now local and generally uncommon to rare (numbers having been depleted in most areas by extensive trapping) in open woodlands and (especially southward) semiopen scrub. Usually conspicuous where it occurs, often perching on top of bushes and small trees; generally in pairs or small groups. Feeds mainly on ground. Has a loud, musical song (which, together with its attractive plumage, accounts for its popularity as a cage bird).
RANGE: Extreme se. Brazil (s. and w. Rio Grande do Sul), Uruguay, and locally across n. Argentina (Salta, Formosa, Corrientes, and s. Misiones south to Río Negro). To about 500 m.

Pheucticus Grosbeaks

Large thick-set finches with massive, heavy bills. All have white wing-patches. Arboreal in semiopen areas, montane and s. South America.

Pheucticus aureoventris

BLACK-BACKED GROSBEAK

PLATE: 25

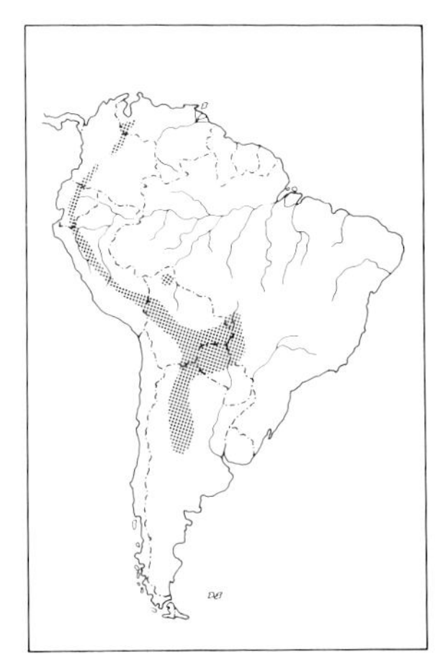

IDENTIFICATION: 22 cm (8½"). *Massive bill,* paler below. Variable. Male of most races (including *uropygialis*) basically *black above and on throat and chest,* yellow on lower underparts, with bold white patches on wings and large white tail corners. *Crissalis* from extreme s. Colombia and Ecuador has entire underparts yellow; it and *uropygialis* of Colombia's Cen. and E. Andes have rump mottled with yellow, while in *meridensis* of Venezuelan Andes rump is entirely yellow. Female like respective male but browner and more mottled with yellow above; below mostly yellow with black speckling except on belly; less white in tail. Female *uropygialis* has more black on throat than females of other races.

SIMILAR SPECIES: Only likely confusion is with Southern Yellow-Grosbeak, which has entire head yellow (male) or essentially yellow (female), never black as in this species.

HABITAT AND BEHAVIOR: Uncommon to fairly common (thinly spread, but usually conspicuous) in semiopen areas with scattered trees and shrubby places, mostly in arid regions (in the Andes) and dry *chaco* woodland. Generally found singly or in dispersed pairs; usually stolid, allowing a close approach, and frequently perches in the open. Has a rich mellow song, rather long and rapidly delivered, and a metallic call, both similar to Rose-breasted Grosbeak's.

RANGE: Locally on Andean slopes in w. Venezuela (Mérida), Colombia (E. and Cen. Andes), w. Ecuador, w. Peru, w. Bolivia (also 1 record from the north in Pando), nw. Argentina (south to San Luis and Córdoba), and east across the n. *chaco* into w. Paraguay (where scarce) to extreme sw. Brazil (w. Mato Grosso). To about 3000 m.

Pheucticus chrysogaster

SOUTHERN YELLOW-GROSBEAK

Other: Golden-bellied Grosbeak, Yellow Grosbeak

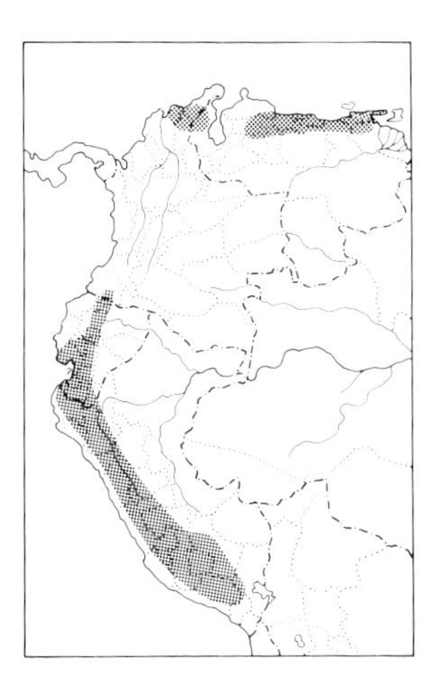

IDENTIFICATION: 21.5 cm (8½"). *Massive bill,* paler below. Resembles Black-backed Grosbeak, but *head and neck golden yellow* (nominate *chrysogaster*) *or yellow mottled olive* (*laubmanni,* of n. Colombia and Venezuela). Above otherwise mostly black, with bold white wing-markings and broad tail corners. Nominate race has pure yellow rump, while in *laubmanni* rump is mottled black and back has yellow streaking. Female duller, streakier, and more olive brown above, but still with *head and neck essentially yellowish.* See C-55, V-39.

SIMILAR SPECIES: Black-backed Grosbeak has entirely (male) or mostly (female) dark head.

HABITAT AND BEHAVIOR: Fairly common at forest borders, in semi-open, partially cultivated areas with scattered trees, and in deciduous woodland and scrub; not as restricted to arid regions as Black-backed Grosbeak. Behavior and voice similar to Black-backed Grosbeak's.

RANGE: Coastal mts. of n. Venezuela (Sucre west to Lara), Perijá Mts. on Colombia-Venezuela border, and Santa Marta Mts. in ne. Colombia; Andes of extreme s. Colombia (Nariño) south through w.

Ecuador (both in Andes and arid sw. lowlands) to s. Peru (south to Arequipa and Puno, mostly on Andean slopes, but in smaller numbers also in coastal lowlands). To about 3000 m (mostly 1500–2500 m in Venezuela and Colombia).

NOTE: We have tentatively followed the 1983 AOU Check-list in considering the S. Am. *chrysogaster* group as a species distinct from Mexican *chrysopeplus* and Middle American *tibialis* but are not entirely convinced that this is the proper course; *chrysopeplus* and *chrysogaster* are *very* similar. The previously suggested English name for this species ("Golden-bellied Grosbeak") is poor, as it implies a difference in this form which does not exist: its belly is no more "golden" than that of any of its relatives. We prefer to emphasize the distributional pattern of the superspecies by calling the S. Am. member the "Southern Yellow-Grosbeak."

Pheucticus ludovicianus

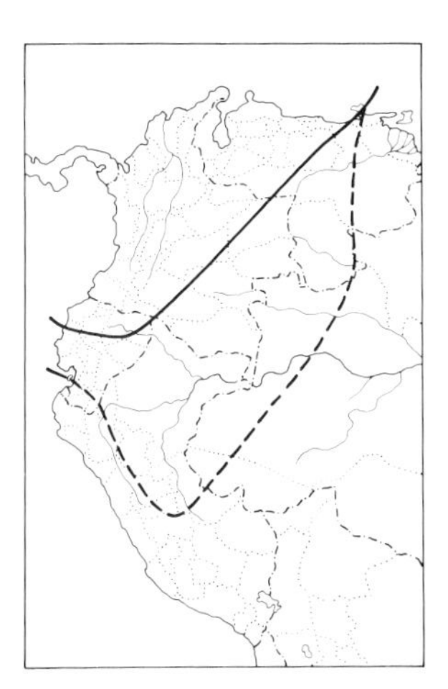

ROSE-BREASTED GROSBEAK

IDENTIFICATION: 18.5 cm (7¼"). *Heavy whitish bill.* Overall form much as in Black-backed Grosbeak but smaller, and color pattern very different. Breeding male (plumage acquired Feb.–Mar.) has *head, upperparts, and throat black,* with *rosy red patch on breast;* rump, lower underparts, and large patches in wing white; under wing-coverts pale rosy red. Nonbreeding male has buff superciliary, coronal stripe, and mottling on upperparts; below tinged buff on throat and breast with some blackish speckling, and *only a hint of the rose* (but some always present). Female resembles nonbreeding male but browner above with *whitish superciliary and coronal stripe* and *brown cheeks;* less white in wings; below whitish lightly streaked brown, never showing any pink; under wing-coverts yellow. See V-39.
SIMILAR SPECIES: Male's rosy breast is unique. Female can be more confusing; look for massive bill and robust overall shape, head striping, and fine streaking below.
HABITAT AND BEHAVIOR: Fairly common n.-winter resident (mostly Nov.–Mar.) in forest borders, lighter woodland, and semiopen areas with scattered trees. Usually found singly or in pairs, most often high in trees, but on migration sometimes in loose, small flocks of its own species. Its distinctive metallic call is often heard (a single "pink"), but even on spring migration males are almost never heard to sing.
RANGE: Nonbreeding visitor to Colombia (not in southeast) and n. Venezuela; more rarely to Ecuador, cen. Peru, s. Venezuela, and Guyana (once). Breeds in North America, wintering from Mexico south; migrates through West Indies. To about 2000 m; most numerous in upper tropical and subtropical zones.

Cyanocompsa and *Cyanoloxia* Grosbeaks

Medium-size blue (males) or brown (females) finches with heavy bills. Favored habitats also vary from humid lowland forest to woodland edge and scrub. Often in pairs and generally inconspicuous except when singing. Some recent authors have advocated merging *Cyano-*

compsa into the bunting genus *Passerina,* but we have opted to maintain it, at least for the present. *Cyanoloxia* is a monotypic genus very like *Cyanocompsa,* but with a smaller bill.

Cyanocompsa brissonii

ULTRAMARINE GROSBEAK

PLATE: 25

IDENTIFICATION: 15–17 cm (6–6¾"). *Heavy bill* with pale base of mandible. *Male mostly dark blue,* with paler, brighter blue forehead, superciliary, malar area, and shoulders. Female cocoa brown above, *paler fulvous-brown below.* N. birds considerably smaller than those from s. part of range (including *sterea*), and male has paler blue rump; see C-55 and V-39.

SIMILAR SPECIES: Respective sexes of Blue-black Grosbeak are very similar, though the 2 tend to separate out by habitat (Blue-black in humid forest) and range. Ultramarine's bill is slightly less heavy and its culmen more curved; n. Ultramarines are notably smaller than Blue-black. Smaller Glaucous-blue Grosbeak has notably stubbier bill, and male is a markedly different shade of blue. Female Ultramarine also easily confused with female Great and Large-billed Seed-Finches, but their bills are proportionately more massive with a straight culmen, and their overall tone of brown is more olivaceous (not as warm); their habitats also usually differ.

HABITAT AND BEHAVIOR: Fairly common in dense thickets in semi-open area, in undergrowth and edge of *chaco* woodland, and (in Colombia and Venezuela) in semiarid scrub. Usually in pairs and often reclusive and hard to more than glimpse; in early morning occasionally perches prominently on top of a bush or low tree. Male of s. race (*sterea*) has fairly loud musical song, slow at first, then characteristically slurring downward.

RANGE: Locally in sw. Colombia (upper Cauca, Patía, and Dagua valleys) and n. Venezuela (Falcón and Lara east to Sucre and south to Guárico); e. Brazil (Piauí, Ceará, and Pernambuco southward) south and west across Paraguay and n. and e. Bolivia (northwest to La Paz) to n. Argentina (south to San Luis, Córdoba, Santa Fe, and Entre Ríos; a few recent sightings from n. Buenos Aires); recent records from w. Uruguay (Río Negro). To about 1500 m.

NOTE: The suggested merger (*Birds of the World,* vol. 13) of the genera *Cyanocompsa, Cyanoloxia,* and *Guiraca* into the N. Am. bunting genus *Passerina* (which we do not follow) has resulted in a change in the species name of the Ultramarine Grosbeak. In order to preserve the name *P. cyanea* for the familiar (to North Americans) Indigo Bunting, the older name for the Ultramarine Grosbeak, *cyanea,* was suppressed in favor of *brissonii.*

Cyanocompsa cyanoides

BLUE-BLACK GROSBEAK

IDENTIFICATION: 16 cm (6¼"). Closely resembles Ultramarine Grosbeak (little or no ecological or range overlap). Build somewhat more robust, with slightly heavier bill with rather straight culmen. Male *rothschildi* of Amazonia very similar, but has *rump concolor with back*

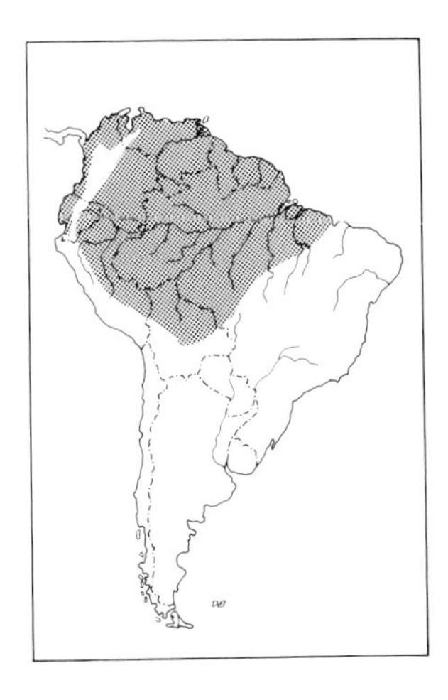

(not brighter, paler blue of n. races of Ultramarine). Nominate race (west of Andes) also dark-rumped and has eyebrow, malar area, and shoulders somewhat less bright blue. See C-55, V-39, P-31. Female rather *uniform deep chocolate brown* (unlike Ultramarine not notably paler brown below).

SIMILAR SPECIES: Cf. Ultramarine Grosbeak. Female quite closely resembles female Great-billed Seed-Finch, but grosbeak's bill is more conical and less massive and overall color is warmer (less olive) brown; habitat and behavior also differ markedly.

HABITAT AND BEHAVIOR: Fairly common to common (though often only as an elusive voice in the forest) in undergrowth of humid forest and forest borders and in more mature secondary woodland. Rarely or never fully in the open. Usually in pairs, which typically forage independently of other birds; shy and furtive. The call is a frequently heard sharp "chink," the song a series of rich notes, first slow, hesitant, and rising, then falling and more jumbled.

RANGE: Widespread in more forested lowlands south to n. Bolivia (south to La Paz, Cochabamba, and w. Santa Cruz) and Amaz. Brazil (south to cen. Mato Grosso and n. Maranhão); on Pacific slope south to sw. Ecuador (El Oro). Also s. Mexico to Panama. Mostly below 1000 m.

Cyanoloxia glaucocaerulea

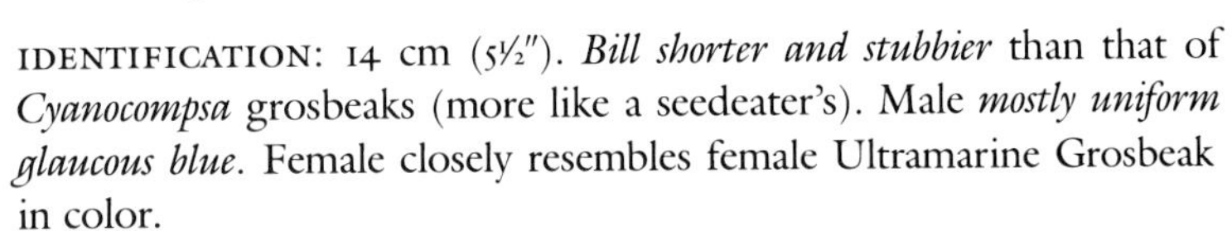

GLAUCOUS-BLUE GROSBEAK

PLATE: 25

Other: Indigo Grosbeak

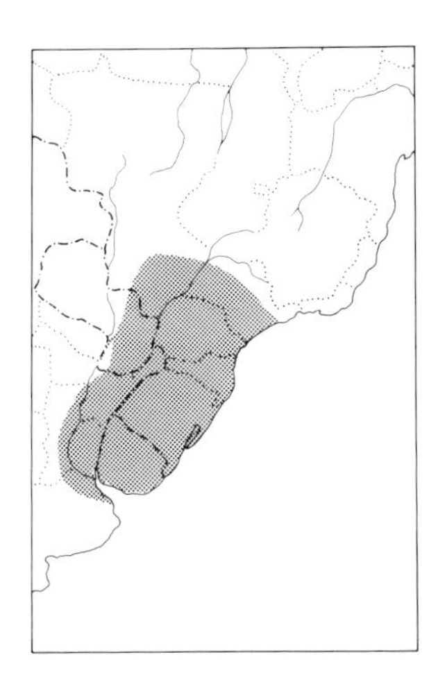

IDENTIFICATION: 14 cm (5½"). *Bill shorter and stubbier* than that of *Cyanocompsa* grosbeaks (more like a seedeater's). Male *mostly uniform glaucous blue*. Female closely resembles female Ultramarine Grosbeak in color.

SIMILAR SPECIES: Ultramarine Grosbeak is notably larger with heavier bill; male is darker blue (lacking this species' grayish suffusion), but female best told by its smaller size and differently proportioned stubbier bill. Blackish-blue Seedeater, especially female, is also similar but has shorter, more conical bill and is brighter tawny-brown; note habitat differences.

HABITAT AND BEHAVIOR: Uncommon in forest and woodland borders and shrubby second growth (infrequent in interior of actual forest). Usually in pairs, and like the other "blue grosbeaks" often shy and retiring. Singing males may mount to an exposed perch (even a wire) to give their rapid, complex, warbling song (without down-slurred effect of Ultramarine's).

RANGE: S. Brazil (north to w. São Paulo and s. Mato Grosso) south through Uruguay and locally in e. Paraguay (few records) to ne. Argentina (south to ne. Buenos Aires and Santa Fe). To about 900 m.

NOTE: The previously suggested name for *C. glaucocaerulea* ("Indigo Grosbeak") is poor as it is even less an indigo blue than is the Indigo Bunting of North America, *P. cyanea*. We suggest a straight translation of its Latin species name, which is an accurate description of the male's color; Hudson (1920:44) called it simply "Glaucous Grosbeak."

Emberizinae

EMBERIZINE FINCHES

A large complex, the Emberizine finches are an imprecisely defined lot whose relations to the tanagers have yet to be resolved (most characters attributed to one or the other group have been shown to break down). In South America its most important genera are the *Sporophila* seedeaters, the *Atlapetes* brush-finches, the *Poospiza* warbling-finches, the *Phrygilus* sierra-finches, and the *Sicalis* yellow-finches, with numerous other smaller genera (including all the sparrows) thrown in. A majority of the S. Am. finches are found in open or semiopen country, though *Atlapetes* and associated genera are mainly birds of forest undergrowth and edge. Most species have relatively heavy conical bills adapted to opening seeds, but there are numerous exceptions, and even those species with large bills are often not entirely granivorous. Many of these finches have loud and often very attractive songs.

Porphyrospiza Finches

A monotypic small finch with distinctive slender yellow bill. Restricted to *cerrado* of Brazil. Its systematic position remains obscure, but we feel that any similarity to the *Passerina*/*Cyanocompsa* complex is purely coincidental and that the merger of *Porphyrospiza* into *Passerina* (*Birds of the World*, vol. 13) is incorrect.

Porphyrospiza caerulescens

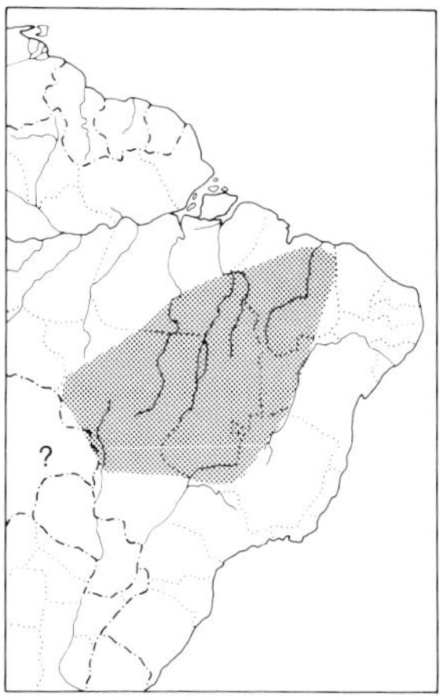

BLUE FINCH

PLATE: 25

IDENTIFICATION: 12.5 cm (5"). *Cerrado of Brazil. Slender bill bright yellow* (with blackish ridge in female); legs dull reddish. Male *uniform bright cobalt blue,* but molting males in fresh plumage have feathers broadly edged rufous-brown (in some almost totally obscuring the blue). Female rufous-brown above, buffy whitish below with *considerable dusky streaking.*

SIMILAR SPECIES: Male is only blue bird in South America with a yellow bill. The streaky brown female can be more confusing, but no similar bird shares the yellow on bill.

HABITAT AND BEHAVIOR: Generally rare to uncommon and local (but perhaps seasonally more numerous?) in open grassy *cerrado* with scattered bushes and low trees. Not very well known, and may be declining due to conversion of much of its *cerrado* habitat for agriculture. In nonbreeding season gathers in small groups which feed on the ground in or near tall grassy cover; seems usually not to associate with other small finches. Breeding males sing a pretty, easily recognized, high, thin "swee-sweeu, swee-sweeu . . ." (usually in pairs of phrases), repeated many times from a perch in a low tree.

RANGE: Interior ne. and cen. Brazil from se. Pará, s. Maranhão, Piauí, and w. Bahia to w. Minas Gerais and s. Mato Grosso; also recorded (questionably?) from se. Bolivia (Cuevo in Chuquisaca). To about 1100 m.

Amaurospiza Seedeaters

A pair of rather rare, inconspicuous seedeaters of lower growth in montane forest. Local (relict?) distribution in n. Andes and se. South America.

Amaurospiza moesta

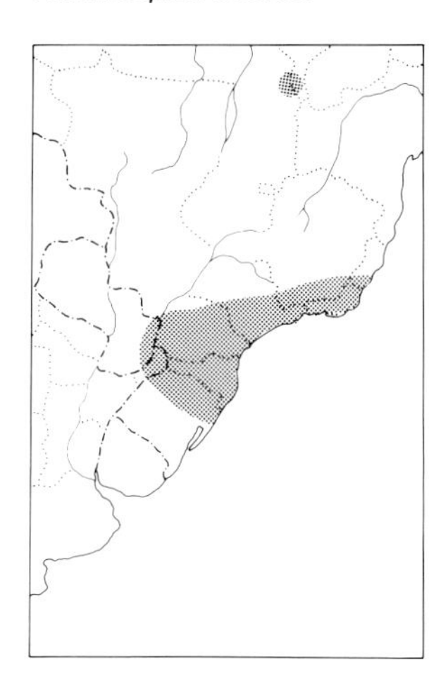

BLACKISH-BLUE SEEDEATER PLATE: 25

IDENTIFICATION: 12.5 cm (5"). *Male uniform slaty blue;* face, throat, and breast slightly more blackish; under wing-coverts white. Female *rather bright tawny-brown,* slightly paler below; under wing-coverts white.
SIMILAR SPECIES: Male recalls male Blue-black Grassquit, but note latter's glossier blue-black plumage and slenderer, more pointed bill. Female quite closely resembles female Lesser Seed-Finch in color, but bill not nearly as heavy. Female tawnier than female of somewhat larger Glaucous-blue Grosbeak. Cf. various female *Sporophila* seedeaters, most of which are smaller and not as bright brown; female Buffy-fronted (with which Blackish-blue is sympatric) has buffy wing-bars.
HABITAT AND BEHAVIOR: Rare (but probably somewhat overlooked) in undergrowth of forest and woodland, particularly where there is extensive growth of bamboo. Not well known and often difficult to observe. Found singly or in pairs, feeding on or near the ground. The song is a bright, spirited, warbling "swee-swee-swi-sweeseeseeu," somewhat variable but always fast in tempo.
RANGE: E. Brazil (1 old record from Tranqueira in s. Maranhão and from Espírito Santo to n. Rio Grande do Sul) and ne. Argentina (Misiones) and se. Paraguay (recorded by Bertoni in early 20th century from Alto Paraná, but not since). To about 1200 m.

Amaurospiza concolor

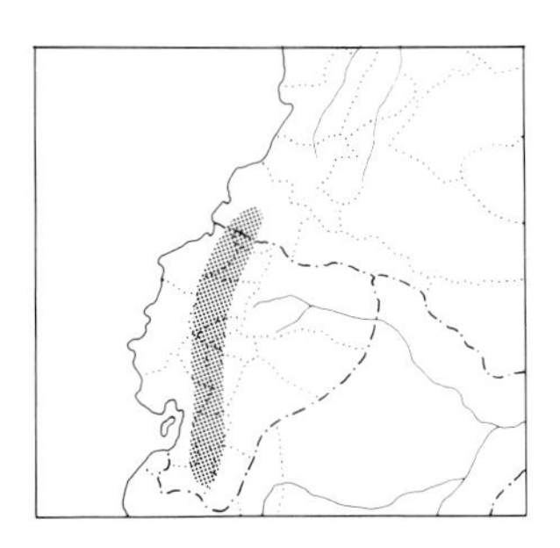

BLUE SEEDEATER

IDENTIFICATION: 12.5 cm (5"). Closely resembles Blackish-blue Seedeater (no overlap). Male slightly brighter, *more dusky blue* than slaty blue, with only foreface slightly more blackish (not entire breast); under wing-coverts brown. Female probably indistinguishable under field conditions, but has brown (not white) under wing-coverts. Some birds (perhaps immature males) are a deeper, more rufous brown. See P-31.
SIMILAR SPECIES: A rather confusing, rare species; often its Andean forest habitat provides the best clue. See discussion under Blackish-blue Seedeater.

HABITAT AND BEHAVIOR: Rare and local in undergrowth of overgrown forest borders and secondary woodland; in Panama usually observed in association with *Chusquea* bamboo, but this has not been noted in South America. Seen singly or in pairs, at times with mixed foraging flocks, usually close to the ground; generally very inconspicuous. Male's song is a short, weak warble, "sweet sweet sweet sa-weet."
RANGE: Pacific slope of sw. Colombia (Nariño) and locally in w. Ecuador (Pichincha, Chimborazo, s. Loja). Also locally s. Mexico to Panama. 1100–2000 m.

Haplospiza Finches

A pair of woodland or forest-inhabiting small finches; usually in montane areas, most often where there is extensive bamboo. Note their sharply pointed bills.

Haplospiza unicolor

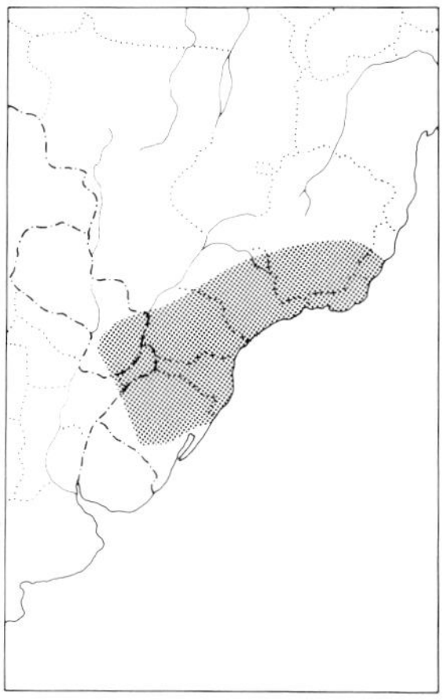

UNIFORM FINCH

PLATE: 25

IDENTIFICATION: 12.5 cm (5″). *Bill conical and sharply pointed.* Male *uniform gray.* Female olive brown above. Below dull whitish tinged yellowish (especially on breast), more olive brown on flanks, with *blurry dusky streaking* (especially across breast).
SIMILAR SPECIES: Male is *only uniformly gray finch in its range.* Cf. Sooty Grassquit. Female is potentially more confusing, but note bill shape and streaking below. Female Blue-black Grassquit smaller with slenderer bill, and its streaking does not reach throat; grassquit usually in open grassy areas.
HABITAT AND BEHAVIOR: Uncommon to fairly common (though can locally, and perhaps temporarily, be much more numerous) in undergrowth of humid forest, especially where there is an extensive growth of bamboo. Often quite difficult to observe, remaining hidden in lower growth, but at other times may be more arboreal and visible. Males have an explosive, buzzy song, somewhat variable but typically "gl-zhwe-e-e-e-e-e."
RANGE: Se. Brazil (north to s. Minas Gerais and Espírito Santo), e. Paraguay, and ne. Argentina (Misiones). To about 1400 m.

Haplospiza rustica

SLATY FINCH

IDENTIFICATION: 12.5 cm (5″). Closely resembles Uniform Finch (but no range overlap). *Bill slightly more slender and sharply pointed.* Male slightly darker and a little more bicolored (darker gray above, slightly paler below). Female has streaking extending only to breast (not to belly) and is somewhat browner (not so olivaceous) overall. Flight feathers narrowly edged rufous. See V-40, C-56.
SIMILAR SPECIES: Cf. Uniform Finch. Male similar in color to male Plumbeous Sierra-Finch, but latter found strictly in *páramo* and *puna*

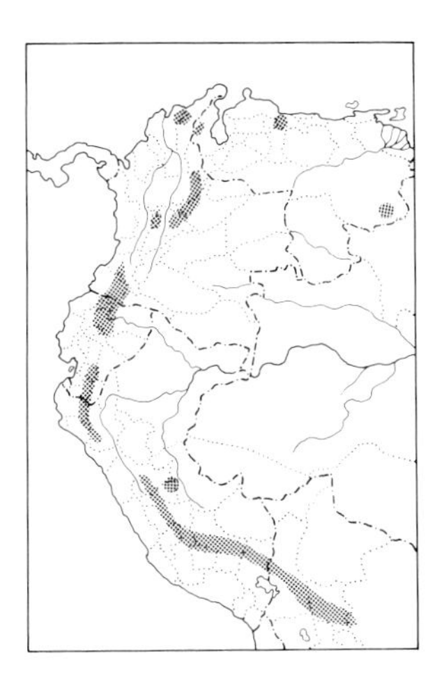

at higher elevations than the finch (mostly in subtropical forests). Female resembles female Blue-black Grassquit but is larger with streaking extending onto throat; note their very different habitats.

HABITAT AND BEHAVIOR: Rare to locally (or temporarily) uncommon in undergrowth and borders of humid montane forest. Usually seen singly or in pairs, but often unobtrusive and hard to see; at times gleans from leaves and branches, but generally remains in lower growth. Song a fast, complex burst of chips, buzzes, and trills, often ending in a buzzy trill.

RANGE: Very locally from n. Venezuela (Distrito Federal) south in Andes through Colombia (including the Perijá and Santa Marta Mts.), Ecuador and Peru (on w. slope south to Lambayeque) to n. Bolivia (south to w. Santa Cruz); s. Venezuela in Bolívar (Cerro Chimantá-tepui). Also s. Mexico to Panama. Mostly 1500–2500 m.

Oryzoborus Seed-Finches

The huge, squared-off bill marks this genus (closely related to *Sporophila,* with which it has fairly often hybridized in the wild, causing some authors to advocate their merger). Males are entirely or mostly black, females rich brown. Unlike *Sporophila,* seed-finches are not gregarious. In some areas, notably Trinidad and Brazil, they have been much persecuted by bird trappers to the point where the larger species especially have been locally extirpated.

LESSER SEED-FINCH

Oryzoborus angolensis

PLATE: 25

Other: Thick-billed or Chestnut-bellied Seed-Finch

IDENTIFICATION: 12.5 cm (5"). *Bill very heavy* (though not as extreme as that of other members of the genus), black. Male mostly glossy black with *contrasting dark chestnut breast and belly* (can look all black in poor light). Small wing-speculum and under wing-linings white. West of Andes *funereus* is *all black* (except for white on wing); see P-31. Some Santa Marta birds show some chestnut on belly, and in upper Magdalena valley *theobromae* has breast and belly chocolate brown. Female dull brown above, fulvous-brown below, with white under wing-linings (but no speculum).

SIMILAR SPECIES: Cf. Large-billed and Great-billed Seed-Finches (both of which have much more massive bills). Female Lessers can be confusing but have notably larger, more squared-off bill than any female *Sporophila* seedeater (most of which are also smaller). They most resemble females of the 2 rare *Amaurospiza* seedeaters; note the seed-finch's markedly heavier bill.

HABITAT AND BEHAVIOR: Fairly common in shrubby or grassy clearings and in low growth at the edge of woodland or forest. Found singly or in pairs, usually independent of other birds, though may join a seedeater flock. Males often sing from a prominent perch, a long series of musical whistled notes, more separated at first, more jumbled and twittery toward the end.

RANGE: Widespread in more humid lowlands south on Pacific slope to sw. Ecuador (El Oro) and to n. Bolivia, e. Paraguay, s. Brazil (south to Rio Grande do Sul, where rare), and ne. Argentina (Corrientes and Misiones). To about 1500 m, but mostly below 1000 m.

NOTE: We concur with most recent authors, most recently S. Olson (*Auk* 98 [2]: 379–381, 1981), in considering the trans-Andean and Middle American *funereus* group (Thick-billed Seed-Finch) as conspecific with *angolensis* of South America east of the Andes (Chestnut-bellied Seed-Finch); the 1983 AOU Check-list, however, retained *funereus* as a full species. Songs from throughout its range appear similar, and seemingly intermediate populations and individuals have been found in w. Colombia (see Olson, op. cit.).

Oryzoborus maximiliani

GREAT-BILLED SEED-FINCH

PLATE: 25

Other: Greater Large-billed Seed-Finch

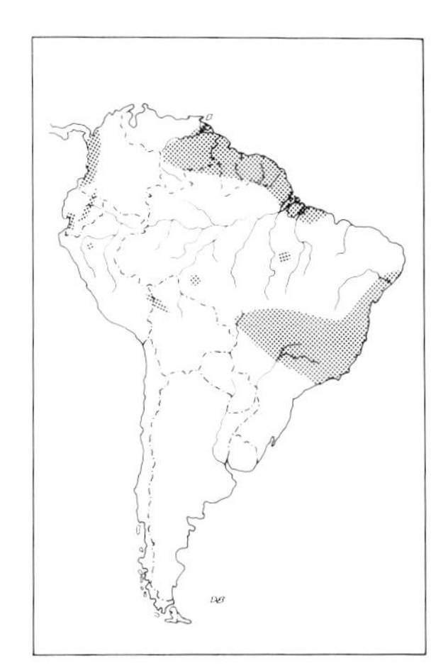

IDENTIFICATION: 14.5–16.5 cm (5¾–6½"). *Bill enormously thick, chalky whitish* in most of range (but *black* in upper Amazonia); black in female. Bill especially thick in *gigantirostris* of se. Colombia and n. Bolivia. Male black, with under wing-coverts white or white mixed with black; nominate race (s. Brazil) and *magnirostris* (ne. South America) also have conspicuous white wing-speculum. Female brown above, buffy brown below; under wing-coverts white (but no speculum).

SIMILAR SPECIES: Large-billed Seed-Finch is *very* similar to the white-billed races of this species (see full discussion under that species). The bill of this species is sufficiently huge to distinguish it from virtually anything else; cf. female Blue-black and Ultramarine Grosbeaks.

HABITAT AND BEHAVIOR: Rare to uncommon (numbers often depleted by cage bird trappers) in damp pastures, moist tall grassy areas along lake and river shorelines, and shrubbier emergent vegetation in marshes. Does not seem to be found far from water. Usually found in well-dispersed pairs. Song in sw. Colombia (*occidentalis*) quite similar to that of *O. crassirostris,* a series of rich, gurgling, whistled notes lasting 4 to 5 seconds, delivered rapidly after some slower introductory calls; the overall effect somewhat Bobolink-like (*Dolichonyx*). In se. Peru the song of *atrirostris* (*gigantirostris*?) is a similar but more leisurely warbling, perhaps even richer in quality.

RANGE: Pacific slope in w. Colombia (Chocó south to Nariño) and sw. Ecuador (Guayas and El Oro); se. Colombia (Putumayo), e. Ecuador (Limoncocha in Napo; also seen south of Sucúa in Morona-Santiago), e. Peru (very local; known only from San Martín and at Tambopata in Madre de Dios), and n. Bolivia (Chatarona in Beni); locally in e. Venezuela (lower Orinoco valley), Guyana (Annai), French Guiana, and lower Amaz. Brazil (Amapá and the Belém region), locally southward in e. and s. Brazil (Alagoas south to São Paulo and west to cen. Mato Grosso; recorded by Sick [1984] from se. Pará and Rondônia as well). To 1100 m.

NOTE: With 1 exception, we have followed the split of *O. maximiliani* from *O. crassirostris* as set forth by Meyer de Schauensee (*Notulae Naturae* 428, 1970).

Novaes (1978: 63) records specimens in breeding condition of both *O. crassirostris* and *O. m. maximiliani* from Amapá, establishing the very strong likelihood that the 2 do in fact breed in virtual (or actual) sympatry. However, we feel that more work on some of the taxa in this complex is still needed. Notably, wing measurements of *occidentalis* (w. Colombia and w. Ecuador) fall with *O. crassirostris* and not with *O. maximiliani* (where Meyer de Schauensee placed it, mainly because of its "wing-tail index," by which it does indeed seem closer to *O. maximiliani* because of its comparatively long tail). The distribution pattern of this form, however, seems to indicate that it must surely be closer to *O. crassirostris;* the 2 forms are parapatric in nw. Colombia, with the other white-billed races of *O. maximiliani* only being found in far-distant e. South America. The black-billed *atrirostris/gigantirostris* races of w. Amazonia pose another problem: they vary in a very unusual way (with extremely large-billed birds at either end of the range [*gigantirostris*] and somewhat smaller-billed birds [*atrirostris*] in the middle), and they differ far more strikingly from the other races of *O. maximiliani* than *O. crassirostris* does. Probably *O. atrirostris* deserves full species status (as Black-billed Seed-Finch), but as with the shift of *occidentalis* we hesitate to do so without a more thorough analysis of the situation. Finally, we have followed the suggestions of F. G. Stiles (*Condor* 86 [2]: 118–122, 1984) regarding the status of *O. nuttingi* (Nicaraguan Seed-Finch) as a full species, though it was considered a race of *O. maximiliani* by Meyer de Schauensee (op. cit.). Again partly on distributional grounds, we suspect that *O. nuttingi* is in fact closer to *O. crassirostris* than it is to *O. maximiliani.*

Oryzoborus crassirostris

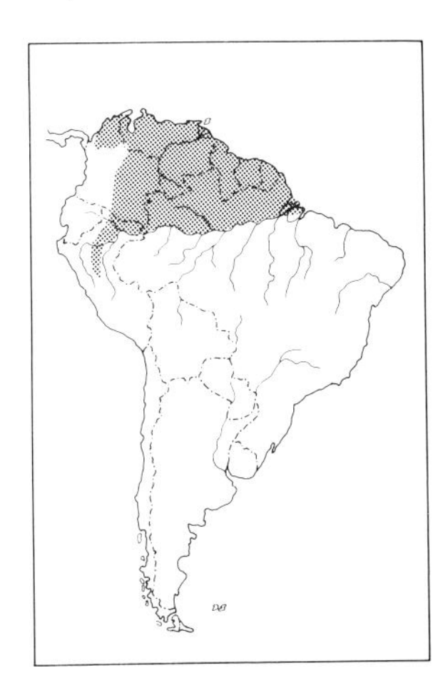

LARGE-BILLED SEED-FINCH

IDENTIFICATION: 13.5–14 cm (5¼–5½"). *Very* closely resembles Great-billed Seed-Finch, and in areas of overlap with pale-billed races of that species (e. Venezuela, Guianas) probably not safely distinguished in the field. *Bill not quite as massive,* always *chalky white in male,* black in female. In ne. Peru (another area of apparent sympatry) bill color would distinguish between males of the 2 species (pale in Large-billed, black in that race of Great-billed). Elsewhere the 2 are virtually identical, but Large-billed is smaller, with proportionately shorter tail (53–59 mm vs. 61–70 mm in *maximiliani*) and smooth and shiny bill texture (not dull and bonelike).

HABITAT AND BEHAVIOR: Similar to Great-billed Seed-Finch's and likewise locally depleted in numbers by the activities of bird trappers. Reportedly quite numerous along the n. bank of the lower Amazon and on Mexiana Is. in its mouth (Sick 1984).

RANGE: Locally in n. and e. Colombia (humid Caribbean lowlands west to Córdoba and locally east of Andes), Venezuela (not recorded from northwest or from most of llanos region), Guianas, Amaz. Brazil (south only to the Amazon), and ne. Peru (Loreto to Huánuco); Trinidad (only formerly?). Mostly below 500 m.

NOTE: See comments under Great-billed Seed-Finch.

Volatinia Grassquits

Abundant small finch of open areas. Often with *Sporophila* seedeaters, but note more slender, pointed bill. Builds a shallow cup-shaped nest (unlike *Tiaris*). It has been suggested (D. Steadman, *Proc. San Diego*

Soc. Nat. Hist. 19 [19]: 279–286, 1982) that the genus *Volatinia* be merged with *Geospiza* (of the Galápagos Is.), but we are not entirely convinced that *Volatinia* is *Geospiza*'s closest mainland relative.

Volatinia jacarina

BLUE-BLACK GRASSQUIT

PLATE: 26

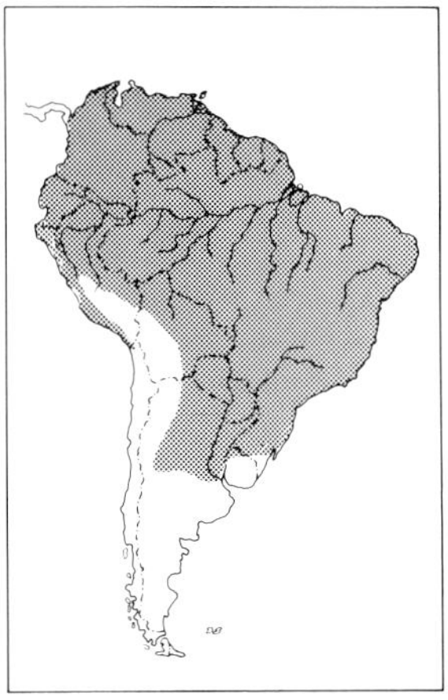

IDENTIFICATION: 10–11 cm (4–4¼"). *Uniform glossy blue-black;* axillars (and sometimes entire wing-lining) white, often visible in flight. Female dull brown above; whitish to pale buff below, with *breast and flanks streaked dusky.* Immature male resembles female but more blackish above; subadults are mottled blackish and brown (at times many more males in this plumage are seen than full adults, and they are known to breed before attaining complete adult plumage).
SIMILAR SPECIES: No *Sporophila* seedeater is entirely black (male) or so prominently streaked below (female); note also difference in bill shape (*narrower and more pointed in this species*). Cf. also male Sooty Grassquit (which can look quite similar, and bill shape is similar) and female Slaty and Uniform Finches.
HABITAT AND BEHAVIOR: Common to locally abundant in all types of open to semiopen country, especially in agricultural areas; also frequent around towns and habitations. One of the more familiar small birds of such situations, usually in scattered pairs when nesting, but at other seasons often gathers in flocks (sometimes large, and regularly in association with various seedeaters). Generally less numerous or even absent from appropriate habitat in extensively forested regions. Males tirelessly repeat a buzzy, explosive "dzee-uu," usually accompanied by a short jump into the air; this is especially frequent during the breeding season but is also given at other times (when usually less persistent).
RANGE: Widespread in lowlands (mostly) south to n. Chile (Arica) and cen. Argentina (to Mendoza and Buenos Aires, though rare at least in the latter), but not recorded from Uruguay and seemingly absent from much of w. Amazonia; Trinidad; has straggled to Curaçao and Bonaire. Also Mexico to Panama, and on Grenada. To about 2000 m, but primarily below 1000 m.

Tiaris Grassquits

Resemble *Sporophila* seedeaters, but bill narrower and more pointed. Both sexes relatively dull-plumaged. Grassy or shrubby areas, where often with seedeaters, though usually not as numerous. The nest is globular or dome-shaped, with a side entrance (very unlike that of *Volatinia* or *Sporophila*).

Tiaris fuliginosa

SOOTY GRASSQUIT

PLATE: 26

IDENTIFICATION: 11.5–12 cm (4½–4¾"). Bill blackish with pink gape in male, duskier with yellowish base in female. Male *uniform sooty blackish,* tinged olive especially on mantle, grayer and often paler below

(especially on belly). Female dull olive brown above, somewhat brighter olive brown below, paling to whitish on center of belly.

SIMILAR SPECIES: Easily confused (especially female). Male Blue-black Grassquit is a glossier, bluer black than this species; even immatures are never as dull or sooty as this species (instead they are mottled brownish). Male Uniform Finch is slightly larger but with proportionately smaller bill and is notably grayer. Female is difficult: it most resembles Black-faced Grassquit but is slightly larger and somewhat darker (probably safely distinguished only when accompanied by males; note also range differences). Also very close to Dull-colored Grassquit (both sexes) but darker brown generally (not pale grayish brown) and without that species' clearly bicolored bill.

HABITAT AND BEHAVIOR: Uncommon and local in clearings and forest or woodland edge situations and also (at least in Colombia) in drier grassy scrub; not well known on the S. Am. mainland (apparently more numerous on Trinidad). Usually found in pairs or small loose groups, foraging in grass independently of other birds. Male's song reported to be a thin, high, wiry "eez-uda-lee," given quickly and run together (S. Hilty).

RANGE: Locally in w. Colombia (upper Magdalena and upper Patía valleys) and n. Venezuela (Perijá Mts.; Carabobo east to Sucre); se. Venezuela (se. Bolívar) and s. Guyana (Bat Mt.); s. Brazil (e. lowlands from Pernambuco south to e. São Paulo and in cen. Mato Grosso); Trinidad. To about 1500 m.

Tiaris bicolor

BLACK-FACED GRASSQUIT

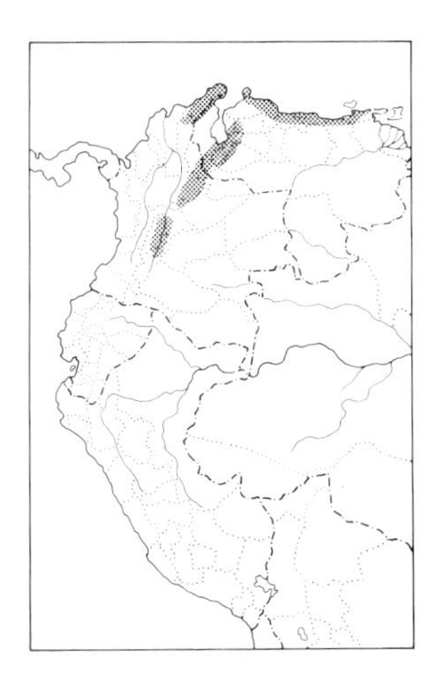

IDENTIFICATION: 10 cm (4"). Closely resembles Sooty Grassquit (females may not always be distinguishable). *Bill blackish with pink gape, somewhat paler in female.* Male dull olive above, becoming *sooty blackish on face and most of underparts;* more olive grayish on belly. Female dull olive above, pale olive grayish below. See C-56, V-40.

SIMILAR SPECIES: Male Sooty Grassquit is uniformly sooty blackish (not so obviously olive above). Female Black-faced and Sooty Grassquits distinguished only with difficulty; Black-faced is slightly smaller, not as brown (more olive grayish). Yellow-faced Grassquit has yellow facial pattern (obvious in male, generally discernible even in young female).

HABITAT AND BEHAVIOR: Uncommon in arid scrub, grassy areas with scattered bushes, and edge of dry woodland; not as numerous or familiar in South America as it is in the West Indies. Usually in pairs or small groups; generally feeds on the ground. Song is a weak, buzzy "tz-tzeeteeeteeeeee," sometimes given in a short, fluttery display flight.

RANGE: Locally in w. Colombia (Guajira Peninsula to s. base of Santa Marta Mts., in Norte de Santander and Boyacá, and in upper Magdalena valley from w. Cundinamarca to Huila) and n. Venezuela (Zulia east to Sucre and inland to Mérida and Portuguesa); Tobago. Also most of West Indies (except Cuba). Mostly below 1000 m.

Tiaris obscura

DULL-COLORED GRASSQUIT

PLATE: 26

Other: Dull-colored Seedeater

IDENTIFICATION: 11 cm (4¼"). *Bill usually bicolored,* dusky above and yellowish below. *A dull, hen-colored grassquit;* sexes virtually alike. Dull grayish olive brown above, paler brownish gray below, lightening to whitish on belly, sometimes with tinge of buff on flanks and crissum. *Haplochroma* (ne. Colombia and w. Venezuela) is browner above and considerably darker grayish brown below, while *pacifica* (coastal s. Peru) is grayer.

SIMILAR SPECIES: Likely to be overlooked as it resembles many female *Sporophila* seedeaters. Seems usually not to associate with them, so watch for singing males, pairs, or groups of seedeaterlike birds which all appear alike. The bicolored bill is a helpful clue, but some female *Sporophila* show this, and shape of this species' bill is not dissimilar. Cf. also female Sooty and Black-faced Grassquits.

HABITAT AND BEHAVIOR: Uncommon and somewhat local in overgrown clearings, woodland borders, and thickly vegetated gardens. Usually occurs singly or in pairs, less often in small groups; rarely with other seedeaters. Generally inconspicuous and not often encountered. Song is an explosive, buzzy "zeetig, zeezeezig" with variations, resembling other grassquits' more than a *Sporophila*'s.

RANGE: Mostly on lower montane and Andean slopes in n. and w. Venezuela (Miranda, Mérida, Táchira, and Perijá Mts.), Colombia (Santa Marta and Perijá Mts., above middle Magdalena valley, lower Cauca valley, and Pacific slope from Valle south), w. Ecuador, w. and s. Peru (south on Pacific slope, including lowlands, to Arequipa), w. Bolivia, and nw. Argentina (south to Tucumán). Mostly 500–2000 m, locally higher and lower.

NOTE: This species was formerly considered a *Sporophila* seedeater. According to a note in *Birds of the World* (vol. 13), P. Schwartz found the nest in Venezuela to be domed (like other *Tiaris*'s) and not cup-shaped (like all known *Sporophila*'s). We further note that its buzzy voice is much more like that of a grassquit than that of a *Sporophila*.

Tiaris olivacea

YELLOW-FACED GRASSQUIT

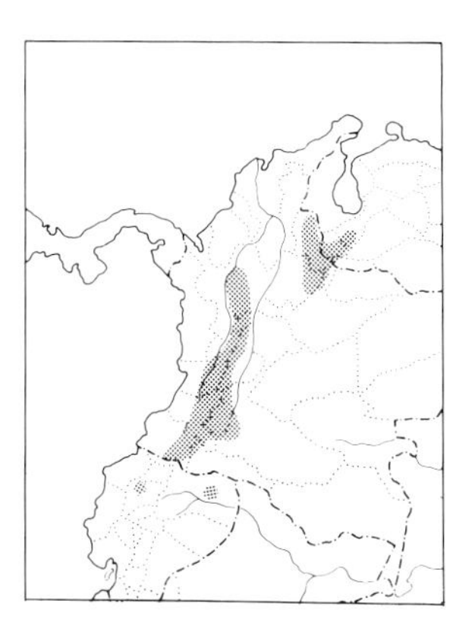

IDENTIFICATION: 10 cm (4"). Olive above except for *black crown, sides of head, and breast,* with *contrasting bright yellow superciliary and throat patch;* black of breast fades into grayish olive on belly. Female much duller, with olive replacing all the black, and yellow face markings are less prominent (but still visible). Immatures resemble female; subadult males have less extensive black on breast than fully adult males. See C-56, V-40.

SIMILAR SPECIES: In form resembles Sooty and Black-faced Grassquits, but easily recognized by yellow facial pattern (even in duller females).

HABITAT AND BEHAVIOR: Common in semiopen and grassy areas, in pastures with bushes and low trees, and along roads. Usually seen as scattered pairs, less often in loose groups in association with various seedeaters. Forages mostly on grass seeds. Its song is a weak, thin trill, "tee-ee-ee-ee."

RANGE: Lower Andean slopes in Colombia (south to Nariño, but not in Santa Marta Mts.) and extreme w. Venezuela (s. Mérida and Táchira); a few records from n. Ecuador (mostly Pichincha, to which it has only recently spread; also once taken in e. lowlands at Limoncocha). Also Mexico to Panama, and Greater Antilles. Mostly 500–2300 m.

Dolospingus Seedeaters

Nearest to a *Sporophila* seedeater but larger with a more conical bill. Locally in semiopen areas of s. Venezuela region.

Dolospingus fringilloides

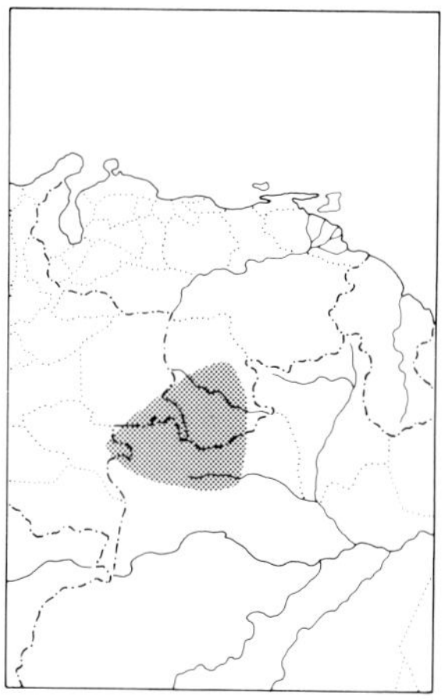

WHITE-NAPED SEEDEATER

PLATE: 26

IDENTIFICATION: 13.5 cm (5¼"). *Large, sharply pointed bill,* apparently pale in male but blackish in female. Mostly black above with narrow white nuchal collar almost meeting on nape and whitish area on rump; *1 broad white wing-bar* and another narrower one, plus a white speculum. Below white, with *black chin* and patch on sides of chest. *Female rather uniform warm brown,* with throat more whitish and *median underparts buffy white.*

SIMILAR SPECIES: Male recalls male Variable Seedeater, but note somewhat larger size, more pointed and pale bill, black chin, and much more white on wing. Somewhat featureless female most resembles female Lesser Seed-Finch (similar rich brown color) but has differently shaped (not nearly as heavy) bill, and seed-finch lacks the whitish below. Compare also to various female seedeaters.

HABITAT AND BEHAVIOR: Little known. Recorded from clearings in low scrubby sandy-belt woodlands and savannas. Voice of 1 male reported to be a loud, fast "ne-ne-ne, te-te-te, ge-ge-ge, jui-jui-jui, tu-e, tu-e tu-e, tu-eé tu-eé tu-eé," the triplets apparently being characteristic (S. Hilty).

RANGE: S. Venezuela (s. Amazonas), extreme nw. Brazil (Rio Xie in upper Rio Negro drainage, Amazonas), and adjacent extreme e. Colombia (sighting near Mitú in e. Vaupés, probably also e. Guainía). To 250 m.

Sporophila Seedeaters

Common and diverse group of small finches ranging virtually everywhere there is a grassy, semiopen country. Most species are found in the lowlands, with greatest diversity being reached in interior s. South America. The thick and stubby bill marks the genus. Males typically are strongly patterned and often colorful, but females usually very difficult, most being some variation on a buffy brown theme, some also with wing-bars; note also bill color.

Male seedeaters of various species are popular cage birds in several regions, notably on Trinidad and in Brazil and Argentina. Intense commercial trapping efforts have led to substantial declines of all Trinidadian species with the exception of the less sought after Ruddy-breasted, but all of these species are numerous on the adjacent S. Am. mainland. The problem is even more serious in ne. Argentina, where heavy trapping pressure has led to serious declines in several species; this has especially affected the group known locally as *capuchinos* (*S. hypoxantha, ruficollis, hypochroma, cinnamomea, palustris,* and *zelichi*), some of which may be severely threatened (*fide* T. Narosky and S. Salvador).

It should be emphasized that the ranges of many seedeaters are still surprisingly imperfectly known. This is especially the case for various species found in s. South America; it is now believed that a number of species may be long-distance migrants. Furthermore, males in identifiable plumage (e.g., fully adult) are often in a distinct minority during the nonbreeding season, making species-level identification difficult.

Note that our Groups are based only on males.

GROUP A

Boldly patterned in black and white (or rusty); bill black.

Sporophila lineola

LINED SEEDEATER

PLATE: 26

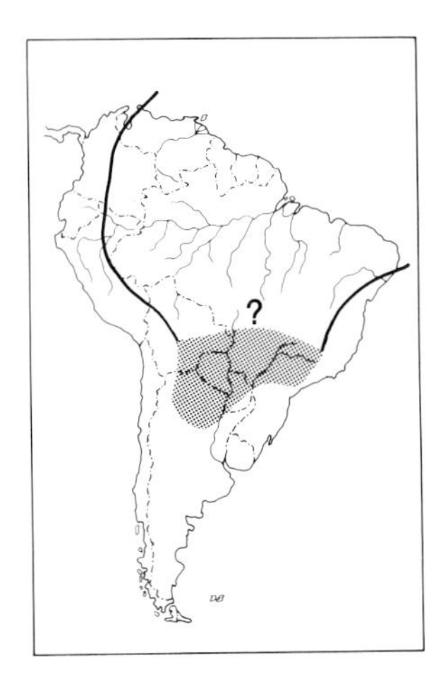

IDENTIFICATION: 11 cm (4¼"). Bill blackish (male) or *mostly yellowish,* especially below (female). Mostly black above with *white coronal stripe, broad malar streak,* rump, and wing-speculum. Throat black, remaining underparts white. Female light olive brown above, buffy to buffy yellowish below.

SIMILAR SPECIES: Cf. Lesson's Seedeater. Male Variable Seedeater is also basically black above and white below but lacks the white crown stripe and malar streak and has more white in wing, a blackish pectoral band, etc. Female Lined difficult to tell from some other female *Sporophila* (e.g., Double-collared, which is not quite so yellowish below; she is indistinguishable from female Lesson's).

HABITAT AND BEHAVIOR: Locally fairly common to common in shrubby clearings and other semiopen areas with tall grass, especially near water. Occurs as scattered pairs when breeding (though can be locally numerous and conspicuous), in small to fairly large groups at other seasons (when often mixes with other seedeaters). Clings to tall grass stems when feeding. Song of breeding male in s. South America a pretty, clear, trilled "tititititi-teé," with final accented note characteristic, often given from a somewhat hidden perch.

RANGE: Apparently an austral migrant, breeding in n. Argentina (south to Tucumán, Santiago del Estero, and Santa Fe), Paraguay (mostly w.), and interior cen. and se. Brazil (at least in s. Mato Grosso, São Paulo, and Paraná, perhaps north locally to Goiás and Bahia), and se. Bolivia (Tarija, perhaps north to Santa Cruz); in austral winter moves north across e. and cen. Amazonia to n. South America (when it occurs sympatrically with then breeding *S. bouvronides*), where recorded from e. Colombia, Venezuela, and the Guianas. Perhaps also a breeding resident north in e. South America to Cayenne, but more information is needed. To about 1200 m.

NOTE: The "vexing" problem of this species (as R. A. Paynter aptly termed it in *Birds of the World*, vol. 13) has yet to be entirely resolved, but we prefer to emphasize the fact that 2 populations do clearly exist and consider them full species. It is now evident, from the work of P. Schwartz (*Ann. Carnegie Mus.* 45: 277–285, 1975) and others subsequently, that both forms are highly migratory, and this contributed significantly to the earlier confusion. *S. bouvronides* breeds only in ne. South America and apparently occurs as a nonbreeding visitant to Amazonia; *S. lineola* apparently breeds in s.-cen. South America (and perhaps also northward in the east to Cayenne) and occurs as an austral winter visitant in ne. South America during the same months when *bouvronides* is nesting. The songs of males of both appear to be similar, and it may be that the main impediment to their interbreeding is the difference in the timing of their nesting cycle.

Sporophila bouvronides

LESSON'S SEEDEATER

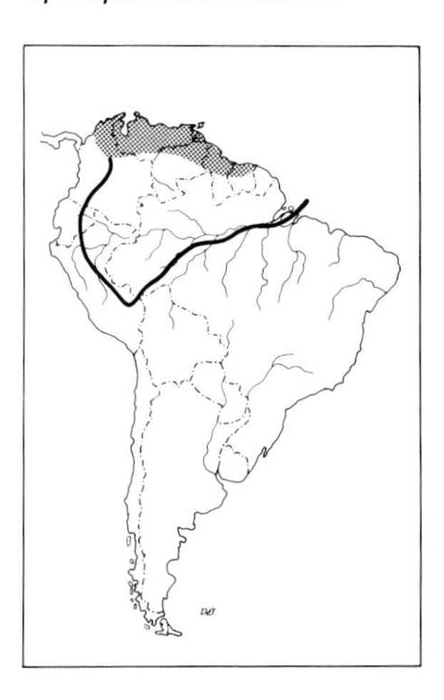

IDENTIFICATION: 11 cm (4¼"). Resembles Lined Seedeater, but *crown entirely black* (no white coronal stripe) and sides and sometimes chest mottled with black. Female indistinguishable from female Lined. See C-56, V-40.

SIMILAR SPECIES: In addition to Lined Seedeater (with which perhaps conspecific), cf. also Variable Seedeater.

HABITAT AND BEHAVIOR: Uncommon to fairly common in shrubby clearings and grassy areas, especially near water. Behavior and song evidently similar to Lined Seedeater's. This species breeds during period when austral migrant Lined Seedeaters are present in n. South America; *bouvronides* also apparently engages in long-distance movements during its nonbreeding season, but details remain to be worked out.

RANGE: Locally in n. and e. Colombia (Caribbean lowlands in lower Magdalena valley; east of Andes in w. Meta, Vaupés, and Amazonas, but probably also elsewhere), Venezuela, and the Guianas south to e. Ecuador, e. Peru, and Amaz. Brazil; Trinidad. Also e. Panama (apparently only as a vagrant). Evidently breeds only in n. part of this range (May–Nov.), occurring southward primarily (or only) as a nonbreeding visitor during the Dec.–June period, but more information is needed. To about 800 m.

NOTE: See comments under Lined Seedeater.

Sporophila americana

VARIABLE SEEDEATER

PLATE: 26

Other: Wing-barred Seedeater

IDENTIFICATION: 11–11.5 cm (4¼–4½"). Mostly black above with whitish to gray rump, white wing-speculum, and some white tipping on wing-coverts (forming indistinct wing-bars). Below whitish, *extending up on sides of neck to form partial nuchal collar;* sides mottled grayish, and has an ill-defined black chest band, often interrupted or even lacking. Nominate *americana* (e. Venezuela to lower Amazon) has more prominent wing-bars than the rather vague wing-bars of *mu-*

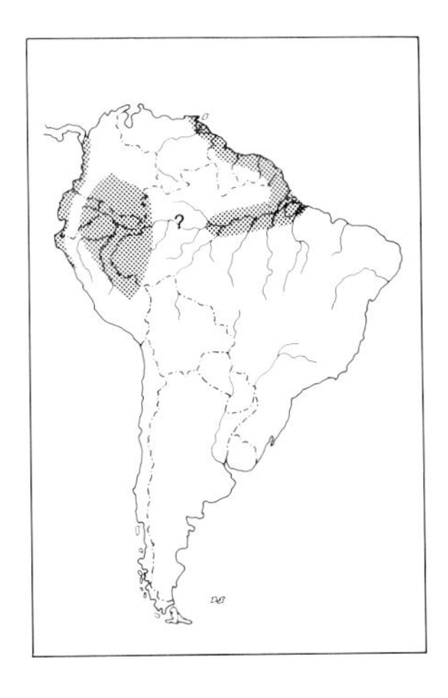

rallae (upper Amazonia). Pacific slope races lack white on wing (aside from the speculum): *hicksi* (Colombia) has broad black chest band, while *ophthalmica* (sw. Colombia to nw. Peru) has neat narrow pectoral band and is whiter below. Female yellowish olive brown above, paler buffy brown below, becoming even paler yellowish buff on belly.

SIMILAR SPECIES: Lined and Lesson's Seedeaters have prominent white malar and lack the nuchal collar. Cf. also White-naped Seedeater. Females closely resemble other female *Sporophila* with dark bills (e.g., Yellow-bellied) and are usually best known by the company they keep; female Lined, Lesson's, and Double-collared Seedeaters all have at least some yellow on bill (especially on lower mandible).

HABITAT AND BEHAVIOR: Common (e.g., on Pacific slope) to rather uncommon (e.g., in upper Amazonia) in grassy and shrubby areas, agricultural areas and roadsides, and towns. Found in pairs or small groups, often accompanying other seedeaters and grassquits; may occur in large flocks when not breeding. The song is a varied musical twittering, usually delivered from a low perch; also frequently calls a sweet "cheeeu."

RANGE: W. Colombia, w. Ecuador, and nw. Peru (south to La Libertad); se. Colombia, e. Ecuador, ne. Peru, and w. Amaz. Brazil; extreme e. Venezuela (Sucre, Monagas, and Delta Amacuro), the Guianas, and lower Amaz. Brazil (Amapá, and upriver along both banks of the Amazon to lower Rio Negro area); Tobago. Also Mexico to Panama. To about 1200 m.

NOTE: We follow most recent authors (but not the 1983 AOU Check-list) in considering the trans-Andean and Middle American *S. aurita* group as conspecific with cis-Andean *americana* and associated races (Wing-barred Seedeater). *Murallae* of upper Amazonia appears to be intermediate, with wing-bars narrow to almost absent, and vocalizations of all populations are similar.

Sporophila collaris

RUSTY-COLLARED SEEDEATER

PLATE: 26

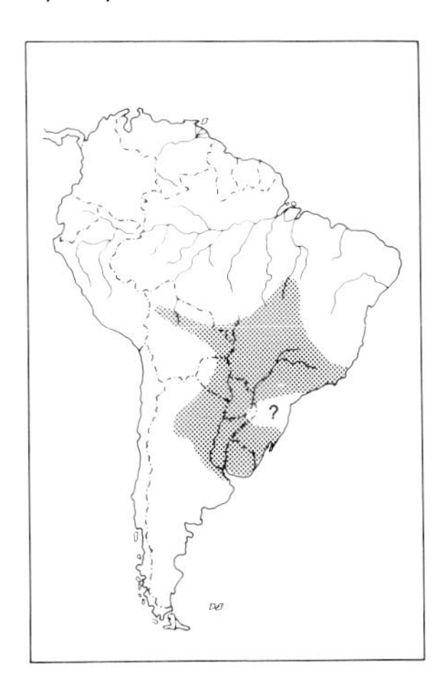

IDENTIFICATION: 12 cm (4¾"). *A boldly marked, rather large seedeater of s. marshlands.* Head black with prominent white patches above lores and below eyes; upperparts black, or blackish mixed with buff or grayish, with *broad tawny-buff collar around neck* and cinnamon buff rump. Throat white, bordered below by conspicuous black pectoral band; lower underparts cinnamon buff. Wings black with 2 bold buff to cinnamon wing-bars and white speculum. N. nominate race (Brazil except in w. Mato Grosso) has *white nuchal collar* only slightly tinged with buff, whitish underparts, and whiter rump. Female rather distinctive for a *Sporophila* as it echoes male's pattern: brown above with at least a trace of buff nuchal collar and with *2 tawny-buff wing-bars* and a buffy whitish speculum. *Throat white, contrasting with buff lower underparts.*

SIMILAR SPECIES: The pretty male is the only seedeater in its range with black pectoral band; Variable Seedeater (not known south of immediate lower Amazon area) has 2 whitish wing-bars, less well defined pectoral band, etc. No other female seedeater combines the distinctly 2-tone underparts with wing-bars.

HABITAT AND BEHAVIOR: Fairly common in grass and shrubbery in marshlands and around lakes and other bodies of water. Usually in pairs or small groups, not associating with other seedeaters. Rarely or never actually on the ground. Readily found in the *pantanal* of sw. Mato Grosso, Brazil, and west of Asunción, Paraguay.

RANGE: N. and e. Bolivia (Beni and Santa Cruz) and s. Brazil (north to cen. Mato Grosso, n. Goiás, and Espírito Santo but absent, apparently, from part of southeast) south through cen. Paraguay to Uruguay and n.-cen. Argentina (south to La Rioja, n. Córdoba, and n. Buenos Aires). To about 500 m.

GROUP B

"*Hooded*" group; *lack* face pattern; *bill bluish.*

Sporophila nigricollis

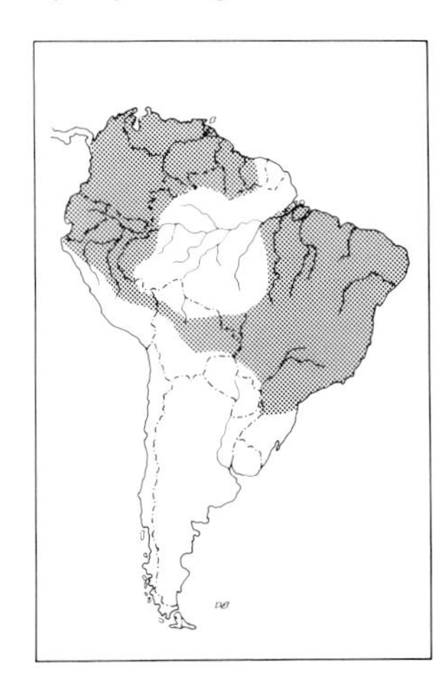

YELLOW-BELLIED SEEDEATER

PLATE: 26

IDENTIFICATION: 11 cm (4¼"). Bill light bluish gray (male) or dark (female). *Crown, sides of head, throat, and chest black* (forming hood), *contrasting with pale yellow lower underparts* (deeper in fresh plumage). Upperparts olive, some birds with a small white wing speculum. *Vivida* (sw. Colombia and w. Ecuador) has somewhat deeper yellow underparts, and some (older?) males are blacker above, extending to entire mantle. *Inconspicua* (Peru) has black restricted to foreface (most of head olive). Female olive brown above, lighter buffy brownish below, becoming yellowish on median belly.

SIMILAR SPECIES: Male distinctive (the only "hooded" seedeater with yellow on underparts), but cf. Black-and-white and Dubois' Seedeater (both with white bellies). Female probably cannot be distinguished from other dark-billed female *Sporophila.*

HABITAT AND BEHAVIOR: Fairly common to common (in some areas numbers vary seasonally) in shrubby or grassy clearings, agricultural areas, and roadsides. Scatters out as pairs when nesting, but in nonbreeding season gathers in flocks, sometimes large, which often associate with other seedeaters. Song short and musical but often ending with 2 buzzier notes, "tsee-tsee-tsee-bseeoo, bzee-bzee" with variations.

RANGE: Widespread in lowlands south to n. and e. Bolivia, s. Brazil (south to Mato Grosso and São Paulo), and ne. Argentina (Misiones); on Pacific slope south only to nw. Peru (Lambayeque and Cajamarca); apparently absent across much of cen. Amazonia from sw. Brazil to French Guiana; Trinidad. Also Costa Rica and Panama, and s. Lesser Antilles. To about 2000 m.

Sporophila ardesiaca

DUBOIS' SEEDEATER

IDENTIFICATION: 11 cm (4¼"). *Se. Brazil.* Closely resembles Yellow-bellied Seedeater (with at least partial range overlap). *Mantle gray* (not

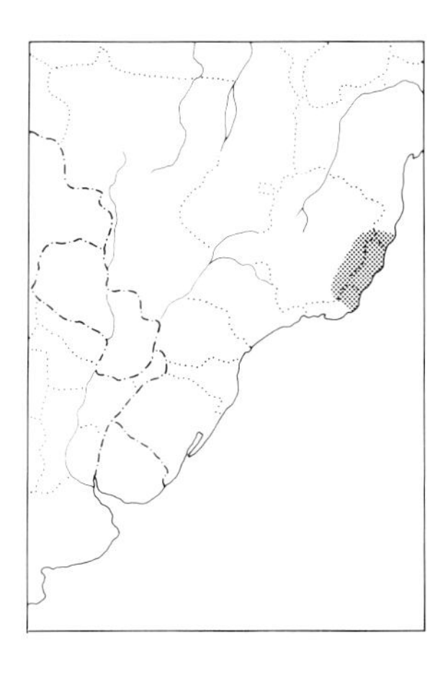

olive) and *lower underparts white* (not pale yellow). Female indistinguishable from female Yellow-bellied (and other dark-billed female *Sporophila*).

HABITAT AND BEHAVIOR: Uncommon in shrubby and grassy clearings and along roads. Usually seen singly, sometimes in association with groups of Yellow-bellied Seedeaters.

RANGE: Se. Brazil in s. Bahia, Espírito Santo, e. Minas Gerais, and Rio de Janeiro. To about 800 m.

NOTE: The status of this form remains in doubt. It may prove to be only a local variant or localized color phase of *S. nigricollis,* and (despite the fact that they seem to be found syntopically) some authors have suggested that *ardesiaca* is only a subspecies of *S. nigricollis*. To date they have not been found breeding together. For more information, see H. Sick, *Bol. do Mus. Nac.*, Nôva Série, Zoologia, No. 235, 1962.

Sporophila melanops

HOODED SEEDEATER

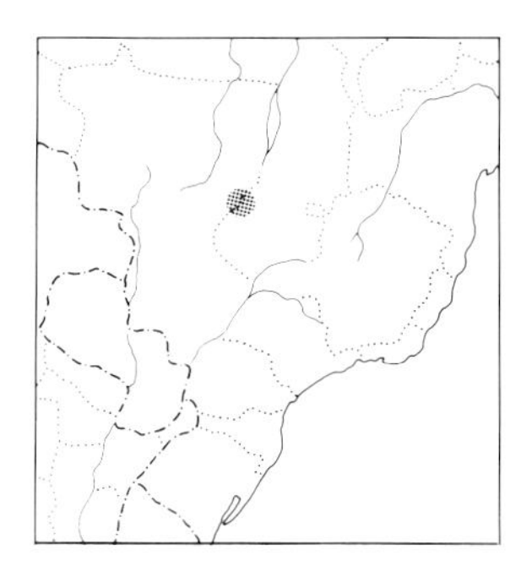

IDENTIFICATION: 11 cm (4¼"). *S.-cen. Brazil.* Resembles Yellow-bellied Seedeater, but pattern of hood somewhat different. Hood black, *contrasting* with olive upperparts and *extending only over throat* (not down over chest). Remaining underparts dingy buff.

HABITAT AND BEHAVIOR: Unknown in life.

RANGE: S.-cen. Brazil in s. Goiás (Porto do Rio Araguaia). Known only from the type specimen, obtained in the 19th century.

NOTE: This species, still known only from the type, seems of dubious validity. Despite described differences in pattern, it seems close to *S. nigricollis,* and it may represent an aberrant individual of that species. A hybrid origin is also possible.

Sporophila luctuosa

BLACK-AND-WHITE SEEDEATER

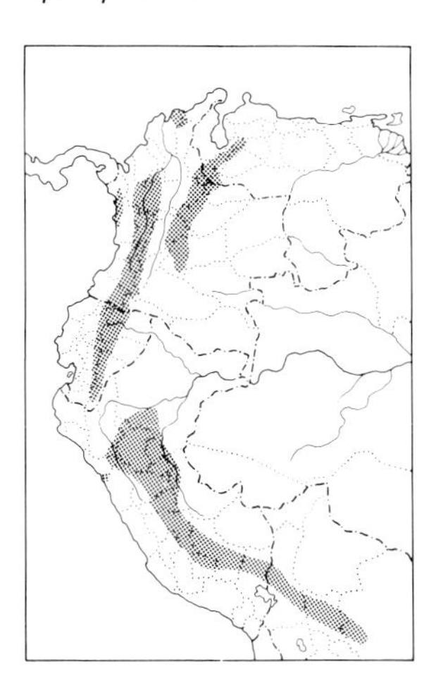

IDENTIFICATION: 11 cm (4¼"). Recalls Yellow-bellied Seedeater. Bill light bluish gray (male) or dark (female). *Black above and on throat and chest,* contrasting strongly with *white lower underparts* and with *conspicuous white wing-speculum.* Female virtually identical to female Yellow-bellied Seedeater. See C-56, V-40.

SIMILAR SPECIES: Yellow-bellied Seedeater has similar pattern but is olive (not black) on mantle and pale yellow (not white) below and usually lacks the wing-speculum. Cf. also other "black and white" seedeaters (Variable, Lined, etc.).

HABITAT AND BEHAVIOR: Uncommon to locally fairly common (especially on Andean slopes) in grassy areas with scattered bushes, in pastures, and along roads in grass and shrubbery. Usually in pairs or small groups, mixing less often with other species than do most seedeaters. Song is unusual and not very melodic, rather blackbirdlike, introduced by 2 especially harsh notes, followed by gurgles of various sorts, some almost shrieking (parrotlike) in quality.

RANGE: Lower Andean slopes (for the most part; smaller numbers also extending into adjacent lowlands) in w. Venezuela (north to Trujillo), Colombia (where rather local; also found on Santa Marta and

Baudó Mts.), Ecuador, Peru, and n. Bolivia (south to w. Santa Cruz). Mostly 1200–2500 m, occasionally much lower (to 1–300 m) or somewhat higher (perhaps engages in seasonal altitudinal movements?).

GROUP C

"*Collared*" group; *gray upperparts; bill yellowish.*

Sporophila caerulescens

DOUBLE-COLLARED SEEDEATER

PLATE: 26

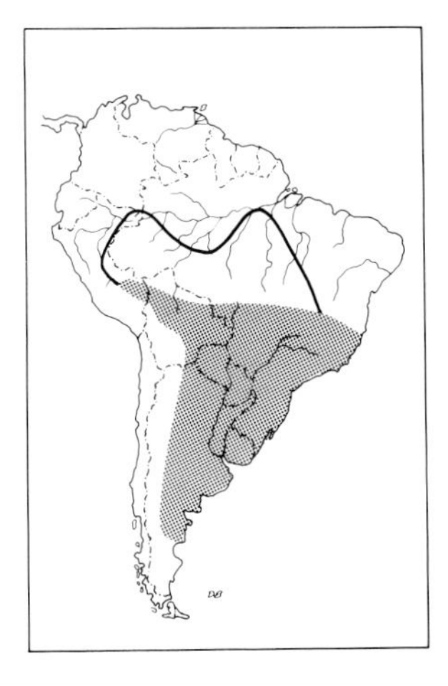

IDENTIFICATION: 11 cm (4¼"). Bill pale greenish yellow (male) or dusky above and yellow below (female). Gray above, tinged olivaceous on back and browner on wings and tail. *Upper throat black,* bordered on sides by *short white malar streak* and below by white band across lower throat, which in turn is bordered below by a *black pectoral band* (forming the "double collar"). Lower underparts white. Female olive brown above, lighter, buffier brown below, becoming whitish on belly.
SIMILAR SPECIES: Male's pattern on throat and breast distinctive within range; no other seedeater is more than vaguely similar (cf. White-throated Seedeater in ne. Brazil). Female can sometimes be known by her bicolored bill, but this is duplicated by female Lined and Lesson's Seedeaters.
HABITAT AND BEHAVIOR: Common and widespread in semiopen and shrubby areas, agricultural regions, and roadsides and lawns. Absent or in relatively low numbers in extensive grasslands; otherwise this is easily the most numerous and familiar seedeater in s. South America. Gathers in large flocks during nonbreeding season, frequently in association with Blue-black Grassquits, or may be joined by other seedeaters. Song is a fast series of musical, jumbled notes, quite variable (e.g., "jew, jitit-jew-jew, jitit").
RANGE: Breeds from n. and e. Bolivia (north to La Paz and Santa Cruz) and cen. Brazil (south of Amazonia) south through much of Paraguay and Uruguay to n. Argentina (to Mendoza, La Pampa, and Buenos Aires); southernmost breeders move northward during austral winter, at which time small numbers are found north into Amazonia (where recorded from e. Peru, extreme se. Colombia at Leticia, and cen. Amaz. Brazil). Status in se. Peru unclear; a few may breed (lingerers have been seen into Nov.), but numbers are greater during austral winter. To about 1500 m (in Bolivia).

Sporophila peruviana

PARROT-BILLED SEEDEATER

PLATE: 26

IDENTIFICATION: 11–12 cm (4¼–4¾"). *Sw. Ecuador and w. Peru. Oversized bill, with very curved culmen, yellowish* (brighter below) in male; duller, more horn-colored in female. Brownish gray above, with prominent white wing-speculum and whitish wing-bars. *Throat and chest black,* often separated from dark head by white streak on sides of neck. Lower underparts white. In nonbreeding season few males are found in this plumage, all or most being at some stage between it and female plumage. Female light brownish above, often with *buff margins to wing-coverts* forming 2 wing-bars. Below buffy whitish.

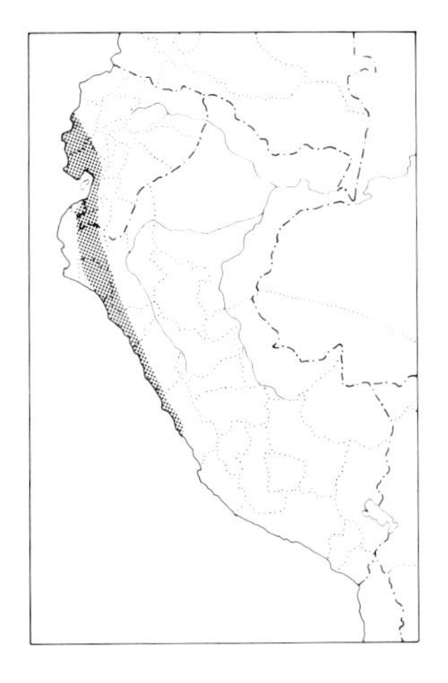

SIMILAR SPECIES: Both sexes can usually be known by their enlarged bill. Female somewhat resembles Drab Seedeater but is larger with much bigger, paler bill. Cf. also Chestnut-throated Seedeater, overall pattern of which is somewhat similar.
HABITAT AND BEHAVIOR: Uncommon to common (with notable seasonal variation in numbers in at least some areas) in arid shrubby areas, agricultural regions with fallow or grassy fields and hedgerows, and shrubbery or low woodland along watercourses. During nonbreeding season regularly in flocks, sometimes quite large and in association with other seedeaters and grassquits. When nesting (which in Ecuador occurs during the Feb.–May rainy season), males sing persistently, moving rapidly from one hidden perch to another, often frustratingly difficult to see except in flight (though they may be singing from all around). The song resembles Double-collared's in quality but is harsher and not as musical and is much shorter, usually only 2 to 3 notes (e.g., "jew-jee-jew" or "jee-jew").
RANGE: Locally on Pacific slope of sw. Ecuador (north to Manabí and inland to s. Loja) and w. Peru (south to Ica). To about 800 m.

Sporophila albogularis

WHITE-THROATED SEEDEATER

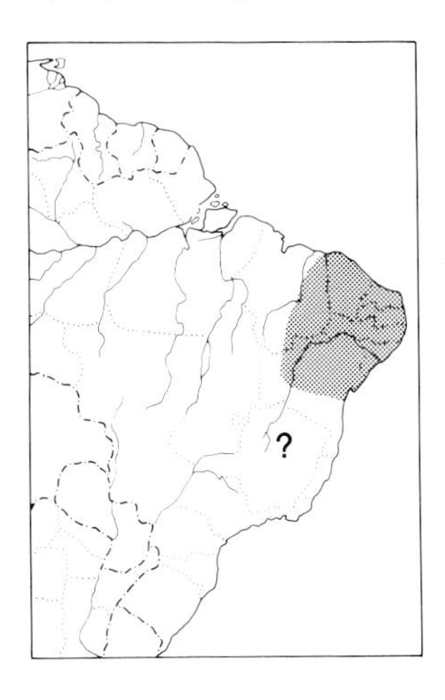

IDENTIFICATION: 11 cm (4¼"). *Ne. Brazil.* Recalls Double-collared Seedeater. Bill orange-yellow (male) or dusky (female). Head blackish, otherwise gray above (without olivaceous tinge). *Entire throat white,* bordered below by *narrow black pectoral band* and extending up on sides of neck to form *partial nuchal collar.* Lower underparts white. Wings and tail blackish, with *prominent white wing-speculum.* Female grayish brown above, buffy whitish below.
SIMILAR SPECIES: No other gray and white seedeater has such a sharp black pectoral band; there are few other seedeaters in its range. Cf. Double-collared Seedeater.
HABITAT AND BEHAVIOR: Uncommon to locally fairly common in dry *caatinga* scrub and shrubby woodland borders. In nonbreeding season found in small groups, sometimes with other seedeaters, feeding on the ground and in grass. Not well known in the wild, though it is not infrequent in captivity.
RANGE: Ne. Brazil from Piauí and Pernambuco to n. Bahia; a few reports from farther south (Minas Gerais, Espírito Santo) may pertain to migrants (or perhaps more likely escapees?). Mostly 500–1200 m.

GROUP D

Predominantly gray to olive; bill yellow or black.

Sporophila frontalis

BUFFY-FRONTED SEEDEATER

PLATE: 26

IDENTIFICATION: 12.5 cm (5"). *A large seedeater of se. forests.* Bill dull yellowish. Mostly olivaceous above, grayer on head; *broad frontal band and narrow, short postocular streak buffy whitish.* Throat whitish, with broad dull olive band across breast (sometimes incomplete) and continuing down on flanks; median lower underparts buffy yellowish.

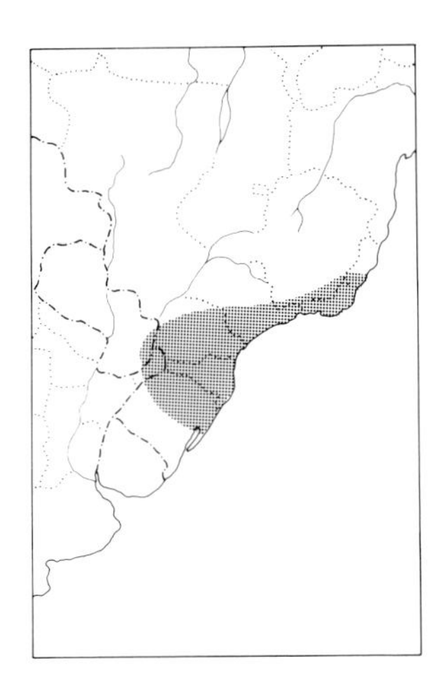

Wings with 2 *buffyish wing-bars* and speculum. Immature male (much more often encountered than full adult) more olivaceous brown above and usually without frontal band. Female browner above with no head markings; whitish on throat, dull buffy olivaceous below, paler on belly; *wing-bars buff,* but no speculum.

SIMILAR SPECIES: The largest seedeater. Smaller Temminck's Seedeater lacks the distinct wing-bars and is never as olivaceous as this species. Cf. also female Uniform Finch (about the same size, but with very different sharply pointed bill, no wing-bars, etc.).

HABITAT AND BEHAVIOR: Rare to uncommon (locally and erratically) in lower growth of forest borders and in shrubby overgrown clearings. Its local distribution may be tied to the brief seeding of bamboo. Notably arboreal for a seedeater, rarely if ever on the ground. Numbers have been reduced not only by habitat destruction but also because it is in such demand as a cage bird. Its local popularity derives from its remarkably loud and explosive (very far-carrying) song, a harsh and rather grating "jeje jét" or "cheh-cheh-chéw!"

RANGE: Se. Brazil (Espírito Santo and se. Minas Gerais south to n. Rio Grande do Sul), extreme e. Paraguay (Alto Paraná; no recent records), and ne. Argentina (Misiones). Now mostly 1000–1500 m (perhaps formerly lower?).

Sporophila schistacea

SLATE-COLORED SEEDEATER

PLATE: 26

IDENTIFICATION: 11 cm (4¼"). *Bill rich yellow* (male) or grayish (female). *Mostly slaty gray,* usually with *small whitish patch on sides of neck* (sometimes hard to see, and may be lacking in younger males). Median breast and belly white. Wings and tail blackish, wings with white speculum and *usually a single upper wing-bar* (narrow to quite broad). Female olive brown above, paler buffy brown below, more creamy whitish on median belly.

SIMILAR SPECIES: Cf. Temminck's Seedeater (no overlap). Gray Seedeater very similar; female probably indistinguishable in the field, but note shallower (less curved) maxilla of Slate-colored as well as habitat differences. Male Gray has more pinkish yellow bill and *never shows whitish wing-bar* (though often does show patch on sides of neck).

HABITAT AND BEHAVIOR: Rare to uncommon (and usually irregular at any single locality; can be briefly more numerous) at forest borders and in regenerating clearings. In some areas its presence seems tied to bamboo or to bamboo flowering, but this is not invariable. *Much more of a woodland bird than Gray Seedeater.* Occurs singly or in scattered pairs, rarely or never in flocks with other seedeaters; often perches quite high in trees. Song is loud but not very musical, a high, fast, sibilant "zit-zit-zee-zee-zee-ze-ze-z-z-z-z," with many variations.

RANGE: Very locally in w. Venezuela (Mérida), Colombia (west of the Andes, east of them on Macarena Mts. in Meta), nw. and e. Ecuador (Pichincha and Napo), e. Peru (Huánuco and Madre de Dios), and nw. Bolivia (Beni and Cochabamba); also in s. Venezuela, and in the Guianas (some confusion with *S. intermedia*?) and extreme ne. Brazil

(Roraima, Amapá, and Pará in the Belém area); Trinidad. Also Mexico, Honduras, Costa Rica, and Panama. Mostly below 1500 m, occasionally higher.

Sporophila intermedia

GRAY SEEDEATER

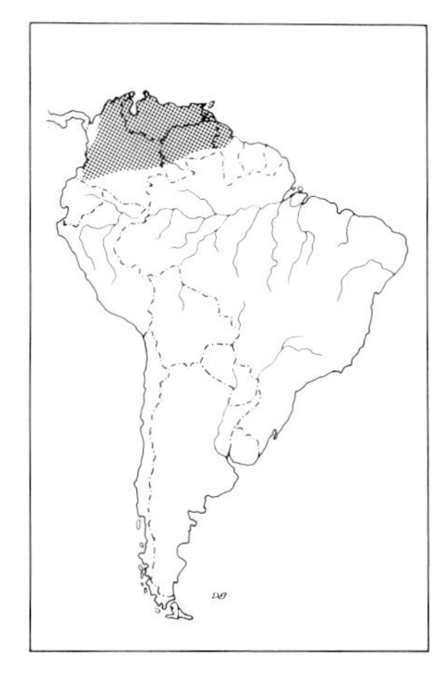

IDENTIFICATION: 11 cm (4¼"). Resembles Slate-colored Seedeater. Male's bill yellowish flesh (not rich yellow), that of female dusky. Male of nominate race (most of range) paler gray below (white belly less demarcated) and lacks submalar mark and wing-bar. See C-56, V-40. In Colombia west of Andes (except Caribbean north) much more similar to Slate-colored, and caution is required; however, the wing-bar is never present. This species tends to be numerous and widespread, which Slate-colored decidedly is not.

HABITAT AND BEHAVIOR: Fairly common to common in semiopen to open grassy or bushy areas and clearings. Often in small groups which regularly associate with other seedeaters. Song is much richer and more musical than Slate-colored Seedeater's (not nearly as sibilant), a varied series of trills, chirps, and twitters; some imitation of other species has been reported.

RANGE: W. Colombia (west of the Andes generally, south to Nariño, and locally in northeast south to Meta and Guainía), Venezuela (south to n. Amazonas and Bolívar), and n. Guyana (Takutu Mts.); Trinidad. Mostly below 2000 m.

Sporophila falcirostris

TEMMINCK'S SEEDEATER

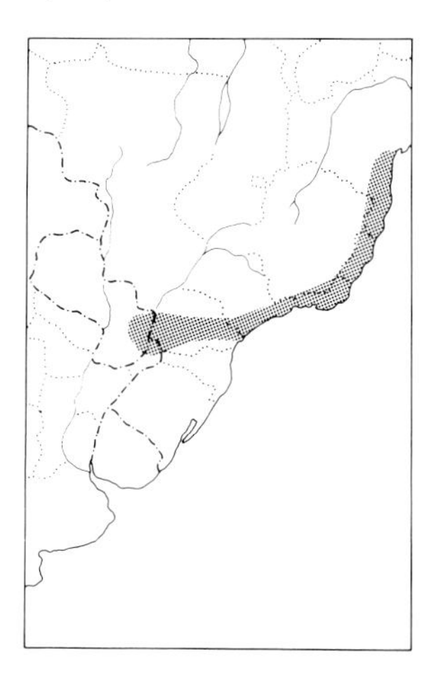

IDENTIFICATION: 11 cm (4¼"). Closely resembles Slate-colored Seedeater and probably is distinguishable only by range. Male somewhat paler gray, with slightly larger white wing-speculum but no white patch on sides of neck. Males sing and evidently breed while still in an immature plumage like females.

SIMILAR SPECIES: Besides Slate-colored, cf. also Buffy-fronted Seedeater (with which sympatric), which is larger and much more olive generally, has 2 buff wing-bars, and (in male) is whitish on head.

HABITAT AND BEHAVIOR: Rare to uncommon (and usually irregular at any single locality) in forest borders and secondary woodland *with much bamboo in understory*. Like Slate-colored, rather an arboreal seedeater. Song recalls Slate-colored's, a high, buzzy, fast trill "ztztztztzt-zt-tzi." Can at times be found at Iguaçu Falls and Itatiaia Nat. Park.

RANGE: Se. Brazil (se. Bahia to Paraná), e. Paraguay (Canendiyu), and ne. Argentina (Misiones). To at least 1200 m.

Sporophila plumbea

PLUMBEOUS SEEDEATER PLATE: 26

IDENTIFICATION: 11 cm (4¼"). *Bill blackish* in both sexes normally, but in se. Brazil some (older?) males apparently have yellow bills. Mostly gray, paler below; white chin and lower median underparts, and fairly

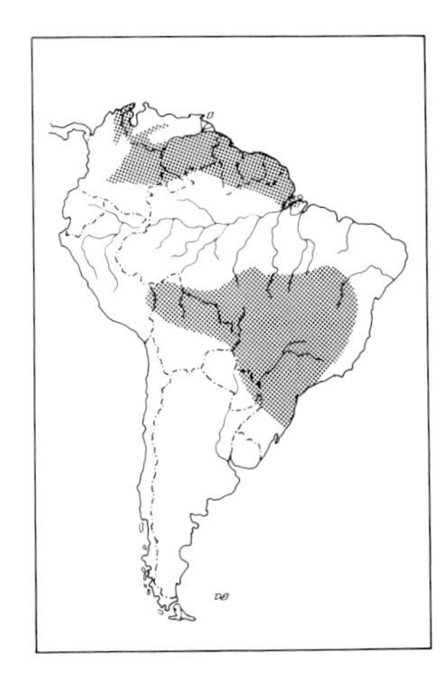

conspicuous white subocular spot. Wings and tail blackish, wings edged with gray with distinct white speculum. Nominate race (south of Amazon) has smaller white subocular spot. Female light brown above, paler brownish buff below, becoming whitish on belly.

SIMILAR SPECIES: *Male is only mostly gray seedeater with a dark bill.* Female probably cannot be distinguished from other dark-billed female seedeaters (but note savanna habitat). In se. Brazil (São Paulo south), where bill is often always yellow, cf. Temminck's Seedeater; their preferred habitats differ, and note white chin and bolder wing-edging of Plumbeous.

HABITAT AND BEHAVIOR: Uncommon to locally fairly common in open savannas with tall grass, frequently near water or damp places. Often in small groups, sometimes associating with other seedeaters. Song reported to be a long series of loud, clear phrases, each often repeated several times, then on to the next; lacks warbling quality of most seedeaters (Hilty and Brown, from P. Schwartz recording).

RANGE: Locally in Colombia (Caribbean lowlands in Cesar Valley and Serranía San Lucas, and in northeast south to Meta and Vichada), Venezuela (Sierra de Perijá, Carabobo, Apure, more widely in Bolívar and n. Amazonas), and the Guianas south to islands in mouth of Amazon in Brazil; se. Peru (Heath Pampas in Madre de Dios), n. Bolivia (La Paz and Beni), interior and s. Brazil (Mato Grosso to Piauí and south to n. Rio Grande do Sul), e. Paraguay, and ne. Argentina (Misiones). May be somewhat migratory from s. part of range. To about 1400 m.

Sporophila simplex

DRAB SEEDEATER

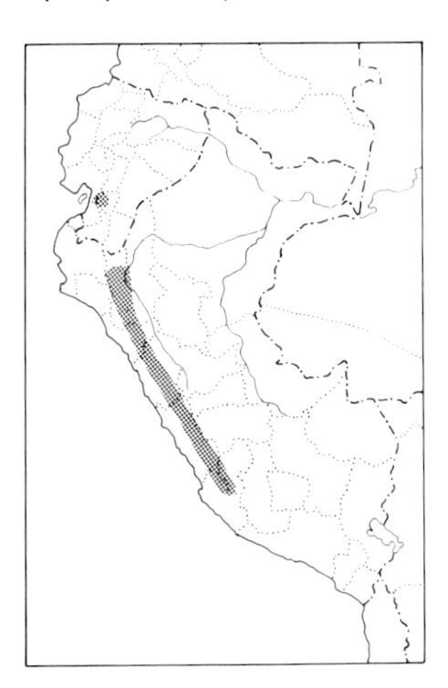

IDENTIFICATION: 11 cm (4¼"). *Sw. Ecuador and w. Peru.* The English name of this *hen-colored* species is appropriate. Bill brownish, paler below. Light grayish brown above; wings with *2 prominent whitish wing-bars* and white speculum. Drab grayish white below, washed with olivaceous across breast. Sexes basically alike, but *female's wing-bars are buffier.*

SIMILAR SPECIES: Despite its drab plumage, in its range quite readily known by its wing-bars. Some female Parrot-billed Seedeaters also show wing-bars, but that species' much larger bill with very curved culmen distinguishes it (Drab's is shaped normally and is darker). Cf. also Dull-colored Grassquit (which lacks wing-bars, etc.).

HABITAT AND BEHAVIOR: Fairly common in shrubby areas, arid scrub, and agricultural regions. Except in breeding season usually in flocks, sometimes large, which often mix with other seedeaters and grassquits. Song reported to be variable, consisting of a series of short, repeated phrases; a frequent motif is "tu-tee-tu-tchay-tchay-tchay" (Koepcke 1970).

RANGE: Locally in arid interior valleys of s. Ecuador (s. Azuay and s. Loja; MCZ specimens) and in w. Peru (upper Marañón drainage and lower Andean slopes from La Libertad to Ica). Mostly below 1500 m, locally somewhat higher.

GROUP E

Sharply bicolored (Bolivian race *black* above); *bill yellow.*

Sporophila leucoptera

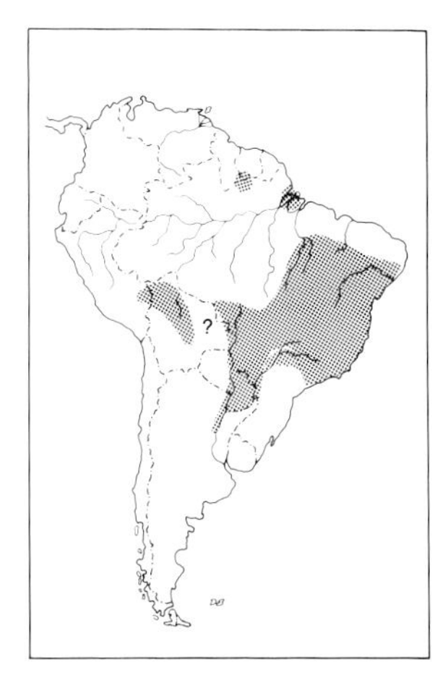

WHITE-BELLIED SEEDEATER

PLATE: 26

IDENTIFICATION: 12 cm (4¾"). The most obviously "bicolored" seedeater. *Bill dull yellow or pinkish yellow. Basically gray above and white below* (*cinereola* of e. Brazil is broadly gray on sides); wings with prominent white speculum, and sometimes shows narrow white rump-band. *Bicolor* of n. Bolivia (Beni and w. Santa Cruz) is *glossy black above.* Female olive brown above, light buffy brown below, whitish on belly.
SIMILAR SPECIES: No other seedeater is so uniformly dark above and white below. Female's pale bill and large size helpful.
HABITAT AND BEHAVIOR: Usually uncommon, found in grassy areas with scattered bushes and thickets, but almost always near water (in marshy areas, margins of rivers or lakes, etc.). Found singly or in scattered pairs, not in flocks, and rarely or never associating with other seedeaters. Song of male of nominate group is a single, clear note, far-carrying and with an odd, ringing quality ("cleeu, cleeu, cleeu, cleeu, cleeu, cleeu").
RANGE: Se. Peru (1 sighting from Tambopata in Madre de Dios), n. and e. Bolivia, cen. and e. Brazil (s. Maranhão, Piauí, and Pernambuco south to n. São Paulo, west through cen. and s. Mato Grosso), e. Paraguay (west to *pantanal* along Paraguay River), and ne. Argentina (south to n. Santa Fe); also isolated populations at mouth of Amazon (Mexiana Is.) and in s. Suriname (Sipaliwini). To about 800 m.

NOTE: Two species may be involved: black-backed *bicolor* in n. Bolivia (east only to w. Santa Cruz) and gray-backed *leucoptera* (with *cinereola*) throughout the remainder of its range. No intermediate specimens have been recorded, but *bicolor* remains poorly known, so we continue to regard them as conspecific. If split, *S. bicolor* might best be called "Black-backed Seedeater" and *S. leucoptera* "Gray-backed Seedeater," for both are "bicolored" and "white-bellied."

GROUP F

Both sexes streaked above and with white at base of tail.

Sporophila telasco

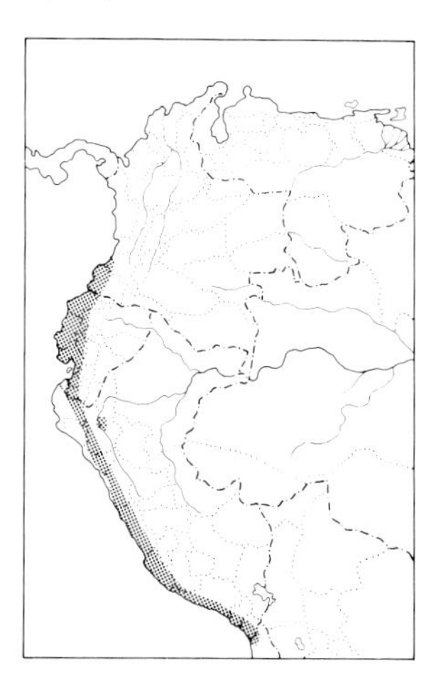

CHESTNUT-THROATED SEEDEATER

PLATE: 26

IDENTIFICATION: 10 cm (4"). *Pacific lowlands from sw. Colombia to n. Chile.* Bill black (male) or pale brown to yellowish (female). Above gray *vaguely streaked dusky on crown and mantle; narrow rump-band white. Upper throat chestnut* (often hard to see), remaining underparts white. Wings and tail blackish, wings with large white speculum, tail basally white. Female light brown above *streaked dusky on crown and mantle;* below whitish with a little vague dusky streaking on breast; small white speculum.
SIMILAR SPECIES: Only other seedeater with streaking above is the rare Tumaco (see below). Male often best known by the white rump-band, especially in flight. Female Parrot-billed Seedeater is larger with strikingly heavier bill, lacks the streaking, and shows 2 wing-bars.
HABITAT AND BEHAVIOR: Common in open grassy areas with scattered bushes and low trees and in agricultural regions generally. May

occur in large flocks when not breeding, and generally is the most numerous and familiar seedeater over much of its range. Its song is a pretty, short warbling delivered from an exposed perch, often a phone wire or fence. Very conspicuous around Guayaquil, Ecuador, especially during the rainy season (Feb.–May) when they are breeding, and singing males are everywhere.
RANGE: Sw. Colombia (Pacific coastal lowlands north to sw. Cauca), w. Ecuador (mostly in more arid areas and inland to s. Loja), w. Peru (mostly Pacific lowlands, but also upper Marañón valley), and extreme n. Chile (Arica). Mostly below 1000 m.

Sporophila insulata

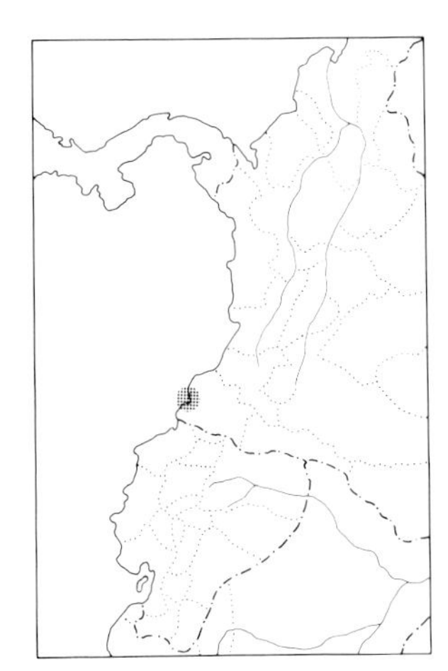

TUMACO SEEDEATER

IDENTIFICATION: 10 cm (4"). *Tumaco Is., sw. Colombia.* Mostly gray above with very narrow band across rump rufous, *base of tail white;* wings blackish with *large white speculum.* Below mostly rufous, but at least some birds (immatures?) have *belly mostly white.* Female olive brown above *vaguely streaked darker on crown and back;* below dull grayish buff, becoming whitish on belly; wings as in male, and also shows a little white at base of tail.
SIMILAR SPECIES: Almost looks like a cross between Chestnut-throated and Ruddy-breasted Seedeaters (and is possibly nothing more than a hybrid population between those 2 species or an isolated race of the former), with wing and tail pattern and female's streaking much like Chestnut-throated's but color of underparts like Ruddy-breasted's.
HABITAT AND BEHAVIOR: Unknown; perhaps extinct (no recent reports). Recorded only on Tumaco Is., but as that island is now so heavily settled it seems doubtful whether it could still survive there. Perhaps will be found on adjacent mainland, although only Chestnut-throated Seedeater has been recently seen there (S. Hilty).
RANGE: Sw. Colombia on Tumaco Is. just off the coast of Nariño.

NOTE: Our conclusion after examining the small series in the AMNH is that *insulata* is almost certainly more nearly allied to *S. telasco* than it is to *S. minuta,* with which it has traditionally been associated. An alternative treatment would be to consider *insulata* a hybrid population between those 2 species. Male *insulata* and *telasco* are alike in having quite conspicuous white at base of rectrices (not shown in any other seedeater); a large white wing-speculum of the same size and shape (while *minuta* shows little or no speculum); and somewhat streaked upperparts (this also in females of both species; *telasco* and *insulata* are the only 2 *Sporophila* showing streaking above). Both AMNH males also show some white on belly, a pattern not found in *S. minuta.*

GROUP G

Small; *cinnamon to chestnut (or black) below.* Many local or rare.

Sporophila bouvreuil

CAPPED SEEDEATER

PLATE: 26

IDENTIFICATION: 10 cm (4"). *Mostly cinnamon* (deeper and more rufescent northward in nominate ssp.; paler in *pileata,* especially below). *Crown contrastingly black;* wings and tail blackish, wing feathers often with buffy margins and white speculum. What is apparently the

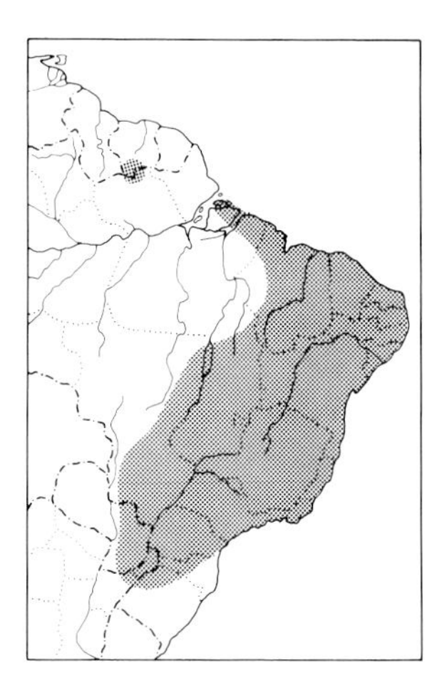

immature male (at least of *pileata*) is white or whitish below, brownish gray on back, with crown, wings, and tail as in male. *Crypta* (of Rio area) evidently breeds while still in an "immature" plumage. *Saturata* (of São Paulo area) is much darker: chestnut rather than cinnamon. Female olive brown above, mostly ochraceous buff below with median lower underparts more yellowish white.

SIMILAR SPECIES: Males are easily recognized by their black caps in striking contrast with their otherwise cinnamon to rufous body plumage. Female probably not safely distinguished from other small female seedeaters.

HABITAT AND BEHAVIOR: Locally fairly common to common in savannas with tall grass and in open *cerrado*. Overall numbers have probably been reduced by overgrazing and excessive burning. During nonbreeding season may occur in quite large groups, often with other seedeaters; this species usually is one of the more numerous members of such flocks.

RANGE: Locally in e. and s. Brazil (islands at mouth of Amazon and from Maranhão south to n. Rio Grande do Sul, inland through Goiás and s. Mato Grosso), e. Paraguay, and ne. Argentina (Misiones and Corrientes); also an isolated population in s. Suriname (Sipaliwini). To about 1100 m.

Sporophila hypoxantha

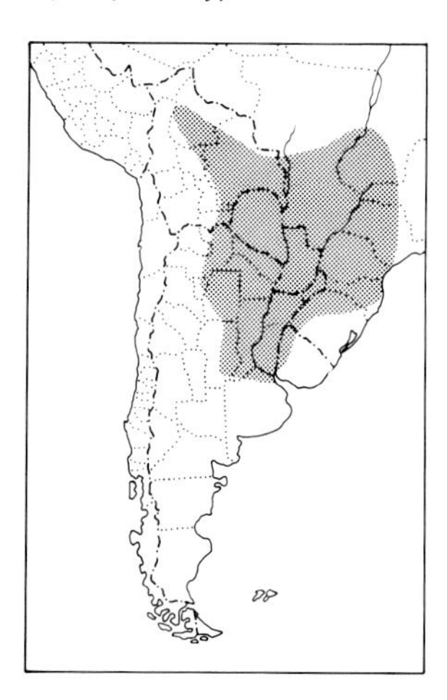

TAWNY-BELLIED SEEDEATER

PLATE: 26

IDENTIFICATION: 10 cm (4"). Mostly brownish gray above; wings and tail duskier, wings with white speculum. *Rump and underparts tawny-rufous* (some birds, perhaps younger males, are paler, more cinnamon tawny), this color extending up over cheeks and ear-coverts, where usually paler. Female brown above, buffy brownish below.

SIMILAR SPECIES: This is generally the commonest and most widespread seedeater of its type in s.-cen. South America; learn it well as the basis of comparison with other less numerous species. Rufous-rumped Seedeater is most similar but has rump and underparts chestnut (not paler tawny-rufous); cf. also Dark-throated and Marsh Seedeaters (all 4, and sometimes others, may occur together during at least the nonbreeding season). Females appear to be indistinguishable (some species even in the hand). See also Ruddy-breasted Seedeater (no overlap).

HABITAT AND BEHAVIOR: Fairly common to common in tall grassy areas, especially near water, and also along roadsides in marshes. During nonbreeding season usually in small groups which sometimes associate with mixed flocks of seedeaters, but also often occurs in groups of its own species. Feeds mostly in tall grass, clinging to the tips of the stems or reaching up from the ground. Song is a simple, clear, whistled "cheeu, cheeu, cheweé, chu" with variations (but never as fast or lively as Ruddy-breasted's).

RANGE: N. and e. Bolivia (Beni, Cochabamba, and Santa Cruz), s. Brazil (s. Mato Grosso and s. Goiás south to n. Rio Grande do Sul), Paraguay, and n. Argentina (south to Tucumán, n. Córdoba, Santa Fe, and Entre Ríos); at least formerly also Uruguay (but not in 20th century). To about 1100 m.

NOTE: We follow L. L. Short (*Wilson Bull.* 81: 216–219, 1969) in considering s. *S. hypoxantha* as a species distinct from *S. minuta*. Besides the morphological differences, their songs do not appear to be that similar.

Sporophila minuta

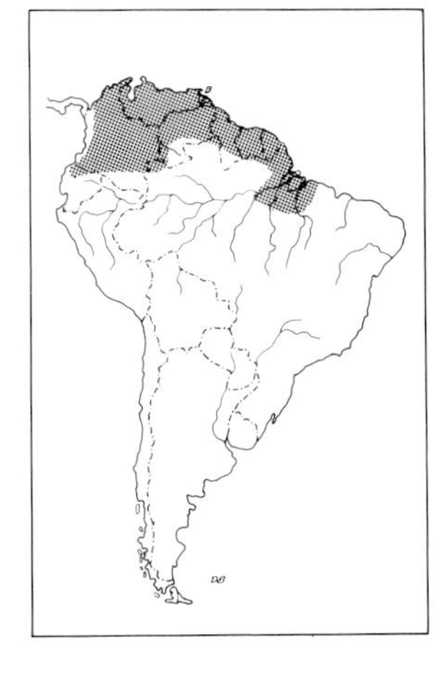

RUDDY-BREASTED SEEDEATER

IDENTIFICATION: 10 cm (4"). Resembles Tawny-bellied Seedeater (no overlap). Male brownish gray above (though color of upperparts can vary from grayish brown to quite pure gray), but has *cheeks and ear-coverts gray* like crown (not cinnamon to tawny); rump and underparts average a deeper rufous, especially on throat; *lacks white wing-speculum* (at most very indistinct). Female buffy brown above, pale dull cinnamon to buffy brown below; wing feathers edged whitish, but has *little or no speculum.*

SIMILAR SPECIES: In its range male likely confused only with Chestnut-bellied Seedeater, which has all-gray upperparts (no rufous rump) and only median underparts chestnut (sides broadly gray). In sw. Colombia, cf. the very rare Tumaco Seedeater.

HABITAT AND BEHAVIOR: Common on open savannas with tall grass and in pastures and along roadsides, especially near water. Separates into scattered pairs while breeding, but most of year found in small groups (at times quite large flocks) which regularly flock with other seedeaters. Its song is pleasant and musical, usually given quite rapidly and from an exposed perch, often commencing with several paired notes.

RANGE: Colombia (west of Andes generally in suitable open country and east of them south to Meta and Guainía), nw. Ecuador (Esmeraldas and Pichincha), Venezuela (generally north of the Orinoco, more local south of it), Guianas, and lower Amaz. Brazil (Amapá and islands of mouth of Amazon upriver to about the lower Rio Tapajós). Also Mexico to Panama. Mostly below 1000 m, locally higher.

NOTE: See comments under Tawny-bellied and Tumaco Seedeaters.

Sporophila hypochroma

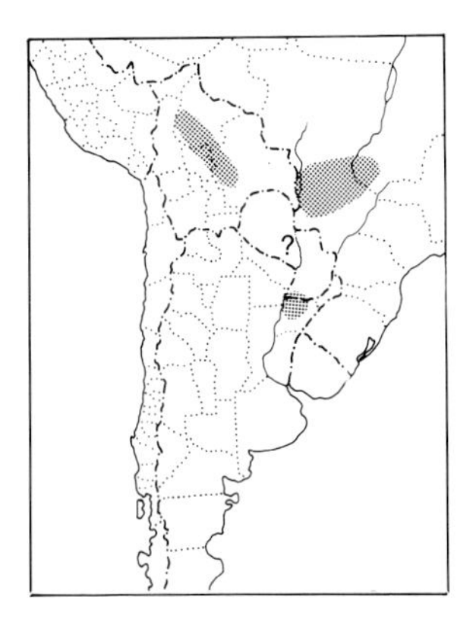

RUFOUS-RUMPED SEEDEATER

PLATE: 26

IDENTIFICATION: 10 cm (4"). *Above mostly gray;* wings and tail blackish with white wing-speculum. *Rump and underparts chestnut,* this color extending up over cheeks and ear-coverts (as in Tawny-bellied Seedeater). Female like female Tawny-bellied Seedeater.

SIMILAR SPECIES: Pattern of male resembles Tawny-bellied Seedeater, but more bluish gray above and color of underparts deep chestnut (not paler tawny-rufous). Chestnut Seedeater has back chestnut (same color as rump), with only the crown gray.

HABITAT AND BEHAVIOR: Poorly known. Birds seen in Oct. 1979 in sw. Mato Grosso, Brazil (RSR), were in mixed flocks of other seedeaters feeding in tall grass near marshy areas; they were always outnumbered and did not differ in behavior from the others.

RANGE: Very locally in n. and e. Bolivia (s. Beni and w. Santa Cruz), sw. Brazil (Emas Nat. Park in extreme s. Goiás and in *pantanal* east of Corumbá, Mato Grosso), and ne. Argentina (east of Ita-Ibate in n. Corrientes); doubtless also Paraguay. Perhaps an austral migrant northward. To about 1100 m.

Sporophila nigrorufa

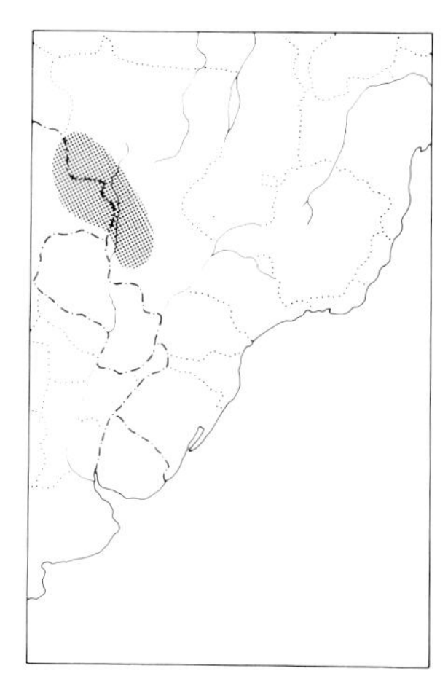

BLACK-AND-TAWNY SEEDEATER

IDENTIFICATION: 10 cm (4″). *Extreme e. Bolivia and w. Mato Grosso, Brazil. Crown, hindneck, and back black;* rump and entire underparts bright cinnamon-rufous. Wings and tail blackish, wings with white speculum and edging to flight feathers. Female does not seem to be definitely known (but probably does not differ appreciably from other small female seedeaters, and probably could not be distinguished in field).

SIMILAR SPECIES: Male's pattern, with black of crown extending down over back, differs from other seedeaters with rufous underparts.

HABITAT AND BEHAVIOR: Rare and virtually unknown in life. Apparently found in grassy *cerrado* and tall grassy areas near water (latter at least when not breeding). One male seen east of Corumbá, Mato Grosso, in Oct. 1979 (RSR) was associating with a mixed flock of nonbreeding seedeaters.

RANGE: Extreme e. Bolivia (Chiquitos in e. Santa Cruz) and extreme sw. Brazil (w. Mato Grosso at Villa Bella de Mato Grosso, Porutí, and east of Corumbá).

Sporophila melanogaster

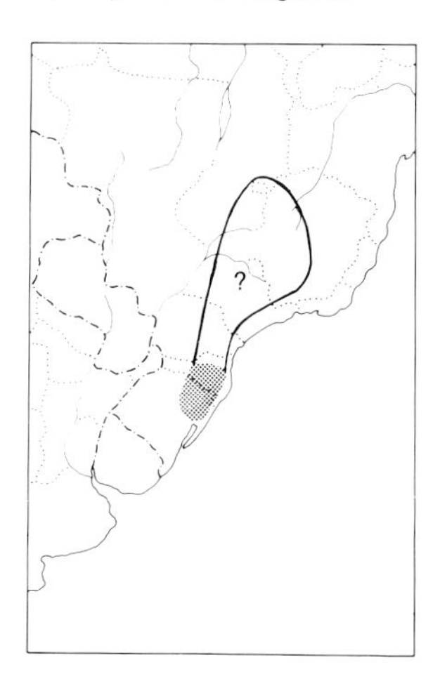

BLACK-BELLIED SEEDEATER

PLATE: 26

IDENTIFICATION: 10 cm (4″). *Se. Brazil.* Mostly rather pale gray, with *black throat and median underparts;* wings and tail blackish, wings with fairly prominent white speculum. Female olive brown above, paler and buffier brown below; wings with whitish speculum as in male.

SIMILAR SPECIES: Dapper gray and black males are unlikely to be confused, but females resemble many others and are probably best identified by their association.

HABITAT AND BEHAVIOR: Locally common in tall grass in and near marshes and boggy depressions. Seasonally numerous (at least Nov.–Mar.) as breeder in highlands of Santa Catarina and ne. Rio Grande do Sul, then occurring in pairs or small groups. Song reported to be a high-pitched simple whistle, followed by a lower trill, then a complex series of wavering warbled notes (W. Belton).

RANGE: Se. Brazil (Rio Grande do Sul north to Minas Gerais and s. Goiás). Apparently an austral migrant to n. part of its range. To about 1000 m.

Sporophila castaneiventris

CHESTNUT-BELLIED SEEDEATER

IDENTIFICATION: 10 cm (4″). Pattern reminiscent of geographically far-distant Black-bellied Seedeater, but *median underparts chestnut* (not

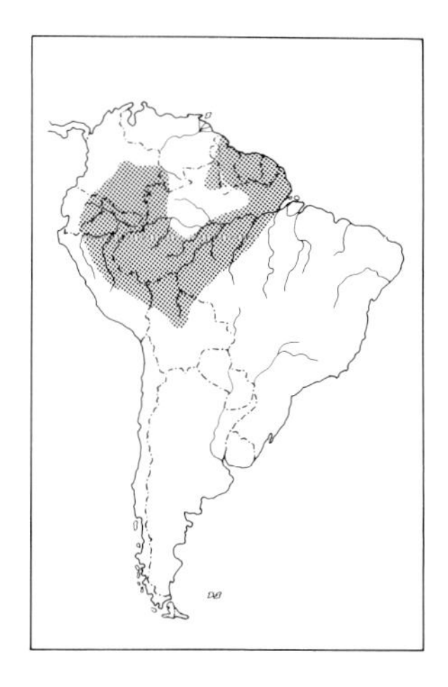

black). *Otherwise bluish gray, including rump;* wings and tail blackish, with speculum faint or lacking. See V-40, C-56. Female mostly olive brown, paler and buffier below (particularly on median underparts).

SIMILAR SPECIES: In its Amaz. range confusion possible only with Ruddy-breasted Seedeater (limited overlap), which has rump and entire underparts rufous (not with sides broadly gray as in this species). Female smaller than other dark-billed seedeaters with which it is sympatric.

HABITAT AND BEHAVIOR: Common in grassy and shrubby clearings, floating vegetation of marshes and lake and river margins, and lawns and gardens in towns and around buildings. Often found in small groups, particularly when not breeding. Generally the most numerous and familiar *Sporophila* in Amazonia.

RANGE: E. Colombia (north to Meta and Guainía) and adjacent sw. Venezuela (sw. Amazonas), e. Ecuador, e. Peru, n. Bolivia (south to La Paz and Cochabamba), Amaz. Brazil (east along Amazon to mouth of Rio Tapajós and Amapá, and in Roraima), and the Guianas. Records from se. Brazil (see Sick 1984, 2: 709) require confirmation. Mostly below 500 m.

Sporophila cinnamomea

CHESTNUT SEEDEATER

PLATE: 26

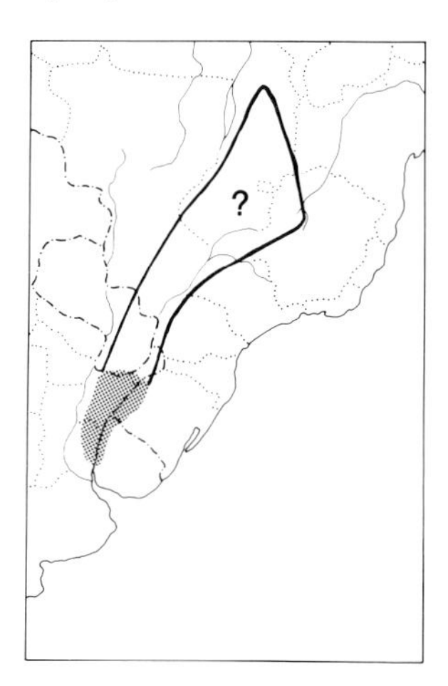

IDENTIFICATION: 10 cm (4"). *Mostly chestnut,* with contrasting *gray crown.* Wings and tail slaty with white wing-speculum. Female like Tawny-bellied Seedeater.

SIMILAR SPECIES: Rufous-rumped Seedeater has similar coloration, but gray of its crown extends to back (which is not chestnut, as in this species). Pattern of this species recalls Capped Seedeater, a much paler bird with a black (not gray) crown.

HABITAT AND BEHAVIOR: Poorly known. Fairly common but very local in tall grass in damp areas or near marshes. Has been found (RSR et al.) to be a regular member of mixed nonbreeding *Sporophila* flocks at Emas Nat. Park, s. Goiás, Brazil (Oct. 1979–1981), where behaviorally it does not differ from the other seedeaters it consorts with. Nesting in some numbers has recently been reported from El Palmar Nat. Park in Entre Ríos, Argentina (T. Narosky and S. Salvador), and it is a fairly widespread nester in Corrientes (T. Parker).

RANGE: Locally in s.-cen. Brazil (recorded from Goiás, se. Mato Grosso, w. Minas Gerais, and w. Rio Grande do Sul), e. Paraguay (1 record from Guaira), and ne. Argentina (Corrientes and Entre Ríos). Apparently only a summer resident (Nov.–Apr.) in s. portion of range; probably a long-distance migrant. To about 1100 m.

Sporophila palustris

MARSH SEEDEATER

PLATE: 26

IDENTIFICATION: 10 cm (4"). Resembles Dark-throated Seedeater (and possibly only a color phase of that species) but *throat and chest white,* contrasting strongly with *rufous-chestnut lower underparts.* Crown and back gray, rump rufous-chestnut. Wings and tail dusky, wings with

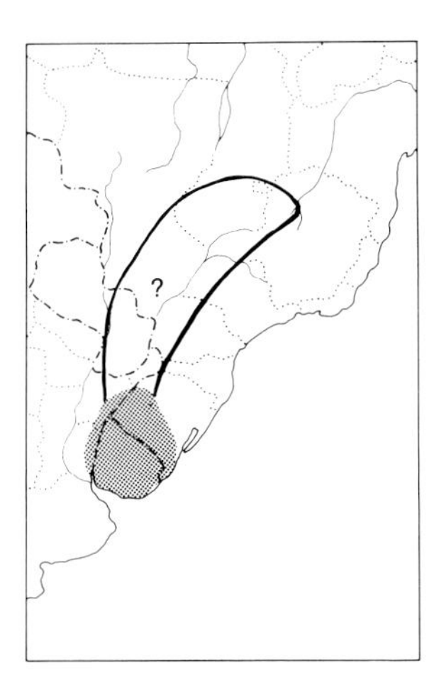

white speculum and whitish edging. Female like female Tawny-bellied Seedeater.

SIMILAR SPECIES: Perhaps the most handsome seedeater and should be quickly recognized. Narosky's (also with white throat and chest) has rufous-chestnut (not gray) back.

HABITAT AND BEHAVIOR: Poorly known. Locally fairly common in tall grass near damp areas or marshes. Has been found to be a regular member of mixed nonbreeding *Sporophila* flocks at Emas Nat. Park, s. Goiás, Brazil (Oct. 1979–1981). Behavior at that season does not differ from that of the various other seedeaters it consorts with, but in ne. Argentina, where it breeds, it has recently been found to be the most strongly marsh-associated seedeater (T. Narosky and S. Salvador). Here it has also declined substantially, a decrease attributed by these observers to heavy trapping for the cage bird market (in which it is considered one of the most desirable species); they suggest that it deserves endangered status.

RANGE: Very locally in s. Brazil (recorded only s. Goiás, s. Mato Grosso, w. Minas Gerais, and Rio Grande do Sul), cen. Paraguay (recorded only from n. *pantanal*), Uruguay (recorded only from Artigas and Rocha), and ne. Argentina (e. Chaco, Corrientes, and Entre Ríos). Migratory status unclear, but apparently only a summer resident (Nov.–Apr.) in s. part of breeding range, and probably an austral migrant, which would account for the scatter of records northward. To about 1100 m.

NOTE: *S. ruficollis* and *S. palustris* are closely related, and the latter may prove to be merely a color phase of the former as Short (1975) has suggested, though we (RSR) have yet to find them at the same locality. We note, however, that *palustris*'s pattern seems to be much less variable than that of *ruficollis*. In our view, further, neither is especially close to *S. hypoxantha*, a considerably more numerous and widespread species.

Sporophila ruficollis

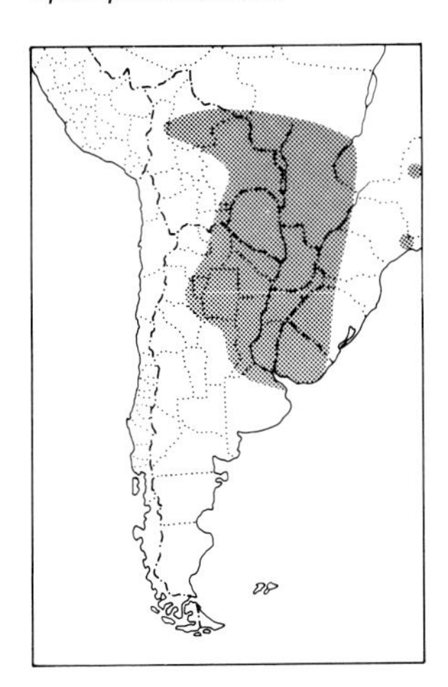

DARK-THROATED SEEDEATER

PLATE: 26

IDENTIFICATION: 11 cm (4¼″). *Crown and back gray* (sometimes more brownish, especially on back); rump cinnamon-rufous. *Throat and chest blackish to dark brown, contrasting with cinnamon-rufous lower underparts;* in some birds only the throat is dark and lower underparts are rather washed out. Wings and tail dusky, wings with white speculum. Female like female Tawny-bellied Seedeater.

SIMILAR SPECIES: Pattern of male, with contrast between dark anterior and lighter posterior underparts, is distinctive.

HABITAT AND BEHAVIOR: Rare to locally uncommon in tall grass near damp areas or in marshes and in partially open shrubby areas. When breeding in ne. Argentina, however, apparently favors less marshy areas than most of its near relatives (T. Narosky and S. Salvador). Usually found singly or in pairs when nesting, but at other seasons may gather in groups with other seedeaters. Has apparently declined substantially in ne. Argentina due to trapping, but not yet to the point where it should be considered threatened.

RANGE: Locally in ne. Bolivia (Beni, Santa Cruz, and Tarija), s. Brazil (recorded s. Mato Grosso, s. Goiás, w. Minas Gerais, w. São Paulo, and Rio Grande do Sul; perhaps more widespread), w. and cen. Paraguay, n. Uruguay (Artigas and Paysandú), and n. Argentina (south to Tucumán, n. Córdoba, Santa Fe, and rarely to n. Buenos Aires). To about 1200 m.

NOTE: See comments under Marsh Seedeater.

Sporophila zelichi

NAROSKY'S SEEDEATER

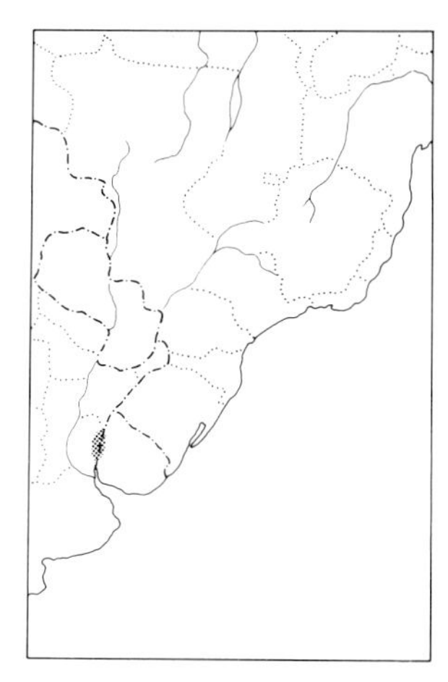

IDENTIFICATION: 10 cm (4"). *Ne. Argentina.* Crown gray, with *broad white nuchal collar, throat, and chest. Back and rump cinnamon chestnut,* as are lower underparts. Wings and tail blackish, wings with white speculum and whitish edging. Female like female Tawny-bellied Seedeater.
SIMILAR SPECIES: This newly described species (if indeed it is not a localized color morph of *S. cinnamomea*) resembles Marsh Seedeater but for the broad white collar (in latter white extends only to sides of neck) and the chestnut (not gray) back.
HABITAT AND BEHAVIOR: Poorly known, but evidently rare and declining as a result of intense trapping efforts by collectors; probably deserves endangered status (T. Narosky and S. Salvador). Reported to occur in semiopen areas near small, clear streams, especially where there are small patches of low woodland, and to gather in mixed flocks with other seedeaters in March, prior to departing for what are presumed to be its winter quarters farther north.
RANGE: Known only from ne. Argentina (Entre Ríos), where apparently only a summer resident (Nov.–Apr.).

NOTE: The status of this recently described species (T. Narosky, *Hornero* 11: 345–348, 1977) remains obscure. It may well prove to be a morph of some known species, and a hybrid origin (*palustris* × *cinnamomea*) has even been suggested. *S. zelichi* resembles *S. cinnamomea* with the addition of a white foreneck and collar—a situation comparable to *S. palustris* and *S. ruficollis*—and thus could be only a localized color phase of *cinnamomea*.

Lysurus Finches

A pair of dark finches which resemble *Atlapetes* brush-finches but are smaller. They skulk in lower growth of humid subtropical forest of Andes and are apparently very local. The nest of the Olive Finch is a domed, moss-covered structure placed in a niche on a damp rockface (T. S. Schulenberg and F. B. Gill, *Condor* 89[3]: 673–674, 1987).

Lysurus castaneiceps

OLIVE FINCH

PLATE: 27

IDENTIFICATION: 15 cm (6"). *A dark, mostly olive green finch* of forest undergrowth. *Crown and nape chestnut; face and throat dark gray.* Otherwise mostly olive, wings and tail duskier.

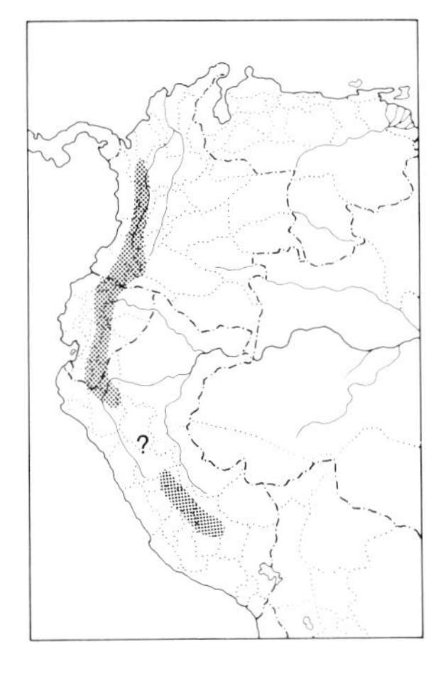

SIMILAR SPECIES: Recalls an *Atlapetes*, but *behavior more reclusive than most*. White-rimmed Brush-Finch most resembles it but is essentially black (not olive) above and has prominent white spectacles; it usually occurs at higher elevations.

HABITAT AND BEHAVIOR: Uncommon and local in undergrowth or on the ground in humid montane forest; also in dense tangles at forest edge or in ravines. The Olive Finch's supposed rarity may more reflect its unobtrusive habits than actual low numbers. Habits poorly known, but its behavior presumably is similar to the better-known (in Central America) Sooty-faced Finch's.

RANGE: Andes in Colombia on Pacific slope from w. Antioquia and s. Chocó south to nw. Ecuador (1 old specimen from Nanegal in Pichincha; no recent records); also locally on e. slope to se. Colombia (se. Nariño; perhaps also north to Caquetá), e. Ecuador, and e. Peru (recorded only from Piura, Amazonas, Pasco, and Cuzco). 700–2200 m, lower on Pacific slope.

Lysurus crassirostris

SOOTY-FACED FINCH

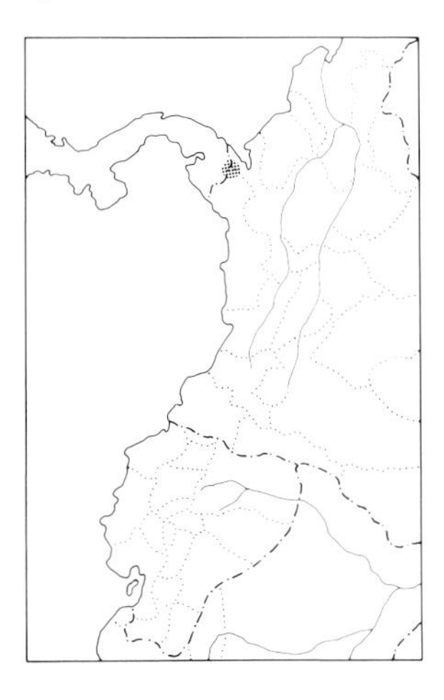

IDENTIFICATION: 15 cm (6"). *Extreme nw. Chocó, Colombia*. Resembles Olive Finch, but has *conspicuous white malar streak* and considerable *yellow on center of breast and belly*. See P-30.

SIMILAR SPECIES: Nothing really similar in range, but in dark light of forest undergrowth could be confused with Chestnut-capped Brush-Finch (though underparts very different).

HABITAT AND BEHAVIOR: Recently recorded from forested slopes on Colombian side of Cerro Tacarcuna (Rodríguez 1982); status there unknown. In Middle America found mostly in pairs, less often in small groups, but rarely joining mixed flocks. Forages near or on the ground in dense forest undergrowth, often near borders or in ravines. Most frequently heard call is a sharp, penetrating "psu-psee."

RANGE: Nw. Chocó in Colombia (Los Kátios Nat. Park). Also Costa Rica and Panama. In Panama mostly 800–1300 m.

Atlapetes Brush-Finches

A large group of mid-size finches of lower growth found mostly at forest edge (though a few are inside forest or out in scrub). They reach their highest diversity in the Andes, though a few are found in montane areas elsewhere. Usually in pairs or small family groups, foraging on or near the ground; most species are bold and relatively easy to see, but a few are much more secretive. Their overall coloration is generally subdued and their patterning simple; head patterns are often most helpful for identification.

GROUP A

Underparts yellow to olive (only a yellow *throat* in one).

Atlapetes leucopis

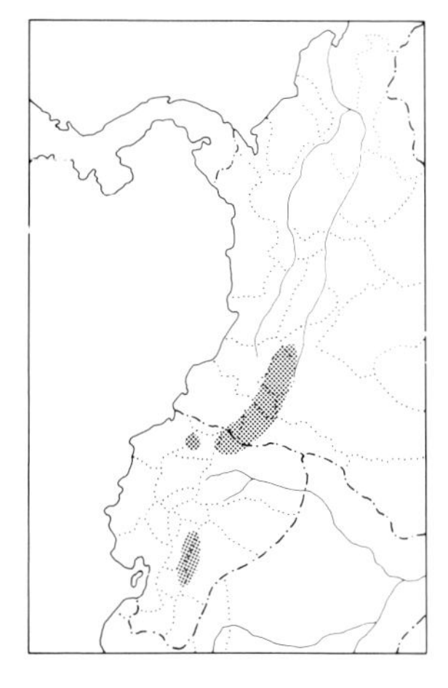

WHITE-RIMMED BRUSH-FINCH

PLATE: 27

IDENTIFICATION: 18 cm (7"). *Locally in sw. Colombia and Ecuador.* A *dark* brush-finch with *conspicuous white eye-ring and short postocular streak.* Crown and nape chestnut, otherwise black above. Below dark olive green, throat and chest mottled dusky.

SIMILAR SPECIES: No other brush-finch has the white spectacles. Otherwise pattern resembles the smaller Olive Finch's.

HABITAT AND BEHAVIOR: Rare to uncommon on the ground and in dense undergrowth of humid forest and at forest borders. Probably often overlooked, with reclusive behavior much like more numerous Chestnut-capped Brush-Finch's. May occur more often with mixed flocks than most brush-finches do.

RANGE: Locally in Andes of Colombia (head of Magdalena valley in Huila and on e. slope in Putumayo and Nariño) and in Ecuador (south on e. slope to Azuay; also a recent record from Imbabura on w. slope, specimen in LSUMZ). 2100–3100 m.

Atlapetes rufinucha

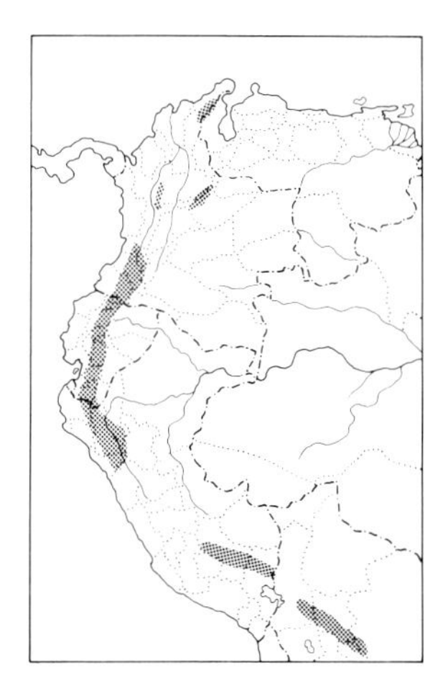

RUFOUS-NAPED BRUSH-FINCH

PLATE: 27

IDENTIFICATION: 16.5–17 cm (6½–6¾"). *Varies strikingly through its Colombia to Bolivia range.* All races have *crown rufous* (paler on hindcrown in nw. Peru) and are *gray to black above* and *mostly yellow below.* White wing-speculum prominent in part of Colombia (in *elaeoproreus* of N.-Cen. Andes and *caucae* of upper Cauca valley) and in se. Ecuador and ne. Peru (*latinuchus*), but small or (usually) absent elsewhere. Upperparts blackest in nominate group (with *melanolaimus* and *carrikeri*) of se. Peru and nw. Bolivia, palest gray in *chugurensis* of nw. Peru. All races have a faint to prominent black malar streak, culminating in the almost entirely black throat of *melanolaimus* of se. Peru. *Phelpsi* of Perijá Mts. has black forehead; it and birds from se. Peru and nw. Bolivia show gray cheeks.

SIMILAR SPECIES: Though the variation is confusing, in most areas this usually common species can be recognized by its combination of rufous crown, slaty back, and bright yellow underparts. Pattern most resembles Tricolored Brush-Finch, but latter's crown much yellower. See also Moustached Brush-Finch (with olive back and white throat) and Santa Marta Brush-Finch.

HABITAT AND BEHAVIOR: Usually common in shrubby forest borders and secondary growth along roads or at edge of pastures; not found in forest undergrowth.

RANGE: Andes of Colombia (locally at n. end of Cen. Andes in Antioquia, in E. Andes from Santander to Cundinamarca, and more widely in W. Andes from Valle south and in Cen. Andes of Cauca), Ecuador (widespread on w. slope and on slopes above cen. valley; on e. slope recorded only from Azuay and Morona-Santiago south), Peru (on w. slope south to Cajamarca, on e. slope south to La Libertad, and in Cuzco and Puno), and nw. Bolivia (south to w. Santa Cruz); Perijá Mts. on Venezuela-Colombia border. Mostly 1500–3000 m, locally somewhat lower and higher.

Atlapetes melanocephalus

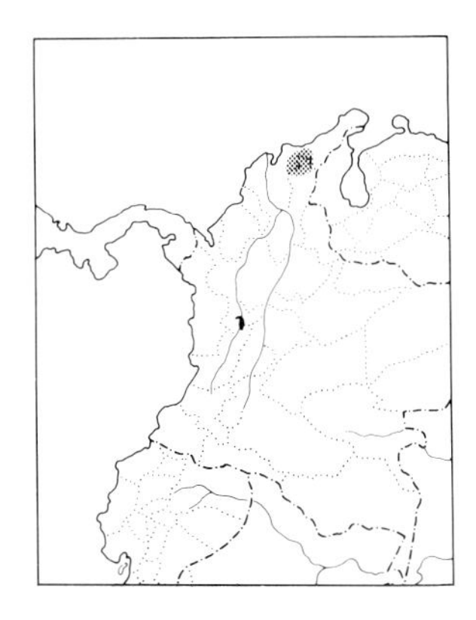

SANTA MARTA BRUSH-FINCH

IDENTIFICATION: 17 cm (6¾"). *Colombia's Santa Marta Mts.* (the only other brush-finch found there is the very different Stripe-headed). Resembles Rufous-naped Brush-Finch (which it replaces in the Santa Martas) but lacks the rufous crown; *entire head, nape, and upper throat black,* with *conspicuous silvery gray patch on ear-coverts.* Upperparts otherwise gray; no white wing-speculum. See C-54.
HABITAT AND BEHAVIOR: Common in shrubby forest borders and dense vegetation along roads and in regenerating areas. One of the more conspicuous and readily seen "Santa Marta endemics" along the San Lorenzo ridge road. Active behavior much like Rufous-naped Brush-Finch's; regularly accompanies mixed foraging flocks.
RANGE: Santa Marta Mts. of n. Colombia. Mostly 1300–3000 m.

Atlapetes albofrenatus

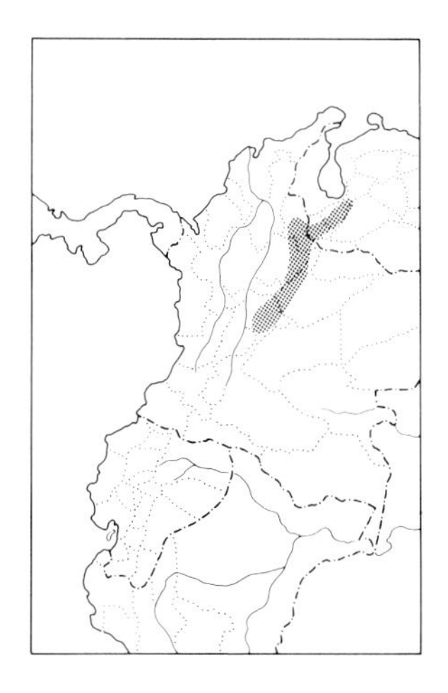

MOUSTACHED BRUSH-FINCH

IDENTIFICATION: 18 cm (7"). *W. Venezuela and ne. Colombia.* Generally resembles Rufous-naped Brush-Finch, but has olive green (not gray) back and no wing-speculum. In area of overlap with Rufous-naped (in ne. Colombia), nominate race of Moustached differs in having a narrow black forehead and *prominent white malar streak and center of throat,* with conspicuous black submalar streak separating the 2 (this area entirely yellow in sympatric Rufous-naped); see C-54. In w. Venezuela *meridae* has all-yellow throat but retains the white malar and black submalar streaks (latter often not very conspicuous) of nominate Moustached; see V-39.
SIMILAR SPECIES: See Rufous-naped Brush-Finch.
HABITAT AND BEHAVIOR: Fairly common at shrubby borders of humid subtropical forest, in thickets along roads, and in regenerating clearings. In some areas also reported to occur in drier lower woodland.
RANGE: Andes of w. Venezuela (Mérida and Táchira) and ne. Colombia (both slopes of E. Andes south to Cundinamarca); Perijá Mts. on Venezuela-Colombia border. Mostly 1500–2500 m, rarely somewhat lower.

Atlapetes tricolor

TRICOLORED BRUSH-FINCH

PLATE: 27

IDENTIFICATION: 16–18 cm (6¼–7"). *Crown rich brownish gold.* Sides of head black and remaining upperparts dark olive. Below yellow, with sides and flanks olive. Nominate race of Peru has crown a brighter gold color and is brighter olive above and yellower below; it is slightly smaller than *crassus* of Colombia and Ecuador.
SIMILAR SPECIES: No other brush-finch has the golden crown, but otherwise somewhat resembles Rufous-naped Brush-Finch (though that species is gray, not olive, above, etc.). Pattern also vaguely recalls Dusky-faced Tanager.

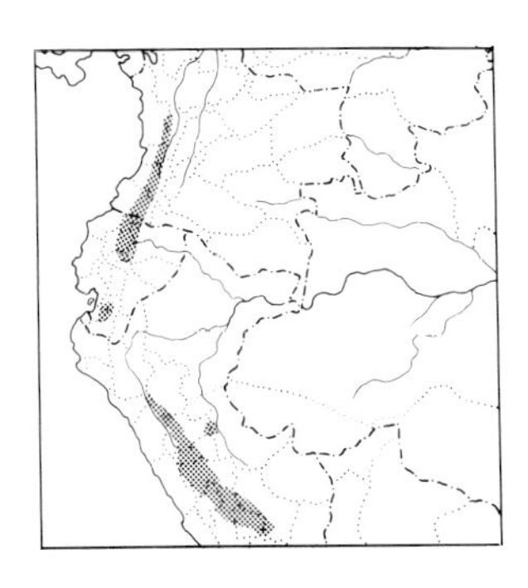

HABITAT AND BEHAVIOR: Uncommon to fairly common and somewhat local at edge of humid forest and in shrubby clearings and forest openings. Probably most numerous in mossy cloud forest areas at appropriate elevations; regular along the road to Mindo in w. Ecuador. The song (in Ecuador) is a fairly high-pitched whistled "tsuwee tsuwee, tsee-tsee, tsí-tsí-tsí" with many variations but always the same quality; a persistent singer.

RANGE: Andes in Colombia on Pacific slope from w. Caldas south locally to w. Ecuador (El Oro); on e. slope in Peru south of the Marañón (La Libertad to Cuzco). 700–1800 m.

Atlapetes pallidinucha

PALE-NAPED BRUSH-FINCH

PLATE: 27

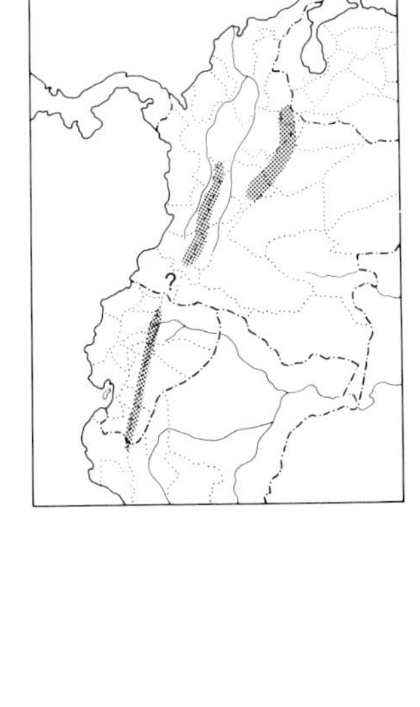

IDENTIFICATION: 18 cm (7"). *Mainly Colombia and Ecuador.* Forecrown cinnamon to yellow, *becoming white on center of hindcrown and nape.* Sides of head and neck black; upperparts otherwise gray. Below yellow, clouded olive especially on sides. Birds from w. Colombia south (*papallactae*) somewhat darker and dingier, with heavy olive suffusion below except on throat.

SIMILAR SPECIES: Rufous-naped Brush-Finch is almost always found at lower (subtropical) elevations and has fully rufous crown; where Rufous-naped's crown color is the palest (in nw. Peru), Pale-naped Brush-Finch does not occur.

HABITAT AND BEHAVIOR: Fairly common in low shrubbery and borders of humid temperate forest, often occurring at or near timberline (and usually most numerous there).

RANGE: Andes of extreme w. Venezuela (Táchira) and E. Andes of ne. Colombia (south to Cundinamarca); Cen. Andes in Colombia from Caldas south on e. slope through Ecuador to extreme n. Peru (Cerro Chinguela on Piura-Cajamarca border). May also occur on w. slope of Ecuadorian Andes, but at best rare or local there (no recent reports). Mostly 2800–3600 m.

Atlapetes gutturalis

YELLOW-THROATED BRUSH-FINCH

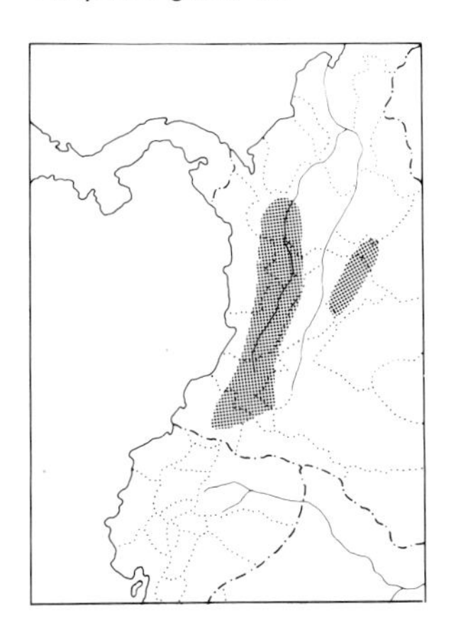

IDENTIFICATION: 18 cm (7"). *Colombia.* Pattern somewhat recalls Pale-naped Brush-Finch, but *only throat yellow;* remaining underparts mostly whitish. Cap mostly black with *white median crown stripe,* but lacking Pale-nape's cinnamon to yellow forecrown. See C-54.

SIMILAR SPECIES: No other brush-finch has the contrasting yellow throat.

HABITAT AND BEHAVIOR: Fairly common in shrubby forest borders and thickets in pastures or along roads; more tolerant of mostly deforested conditions than most other brush-finches.

RANGE: Colombia in W. and Cen. Andes north to Antioquia, in W. Andes south to w. Nariño (ANSP specimen from Yananchá), and in E. Andes mainly in Cundinamarca (probably also southward). Also s. Mexico to Panama. Mostly 1500–2200 m, rarely higher or lower.

NOTE: We here consider *A. gutturalis* as a species distinct from *A. albinucha* (White-naped Brush-Finch) of Mexico; the 2 have by some been considered conspecific (e.g., by Paynter 1978).

Atlapetes fuscoolivaceus

DUSKY-HEADED BRUSH-FINCH

PLATE: 27

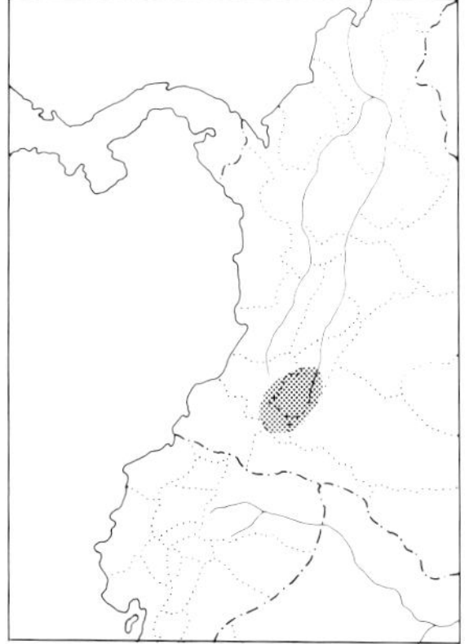

IDENTIFICATION: 17 cm (6¾"). *Colombia's upper Magdalena valley. Head and indistinct malar streak dusky olive brown.* Upperparts otherwise dark olive. Below bright yellow, flanks shaded olive.

SIMILAR SPECIES: No other brush-finch with yellow underparts has an entirely blackish head. Head of immature Dusky-headed is dark olive; this stage thus somewhat resembles Olive-headed Brush-Finch, but it never shows that species' yellow loral area.

HABITAT AND BEHAVIOR: Fairly common at edge of humid forest and in low second-growth and shrubby clearings; rare or absent within actual forest. Numerous at the San Agustín ruins.

RANGE: Andes of Colombia around head of upper Magdalena valley in Huila (both on w. slope of E. Andes and e. slope of Cen. Andes). 1600–2400 m.

Atlapetes flaviceps

OLIVE-HEADED BRUSH-FINCH

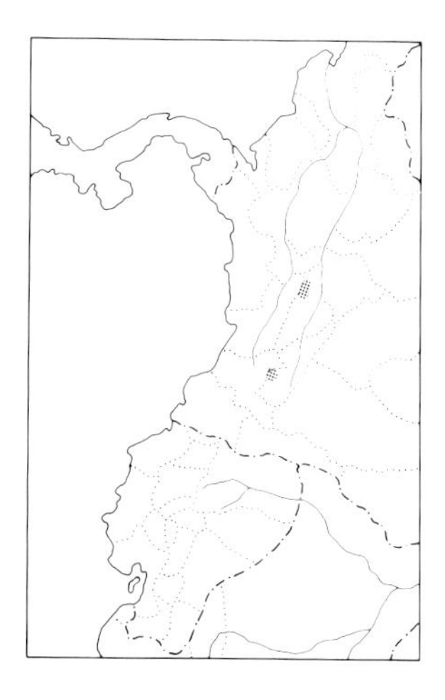

IDENTIFICATION: 17 cm (6¾"). *Colombia's Cen. Andes.* Somewhat resembles Dusky-headed Brush-Finch, but *head mostly yellowish olive,* with *conspicuous yellow lores and narrow eye-ring,* extending back as vague yellowish superciliary. Some birds have considerably yellower heads, while in others the olive feathers are yellow-tipped.

SIMILAR SPECIES: Young Dusky-headed Brush-Finch has head dark olive, but it never shows any yellow.

HABITAT AND BEHAVIOR: Very poorly known; only 1 recent record (1967). We fear that it is endangered by the deforestation which has taken place over most of its range. Its habits are probably much like those of the closely related Dusky-headed Brush-Finch. The original specimens were obtained in "brush covering cleared mountainsides of open valley" (F. Chapman), but 2 subsequently were recorded as having been taken in "forest" (ANSP specimens).

RANGE: Cen. Andes of Colombia on e. slope in Tolima (Río Toche valley, 2000–2200 m) and more recently in Huila (below La Plata Vieja, about 1300 m; Dunning 1982).

NOTE: This rare species, known from only a few specimens, presents a problem. The type (in AMNH) has an olive yellow head and is obviously the bird L. A. Fuertes illustrated in the species' description (F. Chapman, *Bull. Am. Mus. Nat. Hist.* 33: 167–192, 1914), though he slightly exaggerated the brightness of the yellow. The other AMNH specimen has, however, a mostly olive head, with only a scattering of yellow feathers (coming in?) and yellow lores and eye-ring. A bird netted by Dunning (photographs examined) resembles the latter but shows even less yellow. Two ANSP specimens also demonstrate this variation, 1 (a female) having a considerably yellower head than the other (a male) but with intensity of yellow not as great as in the type. We are uncertain which is the more "usual" (i.e., adult); if the type represents the actual adult, then an English name of "Yellow-headed Brush-Finch" would certainly be more appropriate.

Atlapetes citrinellus

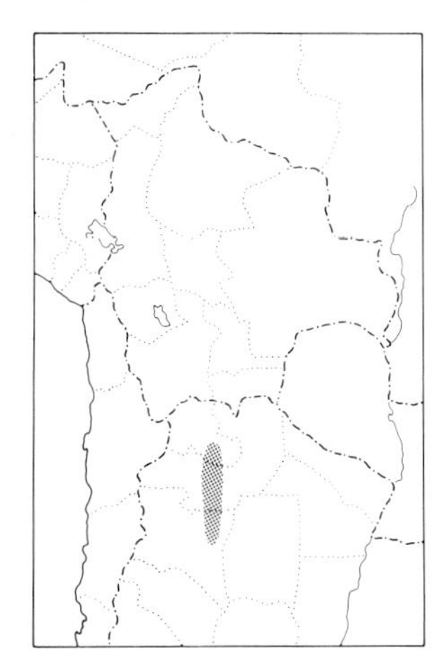

YELLOW-STRIPED BRUSH-FINCH

PLATE: 27

IDENTIFICATION: 17 cm (6¾"). *A striking brush-finch of nw. Argentina with bold blackish and yellow head pattern.* Above mostly olive, with *broad yellow superciliary* extending back from lores and enclosing blackish ocular area and ear-coverts. Below mostly bright yellow, with conspicuous black submalar streak and yellow malar stripe, and chest and sides shaded olive.

SIMILAR SPECIES: In its range likely confused only with Fulvous-headed Brush-Finch (both are yellow below), but Yellow-striped has no rufous on head. Cf. also male Yellow Cardinal (which has a somewhat similar facial pattern but is crested, etc., and occurs in very different scrub habitat).

HABITAT AND BEHAVIOR: Common locally in undergrowth of forest borders and alder woodland. Apparently most numerous in Tucumán (e.g., along upper parts of the road below Tafi del Valle).

RANGE: Andes of nw. Argentina in Jujuy, Salta, and Tucumán (and perhaps also adjacent Catamarca). Recorded also from w. Paraguay, but this record seems very doubtful (Paynter 1978). About 700–2000 m.

Atlapetes semirufus

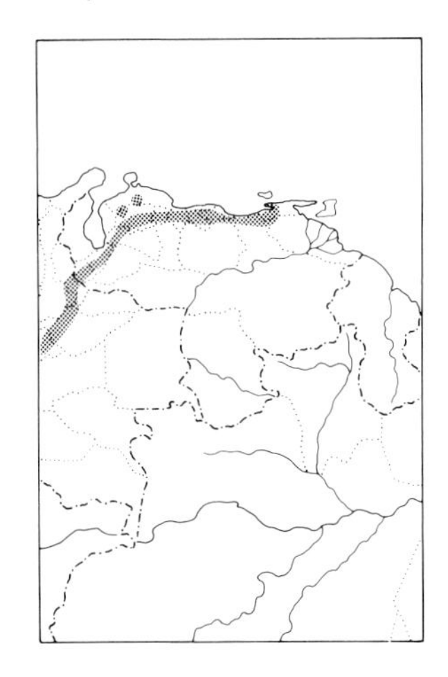

OCHRE-BREASTED BRUSH-FINCH

PLATE: 27

IDENTIFICATION: 17 cm (6¾"). *Venezuela and ne. Colombia. Entire head, throat, and breast cinnamon to cinnamon-rufous.* Above olive green (but more grayish olive in *zimmeri* of extreme ne. Colombia and adjacent Venezuela); lower underparts yellow, more olive on flanks.

SIMILAR SPECIES: The only brush-finch in its range with wholly rufous head.

HABITAT AND BEHAVIOR: Uncommon (in Colombia) to fairly common in shrubbery at forest borders and in regenerating clearings. Seemingly most numerous in coastal cordilleras of n. Venezuela. The song (in Venezuela) is a fast "wheet-wheet, tsu-tsu-tsu" with variations but usually sounding quite like a Yellow-green Grosbeak (the 2 do not overlap).

RANGE: Mts. of n. Venezuela from Paria Peninsula east along coastal cordilleras to Falcón and Lara and in Andes of w. Venezuela and ne. Colombia (E. Andes south to Cundinamarca). In n. Venezuela mostly 1000–2500 m, in w. Venezuela and Colombia mostly 2000–3500 m.

Atlapetes fulviceps

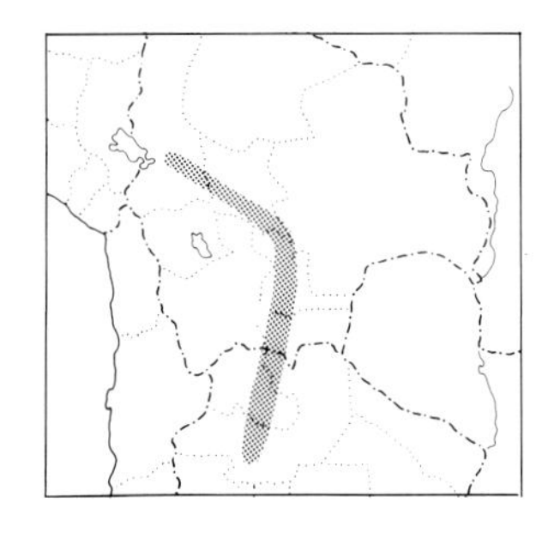

FULVOUS-HEADED BRUSH-FINCH

PLATE: 27

IDENTIFICATION: 17 cm (6¾"). *Bolivia and nw. Argentina. Head and submalar streak rufous.* Otherwise olive above, yellow below (and also on lores and malar stripe).

SIMILAR SPECIES: The only rufous-headed brush-finch in its range. Rust-and-yellow Tanager has similar color pattern but markedly different shape and arboreal behavior.

HABITAT AND BEHAVIOR: At least locally fairly common in woodland undergrowth and at the edge of humid montane forest; seems especially numerous in alder-dominated woodland, at least in Bolivia.

RANGE: Andes of Bolivia (north to La Paz) to nw. Argentina (Jujuy and Salta). About 1500–3000 m.

Atlapetes personatus

TEPUI BRUSH-FINCH

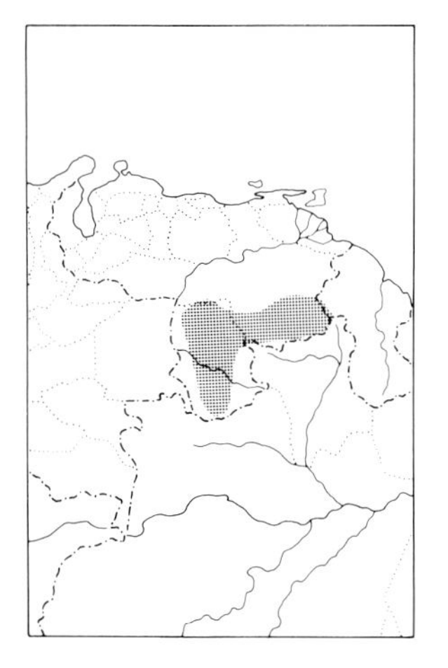

IDENTIFICATION: 18 cm (7″). *Tepui region.* Somewhat resembles Ochre-breasted Brush-Finch (no overlap), but *color of head deeper, more chestnut,* and it varies in its extent; in e. Bolívar (nominate) the chestnut extends only to the chin, while in n. Amazonas (*paraguensis*) it covers the entire throat, and in s. Amazonas and sw. Bolívar (*duidae* group) it extends down over the chest. Above dark gray to black; below mainly yellow, clouded olive on flanks. See V-39.
SIMILAR SPECIES: The only brush-finch in its range.
HABITAT AND BEHAVIOR: At least locally very common (e.g., along the Escalera road in e. Bolívar) in shrubby forest borders and clearings; less numerous in undergrowth of continuous forest.
RANGE: *Tepui* region of s. Venezuela in Bolívar and Amazonas; also adjacent extreme n. Brazil and perhaps also extreme w. Guyana (though not yet recorded). 1000–2500 m.

GROUP B

Underparts white to gray (no yellow).

Atlapetes schistaceus

SLATY BRUSH-FINCH

PLATE: 27

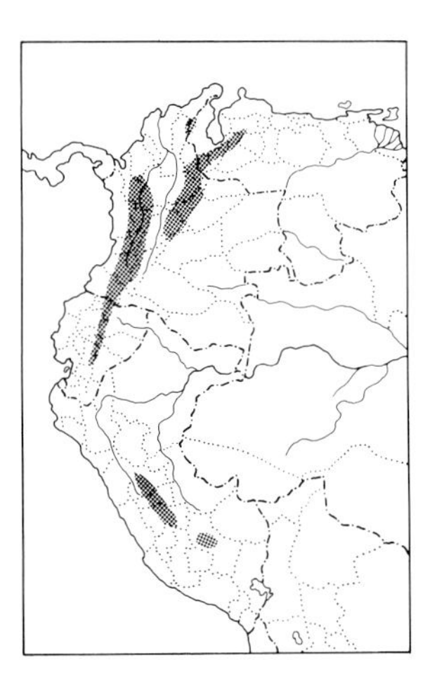

IDENTIFICATION: 18 cm (7″). *Venezuela to Peru. A rather uniform gray brush-finch* with *rufous crown and nape.* Face, sides of neck, and submalar streak black. *Conspicuous malar stripe white;* throat and center of lower underparts whitish, but flanks and crissum gray. Nominate race (most of Colombia and Ecuador) has *much deeper chestnut crown* and a conspicuous white wing-speculum; see C-54. *Canigenis* of e. Peru lacks white malar streak and throat; its face and underparts look very uniform smoky gray.
SIMILAR SPECIES: White-winged Brush-Finch is smaller and much whiter below and has a paler (rufous) crown than the race of Slaty found in Ecuador; the 2 species also differ by habitat and probably never occur together.
HABITAT AND BEHAVIOR: Common at shrubby forest borders and in undergrowth of humid woodland (regularly up to about timberline); usually not found inside continuous forest. Rather arboreal for a brush-finch.
RANGE: Andes of w. Venezuela (north to Trujillo), Colombia, and e. Ecuador (south to Morona-Santiago); e. slope of Andes in cen. Peru (Huánuco to Cuzco); Perijá Mts. on Venezuela-Colombia border. Mostly 2500–3500 m.

Atlapetes leucopterus

WHITE-WINGED BRUSH-FINCH

PLATE: 27

IDENTIFICATION: 15.5–16.5 cm (6–6½″). A *small* gray and white brush-finch. *Crown and nape cinnamon-rufous; conspicuous white wing-speculum.*

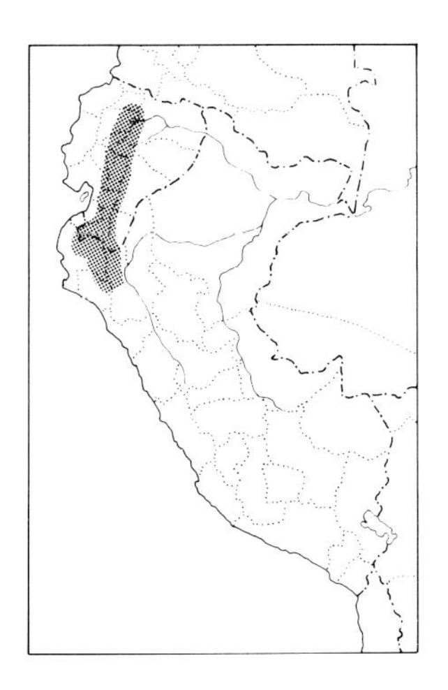

Otherwise gray to brownish gray above, white to buffy whitish below, tinged grayish on flanks. Dresseri of s. Ecuador and nw. Peru is confusingly variable: its forehead is black (extent varies) and it usually (except in extreme north or south of its range) *has at least some white on face* (varying from a narrow eye-ring to birds in which the entire head is virtually white). *Paynteri,* a newly described race from extreme n. Peru (locally in n. Cajamarca and ne. Piura) resembles *dresseri* but has white hindcrown and nape and black sides of head (like nominate race).

SIMILAR SPECIES: See larger and darker Slaty Brush-Finch (no known overlap). *Paynteri,* with its white hindcrown, should be distinctive. *Dresseri,* however, can be confused with Bay-crowned and White-headed Brush-Finches, though neither of these ever has a white wing-patch; head patterns of both also differ in various details.

HABITAT AND BEHAVIOR: Fairly common to common (especially numerous on w. slope of nw. Peru) in dry to fairly humid brushy areas and low woodland. Feeds mostly on or near the ground, hopping in leaf litter in pairs or groups which vary from small to (at least locally) large, with up to 20 to 30 birds per flock in Lambayeque, Peru; in the latter area sometimes accompanied by White-headed and Bay-crowned Brush-Finches.

RANGE: Cen. valley of w. Ecuador (in semiarid regions from Pichincha south to Chimborazo; rarely wandering to w. slope, at least in Pichincha) and from s. Ecuador (Loja and s. El Oro) to nw. Peru (south to Lambayeque and Cajamarca). Mostly 1000–2500 m, locally lower.

NOTE: The newly described form *paynteri* seems very distinct morphologically and apparently differs in its more humid habitat from the more xeric habitats of the other races of *A. leucopterus*. It could prove to be a full species, as acknowledged by its describer (J. W. Fitzpatrick, *Auk* 97 [4]: 883–887, 1980).

Atlapetes seebohmi

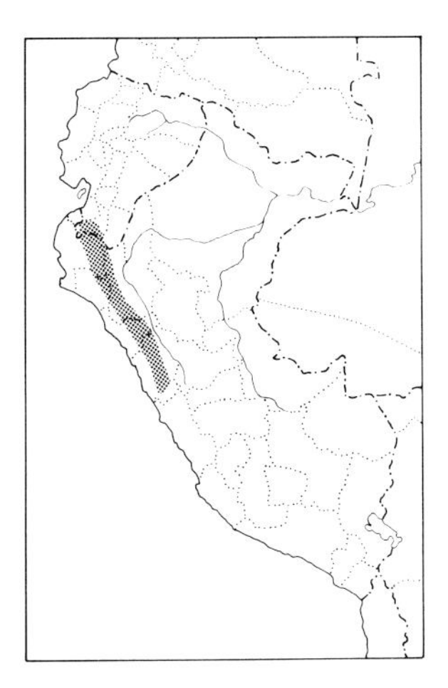

BAY-CROWNED BRUSH-FINCH

IDENTIFICATION: 16–17 cm (6¼–6¾"). *S. Ecuador and nw. Peru.* Resembles White-winged Brush-Finch, *but never shows a white wing-speculum.* Crown and nape rufous, face and sides of neck black (*never shows any white*). Otherwise gray above and on sides and flanks, with white throat and median underparts; rather wide, conspicuous, black submalar streak. Nominate race of w. Peru (La Libertad and Ancash) has black forehead.

SIMILAR SPECIES: Two other gray and white brush-finches are often found with this species: White-winged has white wing-speculum and usually shows white on face, while White-headed has an entirely black and white head (no rufous on crown).

HABITAT AND BEHAVIOR: Uncommon to locally fairly common in undergrowth of low scrubby woodland and on brush-covered hillsides, mostly in rather dry regions.

RANGE: W. slope of Andes from extreme s. Ecuador (Loja) to w. Peru (south to Ancash). 1200–2500 m.

NOTE: Though *A. seebohmi* and *A. nationi* have by some been treated as conspecific (e.g., by Paynter 1972), we prefer to continue to treat them as full species, as did

Meyer de Schauensee (1966, 1970). The 2 races included in *nationi* (nominate and more s. *brunneiceps*) are unique in the genus by virtue of having a cinnamon buff belly. We see no approach to this in *A. s. seebohmi,* the subspecies whose range most closely approaches that of *A. nationi,* and in fact *A. seebohmi's* 3 component subspecies (nominate, *simonsi,* and *celicae*) form a cohesive group showing only minor variation.

Atlapetes albiceps

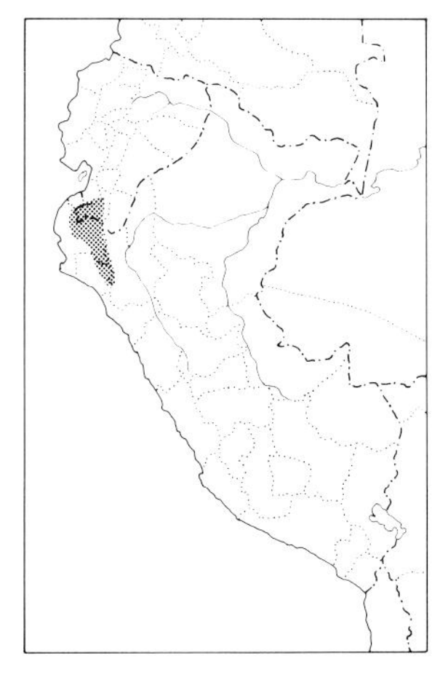

WHITE-HEADED BRUSH-FINCH

PLATE: 27

IDENTIFICATION: 16 cm (6¼"). *S. Ecuador and nw. Peru.* A gray and white brush-finch with *pure white face* and *no rufous on crown. Forehead, entire face, and sides of neck white,* contrasting sharply with *black hindcrown.* Above otherwise pale brownish gray. Below whitish, chest pale gray, flanks and (especially) crissum tinged buff.

SIMILAR SPECIES: No overlap with Pale-headed Brush-Finch. See also White-winged and Bay-crowned Brush Finches (especially sympatric *dresseri* race of the former, some members of which show so much white on face as to resemble this species, though they always have a rufous hindcrown).

HABITAT AND BEHAVIOR: Locally common (at least in Lambayeque, Peru) in undergrowth of low scrubby woodland, especially along streams (to which woodland is restricted in much of its range) in fairly dry regions. Can occur in quite large flocks, often accompanying White-winged Brush-Finches.

RANGE: W. slope of Andes in extreme s. Ecuador (Loja) and nw. Peru (south to Lambayeque). Mostly 400–1300 m.

Atlapetes pallidiceps

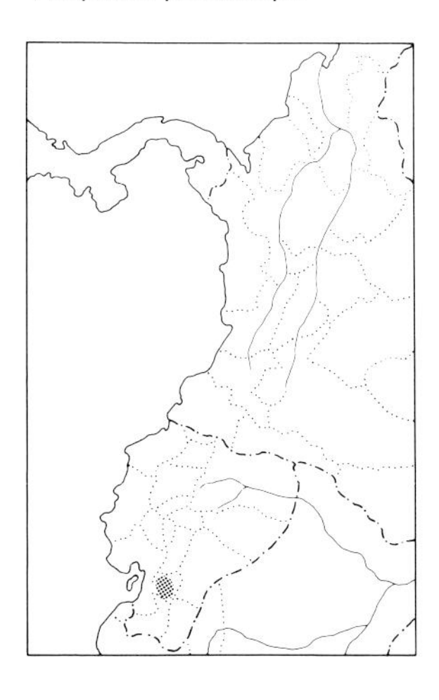

PALE-HEADED BRUSH-FINCH

IDENTIFICATION: 16 cm (6¼"). *S. Ecuador.* Resembles a *faded* White-headed Brush-Finch but has *entire head whitish,* with ill-defined pale brownish stripes on sides of crown and behind eye.

SIMILAR SPECIES: Cf. White-headed Brush-Finch (with crisp black and white head, lacking the dingy effect of this species); note that Pale-headed Brush-Finch is not known to occur sympatrically with any other *Atlapetes.*

HABITAT AND BEHAVIOR: Extremely local, but in the 1960s found to be fairly common in 1 small, isolated patch of low woodland in an otherwise denuded arid region (Paynter 1972). Here mostly arboreal and shy and seen primarily as single individuals. Probably threatened by the near-total removal of natural vegetation in its tiny range; there are no recent sightings.

RANGE: Sw. Ecuador (upper Río Jubones valley in s. Azuay; known only from around Girón and Oña). 1900–2100 m.

Atlapetes nationi

RUSTY-BELLIED BRUSH-FINCH

PLATE: 27

IDENTIFICATION: 17 cm (6¾"). *Sw. Peru. Crown blackish brown,* sometimes with a few scattered white feathers merging into black forehead

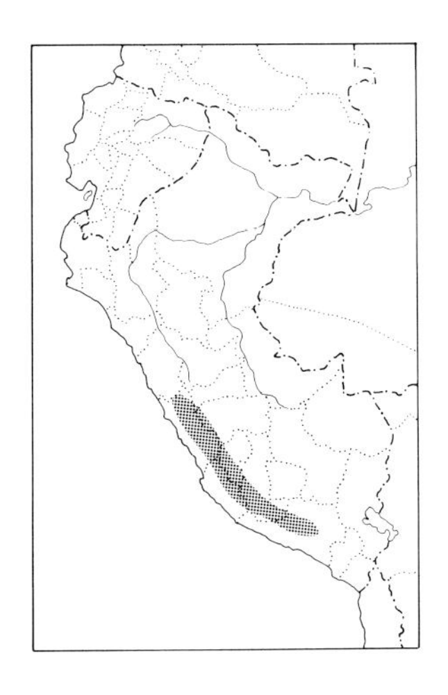

and face. Otherwise brownish gray above. Throat white, with broad black submalar streak. Band across breast and flanks gray, *belly and crissum cinnamon buff. Brunneiceps* (Ica to Arequipa) is paler and has much browner head grizzled whitish on forecrown and face.

SIMILAR SPECIES: The only brush-finch in its range. Bay-crowned Brush-Finch (found just to the north) is smaller, has distinct rufous cap, and lacks buff on belly.

HABITAT AND BEHAVIOR: Uncommon to fairly common but quite local in woodland patches and shrubby areas on Andean slopes. Nominate race can be regularly found above Lima, e.g., along the Santa Eulalia road above Huinco (starting in the shrub zone 1 km past the bridge over the narrow gorge).

RANGE: W. Peru (Lima to Arequipa). 2000–4000 m.

NOTE: See comments under Bay-crowned Brush-Finch.

Atlapetes rufigenis

RUFOUS-EARED BRUSH-FINCH

PLATE: 27

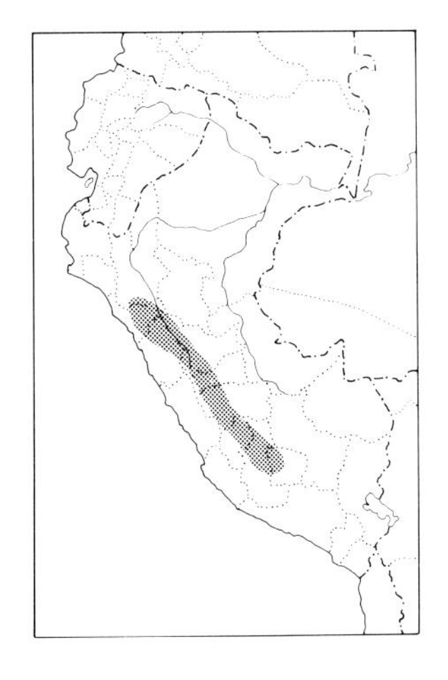

IDENTIFICATION: 19 cm (7½"). *Cen. Peru.* A large brush-finch, *gray above with contrasting rufous head.* Large loral spot and malar stripe conspicuously white, thin submalar streak black. Whitish below, whitest on throat, tinged grayish on sides and flanks. *Forbesi* of Apurímac is darker gray above and richer rufous on its head and has *ocular region conspicuously black.*

SIMILAR SPECIES: Other potential sympatric gray *Atlapetes* have rufous restricted to the crown.

HABITAT AND BEHAVIOR: Uncommon in dense shrubbery and undergrowth of woodland patches (including groves of *Polylepis*). Regularly found in Huascarán Nat. Park in the Cordillera Blanca of Ancash.

RANGE: Andes of w. Peru, mostly in the drainage of the upper Marañón River, from Cajamarca to Huánuco and Ancash; in s. Peru in Apurímac, and recently in Puno (J. Fjeldså). Mostly 3000–4000 m, but *forbesi* recorded at 2750 m.

Atlapetes brunneinucha

CHESTNUT-CAPPED BRUSH-FINCH

PLATE: 27

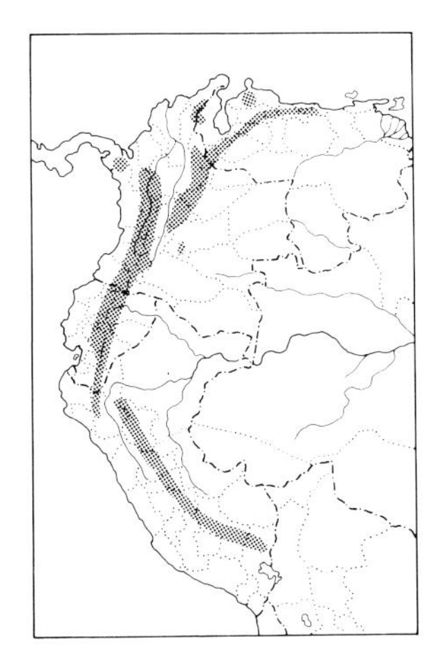

IDENTIFICATION: 19 cm (7½"). *Venezuela to Peru. The only brush-finch with both a chestnut crown* (paler and more golden laterally) *and a black chest band.* Forehead mostly black with 3 inconspicuous white vertical stripes; upperparts otherwise dark olive. *Throat extensively white* (often looks "puffy"); lower underparts whitish, with sides and flanks olive to grayish. Two local races (*allinornatus* of nw. Venezuela and *inornatus* of sw. Ecuador) both lack the black chest band.

SIMILAR SPECIES: Stripe-capped Brush-Finch (many races of which share the chest band) has a very different head pattern with no chestnut on crown.

HABITAT AND BEHAVIOR: Fairly common (but often overlooked) in undergrowth inside humid forest and to a lesser extent at edge. More terrestrial than most *Atlapetes,* feeding by flicking leaves with its bill. Usually not with mixed flocks. The song consists of a few high, thin notes followed by a trill, and it and the bird's *very* high-pitched "pseet" call note (almost inaudible except at close range) are frequently the best indications of the bird's presence.

RANGE: N. Venezuela on coastal mts. from Miranda west to Falcón, in Perijá Mts. on Venezuela-Colombia border, and in Andes of w. Venezuela, Colombia, Ecuador (on w. slope south only to Chimborazo), and e. Peru (south to Puno). Also Mexico to Panama. Mostly 1000–2500 m, locally down to 700 m on Pacific slope of nw. Ecuador.

Atlapetes torquatus

STRIPE-HEADED BRUSH-FINCH

PLATE: 27

IDENTIFICATION: 19–20 cm (7½–7¾"). *Varies strikingly through its Venezuela to Argentina range.* All races are essentially olive above and white below, broadly tinged gray to olive on sides and flanks, and have *white or gray coronal stripe and superciliary* (their widths varying) on an otherwise *black head;* most races also have *conspicuous black chest band.* The only races lacking the chest band are those found from w. Venezuela through Colombian and Ecuadorian Andes to n. Peru (*assimilis* and *nigrifrons*) and again in s. Bolivia and Argentina (*borrelli*). S. races (north to Bolivia) and *phaeopleurus* of the n. coastal range of Venezuela have superciliary white; in all others it is gray.

SIMILAR SPECIES: See Black-headed Brush-Finch (in Colombia). The only other brush-finch with a breast band is the otherwise quite different Chestnut-capped. In n. Bolivia beware the superficially similar local race of Pectoral Sparrow (*nigrirostris*), which has incomplete pectoral band (lacking in local races of brush-finch) and much more conspicuous yellow at band of wing and is notably smaller; it mostly occurs at lower elevations (but they could overlap).

HABITAT AND BEHAVIOR: Generally fairly common (but often overlooked) in undergrowth of humid forest, especially at borders, and in dense secondary growth. Behavior much like Chestnut-capped's, but not as reclusive. The song in Ecuador is short and usually quite thin and squeaky in quality (e.g., "tseek-o-tseé") but with many variations (though the quality is more or less constant).

RANGE: N. Venezuela on coastal mts. from Paria Peninsula west, in Perijá Mts. on Colombia-Venezuela border, and in Andes of w. Venezuela, Colombia, Ecuador, Peru, Bolivia, and nw. Argentina (Jujuy and Salta); on w. slope south only to nw. Peru (to La Libertad); also in Colombia's Santa Marta Mts. Mostly 2000–3000 m (in smaller numbers to timberline), but lower in Venezuela, Bolivia, and Argentina (locally to 1000 m or less).

NOTE: See comments under Black-headed Brush-Finch.

Atlapetes atricapillus

BLACK-HEADED BRUSH-FINCH

IDENTIFICATION: 19 cm (7½"). *Locally in Colombian Andes.* Nominate race of most of Colombian range resembles Stripe-headed Brush-Finch (*without* a black chest band), but *entire head black with no striping;* see

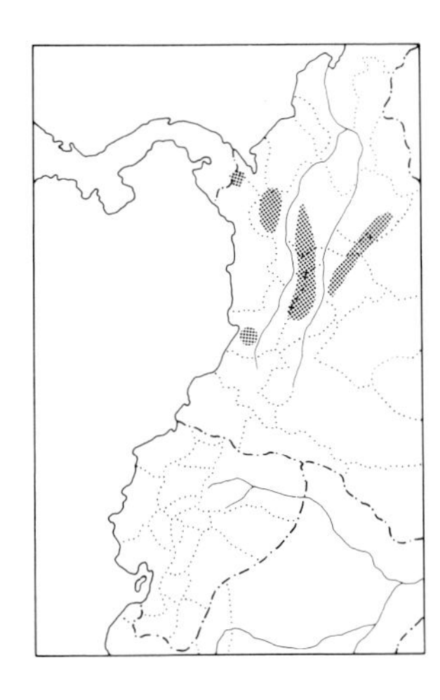

C-54. *Tacarcunae* of Panama border may show trace of gray crown stripe and superciliary, but it is never as obvious as in Stripe-headed Brush-Finch.

HABITAT AND BEHAVIOR: Uncommon and local in undergrowth of humid forest (especially at borders) and in adjacent regenerating areas; perhaps mostly associated with wet, mossy cloud forest situations. Behavior much like Stripe-headed Brush-Finch's.

RANGE: Colombia on slopes of Cerro Tacarcuna on the Panama border and very locally in Andes (W. Andes in Antioquia and Valle; e. slope of Cen. Andes in Caldas and Tolima; w. slope of E. Andes in Santander and Cundinamarca). Also sw. Costa Rica and Panama. Mostly 800–1300 m, locally lower on Pacific slope.

NOTE: The birds found in Costa Rica and w. Panama, *costaricensis,* have been treated either as a race of this species or of *A. torquatus* (maintaining *atricapillus* with *tacarcunae* as a full species; see Paynter 1978) or even as a full species (Gray-striped Brush-Finch). Here we follow Meyer de Schauensee (1966) and the 1983 AOU Check-list in placing *costaricensis* with *A. atricapillus;* as Wetmore, Pasquier, and Olson (1984: 591–592) point out, there does appear to be clinal (or at least stepped) variation in Panama. However, we retain this expanded *A. atricapillus* as a species distinct from *A. torquatus* (*contra* Wetmore, Pasquier, and Olson), for the ranges of these 2 approach each other very closely in Colombia (though habitats may differ) with no evidence of intergradation.

Oreothraupis Finches

A monotypic genus whose form and behavior recall *Atlapetes,* though it is larger.

Oreothraupis arremonops

TANAGER FINCH

PLATE: 27

IDENTIFICATION: 20.5 cm (8"). *A distinctive large rufescent finch of Colombia and Ecuador's w. slope.* Head and chin black with pale *gray coronal stripe and broad superciliary extending to nape.* Upperparts otherwise mostly reddish brown, tail blackish. *Below mostly orange-rufous,* brightest on breast, with center of lower breast and belly gray. Juvenal duller and browner, with head pattern obscure.

HABITAT AND BEHAVIOR: Rare to uncommon and apparently local on or near the ground in humid forest (most often in wet, mossy cloud forest). Much like an *Atlapetes* in behavior. Usually in small groups moving slowly through undergrowth, most often independent of mixed flocks. Not easy to see in most situations, but can be drawn in toward a quiet observer by imitating (or through tape playback) its soft froglike whistled "wert," which is often steadily repeated (but can be overlooked). Most readily found on Cerro Munchique in Cauca, Colombia.

RANGE: Pacific slope of W. Andes of Colombia (very locally from Antioquia south) and in nw. Ecuador (only Pichincha, with no recent reports). 1700–2500 m.

Catamenia Seedeaters

Small finches found mainly at higher elevations in the Andes (lower southward and on Pacific coast). Mainly gray (males) or streaky brownish (females), with stubby pinkish to yellow bills.

Catamenia homochroa

PARAMO SEEDEATER

PLATE: 28

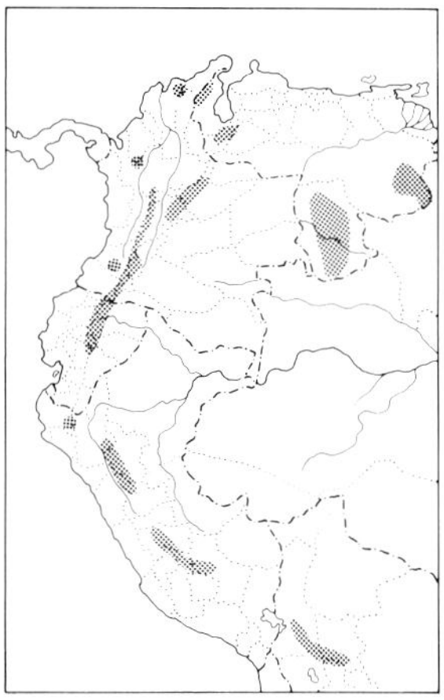

IDENTIFICATION: 13.5 cm (5¼"). Stubby pale pinkish bill (duller in female). Male mostly *dark slaty gray,* with *blackish foreface* and chestnut crissum. *Duncani* of *tepuis* is browner (especially below). Female dark olive brown above streaked with blackish; olive brown below, becoming *more fulvous on belly,* and with chestnut crissum. Immature (both sexes) more coarsely streaked, gradually (in 3 molts) becoming less so and more like full adults.

SIMILAR SPECIES: Both sexes of Plain-colored Seedeater are paler generally and have darker pink bill; male lacks dark foreface, while *female is much more streaked below than female Paramo,* with notably lighter buff (not fulvous) belly. Habitats also usually differ.

HABITAT AND BEHAVIOR: Uncommon and seemingly local in shrubbery at edge of temperate forest and in regenerating clearings; often near timberline, but not in open areas much above it (where Plain-colored is most numerous). Usually in pairs or at most small groups, at times accompanying mixed flocks of high-elevation tanagers, flowerpiercers, etc.

RANGE: Andes from w. Venezuela (Mérida) south locally through Colombia, Ecuador, and Peru to n. Bolivia (La Paz and Cochabamba); Perijá Mts. on Venezuela-Colombia border, Colombia's Santa Marta Mts. (*oreophila*), and on *tepuis* in s. Venezuela (Bolívar and Amazonas) and adjacent n. Brazil (Roraima). Mostly 2500–3500 m, locally a little higher; on *tepuis* 1600–2500 m.

NOTE: L. Zambrano (*Lozania* 23: 1–7, 1977) has shown that *C. oreophila,* originally described and until recently treated as a full species, is merely the Santa Marta subspecies of the widespread *C. homochroa.*

Catamenia inornata

PLAIN-COLORED SEEDEATER

IDENTIFICATION: 13.5–14.5 cm (5¼–5¾"). Bill like Paramo Seedeater's but more brownish pink. Male *rather pale gray,* somewhat tinged brown (especially above); *back streaked blackish;* lower belly tinged buff, and crissum chestnut. Female grayish brown above streaked blackish; yellowish buff below, *streaked dusky on throat and breast;* crissum buff. Extent of female's streaking varies; most conspicuous on immature birds, and may be essentially lost (especially below) on older ones. See C-56, V-40. Nominate race (north to s. Peru) larger than n. *minor.*

SIMILAR SPECIES: Male Paramo Seedeater is much darker and essentially unstreaked above; female Paramo is also darker (especially below, where except in immatures it is unstreaked); see also bill and habitat differences. Cf. also Ash-breasted Sierra-Finch (somewhat smaller, with pointed dusky bill, etc.).

HABITAT AND BEHAVIOR: Common in open grassy areas, *páramo* and *puna,* and pastures with hedgerows and scattered trees and bushes; tends, however, to avoid dense shrubby areas, and unlike Paramo Seedeater rarely at edge of forest or woodland. Forages mostly on ground, often in monospecific flocks (quite large in nonbreeding season), sometimes in association with sierra-finches, etc. Pairs separate when breeding; males then give a short buzzy to musical trill, in some areas introduced by several more musical notes (geographic or individual variation is apparently extensive over this species' wide Andean range).
RANGE: Andes from w. Venezuela (Mérida) south through Colombia (E. and Cen. Andes only), Ecuador, Peru, and w. Bolivia to extreme n. Chile (sightings from Tarapacá) and nw. Argentina (south to Mendoza; also in hills of w. Córdoba). Mostly 2500–3500 m, lower southward.

Catamenia analis

BAND-TAILED SEEDEATER

PLATE: 28

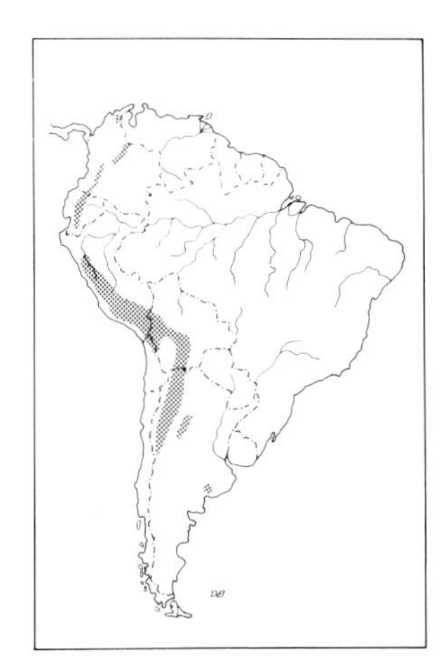

IDENTIFICATION: 12–12.5 cm (4¾–5"). Bill butter yellow. Male *plain gray,* paler below, in s. birds becoming quite white on belly. Foreface blackish (lacking in Colombia), and s. birds (Peru south; nominate race) have a more or less distinct white wing-speculum; crissum chestnut. *Tail blackish with white band across center* (usually visible only in flight or from below). Younger birds apparently more or less streaked black on back. Female rather light grayish brown above streaked blackish; buffy whitish below, streaked blackish except on white belly, with buff crissum; *tail as in male.*
SIMILAR SPECIES: Other *Catamenia* seedeaters lack white in tail (but you often need to flush bird to see it); female paler overall than other female *Catamenia.* Both sexes of Band-tailed Sierra-Finch are quite similar but somewhat larger, with notably longer, more pointed bill.
HABITAT AND BEHAVIOR: Fairly common to locally common in cultivated areas, pastures with hedgerows, and semiopen shrubby and grassy areas. Usually occurs in pairs or small scattered groups (not in large flocks), feeding mostly on ground but perching freely in bushes and small trees. Song is variable, but buzzy trills (sometimes ascending) generally predominate, often interspersed among other, more musical notes; the overall effect is canarylike.
RANGE: Andes in Colombia (E. Andes in Boyacá and Cundinamarca, Cen. Andes locally from Caldas south), Ecuador (Pichincha to Chimborazo), Peru, w. Bolivia, n. Chile (Tarapacá), and n. Argentina (south to Mendoza; also locally in hills of w. Córdoba and sw. Buenos Aires; winters rarely east to Entre Ríos); also Colombia's Santa Marta Mts. Mostly 1000–3000 m, but lower on Pacific slope in Peru.

Idiopsar Finches

A rather large uniform gray finch of very high elevation Andean grasslands. Its long pointed bill separates the genus from the closely related *Phrygilus* sierra-finches.

Idiopsar brachyurus

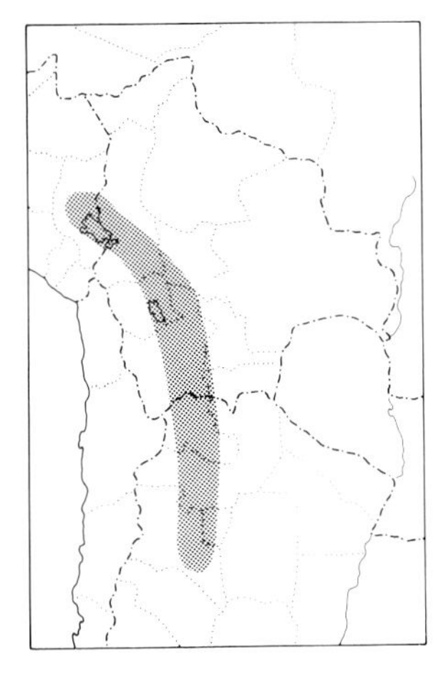

SHORT-TAILED FINCH

PLATE: 28

IDENTIFICATION: 18.5 cm (7¼"). *Bill quite long and thick at base. Entirely leaden gray,* somewhat paler below, with lower face grizzled whitish; wings and rather short tail blackish, primaries edged pale gray. Sexes similar.
SIMILAR SPECIES: Likely confused only with considerably smaller Plumbeous Sierra-Finch; apart from size, note Short-tail's strikingly long bill (which at times gives an almost upturned effect) and the absence of streaky "female-plumaged" birds.
HABITAT AND BEHAVIOR: Uncommon and local in *puna* grasslands, often on steep slopes *with scattered large boulders.* The presence of *Idiopsar* seems in some way tied with boulders, and the species is rarely or never away from them. Found in pairs or small groups, usually not associating with other species; feeds on the ground, but frequently perches on rocks. Often gives a sharp, high-pitched "ziht" call, similar to that of the much commoner White-winged Diuca-Finch.
RANGE: Andes in extreme s. Peru (Puno), n. Bolivia (La Paz and Cochabamba; not recorded southward, though should occur), and nw. Argentina (south to Tucumán and Catamarca). 3300–4500 m.

Phrygilus Sierra-Finches

A fairly large genus of small to mid-size finches found mainly in the Andes at high elevations, but several species occur lower in s. South America. Some shade of gray predominates in all species (though females of some are streakier and brown). They are all mostly terrestrial and often common. Many species are frequently found in groups, typically in open grassy or shrubby terrain.

GROUP A

Mostly plain gray; females brown and streaky.

Phrygilus unicolor

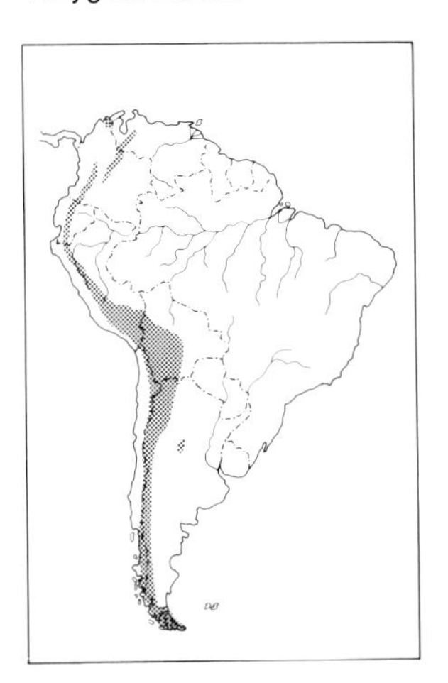

PLUMBEOUS SIERRA-FINCH

PLATE: 28

IDENTIFICATION: 15 cm (6"). Male *uniform leaden gray,* very slightly paler below. N. races (south to Ecuador) have very narrow white eye-ring. Female and juvenal brown above, whitish below, *coarsely streaked dusky throughout;* fully adult females from Peru south are nearly uniform gray (thus resembling males).
SIMILAR SPECIES: Ash-breasted Sierra-Finch is notably smaller; male is paler gray and streaked dusky on back, while female is less coarsely streaked below than female Plumbeous and, like male, has whitish belly. *Catamenia* seedeaters are smaller and have pale bills. Cf. also Slaty Finch (in subtropical forests, not *páramo* or *puna*) and Short-tailed Finch.
HABITAT AND BEHAVIOR: Fairly common in *páramo* and *puna* grasslands and timberline shrubbery. Most often in small groups feeding on the ground, hopping up to perch on rocks or low bushes; regularly

with other species (especially various finches). Often allows a close approach, crouching low to the ground before flushing.

RANGE: Andes from w. Venezuela (north to Trujillo) south through Colombia, Ecuador, Peru, Bolivia, Chile, and Argentina to Tierra del Fuego (where perhaps only casual); hills of w. Córdoba, Argentina, and Santa Marta Mts. of Colombia. Mostly 3000–4500 m, but lower southward.

Phrygilus plebejus

ASH-BREASTED SIERRA-FINCH

PLATE: 28

IDENTIFICATION: 12–12.5 cm (4¾–5″). A drab, *nondescript* sierra-finch, the *smallest of its genus*. Male brownish gray above *streaked dusky on back* and with pale grayish superciliary, grayer on rump. Below paler gray, becoming *whiter on belly* (especially in *ocularis* of Ecuador and extreme n. Peru). Female brown above, coarsely streaked dusky; *whitish below*, finely streaked dusky on throat and breast.

SIMILAR SPECIES: Plain-colored Seedeater, which is actually larger (has proportionately longer tail) and with which it is often found, has stubbier pinkish bill and is buffier below (female) or with chestnut vent (male). Plumbeous Sierra-Finch is larger and male is a darker, more uniform gray (with no back streaking), while female is much more coarsely streaky.

HABITAT AND BEHAVIOR: Common in grassy and shrubby *puna* and *páramo*, especially in more or less sparsely vegetated stony areas. In nonbreeding season often in large flocks, and in general one of the most conspicuous and numerous birds of the high *puna* zone. Also occurs locally, and in smaller numbers, in desert scrub along arid coast of n. Peru and s. Ecuador.

RANGE: Andes of Ecuador (north to Pichincha) south through Peru, Bolivia, n. Chile (south to Antofogasta), and nw. Argentina (south to Mendoza; also in hills of w. Córdoba); also along coast of sw. Ecuador (El Oro) and nw. Peru (Tumbes and Piura). Mostly 2500–4500 m, locally to sea level.

GROUP B

Yellow bill, with *gray to black underparts;* females streaky.

Phrygilus alaudinus

BAND-TAILED SIERRA-FINCH

PLATE: 28

IDENTIFICATION: 14–16.5 cm (5½–6½″). *Rather long, slender, bright yellow bill;* legs also yellow. Male mostly gray, back broadly streaked blackish; *lower breast and belly white, in rather sharp contrast to gray chest;* tail black with *white band across center* (usually visible only in flight or from below). S. birds (north to extreme s. Peru; nominate and *venturii*) are larger and paler gray and have less sharp contrast on underparts. Female brown above streaked blackish; *whitish below,* sharply streaked dusky on throat and chest, tail as in male.

SIMILAR SPECIES: Pattern of both sexes quite like Band-tailed Seedeater's, but note difference in bill shape, seedeater's smaller size (even

in n. part of sierra-finch's range, where it is relatively small), and buffier lower underparts (female) or chestnut crissum (male).

HABITAT AND BEHAVIOR: Fairly common in quite open, often stony or sandy areas with scattered low bushes and sparse grass cover; usually found in arid regions (e.g., rain-shadow valleys). Generally found as scattered pairs (not as gregarious as many other *Phrygilus*). Breeding males may sing during a brief fluttering display flight, mounting to 3 to 5 m above the ground, singing a fairly musical gurgling series of phrases, ending with a long, buzzy "zzhhhhhh," then gliding back to earth. Also sings from a low perch.

RANGE: Andes of Ecuador (north to Carchi), Peru, w. Bolivia, and w. Argentina (south to Catamarca, also in hills of w. Córdoba and adjacent San Luis); also locally on arid coast of sw. Ecuador (Guayas and El Oro), coast and lower slopes of Peru and extreme n. Chile (sighting from Arica), and lowlands and foothills of cen. Chile (Atacama to Valdivia). To about 3500 m.

Phrygilus fruticeti

MOURNING SIERRA-FINCH

PLATE: 28

IDENTIFICATION: 18 cm (7"). *Rather heavy yellow bill* (dusky in female) and yellow legs. Male gray above broadly streaked black (blacker in worn plumage); wings and tail black with 2 bold white wing-bars; *throat and breast black,* flanks gray, center of belly white. In younger males extent of the black bib varies; much grayer and scalier in many birds. Female mostly brown above boldly streaked with blackish, but *rump pure gray;* superciliary and sides of neck grayish, *outlining fulvous-brown cheeks;* wings brown with 2 white bars. Below grayish to whitish, usually streaked dusky on sides of throat and breast (but some, perhaps older, females begin to show some of male's black bib).

SIMILAR SPECIES: This hefty sierra-finch is the largest of its genus; males are readily recognized by their extensive black below, while females can be known by the combination of their size, the outlined brown cheeks, and the prominent wing-bars. Cf. smaller Carbonated Sierra-Finch.

HABITAT AND BEHAVIOR: Common in shrubby areas and cultivated regions with some trees and bushes. Usually in pairs or loose groups (may aggregate even when nesting), but in some areas gathers in flocks during nonbreeding season. Breeding males give a quite loud and explosive "shushglaoww," often introduced or ending with some gurgling notes either from a prominent perch or in a display flight. In general rather confiding and easily seen.

RANGE: Andes from n. Peru (north to Cajamarca) south through w. Bolivia, Chile (south to Llanquihue), and Argentina (south to Chubut); casual south to Tierra del Fuego. Mostly 2000–4000 m, but lower southward (in cen. Chile and s. Argentina sometimes occurs at sea level).

Phrygilus carbonarius

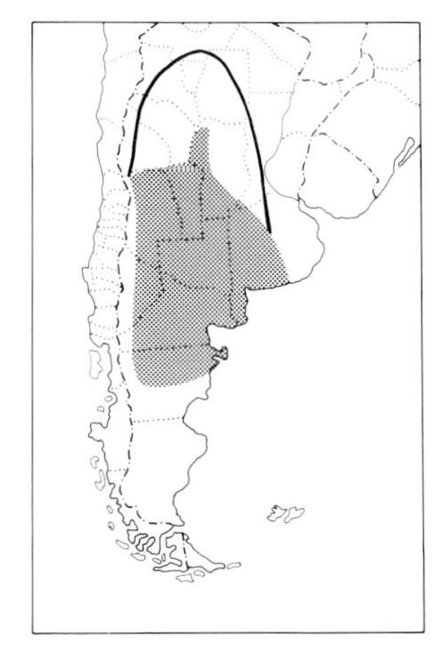

CARBONATED SIERRA-FINCH

IDENTIFICATION: 14.5 cm (5¾"). *Argentina.* In general a *small s. version of better-known Mourning Sierra-Finch.* Bill and legs yellow. Male above gray broadly streaked black, with rump browner; *forecrown, sides of head, and entire underparts* (including center of belly and crissum) *slaty black,* flanks grayer. Female brownish gray above streaked black; whitish below narrowly streaked dusky on breast and sides.

SIMILAR SPECIES: Mourning Sierra-Finch notably larger, and both sexes show prominent wing-bars; furthermore, in male black does not extend down over belly, while female has fulvous cheeks. Female Carbonated rather dull and indistinctive, but lacks any yellow (and thus should not be confused with any *Sicalis* or *Melanodera* finch) and is not as coarsely streaked below as female Plumbeous Sierra-Finch.

HABITAT AND BEHAVIOR: Locally fairly common in semiopen shrubby steppes of Patagonia, migrating north into other semiopen shrubby habitats during austral winter. Breeding males have an attractive display in which they flutter some 8 to 15 m into the air and then sail to a new perch on top of another bush, at the apex of each flight bursting into a fast but musical "treeyee-treeyee-treeyee . . ." (up to 9 to 12 "treeyees"). Numerous in summer on the steppes at the base of the Valdez Peninsula.

RANGE: Breeds on plains of cen. Argentina (Mendoza, Córdoba, and w. Buenos Aires south to Chubut), in winter migrating north to Tucumán and Santiago del Estero.

GROUP C

Hooded effect (gray to black); females duller.

Phrygilus atriceps

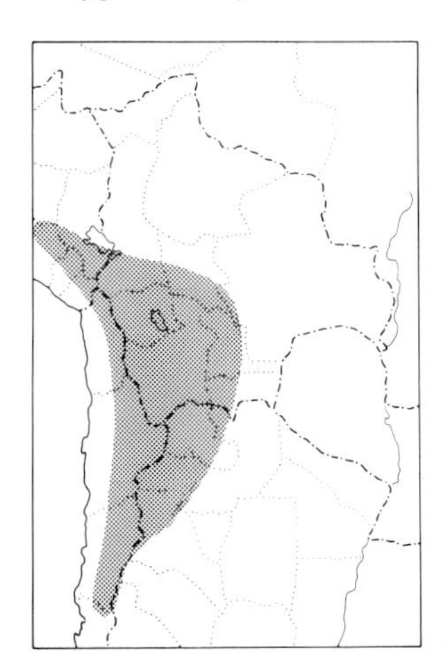

BLACK-HOODED SIERRA-FINCH

PLATE: 28

IDENTIFICATION: 15.5–16 cm (6–6¼"). *High Andes of s. Peru to n. Chile and n. Argentina. Head, throat, and chest black* (forming the "hood") *contrasting with deep ochraceous back and rich yellow lower underparts* and rump; tinged ochraceous especially on breast; wings and tail slaty gray. Female duller than male, but older birds have much the same pattern but with hood slaty gray. Younger females have hood more grayish brown, often show streaking on throat, and may be buffier below.

SIMILAR SPECIES: Males are striking and virtually unmistakable. Female is only female of the "hooded" sierra-finch group with *ochraceous back* (olive or grayish brown in all others), which should serve to distinguish it from Gray-hooded Sierra-Finch in their area of overlap and from Peruvian Sierra-Finch in their area of contact or near-contact in Bolivia.

HABITAT AND BEHAVIOR: Generally common on shrubby slopes and valleys in semiopen areas; often found in areas with considerable cactus growth and generally absent from level grassy altiplano plains. Usually in pairs or small loose groups and generally conspicuous, often feeding in mixed flocks on the ground, frequently with other sierra-finches, yellow-finches, diuca-finches, and certain furnariids. Song consists of a simple, clear repetition of a single note (e.g., "trileé, trileé, trileé . . ." up to 7 to 8 times); usually sings from an exposed perch such as a stone wall or rooftop.

RANGE: High Andes of s. Peru (Arequipa and Tacna), w. Bolivia (except in La Paz), n. Chile (south to Coquimbo), and nw. Argentina (south to Catamarca). Mostly above 3000 m.

Phrygilus gayi

GRAY-HOODED SIERRA-FINCH

PLATE: 28

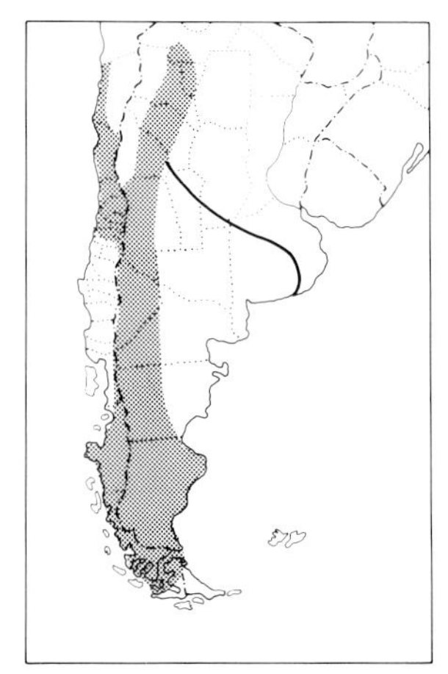

IDENTIFICATION: 15.5–16.5 cm (6–6½"). *Argentina and Chile. Head, throat, and chest steel gray,* contrasting with *olive back* and olive yellow underparts; lower belly and crissum white; wings and tail light gray. Female similar but much duller; hood paler and more brownish gray with some blackish streaking, and *breast usually washed ochraceous;* may show an indistinct black malar streak. Immature browner.

SIMILAR SPECIES: Black-hooded Sierra-Finch has shiny black (not gray) hood and deep ochraceous back (not the pure olive of this species). Duller females and immatures can be very similar, but Gray-hoodeds are more olive on back (no ochraceous). Cf. also Patagonian Sierra-Finch.

HABITAT AND BEHAVIOR: Fairly common to common in shrubby vegetation, particularly in valleys or gulleys where there is water and in Patagonian shrub desert terrain. Particularly numerous and widespread in Chile, where also in woodland borders. Behavior similar to Black-hooded Sierra-Finch's, though in general a bird of thicker vegetation at medium altitudes, rarely or never found at high elevations. Song of male is a somewhat monotonous series of simple, clear notes or phrases, each well enunciated, either as a series (e.g., "sweét, sweét, treelili, treeli, treelili") or a simple repetition of 1 note (e.g., "sweét, sweét, sweét . . .").

RANGE: Andean slopes of Chile (north to Atacama; also a race, *minor,* resident in coastal lowlands from Atacama to Santiago) and Argentina (north to w. Salta) south to n. Tierra del Fuego (where rare, probably only as a summer visitor); some migration across Patagonian lowlands during austral winter, a few reaching sw. Buenos Aires. Mostly 1500–3500 m, lower southward (to near sea level in Tierra del Fuego).

NOTE: We follow the taxonomy of the difficult hooded sierra-finch group proposed by Vuilleumier (1967, and his subsequent unpublished data). Effectively, this splits the formerly disjunct *P. gayi* into 2 full species, the Peruvian *P. punensis* (with *chloronotus*) and the Gray-hooded *P. gayi* (with *minor* and *caniceps*). *P. atriceps* and *P. g. gayi* have been reported to breed sympatrically in n. Chile (both in Atacama and Coquimbo; see Johnson 1967) and may well also do so in nw. Argentina. As these 2 taxa appear to be reproductively isolated, and as *P. punensis* and *P. atriceps* differ in very similar ways and thus likely are also reproductively isolated, we conclude that these latter 2 should also be considered separate species (though they were lumped in *Birds of the World,* vol. 13). Note, further, that adult females of *punensis* and *gayi* differ quite markedly.

Phrygilus punensis

PERUVIAN SIERRA-FINCH

Other: Gray-hooded Sierra-Finch (in part)

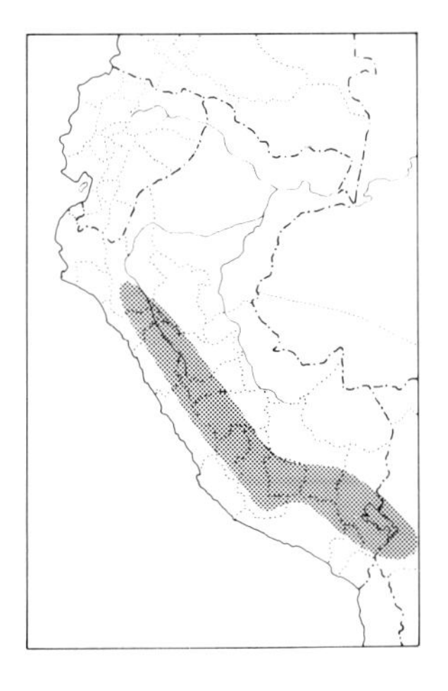

IDENTIFICATION: 15.5–16 cm (6–6¼"). *Andes of Peru and adjacent Bolivia.* Male resembles male Gray-hooded Sierra-Finch in pattern, but back brighter and more yellowish olive, and yellow of breast and belly washed with ochraceous. Female very similar to male but gray of hood somewhat duller and paler. Immature much duller and buffier below. *Punensis* (Puno, Peru, and La Paz, Bolivia) similar but back fairly strongly tinged ochraceous in male.

SIMILAR SPECIES: Much less sexually dimorphic than the Gray-hooded Sierra-Finch of Chile and Argentina. This species is most apt to be confused with Black-hooded Sierra-Finch, especially where their respective ranges approach each other (in extreme s. Peru and nw. Bolivia). Bright male Black-hoods are easy enough, but female Black-hoods quite closely resemble males of nominate race of Peruvian. Females of latter are more or less olive-backed, and as both species usually occur in pairs, this probably provides the best clue; in other words, look for male Black-hoods and female Peruvians. In the late 1960s F. Vuilleumier (pers. comm.) could find no evidence that the 2 forms overlapped or were in contact in La Paz, Bolivia, but further fieldwork may reveal this to be the case (as for *P. atriceps* and *P. gayi*).

HABITAT AND BEHAVIOR: Fairly common on rocky, shrubby slopes and in agricultural areas with scattered bushes and trees. Generally not on flat, open, grassy or scrubby plains of the altiplano. Behavior similar to Black and Gray-hooded Sierra-Finches'. Song of Peruvian resembles Black-hooded's but is somewhat less monotonous and more varied (F. Vuilleumier).

RANGE: Andes of Peru (Cajamarca south to Ayacucho and Puno) and nw. Bolivia (w. La Paz). Mostly 2000–4000 m.

NOTE: See comments under Gray-hooded Sierra-Finch.

Phrygilus patagonicus

PATAGONIAN SIERRA-FINCH

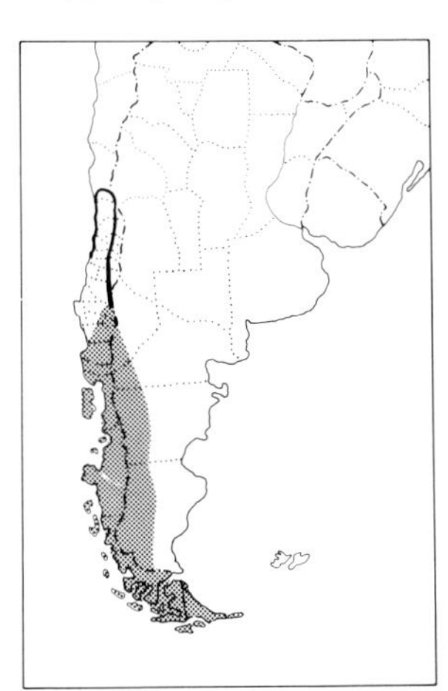

IDENTIFICATION: 14.5–15 cm (5¾–6"). *S. Chile and s. Argentina.* Resembles Gray-hooded Sierra-Finch, but *somewhat smaller;* the 2 do occur together, particularly when not breeding (see below). Male Patagonian easily recognized by *strong tinge of bright ochraceous on back; only its crissum is whitish* (not extending to lower belly). Female Patagonian resembles male but its *back is olive;* it thus resembles male Gray-hooded, the main difference being extent of whitish on lower underparts (to lower belly in Gray-hooded, only crissum in Patagonian). Some immatures can be difficult to separate; note that habitats also usually differ (see below).

HABITAT AND BEHAVIOR: Fairly common in forest borders and openings and shrubby cleared areas, moving into somewhat more open areas during winter. When breeding usually in pairs and more or less arboreal, though occasionally feeding on the ground. Much more of a

forest-based bird when breeding than the Gray-hooded, though the 2 may flock together at other seasons. Male's song is clear and rhythmic, a short series of well-enunciated musical notes (e.g., "cleet-clwett, clweet-weet, cleweet" with variations).
RANGE: S. Chile (north to Ñuble) and Andean slopes of s. Argentina (north to w. Neuquén) south to Tierra del Fuego; during austral winter moves north to cen. Chile (to Aconcagua). To about 1800 m.

GROUP D

Large; *gray above, white below,* with back rufous or gray. Sexes alike.

Phrygilus dorsalis

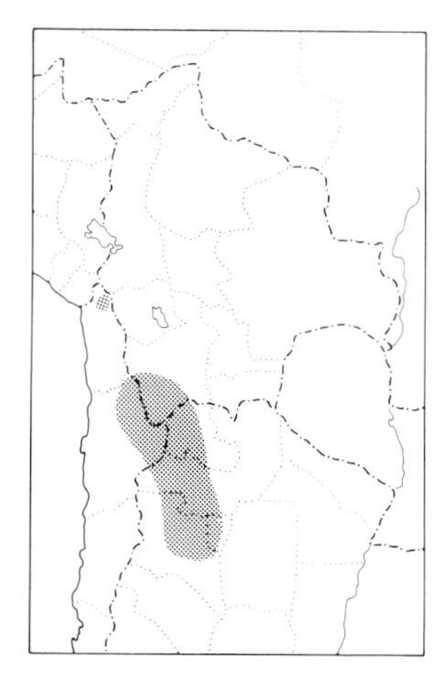

RED-BACKED SIERRA-FINCH

PLATE: 28

IDENTIFICATION: 18 cm (7"). *High Andes of nw. Argentina and adjacent Chile and Bolivia.* Gray above with *back and scapulars contrastingly light rufous;* wings and tail blackish, with *no white;* lores and lower face freckled with white; *center of throat white,* with sides of neck and *breast broadly gray;* belly whitish.
SIMILAR SPECIES: No other mostly gray finch of the high Andes has the pale rufous back. See White-throated Sierra-Finch. Common Diuca-Finch is sometimes tinged brownish above, but that color is never concentrated on back, and it has sharply demarcated gray and white pattern below and white in tail.
HABITAT AND BEHAVIOR: Recorded from very high *puna* grasslands and rocky slopes. Noted in pairs while breeding but in small flocks during winter. Behavior seems similar to White-throated Sierra-Finch's.
RANGE: High Andes of extreme sw. Bolivia (sw. Potosí), n. Chile (Antofogasta, and seen in Dec. 1986 in Arica; RSR et al.) and nw. Argentina (Jujuy to w. Catamarca). Above 4000 m.

NOTE: *P. dorsalis* and *P. erythronotus* are closely related and have nearly allopatric distributions. Their supposed sympatry in Potosí, Bolivia, was based on a series of misidentified ANSP specimens from Oploca. Originally identified as *P. dorsalis,* these are actually *Diuca diuca. P. erythronotus* is thus definitely known to range as far south only as extreme n. Potosí, while *P. dorsalis* has been recorded only from the far southwest of that department (Laguna Colorado, etc.). When the high *puna* zone of w. Bolivia and n. Chile is better known, both may be shown to have more extensive distributions and to occur together. Recent information (D. Scott; RSR et al.) indicates that 1 such area lies in Lauca Nat. Park in extreme n. Chile, where in Dec. 1986 both species were found syntopically, even in the same flock (but *erythronotus* being much commoner), with at least 1 possible hybrid also being seen.

Phrygilus erythronotus

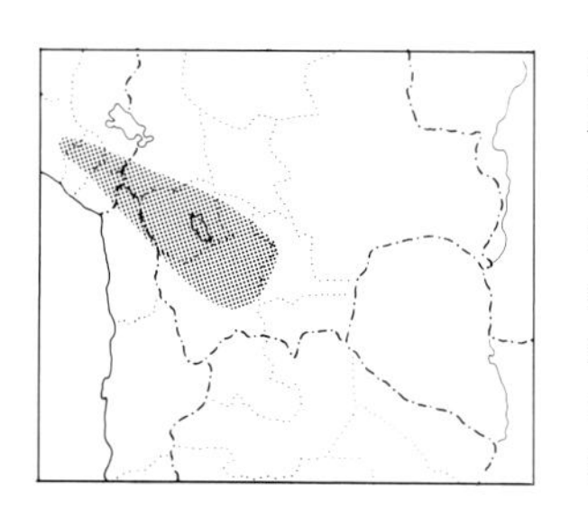

WHITE-THROATED SIERRA-FINCH

IDENTIFICATION: 18 cm (7"). *High Andes of sw. Bolivia and adjacent Peru and Chile.* Resembles Red-backed Sierra-Finch, but *upperparts entirely gray* (no rufous on back).
SIMILAR SPECIES: White-winged Diuca-Finch is superficially quite similar but is somewhat larger and more cleanly patterned with conspicuous white in wing (even when closed) and white crescent below eye. Common Diuca-Finch has much white in tail (lacking in this species), cleaner gray and white pattern below, and rufous on lower flanks.

HABITAT AND BEHAVIOR: Locally fairly common in high *puna* grasslands and on rocky slopes. Occurs mostly in pairs or small groups, feeding on the ground. Usually quite tame, allowing a close approach as they forage or as they perch on stone walls or rocks.
RANGE: High Andes of extreme sw. Peru (Arequipa, Moquegua, and Tacna), extreme n. Chile (Arica and probably Tarapacá), and sw. Bolivia (Oruro and Potosí). Above 4000 m.

Diuca Diuca-Finches

This genus is very similar to *Phrygilus,* but both species show conspicuous white in tail (lacking in all sierra-finches). Both species are gray and white finches found in open country of s. South America, 1 high in the Andes.

Diuca speculifera

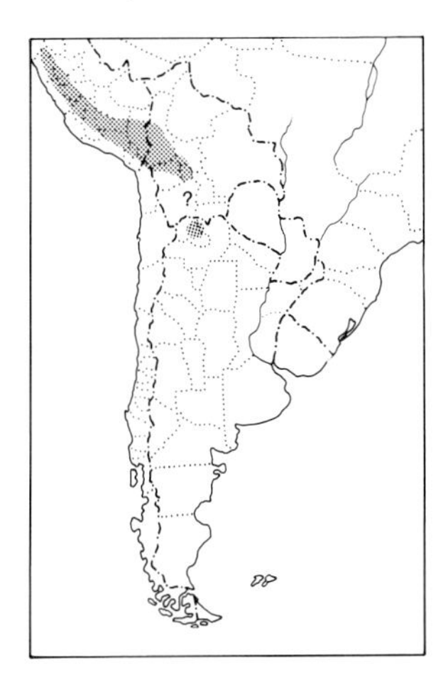

WHITE-WINGED DIUCA-FINCH

PLATE: 28

IDENTIFICATION: 19 cm (7½"). *A clean-cut gray and white finch of the very high Andes.* Above gray with prominent white crescent below eye; wings and tail blackish, with *very conspicuous white patch on primaries* (visible even at rest) and outer web of outer tail feathers also white. Below white with broad pectoral band and flanks gray.
SIMILAR SPECIES: White-throated Sierra-Finch lacks white in wings or tail and its pattern below is much less clearly defined. Cf. also Common Diuca-Finch.
HABITAT AND BEHAVIOR: Locally common in very high *puna* grasslands and rocky slopes. Probably breeds at higher elevations than any other S. Am. passerine (nesting has been recorded at about 5400 m in Bolivia) and has been recorded roosting in glacier crevasses. Usually found in small groups, which may scatter out while foraging on the ground; while feeding birds may crouch low and remain inconspicuous, but in flight they can be seen at great distances. In general comportment will remind N. Am. observers of rosy-finches (*Leucosticte*).
RANGE: High Andes of cen. and s. Peru (north to Ancash), w. Bolivia (La Paz and Cochabamba; also probably southward), extreme n. Chile (Arica), and extreme nw. Argentina (Jujuy). 4000–5500 m, but mostly above 4500 m.

Diuca diuca

COMMON DIUCA-FINCH

IDENTIFICATION: 16–18 cm (6¼–7"). *The common gray and white finch of Chile and s. Argentina.* Gray above (tinged brownish in *minor* of Patagonia, which is also smaller), wings and tail more blackish, with *inner webs of outer tail feathers white* (forming diagonal patch on tail corners, usually flashing in flight); *large throat patch white, sharply demarcated from gray cheeks and pectoral band,* the gray also continuing broadly down flanks; center of breast and belly white; lowermost flanks rufous. Female has gray slightly brownish-tinged.

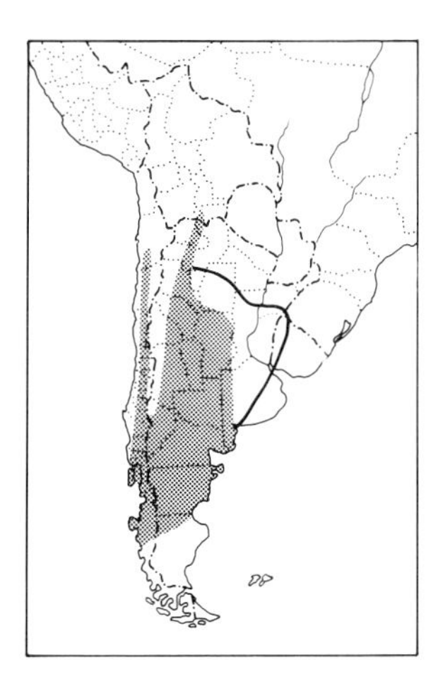

SIMILAR SPECIES: The 2 species with which this finch is most likely to be confused, White-winged Diuca-Finch and White-throated Sierra-Finch, are both high Andean birds; this species is found only in lowlands and lower slopes.

HABITAT AND BEHAVIOR: Common to locally abundant and widespread in shrubby areas, cultivated regions and gardens, Patagonian scrub desert, and even at forest edge (on lower slopes of s. Andes). This dapper finch is often familiar and conspicuous, especially in Chile, where it ranks as one of the most numerous landbirds. It forages mostly on the ground but perches freely in bushes and small trees. The deliberate song in Argentina is loud and musical, recalling Golden-billed Saltator's (e.g., "chit, chuwit, chuwit-chew, chuwit," slowly delivered).

RANGE: S. Bolivia (a series of ANSP specimens from Oploca in Potosí, formerly misidentified as *Phrygilus dorsalis*), Chile (Antofogasta to Aysén, rarely to Magellanes), and Argentina (w. Salta, w. Córdoba, and sw. Buenos Aires south to Santa Cruz); during austral winter *minor* (breeding in Argentina's Patagonian lowlands) moves north in Argentina to Tucumán, Santiago del Estero, and Entre Ríos, and at least formerly it occurred north into Uruguay (Paysandú) and extreme se. Brazil (w. Rio Grande do Sul), but there are no recent records from either of the 2 latter countries. To about 2000 m.

Melanodera Finches

A pair of finches found in Patagonia, males boldly patterned with black throats, females duller. The genus seems closely related to *Phrygilus*, and *Rowettia* of Gough Is. (in South Atlantic) also is very similar.

Melanodera melanodera

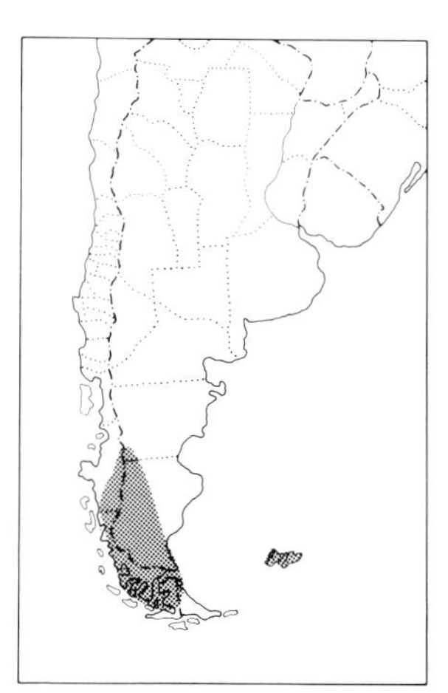

BLACK-THROATED FINCH

PLATE: 28

IDENTIFICATION: 15 cm (6"). *A strongly patterned (male), mostly yellow-winged finch of extreme s. Patagonia.* Crown and hindneck gray; *conspicuous white superciliary and broad malar streak white,* outlining black lores and *large black throat patch.* Back and rump grayish olive. *Wing-coverts and primaries mostly light yellow;* outer tail feathers also pale yellow. Below mostly bright grayish yellow, crissum white. Female much duller: buffier brown above coarsely streaked blackish; whitish below streaked brownish, and with *buffy brown wash across breast;* wings and tail similar to male's but yellow not as bright or extensive. Nominate race of Falkland Is. has much less extensive yellow in wing than *princetoniana* of mainland.

SIMILAR SPECIES: See Yellow-bridled Finch. Even female Black-throated shows so much yellow on wing that confusion is unlikely. Patagonian Yellow-Finch is the yellowest finch occurring with it, and it has brown wings and tail and unstreaked underparts.

HABITAT AND BEHAVIOR: Uncommon to common in open grasslands and around settlements. When nesting usually occurs in pairs or small groups, but during nonbreeding season gathers in large flocks. Particularly numerous on the Falklands, but now much less so in mainland portion of range, with few recent records at least from Argentina (numbers may now be declining due to severe overgrazing by sheep). Forages mostly on the ground. Song of male on the Falklands is a rapidly delivered series of 3 phrases, "cheet, whedít, wheeur; cheet, whedít, wheeur . . ." repeated several times and often commencing with just the last 2 phrases (R. Straneck, tape recording).
RANGE: S. Chile (s. Magellanes) and s. Argentina (s. Santa Cruz) south to Tierra del Fuego; Falkland Is.

Melanodera xanthogramma

YELLOW-BRIDLED FINCH

IDENTIFICATION: 16.5–17 cm (6½–6¾"). Recalls Black-throated Finch; somewhat larger. Male has similar head pattern, but *superciliary and malar streak bright yellow* (not white); *much less yellow on wing* (only some yellowish olive edging on primaries), but tail similar. Color of upperparts varies (2 color phases?), usually bluish gray, but can be yellowish green (in which case underparts are also much yellower). Female closely resembles female Black-throated but shows almost no yellow on wing, is more heavily streaked below, and lacks buffy wash on breast.
SIMILAR SPECIES: Black-throated Finch is more a bird of open regions (this species mostly at forest edge and in openings, at least when breeding). Female could be confused with female Plumbeous Sierra-Finch, but latter more coarsely streaked and lacking any yellow or olive in wings and tail.
HABITAT AND BEHAVIOR: Rare to locally uncommon in low scrub, shrubby forest edges, and open grassy areas; apparently breeds mostly on ridges at and just above timberline, descending to sea level during winter or at any season when higher elevations are briefly covered by snow. Generally in pairs or small groups when breeding, but at other seasons may gather in flocks (at least in s. Chile). Breeding males are reported to have a 2-note song, a long shrill note followed after a pause by a shorter, lower one, "tweet; wheu." Breeds in small numbers in mountains above Ushuaia, Tierra del Fuego (where best seen by walking up from the ski lift).
RANGE: Cen. and s. Chile (north in mts. to Aconcagua) and s. Argentina (north to w. Neuquén) south to Tierra del Fuego; Falkland Is. (at least formerly; no recent records). To about 3000 m (northward near timberline).

Xenospingus Finches

Distinctive slender, yellow-billed finch of arid s. Peru and adjacent Chile.

Xenospingus concolor

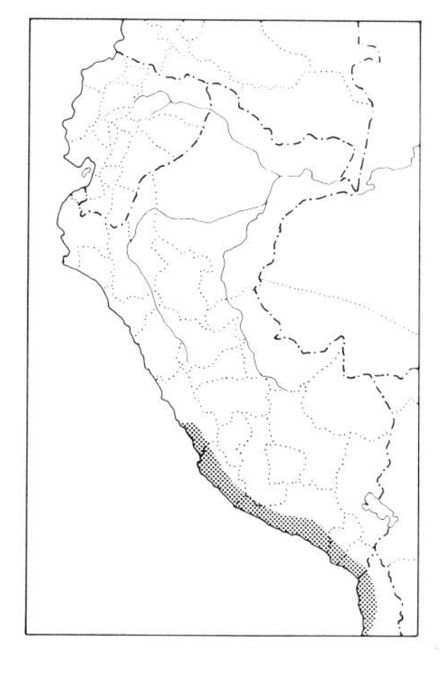

SLENDER-BILLED FINCH

PLATE: 29

IDENTIFICATION: 15 cm (6″). *Bill long and slender, bright yellow;* legs also yellow. Uniform gray above with black loral area; *slightly paler gray below.* Immature olivaceous brown above, wings with 2 vague buffyish wing-bars; yellowish buff below with blurry brownish streaking; bill brownish.
SIMILAR SPECIES: This notably svelte and long-tailed finch should be easily known when adult; immatures slightly recall female Blue-black Grassquit but are much larger and rarely found in open grassy areas.
HABITAT AND BEHAVIOR: Fairly common in shrubby areas and low woodlands, principally along rivers or in irrigated areas. Usually found singly or as scattered pairs (not in flocks, and most often not with other finches), which tend to skulk and perch quietly within foliage, not often remaining long in the open. Forages mostly by gleaning for insects, not on seeds or fruit. The song is a jumbled warble, the call a sharp "zeep."
RANGE: Coastal w. Peru (north to Lima) and n. Chile (Arica and Tarapacá). Mostly below 300 m, but has been found much higher (locally to over 3500 m, at least in Chile).

Piezorhina Finches

Robust pale gray finch with massive yellow bill. Arid coastal nw. Peru.

Piezorhina cinerea

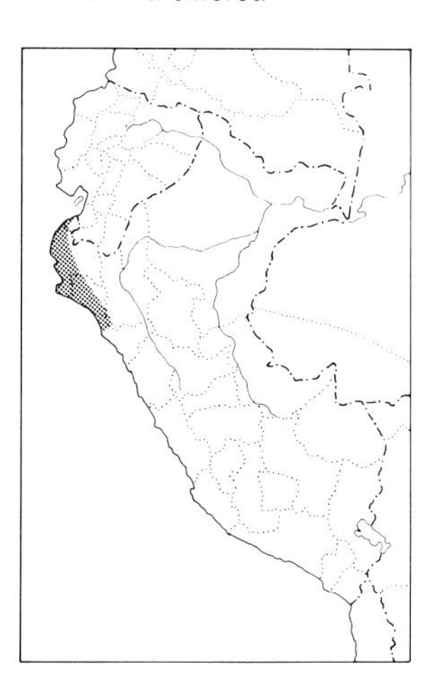

CINEREOUS FINCH

PLATE: 29

IDENTIFICATION: 16.5 cm (6½″). A husky pallid gray finch with a *massive yellow bill* and duller yellow legs. *Pale gray above* with black lores and malar spot and white sides of rump; even paler gray below, white on throat and belly.
SIMILAR SPECIES: Unlikely to be confused in its restricted, desert scrub range. In any case, no other light gray finch has such a heavy bright yellow bill.
HABITAT AND BEHAVIOR: Common on quite open desertlike plains with scattered shrubs and occasional low trees. Conspicuous, often perching fully in the open on the ground, regularly at roadsides (including edge of Pan-American Highway, especially in Piura). Generally singly or in pairs, at most in small loose groups.
RANGE: Coastal nw. Peru from Tumbes south to La Libertad. To about 300 m.

Incaspiza Inca-Finches

Colorful boldly patterned finches of arid scrub (especially where there is cactus) in Andes of w. Peru. Note the conspicuous white outer tail feathers. Females of all 5 species are similar in pattern to their respective males but are somewhat duller, immatures even more so; ridge of their bills is dusky.

GROUP A

Larger species.

Incaspiza pulchra

GREAT INCA-FINCH

PLATE: 29

IDENTIFICATION: 16.5 cm (6½"). *Ancash and Lima, Peru.* Bill and legs orange-yellow. Above mostly brown, with *contrasting rufous inner flight feathers and most of greater wing-coverts* (shoulders are gray, but some rufous on scapulars). Face (including superciliary), sides of neck, and breast gray, outlining *black ocular area and throat patch;* belly buffy whitish. Tail black, but *outer feathers mostly white.*

SIMILAR SPECIES: The only inca-finch with conspicuous rufous on wings. Very similar Rufous-backed Inca-Finch has smaller throat patch and mostly gray wings, with rufous only on scapulars and across back.

HABITAT AND BEHAVIOR: Uncommon and somewhat local on hot, arid slopes and in ravines, especially in areas where large cactus and ground bromeliads are prevalent. Readily found *in early morning* along the Santa Eulalia road above Huinco in Lima (e.g., in shrub zone 1 km past the bridge over the narrow gorge), perching in the open on shrubbery; later in the day they are much less conspicuous.

RANGE: W. Peru (Ancash to s. Lima). Mostly 1500–2100 m, locally somewhat higher and lower.

Incaspiza personata

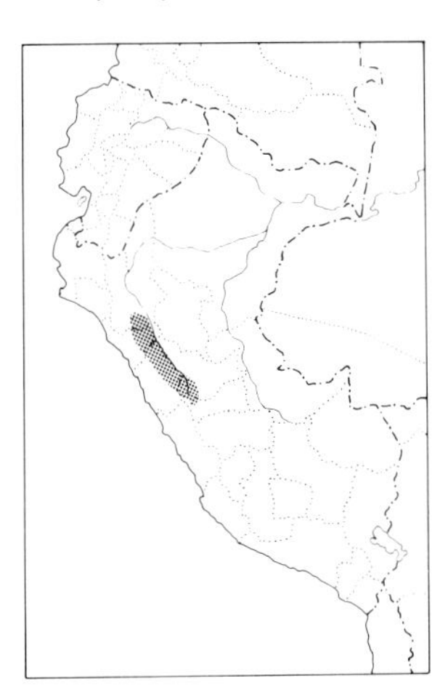

RUFOUS-BACKED INCA-FINCH

PLATE: 29

IDENTIFICATION: 16.5 cm (6½"). *Mostly in upper Marañón drainage, Peru.* Resembles Great Inca-Finch. *Scapulars and entire back rufous* (contrasting with otherwise brown upperparts). Black throat patch smaller (essentially restricted to chin) but extending up onto forehead, forming "shifted up" narrow frontal band. Breast somewhat more olivaceous gray.

SIMILAR SPECIES: Great Inca-Finch has rufous only on wings and none on back; configuration of its black facial area slightly different. Buff-bridled Inca-Finch also has the rufous saddle, but it is smaller with a more extensive black, prominently buff-bordered throat patch.

HABITAT AND BEHAVIOR: Little known. Reported to favor dry mountain slopes where cactus and agave are numerous. Occurs mostly at elevations above where Buff-bridled Inca-Finch is found and has been seen along highway between Huánuco and Lake Junín.

RANGE: W. Peru in upper Marañón River valley (s. Cajamarca, e. La Libertad, e. Ancash, and w. Huánuco), barely crossing onto Pacific slope in La Libertad and Ancash. Mostly 1800–3000 m.

Incaspiza ortizi

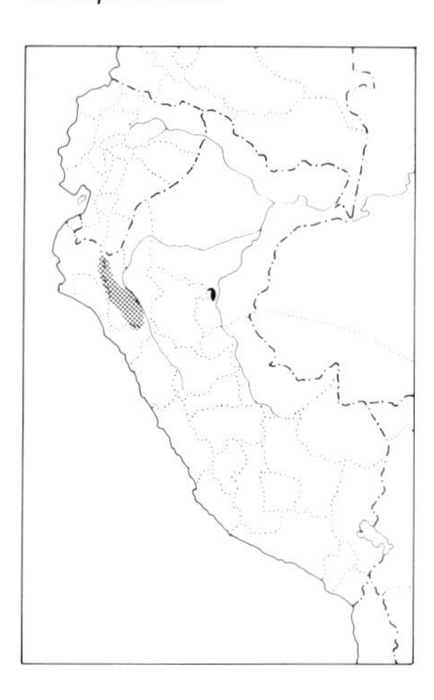

GRAY-WINGED INCA-FINCH

IDENTIFICATION: 16.5 cm (6½"). *Nw. Peru.* Bill orange, legs yellow. Above grayish brown, *streaked with dusky on scapulars and back.* Sides of head and neck gray; narrow forehead, ocular region, and small throat patch black. Breast gray, belly whitish. Tail black, but *outer feathers mostly white.*

SIMILAR SPECIES: Resembles Great and Rufous-backed Inca-Finches (none are known to actually occur together), but is distinguished by *absence of rufous on either back or wings* and streaking on back; its black facial area is shaped like Rufous-backed's.

HABITAT AND BEHAVIOR: Recorded from desert scrub at elevations somewhat lower than Buff-bridled Inca-Finch. Little known.

RANGE: W. Peru in extreme ne. Piura (just east of Huancabamba city) and Cajamarca (locally on w. slope of Marañón River valley, south to Hacienda Limón, east of Celendin). 1800–2300 m.

GROUP B

Smaller species.

Incaspiza laeta

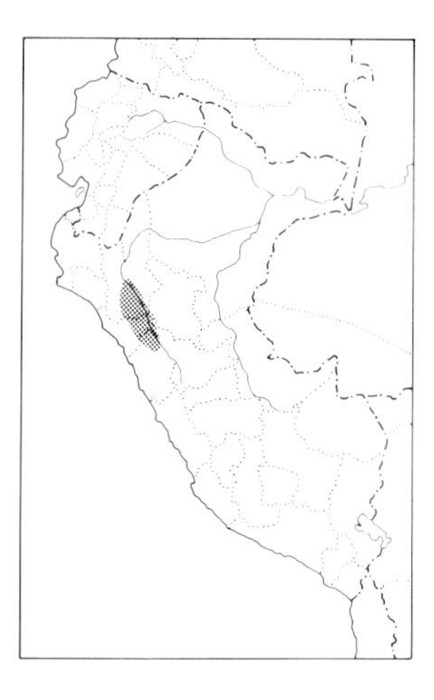

BUFF-BRIDLED INCA-FINCH

PLATE: 29

IDENTIFICATION: 14.5 cm (5¾"). *Upper Marañón drainage, Peru.* Bill and legs orange-yellow. Crown, nape, and rump brownish gray, with *prominent rufous saddle across back and scapulars.* Narrow forehead, ocular area, and throat patch black, the latter bordered by *conspicuous broad pale buff malar streak.* Breast light gray, belly pale ochraceous buff. Tail black, but *outer feathers mostly white.*

SIMILAR SPECIES: This beautifully marked finch should not be confused; it alone among the inca-finches has a malar streak, and only the Rufous-backed also has a rufous saddle (it is larger, with smaller area of black on throat).

HABITAT AND BEHAVIOR: Fairly common in open dry woodland with many *Bombax* trees and a thorny understory. Forages singly or in pairs, on or near the ground, and reported not to be very shy (N. Krabbe).

RANGE: W. Peru in upper Marañón River valley drainage of s. Cajamarca, extreme sw. Amazonas (above Balsas), e. La Libertad, and ne. Ancash (Quiches). Mostly 1500–3000 m.

Incaspiza watkinsi

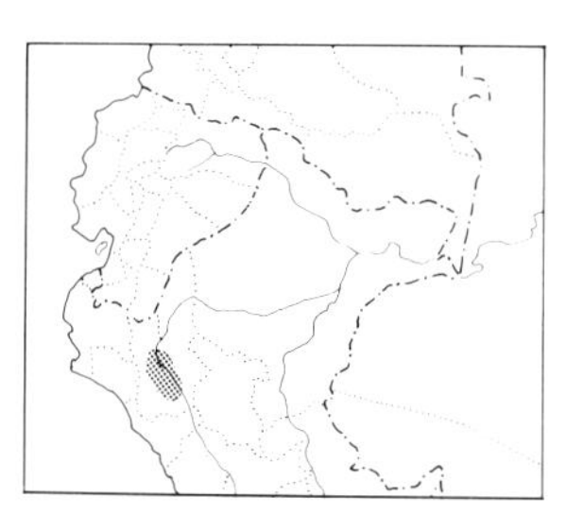

LITTLE INCA-FINCH

IDENTIFICATION: 13 cm (5"). *Middle Marañón valley, Peru. Easily the smallest inca-finch,* but bill almost as long as in other species (and thus proportionately larger). Bill and legs yellow. Head and sides of neck gray, surrounding black foreface and chin (shape like Rufous-backed's); lower throat whitish. Upperparts otherwise mostly brown, back *usually somewhat streaked blackish;* scapulars more rufous, but wing-coverts mainly gray. Breast light gray, belly pale ochraceous buff. Tail black, but *outer feathers mostly white.*

SIMILAR SPECIES: Occurs at lower elevations than any other inca-finch. Resembles a small version of Rufous-backed Inca-Finch, but back not as rufous and usually shows some dark streaking. Gray-winged Inca-Finch also has back streaking, but it is larger, lacks the whitish lower throat, and is whitish (not buff) on belly.
HABITAT AND BEHAVIOR: Recorded from desert scrub, especially where ground bromeliads are numerous. Little known.
RANGE: W. Peru in middle Marañón River drainage in n. Cajamarca and adjacent Amazonas (near Bagua). 700–900 m.

Poospiza Warbling-Finches and Mountain-Finches

Small to mid-size finches found in the Andes and in s. South America, principally in lighter woodland and scrub. Virtually all are boldly patterned and colorful in a subdued way, with rufous, gray, and white predominating. Most are arboreal, gleaning in a warblerlike fashion, often with mixed bird flocks. Despite their name, few or none actually warble!

GROUP A

Lowland and *southeastern* warbling-finches.

Poospiza hispaniolensis

COLLARED WARBLING-FINCH

PLATE: 29

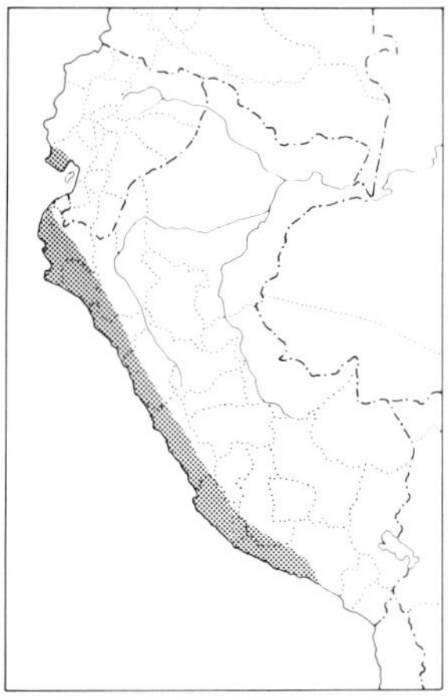

IDENTIFICATION: 13.5 cm (5¼"). *Sw. Ecuador and w. Peru.* Mostly gray above (back tinged brown in many birds), with *long white superciliary and black cheeks;* wing-coverts edged buffy whitish, forming 2 wing-bars. Below mostly white, with *patch on center of chest black,* becoming broadly gray on sides. Rather short tail blackish, but *inner webs of outer feathers mostly white* (conspicuous from below and in flight). Female quite different, more grayish brown above streaked dusky on back, but with *facial pattern similar to male's.* Below whitish, streaked brown across breast; tail as in male.
SIMILAR SPECIES: In its range male liable to be confused only with Black-capped Sparrow; it has a full black pectoral band (no gray on sides), an entirely black crown (no gray), and an all-dark tail. Female less distinctive, but can be known by its facial pattern and the white in tail.
HABITAT AND BEHAVIOR: Uncommon in arid desertlike scrub, dense shrubbery near water, and agricultural areas with scattered trees and low shrubbery. Usually occurs in pairs and most often not associating with other species. The song is a loud, vigorous, ringing "swik, swik-sweéu," reminiscent of Golden-billed Saltator's, delivered from a hidden perch; in Ecuador sings during dry season (at least July–Sept.) when most other birds are silent.
RANGE: Coastal sw. Ecuador (locally in Guayas) and w. Peru (Piura south to Arequipa). To 2500 m, but mostly below 1000 m.

Poospiza torquata

RINGED WARBLING-FINCH

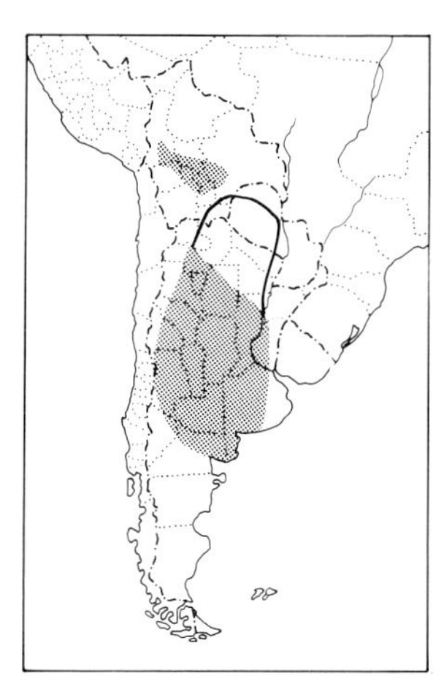

IDENTIFICATION: 13–14 cm (5–5½"). Resembles male Collared Warbling-Finch (no overlap), but with little sexual dimorphism. Somewhat smaller with a considerably slighter bill and a longer tail. Purer *gray above* (no traces of brown), with *broader and more conspicuous white wing-bars. Black pectoral band* more complete (less gray on sides) and with prominent *chestnut crissum.* Outer tail feathers mostly white (flashing conspicuously in flight). Female similar but slightly duller, more brownish on sides. Immature usually lacks black pectoral band, but already has adult's pattern. Nominate race of Bolivia is slightly larger than *pectoralis* of remainder of range and is grayer (not jet black) on cheeks and sides of crown.

SIMILAR SPECIES: Black-capped Warbling-Finch (with which it frequently occurs) lacks the white superciliary and has an unmarked wing and entirely white underparts. Black-crested Finch shows prominent crest and lacks the pectoral band.

HABITAT AND BEHAVIOR: Fairly common in deciduous scrub and woodland. Usually occurs in pairs or small groups which often associate with other birds in mixed foraging flocks, particularly during austral winter. Quite active, gleaning from foliage in somewhat warbler-like fashion.

RANGE: Andean slopes and arid intermontane valleys of Bolivia (Cochabamba, Santa Cruz, and Chuquisaca) and lowlands of much of n. and cen. Argentina south to Mendoza, La Pampa, and Buenos Aires; also occurs, probably only as austral visitant, in w. Paraguay, adjacent n. Argentina (e. Salta, Chaco, and Formosa), and extreme se. Bolivia (Tarija). To about 1800 m (in Bolivia).

Poospiza melanoleuca

BLACK-CAPPED WARBLING-FINCH

PLATE: 29

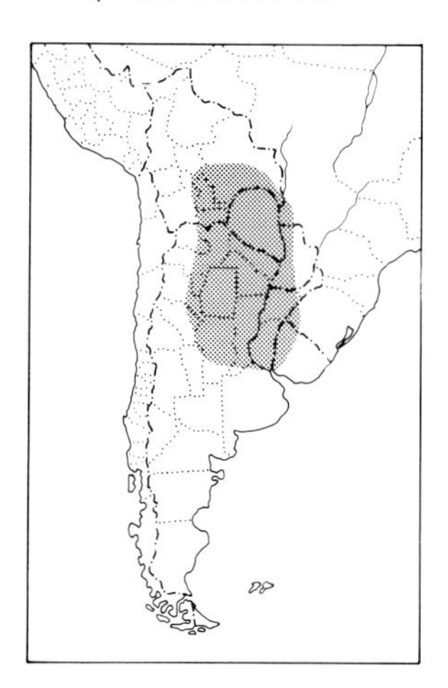

IDENTIFICATION: 13 cm (5"). *Cap glossy black,* contrasting with uniform gray upperparts and *pure white underparts.* Flight feathers edged light gray. Outer tail feathers mostly white (flashing conspicuously in flight). Immature has only sides of head blackish, with crown and hindneck same gray as back.

SIMILAR SPECIES: See Cinereous Warbling-Finch. Ringed Warbling-Finch has white superciliary, white in wing, and obvious pectoral band. Color pattern recalls Masked Gnatcatcher, a frequent associate of this warbling-finch.

HABITAT AND BEHAVIOR: Common and widespread in low woodland and shrubbery, both in arid areas and near water. Often in groups of 4 to 8 birds gleaning actively in foliage, frequently accompanying mixed-species flocks. Rather excitable, often approaching closely in response to squeaking, chipping in an animated fashion.

RANGE: Andean slopes and dry intermontane valleys in Bolivia (Cochabamba and Santa Cruz south to Tarija) and lowlands from se. Bolivia east and south across w. and se. Paraguay (most numerous in *chaco*) to n. Argentina (south to La Rioja, San Luis, Córdoba, and rarely in n. Buenos Aires), w. Uruguay (Artigas to Colonia), and extreme sw. Brazil (both in w. Rio Grande do Sul and in sw. Mato Grosso). To about 1800 m (in Bolivia).

NOTE: Though *P. melanoleuca* and *P. cinerea* have recently been regarded by some as conspecific, we follow Short (1975) and Meyer de Schauensee (1966, 1970) and continue to regard them as full species. We would have been inclined to merge them but for the fact that their bills differ quite markedly; nonetheless they will perhaps eventually be best considered conspecific. In any case, if lumped the species name *melanoleuca* has priority (*contra Birds of the World*, vol. 13).

Poospiza cinerea

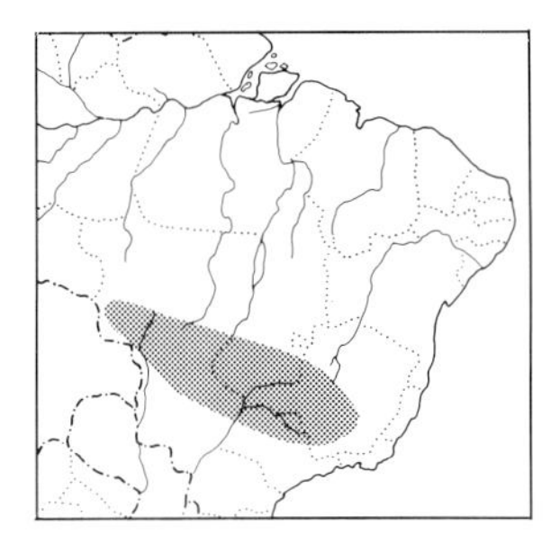

CINEREOUS WARBLING-FINCH

IDENTIFICATION: 13 cm (5"). *Interior s. Brazil.* Resembles Black-capped Warbling-Finch but *lacks black cap in both sexes;* lores and cheeks only slightly sootier than the gray of rest of head and not as black as immature Black-capped's. Bill somewhat longer and more pointed.
HABITAT AND BEHAVIOR: Little known. Recorded from fairly open deciduous woodland and *cerrado.* Behavior probably much like Black-capped Warbling-Finch's; seems inexplicably scarce and local, however.
RANGE: Interior s. Brazil in s. Goiás, w. Minas Gerais, n. São Paulo, and cen. Mato Grosso. Mostly 600–1200 m.

Poospiza lateralis

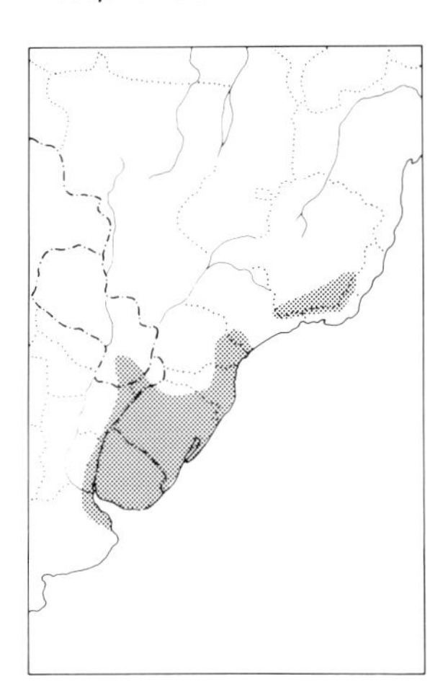

RED-RUMPED WARBLING-FINCH

PLATE: 29

IDENTIFICATION: 15 cm (6"). *Se. Brazil and adjacent areas. The only warbling-finch with contrasting rufous rump. Cabanisi* (most of range) has gray head with *long, narrow, white superciliary,* becoming more olive brown on back. Throat and chest grayish, with center of breast tinged buff on some birds; *sides and flanks rufous,* with center of belly white. Wings and tail blackish; greater coverts and inner flight feathers edged white; corners of tail also white. Nominate race (Itatiaia Mts. and adjacent ne. São Paulo northward) differs in having more yellowish supraloral streak, grayer back, *yellowish buff throat and chest,* sides of breast more broadly rufous (almost meeting at center), and mostly white outer tail feathers.
SIMILAR SPECIES: The smaller Bay-chested Warbling-Finch has *complete* pectoral band, lacks superciliary, etc.
HABITAT AND BEHAVIOR: Fairly common in shrubbery of forest borders and open woodland; seemingly most numerous in the Brazilian portion of its range, particularly in hills and mountains along the coast. Found in pairs or (during austral winter) small groups of up to 4 to 5 birds, often consorting with mixed flocks when not breeding. Its usual song is a repeated, metallic "tzip . . . tzip . . . tzip . . ." (sometimes altered to "tzap"), reminiscent of a hummingbird (e.g., White-vented Violet-ear).
RANGE: Se. Brazil (s. Minas Gerais and Espírito Santo to Rio Grande do Sul), Uruguay, se. Paraguay, and ne. Argentina (Misiones, Entre Ríos, and n. Buenos Aires). To about 1800 m.

NOTE: Two species may be involved, more northern *P. lateralis* (Buff-throated Warbling-Finch) and more southern *P. cabanisi* (Gray-throated Warbling-Finch). As far as we are aware, no intergradation between these 2 rather different forms has been demonstrated.

Poospiza ornata

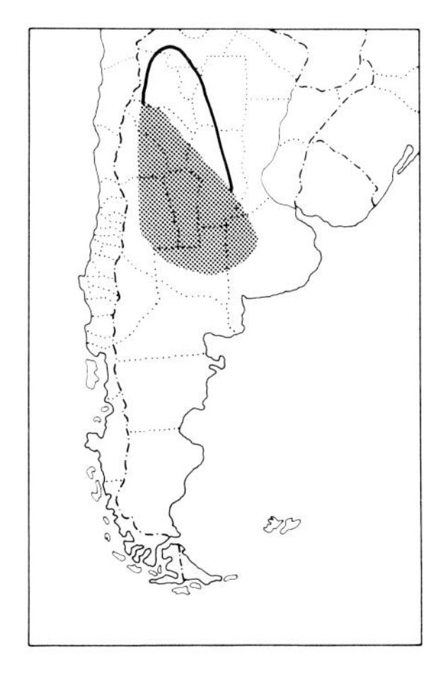

CINNAMON WARBLING-FINCH

PLATE: 29

IDENTIFICATION: 13 cm (5"). *Argentina.* Head and upper back gray, with *long superciliary cinnamon; lower back and rump chestnut.* Throat cinnamon-rufous, *breast rich chestnut,* becoming cinnamon-rufous on belly. Wings and tail dusky, with *2 pronounced white wing-bars,* and outer tail feathers mostly white. Female has similar pattern but is much paler generally, with underparts and wing-bars buffier.
SIMILAR SPECIES: Among other warbling-finches, only the Black-and-chestnut has such rich chestnut; Cinnamon's pattern, in any case, is unique.
HABITAT AND BEHAVIOR: Uncommon in dry woodland and scrub and in pastures and fields with scattered low trees and bushes. Most often in pairs when nesting, in small groups (often with other species) during the nonbreeding season.
RANGE: Argentina, breeding in limited area in San Juan, s. La Rioja, Mendoza, San Luis, La Pampa, and w. Buenos Aires, moving north during austral winter to Salta, Tucumán, and Córdoba. Mostly below 1000 m.

Poospiza thoracica

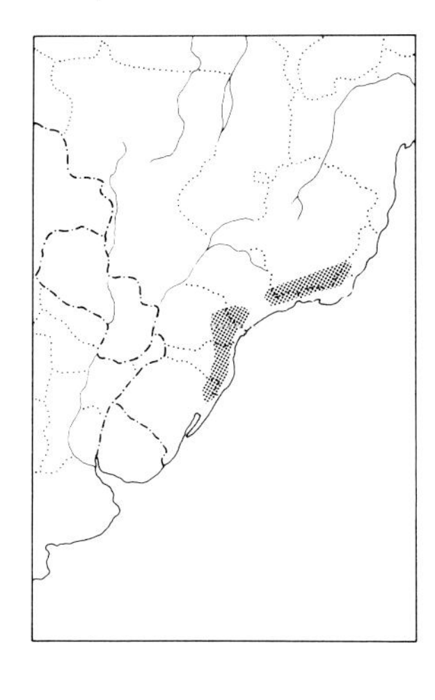

BAY-CHESTED WARBLING-FINCH

PLATE: 29

IDENTIFICATION: 13.5 cm (5¼"). *Coastal mts. of se. Brazil. Above mostly gray,* slightly tinged olivaceous especially on back; white crescent below eye and white edging on primaries. Below white, with *broad pectoral band and sides chestnut.*
SIMILAR SPECIES: The larger Red-rumped Warbling-Finch has a pale superciliary and rufous rump, and shows white in tail.
HABITAT AND BEHAVIOR: Uncommon to fairly common in shrubbery at forest edge, semiopen woodland, and clearings with scattered trees and bushes. Seemingly more associated with montane forest than other s. warbling-finches. Perhaps most numerous in the Itatiaia Mts., but even there outnumbered by Red-rumped. Found in pairs or small groups, often with mixed flocks; gleans warblerlike from foliage and branches, rarely perching for any time in the open.
RANGE: Se. Brazil (e. Minas Gerais to n. Rio Grande do Sul). Mostly 800–1500 m.

Poospiza nigrorufa

BLACK-AND-RUFOUS WARBLING-FINCH

IDENTIFICATION: 15 cm (6"). *Se. South America.* Pattern resembles that of Black-and-chestnut Warbling-Finch of Andes (no overlap; these two were formerly considered conspecific). *Brownish gray* (not slaty) *above;* paler, *more cinnamon-rufous* (not deep chestnut) *below* with less white on center of belly; outer tail feathers only tipped white. Immature lacks superciliary and is whitish below heavily streaked dark brown.

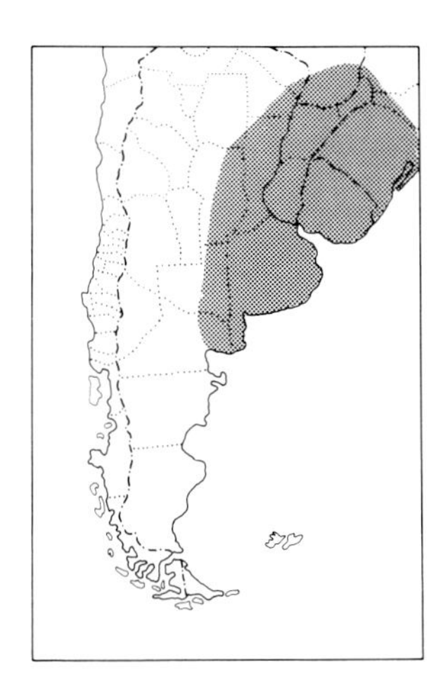

SIMILAR SPECIES: Cinnamon Warbling-Finch has cinnamon (not mostly white) superciliary, chestnut on back, and prominent white on wing (wings uniform brownish in this species). Red-rumped Warbling-Finch has rufous confined to flanks, with grayish throat and breast.

HABITAT AND BEHAVIOR: Fairly common in shrubbery and trees, usually near water but sometimes in gardens. Usually found in pairs, which forage on or near the ground. Singing males often take to an exposed perch, however, and there give a rapidly repeated "swit-swit-cheeu, swit-swit-cheeu . . . ," repeated 3 to 4 times, often ending with a "swit?"

RANGE: Se. Brazil (Santa Catarina and Rio Grande do Sul), Uruguay, se. Paraguay, and e. Argentina (Misiones and e. Chaco south through e. Córdoba and Buenos Aires to La Pampa and e. Río Negro). To about 900 m.

NOTE: We believe se. lowland *nigrorufa* and Andean *whitii* (with Bolivian *wagneri*) represent full species, though they have recently most often been considered conspecific. Color and tail patterns differ (at least as much so as in some other *Poospiza* currently given full species status); furthermore, their primary songs also differ (R. Straneck and pers. obs.).

GROUP B *Andean* warbling-finches and mountain-finches.

Poospiza erythrophrys

RUSTY-BROWED WARBLING-FINCH PLATE: 29

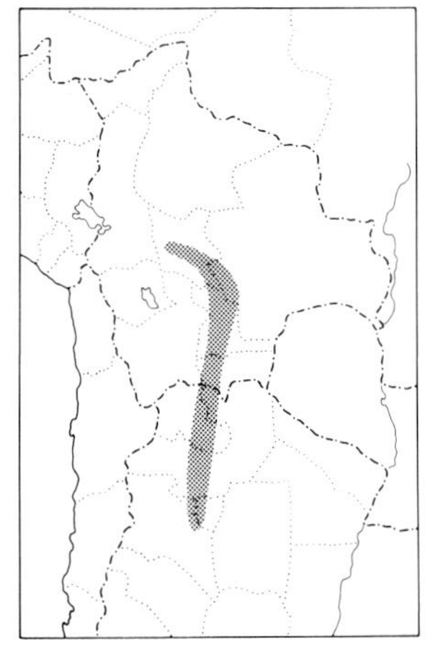

IDENTIFICATION: 14 cm (5½"). *Bolivia and nw. Argentina.* Head and neck gray, with *deep rufous superciliary* and lower eyelid; back olivaceous brown. *Throat and breast deep rufous,* becoming lighter rufous on flanks and whitish on center of belly. Wings and tail dusky, with *greater wing-coverts and primaries edged white,* and outer tail feathers mostly white.

SIMILAR SPECIES: The larger Cochabamba Mountain-Finch lacks white in wings and tail and has a rufous forehead as well as superciliary.

HABITAT AND BEHAVIOR: Locally fairly common in montane woodland, especially where alders are numerous. Usually in pairs or small groups and quite arboreal, gleaning warblerlike from foliage. Often accompanies small flocks.

RANGE: Andes in w. Bolivia (north to Cochabamba) and nw. Argentina (south to Tucumán and n. Catamarca). Mostly 1200–2100 m.

Poospiza rubecula

RUFOUS-BREASTED WARBLING-FINCH

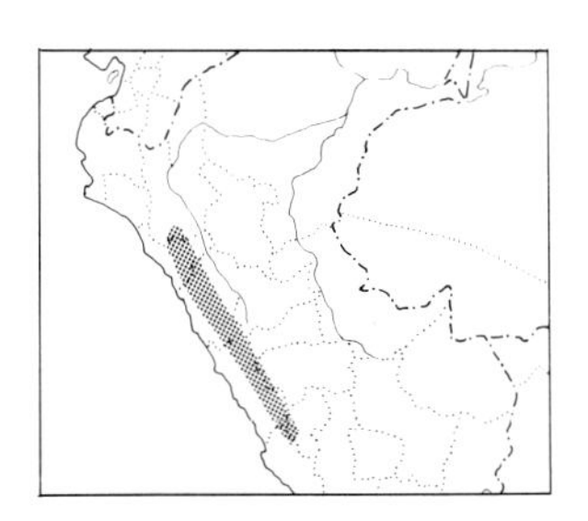

IDENTIFICATION: 16.5 cm (6½"). *W. Peru.* Above slaty gray, tinged olivaceous on back. *Forehead, superciliary, and most of underparts orange-rufous,* becoming whitish on center of belly; *facial area and chin black.* Wings and tail dusky, with *no white.*

SIMILAR SPECIES: Unique within its range; Cochabamba Mountain-Finch (of Bolivia) is rather similar but is larger and lacks black face.

HABITAT AND BEHAVIOR: Rare and with restricted range. Has been seen singly and in pairs in low Andean woodland and adjacent brush,

but the species remains poorly known. Possibly declining as a result of clearing of much woodland in its small range; has been found along the Santa Eulalia road above Huinco in the mountains above Lima.
RANGE: W. slope of Andes in Peru (s. Cajamarca south locally to Ica). 2500–3400 m.

Poospiza baeri

TUCUMAN MOUNTAIN-FINCH

PLATE: 29

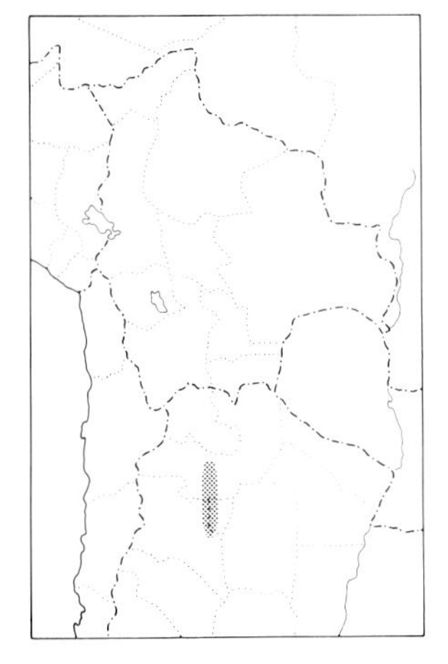

IDENTIFICATION: 18 cm (7"). *Nw. Argentina. Forehead, superciliary, lower eyelid, and broad throat patch orange-rufous.* Otherwise mostly *gray,* somewhat tinged olivaceous on back, and paler on lower belly; crissum also orange-rufous.
SIMILAR SPECIES: No other mostly gray finch has the rufous on the face and throat.
HABITAT AND BEHAVIOR: Uncommon in low woodland, often along streams and especially where alders are frequent. Usually in pairs, which tend to perch rather upright and do not seem to be very active. Can be found in woodland patches and streamside shrubbery above Tafi del Valle in Tucumán.
RANGE: Andes of nw. Argentina in Tucumán and Salta (where seen in Nov. 1984 on road to Cachi; RSR et al.). Mostly 2000–2500 m.

NOTE: This species and the following were formerly separated in the genus *Compsospiza,* but we agree with most recent authors that it is best merged in *Poospiza.* However, we continue to employ "mountain-finch" as a useful subgenus name.

Poospiza garleppi

COCHABAMBA MOUNTAIN-FINCH

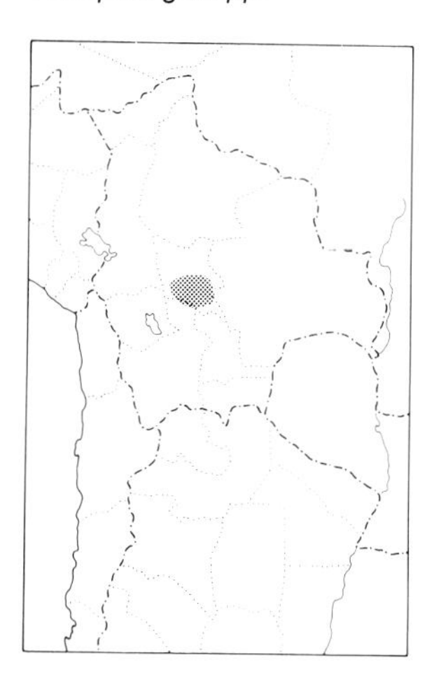

IDENTIFICATION: 18 cm (7"). *Cochabamba, Bolivia.* Mostly gray above; *orange-rufous forehead, superciliary, lower eyelid, and most of underparts,* somewhat paler on center of belly.
SIMILAR SPECIES: Overall pattern reminiscent of Tucuman Mountain-Finch, but entire underparts are orange-rufous and purer gray above. Rusty-browned Warbling-Finch also somewhat similar but is smaller with white in wings and tail and has all-gray crown.
HABITAT AND BEHAVIOR: Uncommon in patches of low woodland (dominated by *Polylepis* and alders), usually near or just below timberline. Usually in pairs and most often with small bird parties; generally perches low, but fairly tame and conspicuous. Like the Tucuman Mountain-Finch, tends to perch rather upright and not to move about as actively as most other warbling-finches. May be threatened by cutting of trees for firewood and clearing for agriculture; its total range is very small, and even its original habitat is limited.
RANGE: Andes of w. Bolivia in Cochabamba. Mostly 3000–3500 m.

NOTE: See comments under Tucuman Mountain-Finch.

Poospiza hypochondria

RUFOUS-SIDED WARBLING-FINCH

PLATE: 29

IDENTIFICATION: 16–16.5 cm (6¼–6½"). *Bolivia and nw. Argentina.* Brownish gray above (brownest on back), with *white superciliary* and

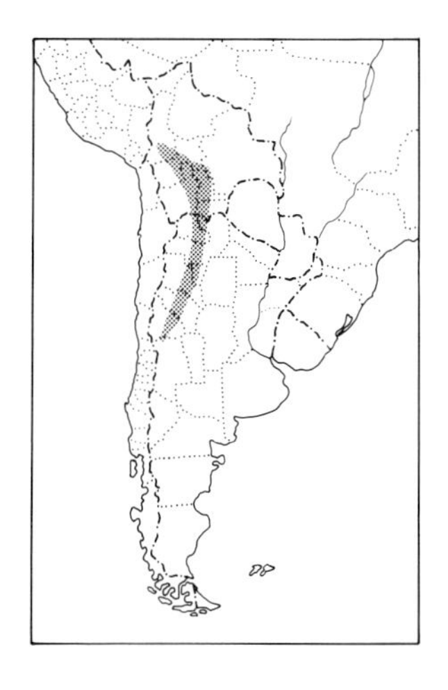

subocular spot. Throat whitish, accented by black malar streak; *breast dingy grayish;* belly buffy whitish, with *rufous flanks.* Wings and tail dusky, with wing-coverts and inner flight feathers edged buff, and outer tail feathers mostly white. *Affinis* (Argentinian portion of range) slightly larger, buffier on belly and grayer on rump, and with less white on tail (only terminal half).

SIMILAR SPECIES: Bolivian Warbling-Finch (wholly sympatric) is similar but has rufous (not grayish) across breast and lacks the malar streak. Cf. also Plain-tailed Warbling-Finch (of Peru).

HABITAT AND BEHAVIOR: Common on shrubby hillsides and in ravines, principally in more arid regions. Often in small groups, foraging actively in shrubbery and low trees, but seems rarely to actually go to the ground.

RANGE: Andes of w. Bolivia (north to La Paz) and nw. Argentina (south to n. Mendoza). Mostly 2500–4000 m.

Poospiza alticola

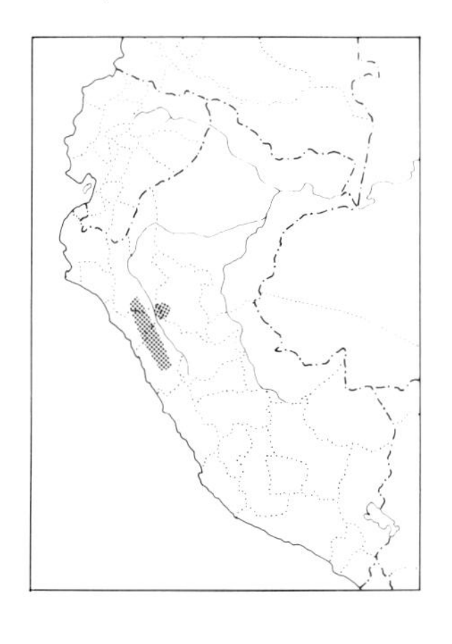

PLAIN-TAILED WARBLING-FINCH

IDENTIFICATION: 16 cm (6¼″). *W. Peru.* Closely resembles Rufous-sided Warbling-Finch. Browner above, particularly on crown. Underparts mostly dingy whitish (no gray across breast), with *rufous on sides of breast, becoming pale cinnamon buff on flanks.* Wing-coverts and flight feathers edged whitish (not buff), but *no white in tail.*

HABITAT AND BEHAVIOR: Locally fairly common in shrubbery and low woodlands, usually just below the *Polylepis* zone, and especially in or near ravines. Usually in pairs, which forage actively through foliage, generally near the ground. Can be found above Yungay in Huascarán Nat. Park, Ancash (M. Robbins).

RANGE: Andes of w. Peru in s. Cajamarca, La Libertad, and Ancash; mostly west of the Marañón River, locally east of it in La Libertad. 2900–3600 m.

Poospiza boliviana

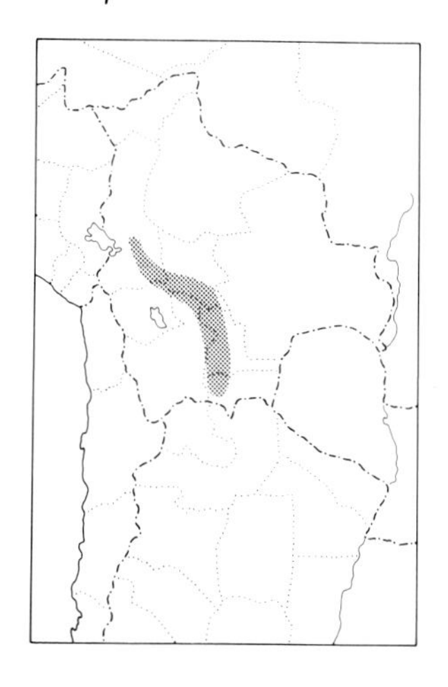

BOLIVIAN WARBLING-FINCH

IDENTIFICATION: 16 cm (6¼″). Somewhat resembles Rufous-sided Warbling Finch. Brown above, somewhat grayer on sides of head, with *long white superciliary.* Throat white (with *no malar streak*), bordered below by *broad rufous pectoral band,* the *rufous extending down sides to flanks;* center of belly white. Wings and tail dusky, inner flight feathers edged cinnamon, and outer tail feathers mostly white.

SIMILAR SPECIES: Rufous-sided Warbling-Finch lacks the rufous across breast (has it only on sides), has a black malar streak, and is grayer (not so brown) above. Cf. also Rusty-browed Warbling-Finch.

HABITAT AND BEHAVIOR: Fairly common in scrub and low woodland. Found in pairs or small groups, in some places with Rufous-sided Warbling-Finch (though tends to occur lower than that species). Can be found in numbers on shrubby hillsides outside of Cochabamba city.

RANGE: Andes of Bolivia in La Paz, Cochabamba, w. Chuquisaca, and w. Tarija. Mostly 1600–3000 m.

Poospiza whitii

BLACK-AND-CHESTNUT WARBLING-FINCH

PLATE: 29

Other: Chestnut Warbling-Finch, Black-and-rufous Warbling-Finch (in part)

IDENTIFICATION: 15 cm (6"). *Bolivia and nw. Argentina. Mostly slaty gray above,* with long white superciliary becoming deep chestnut behind; cheeks black and malar streak white; *below mostly deep chestnut,* center of belly contrastingly white; tail blackish, terminal half of outer tail feathers white. Immature duller.

SIMILAR SPECIES: See Black-and-rufous Warbling-Finch (no overlap; formerly considered conspecific). No other warbling-finch in the Andes has the deep chestnut (almost maroon) on underparts.

HABITAT AND BEHAVIOR: Fairly common in woodland borders, low shrubbery, and agricultural areas with scattered trees and hedgerows. Usually singly or in pairs, perching in shrubbery. Male's vigorous song is rich and gurgling in quality, a spirited series of phrases quite grosbeaklike (*Pheucticus*) overall, and strikingly different from the measured cadence of Gray-and-rufous Warbling-Finch.

RANGE: Andean slopes in w. Bolivia (La Paz, Cochabamba, and w. Santa Cruz south) and nw. Argentina (Jujuy and Salta south to n. La Rioja; also hills of w. Córdoba and n. San Luis). Mostly 600–2500 m, locally a little higher.

NOTE: Here Andean *P. whitii* is considered a species distinct from lowland-inhabiting *P. nigrorufa;* see comments under that species. We believe that the previously suggested name for *whitii* ("Chestnut Warbling-Finch") fails to convey its close relationship with *nigrorufa* and hence add the "Black."

Poospiza caesar

CHESTNUT-BREASTED MOUNTAIN-FINCH

PLATE: 29

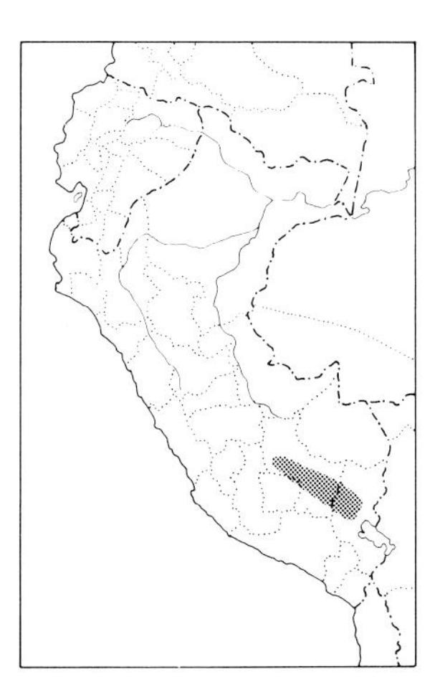

IDENTIFICATION: 18.5 cm (7¼"). *S. Peru.* Gray above, with black forecrown and sides of head and long white superciliary. Throat white, contrasting with *broad rufous-chestnut pectoral band;* flanks broadly gray, center of belly white, with rufous-chestnut crissum.

SIMILAR SPECIES: Nothing really similar in range; notably larger than most other warbling-finches and more terrestrial than any. Overall pattern similar to that of Bolivian Warbling-Finch, but much grayer above and with gray (not rufous) on flanks.

HABITAT AND BEHAVIOR: Locally fairly common in scrub and low woodland. Usually in pairs, which often skulk close to the ground. Often feeds on the ground somewhat *Atlapetes*-like. Readily found around Cuzco (e.g., along the road south to the Huaparcay Lakes and at the Peñas ruins above Ollantaytambo); the species might almost better be called the "Cuzco Mountain-Finch."

RANGE: Andes of s. Peru in Cuzco and Puno. Mostly 2500–3500 m.

NOTE: We believe that this species, formerly separated in the monotypic genus *Poospizopis,* is (like *Compsospiza*) better included in *Poospiza.* It is, however, more divergent, being notably larger and rather more terrestrial than any other *Poospiza.*

Arremon Sparrows

Fairly small finches (otherwise somewhat similar *Atlapetes* are larger), typically with striped head, olive or gray upperparts, and white underparts usually with a pectoral band. Some have brightly colored bills. Lower growth, often foraging on ground, in woodland and forest borders; unlike *Atlapetes,* they range mostly in the lowlands. Nests are bulky and roofed with a side entrance, placed on or near the ground.

Arremon taciturnus

PECTORAL SPARROW

PLATE: 30

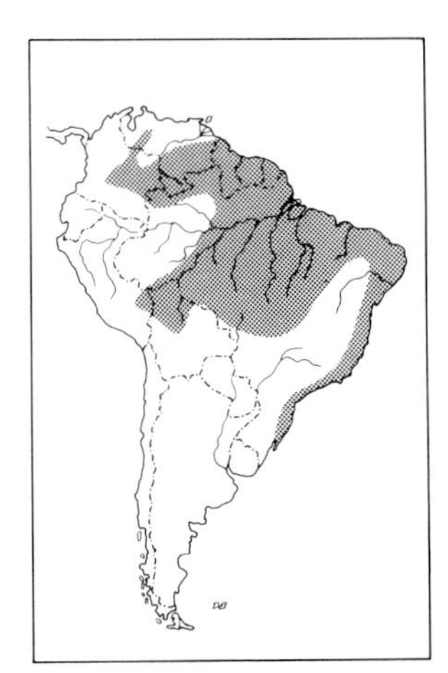

IDENTIFICATION: 15 cm (6"). *Bill black* (with *yellow mandible* in w. Venezuela and ne. Colombia along base of Andes and in se. Brazil). Cap black with gray coronal streak and white superciliary extending back from lores. Above olive, with prominent yellow on shoulders. Below white with *black pectoral band.* Pectoral band incomplete (only on sides of chest) in w. Venezuela and adjacent Colombia (*axillaris,* which also has superciliary starting only above eyes), in se. Peru and ne. Bolivia (*nigrirostris*), and in se. Brazil (*semitorquatus,* which also lacks the yellow shoulders). Females echo pattern of male but are duller and much *buffier below.*

SIMILAR SPECIES: The only *Arremon* in most of Amazon basin; the Pectoral apparently does not occur sympatrically with any of its congeners (though Saffron-billed comes close in Bolivia and cen. Brazil). Orange-billed and Saffron-billed Sparrows both have yellow-orange bills and complete pectoral bands. In Bolivia cf. also Stripe-headed Brush-Finch.

HABITAT AND BEHAVIOR: Fairly common (but inconspicuous) in undergrowth of humid forest and mature second growth. Has an exceptionally high, simple song with an almost hissing quality, "zitip, zeeee-zeeee-zeeee."

RANGE: W. Venezuela (Portuguesa to Táchira and Apure) south along e. base of Andes in ne. Colombia (to w. Meta); extreme e. Colombia (e. Vichada, Guainía, and Vaupés), s. Venezuela (Amazonas and Bolívar), and the Guianas south through much of cen. and e. Brazil (south along e. coast to n. Rio Grande do Sul, in interior south locally to s. Goiás and cen. Mato Grosso), and in n. Bolivia (south to La Paz and Cochabamba) and se. Peru (Cuzco, Madre de Dios, and Puno). Mostly below 1000 m, but locally higher on Andean slopes.

Arremon flavirostris

SAFFRON-BILLED SPARROW

PLATE: 30

IDENTIFICATION: 15–16.5 cm (6–6½"). *S. South America. Bill mostly orange-yellow. Polionotus* (e. Paraguay, ne. Argentina) has cap black with bold white postocular stripe and trace of gray coronal stripe. *Above gray,* with some yellow on bend of wing. Below mostly white with *black pectoral band.* N. (nominate) and w. (*dorbignii*) races have mantle mostly olive green, with only upper back gray. *Dorbignii* also has broad gray coronal streak, and its superciliary starts from lores. *Devillii* (sw.

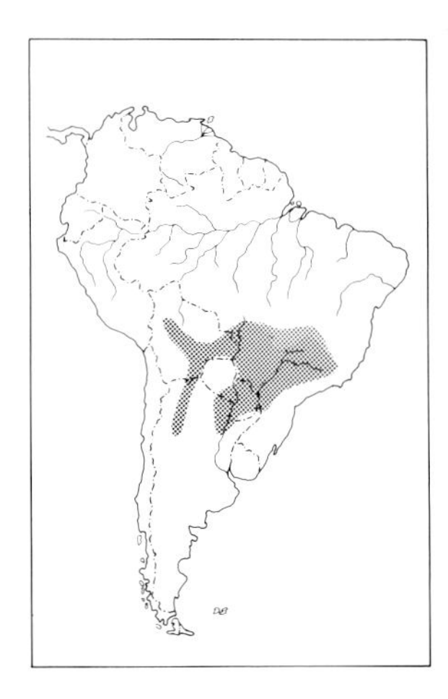

Brazil, adjacent Bolivia) has olive only on wing-coverts. Female resembles male but is somewhat duller and washed with buff below.

SIMILAR SPECIES: In or near its range this is the only *Arremon* sparrow with a yellow bill. *Dorbignii* (of Andean Bolivia and nw. Argentina) resembles Pectoral Sparrow in pattern above but differs in bill color and in presence of pectoral band (absent in the closest race of Pectoral, *nigrirostris*).

HABITAT AND BEHAVIOR: Locally fairly common in undergrowth of deciduous woodland and shrubby areas, mostly in more arid regions (exception in Argentina and e. Paraguay, where also in more humid woodland). The song is a short phrase of high, thin, spitting notes, "tsit, tsee-tsi-tsi, tseép-seép-tseép."

RANGE: E.-cen. Bolivia (Cochabamba) south along e. slope of Andes to nw. Argentina (to Catamarca); e. Bolivia (e. Santa Cruz) east across sw. and interior s.-cen. Brazil to w. Bahia and w. Minas Gerais, and south through w. São Paulo and w. Paraná to e. Paraguay and ne. Argentina (south to ne. Santa Fe, Corrientes, and Misiones). To 1400 m (in Bolivia).

Arremon aurantiirostris

ORANGE-BILLED SPARROW

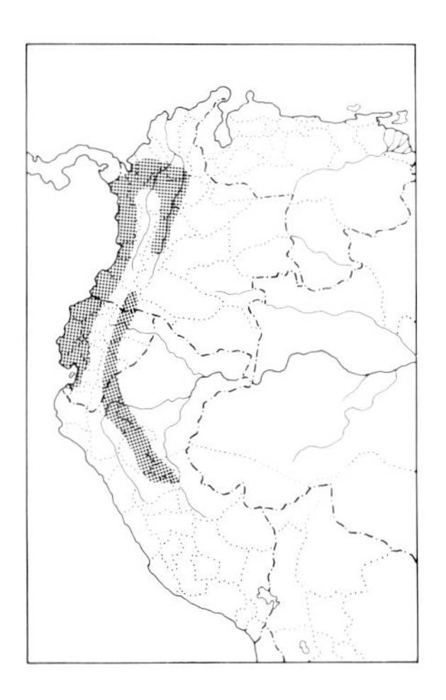

IDENTIFICATION: 15 cm (6"). *Colombia to n. Peru.* Resembles Saffron-billed Sparrow. *Bill entirely bright orange.* White superciliary starting at lores and entirely olive mantle (a little gray only on nape). Bend of wing usually shows more yellow. Black pectoral band much wider in *stictocollaris* of extreme nw. Colombia (Chocó), same as Saffron-billed in all other races. See C-55, P-30.

SIMILAR SPECIES: Golden-winged Sparrow has all-black head and inhabits more arid regions. Range of this species may meet Pectoral Sparrow's along e. base of Colombian Andes; that race of the latter (*axillaris*), however, has bicolored bill (black above, yellow below) and an incomplete pectoral band.

HABITAT AND BEHAVIOR: Fairly common (but unobtrusive) in undergrowth of humid forest and advanced secondary woodland. Song is very high and sibilant, a variable phrase which at times resembles the song of the American Brown Creeper (*Certhia americana*); it is usually given from a low perch, often from a log.

RANGE: Pacific w. Colombia, w. Ecuador, and extreme nw. Peru (Tumbes); also in more humid regions of n. Colombia and south in middle Magdalena valley to Cundinamarca, and from se. Colombia (Putumayo) south through e. Ecuador to ne. Peru (south to e. Huánuco in Tingo María area), mostly close to e. base of Andes. Also Mexico to Panama. To about 1200 m (rarely higher).

Arremon schlegeli

GOLDEN-WINGED SPARROW

PLATE: 30

IDENTIFICATION: 16 cm (6¼"). *Ne. Colombia and n. Venezuela. Bill mostly golden yellow* (black in immature). Cap black (*no striping*) extend-

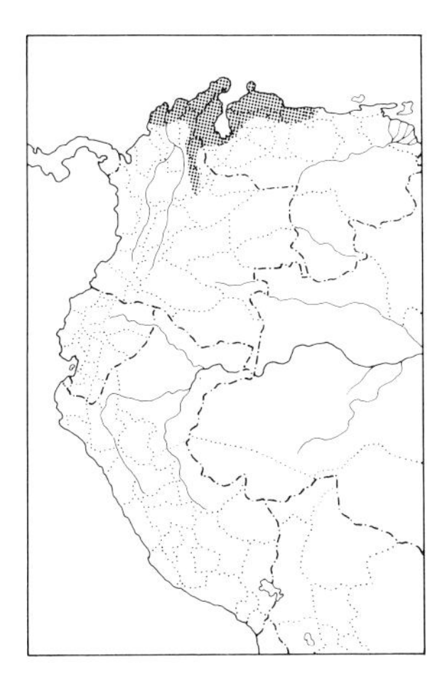

ing down on sides of neck to partial collar on sides of chest. Above mostly gray, with *band from wing-coverts across lower back bright yellowish olive;* prominent yellow on shoulders. Below white, clouded gray on sides. Birds from Santander and Boyacá in Colombia (*canidorsum*) are golden olive only on wing-coverts (back essentially gray).

SIMILAR SPECIES: This handsomest of the *Arremon* sparrows is the only member of its group *without* head striping.

HABITAT AND BEHAVIOR: Fairly common (but usually inconspicuous) in deciduous woodland and adjacent shrubby clearings; mostly in arid regions. Song is a high, thin "soot-soot-soot-see?" given from a low, usually hidden perch.

RANGE: N. Colombia (Bolívar east to the Guajira Peninsula and locally above Magdalena valley in Santander and Boyacá) and n. Venezuela (Zulia east locally through Falcón and Lara to Distrito Federal). To 1400 m.

Arremon abeillei

BLACK-CAPPED SPARROW

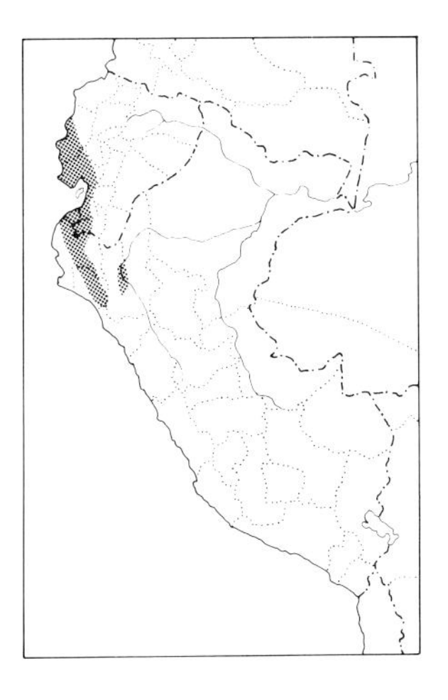

IDENTIFICATION: 15 cm (6"). *Sw. Ecuador and nw. Peru.* Bill black, but otherwise most resembles gray-backed races of Saffron-billed Sparrow. *Head and neck black* with *white superciliary starting above eyes.* Upperparts bluish gray, with single faint white wing-bar, but no yellow at bend of wing. Below white with black pectoral band. *Nigriceps* of Marañón valley of Peru has white superciliary commencing at lores and olive back and rump (thus more like Pectoral Sparrow). Females resemble males but tinged buff on sides.

SIMILAR SPECIES: Only other potentially sympatric *Arremon,* the Orange-billed Sparrow, has an orange bill and is entirely bright olive above with yellow on bend of wing; it is restricted to more humid regions (with Black-capped in more arid areas). See also male Collared Warbling-Finch (with more white in wing and tail, very different behavior, etc.).

HABITAT AND BEHAVIOR: Fairly common in undergrowth of deciduous forest and woodland. During rainy season (Jan.–May) Ecuadorian birds often sing a high, thin "tsee, tsew, tee-tee, ti-i-i-i-i" with variations, repeated over and over.

RANGE: Sw. Ecuador (north to Manabí) and nw. Peru (south on Pacific slope to Cajamarca, also in Marañón valley in e. Cajamarca). To 700 m.

Arremonops Sparrows

A duller version of *Arremon.* Found in shrubbery in n. South America, with 2 more species in Middle America. The nest is a bulky domed structure with a large side entrance (A. Skutch).

Arremonops conirostris

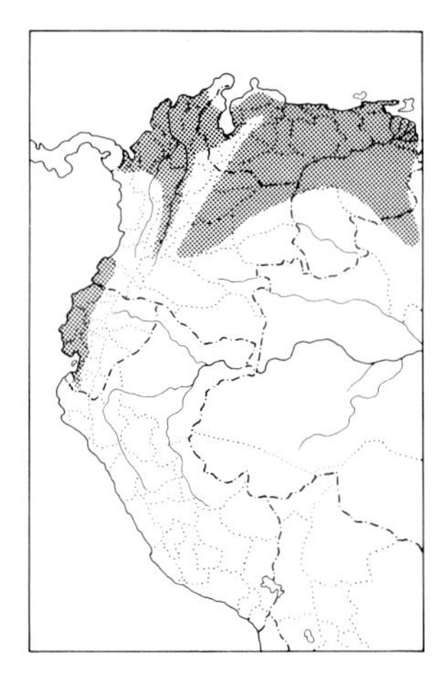

BLACK-STRIPED SPARROW

PLATE: 30

IDENTIFICATION: 15–18 cm (6–7"). *Head gray with bold black striping.* Above olive green with yellow bend of wing. Below pale gray, whitest on throat and center of belly. Birds from Pacific slope (*striaticeps*) are larger and brighter olive above. Birds from Maracaibo basin (*umbrinus*) and upper Magdalena valley (*inexpectatus*) are smaller and dingier, particularly the latter.

SIMILAR SPECIES: See Tocuyo Sparrow. *Arremon* sparrows are all more boldly patterned and have at least a partial black pectoral band.

HABITAT AND BEHAVIOR: Fairly common (often not very conspicuous except when singing, when it may mount to the top of a bush) in low deciduous woodland, shrubby clearings, and thick hedgerows. Usually seen singly or in pairs, on or near the ground, and is often quite shy. Best known from its far-carrying and distinctive song, a 2-part series: first several slow, well-spaced notes and then an accelerating stutter, something like an engine starting up: "cho; cho; cho, cho, cho-cho-cho-chochochochch." Also gives a sharply accented whistled call: "ho-wheet."

RANGE: Pacific sw. Colombia (north to sw. Cauca) south locally (absent from very humid regions) to w. Ecuador (to El Oro and w. Loja); Caribbean n. Colombia (n. Chocó east and south in Magdalena valley to Huila) and ne. Colombia (south to w. Meta) east across Venezuela north of the Orinoco to Delta Amacuro, south of it in Bolívar, and extending to extreme n. Brazil (Roraima). Also Honduras to Panama. To 1500 m.

Arremonops tocuyensis

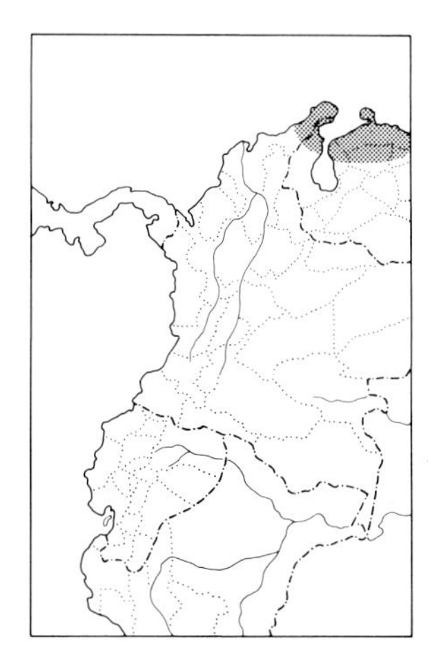

TOCUYO SPARROW

IDENTIFICATION: 14 cm (5½"). *Closely* resembles Black-striped Sparrow. *Smaller* than even the smallest race of that species (and occurs only with the larger nominate race) and *paler and more washed out,* particularly on the head, which is more clay-colored. See C-55, V-39.

HABITAT AND BEHAVIOR: Uncommon in scrubby areas and thickets in low deciduous woodland in arid regions. Black-striped Sparrow tends to occur in somewhat more humid areas in the zone of overlap, but the 2 species are reported to occur sympatrically at several sites. Behavior resembles Black-striped Sparrow's, but song reported to be thinner and sweeter (S. Hilty, from P. Schwartz recording), though overall pattern is somewhat similar.

RANGE: Locally in ne. Colombia (Guajira Peninsula from Riohacha area east) and nw. Venezuela (arid areas in n. Zulia, Falcón, and Lara). To 1100 m (in Colombia only near sea level).

Zonotrichia Sparrows

Well-known sparrow with obvious rufous collar. The southernmost representative of a common N. Am. genus, it is widespread in semi-open areas.

Zonotrichia capensis

RUFOUS-COLLARED SPARROW

PLATE: 30

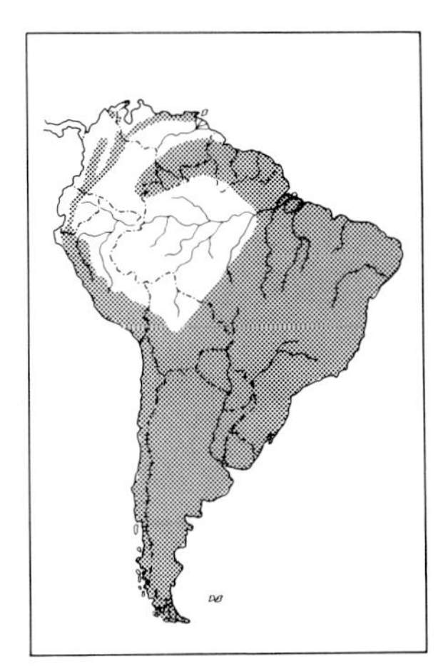

IDENTIFICATION: 14–15 cm (5½–6″). *Slightly bushy-crested. Head gray with black striping, contrasting with conspicuous rufous collar on hindneck.* Above brown to rufous streaked dusky. Below whitish with black patch on sides of chest. *Australis* of Patagonia (migrating north to s. Bolivia) lacks black head striping and thus appears mostly gray-headed. Races found on *tepuis* are dark and dingy, quite blackish on head. Juvenal much duller and streakier (especially below), but usually showing at least a trace of rufous collar; see C-55, V-39.

SIMILAR SPECIES: Perky adults are readily recognized (and will soon be familiar). Juvenals resemble many other "streaky" sparrows and other finches (e.g., female Plumbeous Sierra-Finch) but can usually be known by their habits or the company they keep.

HABITAT AND BEHAVIOR: Usually common to abundant and found in open or semiopen areas (both natural and artificial) almost throughout; numerous in mountains and s. South America. Tame and conspicuous, it frequently occurs around habitations and in agricultural areas; remains numerous in cities still without House Sparrows. Its song is well known and pretty, typically 1 or 2 long, slurred whistles followed by a trill (e.g., "tee-teeoo, t-e-e-e-e") but is variable. The "chink" call note also varies geographically.

RANGE: Widespread in Andes and other montane areas and across much of semiopen s. and e. South America (south to Tierra del Fuego and north in Brazil to Mato Grosso, se. Pará, and s. Maranhão); largely absent from the Orinoco and Amazon basins, but found widely on the *tepuis* and more locally on natural savannas in e. Colombia, nw. Brazil, and s. Venezuela and in the Guianas and near the mouth of the Amazon in Brazil; Netherlands Antilles (Curaçao and Aruba); southernmost breeders apparently migrate northward during austral winter (actual extent uncertain). Also s. Mexico to Panama, and Hispaniola. To 3500 m (sometimes higher).

Ammodramus Sparrows

Small, plain sparrows of grassy areas, always with some yellow on lores or face. The genus *Myospiza,* in which 2 of the S. Am. species were formerly placed, is now usually merged into *Ammodramus.*

Ammodramus humeralis

GRASSLAND SPARROW

PLATE: 30

IDENTIFICATION: 13 cm (5″). A small sparrow of *tall grasslands.* Brownish gray above with blackish streaking (streaks on the mantle edged chestnut), and also with chestnut edging on inner flight feathers. Bend of wing and *lores (only) yellow,* narrow eye-ring white. Whitish below, chest and sides tinged pale grayish buff. In parts of range pinkish buff staining below frequent. Juvenal has dusky streaking across breast.

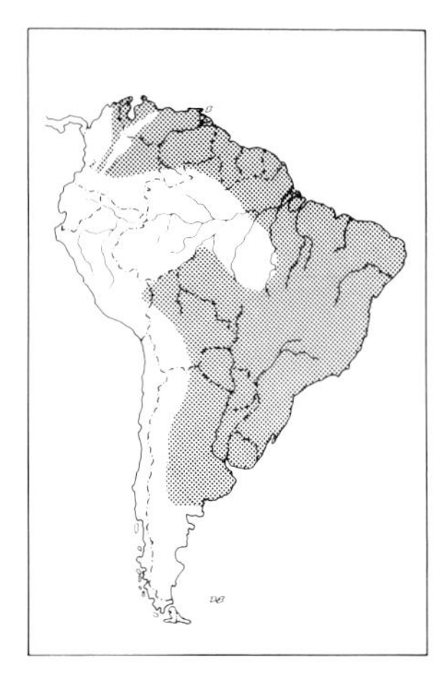

SIMILAR SPECIES: Yellow-browed Sparrow has much more yellow on face, different habitat and song. See also Grasshopper Sparrow. Other streaked finches are considerably larger.

HABITAT AND BEHAVIOR: Locally common in tall grassy savannas and *campos* and in *cerrado* (if not too heavily grazed or recently burned). Often quite secretive, feeding on the ground and flushing only short distances; upon alighting generally disappears into heavy cover. More conspicuous when singing; also may feed more in the open early and late in the day. Song is high and thin but often quite musical: "eee telee, teeeee," with many variations; *it totally lacks Yellow-browed's buzzy quality.*

RANGE: Locally in w. and n. Colombia (south to the Cauca and Magdalena valleys and in upper Patía valley) and in llanos of northeast (south to Meta and w. Vaupés), virtually throughout Venezuela, locally in the Guianas and e. Amaz. Brazil (Amapá and along the lower Amazon, upriver to the lower Rio Tapajós), south across e. and s. Brazil through Uruguay and Paraguay to n. and e. Argentina (south to n. Río Negro), and west across e. and n. Bolivia to extreme se. Peru (Heath Pampas in Madre de Dios). To about 1100 m (locally higher in s. Venezuela and perhaps elsewhere).

Ammodramus aurifrons

YELLOW-BROWED SPARROW

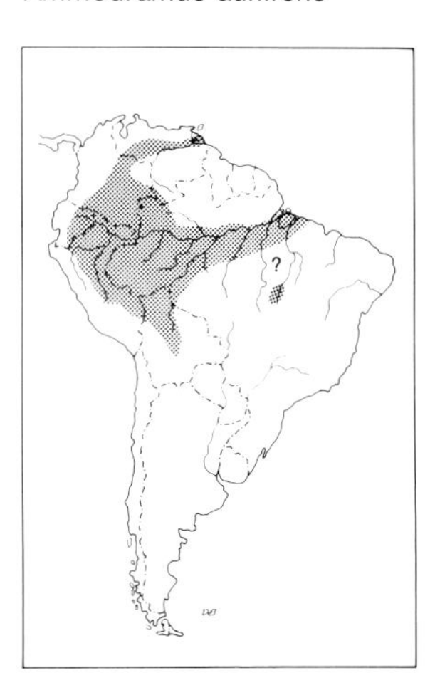

IDENTIFICATION: 13.5 cm (5¼"). A somewhat pallid version of the Grassland Sparrow. Upperparts less boldly streaked and with no chestnut edging. *Face conspicuously yellow* (lores, superciliary, eye-ring, and cheeks). See C-56, V-40. *Cherriei* of w. Meta in Colombia lacks yellow on cheeks.

SIMILAR SPECIES: See Grassland Sparrow (which normally has much less yellow on face, though some individuals can be confusing—then go by habitat and, if possible, voice).

HABITAT AND BEHAVIOR: Common and widespread in grassy areas of roadsides, towns, riparian areas, and agricultural regions. Unlike Grassland Sparrow, this is a conspicuous species which hops about in the open and quickly becomes familiar. Its song likewise will rapidly become known, a buzzing "tic, tzzz-tzzzzz," repeated endlessly from a low perch, even in the heat of the day.

RANGE: E. Colombia and locally in s. Venezuela (Barinas, Apure, and sw. Amazonas south) south through e. Ecuador, e. Peru, and Amaz. Brazil (north of the Amazon east only to Rio Jamunda, south of it east to e. Pará and south to n. Goiás and ne. Mato Grosso) to n. Bolivia (to La Paz, Cochabamba, and Santa Cruz). To about 1000 m (locally along base of Andes).

Ammodramus savannarum

GRASSHOPPER SPARROW

IDENTIFICATION: 12 cm (4¾"). *Local in Colombia and nw. Ecuador.* Resembles Grassland Sparrow but is slightly smaller and has narrower,

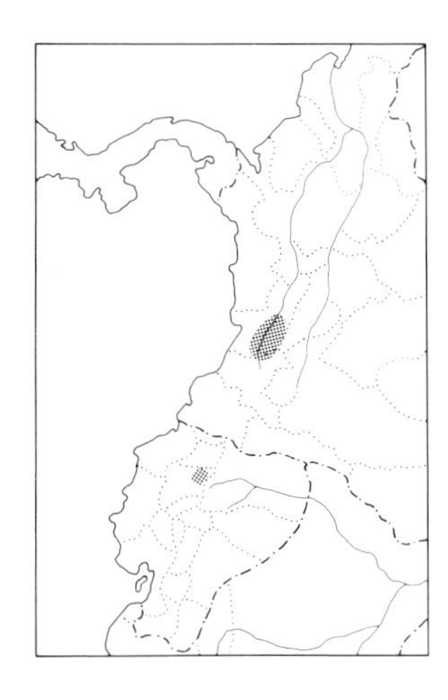

shorter, and more pointed tail; *narrow pale buff coronal streak.* Superciliary yellowish buff, yellowest in front of eye. *Below buff,* becoming whitish on belly. Immature has dusky streaking on breast.

SIMILAR SPECIES: Grassland and Yellow-browed Sparrows lack the coronal streak.

HABITAT AND BEHAVIOR: Very local in tall grasslands; known to be not uncommon at 1 site in Valle, Colombia. Difficult to see unless singing; forages mostly on ground, well hidden in tall grass, and not easy to flush. Singing males sometimes perch in the open on a fence or low bush; song is a buzzy trill, "pi-tup-tzzzzzzzzzz," which, though not strong, does carry well. S. Am. populations may be threatened by overgrazing and the conversion of natural grasslands in its limited range.

RANGE: Locally in w. Colombia (Cauca valley, apparently only in Valle) and nw. Ecuador (Pichincha; no recent records). Also North America south locally to Panama and Greater Antilles. To 3000 m (once), mostly below 1000 m.

Aimophila Sparrows

The 2 S. Am. species are larger and longer-tailed than *Ammodramus* sparrows, with more complex facial patterns. As with *Zonotrichia,* these are the southernmost representatives of a diverse N. Am. genus.

Aimophila strigiceps

STRIPE-CAPPED SPARROW

PLATE: 30

IDENTIFICATION: 16–17 cm (6¼–6¾"). *Argentinian chaco* (mostly). Mandible mostly pale. *Head and nape gray with conspicuous brown striping.* Above grayish brown to brown, streaked blackish except on rump; tail brown, fairly long. Throat white with black submalar streak. Breast light gray, belly whitish. *Dabbenei* (w. part of range) is larger and rustier dorsally and has distinct blackish preocular area.

SIMILAR SPECIES: Nothing really similar in its range; Rufous-collared Sparrow has rufous collar and black (not brown) head striping and lacks malar streak.

HABITAT AND BEHAVIOR: Fairly common but seemingly somewhat local in low semiopen woodland interspersed with grassy areas. Usually occurs in pairs or small groups, feeding on the ground but perching readily in shrubbery and low trees. Generally quite conspicuous where it is found; *dabbenei* is, for instance, easily seen north of the city of Salta, in Salta, Argentina (though the nominate race of the *chaco* seems scarcer). Nominate male's song is loud and ringing, somewhat variable, but typically a repetition of a single note followed by a trill (e.g., "chee-chee-chee-chee, trrr"; R. Straneck recording).

RANGE: N. Argentina along base of Andes from Jujuy to Tucumán; from ne. Buenos Aires north to Chaco (and doubtless also Formosa) and sw. Paraguay (known only from Lichtenau, in Presidente Hayes). To about 1000 m.

Aimophila stolzmanni

TUMBES SPARROW

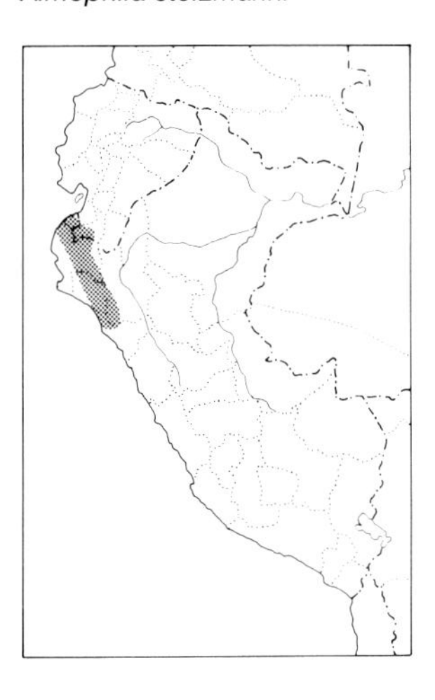

IDENTIFICATION: 14.5 cm (5¾"). *Sw. Ecuador and nw. Peru. Bill heavy,* lower mandible whitish. Resembles Stripe-capped Sparrow in its *striped head* and streaked upperparts, but larger bill and proportionately shorter tail give it a relatively bull-headed and squat appearance. *Shoulder quite conspicuously chestnut* and bend of wing (often hidden) yellow. Underparts essentially like Stripe-capped Sparrow's.

SIMILAR SPECIES: Nothing really similar in its range; see Rufous-collared Sparrow.

HABITAT AND BEHAVIOR: Fairly common in scrubby desert country with scattered bushes, cactus, and low trees, but often only sparse grass cover. Usually hugs the ground, and though remaining close to cover not shy or particularly hard to see. Not in flocks.

RANGE: S. Ecuador (known only from w. Loja) and Pacific slope of nw. Peru from Tumbes south to sw. Cajamarca and n. La Libertad. To about 1400 m.

NOTE: Formerly placed in the monotypic genus *Rhynchospiza,* which is now usually merged into *Aimophila* (see R. A. Paynter, Jr., *Breviora* 278, 1967; also M. D. Williams, *Condor* 1983 [1]: 83–84, 1981).

Spiza Dickcissels

Locally abundant winter visitant from North America to ricefields in (especially) Venezuela. Plumages vary, but with some yellow on breast, rusty on shoulders. Its taxonomic affinities remain uncertain; some authors place it with the Cardinalinae, while others put it in the Icterinae (e.g., Wetmore, Pasquier, and Olson 1984: 332); we place it here because of its superficial similarity to various sparrows.

Spiza americana

DICKCISSEL

IDENTIFICATION: 15 cm (6"). *Sparrowlike.* Breeding male has gray head with *yellow superciliary* and malar streak. Above otherwise light brown streaked dusky, with *usually prominent rusty shoulders.* Upper throat white, with *V-shaped black bib on lower throat and chest. Breast yellow* fading to whitish on belly. Nonbreeding male duller with less yellow; bib veiled. Female like nonbreeding male but less rusty on shoulders; below mostly whitish with some streaking (especially on breast and sides); retains *band of pale yellow across chest.* Immature like female but even duller, usually with some dusky streaking on breast. See V-40, C-56.

SIMILAR SPECIES: Pattern of immature Saffron Finch recalls female, but it lacks yellow superciliary and chestnut shoulders and has yellow crissum. Cf. female House Sparrow and Bobolink as well as much smaller Yellow-browed and Grassland Sparrows.

HABITAT AND BEHAVIOR: Locally common (at times abundant on the Venezuelan llanos), but erratic, n.-winter visitor (mostly Oct.–Apr.) to open country and agricultural areas. Almost always in flocks, sometimes huge (thousands or more), which often concentrate on ricefields. *Flying flocks are often very dense* (like shorebirds). Calls frequently, a raspy "drrt" and assorted twitters; a large flock can generate much noise.

RANGE: Winters in Colombia (Caribbean lowlands and llanos of northeast) and Venezuela (not the far south) and in much smaller numbers in the Guianas and extreme n. Brazil (Roraima); Trinidad (where irregular and not present at all in some years). Breeds cen. North America, migrating through Middle America. Mostly below 500 m.

Embernagra Pampa-Finches

A pair of large, heavy-set finches of s. grasslands. Dark and olive, with orange bill. The genus is probably close to *Emberizoides,* which it resembles in overall color pattern and proportions as well as bill.

Embernagra platensis

GREAT PAMPA-FINCH

PLATE: 30

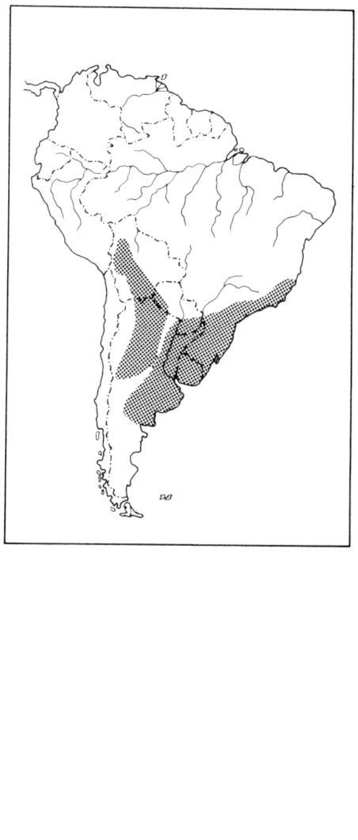

IDENTIFICATION: 20.5–23 cm (8–9"). A large olive finch of damp grasslands; *bill yellowish orange* (deeper orange in west) with blackish ridge. *Head olive gray, duskier on face,* becoming purer olive on rest of upperparts; back streaked blackish and wing-edging yellowish olive. *Below pale grayish,* becoming buffy whitish on belly. *Olivascens* (Bolivia to w. Argentina) is unstreaked above. Juvenal more coarsely streaked, especially above but also on breast.

SIMILAR SPECIES: See Pale-throated Serra-Finch. *Emberizoides* grassfinches have longer and more pointed tails and are more streaked above and not so gray on head and breast.

HABITAT AND BEHAVIOR: Generally common in tall grassy areas, often with scattered shrubs, including roadsides; in many regions prefers damp places. Usually quite conspicuous (more so than *Emberizoides*), often perching on fence posts or on top of low bushes. Flight rather weak and jerky, low over the grass. The song of the nominate race is a brief, gurgling, somewhat musical "gledit, gledit, gleeu" with many variations, including often a more sputtering longer ending.

RANGE: E. Bolivia (north to Beni) south to cen. Argentina (south to La Pampa and Río Negro), most of Paraguay (local in dry *chaco*), Uruguay, and se. Brazil (north to Minas Gerais and Espírito Santo). To about 2500 m (in Bolivia and higher mts. of se. Brazil).

NOTE: The reported (Nores, Yzurieta, and Miatello 1983: 96) overlap of the 2 supposed races of this species in Córdoba deserves further investigation; formerly these 2 forms were considered full species, and this may prove to be correct. If split, *E. olivascens* could be called Olive Pampa-Finch.

Embernagra longicauda

PALE-THROATED SERRA-FINCH

Other: Buff-throated Pampa-Finch

IDENTIFICATION: 21.5 cm (8½"). Very local in e. Brazil. Resembles Great Pampa-Finch (its geographically distant unstreaked w. form),

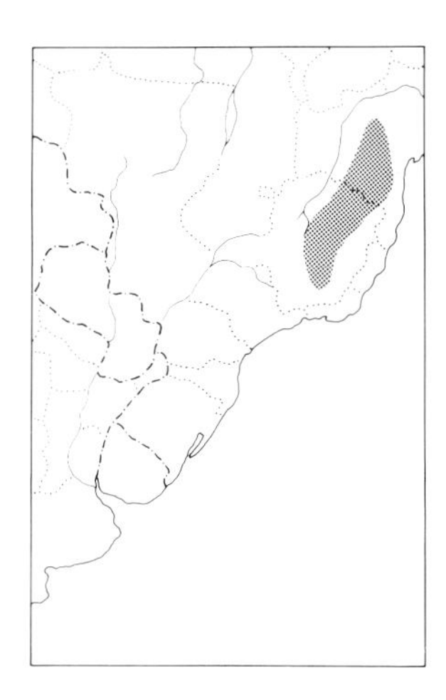

but has longer and narrower tail. *Head purer gray,* contrasting more with back, with white supraloral streak and lower eyelid; primaries and tail brighter olive; *median throat white* outlined by dusky malar area.
HABITAT AND BEHAVIOR: Not well known. Recently found in dry savannas and fields with scattered palms and ground bromeliads, mainly in high plateau country and on *serras* (G. Mattos and H. Sick); the closely related Great Pampa-Finch occurs in virtual sympatry with it, but always (in the area of overlap) at lower elevations and in damper areas. Behavior is similar to that of Great Pampa-Finch, but seems shyer. Its song is a loud and penetrating, "tsi, tsoweeé," repeated steadily at 4- to 5-second intervals, mostly during the early morning and late afternoon; may sing from an exposed perch, but at other times this can be a hard bird to locate. Has recently been found in some numbers at various points on the Serra do Espinaço in Minas Gerais (e.g., on the Serra do Cipó near Belo Horizonte), and also still found near the summit of the Morro do Chapeu in Bahia.
RANGE: Locally on *serras* in e. Brazil in interior cen. Bahia and Minas Gerais. 700–1300 m.

NOTE: The throat of this species (as seen in 3 AMNH specimens, and in life) is white and not buff, hence its previous descriptive English name ("Buff-throated") is inaccurate. Further, calling this bird a "pampa-finch" is grossly misleading, as (unlike its congener the Great Pampa-Finch) the species is not found in "pampas" but rather is typical of the serras of e. Brazil. We thus propose naming this interesting and scarce bird the Pale-throated Serra-Finch; we would have called it the White-throated Serra-Finch (in fact slightly more accurate) but for the possibility of confusion with the White-throated Sierra-Finch (*Phrygilus erythronotus*) of the Andes. For more details on the distribution and ecology of this species, until recently unknown in life, see G. T. Mattos and H. Sick (*Rev. Brasil. Biol.* 45 [3]: 201–206, 1985).

Emberizoides Grass-Finches

Fairly large finches with very long pointed tails. Tall grass and shrubbery in open country. We follow E. Eisenmann and L. L. Short (*Am. Mus. Novitates* 2740, 1982) in considering both *E. ypiranganus* and the little-known *E. duidae* as full species, though the status of the latter remains somewhat in doubt.

Emberizoides herbicola

WEDGE-TAILED GRASS-FINCH

PLATE: 30

IDENTIFICATION: 18–20 cm (7–8"). Bill mostly yellow, with black ridge. *Very long tail, graduated and extremely pointed.* Above light olive brown streaked black, wings more olive with bend of wing yellow (often concealed). Lores and conspicuous eye-ring whitish. Below whitish, tinged buffy on breast and sides. Nominate race (south of Amazon) slightly larger than n. *sphenurus* group; *apurensis* (of extreme e. Colombia and adjacent Venezuela) is the palest race above.
SIMILAR SPECIES: See Lesser and Duida Grass-Finches. Great Pampa-Finch is larger and has orange bill, grayer head and chest, less streaking above, and "normal" tail.

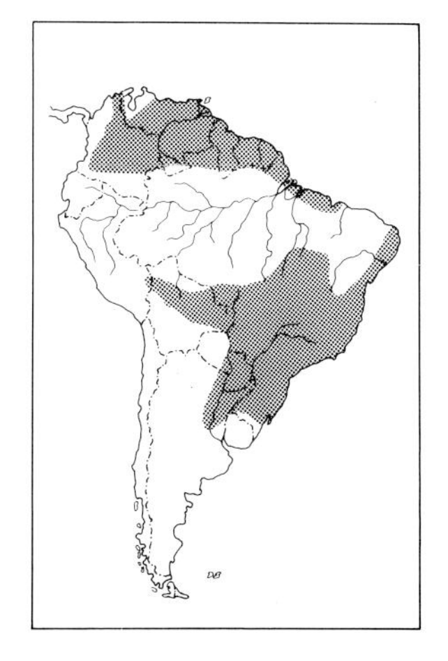

HABITAT AND BEHAVIOR: Uncommon to locally common (more numerous and widespread south of the Amazon) in taller grasslands, usually with scattered bushes; sometimes also in roadside verges (where the grass is tall). Unless singing usually not very conspicuous, remaining inside cover or on ground. Singing males often perch prominently on fences or low bushes; the song is a rather musical "tee, teedelee" with variations, and males also give more chattering calls.

RANGE: Colombia (locally in Santa Marta area, in Magdalena and Cauca valleys, and east of Andes south to Meta and Vaupés), Venezuela (local, but including grasslands on the Gran Sabana in Bolívar), the Guianas, and extreme e. Amaz. Brazil (Amapá, islands in the mouth of the Amazon, and n. Maranhão), and in e. and s. Brazil (south to Rio Grande do Sul), e. Paraguay, ne. Argentina (south to Santa Fe and Entre Ríos), n. and e. Bolivia, and extreme se. Peru (Heath Pampas in Madre de Dios). Also Costa Rica and Panama. Mostly below 1500 m.

Emberizoides ypiranganus

LESSER GRASS-FINCH

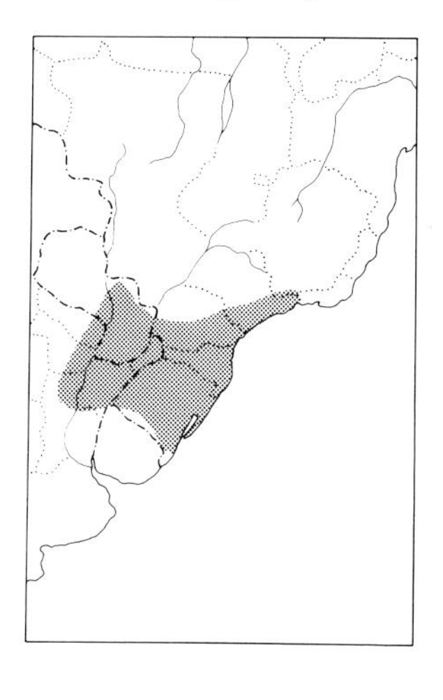

IDENTIFICATION: 18 cm (7″). Resembles Wedge-tailed Grass-Finch but *smaller* (especially than the locally sympatric nominate race of that species). More contrasty above, with heavier black streaking on a paler ground color. *Cheeks decidedly darker and grayer, contrasting with white throat.*

HABITAT AND BEHAVIOR: Uncommon and somewhat local in marshes and damp grassy areas; not found in drier, better-drained grasslands where Wedge-tailed Grass-Finch takes over. Behavior similar to that species'; usually rather furtive, but does respond to squeaking, when it may perch in the open. The song is a fast, chattering "ch, ch, ch-ch-ch-ch-ch-ch-ch-ch," very different from the more musical song of the Wedge-tailed Grass-Finch.

RANGE: Locally in se. Brazil (s. São Paulo south to Rio Grande do Sul), e. Paraguay, and ne. Argentina (Corrientes, ne. Santa Fe, and e. Formosa). To 900 m.

Emberizoides duidae

DUIDA GRASS-FINCH

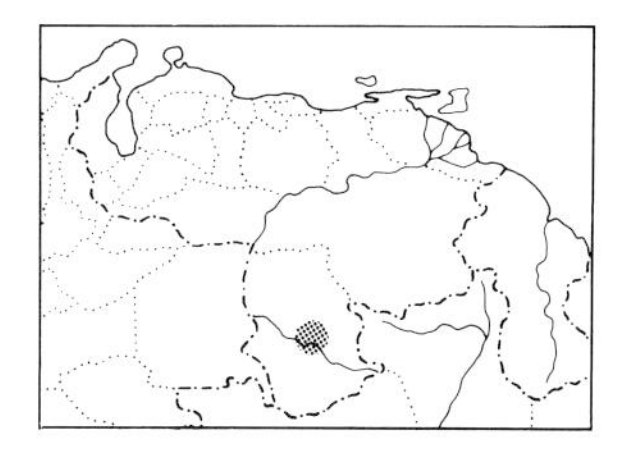

IDENTIFICATION: 21 cm (8¼″). *Amazonas, Venezuela.* Resembles Wedge-tailed Grass-Finch, but *larger with proportionately longer tail;* darker above, especially on crown and upperside of tail, with duskier cheeks.

HABITAT AND BEHAVIOR: Nothing recorded. Presumably similar to Wedge-tailed Grass-Finch.

RANGE: *Tepuis* of s. Venezuela (Cerro Duida in Amazonas). 1300–2100 m.

Coryphaspiza Finches

A boldly patterned finch with white in graduated tail. Local in less disturbed savannas with tall grass in s. and e. South America.

Coryphaspiza melanotis

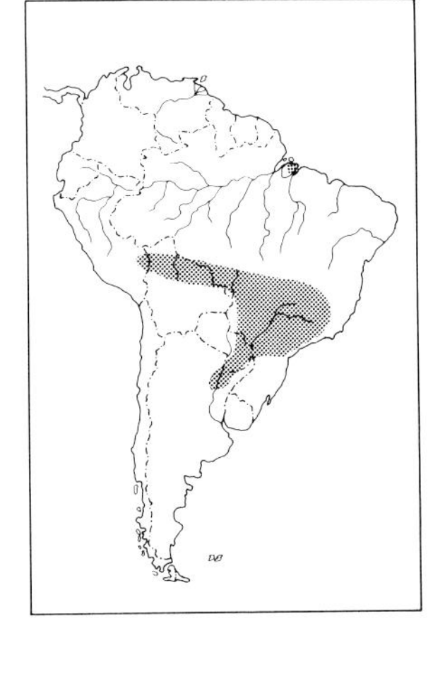

BLACK-MASKED FINCH

PLATE: 30

IDENTIFICATION: 13.5 cm (5¼"). *Bill conspicuously bicolored,* mandible yellow. *Crown and sides of head black with long white superciliary* and some gray on nape. Back olive brown broadly streaked chestnut; yellow shoulder and yellowish olive wing-edging. Below white, with black patch on either side of chest (usually mostly hidden). Tail *rather long and graduated,* outer feather *broadly tipped white* (often flashed in flight). Female has head more grayish (without male's black) and is dingier generally except for more prominent greenish yellow wing-edging, with less white in tail.

SIMILAR SPECIES: Male's snappy head pattern renders it virtually unmistakable; female can be more difficult, but should be known by overall shape and white in tail (and rarely is it found alone).

HABITAT AND BEHAVIOR: Locally common in tall grasslands, but now very local due to overgrazing and burning (overall numbers have doubtless declined considerably). Grasslands may have some scattered low bushes, but not many. Usually feeds on the ground under cover of tall grass, but sometimes perches in open (especially early or late in the day and when singing). The song is a very weak, thin, spiritless "tsees-lee," quite insectlike. Gathers in loose groups during nonbreeding season, then often associating with other grassland birds (e.g., Cock-tailed Tyrant, Grassland Sparrow).

RANGE: Locally from extreme se. Peru (Heath Pampas in Madre de Dios) east across n. Bolivia to interior cen. and s. Brazil (Mato Grosso, Goiás, and Minas Gerais south to São Paulo), and south through e. Paraguay to ne. Argentina (south to e. Chaco and ne. Santa Fe); also Marajó Is. in mouth of Amazon, Brazil. To about 1000 m.

Donacospiza Reed-Finches

A small buffy finch found in and near marshes in s. South America. Rather long slender tail. Probably most closely related to the *Poospiza* warbling-finches.

Donacospiza albifrons

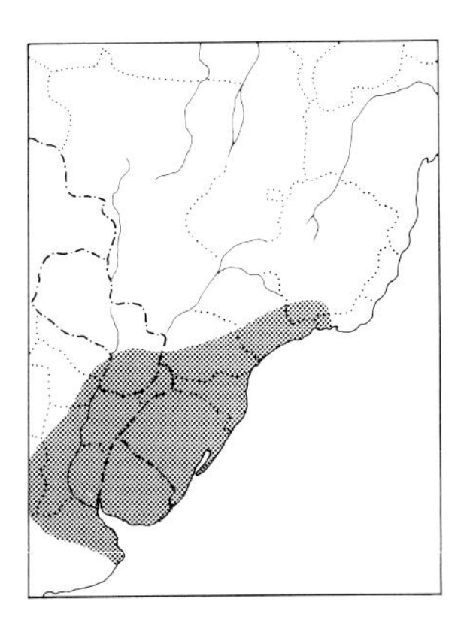

LONG-TAILED REED-FINCH

PLATE: 30

IDENTIFICATION: 15 cm (6"). *A mostly buffy brown, rather long-tailed finch of reedbeds and shrubbery near water. Short superciliary and half-moon under eye whitish.* Head grayish brown to gray, becoming rufescent brown streaked dusky on back. Shoulders bluish gray. *Underparts tawny-buff.* Female more broadly streaked above.

SIMILAR SPECIES: Grass-finches have much more prominent streaking above and conspicuous yellowish olive on wings, and they lack the warm buff below. Pattern is somewhat reminiscent of several warbling-finches, but relatively long tail and habitat are not.

HABITAT AND BEHAVIOR: Fairly common in reedbeds and tall grass and shrubbery in and near water. Regularly perches in the open on a grass stem or on top of a bush. Flight is relatively weak, low over reeds

or grass. Usually in pairs. Song is spritely and fast, a series of 2 simple notes seemingly repeated in an almost random fashion; overall effect is strongly reminiscent of some warbling-finches.
RANGE: Se. Brazil (north to s. Minas Gerais and Rio de Janeiro), se. Paraguay, Uruguay, and ne. Argentina (south to Buenos Aires and e. Córdoba). To about 900 m.

Saltatricula Chaco-Finches

A colorful, boldly patterned finch of scrub in the *chaco* of s.-cen. South America.

Saltatricula multicolor

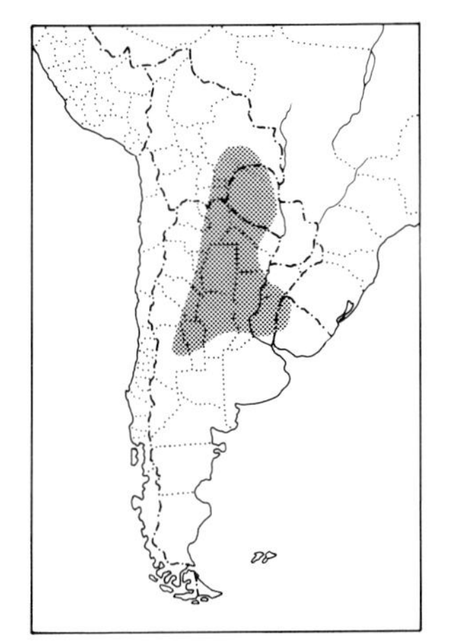

MANY-COLORED CHACO-FINCH PLATE: 30

IDENTIFICATION: 18 cm (7″). *N. Argentina, w. Paraguay, and adjacent areas. Mandible yellow.* Above mostly light sandy brown, with *prominent white postocular stripe* and *black mask from lores across face to sides of neck.* Center of throat white, band across chest pale gray. Breast and sides pinkish buff; center of belly white. Rather long, graduated tail dusky with *outer feathers broadly tipped white.*
SIMILAR SPECIES: Might conceivably be confused with female of larger Golden-billed Saltator.
HABITAT AND BEHAVIOR: Common in grassy and shrubby edges of dry *chaco* woodland. Found mostly in pairs or small groups of up to 6 to 8 birds, often in association with Red Pileated-Finches and other finches. Forages mostly on the ground, flushing up into low bushes when disturbed, where it often perches fully in the open; white in the tail is prominent, more so than any other sympatric finch. Male's song is a fast but spiritless "weeaweeawee," usually delivered from a low, partially hidden perch, often through the heat of the day.
RANGE: Se. Bolivia (north to s. Santa Cruz), w. Paraguay (only well west of Paraguay River), n. Argentina (south to Mendoza, San Luis, Córdoba, Santa Fe, and Entre Ríos), and nw. Uruguay (Paysandú). To 400 m.

Lophospingus Finches

A pair of mostly gray finches with obvious upstanding crests. Local in arid scrub of interior s. South America.

Lophospingus pusillus

BLACK-CRESTED FINCH PLATE: 30

IDENTIFICATION: 14 cm (5½″). Mandible yellow flesh. *Conspicuous upstanding black crest. Broad superciliary white,* sides of head black. Upperparts otherwise gray. *Center of throat black,* its sides broadly white. Underparts otherwise pale gray, center of belly and crissum white. Wing-coverts broadly tipped whitish; tail dusky, *with large white cor-*

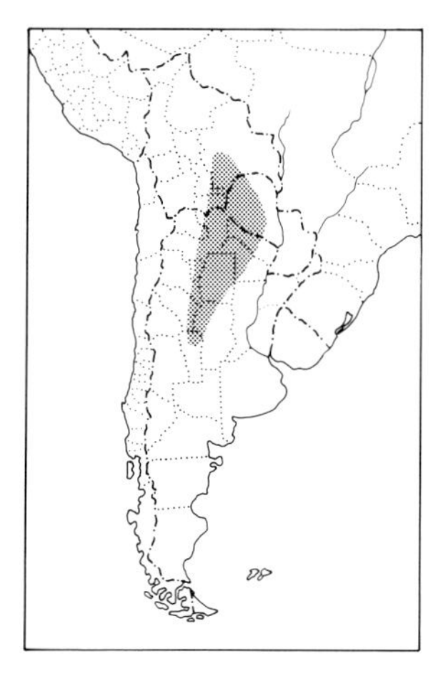

ners (conspicuous in flight). Female has less crisp pattern, with gray tinged brownish (especially above); it retains the *crest* but usually lacks the black throat patch.

SIMILAR SPECIES: Dapper males are virtually unmistakable, and females are sufficiently similar to be easily recognized. See Ringed Warbling-Finch (lacking the crest) and Gray-crested Finch (lacking the pattern and not known to be sympatric).

HABITAT AND BEHAVIOR: Generally uncommon and rather local in semiopen arid *chaco* scrub, often in association with sandy open areas; can be locally somewhat more numerous. Usually in small flocks (especially when not breeding) which feed on the ground or at roadsides, flushing when disturbed into low trees and shrubbery. At times consorts with other commoner *chaco* birds (e.g., Red Pileated-Finch, Red-crested Cardinal, etc.).

RANGE: Se. Bolivia (north to sw. Santa Cruz) to w. Paraguay (only well west of Paraguay River) and nw. Argentina (south to La Rioja, San Luis, and Córdoba; east to w. Formosa and Chaco). To 2200 m (in se. Bolivia), but mostly below 1000 m.

Lophospingus griseocristatus

GRAY-CRESTED FINCH

PLATE: 30

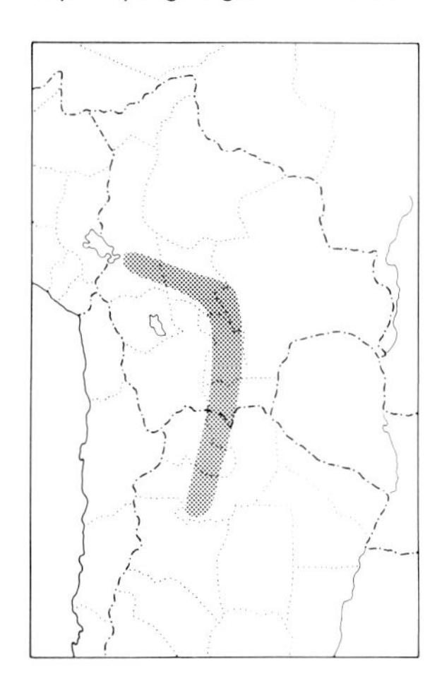

IDENTIFICATION: 14 cm (5½"). Mandible yellowish flesh. *Conspicuous upstanding crest. Mostly gray,* somewhat paler below, with white center of belly and crissum. Tail dusky, with *large white corners* (conspicuous in flight). Immature slightly browner.

SIMILAR SPECIES: A number of other finches are mostly gray, but this is the only one with a spikelike crest. See Black-crested Finch.

HABITAT AND BEHAVIOR: Common locally in arid intermontane valleys with xeric desertlike vegetation and cultivated fields along watercourses. Forages mostly on the ground, usually in the open (often at roadsides); generally in small flocks. Very numerous in w. Santa Cruz (Bolivia) along Santa Cruz–Cochabamba highway between Samaipata and Comarapa.

RANGE: W. Bolivia (from e. La Paz locally through Cochabamba and w. Santa Cruz and south to w. Tarija) and extreme nw. Argentina (locally in Salta and Jujuy). Mostly 1000–2500 m, locally lower (in s. Bolivia).

Charitospiza Finches

A distinctive finch of Brazilian *cerrado,* with usually laid-back crest and unmistakable pattern in male.

Charitospiza eucosma

COAL-CRESTED FINCH

PLATE: 30

IDENTIFICATION: 11.5 cm (4½"). *A boldly patterned small finch of the cerrado. Crown and crest black* (but crest usually held laid back), *contrasting with white ear-coverts and cheeks.* Back pale gray, wings and tail

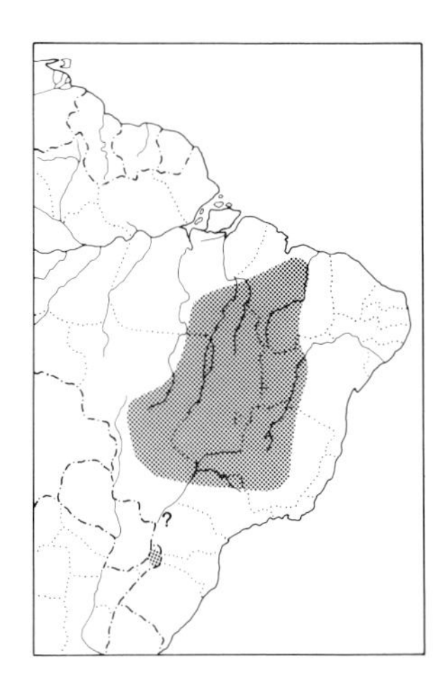

black with whitish wing-coverts, and *basal half of outer tail feathers white* (prominent in flight). *Throat and bib extending over center of chest black,* contrasting with *cinnamon buff lower underparts* (deeper chestnut on center of breast). Female quite different, duller and browner, with facial pattern only faintly echoed; below dull buffy, lacking black on throat and chest.

SIMILAR SPECIES: Male with its titlike pattern and laid-back crest is unique. Female known by its crest and white in tail (both as in male).

HABITAT AND BEHAVIOR: Uncommon to fairly common (but seemingly very local and erratic) in grassy *cerrado* with scattered bushes and low trees. Generally found in small groups of its own species, feeding on bare ground near grassy cover, flushing to perches in shrubbery. Possibly most numerous in *cerrado* which has been recently burned; local movements may be associated with the onset of the rainy season. May perhaps be at risk due to degradation and conversion of much of its *cerrado* habitat.

RANGE: Interior ne. and cen. Brazil from cen. Piauí, s. Maranhão, and se. Pará south through Goiás, w. Bahia, and w. Minas Gerais to se. Mato Grosso and nw. São Paulo; also recorded from Misiones in ne. Argentina. To about 1200 m.

Coryphospingus Pileated-Finches

A pair of small finches with distinctive flat red, black-bordered crests in males; otherwise plain red or mostly gray. Found in scrubby areas and lighter woodland. As the crests of both species are identical and characteristic of the genus, we feel it worthwhile to emphasize this by giving both species the group name "pileated-finch," preceding this by each species' dominant color. The two are known to hybridize where they come into contact with each other in Brazil (Sick 1984), but we prefer to maintain them as full species until more is known about the situation.

Coryphospingus cucullatus

RED PILEATED-FINCH

PLATE: 30

Other: Red-crested Finch

IDENTIFICATION: 13.5 cm (5¼"). *Crown mostly black,* usually at least partially concealing a *scarlet coronal stripe* (scarlet can also be exposed as bushy crest, with black spread to sides). *Above dark vinous red* (some, perhaps younger, birds browner) with rump more crimson; narrow eye-ring white. *Below dull crimson* (perhaps brighter in older birds). Female lacks black and red on crown and is browner above, but still with dull crimson rump and *narrow white eye-ring*. Throat whitish, *below rosy pink* (varying in brightness).

SIMILAR SPECIES: No other small finch is so uniformly reddish; for duller, browner females often the eye-ring is the best clue.

HABITAT AND BEHAVIOR: Common to sometimes locally abundant (especially in the *chaco*) in arid scrub, drier woodland, and agricultural

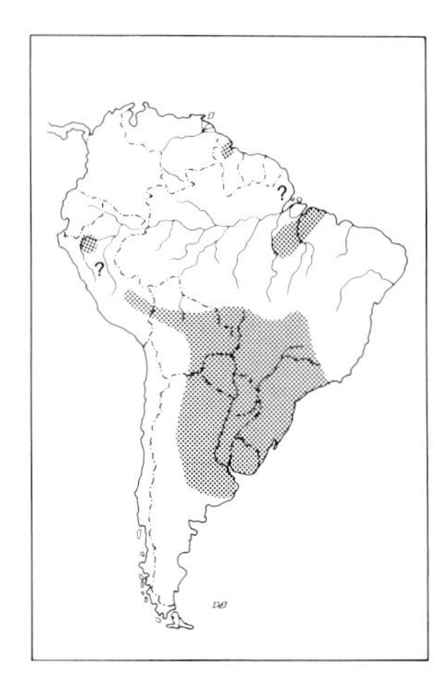

areas. Though widespread in semiopen and more arid parts of s. South America, the Red Pileated-Finch usually does not occur around habitations and thus is rarely as familiar as the Rufous-collared Sparrow. In nonbreeding season often gathers in large loose flocks, with which other finches often associate in smaller numbers. Usually forages on ground, especially at grassy borders but also within tangles and inside woodland. Male's song consists of a simple phrase repeated 3 to 6 times, with a short introductory and final note (e.g., "chewit, weet-chewit, weet-chewit, weet-chewit . . ."); it is given mostly soon after dawn.

RANGE: Locally in Guyana and ne. Brazil (along the lower Amazon upriver to the Rio Tocantins); much more widely in interior s. Brazil (north to cen. Mato Grosso, s. Goiás, w. Minas Gerais, and Rio de Janeiro) south and west through Paraguay and Uruguay to n. Argentina (south to ne. Buenos Aires, Córdoba, and La Rioja), n. and e. Bolivia, and locally north in arid intermontane valleys to n. Peru (recorded north to Marañón valley in n. Cajamarca, where resident). Mostly below 1500 m, locally higher in Andean valleys.

Coryphospingus pileatus

GRAY PILEATED-FINCH

PLATE: 30

Other: Pileated Finch

IDENTIFICATION: 13 cm (5"). *Crown and crest as in Red Pileated-Finch. Above mostly light gray* with narrow white eye-ring and whitish lores. Below whitish, breast and sides tinged light gray. Female lacks black and red on crown, but otherwise resembles male (*including the eye-ring*); some blurry grayish streaking below.

SIMILAR SPECIES: Female quite nondescript, but no other essentially grayish finch is quite so plain (and none has the eye-ring). Female seedeaters are buffier and browner and have thicker bills.

HABITAT AND BEHAVIOR: Fairly common to common in arid scrub and low woodland and in shrubby borders of more humid woodland; principally in grass at edges, such as roadsides. Often in small groups, especially when not breeding, but most often not associating with other species. Usually forages on ground, at edge or inside thickets, but frequently perches in the open.

RANGE: N. Colombia (Guajira Peninsula west locally to Santa Marta area and in upper Magdalena River valley) and most of n. Venezuela (south to n. Bolívar); e. Brazil (s. Maranhão and Ceará to cen. Mato Grosso, s. Goiás, Minas Gerais, and Rio de Janeiro). To 1000 m.

Sicalis Yellow-Finches

Small to mid-size, predominantly yellow finches of open country, where they feed on the ground, often in very large flocks. Identification is difficult, especially in the Andes, where several species are frequently sympatric or nearly so (though they rarely actually flock together).

GROUP A

Yellow-finches of the *arid Pacific slope* or *widespread in lowlands* (some locally higher).

Sicalis taczanowskii

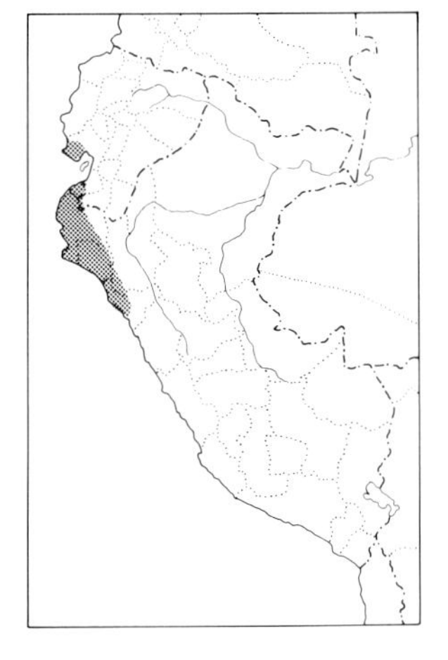

SULPHUR-THROATED FINCH

PLATE: 31

IDENTIFICATION: 12 cm (4¾"). *Sw. Ecuador and nw. Peru. Very heavy horn-colored bill.* Light grayish brown above, streaked darker brown on back. *Superciliary, malar streak, and upper throat pale yellow.* Below dull whitish, tinged yellow on crissum. Primaries edged pale yellow.
SIMILAR SPECIES: This nondescript finch looks sparrowlike, but nothing similar combines the stout bill with the yellow face markings (the yellow throat can be hard to discern, however). Appears short-tailed in the field. Note also its *characteristic habit of moving in large flocks.*
HABITAT AND BEHAVIOR: Seasonally common in open desertlike areas with scattered bushes and low trees; also in barren rocky regions almost devoid of vegetation (e.g., w. part of Ecuador's Santa Elena Peninsula), but not in absolute sand desert of Peru. Mostly near coast. Notably erratic, but can occur in flocks of several hundred or more birds. Flying flocks often very compact, but they usually spread out when feeding on the ground.
RANGE: Sw. Ecuador (Guayas, and 1 recent sighting from El Oro) and nw. Peru (Tumbes to La Libertad). Mostly below 200 m.

NOTE: Most recent authors have favored the inclusion of monotypic *Gnathospiza* in *Sicalis*. We agree, though its massive bill clearly sets it apart; its behavior is typical of the genus.

Sicalis flaveola

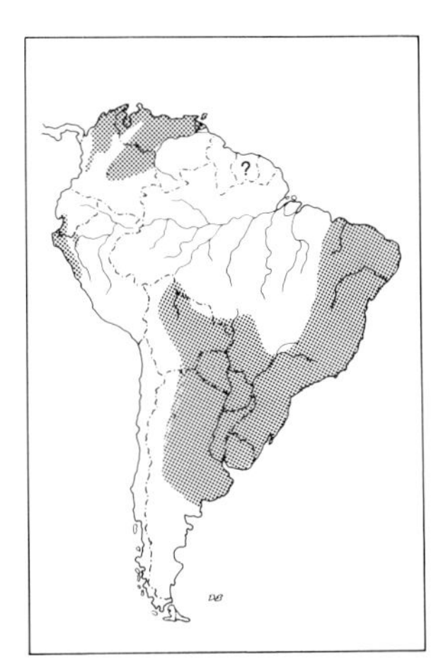

SAFFRON FINCH

PLATE: 31

IDENTIFICATION: 13.5–14 cm (5¼–5½"). The most *widespread and familiar* yellow-finch in the lowlands. *Mostly bright yellow,* becoming *orange on forecrown;* upperparts somewhat more olive, faintly streaked dusky on back. Female somewhat duller, with less orange on crown. Juvenal very streaky: brownish above streaked dusky, whitish below streaked brownish. More frequently seen are immatures; these are like juvenals but more yellowish above and more or less unstreaked below, with *yellow pectoral band* often extending up around nape as a nuchal collar. Male of s. race *pelzelni* (north to s. Brazil and e. Bolivia) is duller with less orange on crown, more olive and heavily streaked upperparts, and an olive wash across breast. Female *pelzelni* resembles streaky juvenals of n. races.
SIMILAR SPECIES: Learn this *Sicalis* well as a basis for comparison with other yellow-finches. Orange-fronted is the most similar but is notably smaller, etc.
HABITAT AND BEHAVIOR: Common to abundant in semiopen areas with scattered bushes and trees, agricultural regions, and many towns and cities (where it often feeds on lawns). Most numerous in drier regions. Usually forages on the ground; sometimes in large flocks. A popular cage bird, it has been introduced (accidentally?) to many places where it did not naturally occur. The song (apparently that of all

races is similar) is bright and lively but somewhat repetitious, an often long-continued series of musical notes and phrases (e.g., "chididi, tsee; chididi, tsee; chididi, tsee . . ." with variations).
RANGE: Three disjunct populations: n. Colombia (Caribbean lowlands and south in Cauca valley; also in ne. llanos south to Meta and Vichida), n. Venezuela (south to n. Bolívar), and the Guianas (where present status needs to be clarified, with few or no recent records); sw. Ecuador (north to Guayas) and nw. Peru (south to Ancash); n. and e. Bolivia (north to s. Beni) east across much of s. and e. Brazil (not in Amazon basin), and south through Paraguay and Uruguay to n. and cen. Argentina (south to Mendoza, La Pampa, and Buenos Aires); Trinidad. Also introduced to Panama, Puerto Rico, and Jamaica. To 2000 m (in Bolivia), but mostly below 1000 m.

Sicalis columbiana

ORANGE-FRONTED YELLOW-FINCH

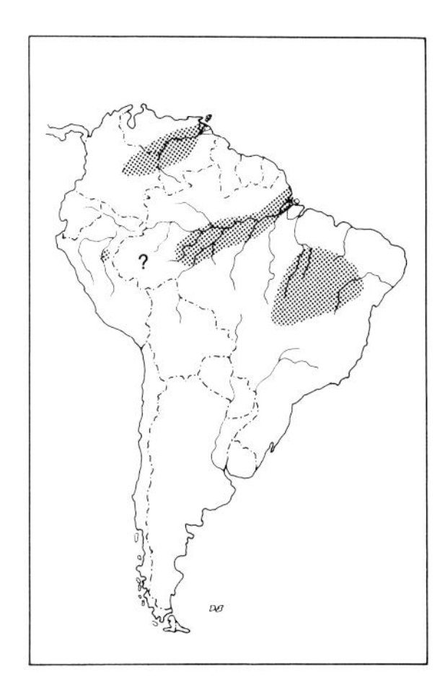

IDENTIFICATION: 11.5 cm (4½"). Resembles Saffron Finch, but *notably smaller* and with *virtually no streaking* (even in female or immature). Male mostly bright yellow, but upperparts bright olive yellow with orange forecrown and *dusky lores* (yellow in Saffron). Some birds (especially *goeldi* of Amazon basin) also have orange tinge on breast. Female olive brown above, whitish below with grayish buff tinge on sides of breast. See C-56, V-40. Some birds (immature males?) resemble immature Saffron Finch but are brighter olive yellow and unstreaked above.
HABITAT AND BEHAVIOR: Fairly common to common (but somewhat local) in shrubby riparian growth and around habitations; presence seems closely tied to water (rivers, lakes, and stock ponds in ranching country) and never seems to be found in towns like Saffron Finch. Usually in pairs or small flocks, foraging on the ground but perching freely in small trees or on ranch buildings.
RANGE: Locally in Orinoco River basin in ne. Colombia (south to Meta and Vichada) east across cen. and e. Venezuela (Apure, s. Guárico, and n. Bolívar east to Delta Amacuro); cen. Amazon basin in Brazil (lower Rio Tapajós area upriver to lower Rio Purús) and ne. Peru (only 1 old record from Loreto); interior e. Brazil (n. Goiás, w. Bahia, s. Piauí, and s. Maranhão; records from farther south in e. Brazil require confirmation); Trinidad (1 old record). Below 300 m.

Sicalis citrina

STRIPE-TAILED YELLOW-FINCH

PLATE: 31

IDENTIFICATION: 12 cm (4¾"). Male olive above with *forecrown contrastingly citrine* and streaked dusky on mantle; *inner web of outer 2 tail feathers mostly white on terminal half*; mostly bright yellow below, breast clouded olive. Female much duller: above more brownish and conspicuously streaked (including crown); below pale yellowish, *streaked dusky especially on breast;* tail as in male, but white patches somewhat smaller.
SIMILAR SPECIES: Resembles more widespread Grassland Yellow-Finch (and locally sympatric with it); the white in tail is visible mainly

from below and thus is hard to see even when flying. Better marks are Stripe-tailed's plain olive face (lacking Grassland's yellow ocular area) and its citrine forecrown (olive streaked dusky in Grassland). Female Grassland *lacks any streaking below;* female Stripe-tailed is the only yellow-finch with streaking below on a *yellow* ground color.

HABITAT AND BEHAVIOR: Fairly common but quite local in grasslands, open cultivated areas, sometimes (but not always) near water. Occurs in flocks, sometimes fairly large, particularly when not breeding; these forage on the ground. The song in Brazil is a fast sputtering which can be paraphrased as "switchity, switch-you, switch-you"; it is usually given from an elevated perch.

RANGE: Disjunctly in highlands of n. and w. Colombia (Santa Marta Mts., n. end of Cen. Andes in Antioquia, upper end of Magdalena valley, E. Andes on Bogotá plateau and on w. slope in Norte de Santander, and upper Patía valley in s. Cauca), Venezuela (Perijá Mts. in Zulia, coastal cordilleras from Carabobo to Miranda, and *tepuis* in cen. Amazonas and s. Bolívar), interior Guyana and Suriname, and extreme n. Brazil (Roraima); Andes in extreme s. Peru (Oconoque in Puno) and nw. Argentina (Tucumán); and locally (perhaps partially migratory) in interior Brazil (s. Pará and Piauí south to Mato Grosso and Paraná). Mostly 1000–3000 m.

Sicalis luteola

GRASSLAND YELLOW-FINCH

PLATE: 31

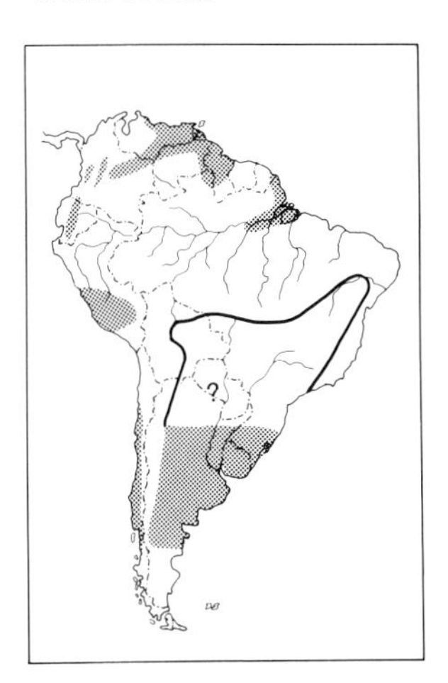

IDENTIFICATION: 11.5–12.5 cm (4½–5"). Male olive brownish above, streaked with dusky on crown and back, rump plain yellowish olive; *lores and ocular area bright yellow* (less prominent in *bogotensis* of Andes from Colombia to Peru); wings and tail brownish edged pale buff. Below bright yellow, tinged olive on breast (darker and more pronounced grayish olive on breast in larger *luteiventris* of s. South America). Female similar but browner above; throat, breast, and flanks light buffy brownish, belly yellow.

SIMILAR SPECIES: See Stripe-tailed and Raimondi's Yellow-Finches. Grassland's contrasting yellow on face is unique among *Sicalis* (less obvious in female *but still present*).

HABITAT AND BEHAVIOR: Locally common in tall grasslands (where they usually breed, sometimes loosely colonial), fields, and the edges of marshes. Often occur in large flocks, feeding mostly on the ground or in tall grass, flushing abruptly (sometimes en masse). Breeding males' song is a series of trills, some buzzy and others melodic, given both from low perches and in hovering display flights. Particularly numerous during austral summer on the pampas of Buenos Aires province, Argentina. Here males giving their brief display flight with fluttering wings and raised rump feathers can be seen on almost every roadside.

RANGE: Locally in lowlands of n. Colombia (south in Magdalena and Cauca valleys and in ne. llanos south to Meta and Vichada), much of lowland Venezuela, Guyana, and French Guiana (status in Guianas requires clarification) and ne. Brazil (along lower Amazon upriver to mouth of Rio Tapajós, perhaps also in Amapá); locally in highlands of

w. Venezuela (Mérida), Colombia (E. Andes in Boyacá and Cundinamarca, and in Nariño), Ecuador, and Peru (where also locally in Pacific lowlands); s. Brazil (Rio Grande do Sul), Uruguay, and n. and cen. Argentina (Corrientes and Santiago del Estero south to Río Negro), and in cen. Chile (Atacama south to Aysén). S. breeders apparently migrate northward during austral winter, reaching Paraguay, e. Bolivia, and e. Brazil (north to Mato Grosso and Bahia), but details not known, and may breed at least locally farther north. Also locally in Middle America and on some Lesser Antilles (introduced). To about 3000 m (Andean populations mostly 2000–3000 m).

NOTE: It has been suggested that more than 1 species may be involved. The range of the Andean form, *bogotensis* (Montane Yellow-Finch), lies entirely above that of true *luteola* of the lowlands of n. South America. No intergradation between the 2 has been shown (they seem unlikely to come into contact). However, the 2 forms are very close in plumage, and in the absence of information concerning differences in behavior or vocalizations we continue to regard them as conspecific. Geographically far-removed *luteiventris* (Misto Yellow-Finch) of lowlands of s. South America may also deserve full species rank, but again the evidence is inconclusive, and the little-known forms of the lower Amazon region (which may be intermediate) remain a complicating factor.

Sicalis raimondii

RAIMONDI'S YELLOW-FINCH

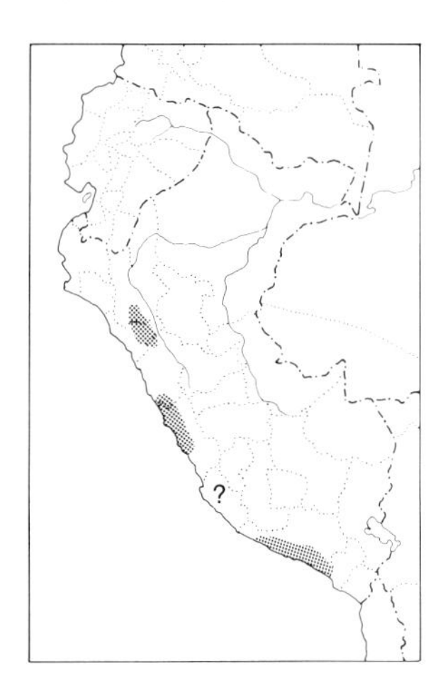

IDENTIFICATION: 11.5 cm (4½"). *W. Peru.* Closely resembles Grassland Yellow-Finch. More grayish brown above with finer dusky streaking. *No yellow on lores or ocular area.* Forehead, sides of crown, and nape yellowish, but *center of crown and ear-coverts gray.* Below yellow with gray on flanks. Female very dingy, essentially dull grayish with dusky streaking above.

SIMILAR SPECIES: Raimondi's Yellow-Finch is a confusing bird, long considered a subspecies of Grassland until they were shown to be sympatric in w. Peru. Most likely confused with that species, but see also Greenish Yellow-Finch (larger and more uniform in coloration, etc.).

HABITAT AND BEHAVIOR: Fairly common but very local and erratic on rocky slopes, often where there is little or no vegetation, sometimes in areas with sparse loma ground cover (scattered shrubs or grass), but rarely or never in agricultural land. At certain seasons may gather in flocks of hundreds of birds. Male's voice similar to Grassland Yellow-Finch's, but reportedly does not engage in display flights (M. Koepcke).

RANGE: W. Peru (locally from Cajamarca to Arequipa). Mostly 500–2000 m, but sometimes descends to coast.

GROUP B

Andes (Peru south) and/or *Patagonia.*

Sicalis uropygialis

BRIGHT-RUMPED YELLOW-FINCH

PLATE: 31

IDENTIFICATION: 14 cm (5½"). Crown and rump olive yellow, with *contrasting brownish gray back,* indistinctly streaked dusky; *lores and large cheek area gray.* Below bright yellow, breast clouded olive; sides and flanks gray. Wings and tail dusky, edged paler. Female similar but

duller; somewhat browner above and paler yellow below. *Sharpei* of n. and cen. Peru has gray cheek less well defined and lacks gray on sides and flanks.

SIMILAR SPECIES: Among high Andean *Sicalis,* only Citron-headed (of Bolivia) is as gray above and has as much gray on sides and flanks; it differs in having an entirely olive yellow head, with no gray on cheeks. In most of its range the gray cheeks are an excellent field mark, not shared by any other yellow-finch.

HABITAT AND BEHAVIOR: Fairly common to common in high *puna* grasslands, often on rocky slopes with interspersed patches of grass and open ground; sometimes found around habitations. Usually in small groups which forage on the ground, often with other species (especially various finches and furnariids) feeding nearby.

RANGE: Andes of Peru (north to Cajamarca), w. Bolivia (La Paz and Cochabamba to Potosí), n. Chile (south to Antofogasta), and nw. Argentina (Jujuy and Tucumán). Mostly 4000–4800 m northward, but lower in Chile (down to 2500 m).

Sicalis luteocephala

CITRON-HEADED YELLOW-FINCH

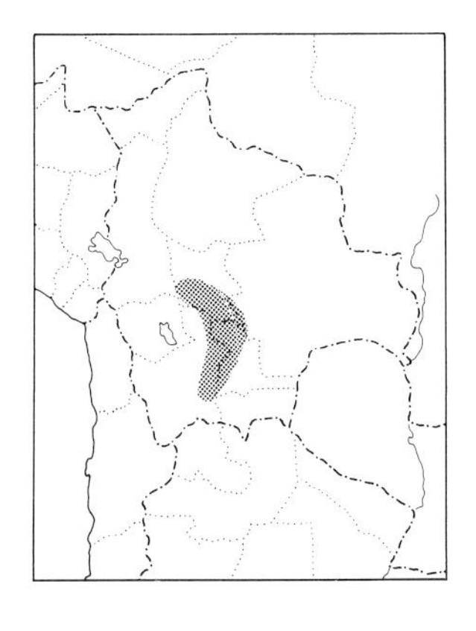

IDENTIFICATION: 14 cm (5½"). *Bolivia.* Recalls Bright-rumped Yellow-Finch but *entire face citron yellow* (no gray on cheeks), with brownish gray of back coming up onto nape and *extending down over rump.* Median underparts bright yellow, but *sides and flanks broadly gray.* Lower belly white. Female slightly duller and browner.

HABITAT AND BEHAVIOR: Locally fairly common on steep shrubby slopes with scattered fields and open grassy areas. Not found in *puna* grasslands of Bright-rumped Yellow-Finch. Usually in small groups, foraging on the ground in the open. Can be seen along the Cochabamba–Santa Cruz highway about 100 km east of the former.

RANGE: Highlands of w. Bolivia in Cochabamba, w. Santa Cruz, Chuquisaca, and Potosí. Mostly 2800–3500 m.

Sicalis lutea

PUNA YELLOW-FINCH

PLATE: 31

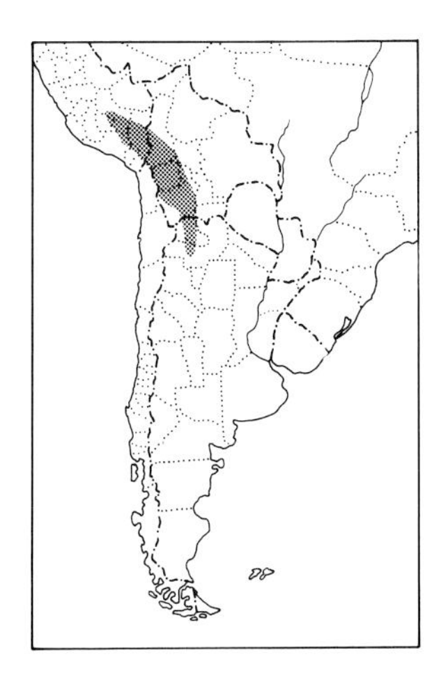

IDENTIFICATION: 13.5 cm (5¼"). *The brightest and most uniform of the high Andean yellow-finches. Above bright yellowish olive,* yellower on rump, but with wings and tail duskier. *Below bright yellow.* Female very similar to male (perhaps averaging a little more olive-tinged below).

SIMILAR SPECIES: Closely resembles male Greenish Yellow-Finch but yellower above. Female Greenish is much browner above; thus in a flock of Punas all birds (both sexes) look about the same, whereas in a flock of Greenish there are 2 plumages (male and female). There also are elevation and habitat differences.

HABITAT AND BEHAVIOR: Common in high *puna* grasslands and adjacent shrubby areas, usually on relatively level plains (not on steep slopes); unlike Bright-rumped and Greenish Yellow-Finches this species is rarely found in or around towns. Particularly numerous in the Titicaca basin, less so southward (quite scarce in Argentina). Usually

in small flocks, feeding on the ground, sometimes on roadsides.
RANGE: Andes of s. Peru (Cuzco and Arequipa south through Puno, Moquegua, and Tacna), w. Bolivia (recorded only Oruro and Potosí, but perhaps more widespread), and nw. Argentina (Jujuy and Salta). Mostly 3500–4300 m.

Sicalis olivascens

GREENISH YELLOW-FINCH

PLATE: 31

IDENTIFICATION: 14 cm (5½"). *Above rather dull olive,* but somewhat brighter on rump. *Below mostly olive yellow,* becoming purer yellow in center of belly. Female *considerably browner above* and often mostly rather drab grayish buff below (only throat and belly being yellow).
SIMILAR SPECIES: Males are more uniformly olive above than other *Sicalis*. Puna Yellow-Finch is similar but is much yellower generally and *its sexes are more or less alike*. Cf. also Greater Yellow-Finch.
HABITAT AND BEHAVIOR: Locally common in semiopen shrubby areas, often interspersed with scattered fields; usually not found in pure *puna* grasslands with no bushes. In Argentina cacti are often frequent where this bird is found. Can occur in large flocks, at times restless and wary, at others feeding unconcernedly on roadsides.
RANGE: Andes of Peru (north to upper Marañón valley in Huánuco and La Libertad, on w. slope from Ancash and Lima south, and in Cuzco), w. Bolivia (La Paz and Cochabamba to Potosí), n. Chile (south to Coquimbo), and w. Argentina (south to Mendoza and San Luis). Mostly 2500–3800 m.

Sicalis auriventris

GREATER YELLOW-FINCH

IDENTIFICATION: 15 cm (6"). *Closely* resembles Greenish Yellow-Finch and is only slightly larger than that species. Head and underparts rather bright yellow, *richer and brassier* than in Greenish. Back olive, vaguely and lightly streaked dusky. Wings and tail dusky, with *coverts and flight feathers broadly edged gray.* Female somewhat browner.
SIMILAR SPECIES: Besides Greenish Yellow-Finch (overlapping primarily in Chile), see also Bright-rumped Yellow-Finch (also only overlapping in Chile, and here the Bright-rumped has distinctive gray flanks, cheeks, and back).
HABITAT AND BEHAVIOR: Common in open shrubby and grassy areas on Andean slopes, apparently with some downward movement during winter. Also often seen around habitations, frequently nesting in buildings, stone walls, etc. Like the other yellow-finches, forages mostly on the ground, gathering in considerable flocks during nonbreeding season. Numerous in the Andes of Santiago, Chile, during the Nov.–Jan. nesting season.
RANGE: Andes of n. and cen. Chile (Antofogasta to Talca) and s.-cen. Argentina (Mendoza and w. Neuquén). Mostly 1800–2500 m, lower in winter.

Sicalis lebruni

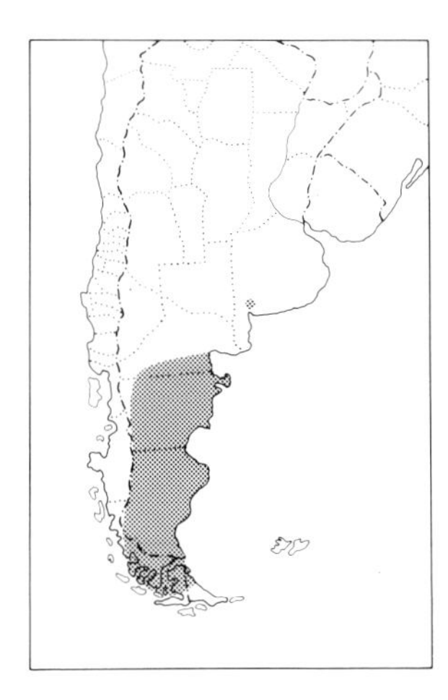

PATAGONIAN YELLOW-FINCH

IDENTIFICATION: 14 cm (5½"). *Patagonia.* Resembles Greenish Yellow-Finch (no overlap), but olive of *upperparts strongly suffused with gray* (sometimes giving streaked effect). Wing-edging broader and grayer (not olive yellow). Sides usually washed with gray (with little or no olive below). Female paler and duller.
SIMILAR SPECIES: The only yellow-finch in its far-southern range. Female *Melanodera* finches are generally streaked.
HABITAT AND BEHAVIOR: Uncommon on open grassy plains with low shrubs and in more sheltered areas with shrubs and low trees (at least when nesting). Occurs in pairs or small flocks, the latter principally when not breeding; forages entirely on the ground. Can usually be found in the Valdés Peninsula region (where it is the only *Sicalis* present).
RANGE: Extreme s. Chile (s. Magellanes) and s. Argentina (north to Río Negro and sw. Buenos Aires).

Rhodospingus Finches

Male's *red and black* plumage unmistakable, but nondescript female easily confused; note rather slender, pointed bill. Limited range mainly in w. Ecuador.

Rhodospingus cruentus

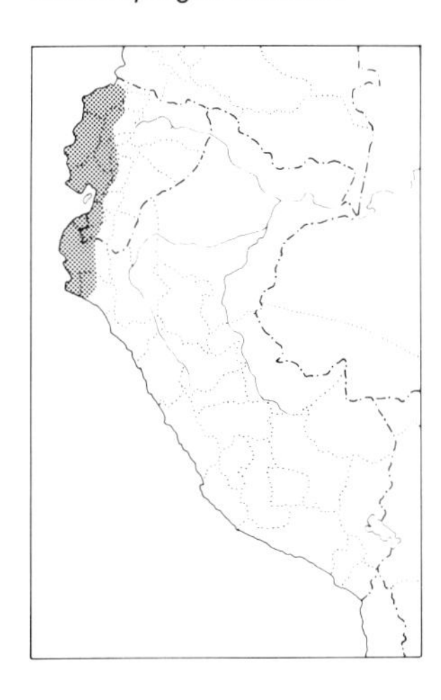

CRIMSON FINCH

PLATE: 31

IDENTIFICATION: 11 cm (4¼"). *Bill rather slender, sharply pointed.* Male unmistakable: *black above, pinkish scarlet below,* fading to buffy yellowish on lower belly; center of crown also pinkish scarlet. Birds in fresh plumage have pale edging to feathers of upperparts. Female light brown above; yellowish buff below, washed with light brown on sides and flanks. Immature male like female but with orange wash across breast (many are seen in this or a similar plumage).
SIMILAR SPECIES: Female resembles a female *Sporophila* seedeater, but note the sharply attenuated bill (not short and conical). Female Guira Tanager is larger, yellower below, and essentially arboreal.
HABITAT AND BEHAVIOR: Fairly common to locally common in low woodland and scrub, usually with rather dense grass cover, and usually in more arid regions. More arboreal when nesting (during Jan.–May rainy season), but at other seasons often gathers in large flocks which forage like, and often with, seedeaters. In these nonbreeding flocks males in good plumage are scarce or absent. Breeding males give a sibilant "tsee-tzztzz," rapidly repeated over and over, with quality similar to that of Blue-black Grassquit's. Much more numerous in Ecuador (especially Guayas and El Oro) than in Peru.
RANGE: W. Ecuador (w. Pichincha and coastal Esmeraldas south) and nw. Peru (south to Piura, rarely to Lambayeque). To about 300 m.

Fringillidae

CARDUELINE FINCHES

THE S. Am. members of this family (all in the genus *Carduelis*) are now considered to belong in the Cardueline subfamily in the Fringillidae, a family of finches which reaches its greatest diversity in Eurasia. They are apparently only distantly related to the predominantly New World Emberizidae. In South America they are found in semiopen habitats mainly in the Andes; 2 European species have been introduced into the southeast.

Carduelis Siskins and Goldfinches

Small finches, with males typically boldly patterned in black, yellow, and olive. Females often much duller, with some species difficult to identify. Very gregarious in semiopen to open terrain, with greatest diversity attained on the slopes of the Andes. Feed mostly in trees and shrubbery, sometimes dropping to the ground; note their distinctive bounding flight. We follow the 1983 AOU Check-list and most other recent authors in merging the genus *Spinus* into *Carduelis;* this has necessitated some alterations to the species name endings.

GROUP A "*Hooded*" siskins.

Carduelis magellanica

HOODED SISKIN PLATE: 31

Other: Santa Cruz Siskin

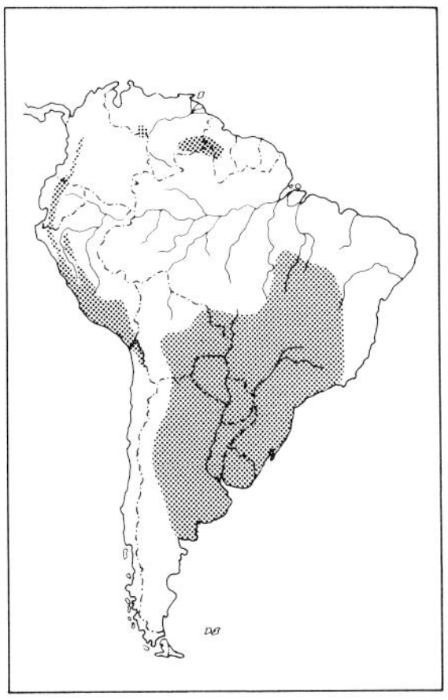

IDENTIFICATION: 11–12 cm (4¼–4¾"). *The most widespread and generally the commonest S. Am. siskin. Hood black* contrasting strongly with olive back and bright yellow underparts. Rump usually yellow, but more or less concolor with back in *capitalis* of n. Andes (south to n. Peru). Olive back often spotted with dusky or black (this most marked in s. Andes). Wings black with *yellow band on greater coverts and another across base of flight feathers* and some white edging on tertials; tail black with yellow at base. Female much duller, with no black on head: essentially olive above but more yellowish on rump; olive yellowish below, becoming brighter on belly (but *more grayish white below* in females from Colombia, Ecuador, and most of Peru).

SIMILAR SPECIES: Learn this species well as a basis of comparison with other less numerous siskins. Female Andean Siskin is similar but in narrow area of overlap (s. Colombia, n. Ecuador) Hooded has whitish (not yellowish) underparts.

HABITAT AND BEHAVIOR: Common and widespread in semiopen or cultivated areas with scattered trees; also often in gardens around houses and in city parks. Most often seen in small groups, foraging at all levels. Often quite tame.

RANGE: Andes in sw. Colombia (mostly Nariño, a few records north in Cen. Andes to Caldas), Ecuador, Peru (also in Pacific lowlands from sw. Ecuador to Arequipa, Peru), and n. Chile (Arica and Tarapacá); lowlands and Andean slopes in e. and s. Bolivia (north to Cochabamba and Santa Cruz), s. and cen. Brazil (north to se. Pará and Piauí), Paraguay, Uruguay, and n. and cen. Argentina (south to n. Río Negro); *tepuis* of s. Venezuela (Bolívar and Amazonas) and adjacent Guyana and extreme n. Brazil (Roraima). Locally to about 4000 m.

NOTE: We follow Short (1975) in considering *santaecrucis* (Santa Cruz Siskin) as a race of *C. magellanica*. It is not very different from neighboring subspecies, and as presently understood its distribution does not coincide with many other taxa having achieved full species status. On the other hand, *C. siemeradzkii* does appear distinct, as it differs quite markedly from the geographically closest races of *C. magellanica* (*capitalis* and *paula*) and no intermediates are known. The case of *C. olivacea* remains less clear: morphologically it is not very different from *magellanica*, but its range and habitat are divergent. We thus tentatively continue to regard it as a species.

Carduelis siemiradzkii

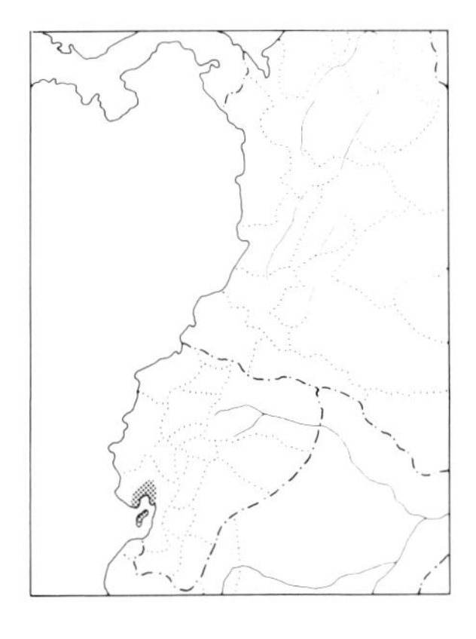

SAFFRON SISKIN

IDENTIFICATION: 11 cm (4¼"). Sw. Ecuador. Closely resembles Hooded Siskin, but *both sexes are much yellower*. Male has *back clear olive yellow* (not olive, and without the blackish mottling of the 2 Hooded Siskin races with which it might come into contact, *capitalis* of the highlands and *paula* to the south).

HABITAT AND BEHAVIOR: Rather rare in arid scrub and deciduous woodland. Not often found in recent years, but has been seen in Chongon Hills west of Guayaquil. Usually in pairs. Behavior virtually identical to Hooded Siskin's.

RANGE: Sw. Ecuador (Guayas and Puna Is.; possibly more widespread?). To only 100 m.

NOTE: See comments under Hooded Siskin.

Carduelis olivacea

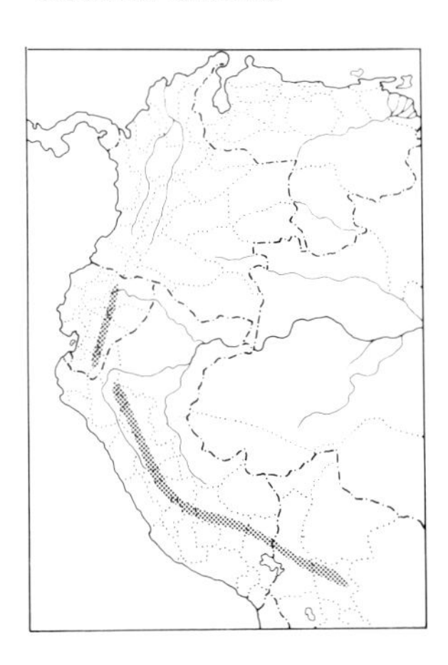

OLIVACEOUS SISKIN

IDENTIFICATION: 11 cm (4¼"). *Very* closely resembles Hooded Siskin (identifiable mostly by range and habitat). Both sexes have yellow underparts quite *strongly tinged olive* and are *slightly* smaller than the various races of Hooded Siskin found at higher elevations above the range of this species.

HABITAT AND BEHAVIOR: Uncommon and somewhat local in clearings with scattered trees and at edge of subtropical forest. Other than the habitat difference, behavior virtually identical to Hooded Siskin's (of which it may be only a race).

RANGE: E. slope of Andes in Ecuador (north to Napo; likely also in extreme se. Colombia), Peru, and nw. Bolivia (south to Cochabamba). Mostly 1200–3000 m.

NOTE: See comments under Hooded Siskin.

Carduelis crassirostris

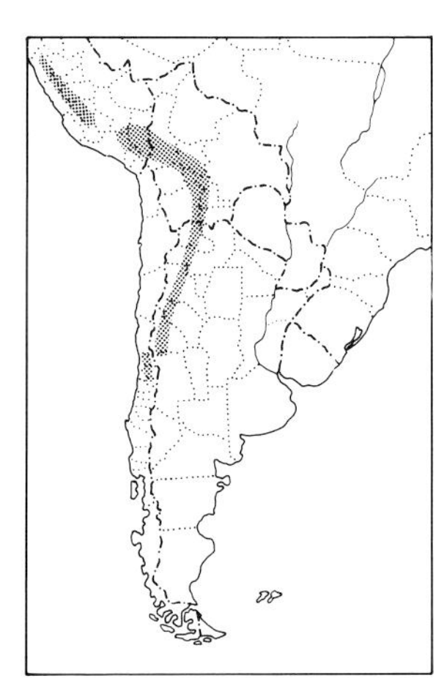

THICK-BILLED SISKIN

PLATE: 31

IDENTIFICATION: 13.5–14 cm (5¼–5½"). *Bill much thicker than in other siskins;* usually recognizable on that basis alone. Plumage resembles Hooded Siskin's. Male has *center of belly white.* Female very dingy and grayish, with whitish belly. Both sexes of slightly larger *amadoni* (sw. Peru) are even duller, male with less yellow (dull olive above including rump, *more grayish white below*), female even grayer (less olive) above.
SIMILAR SPECIES: Good-plumaged males seem infrequent (few are as bright as a typical male Hooded). Hooded Siskin is sympatric in some areas but has much slighter bill. Yellow-rumped Siskin not only has smaller bill but also blackish (not olive) back and black chest.
HABITAT AND BEHAVIOR: Uncommon to rare and apparently local in patches of sparse *Polylepis* woodland (*amadoni*) and also brushy slopes and ravines. Usually in pairs or small groups.
RANGE: Andes of w. and s. Peru (locally from Ancash and Pasco south to Arequipa, Tacna, and Puno), w. Bolivia (recorded only from Cochabamba and Potosí, but probably more widespread), w. Argentina (Jujuy south locally to Mendoza), and locally in cen. Chile (Aconcagua and Santiago); *amadoni* may also occur in extreme n. Chile and adjacent w. Bolivia. Mostly 3000–4000 m, lower southward.

Carduelis cucullata

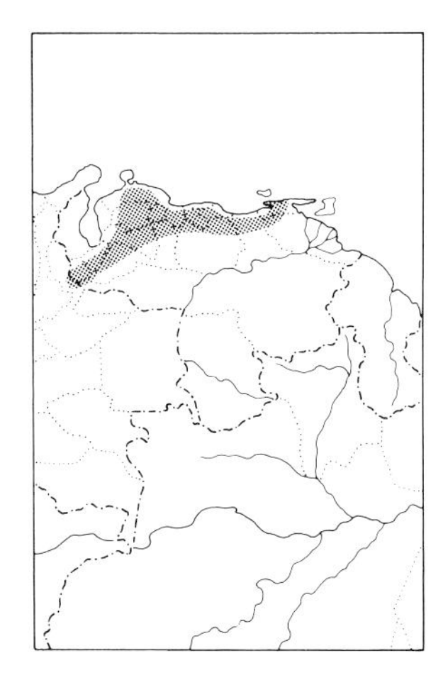

RED SISKIN

PLATE: 31

IDENTIFICATION: 10 cm (4"). *Venezuela and ne. Colombia.* Male unique: *mostly bright orange-vermilion,* somewhat darker on back; hood, wings, and tail black, with band across base of primaries pale orange-vermilion. See V-40, C-56. Female brown above, redder on rump; grayish white below, with *salmon red wash on sides of breast* (the red sometimes more extensive); wings and tail dusky, *wings marked as in male but paler.*
SIMILAR SPECIES: Pattern of male identical to Hooded Siskin, but color renders confusion impossible. Female shows enough salmon red (same color as female Vermilion Flycatcher) to be readily recognizable.
HABITAT AND BEHAVIOR: Rare in semihumid woodland and forest borders when breeding, at other seasons wandering through open grassy areas with scattered bushes and low trees. Usually in small flocks. Numbers in most areas much depleted by trapping; they are highly popular cage birds and are also greatly desired for interbreeding with domestic canaries. Trapping activities have caused the species to be given endangered status.
RANGE: Locally in n. Venezuela (Sucre and Monagas east to Falcón, Lara, Portuguesa, and Mérida, mostly on slopes of coastal cordilleras) and extreme ne. Colombia (Norte de Santander); Trinidad (apparently extirpated). Puerto Rico (introduced). Mostly 300–1200 m.

GROUP B "*Capped*" and "*black-mantled*" siskins.

Carduelis yarrellii

YELLOW-FACED SISKIN

PLATE: 31

IDENTIFICATION: 10 cm (4"). *Mostly ne. Brazil. A very yellow siskin.* Male has *black crown,* contrasting with *bright yellow face and underparts;* back olive yellow with some blackish spotting, but rump yellow. Wings black with yellow markings more or less as in other siskins. Female rather bright yellowish olive above, rump notably yellower; below quite bright yellow; wings as in male.

SIMILAR SPECIES: Limited or no range overlap with other siskins. Most resembles Andean Siskin (no overlap), but that is a much more olive bird generally.

HABITAT AND BEHAVIOR: Little known. Specimens collected by E. Kaempfer in the 1920s were taken in regions with "fine old forests and coffee plantations." It has been suggested that the species may be at risk from cage bird trappers.

RANGE: Interior ne. Brazil (Ceará and Paraíba to n. Bahia); n. Venezuela (specimens from 2 localities, Hacienda El Trompillo and Pira-Pira, Hacienda La Araguata, both in se. Carabobo; A. Altman has suggested that Venezuelan records may refer to escaped cage birds).

Carduelis spinescens

ANDEAN SISKIN

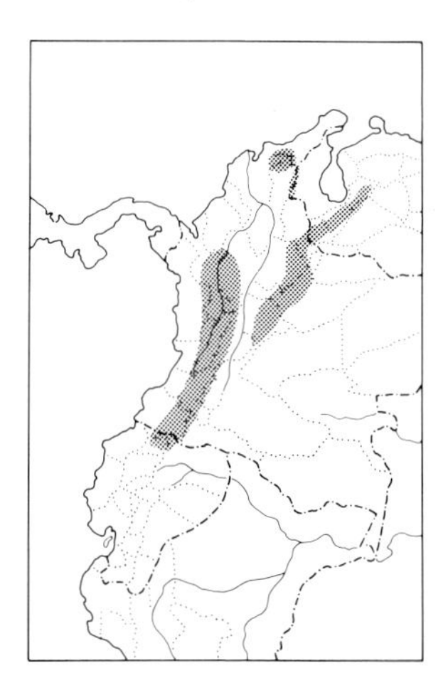

IDENTIFICATION: 10 cm (4"). *Mainly mts. of Venezuela and Colombia.* Resembles Yellow-faced Siskin in pattern, but *much more olive generally* (with less yellow). Male has *black crown;* otherwise essentially dark olive above, yellowish olive below; rump not yellower than back. Yellow wing- and tail-markings as in other siskins except that *nigricauda* (Colombia's W. and N.-Cen. Andes) has all-black tail. Female lacks black cap and is duller, with whitish mid-belly. See C-56, V-40.

SIMILAR SPECIES: Hooded Siskin has entirely black hood (not just crown). Female Hooded is similar to female Andean but in area of overlap (s. Colombia and n. Ecuador) has grayish white (not mostly yellowish olive) underparts. Female Yellow-bellied Siskin also resembles female Andean but is slightly larger with more 2-tone underparts (olive throat and breast, yellow belly). Female siskins usually occur with males, which reduces potential confusion.

HABITAT AND BEHAVIOR: Fairly common in open areas with scattered trees, agricultural regions, forest borders, and *páramo* with *Espeletia* sp. (often feeding on the latter). Generally feeds near ground, but may perch high. Most often in small flocks.

RANGE: Coastal mts. of n. Venezuela (ne. Aragua at Colonia Tovar) and Andes of w. Venezuela (Trujillo to Táchira), Colombia, and extreme n. Ecuador (photographed in Carchi in Jan. 1982 by S. Greenfield et al.); Santa Marta Mts. of Colombia and Perijá Mts. on Colombia-Venezuela border. Mostly 1800–3700 m.

Carduelis barbata

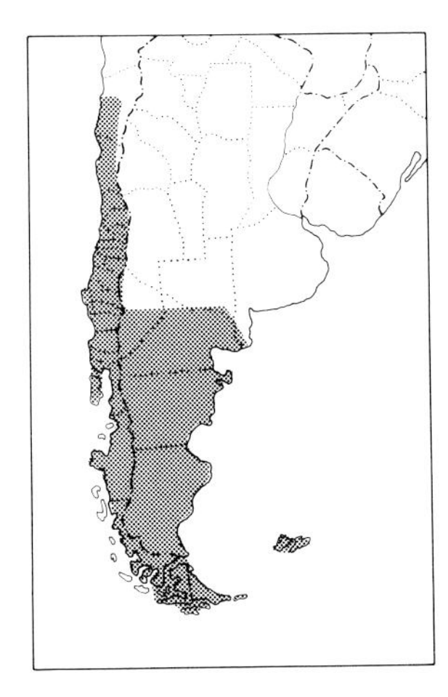

BLACK-CHINNED SISKIN

PLATE: 31

IDENTIFICATION: 13 cm (5"). *Chile and s. Argentina.* Somewhat heavier bill than other siskins' (except Thick-billed). Male has *crown and center of throat black.* Otherwise yellowish olive above with blackish streaking; rump somewhat yellower. Mostly bright olive yellow below, but lower belly whitish. Wings black with 2 yellowish bars and yellow patch on primaries. Female lacks black on crown and throat. Above like male but duller and more grayish olive, with vague yellowish superciliary; below pale yellowish olive, becoming whiter on belly.
SIMILAR SPECIES: *Occurs farther south than any other siskin.* Confusion most likely with female Hooded Siskin, which is somewhat smaller and shows more or less distinct yellow rump; male has full black hood. Cf. also Yellow-rumped Siskin.
HABITAT AND BEHAVIOR: Common in forest borders, lightly wooded areas, and trees and bushes around habitations and in towns. Most often seen in flocks, sometimes large (especially in winter); regularly forages on the ground.
RANGE: Chile (Atacama south) and s. Argentina (Neuquén and Río Negro south) to Tierra del Fuego; Falkland Is. To about 1500 m.

Carduelis uropygialis

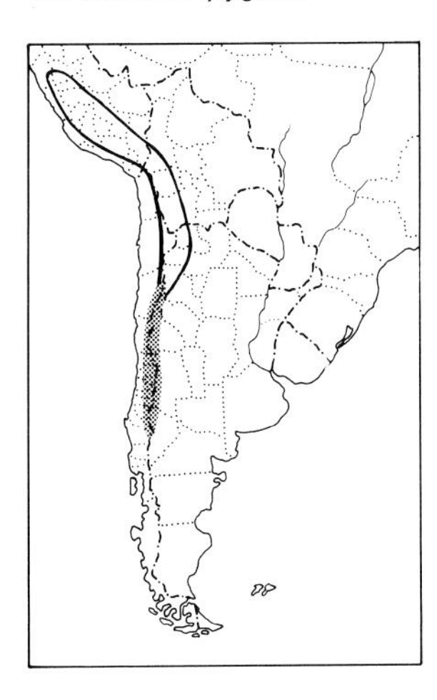

YELLOW-RUMPED SISKIN

PLATE: 31

IDENTIFICATION: 13 cm (5"). *Mainly Chile and Argentina,* a few migrating northward. *Mostly sooty black,* but *rump, breast, and belly bright yellow.* Band across base of flight feathers and basal half of tail also yellow. Sexes nearly alike, male being somewhat deeper black and brighter yellow; on most birds back feathers are somewhat edged olive, giving mottled effect.
SIMILAR SPECIES: Black Siskin lacks yellow rump and has more extensive black on underparts (only lower belly is yellow). Yellow-bellied Siskin male somewhat similar but also lacks yellow rump; it is found at lower elevations in humid forest border situations (not in high, semi-open shrubby areas).
HABITAT AND BEHAVIOR: Rare (in Peru and Bolivia) to quite common (in Chile) on open shrubby slopes and in ravines. Usually not found sympatrically with Black Siskin, which is found mostly at higher elevations, in *puna* grasslands on the altiplano. Generally in small groups, feeding in bushes or on the ground.
RANGE: Andes of cen. Peru (north to Lima and Huancavelica) south locally through w. Bolivia (recorded only La Paz and Potosí, but probably occurs elsewhere in altiplano) to cen. Chile (south to Bio-Bio) and cen. Argentina (south to Mendoza). In Peru and Bolivia apparently an austral migrant, recorded only Apr.–Oct. Mostly 2000–4000 m, but once (Aug. 1968) seen near sea level at Mollendo, Peru (R. Hughes).

Carduelis xanthogastra

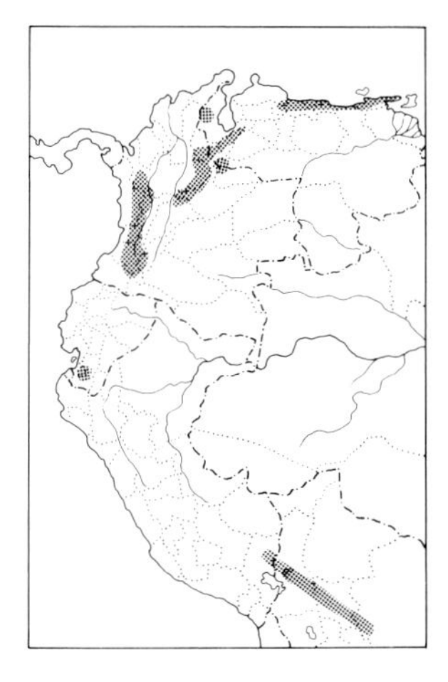

YELLOW-BELLIED SISKIN

IDENTIFICATION: 11.5 cm (4½"). *Locally in subtropics from Venezuela to Bolivia.* Male *mostly black* with sharply contrasting *yellow breast and belly.* Bold yellow band at base of flight feathers and at base of tail. Female has olive replacing male's black but echoes pattern of male in having yellow breast and belly; center of lower belly white; wings and tail as in male. See C-56, P-31.

SIMILAR SPECIES: Occurs lower and in different (forested) habitat from Yellow-rumped Siskin; males are superficially similar but lack yellow on rump. Male Lesser Goldfinch is all yellow below. Female Yellow-bellied resembles female Hooded and Andean Siskins but is darker olive above, and usually shows more contrast between olive and yellow on upperparts.

HABITAT AND BEHAVIOR: Fairly common in forest borders and in clearings with scattered trees and bushes; usually not in mostly open or deforested country. Often somewhat local and erratic. Usually in small groups, which generally perch and forage well up in trees, not often coming to ground.

RANGE: Coastal mts. of n. Venezuela (Paria Peninsula west to Yaracuy) and in Andes of w. Venezuela (Mérida south), locally in Colombia (not Nariño), sw. Ecuador (only recorded from El Oro, but perhaps more widespread), se. Peru, and nw. Bolivia (La Paz to Santa Cruz); Perijá Mts. on Colombia-Venezuela border. Also Costa Rica and w. Panama. Mostly 1500–2500 m.

Carduelis atrata

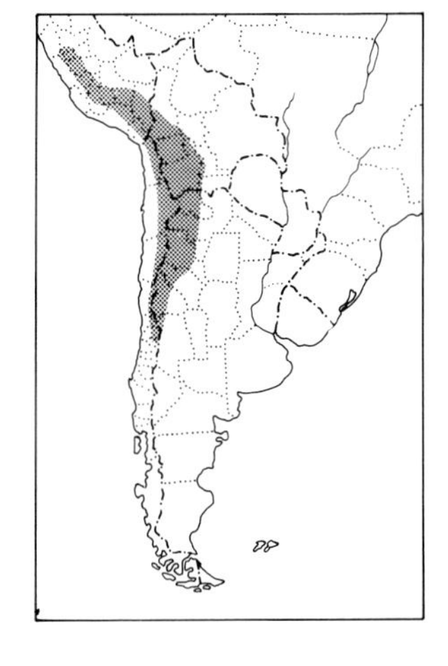

BLACK SISKIN

PLATE: 31

IDENTIFICATION: 13 cm (5"). *High Andes* of Peru to nw. Argentina and n. Chile. *Mostly glossy black* with bright yellow lower belly and crissum; band across base of flight feathers and basal half of tail also yellow. Sexes alike in pattern, but female a browner, sootier black.

SIMILAR SPECIES: This striking siskin is liable to be confused only with Yellow-rumped (which has yellow rump and shows much more yellow below); beware all the yellow in Black's wings and basal tail, which in fluttering flight can look deceptively like a yellow "rump."

HABITAT AND BEHAVIOR: Fairly common in high *puna* grasslands, on rocky slopes, and in ravines. Also quite frequent around habitations. Usually in pairs or small flocks, often foraging on the ground.

RANGE: Andes of cen. Peru (north to Huánuco and Lima), w. Bolivia (La Paz to Potosí), n. Chile (south to Antofogasta, rarely to Santiago), and w. Argentina (south to Mendoza). Mostly 3500–4500 m.

Carduelis psaltria

LESSER GOLDFINCH

Other: Dark-backed Goldfinch

IDENTIFICATION: 10 cm (4"). *Male glossy black above and bright yellow below.* Wings with *white* patch at base of primaries and edging on ter-

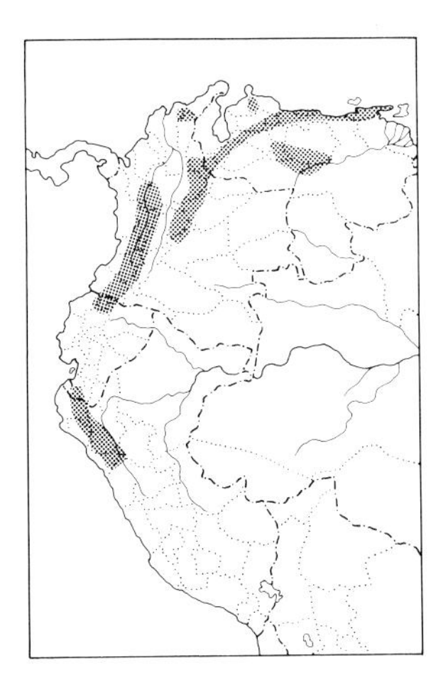

tials. Female olive above, wings brownish dusky but marked as in male; dull yellow below. See C-56, V-40.

SIMILAR SPECIES: All other S. Am. *Carduelis* have yellow (not white) in wing. Pattern of several male *Euphonia* is reminiscent of male, but all have yellow on forecrown.

HABITAT AND BEHAVIOR: Fairly common to common in semiopen or agricultural areas and around towns and gardens. Usually in pairs or small groups and generally conspicuous and tame. Has a somewhat disjointed, musical, twittery song, but more often heard are its clear "kleeu" call and a rather grating "ch-ch-ch-ch," most often given in flight.

RANGE: N. Venezuela (widely north of Orinoco River, locally south of it in n. Bolívar), w. Colombia (not east of E. Andes), locally in w. Ecuador (south to Pichincha and in El Oro), and nw. Peru (south to La Libertad, both on w. slope and in Río Marañón drainage). Also w. United States to Panama. Mostly 500–2500 m.

Introduced Waxbills, Sparrows, and Finches

A group of 4 species, all of them introduced into e. South America in the late 19th and early 20th centuries. The waxbill is a member of the Estrildidae family, native to tropical parts of the Old World, while the sparrow is classified in the Passeridae, found widely in the Old World. Both species remain closely associated with humans, the waxbill in parks, the sparrow around buildings in towns and cities. The 2 finches are both in the same genus as the native S. Am. siskins, all in the Cardueline subfamily of the Fringillidae.

Estrilda astrild

COMMON WAXBILL

PLATE: 31

IDENTIFICATION: 11.5 cm (4½"). *Locally in e. Brazil. Vivid red bill* and *red streak through eye onto ear-coverts.* Above brown, faintly vermiculated darker. Below lighter brown, also faintly vermiculated darker (except on whitish throat), with crimson stain on mid-belly; crissum black.

SIMILAR SPECIES: No native S. Am. bird has the red bill and streak through the eye.

HABITAT AND BEHAVIOR: Fairly common in parks and gardens, usually in urban areas. Usually quite unobtrusive and generally seen in groups feeding quietly on lawns, sometimes near *Sporophila* seedeaters and other finches.

RANGE: Locally in Brazil from Amazonia (Manaus and Belém) south mostly along e. coast in various cities to Porto Alegre, Rio Grande do Sul; also in Brasilia, and in other cities in s. Mato Grosso, Minas Gerais, etc. Apparently introduced into São Paulo around 1870 (possibly earlier?), with most other populations probably founded by accidentally escaped cage birds (H. Sick). Native to Africa; introduced locally elsewhere.

Passer domesticus

HOUSE SPARROW

IDENTIFICATION: 15 cm (6"). *"The" sparrow;* to most, so familiar as to almost obviate the need for a description. Male with *gray crown, chestnut nape, white cheeks and sides of neck,* and *black bib on throat and chest;* otherwise mostly brownish above streaked black, but rump gray; below whitish. Female nondescript: light grayish brown above with back streaked dusky, and *whitish superciliary;* below dingy whitish. Both sexes show 1 white wing-bar, more indistinct in female.
SIMILAR SPECIES: In South America can almost always be known by its commensal behavior with humans and their buildings. Males, which actually can be quite dapper, should be easily recognized, while the drab females can usually be known by their company and their surroundings.
HABITAT AND BEHAVIOR: Locally common around towns and cities; in South America usually not in purely agricultural regions. Is currently colonizing many urban areas from which it was not previously known, a process in which it is apparently not now being directly assisted by humans (i.e., it is dispersing on its own); seems likely to continue to spread. Quite noisy and aggressive: when House Sparrows arrive, Rufous-collared Sparrows usually decline.
RANGE: Locally in w. Colombia (Buenaventura, Valle, Guapi, Cauca), w. Ecuador (Esmeraldas, Quevedo, and cities on Santa Elena Peninsula; a few also in Cuenca and Guayaquil), w. Peru, and more widespread from Bolivia (north to Beni) and Chile east across most of Argentina, Uruguay, and Paraguay to e. Brazil (where has recently even begun to invade e. Amazonia, in e. Pará, apparently along Trans-Amazon highway). Introduced first (in late 19th and early 20th centuries) into Chile, Argentina, and s. Brazil. Has tended not to become established in humid, hot areas, but this may now be changing. Native to Eurasia, but widely introduced. Locally to about 3000 m.

Carduelis carduelis

EUROPEAN GOLDFINCH

IDENTIFICATION: 12 cm (4¾"). *Uruguay* (mostly). An unmistakable, colorful small finch. *Foreface and throat bright scarlet,* contrasting with *black hindcrown* and white ear-coverts, sides of neck, and underparts. *Above and on sides of breast brown;* wings and tail black, wings with broad yellow band across base of primaries, tail with white patches. Sexes similar; immatures duller and more grayish, but pattern usually apparent.
SIMILAR SPECIES: This gaudy little finch resembles native siskins only in its yellow on wing.
HABITAT AND BEHAVIOR: Locally common in gardens, agricultural areas with scattered trees and shrubbery, and light open woodland. Introduced in the early 20th century, and apparently well established in Uruguay, though it has yet to show any definite signs of spreading.
RANGE: Coastal s. Uruguay (Colonia east to Maldonado); the few reports from ne. Argentina (Buenos Aires) have been regarded as iso-

lated releases, though natural dispersal would not seem out of the question. Native to Eurasia.

Carduelis chloris

EUROPEAN GREENFINCH

IDENTIFICATION: 15 cm (6"). *Uruguay* (mostly). A *mostly green* finch with *conspicuous yellow in wings and tail.* Rather heavy pale bill. Male dusky olive above, purer olive on rump; below yellowish olive, whitish on lower belly. *Primaries with prominent flash of canary yellow* toward their base, and basal half of tail also canary yellow. Female and immature duller and grayer, some with ill-defined dusky streaking above, but *always showing at least some yellow in wings and tail.*
SIMILAR SPECIES: In its limited S. Am. range likely confused only with various *Sicalis* finches, all of which are considerably smaller and yellower or streakier; cf. especially Saffron Finch.
HABITAT AND BEHAVIOR: Locally common in gardens and patches of open woodland or scattered trees in agricultural regions. Introduced in the late 19th century and apparently well established locally, though it has been slow to spread.
RANGE: Coastal s. Uruguay (Colonia east to Rocha) and also recorded (1978) in ne. Argentina (Villa Gessell in Buenos Aires; H. Montaldo, *Hornero* 12 [1]: 57–58, 1979). Native to Eurasia.

NOTE: The genus *Chloris,* in which the species was formerly placed, is now generally included in *Carduelis.*

Appendix

CASUAL MIGRANTS FROM NORTH AMERICA

OF the numerous migrants from North America which reach South America, a proportion have been recorded so infrequently that we felt it was preferable not to accord them full species treatment but rather to place them in an appendix, with merely a summary of their occurrences here. None of the following 24 species appears to occur annually, none has ever been found other than in the n. part of the continent, and several are surely "accidental," having occurred but once and perhaps not likely to do so again. We are aware that there exists a gray area involving species which we consider slightly more regular and numerous: these we have placed in the main text. For example, the Cape May Warbler is placed there, whereas the Townsend's Warbler will be found here. Quite apart from the difference in the actual number of S. Am. records of these 2 species, we take cognizance of the fact that, for example, the Cape May is now annual in Panama (albeit in small numbers), whereas the Townsend's has occurred even there only on 2 or 3 occasions and only in extreme w. Panama: it thus surely will never be more than an accidental vagrant to Colombia. For identification details, we suggest that readers refer to the full treatments in any of the N. Am. field guides or to Hilty and Brown's *Guide to the Birds of Colombia* (1986).

CUBAN MARTIN, *Progne cryptoleuca*
Specimens collected on Curaçao; Sept.–Oct. Here regarded as a full species, following 1983 AOU Check-list, but identification in the field problematic at best; see main text under Caribbean Martin.

VIOLET-GREEN SWALLOW, *Tachycineta thalassina*
One sighting from the Santa Marta Mts., Colombia; Jan.

NORTHERN ROUGH-WINGED SWALLOW, *Stelgidopteryx serripennis*
One sighting from Bonaire; Oct. Also a specimen possibly from Brazil. Seems likely to have been overlooked, but there are no other records. The species was recently split from s. birds, *S. ruficollis,* which occur widely in South America. See main text under Southern Rough-winged Swallow.

CAVE SWALLOW, *Hirundo fulva*
One specimen collected on Curaçao; Oct. N. populations are known to be migratory and presumably pass the nonbreeding season somewhere in South America, but there are no actual records. Populations resident in South America are here regarded as a full species, *H. rufocollaris.* See main text under Chestnut-collared Swallow.

NORTHERN WHEATEAR, *Oenanthe oenanthe*
Only on Netherlands Antilles, with 1 sighting from Curaçao and another from Bonaire (the latter photographed); Nov., Dec.

WOOD THRUSH, *Hylocichla mustelina*
One specimen collected in Chocó, Colombia, and a sighting on Curaçao; Oct., Dec.

GRAY CATBIRD, *Dumetella carolinensis*
Three records, including 2 specimens collected from n. Colombia, in Chocó and the Santa Marta area; Jan., Mar. This is conceivably of regular occurrence.

BROWN THRASHER, *Toxostoma rufum*
One specimen collected on Curaçao; Oct.

WHITE WAGTAIL, *Motacilla alba*
One recent sighting from Trinidad; Dec.

CEDAR WAXWING, *Bombycilla cedrorum*
Two records (1 specimen collected) from w. Colombia in Chocó and Valle, and 1 from w. Venezuela in Zulia (also collected); Jan., Feb. Irregular even in Panama.

EUROPEAN STARLING, *Sturnus vulgaris*
Three records from Netherlands Antilles, 1 involving 2 birds (photographed in 2 instances); Nov.–Jan. There is also a recent observation in Panama.

PHILADELPHIA VIREO, *Vireo philadelphicus*
Three records from n. Colombia, twice in Chocó and once in Cundinamarca; Oct., Nov. As it has never been found in e. Panama, there seems little reason to believe that it is regular in South America.

BLUE-WINGED WARBLER, *Vermivora pinus*
One specimen collected in the Santa Marta area, Colombia; Mar.

NORTHERN PARULA, *Parula americana*
A number of records from the Netherlands Antilles and 1 from Los Roques off n. Venezuela, but none from S. Am. mainland; Oct.–Mar.

MAGNOLIA WARBLER, *Dendroica magnolia*
Several records (including 1 specimen taken) from n. Colombia in Boyacá and the Santa Marta/Guajira area, 1 sighting from n. Venezuela in Aragua (*fide* A. Altman), and at least 2 sightings from Trinidad/Tobago; Dec.–Mar.

YELLOW-RUMPED WARBLER, *Dendroica coronata*
Two records (including 1 specimen taken) from the Santa Marta area of n. Colombia, one from e. Venezuela in Delta Amacuro, and 4 records from Netherlands Antilles; Dec.–Apr.

TOWNSEND'S WARBLER, *Dendroica townsendi*
One specimen collected in n. Colombia on Guajira Peninsula; Jan.

YELLOW-THROATED WARBLER, *Dendroica dominica*
Two sightings from n. Colombia, both in Santa Marta area; Oct.–Dec.

PRAIRIE WARBLER, *Dendroica discolor*
One specimen collected in n. Colombia in Córdoba, 1 known sighting from Trinidad, and several reports from Netherlands Antilles; Aug.–Apr.

PALM WARBLER, *Dendroica palmarum*
A few records from Netherlands Antilles, all during one n. winter (1956–57), with 2 birds collected; Nov.–Mar.

WORM-EATING WARBLER, *Helmitheros vermivorus*
One recent record from Venezuela.

BLACK-HEADED GROSBEAK, *Pheucticus melanocephalus*
One sighting from Curaçao; Dec.

INDIGO BUNTING, *Passerina cyanea*
Several specimens and sightings from n. Colombia in Chocó and the Santa Marta area, 1 record from w. Venezuela (Zulia), 1 sighting from Trinidad, and a number of reports from Netherlands Antilles (including 1 specimen); Nov.–May. Does not appear to be regular on the S. Am. mainland, but possibly annual on Netherlands Antilles.

BLUE GROSBEAK, *Guiraca caerulea*
Two records from Colombia (1 specimen collected), another collected in e. Ecuador; Dec., Apr.

BIBLIOGRAPHY

American Ornithologists' Union. 1983. *Check-list of North American Birds*, 6th ed.

Belton, W. 1985. Birds of Rio Grande do Sul, Brazil. Part 2. Formicariidae through Corvidae. *Bull. Am. Mus. Nat. Hist.* 180 (1): 1–241.

Bond, J. 1951. Notes on Peruvian Fringillidae. *Proc. Acad. Nat. Sci. Phil.* 103: 65–84.

———. 1953. Notes on Peruvian Icteridae, Vireonidae, and Parulidae. *Not. Nat. (Phil.)* 255: 1–15.

———. 1955. Notes on Peruvian Coerebidae and Thraupidae. *Proc. Acad. Nat. Sci. Phil.* 107: 35–55.

———. 1956. Additional notes on Peruvian birds II. *Proc. Acad. Nat. Sci. Phil.* 108: 227–247.

Bond, J., and R. Meyer de Schauensee. 1942. The birds of Bolivia. Part 1. *Proc. Acad. Nat. Sci. Phil.* 94: 307–391.

Chapman, F. M. 1917. The distribution of bird life in Colombia. *Bull. Am. Mus. Nat. Hist.* 36: 1–729.

———. 1926. The distribution of bird life in Ecuador. *Bull. Am. Mus. Nat. Hist.* 55: 1–784.

Cracraft, J. 1985. Historical biogeography and patterns of differentiation within the South American avifauna: Areas of endemism. In P. A. Buckley et al. (eds.), *Neotropical Ornithology*. Ornithol. Monogr. no. 36: 49–84.

Cuello, J., and E. Gerzenstein. 1962. Las aves del Uruguay (lista sistemática, distribución y notas). *Comun. Zool. Mus. Hist. Nat. Montev.* 93 (6).

Davis, T. J. 1986. Distribution and natural history of some birds from the departments of San Martín and Amazonas, northern Peru. *Condor* 88 (1): 50–56.

Donahue, P. K., and J. E. Pierson. 1982. *Birds of Suriname: An Annotated Checklist*. Austin, Tex.: Victor Emanuel Nature Tours.

Dunning, J. S. 1982. *South American Landbirds: A Photographic Guide to Identification*. Newtown Square, Pa.: Harrowood Books.

ffrench, R. 1973. *A Guide to the Birds of Trinidad and Tobago*. Wynnewood, Pa.: Livingston Press.

Fjeldså, J. 1987. *Birds of Relict Forests in the High Andes of Peru and Bolivia*. Technical Report, Zoological Museum, University of Copenhagen, Denmark.

Fjeldså, J., and N. Krabbe. 1986. Some range extensions and other unusual records of Andean birds. *Bull. B.O.C.* 106 (3): 115–124.

Friedmann, H. 1948. Birds collected by the National Geographic Society's expeditions to northern Brazil and southern Venezuela. *Proc. U.S. Natl. Mus.* 97: 373–570.

Frisch, J. D. 1981. *Aves Brasileiras.* Vol. I. Verona: Mondadori.

Fry, C. H. 1970. Ecological distribution of birds in northeastern Mato Grosso State, Brazil. *Ann. Acad. Bras.* 42: 275–318.

Goodwin, D. 1976. *Crows of the World.* Ithaca, N.Y.: Cornell University Press.

Gore, M. E. J., and A. R. M. Gepp. 1978. *Las aves del Uruguay.* Montevideo: Mosca Hnos., S.A.

Graham, G. L., G. R. Graves, T. S. Schulenberg, and J. P. O'Neill. 1980. Seventeen bird species new to Peru from the Pampas of Heath. *Auk* 97 (2): 366–370.

Gyldenstolpe, N. 1945a. The bird fauna of Rio Juruá in western Brazil. *K. Sven. Vetenskapsakad. Handl. Ser. 3* 22 (3): 1–388.

———. 1945b. A contribution to the ornithology of northern Bolivia. *K. Sven. Vetenskapsakad. Handl. Ser. 3* 23 (1): 1–300.

———. 1951. The ornithology of the Rio Purús region in western Brazil. *Ark. Zool. Ser. 2* 2: 1–320.

Haffer, J. 1974. *Avian Speciation in Tropical South America.* Publ. Nuttall Ornithol. Club no. 14. Cambridge, Mass.

———. 1975. *Avifauna of Northwestern Colombia, South America.* Bonn. Zool. Monogr. no. 7. Bonn.

Haverschmidt, F. 1968. *Birds of Surinam.* Edinburgh and London: Oliver and Boyd.

Hellmayr, C. E. 1918–1938. *Catalogue of Birds of the Americas.* Field Mus. Nat. Hist. Publ. Zool. Ser., vol. 13, parts 1 (nos. 1–4), 2 (nos. 1–2), 3–11.

———. 1929. A contribution to the ornithology of northeastern Brazil. *Field Mus. Nat. Hist. Zool. Ser.* 12: 235–501.

Hilty, S. L., and W. L. Brown. 1983. Range extensions of Colombian birds as indicated by the M. A. Carriker, Jr., collection at the National Museum of Natural History, Smithsonian Institution. *Bull. B.O.C.* 103: 5–17.

———. 1986. *A Guide to the Birds of Colombia.* Princeton, N.J.: Princeton University Press.

Holt, E. G. 1928. An ornithological survey of the Serra do Itatiaya, Brazil. *Bull. Am. Mus. Nat. Hist.* 57: 251–326.

Hudson, W. H. 1920. *Birds of La Plata.* Vol. I. New York: E. P. Dutton and Co.

Humphrey, P. S., D. Bridge, P. W. Reynolds, and R. T. Peterson. 1970. *Birds of Isla Grande (Tierra del Fuego), Preliminary Smithsonian Manual.* Univ. Kans. Publ. Mus. Nat. Hist.

Johnson, A. W. 1967. *The Birds of Chile and Adjacent Regions of Argentina, Bolivia and Peru.* Vol. 2. Buenos Aires: Platt Estab. Gráficos.

———. 1972. *Supplement to the Birds of Chile and Adjacent Regions of Argentina, Bolivia and Peru.* Buenos Aires: Platt Estab. Gráficos.

Koepcke, M. 1961. Birds of the western slope of the Andes of Peru. *Am. Mus. Novitates* 2028: 1–31.

———. 1970. *The Birds of the Department of Lima, Peru*. Wynnewood, Pa.: Livingston Publishing Co.

Laubmann, A. 1939. *Wissenschaftliche Ergebnisse der deutschen Grand Chaco Expedition*. Part 2. *Die Vogel von Paraguay*. Stuttgart: Strecker and Schroder.

Marchant, S. 1958. The birds of the Santa Elena Peninsula, S.W. Ecuador. *Ibis* 100: 349–387.

Mayr, E., and W. H. Phelps, Jr. 1967. The origin of the bird fauna of the south Venezuelan highlands. *Bull. Am. Mus. Nat. Hist.* 136 (5): 269–328.

Meyer de Schauensee, R. 1948–1952. The birds of the Republic of Colombia. Parts 1–5. *Caldasia* 22–26: 251–1212.

———. 1959. Additions to the birds of the Republic of Colombia. *Proc. Acad. Nat. Sci. Phil.* 111: 53–75.

———. 1964. *The Birds of Colombia*. Narberth, Pa.: Livingston Publishing Co.

———. 1966. *The Species of Birds of South America with Their Distribution*. Narberth, Pa.: Livingston Publishing Co.

———. 1970. *A Guide to the Birds of South America*. Wynnewood, Pa.: Livingston Publishing Co. (Reprinted by International Council for Bird Preservation with new addenda, 1982.)

Meyer de Schauensee, R., and W. H. Phelps, Jr. 1978. *A Guide to the Birds of Venezuela*. Princeton, N.J.: Princeton University Press.

Mitchell, M. H. 1957. *Observations on Birds of Southeastern Brazil*. Toronto, Ontario: University of Toronto Press.

Moscovits, D., J. W. Fitzpatrick, and D. E. Willard. 1985. Lista preliminar das aves da Estação Ecológica de Macará, Território de Roraima, Brasil, e áreas adjacentes. *Pap. Avuls., Zool.* (São Paulo) 36 (6): 51–68.

Naumburg, E. M. B. 1930. The birds of Matto Grosso, Brazil. *Bull. Am. Mus. Nat. Hist.* 60: 1–432.

Nores, M., and D. Yzurieta. 1982. Nuevas localidades para aves Argentinas. Part 2. *Historia Natural* 2 (13): 101–104.

Nores, M., D. Yzurieta, and R. Miatello. 1983. Lista y distribución de las aves de Córdoba, Arg. *Bol. Acad. Nac. Cienc.* (Córdoba) 56 (1–2).

Novaes, F. C. 1957. Contribuição a ornitologia do noroeste do Acre. *Bol. Mus. Par. Emi. Goeldi Nova Ser. Zool.* 9: 1–30.

———. 1960. Sobre uma coleção de aves do sudeste do estado do Pará. *Arq. Zool.* (São Paulo) 11: 133–146.

———. 1974, 1978. Ornitologia da territorio do Amapá I & II. *Publ. Avul. Mus. Goeldi* 25: 1–121; 29: 1–75.

———. 1976. As aves do rio Aripuañá, estados de Mato Grosso e Amazonas. *Acta Amazônica* 6 (4). Suplemento: 61–85.

Olivares, A. 1969. *Aves de Cundinamarca*. Bogotá: Universidad Nacional de Colombia.

Olrog, C. C. 1979. Nueva lista de la avifauna Argentina. *Opera Lilloana* 27: 1–297.

———. 1984. *Las aves argentinas "Una nueva guia de campo."* Buenos Aires: Administración de Parques Nacionales.

O'Neill, J. P. 1974. The birds of Balta, a Peruvian dry tropical forest locality, with an analysis of their origins and ecological relationships. Ph.D. dissertation, Louisiana State University.

O'Neill, J. P., and D. L. Pearson. 1974. Estudio preliminar de las aves de Yarinacocha, Dept. de Loreto, Peru. *Publ. Mus. Hist. Nat. "Javier Prado" Ser. A Zool.* 25: 1–13.

Parker, T. A., III. 1982. Observations of some unusual rainforest and marsh birds in southeastern Peru. *Wilson Bull.* 94 (1): 477–493.

Parker, T. A., III, and J. P. O'Neill. 1980. Notes on little known birds of the upper Urubamba Valley, southern Peru. *Auk* 97 (2): 167–176.

Parker, T. A., III, and S. A. Parker. 1982. Behavioral and distributional notes on some unusual birds of a lower montane cloud forest in Peru. *Bull. B.O.C.* 102: 63–70.

Parker, T. A., III, S. A. Parker, and M. A. Plenge. 1982. *An Annotated Checklist of Peruvian Birds.* Vermillion, S.D.: Buteo Books.

Parker, T. A., III, T. S. Schulenberg, G. R. Graves, and M. J. Braun. 1985. The avifauna of the Huancabamba region, northern Peru. In P. A. Buckley et al. (eds.), *Neotropical Ornithology.* Ornithol. Monogr. no. 36: 169–197.

Partridge, W. H. 1953. Observaciones sobre aves de las provincias de Córdoba y San Luis. *Hornero* 10: 23–73.

Paynter, R. A., Jr. 1972. Biology and evolution of the *Atlapetes schistaceus* species-group (Aves: Emberizinae). *Bull. Mus. Comp. Zool.* 143 (4): 297–320.

———. 1978. Biology and evolution of the avian genus *Atlapetes* (Emberizinae). *Bull. Mus. Comp. Zool.* 148 (7): 323–369.

Pearson, D. L., D. Tallman, and E. Tallman. 1977. *The Birds of Limoncocha, Napo Prov., Ecuador.* Quito, Ecuador: Instituto Lingüístico de Verano.

Peña, L. E. 1961. Explorations in the Antofogasta ranges of Chile and Bolivia. *Postilla* 49: 3–42.

Peters, J. L. 1931–1986. *Check-list of the Birds of the World.* Cambridge, Mass.: Harvard University Press.

Phelps, W. H., and W. H. Phelps, Jr. 1963. Lista de las aves de Venezuela y su distribución (2nd ed.). Vol. 1, part 2, Passeriformes. *Bol. Soc. Venez. Cien. Nat.* 24 (104, 105): 1–479.

Pinto, O. M. de O. 1932. Resultados ornitológicos de uma excursão pelo oeste de São Paulo & sul de Matto-Grosso. *Rev. Mus. Paulista* 17 (2): 641–708.

———. 1937. Catálogo das aves do Brasil (1a parte). *Rev. Mus. Paulista* 22: 1–566.

———. 1944. *Catálogo das aves do Brasil (2a parte).* São Paulo: Secretário Agricultura.

———. 1952. Sumula histórica e sistemática da ornitologia de Minas Gerais. *Arq. Zool.* (São Paulo) 8: 1–52.

———. 1954. Resultados ornitológicos de duas viagems científicas ao Estado de Alagoas. *Pap. Avuls. Zool.* (São Paulo) 12: 1–98.

Pinto, O. M. de O., and E. A. de Camargo. 1954. Resultados ornitológicos de uma expedição ao Territorio do Acre pelo Departamento de Zoologia. *Pap. Avuls. Zool.* (São Paulo) 11: 371–418.

———. 1957. Sobre uma coleção de aves da região de Cachimbo (Sul do Estado do Pará). *Pap. Avuls. Zool.* (São Paulo) 13: 51–69.

———. 1961. Resultados ornitológicos de quatro recentes expedições do Departamento de Zoologia ao nordeste de Brasil, com a descrição de seis novas subespecies. *Arq. Zool.* (São Paulo) 11: 193–284.

Remsen, J. V., Jr. 1984. Natural history notes on some poorly known Bolivian birds. Part 2. *Gerfaut* 74: 163–179.

———. 1986. Aves de una localidad en la sabana húmeda del norte de Bolivia. *Ecol. en Bol.* 8: 21–35.

Remson, J. V., Jr., T. A. Parker, III, and R. S. Ridgely. 1982. Natural history notes on some poorly known Bolivian birds. *Gerfaut* 72 (1): 77–87.

Remson, J. V., Jr., and R. S. Ridgely. 1980. Additions to the avifauna of Bolivia. *Condor* 82 (1): 69–75.

Remson, J. V., Jr., C. G. Schmitt, and D. C. Schmitt. 1988. Natural history notes on some poorly known Bolivian birds. Part 3. *Gerfaut* 78: 363–381.

Ridgely, R. S. 1976. *A Guide to the Birds of Panama.* Princeton, N.J.: Princeton University Press.

———. 1980. Notes on some rare or previously unrecorded birds in Ecuador. *Am. Birds* 34 (3): 242–248.

Ridgely, R. S., and S. J. C. Gaulin. 1980. The birds of Finca Merenberg, Huila Department, Colombia. *Condor* 82 (4): 379–391.

Rodríguez, J. V. 1982. *Aves del Parque Nacional Los Katíos.* Bogotá: INDERENA.

Ruschi, A. 1979. *Aves do Brasil.* São Paulo: Editora Rios.

Schulenberg, T. S. 1986. Adiciones a la avifauna de Pampa Galeras. *Bol. de Lima* 48: 89–90.

———. 1987. New records of birds from western Peru. *Bull. B.O.C.* 107 (4): 184–189.

Schulenberg, T. S., S. E. Allen, D. F. Stotz, and D. A. Wiedenfeld. 1984. Distributional records from the Cordillera Yanachaga, central Peru. *Gerfaut* 74: 57–70.

Scott, D. A., and M. de L. Brooke. 1985. The endangered avifauna of southeastern Brazil: A report on the BOU/WWF expeditions of 1980/1981 and 1981/1982. In A. W. Diamond and T. E. Lovejoy (eds.), *Conservation of Tropical Forest Birds.* Cambridge: ICBP.

Short, L. S. 1968. Sympatry of Red-breasted Meadowlarks in Argentina, and the taxonomy of meadowlarks (Aves: *Leistes, Pezites,* and *Sturnella*). *Am. Mus. Novitates* 2349: 1–30.

———. 1971. Aves nuevas o poco comunes de Corrientes, Republica Argentina. *Rev. Mus. Argent. Cienc. Nat. "Bernardino Rivadavia" Inst. Nac. Invest. Cienc. Nat. Zool.* 9: 283–309.

———. 1975. A zoogeographic analysis of the South American Chaco avifauna. *Bull. Am. Mus. Nat. Hist.* 154 (3): 165–352.

Sick, H. 1955. O aspecto fitofisionómico da paisagem do medio Rio das Mortes, Mato Grosso e a avifauna do região. *Arq. Mus. Nac. Rio de Jan.* 42: 541–576.

———. 1979. Notes on some Brazilian birds. *Bull. B.O.C.* 99 (4): 115–120.

———. 1985. *Ornitologia brasileira, uma introdução.* Vols. 1 and 2. Brasilia: Editora Universidade de Brasilia.

Sick, H., L. Aldo de Rosario, and T. Rauh de Azevado. 1981. Aves do Estado de Santa Catarina. *Sellowía,* Ser. Zool. no. 1.

Snethlage, E. 1914. Catálogo das aves amazônicas. *Bol. Mus. Par. Emilio Goeldi* 8: 1–534.

Snyder, D. E. 1966. *The birds of Guyana.* Salem, Mass.: Peabody Museum.

Steinbacher, J. 1962. Beitrage zur Kenntnis der Vogel von Paraguay. *Abh. Senckenk. Naturf. Ges.* 502: 1–106.

———. 1968. Weitere Beitrage über Vogel von Paraguay. *Senckenb. Biol.* 49: 317–365.

Stiles, F. G., and S. M. Smith. 1980. Notes on bird distribution in Costa Rica. *Brenesia* 17: 137–156.

Taczanowski, M. L. 1884–1886. *Ornithologie du Pérou.* 3 vols. Berlin: R. Friedlander and John.

Teixera, D. M., J. B. Nacinovic, and M. S. Tavares. 1986. Notes on some birds of northeastern Brazil. *Bull. B.O.C.* 106 (2): 70–74.

Terborgh, J. W., J. W. Fitzpatrick, and L. Emmons. 1984. Annotated checklist of bird and mammal species of Cocha Cashu Biological Station, Manu National Park, Peru. *Fieldiana Zool. Mus. Nat. Hist.* n.s. 21: 1–29.

Todd, W. E. C., and M. A. Carriker, Jr. 1922. The birds of the Santa Marta region of Colombia: A study in altitudinal distribution. *Ann. Carnegie Mus.* 14.

Traylor, M. A., Jr. 1958. Birds of northeastern Peru. *Fieldiana Zool.* 35: 87–141.

Voous, K. H. 1983. *Birds of the Netherlands Antilles.* Curaçao: De Walburg Pers.

Vuilleumier, F. 1967. Speciation in high Andean birds. Ph.D. dissertation, Harvard University. 444 pp.

———. 1969. Systematics and evolution in *Diglossa* (Aves, Coerebidae). *Am. Mus. Novitates* 2381: 1–43.

Weske, J. 1972. The distribution of the avifauna in the Apurímac Valley of Peru with respect to environmental gradients, habitat, and related species. Ph.D. dissertation, University of Oklahoma.

Wetmore, A. 1926. Observations on the birds of Argentina, Paraguay, Uruguay, and Chile. *Bull. U.S. Natl. Mus.* 133: 1–448.

Wetmore, A., R. F. Pasquier, and S. L. Olson. 1984. *The birds of the Republic of Panama, Part 4, Passeriformes: Hirundinidae (Swallows) to Fringillidae (Finches).* Washington, D.C.: Smithsonian Institution Press.

Wiedenfeld, D. A., T. S. Schulenberg, and M. B. Robbins. 1985. Birds of a tropical deciduous forest in extreme northwestern Peru. In P. A. Buckley et al. (eds.), *Neotropical Ornithology*. Ornithol. Monogr. no. 36: 305–315.

Willis, E. O. 1976. Effects of a cold wave on an Amazonian avifauna in the upper Paraguay drainage, western Mato Grosso, and suggestions on Oscine-Suboscine relationships. *Acta Amazônica* 6 (3): 379–394.

———. 1977. Lista preliminar das aves da parte noroeste e areas vizinhas da Reserva Ducke, Amazonas, Brasil. *Rev. Brasil. Biol.* 37 (3): 585–601.

Willis, E. O., and Y. Oniki. 1981. Levantamento preliminar de aves em treze áreas do Estado de São Paulo. *Rev. Brasil. Biol.* 41 (1): 121–135.

———. 1985. Bird specimens new for the State of São Paulo. *Rev. Brasil. Biol.* 45 (1/2): 105–108.

Zimmer, J. T. 1931–1955. Studies of Peruvian birds, 1–66. *Am. Mus. Novitates.*

INDEX TO ENGLISH NAMES

INDEX TO SCIENTIFIC NAMES

PARAGUAY
MATO GROSSO
SÃO PAULO
Rio Paraná
Rio Paraguay
BRAZIL
PARANÁ
ANTOFAGASTA
JUJUY
Salta
SALTA
FORMOSA
Asunción
ATACAMA
Tucumán
TUCUMAN
CHACO
MISIONES
SANTA CATARINA
CATAMARCA
SANTIAGO DEL ESTERO
CHILE
CORRIENTES
LA RIOJA
SANTA FE
RIO GRANDE DO SUL
COQUIMBO
SAN JUAN
Córdoba
Río Paraná
ENTRE RÍOS
ACONCAGUA
SAN LUIS
CÓRDOBA
URUGUAY
SANTIAGO
O'HIGGINS
COLCHAGUA
MENDOZA
Buenos Aires
CURICÓ
TALCA
MAULE
LINARES
ARGENTINA
ÑUBLE
CONCEPCIÓN
LA PAMPA
BUENOS AIRES
ARAUCO
BÍO-BÍO
MALLECO
NEUQUÉN
CAUTÍN
VALDIVIA
RÍO NEGRO
OSORNO
LLANQUIHUE
Valdez Peninsula
CHILOÉ
CHUBUT
AYSÉN
SANTA CRUZ
FALKLAND ISLANDS
MAGALLANES
TIERRA DEL FUEGO